Pierre Léna • Daniel Rouan
François Lebrun • François Mignard
Didier Pelat

In collaboration with Laurent Mugnier

Observational Astrophysics

Translated by S. Lyle

Third Edition

Pierre Léna
Université Paris Diderot &
Observatoire de Paris (LESIA)
92195 Meudon
France
pierre.lena@obspm.fr

Daniel Rouan
Didier Pelat
Observatoire de Paris (LESIA)
92195 Meudon
France
daniel.rouan@obspm.fr
pelat@obspm.fr

François Lebrun
Laboratoire Astroparticules et Cosmologie
Université Paris Diderot & Commissariat à l'énergie atomique et aux énergies alternatives
10 rue A. Domon & L. Duquet, 75205 Paris Cx13
France
francois.lebrun@apc.univ-paris7.fr

François Mignard
Observatoire de la Côte d'Azur
Laboratoire Cassiopée
Bd de l'Observatoire
06304 Nice Cedex 4
France
francois.mignard@oca.eu

Translation from the French language edition of *L'observation en astrophysique*.

ISSN 0941-7834
ISBN 978-3-642-21814-9 e-ISBN 978-3-642-21815-6
DOI 10.1007/978-3-642-21815-6
Springer Heidelberg Dordrecht London New York

Library of Congress Control Number: 2011942601

Cover figure: The mobile auxiliary telescopes (1.8-m in diameter) of the European *Very Large Telescope* (VLT), contributing to the interferometric mode (VLTI) of the VLT (see Chap. 6). The sky background is added from a real photograph.

Credit: Pierre Kervella, Observatoire de Paris

Printed on acid-free paper

Springer is part of Springer Science+Business Media (www.springer.com)

ASTRONOMY AND ASTROPHYSICS LIBRARY

For further volumes:
http://www.springer.com/series/848

Preface

Up until the end of World War II, almost the only tools available for astronomical observation were telescopes, spectrometers, and photographic plates, limited to the visible range of the electromagnetic spectrum. This was a relatively simple technology, but carried to a high level of performance by the combined efforts of opticians and astronomers. Then in the 1950s came radioastronomy, followed by infrared, ultraviolet, X-ray, and γ-ray astronomy, the birth and growth of space-based observation, *in situ* observation of the Solar System, and the advent of computing with the massive improvement in data processing capacity that resulted from it, so many factors leading to an unprecedented explosion in astrophysical activity. The first French edition of this book appeared in 1986, after three decades of new observational developments, followed in 1988 by the English translation *Observational Astrophysics*, published by Springer. And yet, ten years later, this first edition had already given way to a second: CCD detectors had replaced photography and a new generation of giant optical telescopes was coming into being on the Earth's surface, while the first cosmic neutrinos had been detected and the existence of gravitational waves indirectly demonstrated. The world astronomical community had also evolved since, apart from the English translation of the second edition in 1998, a slightly updated Chinese version appeared in Taiwan in 2004.

But after a further decade of astonishing developments in astronomical observation, the book must yet again be reworked. Adaptive optics has opened up entirely novel prospects for Earth-based optical telescopes, while interferometry can achieve angular resolutions on the ground today, and soon in space, that were previously only obtained at radiofrequencies. Meanwhile an assortment of new Earth- and space-based instruments are being developed today to explore the submillimeter range, still virtually uncharted, to observe objects with very high spectral shifts and the cosmological background radiation. The discovery of an ever-increasing number of exoplanets has led to many refinements of older techniques, such as coronography, while opening a new and fascinating chapter in the history of astronomy — the search for life in the Universe — in which physics, chemistry, and biology each play their role. There are new and more refined neutrino telescopes, while those developed to seek out gravitational waves are gradually being brought into service. And space exploration of the Solar System is still an issue, as more and more probes and *in situ* experiments go out to Mars, Titan, and cometary nuclei. The temporal and spatial reference frames used by astronomers, but others too, e.g., for detailed study of continental drift, are becoming increasingly accurate.

With the help of several new authors, we have therefore rewritten the book, reorganising and extending the material used in the previous editions. The work started out as lecture notes for a course one of us (PL) delivered to graduate students of astrophysics at the Denis Diderot University (Paris VII). The original section on methodology has been maintained. Apart from surveying the broad range of techniques specific to each wavelength, the details of which can be found in more specialised sources, our aim has been to present the *physical* foundations for the various types of instrumentation: telescopes gathering data, spectrometers analysing it, and detectors converting it into a signal. After the first four chapters, which deal with information carriers (Chap. 1), the effects of the Earth atmosphere (Chap. 2), basic photometry (Chap. 3), and spatial and temporal reference systems (Chap. 4), there follow a chapter each on telescopes, detectors, and spectrometers (Chaps. 5–8), with some emphasis of course on image formation.

The idea has been to bring out the main principles, describing levels of performance or the ultimate limits allowed by the laws of physics. So the guiding thread here lies in the properties of the photon (or the electromagnetic wave), since this remains the main information carrier in astrophysics. Acquisition, measurement, and quantitative techniques for analysing data constitute the theme of this book, and the choices made here reflect this objective. Such an approach necessarily limits what can be covered, and we make no pretence to exhaust all observational methods, nor to provide a complete and systematic presentation of the corresponding tools.

The increasing complexity, development timescales, and costs involved in today's instrumentation have radically changed the way this kind of work is now organised, and indeed the whole profession. Very often, too often perhaps, those who design and build an instrument are not the same as those who use it and who interpret the observations. The present book will have achieved its aim if it provides some with the means to advance the pursuit of data, and others with the lights to understand the 'black boxes' that constitute contemporary observational equipment.

There have been two major additions to the new edition. One is a more detailed discussion of signal processing in Chap. 9, stressing the universal digitization of data and the power of computational tools which have revolutionized the way information is processed. This chapter is inevitably rather mathematical and stands out from the rest of the book, but we have no doubt that it will be of great interest to readers. Apart from this, Chap. 10 is entirely new, describing the way modern instruments gather huge volumes of data, making them available to all in data banks. This leads to the idea of the *virtual observatory*, something that has transformed the everyday life of the astrophysicist. Finally, the essential mathematical tools, such as the Fourier transform and an introduction to probability and statistics, can be found in the appendices. We have kept the exercises included in the earlier editions without modification or addition. Despite their sometimes rather simple or even dated nature, students have found them of some use, at least at the elementary level.

The rich supply of information, images, and up-to-date news available on the Internet might make it seem pointless to try to catch all this knowledge in long-lasting written form. Naturally, the book includes a detailed webography, wherein the reader may find updates for all the subjects treated here. However, efficient use of the web can only be achieved within the kind of framework we hope to provide through this book. This has been the underlying idea that guided us while we were writing it.

Since the aim has been to produce a reference book, we have chosen to remove bibliographical references from the text as far as possible. We have simply put together a short bibliography at the end, not intended to be exhaustive. The reference books that seem to us to be potentially the most useful to the student, researcher, or teacher have been organised according to theme.

We could not possibly name or thank all colleagues or students, often later to become colleagues, who have contributed to the two first editions and provided illustrations. We would just like to thank Mme Claude Audy, who prepared the final version of the manuscript, and Mme Hélène de Castilla of InterEditions (Paris), together with Eric Gendron, who carefully copy-edited. The current edition is indebted to Laurent Mugnier, who wrote part of Chap. 9, and Marc Huertas, who put together the webography. We are also grateful to Laurent Pagani for radiofrequencies, Michel Cribier for neutrinos, Philippe Laurent for gravitational waves, Jean Ballet for X-ray astronomy, Philippe Goret for ground-based γ-ray astronomy, and Claude Pigot, who accepted to write or proofread parts of the text. The *Fondation des Treilles* generously hosted one of us (PL) in Provence (France) while the book was being finalised. We thank them for that, and also Michèle Leduc for her tireless supervision of the *Savoirs actuels* series.

We have not forgotten that the two previous editions of this book were dedicated to the memory of the astronomer and physicist Philippe Delache (1937–1996). We hope that the present edition, following his example, will excite the enthusiasm of many new generations of students, attracted into this most wonderful of sciences — astronomy.

Paris

Pierre Léna
Daniel Rouan
François Lebrun
François Mignard
Didier Pelat

A detailed bibliography is given at the end of the book. Only a few specific references are given in the course of the chapters, in the text or in footnote. Beside the classical names of journals, some specific abbreviations are used for frequent quotations of documents detailed in the bibliography, namely:

- AF for the book Astrophysical Formulae.
- AQ for the book Astrophysical Quantities
- ARAA for Annual Review of Astronomy and Astrophysics.

Contents

Part I Foundations

1 Astrophysical Information ... 3
- 1.1 Carriers of Information ... 4
  - 1.1.1 Electromagnetic Radiation ... 4
  - 1.1.2 Matter: From Electrons and Nuclei to Meteorites ... 5
  - 1.1.3 Neutrinos ... 6
  - 1.1.4 Gravitational Waves ... 9
  - 1.1.5 In Situ Observation ... 10
- 1.2 Data Acquisition ... 12
  - 1.2.1 The Main Characteristics of Photons ... 12
  - 1.2.2 Observing Systems ... 12
  - 1.2.3 Reaching a Systematic Description of Observation ... 27
- 1.3 Global Organisation of Astronomy ... 28
  - 1.3.1 People ... 29
  - 1.3.2 Research Policies and Institutions ... 31
  - 1.3.3 Publications ... 34

2 The Earth Atmosphere and Space ... 39
- 2.1 Physical and Chemical Structure of the Atmosphere ... 40
  - 2.1.1 Vertical Structure ... 40
  - 2.1.2 Constituents of the Atmosphere ... 41
- 2.2 Absorption of Radiation ... 45
- 2.3 Atmospheric Emission ... 50
  - 2.3.1 Fluorescent Emission ... 50
  - 2.3.2 Thermal Emission ... 55
  - 2.3.3 Differential Measurement Techniques ... 56
- 2.4 Scattering of Radiation ... 58
- 2.5 Atmospheric Refraction and Dispersion ... 61

2.6 Turbulence Structure of the Earth Atmosphere 62
2.6.1 Turbulence in the Lower and Middle Atmosphere 63
2.6.2 Ionospheric Turbulence 70
2.7 The Atmosphere as Radiation Converter 70
2.7.1 Ground-Based Gamma-Ray Astronomy 70
2.7.2 Air Showers and Cosmic Rays 71
2.8 Terrestrial Observing Sites 71
2.8.1 Visible, Infrared, and Millimetre Observations 72
2.8.2 Centimetre and Metre Wave Radioastronomy 74
2.8.3 Very High Energy Gamma-Ray Astronomy 75
2.8.4 Very High Energy Cosmic Radiation 75
2.8.5 Man-Made Pollution and Interference 75
2.8.6 The Antarctic 76
2.9 Observation from Space 77
2.9.1 The Advantages of Observation from Space 79
2.9.2 Sources of Perturbation 79
2.9.3 Choice of Orbits 86
2.10 The Moon as an Astronomical Site 87
Problems 89

3 Radiation and Photometry 93
3.1 Radiometry 94
3.2 Aspects of Radiation 99
3.2.1 Blackbody Radiation 99
3.2.2 Coherence 100
3.3 Magnitudes 104
3.4 Photometry Through the Atmosphere 109
3.5 Calibration and Intensity Standards 110
3.5.1 Radiofrequencies 110
3.5.2 Submillimetre, Infrared, and Visible 112
3.5.3 Ultraviolet and X Rays 117
3.5.4 Gamma-Ray Radiation 120
3.5.5 Some Examples of Spectrophotometry 120
3.6 Calibration of Angular Dimensions 123
Problems 124

4 Space–Time Reference Frames 127
4.1 Spatial Reference Systems 129
4.1.1 Definitions of Spatial Frames 129
4.1.2 Astronomical Reference Frames 131
4.1.3 Change of Frame 138
4.2 Practical Realisation of Spatial Frames 144
4.2.1 Celestial Reference Systems 144
4.2.2 Fundamental Catalogues 145
4.2.3 The Extragalactic System 147

4.2.4 The Hipparcos Frame 151
4.2.5 The Near Future: The Gaia Mission 155
4.3 Temporal Reference Systems 157
4.3.1 Time Scales 157
4.3.2 Atomic Time 161
4.3.3 Coordinated Universal Time (CUT or UTC) 164
4.3.4 GPS Time 166
4.3.5 Dynamical Time Scales 167
4.3.6 Dates and Epochs. Dealing with Long Periods 169

Part II Data Collection

5 Telescopes and Images 175
5.1 Image and Object in Astronomy 176
5.1.1 The Telescope and Geometrical Optics 177
5.1.2 Gravitational Optics 183
5.2 Telescopes 184
5.2.1 Radiotelescopes 185
5.2.2 Ground-Based Optical Telescopes: Visible and Near Infrared 189
5.2.3 Space Telescopes, from Ultraviolet to Submillimetre 194
5.2.4 X-Ray Telescopes 199
5.2.5 Gamma-Ray Telescopes 201

6 Diffraction and Image Formation 209
6.1 Diffraction by an Arbitrary Aperture 210
6.1.1 The Zernike Theorem 211
6.1.2 Coherence Etendue 214
6.1.3 Diffraction at Infinity 216
6.1.4 Spatial Filtering by a Pupil 221
6.2 The Earth Atmosphere and Coherence Losses 228
6.2.1 Perturbations of the Wavefront 229
6.2.2 The Perturbed Image 232
6.2.3 Effect of the Atmosphere on Interferometry 238
6.3 Adaptive Optics 240
6.3.1 Wavefront Measurement 241
6.3.2 Phase Correction Devices 245
6.3.3 The Final Image 246
6.3.4 Sensitivity and Reference Sources 248
6.3.5 New Concepts 252
6.4 Astronomical Interferometry 256
6.4.1 Obtaining an Interferometer Signal 257
6.4.2 Light Transfer 262
6.4.3 Temporal Coherence 264
6.4.4 Loss of Spatial Coherence 264

6.4.5 Calibrating the Instrumental MTF 268
6.4.6 Phase Closure 271
6.5 Astronomical Interferometers........ 274
6.5.1 Radiotelescope Arrays........ 274
6.5.2 Ground-Based Optical Arrays 286
6.5.3 Space-Based Optical Interferometry........ 294
6.6 High Dynamic Range Imaging (HDRI) 298
6.6.1 Coronagraphy and Apodisation 299
6.6.2 Nulling Interferometry 311
Problems 316

7 Detectors........ 323
7.1 General Properties........ 324
7.1.1 Amplitude Detectors. Quadratic Detectors 325
7.1.2 Spatial Structure of Detectors........ 326
7.1.3 Temporal Response........ 329
7.1.4 Noise 330
7.1.5 Characterisation of Detectors 331
7.2 Fundamental Fluctuations 332
7.2.1 Quantum Noise 336
7.2.2 Thermal Noise 340
7.3 Physical Principles of the Detection of Electromagnetic Radiation........ 343
7.3.1 Detection of Quanta 344
7.3.2 Detection of the Electromagnetic Field........ 355
7.4 Astronomical Detectors from X Ray to Submillimetre........ 355
7.4.1 Noise Performance 356
7.4.2 Photographic Plates........ 357
7.4.3 Photomultipliers and Classical Cameras: X Ray, UV, and Visible........ 359
7.4.4 X-Ray Detection (0.1–10 keV) 364
7.4.5 Solid-State Imagers........ 365
7.4.6 Charge Coupled Device (CCD) 366
7.4.7 The Hybrid CMOS Detector 373
7.4.8 Observing Conditions in the Infrared........ 380
7.4.9 Development of Solid-State Imaging Arrays........ 381
7.4.10 Bolometers........ 383
7.5 Astronomical Detectors: Radiofrequencies 387
7.5.1 General Features........ 388
7.5.2 Heterodyne Detection 393
7.5.3 The Diversity of Radioastronomy 403
7.6 Observing Systems for Gamma-Ray Astronomy........ 404
7.6.1 Spatial Resolution of Gamma-Ray Sources........ 407
7.6.2 Spectral Analysis of Gamma-Ray Sources........ 412

7.7 Neutrino Observing Systems 420
7.7.1 Radiochemical Detection of Solar Neutrinos 421
7.7.2 Neutrino Detection by Cherenkov Radiation 424
7.7.3 High Energy Neutrino Astronomy 425
7.8 Gravitational Wave Detection 431
Problems 437

8 Spectral Analysis 441
8.1 Astrophysical Spectra 442
8.1.1 Formation of Spectra 442
8.1.2 Information in Spectrometry 448
8.2 Spectrometers and Their Properties 455
8.2.1 Quantities Characterising a Spectrometer 456
8.2.2 Spectral Discrimination 459
8.2.3 The Modes of a Spectrometer 460
8.3 Interferometric Spectrometers 462
8.3.1 General Criteria 462
8.3.2 Interference Filters 463
8.3.3 Grating Spectrometers 463
8.3.4 Fourier Transform Spectrometer 481
8.3.5 The Fabry–Perot Spectrometer 489
8.3.6 The Bragg Crystal Spectrometer (X-Ray Region) 491
8.4 Radiofrequency Spectrometry 494
8.4.1 Spectral Discrimination Methods 495
8.4.2 Submillimetre Spectroscopy 501
8.5 Resonance Spectrometers 503
Problems 504

Part III Data Analysis

9 The Signal in Astronomy 509
9.1 The Signal and Its Fluctuations 510
9.1.1 Observing System and Signal 510
9.1.2 Signal and Fluctuations. Noise 511
9.1.3 Elementary Signal Processing 519
9.1.4 A Specific Example of Data Processing 528
9.2 Complete Model of an Observing System 529
9.3 Overall Performance of an Observing System 532
9.3.1 Observing with the IRAM Millimetre Interferometer 533
9.3.2 Observing with NAOS Adaptive Optics 536
9.3.3 Observing with the Photometric Satellite COROT 538
9.3.4 Observing with a Coded Mask Gamma-Ray Instrument 541
9.4 Removing Instrumental Signatures 544
9.4.1 Intrinsic Emission from the Instrument 545
9.4.2 Dark Current 545

9.4.3 Non-Linearity Defects 546
9.4.4 Bias 547
9.4.5 Light Interference 547
9.4.6 Flat Field Corrections 548
9.4.7 Defective Pixels 549
9.4.8 Effects of High Energy Particle Impacts 549
9.5 The Problem of Estimation 550
9.5.1 Samples and Statistics 550
9.5.2 Point Estimation 551
9.5.3 Elements of Decision Theory 551
9.5.4 Properties of Estimators 554
9.5.5 Fréchet or Rao–Cramér Inequality 564
9.5.6 Efficient Estimators 566
9.5.7 Efficiency of an Estimator 568
9.5.8 Biased Estimators 568
9.5.9 Minimum Variance Bound and Fisher Information 570
9.5.10 Multidimensional Case 570
9.5.11 Robust Estimators 571
9.5.12 Some Classic Methods 573
9.6 From Data to Object: the Inverse Problem 575
9.6.1 Posing the Problem 576
9.6.2 Well-Posed Problems 579
9.6.3 Conventional Inversion Methods 581
9.6.4 Inversion Methods with Regularisation 587
9.6.5 Application to Adaptive Optics Imaging 592
9.6.6 Application to Nulling Interferometry 595
Problems 597

10 Sky Surveys and Virtual Observatories 605
10.1 Statistical Astrophysics 605
10.2 Large Sky Surveys 608
10.2.1 Sky Surveys at Visible Wavelengths 610
10.2.2 Infrared Sky Surveys 614
10.3 A Virtual Observatory 615

A Fourier Transforms 619
A.1 Definitions and Properties 619
A.1.1 Definitions 619
A.1.2 Some Properties 620
A.1.3 Important Special Cases in One Dimension 622
A.1.4 Important Special Cases in Two Dimensions 625
A.1.5 Important Theorems 626
A.2 Physical Quantities and Fourier Transforms 631
A.3 Wavelets 635

B Random Processes and Variables 637
B.1 Random Variables 637
B.2 Random or Stochastic Processes 644
B.3 Physical Measurements and Estimates 653
B.3.1 An Example of Estimation: The Law of Large Numbers 654
B.3.2 Estimating the Moments of a Process 655

C Physical and Astronomical Constants 659

D Tables of Space Missions 661

E Webography 663
E.1 Main Earth-Based Telescopes 663
E.2 Recent Space Missions 667
E.3 Databases 669
E.4 Journals 672
E.5 Bibliographical Research 673
E.6 Image Sources 673
E.7 Education 675
E.8 Computing and Astronomy 676
E.9 Resources 677

F Acronyms 679

Bibliography 687

Index 705

Part I
Foundations

Chapter 1
Astrophysical Information

The aim of astrophysics is to describe, to understand and to predict the physical phenomena that occur in the Universe. The physical content of the Universe — dense or rarefied, hot or cold, stable or unstable — can usually be classified into categories, such as planets, stars, galaxies, and so on. The *information* received by observers and transformed into *signals* is the basis for these classifications, for the physical models and for the predictions, which together make up the science of astrophysics.

The aim of *observation* is to work out a strategy for collecting this information, and to order the various variables or physical parameters measured; to analyse this information in such a way that it is neither over-interpreted nor wasted; and to store it for later investigations, possibly by future generations. In this introductory chapter, we look first at the carriers of this information. These carriers, originating in the object under study, are generally modified during their journey through space, and then collected by observers and their instruments. The main carriers are electromagnetic waves, but there are others, such as gravitational waves and neutrinos, not to mention elementary particles like electrons, protons, nuclei, and atoms, or dust grains of various sizes, such as meteorites. The carrying of information results from a carrying of *energy*, in whatever form, from the source to the observer.

It is quite clear, by considering how this information is *collected* when it reaches the Earth, that it is impossible to gather and measure simultaneously every one of its components. Each observational technique acts as an *information filter*, selecting usually only a small fraction of the wealth of information available at any instant. The great diversity of such filters, producing images, spectra, light curves, and so on, is closely linked to the technology and physical tools available at any given time. Finally, there is no point in collecting the information if it cannot be stored, handled or refined, should the occasion arise.

Astronomical information can also be considered as an *economic commodity*; it is costly to acquire and to analyse, so decisions must often be made at the political level or, at least, on the level of the science research budget. We will describe some of the orders of magnitude involved.

P. Léna et al., *Observational Astrophysics*, Astronomy and Astrophysics Library,
DOI 10.1007/978-3-642-21815-6_1, © Springer-Verlag Berlin Heidelberg 2012

1.1 Carriers of Information

Here we shall discuss the different information carriers that make their way to our astronomical instruments from the depths of space, allowing us to observe the Universe from a distance: electromagnetic radiation, of course, but also matter particles and other signals, such as neutrinos and gravitational waves, associated with other interactions than the electromagnetic interaction.

1.1.1 Electromagnetic Radiation

Electromagnetic radiation plays a crucial role, carrying virtually all the information which constitutes our knowledge of the Universe, and upon which modern astrophysics is built. This is certainly not a consequence of any fundamental physical law – neutrinos propagate more readily in space than photons – but is rather a consequence of the historical role played by the human eye. This organ is far richer and more elaborate than the organs of the other senses. Without any technical aids, the *naked eye* can gather, and transcribe for the brain, information originating from objects as distant as the Andromeda Galaxy (M31), three million light-years (about 1 Megaparsec) away.

Production of electromagnetic (EM) radiation is very directly related to the whole range of physical conditions prevailing in the emitter, including the nature and motion of particles, atoms, molecules or dust grains, and also the temperature, pressure, presence of a magnetic field, and so on. No electromagnetic wavelength is privileged, each energy domain being associated with one or several classes of object. Electromagnetic radiation will be studied in greater detail in Chap. 3.

The terms describing the different regions of the EM spectrum, and given in Table 1.1, have arisen through practical usage. They are rather arbitrary, being linked historically to the various methods of detection, and in particular, have no well-defined limits. Moreover, it is common practice to describe high-energy photons ($\gtrsim$100 eV), such as γ- and X-ray photons, in terms of their energy (in eV), those

Table 1.1 The electromagnetic spectrum. *Note:* The main divisions illustrated here have been further subdivided, once again in a somewhat arbitrary way: soft X-rays (1–10 nm), extreme ultraviolet (10–90 nm), far ultraviolet (90–200 nm), near ultraviolet (200–300 nm), near infrared (0.8–15 μm), far infrared (15–200 μm), submillimetre (0.2–1 mm), then radiofrequencies by wavelength (millimetre, centimetre, . . . , kilometre) or by frequency (UHF from 3 GHz to 300 MHz, VHF from 300 MHz to 30 MHz, . . . , down to ULF from 3 000 Hz to 300 Hz)

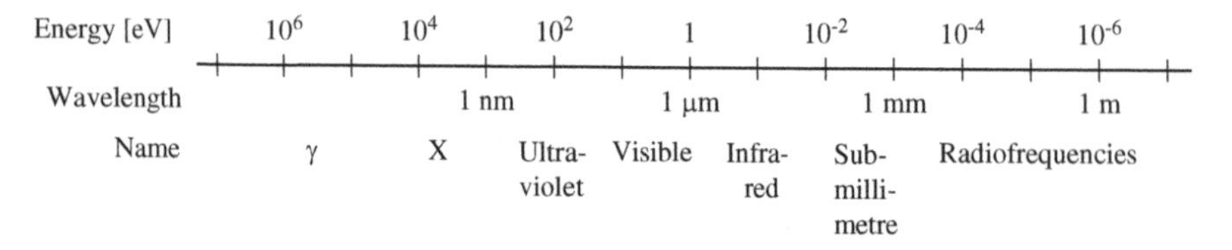

of intermediate energy in terms of their wavelength, and those of long wavelength ($\gtrsim$0.5 mm) in terms of their frequency. The term 'optical' is still widely used. It refers to wavelengths in the visible (320–700 nm) and also in the near infrared (up to about 5–12 μm), which are observable from the Earth's surface, although it is sometimes restricted to visible light.

The *propagation* of the information carried by electromagnetic waves is affected by conditions along its path: the trajectories it follows depend on the local curvature of the universe, and thus on the local distribution of matter (*gravitational lensing*); different wavelengths are affected unequally by *extinction*; hydrogen, in neutral atomic form H, absorbs all radiation below the Lyman limit (91.3 nm); and absorption and scattering by interstellar dusts are greater as the wavelength is shorter.

Interstellar plasma absorbs radio wavelengths of one kilometre or more. The inverse Compton effect raises low energy (millimetre) photons to higher energies through collisions with relativistic electrons, while γ- and X-ray photons lose energy by the direct Compton effect, inelastic scattering of light by a particle of matter. The radiation reaching the observer thus bears the imprint both of the source and of the accidents of its passage through space. It is the task of *interpretation* to separate the one from the other, in the global mix of information received.

The Earth's atmosphere presents the final obstacle between source and observer, confining observation to the narrow window of optical and radio waves until as recently as the middle of the twentieth century. Observation from space has put all wavelengths back on the same footing, the choice of observing method being determined only by the object under study, and, to some extent, economic considerations.

1.1.2 Matter: From Electrons and Nuclei to Meteorites

The Earth is incessantly bombarded by matter, whose information content should not be neglected.

Cosmic rays (so-called through an historical confusion with EM γ rays, at the beginning of the twentieth century) consist of electrons and also of atomic nuclei, from protons right up to heavy nuclei like iron, or heavier (see Fig. 1.1), most (if not all) of which originate in the high-energy processes of the Galaxy, such as supernova explosions. These charged particles interact with the galactic magnetic field, and this gives them a highly isotropic spatial distribution, to such a degree that no direction is distinguished when they are observed from the Earth. This is not at all the case at very high energies (above 10^{18} eV), where deflection is so small that particles can retain some memory of the direction to their source. In contrast, matter particles originating in the Sun when calm (*solar wind*) or from solar flares, are strongly deflected by the Earth's magnetic dipole field, and either contribute to the magnetosphere or reach the surface of the Earth.

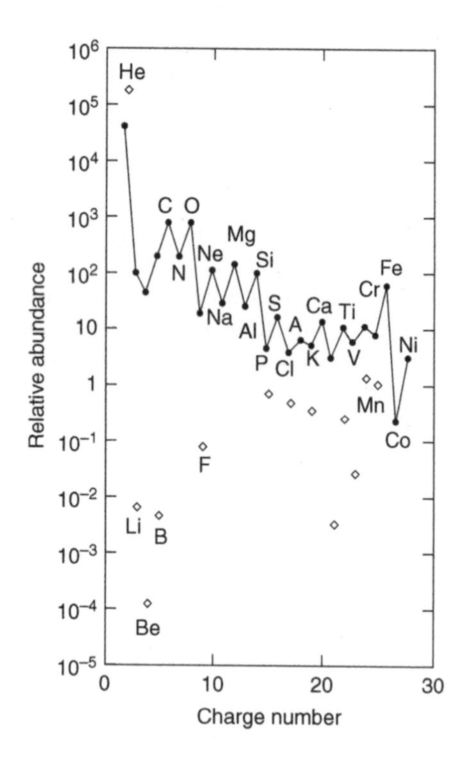

Fig. 1.1 Cosmic rays. (•) Abundance of the elements relative to silicon (Si = 100) in low-energy cosmic rays (70–280 MeV per nucleon). The composition at high energies (1–2 GeV per nucleon) is very similar. (◇) Abundance of the elements in the Solar System: only those elements whose abundances differ greatly from those in cosmic rays are shown[1]

Matter particles known as *meteorites*, ranging in size from microscopic to a mass of several tons, also rain down upon the Earth. This flow of extraterrestrial matter gives direct information about the *abundance* of the various elements at the time and place where it was produced, whether it be the present (solar wind), or the past (meteorites) of the Solar System, high-energy reactions on the surface of stars (explosive nucleosynthesis), or the early Universe (helium abundance in cosmic rays).

Space exploration is gradually allowing us to gather this information directly from samples taken outside the Earth. This sampling can now be made from spacecraft in low orbit (300–500 km) about the Earth, on the surface of the Moon or of Mars, in the atmospheres of Venus or of Saturn, and even in the nuclei of comets.

1.1.3 Neutrinos

The existence of the neutrino was first suggested by Wolfgang Pauli in 1930 to explain β decay, although Fermi provided the name, but it was not actually observed

[1]After J.A. Simpson in: *Composition and Origin of Cosmic Rays*, ed. by M. Shapiro (Reidel, Dordrecht 1983) pp.1–24. With the kind permission of D. Reidel Publishing Company.

until 1956 by Frederic Reines (1918–1998, American physicist, Nobel Prize for Physics with Martin Perl in 1995 for the detection of the neutrino and for his other work on leptons) and Cowan, who used the first nuclear reactors as source. The simplest β decay reaction is the weak decay of the neutron: $n \rightarrow p + e^- + \overline{\nu}_e$.

For the physicist, neutrinos are truly elementary particles like electrons and quarks. They belong to the set of 12 fundamental building blocks of matter. Our understanding of neutrinos has made much progress recently. It is now known that they are not massless, but have a much lower mass than the other elementary particles. The neutrino is a lepton, a class of particles which also contains the electron, the muon, and the tauon. The muon and tauon have very similar properties to the electron, but are 200 and 3 500 times more massive, respectively, and are unstable. The leptons thus fall into three families, each of which has its own neutrino: (e, ν_e), (μ, ν_μ), and (τ, ν_τ). Each family is characterised by its *flavour*. To each lepton there also corresponds an antilepton.

The Standard Model of particle physics, whose predictions are accurately measured in particle accelerators, was built on the assumption that neutrinos were massless. Over the past few years, however, experimental work on solar and atmospheric neutrinos, confirmed by terrestrial neutrino sources (nuclear reactors and accelerator beams), has shown that this hypothesis cannot be upheld. Although very small, at least 250 000 times less than the electron mass, the neutrino masses have one very important consequence for astronomical observation: neutrinos can oscillate from one species to another while propagating from source to detector. It is naturally important to take this effect into account, since it can greatly modify the detected signal with respect to the one emitted by the neutrino source.

The neutrino is, by a wide margin, the elementary particle that interacts the least with the matter it encounters on its path. Since it has no electric charge, it can only interact via the weak interaction. Out of 100 000 billion solar neutrinos passing through the Earth, only one will actually be stopped, so small is the neutrino's effective interaction cross-section. But it is precisely the rarity of neutrino interactions that so drastically complicates research into the properties of these extraordinary long-haul astronomical messengers. Like photons, neutrinos traverse spacetime geodesics in vacuum, thereby retaining information about the direction of the source which produced them and also their initial energies. Moreover, because they have a very small effective cross-section in their interaction with matter, the Universe is far more transparent to them than to photons, and they can travel very great distances. Consequently, they give direct access to very dense regions such as the cores of stars, or past states of the Universe when the temperature was greater than $T = 10^{10}$ K. Released from nuclear reactions, they can inform us directly about those reactions, e.g., solar neutrino measurements. The charged particles in cosmic rays have been observed, but not the neutrinos which they must necessarily contain. However, neutrinos from the Sun's core are observable. In addition, in 1987, during observation of the supernova 1987a of a star in the Large Magellanic Cloud, neutrinos coming from outside our Galaxy were detected for the first time.

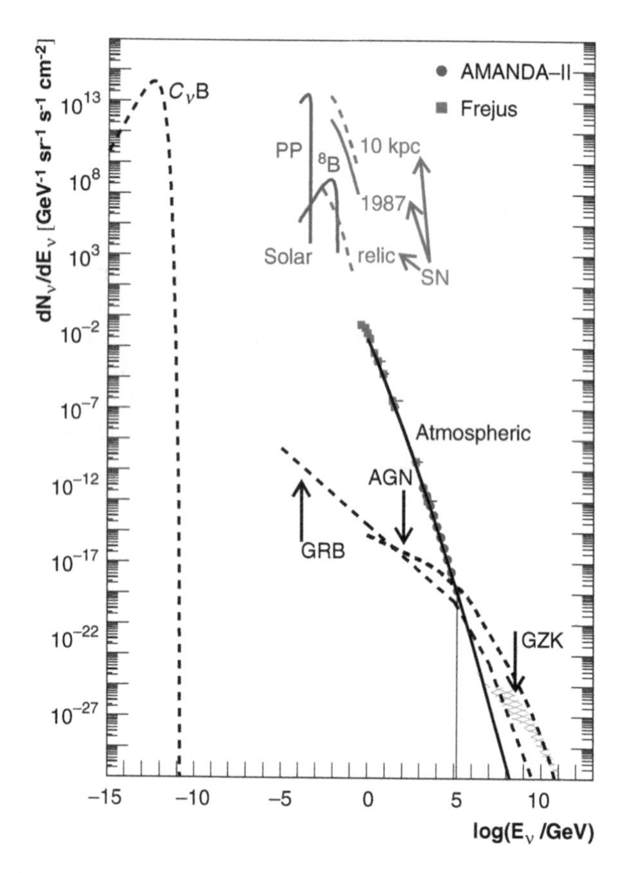

Fig. 1.2 Neutrinos reaching the Earth over a broad energy range. The number of neutrinos N_{ν} per unit energy range is plotted versus the neutrino energy E_{ν}. *Blue*: Solar origin. *Red*: Supernova contributions at 10 kpc, from the 1987 supernova, and from relic supernovas. Expected contributions from active galactic nuclei (AGN), gamma-ray bursts (GRB), interaction of cosmic rays with cosmological background photons (Greisen–Kuz'min–Zatsepin GZK), and cosmic neutrino background (CνB) are indicated. Terrestrial secondary atmospheric neutrinos limit the sensitivity of Earth-based detection: on this curve, *blue circles* are data from AMANDA (see Chap. 7) and *red squares* are from Frejus (France–Italy) detectors. Source Becker J.K., Phys. Rept. **458**, 173–248, 2008

We shall not discuss in detail the way these particles are produced in the Universe.[2] Many high-energy processes (>100 MeV) result in their emission, including de-excitation of nuclei, electron capture by nuclei, and electron–positron annihilation. The astrophysical sources are correspondingly diverse (see Fig. 1.2).

This new type of messenger thus brings important information, quite complementary to that delivered by the photon.

---

[2]See, e.g., Longair, M.S. (1981): High Energy Astrophysics (Cambridge University Press); see also AF, Sect. 4.3.

1.1.4 Gravitational Waves

If the spatial mass distribution of a system changes in time, the resulting perturbation of the gravitational field propagates at a finite speed, in conformity with the theory of relativity. The field propagates as a *gravitational wave*.

Now, an electromagnetic wave is described by a wave vector $\boldsymbol{k}$, and by a vector potential $\boldsymbol{A}$ which can have one of two *helicities* (± 1), corresponding to two opposite circular polarisations. A gravitational wave is described by a tensor $\boldsymbol{K}$ (transverse, symmetric and traceless), with helicity ± 2. This means that a circular ring of particles, in free fall parallel to the plane of the wave, will become elliptical under the effect of the wave. The wave amplitude is characterised by the dimensionless quantity

$$h(t) = \frac{\delta L}{L}$$

where $\delta L/L$ represents, in the example above, the relative deformation of the ring.

In vacuum, gravitational waves propagate at the speed of light c (according to general relativity). In matter, the dispersion and absorption of gravitational waves is negligible because of the weakness of gravity (about 10^{-38} of the strength of the electromagnetic interaction). Thus the Universe is completely transparent to gravitational waves, including regions where it is almost opaque to electromagnetic waves (for example, galactic nuclei, dense globular clusters, and compact objects such as neutron stars). The gravitational *cosmological horizon* thus lies much further back in time than the electromagnetic horizon. However, it is likely to remain inaccessible to our instruments for a long time yet, since they are not nearly sensitive enough to pick up this tiny interaction.

For observation of these waves, the relevant astronomical objects are those in which the masses and speeds are suitable to create a potentially detectable wave. Figure 1.3 summarises some of the theoretical predictions about such objects, to which we shall return in Chap. 7.

The following sources are distinguished:

- Periodic sources (binary stars, pulsars, etc.) generating quasi-sinusoidal waves of very low amplitude due to their very low frequency (10^{-4}–10^{-1} Hz). There are many such sources and they are not distant.
- Low-frequency impulsive (or burst) sources connected with the presence of black holes in massive objects (10^5–$10^9\,M_\odot$, where $M_\odot$ is the mass of the Sun), such as galactic nuclei and quasars. These events occur only at very low rates (one every ten years, on average).
- Higher frequency impulsive (or burst) sources (10–10^4 Hz). These waves are emitted in the gravitational collapse of stars of masses $M = 1$–$10\,M_\odot$ during a supernova event, when the core contracts to form a dense object. The occurrence rate is about 3 to 10 times per century in the Galaxy ($h \sim 10^{-18}$, where h is the dimensionless parameter defined above), and from 1 to 3 times per month in

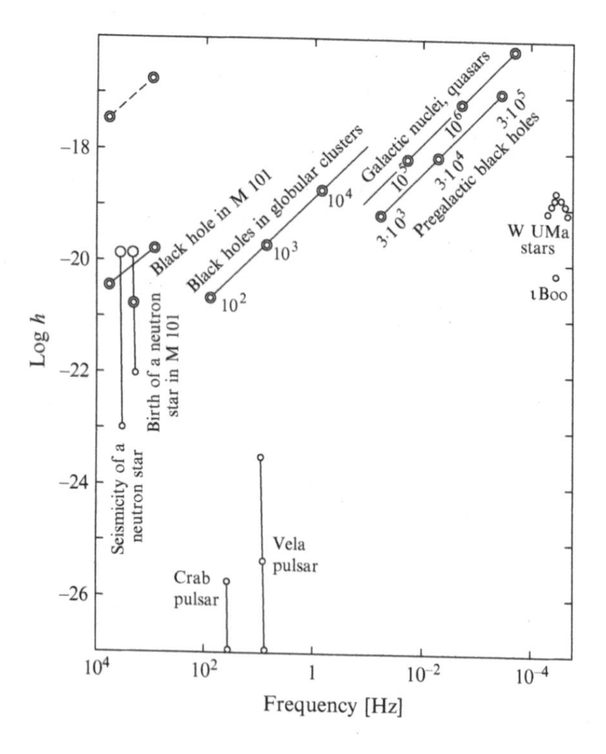

Fig. 1.3 Theoretical predictions of dimensionless amplitude h for various potential sources of gravitational waves, as a function of their frequency. Periodic sources are denoted by ∘ and burst sources by ⊚. From Brillet A., Damour T., Tourenc C., Ann. Phys. **10**, 201 (1985). With the kind permission of Editions de Physique

the Virgo cluster ($h \sim 10^{-21}$). The latter presents the most interesting prospects, being sufficiently intense, occurring sufficiently often, and promising important insights into the collapse process. It should be characterised by a fairly flat power spectrum at low frequencies, the square root of the spectral density being of the order of $10^{-23}\,\mathrm{Hz}^{-1/2}$ (see Appendix B), with a sharp spike at the oscillation frequency of the star.

Although they have only been detected indirectly (Fig. 1.4), gravitational waves are of sufficient fundamental importance for their study to constitute a new branch of astrophysics. We shall therefore discuss methods of detecting these waves in greater detail in Chap. 7.

1.1.5 In Situ Observation

Before the advent of space exploration, which began in the period between 1945 and 1959 (see Sect. 2.9), all our knowledge of the terrestrial environment, the Solar System, and the farthest confines of the Universe came from the collection of information transmitted in the form of light or of matter particles. The recently coined term *remote sensing* refers to observation of the Earth from a distance, in particular, from space, but it could just as well have been applied to the whole of astronomical observation before this period.

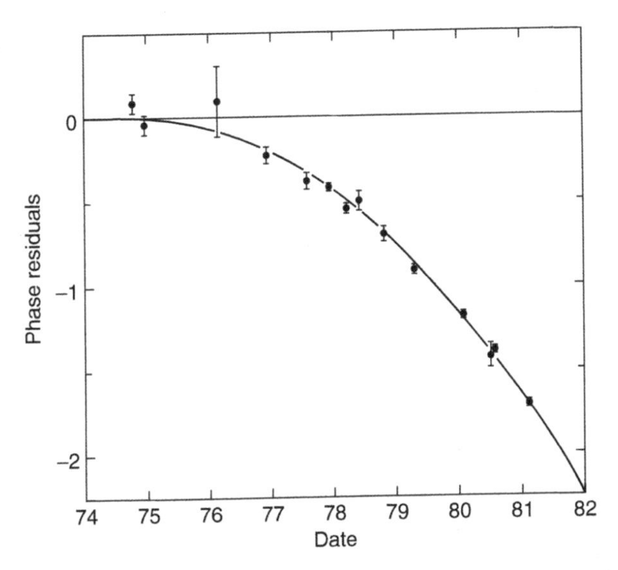

Fig. 1.4 Indirect observational proof of the existence of gravitational waves. The orbital period of the binary pulsar PSR 1913+16 decreases in time. The graph shows the variation of this period, measured between 1974 and 1982 using the emitted EM radiation (430 and 1 410 MHz). The observed values are plotted with error bars. The parabola describes the variation of period predicted by general relativity and corresponds to the emission of quadrupole gravitational waves. The timing of the object has been achieved to extraordinary accuracy (see Chap. 8). The period is $P = 59.029\,997\,929\,883(7)$ ms and its time derivative $\mathrm{d}P/\mathrm{d}t = 8.626\,29(8) \times 10^{-18}$ s s^{-1}. From Taylor J.H., Weisberg J.M., Ap. J. **253**, 908 (1982). With the kind permission of the Astrophysical Journal

Today, however, a considerable amount of information can be acquired from measurements made *in situ* right across the Solar System. Some such measurements are still remote sensing, using the close approach of a space probe to obtain detailed information or to study regions hidden from an observer on the Earth. The Voyager mission to the outer planets (in the decade 1980–1990), the Ulysses mission (1990) passing over the poles of the Sun, the Clementine mission (1994) procuring images of the whole surface of the Moon, and the Mars missions in the 2000s are good examples of remote sensing, in which the main aim was to increase angular resolution of images and pick up more energy. Most of the observation methods described in this book are relevant mainly to such missions.

But space exploration also allows *local measurements* of electric fields, magnetic fields, gravitational fields, temperature and density of ions or electrons, chemical composition, taking of samples, mineralogical and even biological analysis, and so on. A good example is the study of the atmosphere of Titan, one of Saturn's moons, using mass spectrometers on the Huygens probe (Cassini–Huygens mission, in place 2004–5).

Indeed, it has become possible to *experiment* in the same way as a physicist, a chemist, or a biologist. Consider the following examples of experiments which clearly go beyond classical remote sensing: study of the Moon's internal structure by conventional seismic methods (using explosives) or the effect of probes impacting on the surface; biological study of samples taken on Mars by the Viking probe (1976) or by the robot Opportunity (2004); local emission of radiofrequencies for radar sounding of the surface of a planet (or lidar sounding of the atmosphere) or to perturb the magnetospheric plasma.

The subject of the last two paragraphs, developing so successfully today, will not be treated here. In parallel to this book, which deals with 'astronomical remote sensing', another text could be written, devoted entirely to *methods of in situ observation*.

1.2 Data Acquisition

There is no privileged messenger in observational astrophysics: measuring the flux of neutrinos emitted from the centre of the Sun may have just as important cosmological consequences as observing the relict 2.7 K radiation. Today, astrophysics is still in essence built on the information carried by photons, but neutrinos and gravitational waves are bound to play an increasing role. The main observational strategies in astronomy revolve around the collection of this information, and it is therefore to photons that the following refers, even though much of the method could apply just as well to other information carriers, such as those mentioned above.

1.2.1 The Main Characteristics of Photons

Table 1.2 summarises the properties of all electromagnetic radiation. Each of these properties gives a certain type of information which cannot be supplied by the others, and therefore an observational strategy must be devised specifically for each.

1.2.2 Observing Systems

Figure 1.5 shows schematically the basic setup of an astronomical instrument, or rather what we call the *observing system*, defined as the arrangement of physical devices (*subsystems*) between the collection of radiation and its processing and storage.

The function of each of the subsystems, whose arrangement constitutes the observing system, can be identified in the diagram:

- The photon flux from the source is collected by a surface A, called the *aperture* of the detector, generally a mirror, the *primary mirror*. The area of aperture varies enormously, from several thousand square metres for a radiotelescope down to just a few tens of centimetres for an X-ray telescope.
- The photons are collected in a solid angle $\Delta\Omega$ called the *field of view* of the system. This also varies widely, with a value of several square degrees for a Schmidt telescope or even a few steradians for a gamma-ray telescope.

Table 1.2 Properties of photons

Photon property	Observational strategy	Chapter
Energy, wavelength, frequency	Spectral coverage	8
	Transmission through Earth atmosphere	2
	Choice of appropriate detector	7
Number of photons received (flux)	Size of collecting area (telescopes)	5
Radiation intensity	Detector sensitivity	7, 9
	Photometry	3
Time dependence and temporal coherence ($t \gtrsim 1/\nu$)	Spectral analysis Spectral resolution	8
Time dependence ($t \gg 1/\nu$)	Time resolution Rapid photometry ($t \lesssim 1$ s)	3
Spatial (angular) dependence	Mapping, imaging, spatial (angular) resolution	6
Spin	Polarimetry	3

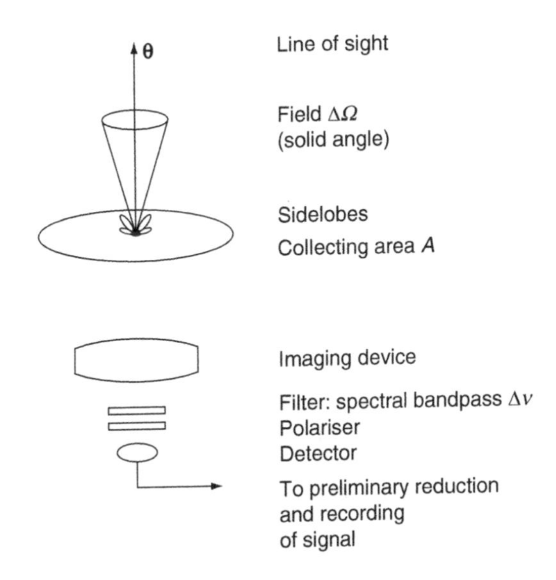

Fig. 1.5 Schematic diagram of an astronomical observing system

- There may be parasitic *sidelobes* which take energy from directions other than the principal direction of observation. Neither the principal lobe nor the sidelobes necessarily show axial symmetry about the direction of observation.
- An *optical system*, a set of several mirrors, lenses, or a combination thereof, concentrates the received energy and forms an image, usually plane, in the *image plane* or *focal plane*. The field of view $\Delta\Omega$ is thus decomposed into image elements or *pixels*, each one subtended by a solid angle $\delta\omega$ of the source.
- A device for *spectral selection* isolates a particular frequency domain $\Delta\nu$ in the incident radiation. This selection may result simply from the spectral selectivity imposed by the physical characteristics of the optical system or the detector itself,

or from deliberate filtering of the radiation. A set of different frequency domains, of width $\delta\nu$, contained within $\Delta\nu$ can be sampled at the same time, should the need arise.

- The polarisation of the radiation can be determined by means of a *polarising filter*, which selects a particular polarisation (linear or circular) in the incident radiation. Once again, certain detectors (at radiofrequencies) are only sensitive to a specific polarisation, thus combining detection and measurement of polarisation.
- The incident EM signal is transformed by the *detector* or *receiver* into a physical quantity which can be measured and stored, such as current, voltage, chemical transformation (photographic plate), and so on.
- The detector is generally followed by a set of electronic devices which make up the *acquisition system* for *analysing* and *recording* the signal. If the receiver is a photographic plate, this analysis is not done simultaneously; but if the detector produces an electric signal which can be directly manipulated (for example, a photoelectric analogue device), this analysis may be done in real time, at least partially. These days, the latter is by far the most usual case.

In the γ-ray part of the spectrum, the observing system can still select a specific direction and energy, but it is often impossible to isolate the various subsystems, as can be done in the other spectral ranges. For this reason, we shall include γ-ray observation systems in Chap. 7, alongside those for neutrinos and gravitational waves.

Starting from this basic setup, very different observing strategies are possible: the spectrum of an object can be measured with or without spatial or temporal resolution; a region can be mapped with or without sensitivity to faint photon fluxes; and so on. Even though the various aspects are separated here, for the purposes of presentation, observation never simply picks out just one of the independent parameters and characteristics of the radiation under study. The discovery of new phenomena and the detailed physical understanding of the objects studied are very much dependent on the drive to improve the performance of each property of the observing system. We shall now illustrate the development of astronomical observing systems, from a rather general standpoint, but concentrating on the main photon properties as presented in Table 1.2.

Coverage of the Electromagnetic Spectrum

Figure 1.6 shows the remarkable historical development of access to the whole electromagnetic spectrum. Although complete, this coverage was initially very variable with regard to performance in the domains of sensitivity, spectral resolution, or angular resolution. These levels of performance, which depend on the wavelength and the technology of the instruments, have nevertheless been improving constantly since 1960, both on the ground and in space. Space-based observation, which has removed the problem of absorption by the Earth's atmosphere, described in detail in Chap. 2, was a determining factor, along with the development of ever more

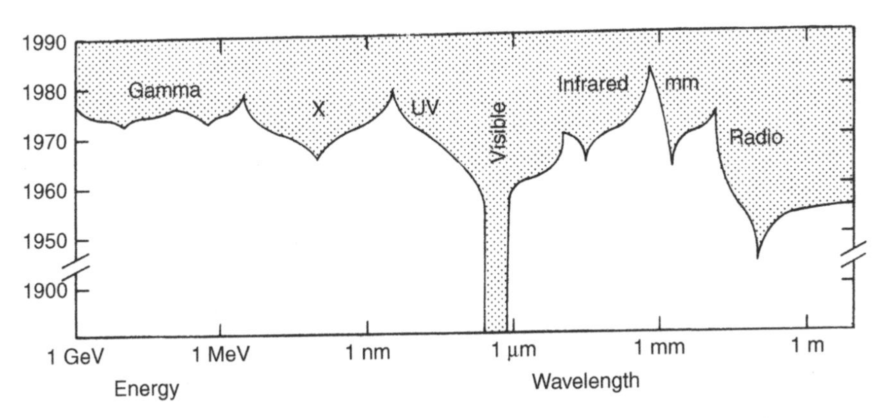

Fig. 1.6 The progressive coverage of the electromagnetic spectrum by observation. This coverage was complete up to GeV by the 1990s, and now extends to TeV

Fig. 1.7 The Galaxy viewed at different wavelengths, from radiofrequencies to γ rays. Space-based and ground-based surveys were carried out to obtain these images, which cover almost the whole sky. Courtesy of NASA-GSFC, mwmw.gsfc.nasa.gov/mmw-sci.html

sensitive and better designed detectors for the relevant wavelengths, as we shall see in Chap 7.

The importance of wide spectral coverage is to provide a full and quantitative view of the relevant object, as shown by the example in Fig. 1.7. The photons observed in the various spectral regions may be emitted from vastly different sources; these images of our Galaxy correspond to emission from very hot regions (UV), stars (visible), dust or cold gases, depending on the wavelength. Clearly, in order to make this kind of comparison, the angular resolution or the level of sensitivity of the various observations should be comparable. The synthesised view thus obtained is extremely rich and hence, for any new observational projects in space or on the Earth, every attempt is made to use *observatories* of comparable performance in terms of sensitivity, and spectral, angular, and temporal resolution, in the relevant wavelength ranges. The data are collected in a worldwide database to be used and conveniently compared in what is now known as a *virtual observatory*, discussed in Chap. 10.

Measurement of Intensity

The number of photons received per unit time depends, all things being equal, on the collecting area used. The *telescope* has thus the double function of collecting the radiation and of forming an image (angular resolution). It is mainly thermal and mechanical effects, or the maximum payload possible in space, which limit the collecting power of telescopes, a power in constant growth, historically to begin with in the visible range, as shown in Fig. 1.8, then at all wavelengths (see Fig. 1.9).

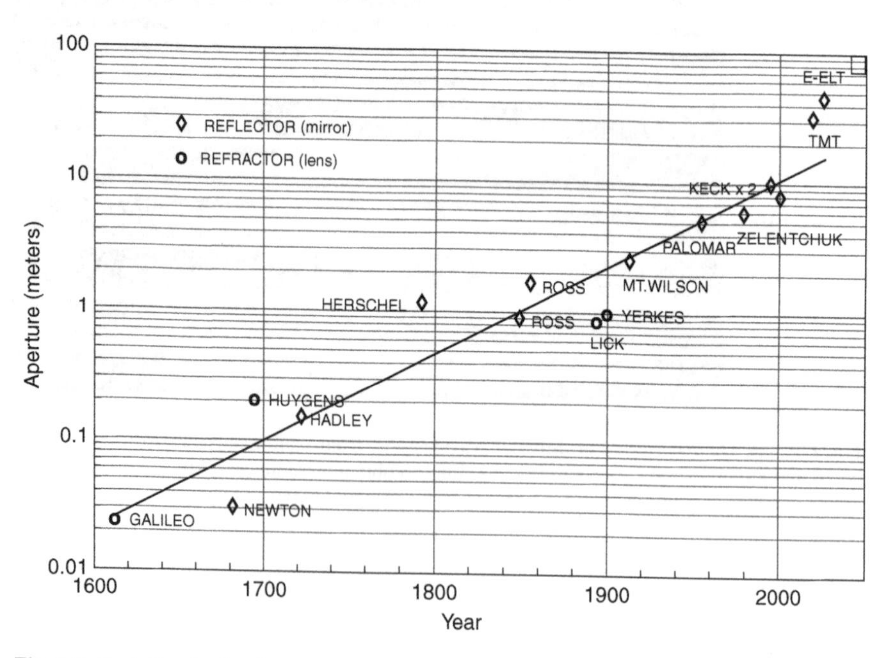

Fig. 1.8 Evolution in the diameters of astronomical instruments in the optical range. Note the gradual disappearance of refractors (○) in favour of reflectors (◇). Only the main stages have been shown here. The Keck I and II telescopes and the European Very Large Telescope (VLT) represent a whole family of instruments in the range of 8–10 m, not all shown here (see Sect. 5.2), that came into service around the year 2000. The Thirty Meter Telescope (TMT) in California and the European Extremely Large Telescope (E-ELT) are shown with projected completion dates. (Figure adapted by P. Bely)

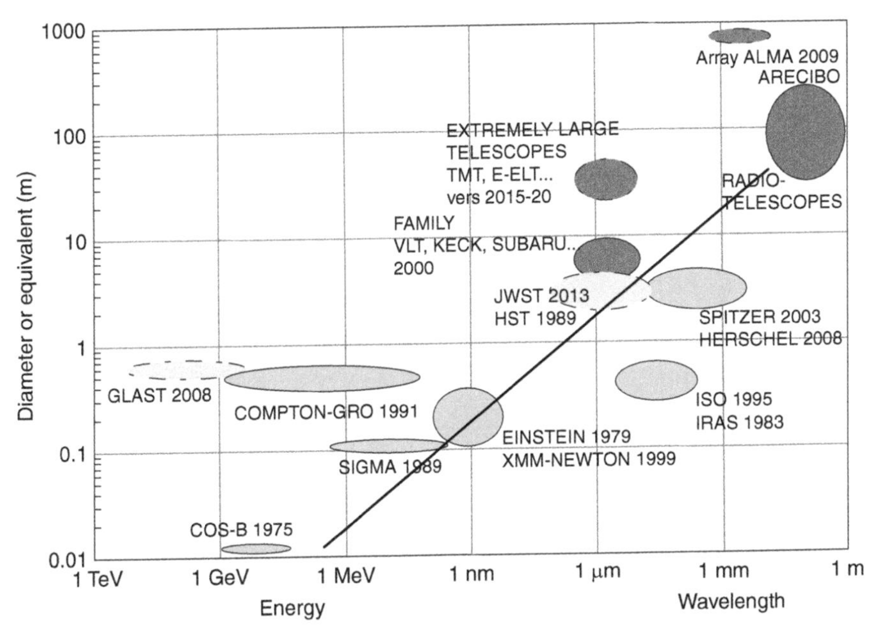

Fig. 1.9 Progress in collection capacity of electromagnetic radiation by instruments prior to 2007 (*continuous line*) or in the future (*dashed line*), on the ground (*dark grey*) and in space (*light grey*), as a function of operating wavelength. The diameters indicated on the *ordinate* are those of the instruments, or those deduced from the total area in the case of an array. The ranges shown are typical of an instrument or family of instruments, and show the evolution over several decades. Note the empirical dependence $D \propto \lambda^{1/3}$ below MeV. (See the table at the end of the book for a list of acronyms and space missions)

The received radiation intensity is determined by *photometry*, not relative to reference objects such as standard galaxies or stars, for example, but in an absolute manner and expressed in fundamental physical units. This approach to observation, essential but difficult, uses *absolute calibration techniques* (see Sect. 3.5).

The performance of the detector determines both the *precision* and the *ultimate sensitivity* which can be attained. This performance depends on the technology available, but also on fundamental physical limitations, which impose a limit on possible progress in detector sensitivity. Thus statistical fluctuations in the arrival of photons limit the photometric precision of a given instrument for a given observation time. These limitations are considered in detail in Chap. 7.

The astronomical knowledge accumulated in different civilisations — Mediterranean, but also Indian, Chinese, and Maya — using observations made exclusively with the naked eye, with no other instruments than sighting tools, is impressive. The accuracy of Tycho Brahe's visual observations of Mars, which allowed Kepler to assign an elliptical orbit to this planet, is also quite remarkable. Figure 1.10 shows the gain in sensitivity obtained over the past four centuries by the use of refracting and reflecting telescopes of ever increasing diameter.

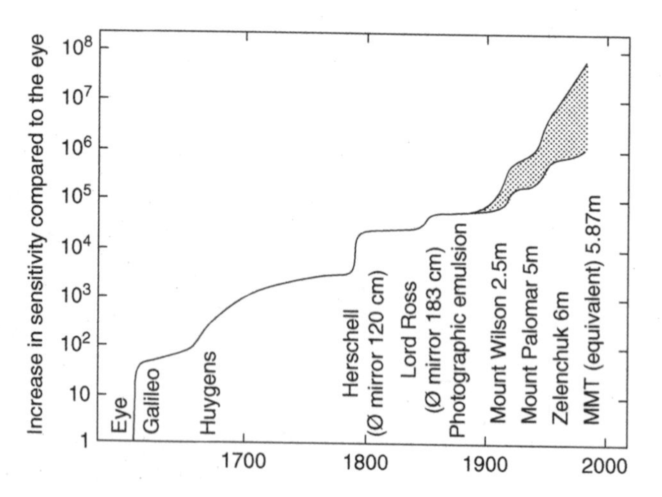

Fig. 1.10 Progress in sensitivity of detection in the visible over the last four centuries, by comparison with the eye. The *shaded area* shows the gain due to the integrating capacity of the photographic plate, and then, later, due to the quantum efficiency of the photoelectric detectors (Sect. 7.4). From Harwit M. (1981), *Cosmic Discovery*, Harvester, Brighton. With the kind permission of the publisher

The modelling and understanding of astronomical objects is greatly assisted by having access to observations made across the whole wavelength range. It is then useful to obtain uniform levels of performance with regard to sensitivity, but also spatial and angular resolution, at all frequencies of the electromagnetic spectrum. But what do we mean by *uniform* in this context? Figure 1.11 will help the reader to grasp the basic strategy wherein instruments are designed to cover the whole spectrum in a harmonious way. It is easy to appreciate the sensitivity achieved by several major instruments across the various spectral ranges in the decades 1980–2000. The *uniformity criterion* is obtained by considering the spectral distribution of a source emitting a constant number of photons per frequency interval. This *spectral energy distribution* (SED) is close to the characteristic emission of a quasar, observed over the whole of the EM spectrum.

Spectral Analysis

It is safe to say that astrophysics is built upon the achievements of spectroscopy, given the results made possible using *spectral analysis*; these concern chemical and isotopic composition, velocity fields, turbulence, temperature, pressure, magnetic fields, gravity, and innumerable others. Thus great efforts have gone into developing the techniques (see Chap. 8), as illustrated in Fig. 1.12.

The *spectral resolving power* shown in this figure is defined as the ability of an instrument to measure independently two emissions of distinct frequency. In principle it depends only on the device (*spectrograph*) used to analyse the photons.

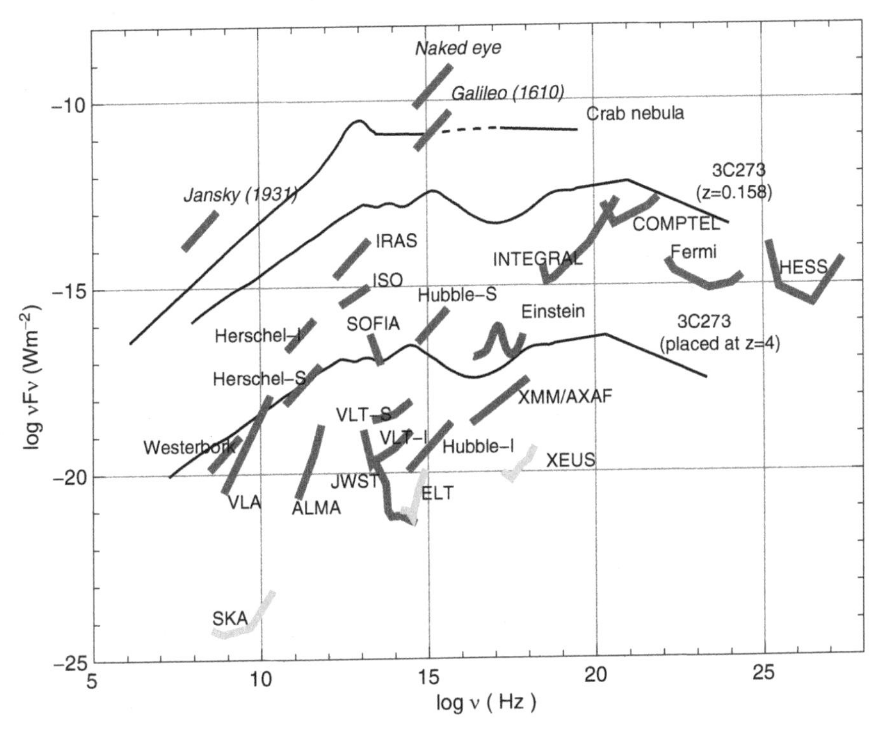

Fig. 1.11 Sensitivity of various instruments and/or observatories, covering the whole electromagnetic range, with the existing instruments for the period 1980–2010, and some others in the near future. *Ordinate*: The quantity νF_ν, F_ν being the spectral illumination. *Abcissa*: The $\log_{10}$ of the frequency ν, on the whole range of current astrophysical observations. The plotted value of νF_ν corresponds to the detectable astronomical signal by the considered instrument with the integration time (minutes, hours, or days) made possible by this instrument. Values are orders of magnitude only. Also plotted for reference are the spectrum of a supernova remnant (Crab nebula in the Galaxy), the quasar 3C273, the same quasar placed at a much larger distance (redshift $z = 4$). S = spectrographic mode (resolution $R = 10^3$ to 10^4). I = wide band imaging mode (resolution $R = 1$ to 10). For specific data on the instruments, see the table at the end of the book)

However, in practice, the ability to obtain the ultimate resolution also depends on the number of photons available, and hence the collecting area and measurement time, as well as the sensitivity of the detectors being used. In almost every spectral range, except for those at very high energies (>1 keV), spectrographs have achieved resolving powers greater than those required to analyse the relevant emissions. Indeed, the proper width of the spectral lines produced by a given object, caused by internal motions within the source and the resulting Doppler shift, also limit the useful resolution (velocity scale on the left of Fig. 1.12). In the end, it is this broadening of the lines when they are actually produced that makes it unnecessary to seek even higher spectral resolutions, except in certain very special cases, e.g., asteroseismology measurements or the search for exoplanets.

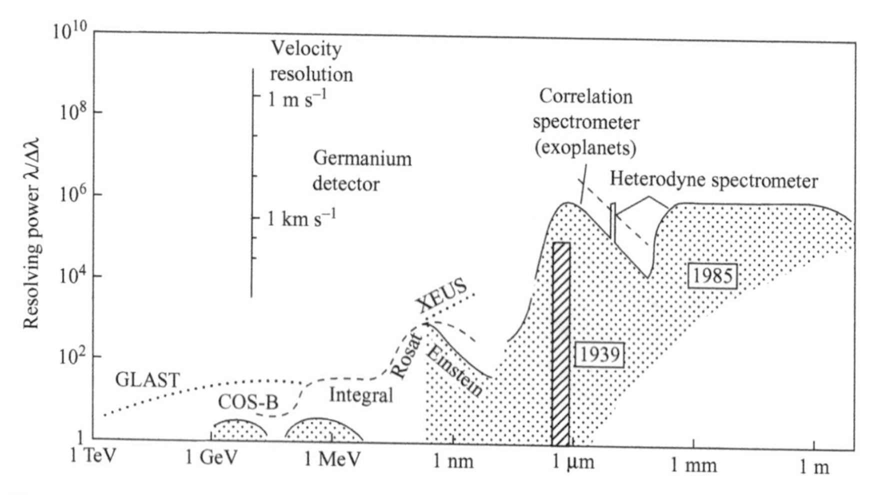

Fig. 1.12 Progress in relative spectral resolution. *Shaded*: Performance levels in 1985. *Dashed*: Performance levels in 2005. *Dotted*: Performance levels specified for future missions (2010–2020). Note the exceptional performance of the spectrometers used in the search for exoplanets (see Sect. 8.3). Updated from Harwit M. (1981), *Cosmic Discovery* Harvester, Brighton

Nowadays, progress in spectroscopy increasingly involves the ability to measure simultaneously other characteristics of the radiation:

- Rapid spectroscopy and photometry, in which successive spectra are obtained at very short intervals, sometimes less than one millisecond (solar flares, eruptive variable stars, accretion phenomena, X-ray sources).
- Spectroscopy and imaging, in which images at different wavelengths are obtained simultaneously (spectroheliograms of solar activity, X-ray mapping of the solar corona in emission lines of different excitations, maps of the hydrogen velocity distribution in a galaxy using the Doppler shift of the 21 cm line).

Time Variability

Variable stars, which have a slow time variation, have been known since ancient times (Mira Ceti), but the first *pulsar*, of period 1.377 ms, was not discovered until 1968. Since then, the study of very rapidly varying phenomena has developed as a consequence of progress in detector sensitivity. A good example is provided by γ ray bursts, discovered in the 1970s and only understood much later.

Observation of such phenomena is a rich source of information, as can be seen from Table 1.3, in which the characteristic time scales are intended as an indication of the order of magnitude.

The possibility of observing these phenomena is closely linked to detector sensitivity, since here the measurement time is externally imposed. We shall see in Chaps. 7 and 9 how sensitivity improves with increased measurement time. Figure 1.13 shows the progress in time resolution.

Table 1.3 Time variability of astronomical sources

Phenomenon	Wavelength region	Characteristic time scale or range [s]
Accretion in binary systems	All	$>10^{-1}$
Black holes	X-ray, visible	$>10^{-3}$
Gamma-ray bursts (supernovae, neutron star coalescence)	γ-ray, radio	10^{-4}–10^{1}
Interstellar scintillation	Radio, X-ray	$>10^{-3}$
Pulsars	All	10^{-3}–10
Solar flares	Visible, IR, radio	10^{-6}–10^{3}
Variable stars	All	10^{1}–10^{7}
Supernovæ	Visible	10^{2}–10^{6}
Solar and stellar oscillations	Visible, UV (gravity)	10^{2}–10^{4}
Stellar variability (convection, magnetic fields)	Radio, IR, UV, visible	10^{5}–10^{10}
Variability of galactic nuclei, quasars	All	$>10^{6}$

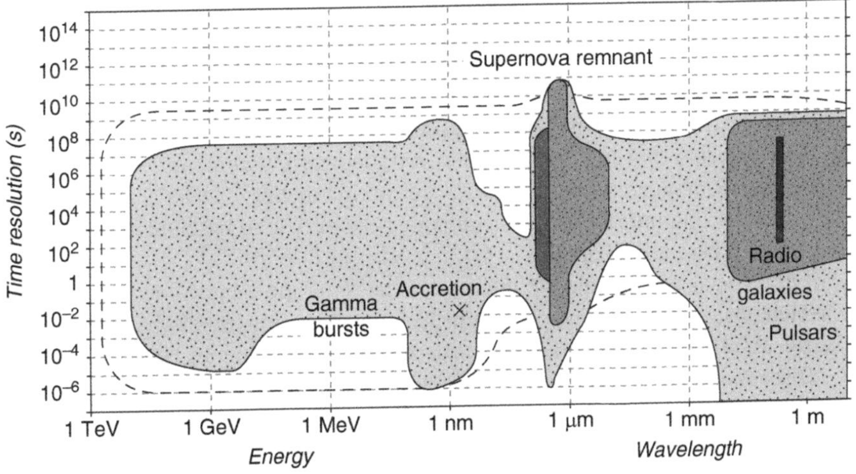

Fig. 1.13 Progress in time resolution. The region covered by observations is shown progressively *shaded* for the years 1939, 1959, 1979, and 1989. *Dashed line*: Performance achieved by 2000. The upper part of the figure corresponds to observations covering decades or centuries, such as observation of the Crab supernova (1054). Some important discoveries of variable objects are shown. From Harwit M. (1981), *Cosmic Discovery* Harvester, Brighton and 2006, personal communication

Imaging

The aim here is to distinguish between rays coming from different directions in space, and the capacity of a given observation device to do this is called its *spatial* (or *angular*) *resolving power* (see Chap. 6). In practice, any observation generally

Table 1.4 Angular resolution and imaging

Resolution	Number of pixels to cover 4π sr	Information available on this scale (2007)	Spectral region
0.1 degree	4×10^6	Cosmic microwave chart	Millimeter, submillimeter
		Sky survey	Gamma-Ray
1 arcmin	1.5×10^8	Sky survey	Infrared (10–100 μm)
1 arcsec	5.4×10^{11}	Sky survey	Visible
		Specific objects (restricted field)	mm, IR, UV, X-ray
10^{-3} arcsec	5.4×10^{17}	Specific objects	IR, visible
		Specific objects	Radio (mm, cm)
10^{-6} arcsec	5.4×10^{21}	Specific objects	Radio (cm, mm)

associates an angular resolution with a particular spectral resolution and a particular level of sensitivity, as for the other parameters already described.

Table 1.4 shows how many independent image elements (or *pixels*) would be required to cover the celestial sphere (4π steradians) at various angular resolutions.

There is a great difference in resolution between the *global sky surveys* available in the various spectral regions, varying from a few minutes of arc at radio frequencies, one second of arc or better in the visible, to at best one arc minute in the gamma-ray region. These charts are obtained by juxtaposition of successive fields of view produced by the same detector, or by pixel-by-pixel scanning at radiofrequencies. Nevertheless, a more restricted region, containing some specific object, can be imaged with a far greater resolution, better than 10^{-3} arcsec or even 10^{-6} arcsec using interferometric techniques and very long baseline interferometry (VLBI), discussed in Chap. 6.

It is crucial to be able to interpret images of the same object at different wavelengths, whence the importance of locating these images relative to each other (*justification*) and of having access to adequate reference frames (see Chap. 4).

Three factors determine the angular resolution which can be attained: first, the size of the instrument, this limit being fixed by diffraction and the wavelike nature of radiation; second, the wavelength of the radiation, as a consequence of the first; and finally, for ground-based observation, the effects of atmospheric turbulence on the formation of images. Figure 1.14 shows the progress made, but also the limits on instruments available in the period 1980–2010.

Progress in angular resolution has been rapid (Fig. 1.15), resulting mainly from the increasing size of the instruments, which is not even limited by the size of the Earth today (2007), thanks to spaceborne very long baseline interferometry, but resulting also from the development of observational tools like adaptive optics (see Sect. 6.3), which minimise the disruptive effect of the atmosphere. The latter constraint is removed by putting instruments in space, and the only remaining limit imposed on angular resolution is the number of available photons as the angular size of the observed region is reduced.

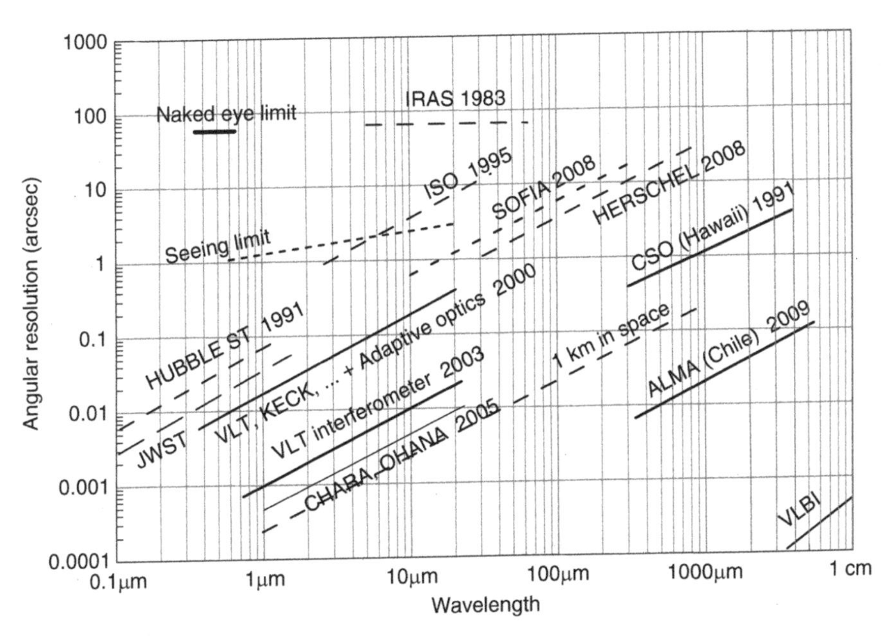

Fig. 1.14 Progress in angular resolution in the wavelength range 100 nm to 1 cm. All straight lines of unit slope correspond to the diffraction limit. The atmospheric seeing limit is indicated. *Continuous lines*: Ground-based instruments. *Dashed lines*: Space-based instruments. See the table at the end of the book for space missions, acronyms, and websites of ground-based instruments

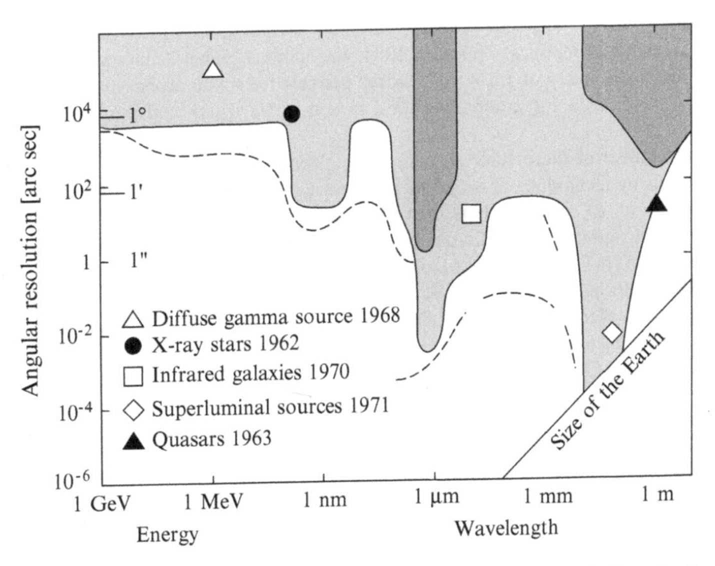

Fig. 1.15 Progress in angular resolution between 1959 and 1990 (*heavy to lighter shading*). Some important discoveries due to this progress are indicated, the resolution attained making possible an unambiguous identification of the source. From Harwit M. (1981), *Cosmic Discovery* Harvester, Brighton, completed with a *dashed line* for the period 1985–2000

Polarisation

The information carried by the polarisation of EM radiation is important. It is usually very characteristic both of the physical conditions relating to its emission, such as scattering, presence of magnetic fields, or bremsstrahlung; and physical conditions encountered by the radiation on its path, for example, the presence of a macroscopically oriented anisotropic medium, such as interstellar grains.

Measurement of polarisation, or *polarimetry*, is briefly discussed in Sect. 3.1. It is sometimes an intrinsic feature of the detector (radiofrequencies, Compton telescope, pair creation telescope).

Space–Time Reference Frames

It is essential to be able to refer astronomical events to some *space–time reference system* in order to be able to use correctly the information transported by photons. Even until quite recently, knowledge of the Earth's motion was the basis for temporal reference. *Location* of objects in some well-defined spatial frame is also essential, whether it be to chart out their proper motion over some period of time, or to accurately locate images of a single object observed in several regions of the spectrum. Chapter 4 deals with the way these frames are determined. The aim is to possess a frame as accurate as the best available angular resolutions, which is to say 0.001 arcsec, and this at whatever the wavelength.

Catalogues are available, giving the positions and proper motions of stars for a relatively dense coverage of the sky and with the required accuracy. The FK5 catalogue, adopted by the International Astronomical Union in 1984, maintains a global accuracy of $0.02''$ for stellar positions, and an accuracy of 1.5×10^{-3} arcsec per year for stellar proper motions (the two levels of accuracy therefore being compatible for a period of 20 years), and this for several thousand stars.

A small number of these stars also emit in the centimetre range, which means that they can be observed using techniques of very long baseline radio interferometry (VLBI, see Sect. 6.5.1), to an accuracy of 10^{-4} arcsec. It is then possible to relate the frames used in the visible, IR, and radio wavelengths. The Aladin database (Centre de Données Astronomiques de Strasbourg, France), discussed in Chap. 10, is used to cross-identify objects known by their position at a certain wavelength and locate them on Schmidt reference plates. Observing from space eliminates the perturbing effects of the Earth's atmosphere (seeing disks) and allows important gains in accuracy, such as those obtained by the Hipparcos satellite (1989–1993) and expected from the Gaia mission. This is discussed in Chap. 4.

Processing and Storage of Information

Astrophysical information is taken here to mean *data*, gathered across the whole range of observed sources. This data, whilst always being susceptible to improvement as regards accuracy, has great *historical* value, in the sense that any variation in the course of time is itself useful information: the change in position of a star may

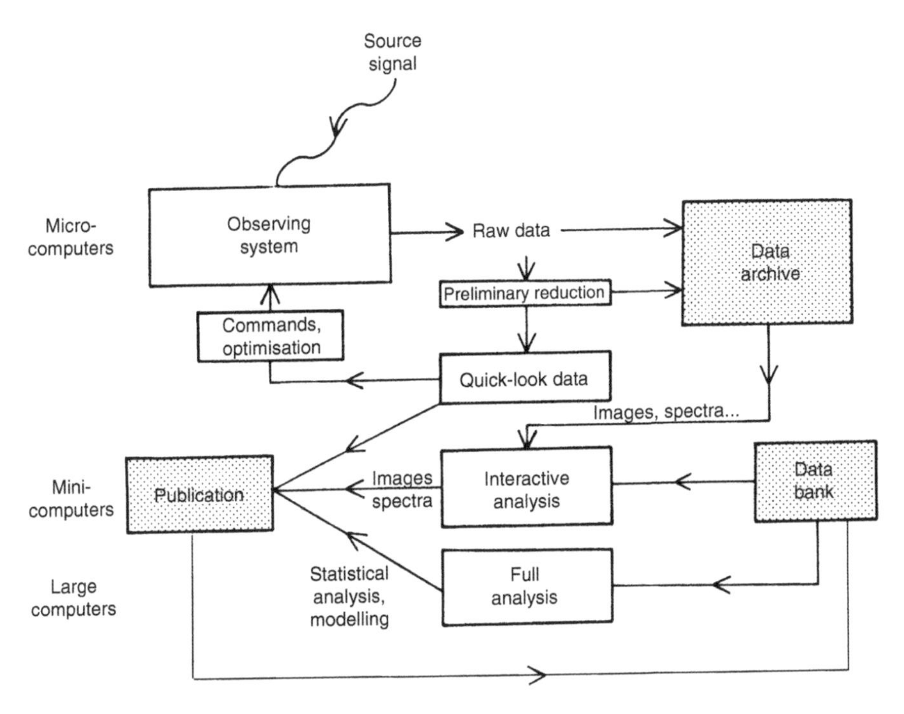

Fig. 1.16 Acquisition and processing of astronomical data

reveal the presence of an unseen companion, the variability of a quasar characterises the physical processes in its nucleus, the identification of the star which has become a supernova is essential, and so on.

Figure 1.16 shows the different stages in the processing of astronomical data, as well as the role played by computer systems.

The increase in *volume* of data gathered is enormous. About 10^4 stars visible to the naked eye had been catalogued by the seventeenth century, compared with around 10^7 in 2007 (position, magnitude, and colour, etc.), while the number of galaxies now catalogued is much higher.

More examples of this increase can be given. In space radioastronomy, a single satellite designed to study solar variability and solar wind produced 10^8 bits per year in 1972 and 10^{10} bits per year in 1982. Between 1970 and 1985, the production of radioastronomical data from space increased ten-fold every five years! The satellite International Ultraviolet Explorer (IUE), designed to study ultraviolet emission from stars, produces 2 000 spectra per year, i.e., around 10^{10} bits per year. The Cassini–Huygens mission to Saturn, which finally arrived in 2004, produces around 10^{10} bits per month.

In addition, there is a considerable volume of data, acquired before about 1980 on photographic plates, which is currently being digitised, and is thus becoming easier to access (see Chap. 10). Digitising all these photographic plates with a resolution of 10 arcsec will yield about 10^{11} bits of information.

The *acquisition of information in real time* is achieved by coupling directly to the measurement system (telescope, spectrographs, detectors, and so on). The so-called *quick look* optimises an observation in order, for example, to choose the integration time required to attain the desired signal-to-noise ratio, as will be exemplified by several cases of actual observations in Chap. 9 (see Sect. 9.3).

The *acquisition rate* can vary widely, from the detection of individual photons from a weak source (one or two per hour or per day), to the production of instantaneous images using adaptive optics in the visible, with an imaging spectrophotometer capable of recording 10^8 or 10^9 bits per second.

Real-time data handling (data compression and filtering) is often included as an early stage of the acquisition process, reducing the volume of the *raw data* and facilitating permanent storage.

Since the beginning of the 1990s, *data compression* has been developing. Its aim is to reduce the volume of transmitted data without significant loss of information, using some other available information concerning the object. A simple example shows what kind of gains are possible. Consider a 1000×1000 pixel CCD image of a region containing a few hundred stars; the useful fraction of the values representing the signal is only 10^{-3} or 10^{-4} and the rest need not be transmitted. Data compression techniques are particularly useful for long-distance transmission (satellite to ground, distant ground telescope to laboratory, laboratory to laboratory). The compression factor varies from a small multiple of 1 to a small multiple of 10 depending on the type of data transmitted (see Sect. 9.1.3).

The rate of production of raw or reduced data is increasing rapidly. The information output of the Hubble Space Telescope (launched in 1989) is 4×10^3 bits per second for 80% of the time and 10^6 bits per second for 20% of the time, which means around 10^{13} bits per year. The output of space probe missions observing the Earth or other planets is of the same order. To put this into perspective, the storage capacity of an *optical disk* is 10^{11} bits in 2007.

Subsequent analysis is generally *interactive* in the sense that it involves some interaction between scientist, computer, and data. This interaction includes, for example, selection of the best data, *optimal filtering* of noise (see Chap. 9), various corrections for properties of the observing system (variations in sensitivity, atmospheric perturbations, pointing drift, and others), and calibration by comparison with the standard sources (see Sect. 3.5).

In the mid-1990s, this interactive processing is handled by work stations, usually UNIX. Based on 32 or 64 bit systems (RISC systems), their operating speed can reach several tens of Mflops (number of floating point operations per second). The power of these work stations is increased by increasing their clock speed, which can reach several GHz, and by the advent of parallel computing.

There are at least three categories of data-processing software: first, a set of algorithms for thematic *extraction* and *analysis* of information, exemplified by IRAF, MIDAS, IDL, AIPS (radioastronomy), and so on; second, a set of programs, often called *calibration* programs, close to the instruments and designed to optimise the working of some spectrometer, some camera, and so on; and third, *viewing* programs to simplify the interface between calculation tools and the user, allowing, for example, the three-dimensional manipulation of objects.

Processed data includes error estimates, that is to say, the signal-to-noise ratio after processing, without which it would be difficult to use! This data can be relayed to the scientific community, either directly as a publication, or by making it available as part of a *data bank*.

Accordingly, for each year of the decade 1980–1990, around 50 000 stars have found their way into some publication through one or other of their properties (spectrum, position, photometry, etc.). To give a further example, during the period 1976–1990, the IUE satellite produced thousands of spectra. Similarly, observations from IRAS (InfraRed Astronomical Satellite, 1983) led to publication of a catalogue (*Point Source Catalogue*) featuring 250 000 objects, and each of those in terms of twelve different attributes (position, photometry in four spectral ranges, etc.). In Chap. 10, we shall investigate some of the consequences of this data explosion, which has continued to progress during the decade up to 2010.

It is crucial to address the question of *standardising* these data banks. Indeed, every observing device (X-ray satellite, Schmidt telescope, aperture-synthesis radiotelescope, and so on) produces data in its own form. A single physically homogeneous source will thus appear under different names in the various data bases. The data will differ in field of view (from one arcmin down to 10^{-2} arcsec, or even less), as well as in their precision. Thus, a cross-identification catalogue (Aladin) has been made available in the Centre de Données Astronomiques de Strasbourg, France, for the 550 000 objects in its data base; cross-references can be made between all the recorded properties of a given object. The advent of space observatories with very high data output, in the decade 1990–2000 (Hubble Space Telescope, X-ray, γ-ray, and infrared observatories, and others), has made it essential to create a global astronomical data bank, described as a *virtual observatory* (see Sect. 10.3).

Users link in to data bases thanks to the very rapid development of networks since the 1990s: local connections within a site or campus, operating under Internet protocol, have transmission rates of 1–10 Gbits s^{-1}; those between sites on a nationwide scale reach 2.5 Gbits s^{-1}; and intercontinental links reach 10 Gbits s^{-1} in 2007 (research networks). Compare this with the two transatlantic lines of capacity 2 Mbits s^{-1} which linked Europe to the American continent in the 1990s.

The last stage is the *detailed treatment*, in which a large volume of data or properties of sources is extracted from the data bank to be handled by powerful machines (work stations), with a view to revealing any statistical features, or for sophisticated physical modelling.

In the mid-1990s, machines like the CRAY C98 possessed 8 processors of 2 Gflops, 4 Gbytes of central memory and more than 100 Gbytes of disk space. (The *floating point operation* or *flop* corresponds to one elementary operation such as an addition or multiplication.) The highly parallel *Connexion Machine* included 32 000 processors. In 2007, these values would only be worthy of an individual processing station. Computational capacities of several tens of teraflops (10^{12} flops) can be obtained by interconnecting machines distributed over several sites. For example, Europe has a capacity of 30 Tflops by combining IDRIS (CNRS, France), CINECA (Italy), and Garching (Max-Planck Gesellschaft, Germany).

All the above operations take place in the fast-changing context of computing and the associated algorithms.

1.2.3 Reaching a Systematic Description of Observation

The great difficulty in presenting the physical methods which underlie observation in astronomy is the high degree of interrelationship between all the elements of the observing system, the properties of the incident radiation and those of the source being studied. The sensitivity and imaging capabilities of the detector, the size of collector, spectral properties of materials used to reflect, transmit or detect the radiation are some of the many inseparable factors. We have tried to carry out this analysis, and will continue to do so in the rest of the book.

Before reaching the observing system, the electromagnetic wave leaving the source may have undergone perturbation. The effects of passage through the Earth's atmosphere pose particular problems, which deserve a special treatment in Chap. 2. Observation conditions at ground level or in space are compared (*The Earth's Atmosphere and Space*).

A few general properties of EM radiation and its measurement are then presented in Chap. 3 (*Radiation and Photometry*). Chapter 4 (*Space–Time Reference Frames*) shows how astronomical observation can be related to space and time reference frames, the only way to bestow it with a proper physical meaning and to compare measurements. Chapter 5 (*Telescopes*) deals with energy gathering and image formation by telescopes, but discusses only the relevant features of geometric optics, while Chap. 6 (*Image Formation and Diffraction*) goes into the analysis of images and the kind of resolutions that can be obtained, which requires some understanding of the coherence of radiation. Chapter 7 (*Detectors*) investigates the physical principles underlying the detectors placed at the focal points of telescopes and describes several applications. The principles of spectroscopy and spectrometer families are discussed in Chap. 8 (*Spectral Analysis*). The development of computers has led to considerable progress in the processing and storage of data, and this is presented in Chap. 9 (*The Signal in Astronomy*). The capacity of today's observational tools and the constitution of huge data bases has led to the idea of a *virtual observatory*, and all this is discussed in Chap. 10 (*Large-Scale Surveys and Virtual Observatories*).

Finally, Appendices A and B consist of mathematical supplements that will be useful for the book as a whole, one on the *Fourier Transform* and the other on *Random Processes and Variables*.

1.3 Global Organisation of Astronomy

In the earliest civilisations, astronomical knowledge was a valuable commodity, around the Mediterranean, along the silk roads, between India and the West, and between the Arabic and Latin worlds. Since the Renaissance, astronomers have built up a genuine international community which, with the globalisation of trade and information, is ready to tackle the twenty-first century with enthusiasm, inspired by the volley of discoveries recently brought by progress with instrumentation in the twentieth century, combined with the ground-breaking interpretive tools provided by modern physics. The life and development of this community rests primarily on its members, but also on research policies, communication of results through publications, and the sharing of accumulated knowledge across the whole of society.

The first three of these four points will be discussed briefly here, as they appear just before the beginning of 2009, celebrated by the United Nations Organisation as World Astronomy Year.[3]

1.3.1 People

Research activities require personnel with a wide range of skills and knowhow, including researchers, engineers, technicians, and administrators. These personnel are usually employed by research institutes or universities depending on state financing. An interesting figure is the ratio of supporting staff to researchers, which is of the order of 1 or 2 in a laboratory, but may reach a value of 10 in an operational observatory. There is a wide range of different lines of work, even among those who could be described as astronomers. While training in physics was primordial up to the middle of the twentieth century, several factors have transformed the traditional profile of the astronomer, including the renewal of celestial mechanics, the birth of astrochemistry and astrobiology (also called bioastronomy), the increased role of signal processing accompanying the generalised digitisation of data, and the technological complexity of contemporary instrumentation. It is also the complexity of the technology involved in today's observational equipment that has led to an unprecedented collaboration with engineers and industry in the fields of optics, electronics, computing, robotics, and so on.

Figure 1.17 shows the geographical spread of International Astronomical Union members across the world. Table 1.5 complements the figure by specifying the size of the astronomical community and the number of young people preparing doctorates in this field in several different countries, in particular in Europe. The ratios of doctoral students to researchers illustrate the rather different policies implemented in different countries. In countries like France, the number of doctoral students is more or less adapted to suit the number of professional posts available for them later on. Other countries like Germany aim to produce a large number of doctoral students, thereby populating their research institutes with a young, creative, and relatively cheap work force, able to take on the huge availability of observational data and the wide variety of interesting research subjects.

It would be a mistake to end this quick overview without mentioning *amateur astronomers*. In every country, these enthusiasts build their own equipment, often quite sophisticated, sometimes following the professional literature, or even contributing to some original piece of research through programmes requiring time and availability: observations of variable stars, meteorites, or comets, or even transits

[3]Three outstanding works provide an in-depth panorama of the four points mentioned: Woltjer, L., *Europe's Quest for the Universe*, EDP Sciences, Paris (2006); *Organizations and Strategies in Astronomy*, Heck, A., Ed., Vol.7, Springer (2006); *Future Professional Communication in Astronomy*, Heck, A. & Houziaux, L., Eds., Académie royale de Belgique, xxviii, 2047 (2007), from which several parts of the following discussion have been extracted.

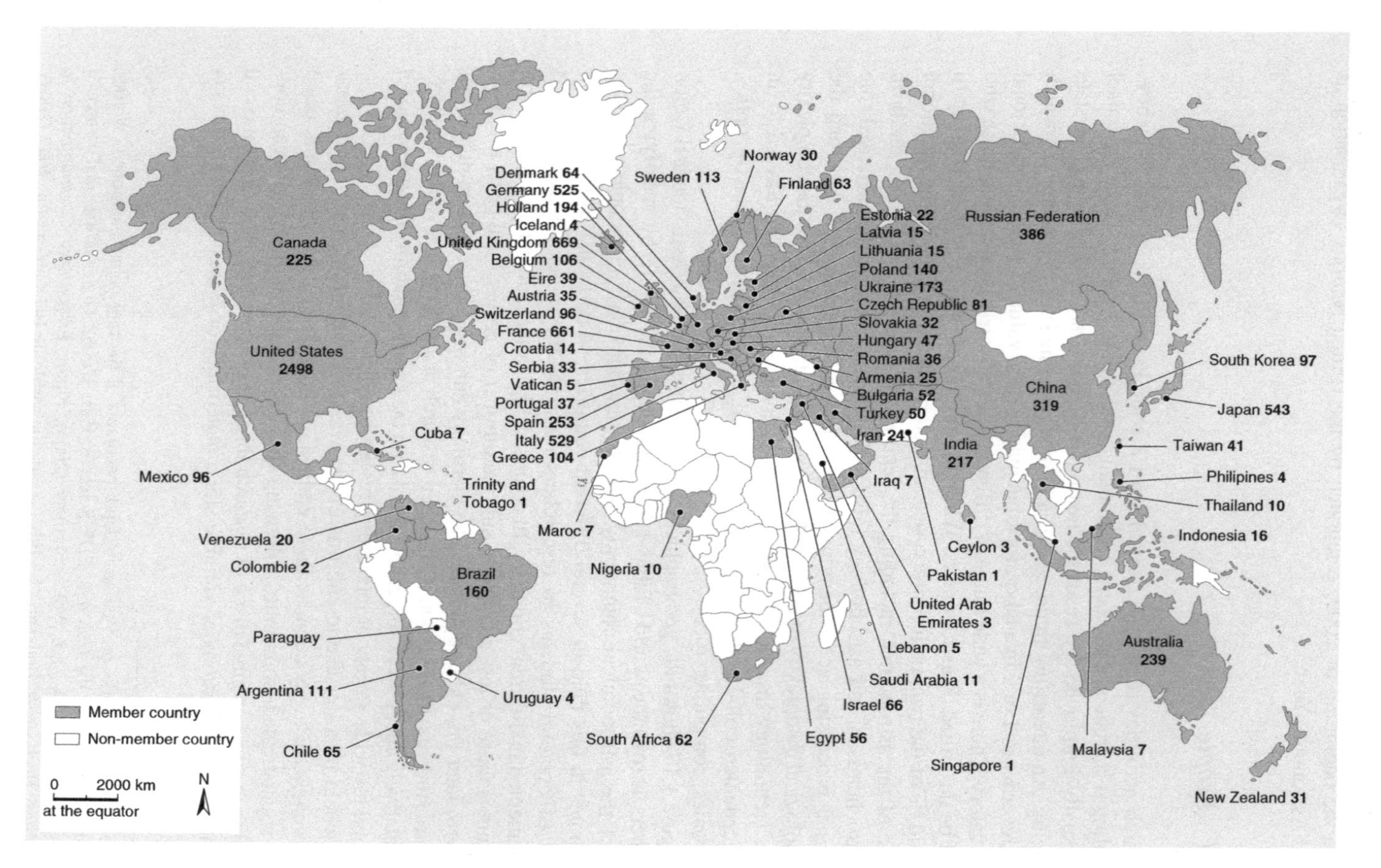

Fig. 1.17 Geographical distribution of International Astronomical Union members in 2007. The name of each country is followed by the number of members, if there are more than five. *Shaded countries* belong to the Union, while there may be private members in others. Kindly provided by the International Astronomical Union (UAI), www.iau.org

Table 1.5 Research scientists working in astronomy around the world. Columns: (**1**) Country. (**2**) Number of researchers in astronomy. (**3**) Number of young people preparing a doctorate related in some way to astronomy. (**4**) Fraction of these researchers in relation to the whole research community of the country. (**5**) Average number of pages published per year and per researcher in astronomy. (**6**) Number of astronomers per million inhabitants in the country. (**7**) Average growth in the number of members of the International Astronomical Union over the period 1992–2004. UE* European Union before 2004, to which Norway and Switzerland have been added here. (These estimates are taken from Woltjer, L., *Europe's Quest for the Universe*, EDP, Paris, 2006, with the kind permission of the author)

Country	Researchers	Doctorates	% researchers	Pages per astron.	Astronomers per 10^6	Δ IAU (%)
France	800	200	0.65	4.0	17	1.0
Germany	850	500	0.58	4.0	16	1.1
Italy	950	100	1.39	4.4	18	1.4
UK	800	500	0.91	6.2	22	1.6
Austria	50	50	0.78	2.8	12	1.3
Belgium	(191)		(0.83)	(3.2)	(19)	1.9
Denmark	45	13	0.35	5.7	11	1.4
Finland	75	46	0.72	4.5	23	5.3
Greece	130	30	2.0	2.8	15	1.4
Iceland	4	1	0.44	4.2	18	2.4
Eire	32	28	0.72	1.8	16	2.5
Holland	125	80	0.65	9.0	13	0.9
Norway	(42)		(0.26)	(2.9)	(9)	0.8
Portugal	(65)		0.56	2.7	7	6.7
Spain	315	133	0.86	4.1	11	3.6
Sweden	100	70	0.50	5.0	19	2.1
Switzerland	100	55	0.72	5.2	14	4.1
UE*	4 200	1 900	0.71	4.9	16	1.6
US	4 500	(900)	0.56	5.6	20	1.4

of exoplanets, these being tasks that do not necessarily need very large modern telescopes, but require a great deal of tenacity and care. The contribution made to science by these amateurs is far from being negligible, and they often provide the bridge between professional science and the general public, mentioned again below. As an example, in 2006, the United Kingdom numbered some 10 000 amateur astronomers.

1.3.2 Research Policies and Institutions

Most institutes involved in research depend on public financing, but sometimes also private initiative, with a broad range of different types of organisation: dedicated research institutions like the *Max-Planck Gesellschaft* in Germany and the *Centre national de la recherche scientifique* (CNRS) in France, academies of science based on the Soviet model in Russia and China, and state-run or sometimes private universities like the *Pontificia Universidad Católica* in Chile and the *California Institute of Technology* in the United States. Whereas at the beginning of the twentieth century, observing instruments were generally located close to urban centers or were themselves research centers, e.g., Meudon in France, Yerkes in Chicago, observatories rapidly moved away to better, high altitude sites (see Sect. 2.8), while maintaining their links with one or more research centers.

The recent trend toward very large observing systems, both on the ground and in space, has required international collaborations for their construction, e.g., in Europe (European Space Agency founded in 1975, European Southern Observatory founded in 1962) or even worldwide (ALMA submillimeter telescope in Chile). This kind of collaboration concerns not only observatories in the strict sense of the term, but also the wide range of costly instrumentation required to equip them, both on the ground and in space.

Observational astrophysics requires:

- Observing apparatus, telescopes, on the ground or in space.
- Space probes, exploring the Solar System *in situ* (the Moon, the planets, planetary satellites such as Titan, asteroids and comets) and sometimes bringing samples back to Earth.
- Sophisticated instrumentation working in relation with these telescopes and probes: cameras, detectors, spectrographs, and the like.
- The means of processing the data: computers, data bases, etc.
- The means to place equipment in space, including rockets, launch bases, tracking antennas, and so on, shared by all those involved in the space adventure.

A whole range of technology, including optoelectronics, mechanics, robotics, systems science, artificial intelligence, and others, is brought into play for the various tools required.

From another angle, theoretical astrophysics uses not only the tools of mathematics, physics, and chemistry, but also large-scale numerical calculation, which demands computers of the highest level of performance in speed and storage capacity.

In all these cases, every country has a national research policy.[4] Research programmes and new ideas are examined, resources attributed, collaborations organised, and the ever more necessary international agreements negotiated. The interplay between the free initiative of individuals and institutions on the one hand and state regulations on the other, depends to some extent on political traditions, in particular the degree to which they are centralised, and the wealth of the given

[4] See, for example, Paul Murdin's contribution to *Organizations and Strategies in Astronomy*, Heck, A., Ed., Vol.7, Springer (2006) for a description of the state of astronomy in the United Kingdom, with an interesting historical perspective.

state. But since astronomy is a scientific discipline which has always had the goal of exploration and understanding, rather than producing immediately useful applications, the role of the state institutions tends to be strengthened here.[5]

How can one keep up with the accelerating rate of technological change by investing in observing systems and computers? What collaborations should be advocated? What new, unexplored areas should be charted out (gravitational waves, bioastronomy, etc.)? In most developed countries, these questions stimulate regular publication of *prospect reports* which, after broad consultations on the national and international level, provide a regular update of the orientations to be adopted in the years and decades to come: *Denkschrift für Astronomie* in Germany, *Rapports de prospective* by the *Institut national des sciences de l'univers* (INSU) in France, white papers or analogous documents elsewhere in Europe, the *US Decade Report* in the United States, and the *Megascience Forum* for the OECD, etc.

The European investment in astronomical research can be illustrated by some figures. Within Europe, astronomers have organised collaborative efforts between countries, mainly in the context of the European Space Agency (ESA) and the European Southern Observatory (ESO). Direct or indirect spending on the projects and instruments carried by these two institutions represents some 42% of European expenditure in astronomy.[6] We see here the need for broad collaboration in order to procure ever more ambitious and sophisticated means of observation.

The unit costs of equipment and projects rise steadily with increasing size and complexity. Thus the cost of traditional ground-based optical telescopes was quite well described at the beginning of the 1980s by the formula

$$\text{cost (M\$ 1980)} = 0.45D^{2.6} ,$$

where D is the diameter in metres. On the other hand, technological progress leads to reduction in costs (see Chap. 5). The above power law would have implied a cost of about 350 M$ (1995) for the Keck telescope, whereas its actual cost was around 100 M$ (1995). The European Very Large Telescope VLT, opened at the end of the 1990s, cost around 260 MEuro (2004), not including the cost of developing the Paranal site in Chile, and the cost of its instrumentation and interferometric mode, all factors ignored in the above formula. However, this formula would have implied a cost of 760 M$ (1995), or around 730 MEuro (2004), rather than the actual cost of 260 MEuro. We see how the advancement of technology — segmented primary mirror for the Keck, or thin and active for the VLT — has succeeded in limiting the growth in costs and making it possible to construct these major telescope facilities.

Notwithstanding, the cost of the next generation of optical giants known as *extremely large telescopes* (ELT) is not yet firmly established in 2007. The *Thirty Meter Telescope* (TMT), with a diameter of 30 m, currently under study in the United States, is expected to cost some 400 M$ (2006), comparable to the cost of the future *European Extremely Large Telescope* which would

[5]Not everyone shares this view. For example, Jacques-E. Blamont, in his book *Le Chiffre et le songe. Histoire politique de la découverte* (O. Jacob, Paris, 1993), discerns a strong connection between the historical advancement of astronomy and the level of royal concern.

[6]This figure is given for the beginning of the decade 2000–2010, L. Woltjer, *Europe's Quest for the Universe*, EDP Sciences, Paris (2006). The definition of what constitutes Europe is somewhat fuzzy here, since it does not correspond exactly to the European Union, which is itself evolving, and since these two perimeters involve a considerable geographic diversity of members, also growing.

have a diameter of 42 m, although the feasibility of building such a thing within this budget remains to be proven.

The costs of space missions are an order of magnitude larger, although this may decrease slowly as access to space becomes more commonplace.

To give an example, the cost of the Hubble Space Telescope mission (NASA + 15% Europe) was estimated at 4 G$ (1995) for ten years of observation in orbit.

The European Space Agency (ESA) recognises three categories of mission for the period up to 2005:

- Minor missions (small Earth-orbiting satellites), costing up to 200 MEuro.
- Major missions, such as observatories (infrared, extreme UV, submillimetre, etc.), costing around 700 MEuro (2005).
- Missions for exploration and/or return of samples, of very high cost, such as a Mars exploration vehicle, a probe in the atmosphere of Titan (satellite of Saturn), or gathering and return of samples from comets. Any one of these missions would cost between 1 and 2 GEuro (2005), and they are only conceivable in a context of international cooperation.

The progress in *productivity* has been spectacular, whether for ground-based instruments or for space missions. A careful choice of sites or orbits, improvement in detectors, better analysis of available photons, optimised signal processing, and expert systems improving real-time decision capability and optimising the use of available observation time, are all factors contributing to a higher yield of information per hour from the instruments used.

Unfortunately, despite the wealth of quantitative data given above, there still does not exist any reasonably sure way of quantifying the yield in terms of discovery of a research instrument; neither the volume of publications, nor the number of citations of those publications, would be sufficient in themselves for this purpose. For the moment, we are unable to evaluate the intrinsic worth of a discovery or an observation, and its longer term importance!

1.3.3 Publications

Publication is in integral part of research, and the growth in the means of observation, and the capacities of those means, is accompanied by a parallel growth in the exchange of acquired and processed information. There are two reasons why publication should be considered so important: for one thing it allows the gradual authentication of new scientific results, and for another, it plays an essential, although often excessive, role in the professional recognition and reputation of the scientists involved.[7]

In Chap. 10, we shall see how this *primary* information, so-called because it arrives almost directly from the observing system, is stored and made available

[7] See *Future Professional Communication in Astronomy*, Heck, A. & Houziaux, L., Eds., Académie royale de Belgique, XXVIII, 2047 (2007).

to potential users in the form of data bases, and in fact, mainly in the form of *virtual observatories*. Most professional publications in astrophysics are accounts of observations, models, theoretical predictions, proposal or validation of methodologies, and descriptions of instruments (see Fig. 1.18). Their volume is increasing all the time, while the means of communication between researchers is also evolving very fast, both in volume and speed, and at the same time diversifying thanks to the digital revolution underway since the 1980s. It is easy to gauge the close link maintained between publication and observation by glancing at Fig. 1.19. This

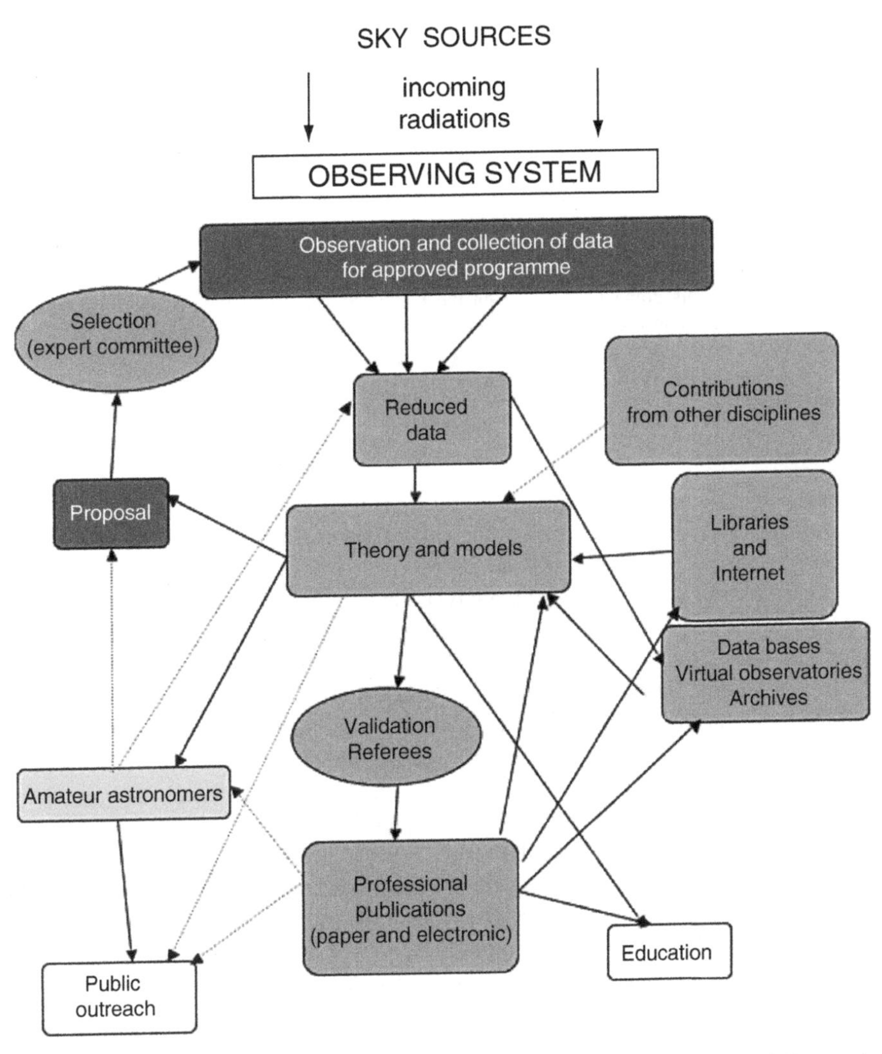

Fig. 1.18 Paths of data transfer in astronomy. From Heck, A., in *Future Professional Communication in Astronomy*, Heck, A. & Houziaux, L., Eds., Académie royale de Belgique, XXVIII, 2047 (2007), with the kind permission of the author and publisher

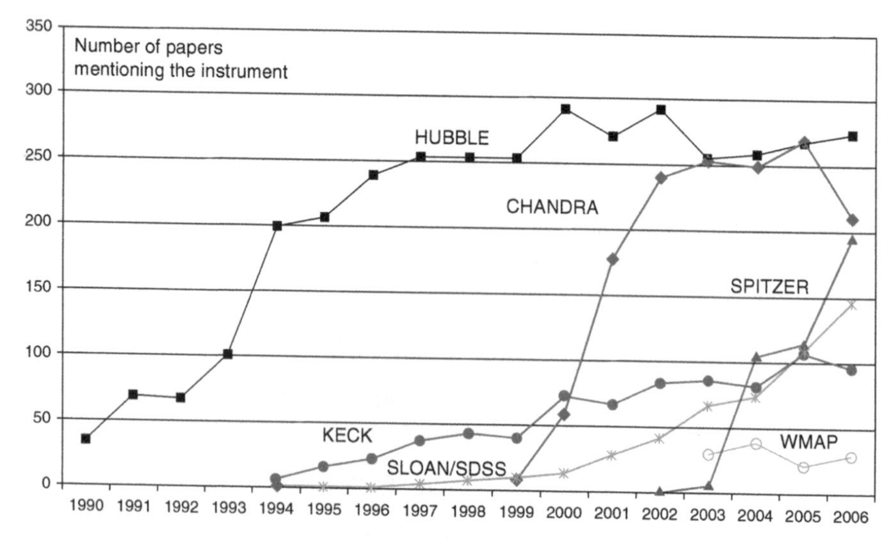

Fig. 1.19 Relation between publication and observing system. *Curves* give the number of papers in the *Astrophysical Journal* (including *Letters* and *Supplements*), whose titles mention one of the main instruments going into service in the United States during the period considered. In 2006, these papers constituted roughly a quarter of all articles published in the journal. From Burton, W.B., in *Future Professional Communication in Astronomy*, Heck, A. & Houziaux, L., Eds., Académie royale de Belgique, XXVIII, 2047 (2007), with the kind permission of the author and publisher

figure concerns one of the most important professional journals in this field, the *Astrophysical Journal*, published in the United States, and examines the *titles* of the papers published between 1990 and 2006. These titles may or may not mention a specific observing instrument: the analysis retains the main instruments set up by NASA over this period, and the figure clearly shows the impact on contemporary publications.

In this digital era, the evolution of communication tools has gradually transformed the means of communication used for research. The *lingua franca* of science is English. In addition to the well established and commercial system of professional journals, with their in principle rigorously ethical committees of referees, guaranteeing an equitable treatment for all authors, peer assessment, competence, and an open mind toward new or heterodox ideas, we now find other means of publication using digital electronic technology (Internet). While the conventional journals are gradually adapting to the paradigm of electronic publication, and can now, for example, instantaneously provide a complete listing or archives online, new tools are appearing. The latter are more flexible and aim to answer the criticisms made of the more traditional methods, e.g., editorial bias, high cost, and sluggish response.

Going by the name of *open acess*, these new approaches can be very varied, from authors publishing without validation by referees, exposing their articles to criticism under their own responsibility, to professional reviews freely circulating recent papers for some limited period, and then again after some embargo period, or in some cases allowing authors to disseminate their own manuscripts after publication. Many different possibilities have been under trial during this decade, the problem being to reconcile the economic viability of the system with a rigorous approach to validation.

One of the most significant attempts to provide semi-open access to publications has been undertaken jointly by NASA and CDS in Strasbourg since 1992. It goes by the name of the *Astrophysics Data System* (ADS). The system is accessible by Internet, and articles are registered and fully accessible upon subscription, while their title, author(s), and abstracts are freely accessible to all. This is an exceptionally useful tool for the bibliographical aspects of any kind of research in astronomy.

Needless to say, *ethical considerations* are receiving renewed attention today, in particular with regard to research publications. We may cite the opinion recently reached by an ethical committee in France[8], which recommends the following code of conduct:

- To identify and support accessible, high quality publication systems, in order to ensure as broad a dissemination of knowledge as possible.
- To raise the awareness of research scientists regarding the means available to them to communicate their results.
- To correct for abuses facilitated by dominant position or monopoly in publications.
- To find ways to counterbalance the predominance of English in systems for disseminating knowledge, in particular by favouring bilingualism, and in some disciplines, computer assisted translation.
- To raise the awareness of ethical responsibility in those generating knowledge. The act of publication must be reasoned, and the contribution appropriate, balanced, and justified by more than just the career prospects of the authors or their status with regard to international competition.
- To base evaluative judgements on a range of indicators, taking into account the originality and inventivity of results, and never just purely quantitative criteria such as impact factors of reviews or citation indexes.

We shall not discuss the last point of these recommendations, that is, ways of measuring the impact of a publication, an author, or an institution through *citations*. It is a vast and important subject that goes beyond the scope of this book.

[8]Comité d'éthique of the CNRS (COMETS, March 2007), www.cnrs.fr/fr/organisme/ethique/comets; see also the *Universal Code of Ethics for Scientists* published by the UK Council for Science and Technology

Chapter 2
The Earth Atmosphere and Space

The Earth's atmosphere has always acted as a screen between the observer and the rest of the Universe. The pre-Copernicans regarded it as the seat of the volatile elements because of its mobility, separating as it did the sublunar world from the world of the stars. From the time of Galileo, and up until the conquest of space, observations of photons were limited to the narrow window of the visible, and this range was extended only recently by the addition of radio frequencies.[1] Despite the recent development of observation from space, ground-based observation retains considerable advantages in terms of both access and cost. The global strategy of observational astronomy therefore requires an exact knowledge of the properties of the Earth's atmosphere. With such a knowledge, the potential or the limits of ground-based observation can be defined, and, for each wavelength of the spectrum, the best altitude can be determined and the best sites chosen for new instruments. The choice of site is crucial. Many factors must be taken into account, and we shall describe them here. The Antarctic continent is now accessible for astronomy and will no doubt provide many important opportunities.

In this chapter, we examine one by one the physical properties and composition of the atmosphere, which lead to its *opacity*, that is, to its capacity to absorb radiation. *Atmospheric scattering* prevents daytime observation in the visible, and causes light pollution at night. *Refraction* and *dispersion* deviate the apparent direction of a star chromatically, i.e., in a wavelength-dependent way, from its true direction. The *thermal emission* of the atmosphere perturbs infrared and millimetre observation during both the day and the night. *Atmospheric turbulence* degrades images and causes phase fluctuations which affect the working of telescopes and interferometers. *Ionisation* of the upper atmosphere creates a plasma which modifies the propagation of radio waves. Furthermore, all these phenomena vary with time and, for the most part, are strongly dependent on geographical location.

---

[1]The first radioastronomical observation was made by American astronomer Karl Jansky (1905–1950), who observed the Sun in 1933, using a telecommunications antenna.

P. Léna et al., *Observational Astrophysics*, Astronomy and Astrophysics Library,
DOI 10.1007/978-3-642-21815-6_2, © Springer-Verlag Berlin Heidelberg 2012

During the twentieth century, development of observation from space has made the *whole electromagnetic spectrum* available to observation.[2] However, even in space, there remain parasitic phenomena which perturb observation, in the same way as the atmosphere for Earth-based observation, and it will not be out of place also to tackle these questions in the present chapter, for the sake of completeness.

2.1 Physical and Chemical Structure of the Atmosphere

In this section we discuss the highly variable properties of the Earth atmosphere as a function of altitude. This is important when deciding whether to situate an observatory at sea level, on a mountain top, aboard an aircraft, or in a stratospheric balloon, depending on the properties required for observing the relevant wavelength.

2.1.1 *Vertical Structure*

To a first approximation, the Earth's atmosphere is in radiative equilibrium with its surroundings. The overall balance, over the whole Earth, of the flux received from the Sun and that which is re-radiated into space leads to a more or less stationary distribution of temperatures, pressures, and so on, with respect to time, although daily, annual, and even secular cycles are superposed.

The average structure of the atmosphere is described as a function of altitude z by temperature and density distributions, $T(z)$ and $\rho(z)$, respectively (Fig. 2.1). Note the *troposphere* ($\partial T/\partial z < 0$, in general), which is separated from the *stratosphere* ($\partial T/\partial z > 0$, in general) by the *tropopause*. The height of the tropopause is strongly dependent on the latitude, almost reaching ground level over the Antarctic continent (altitude of the Southern Polar cap ~ 3.4 km). At all latitudes, significant deviations from the average distribution are observed near ground level. In particular, there are *inversion layers* where the temperature gradient changes sign through a certain vertical distance, which can be more than one kilometre.

From 0 to 90 km, the composition of the air can be considered as constant, and the pressure is approximately described by an exponential law:

$$P(z) = P_0 \exp(-z/H), \quad H = \frac{R}{M_0}\frac{T_\mathrm{m}}{g},$$

where H is the scale height, M_0 the mean molecular mass of the air, and T_m a mean temperature. Taking $R = 8.32\,\mathrm{J\,K^{-1}\,mole^{-1}}$, $M_0 = 0.029\,\mathrm{kg}$, gives the value $H = 7\,998\,\mathrm{m} \simeq 8\,\mathrm{km}$ near ground level.

[2]The first balloons equipped with detectors date from about 1910, the first scientific rocket launches from 1946 (using the German V-2 rocket), and the first satellites in orbit from 1960.

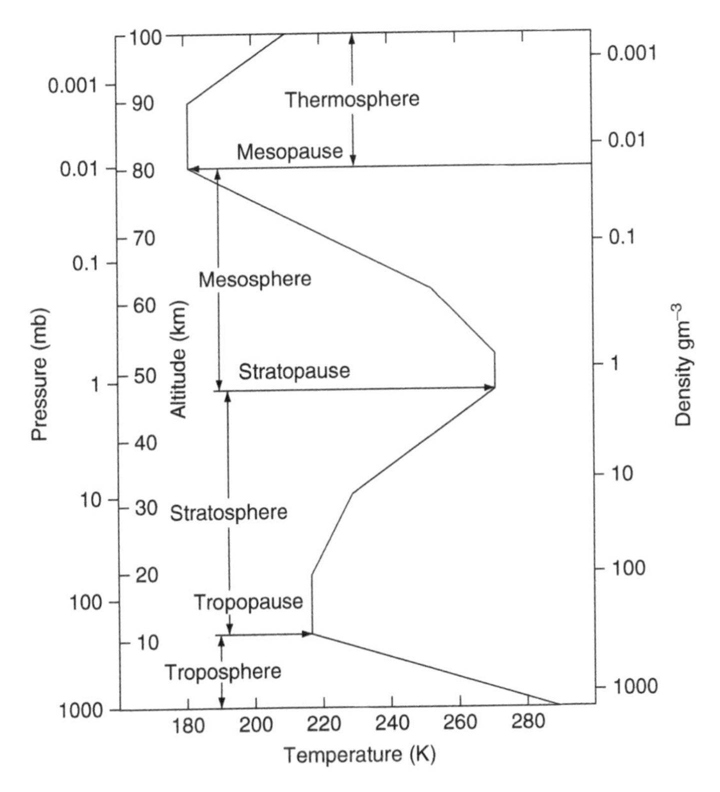

Fig. 2.1 Temperature profile $T(z)$, density $\rho(z)$, and pressure $P(z)$ in the mean atmosphere. The names of the layers and their boundaries are those adopted by the World Meteorological Organisation. (After U.S. Standard Atmosphere)

The adiabatic gradient of dry air has value around $0.01°\mathrm{m}^{-1}$ near ground level, and is given by the formula

$$\left(\frac{\partial T}{\partial z}\right)_{\mathrm{ad}} = -\frac{g}{R} M_0 \frac{C_p - C_v}{C_p} \qquad (C_p,\ C_v \text{ specific heats}).$$

Any gradient larger than this in absolute value would imply *convective instability* of the atmosphere, and hence vertical currents.

Space observatories are placed in orbits above 300 km, an altitude at which the residual amount of atmosphere is not totally negligible (Fig. 2.2).

2.1.2 Constituents of the Atmosphere

The principal constituents are O_2 and N_2, whose relative proportions are constant between 0 and 100 km. Study of the *minor atmospheric constituents* in these layers

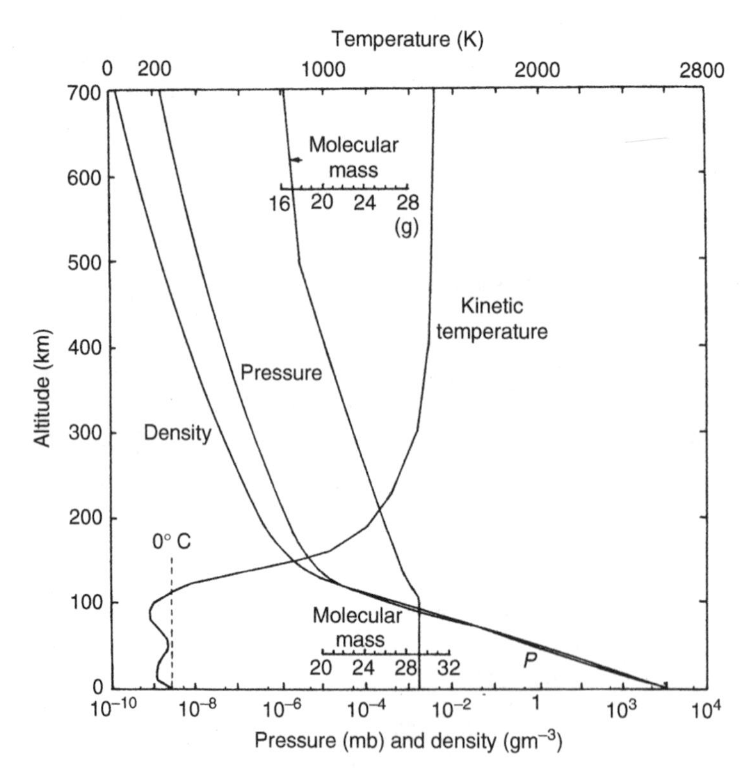

Fig. 2.2 Temperature, density, and molecular mass profiles between 0 and 700 km altitude. There are significant variations with latitude or solar activity in these average profiles. (U.S. Standard Atmosphere)

is an active area of research, considering their role in the maintenance of physical conditions at the Earth's surface (radiative balance, ultraviolet flux), and the possible disturbance of the natural equilibrium by human activity. Some of these constituents also play an important role in astronomical observations, because of the strong absorption they cause in certain spectral bands. In particular, water vapour, carbon dioxide, and ozone can be cited.

Water Vapour

The *mixing ratio* or *fractional content* is defined locally as

$$r = \frac{\text{mass of H}_2\text{O per m}^3}{\text{mass of air per m}^3},$$

which is usually measured in $\mathrm{g\,kg^{-1}}$. This ratio varies between 0 and a maximum value $r_s(T)$ which is characteristic of saturation, and is also a very rapidly varying

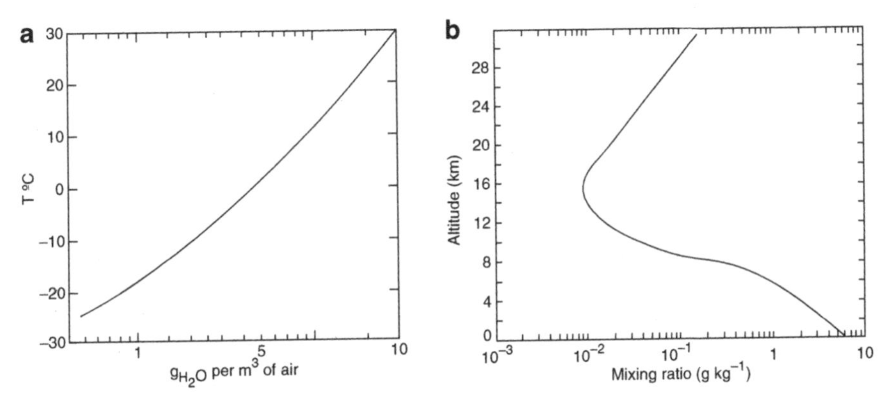

Fig. 2.3 (**a**) Mass concentration of water vapour in saturated air, at normal pressure, as a function of temperature. (**b**) Mixing ratio $r(z)$ of water vapour in air as a function of altitude for an average atmosphere. $r(z)$ is measured in g kg^{-1}, so that it is close to unity for the most frequently encountered meteorological conditions near ground level

function of air temperature (Fig. 2.3a). The mixing ratio is a function of altitude, and is also strongly dependent on both time and latitude (Fig. 2.3b).

The *quantity of precipitable water* above altitude z_0 is defined as

$$w(z_0) = \int_{z_0}^{\infty} N_{H_2O}\, dz,$$

where $N_{H_2O}(z)$ the number of molecules per unit volume. For normal pressure and temperature P_0 and T_0, respectively,

$$N_{H_2O}\ [m^{-3}] = 4.3 \times 10^{25} \frac{P}{P_0} \frac{T}{T_0} r(z).$$

This quantity can also be expressed as a column of precipitable water

$$h_{H_2O}\ [cm] = \rho_0\ [g\, cm^{-3}] \int_{z_0}^{\infty} r(z) e^{-z/H}\, dz,$$

where ρ_0 is the air density at altitude z_0, and z is given in centimetres.

Because of the rapid variation of $r(z)$ with height, the scale height of water vapour is considerably less than that of dry air H, being something like 3 km in the troposphere. Thus, siting an observatory on a high mountain, at an altitude of several kilometres, significantly improves the quality of observation, especially in the infrared and at millimetre wavelengths. The Antarctic plateau, at an altitude of around 3 000 m, also provides favourable observing conditions (see Sect. 2.8.6).

Ozone

The vertical distribution of ozone depends on the latitude and the season. This layer has been the subject of much interest since the detection of perturbations which may be due to human activity, notably the destructive role of certain industrial products such as fluorocarbons in the 1980s. Referred to normal conditions, the integrated quantity of ozone in the whole atmosphere varies from a column of 0.24 cm to a column of 0.38 cm STP (Standard Temperature and Pressure) in going from low to high altitudes. The maximum concentration occurs at about 16 km, although ozone is present even at 80 km. It absorbs mainly in the ultraviolet ($\lambda \lesssim 300$ nm).

Carbon Dioxide

This constituent is an important source of infrared absorption. Its vertical distribution is similar to those of O_2 and N_2, and its mixing ratio is independent of altitude. It absorbs mainly in the mid-infrared.

Ions

The atmosphere becomes increasingly ionised above 60 km, because of the Sun's ultraviolet radiation, which cause a series of photochemical reactions of the type

$$O_2 + h\nu \longrightarrow O_2^{+*} + e^-,$$

$$O_2 + h\nu \longrightarrow O^+ + O + e^-,$$

where O_2^{+*} denotes an excited state of O_2. The inverse reactions of recombination, and of radiative or collisional de-excitation, also occur, in such a way that the electron density is not constant at a given altitude. It varies considerably with altitude and solar illumination (the *circadian* or day–night cycle), as well as with solar activity. Solar flares (see Fig. 6.37) increase the ultraviolet flux, and cause electron showers which are channelled along the Earth's magnetic field lines and which ionise the atmosphere at high latitudes (auroral zones).

Several ionospheric layers are distinguished, corresponding to local maxima of electron density N_e, as shown in Table 2.1. Beyond these layers, the level of ionisation remains effectively constant up to about 2 000 km, with $N_e \sim 10^4\ \mathrm{cm}^{-3}$.

Table 2.1 Ionospheric layers

Layer	Altitude [km]	Electron density N_e [cm^{-3}]
D	60	10^3
E	100	10^5
F	150–300	2×10^6

2.2 Absorption of Radiation

Absorption of radiation by the constituents of the atmosphere can be either total or partial. If it is total, *transmission windows* can be defined at a given altitude, and the minimum altitude at which an observation becomes possible can be determined. If it is partial, it will modify the spectra of the sources under observation; these spectra will be affected by *telluric absorption bands*, whose position, intensity, and equivalent width must be identified (see Sect. 8.1).

Atomic and Molecular Transitions

These transitions cause absorption at discrete wavelengths. The types of transition are:

- *pure rotational* molecular transitions, e.g., H_2O, CO_2, O_3,
- *rotational–vibrational* molecular transitions, e.g., CO_2, NO, CO,
- *electronic* molecular transitions, e.g., CH_4, CO, H_2O, O_2, O_3, or radicals such as OH,
- *electronic* atomic transitions, e.g., O, N.

Atomic and molecular physics allow calculation of the *absorption coefficients* $\kappa_i(\lambda)$ (with units $\mathrm{cm^2\,g^{-1}}$) and *cross-sections* (with units $\mathrm{cm^2}$) for the various transitions of a given constituent, as a function of wavelength.

Atomic physics gives the cross-sections σ [$\mathrm{cm^2}$] and the physical conditions of the gas determine the population n_i [$\mathrm{cm^{-3}}$] of a given energy level. The mass absorption coefficient is then

$$\kappa_i = \frac{\sigma n_i}{r_i \rho_0}.$$

These coefficients are often given for $P = 1$ atm and $T = 273$ K, so that a correction is therefore required to take into account variations of altitude. At an altitude z_0, the *optical depth* along a vertical line of a constituent i with mixing ratio $r_i(z)$ is given by the expression

$$\tau_i(\lambda, z_0) = \int_{z_0}^{\infty} r_i(z)\rho_0(z)\kappa_i(\lambda)\mathrm{d}z,$$

where $\rho_0(z)$ is the mass density of the air.

The attenuation of an incident ray of intensity I_0, received at altitude z_0, and making an angle θ to the zenith (*zenith distance*), is

$$\frac{I(z_0)}{I_0(\infty)} = \exp\left[-\frac{1}{\cos\theta}\sum_i \tau_i(\lambda, z_0)\right],$$

where the sum is over all absorbing species.

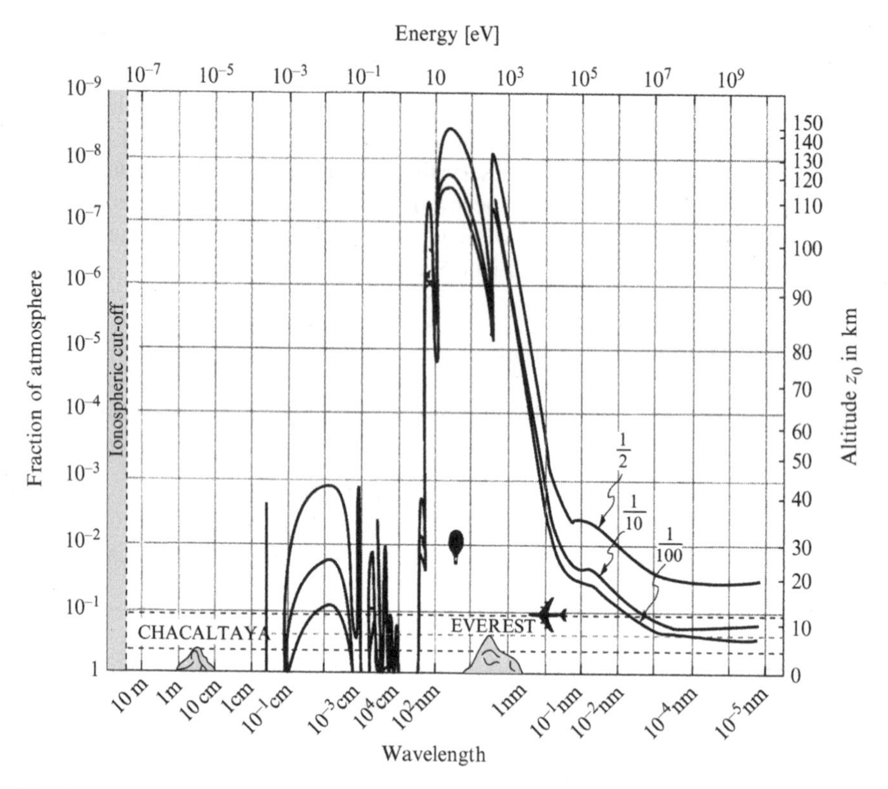

Fig. 2.4 Attenuation of electromagnetic radiation by the atmosphere. Curves give the altitude z_0 (*right-hand scale*) or the residual fraction of the atmosphere, in mass, above z_0 (*left-hand scale*), for three values of the ratio $I(z_0, \lambda, \theta = 0°)/I_0(\infty, \lambda)$. Chacaltaya is a site in the Andes (altitude ~6 000 m)

Figure 2.4 gives, as a function of wavelength, the altitudes z_0 at which the ratio $I(z_0, \lambda, \theta = 0°)/I_0(\infty, \lambda)$ takes the values 0.5, 0.1, 0.01, respectively, which correspond to optical depths of 0.69 (transparent), 2.3, and 4.61 (opaque), again respectively. This figure covers the whole electromagnetic spectrum, from the cutoff at $\lambda = 23$ m (radiofrequencies), caused by the ionospheric plasma, right through to γ radiation at several GeV.

The atmosphere is totally opaque if $\tau_0 = 10$, and astronomical observation can be regarded as feasible when $\tau(\lambda, z) < 0.5$, implying transmission greater than 61%. Qualitatively, millimetre wavelengths are dominated by pure rotational bands of H_2O and O_2; the infrared and submillimetre region by rotational bands of H_2O and CO_2, as well as by the rotational–vibrational bands of these constituents; and the near ultraviolet by electronic transitions of O_2 and O_3. In the near ultraviolet, there appears the continuous absorption band of oxygen O_2, which is characteristic of the ionisation of this molecule in the ionosphere. The absorption continuum of N_2 dominates the far ultraviolet ($\lambda < 20$ nm).

At wavelengths less than 10 nm, molecular ionisation is complete and the absorption coefficient is effectively constant. The optical depth of a layer of thickness l [cm] is then given by

$$\tau_\lambda(\lambda \lesssim 10\,\text{nm}) \sim 30\,l\,[\text{cm}]\frac{P}{P_0}\frac{T_0}{T}.$$

For these wavelengths, even at the very low pressures encountered by stratospheric balloons ($z = 30$ km, $P = 10^{-3} P_0$), we obtain $\tau = 1$ (transmission 37%) for a very short path, of length $l = 30$ cm.

Figure 2.4 can be used to define the domains of ground-based and space-based astronomical observatories:

- *Ground-based astronomy* is limited to the visible, the near infrared ($\lambda < 25\,\mu$m), where there remain, nevertheless, many absorption bands, millimetre wavelengths ($\lambda > 0.35$ mm), subject also to some absorption, and then centimetre wavelengths and beyond.
- *Space-based astronomy* covers all the rest of the electromagnetic spectrum, including γ-ray, X-ray, ultraviolet (UV) and infrared (IR). It should be noted, however, that the 50% transmission level occurs at quite different altitudes for different wavelengths. X-ray and γ-ray astronomy would just be possible at altitudes attainable by balloons (30 to 40 km), as would observation in the near UV ($\lambda > 200$ nm). Infrared and submillimetre observation is possible from an altitude of 12 km, which can be reached without difficulty by commercial aircraft (Airbus, B747), or on the polar ice caps of the Antarctic plateau.

The relatively low cost of these intermediate observational platforms (aircraft and balloons) led to their intensive use in the period 1960 to 1980, although this use gradually decreased between 1980 and 1990, with the development of powerful space observatories. A good example is NASA's Kuiper Airborne Observatory (KAO), which observes in the infrared and submillimetre ranges using a 0.9 m telescope carried aboard a large transport aircraft; and a 2.5 m replacement for this telescope (SOFIA) is being prepared for launch. Another example is provided by a very large stratospheric telescope (2 m), flown by the CNES in the 1990s on a balloon going above 30 km altitude (PRONAOS). In the future, such platforms will probably be reserved for very specific missions, since most observations are now made from genuine orbiting observatories, each one specialised in some wavelength range and having its orbit optimised accordingly.

The complete control of observing conditions now within our reach has totally freed astronomical observation of electromagnetic radiation from atmospheric absorption and opened up an energy range extending over sixteen powers of ten. This is a considerable improvement on the octave of traditional visible wavelengths (350–800 nm).

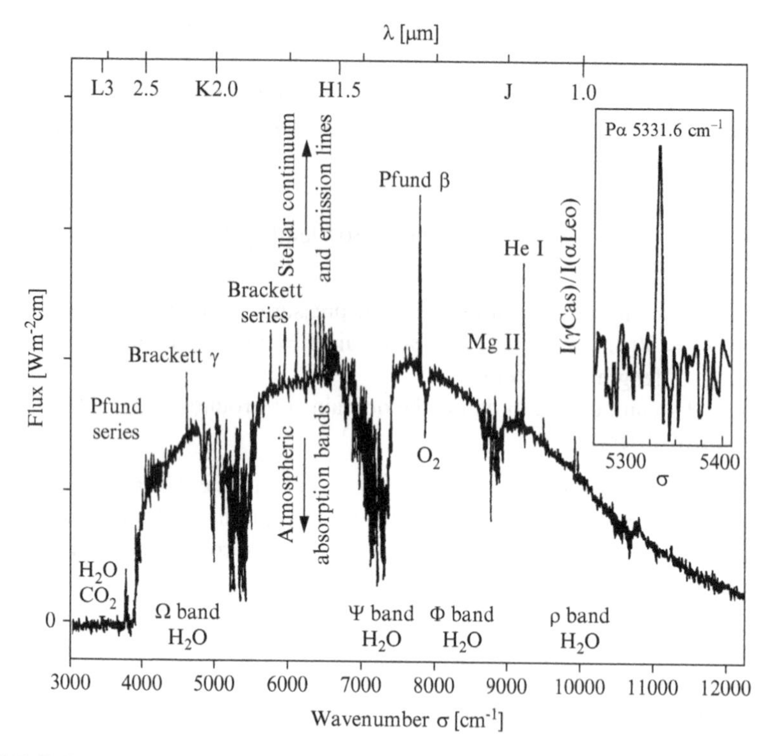

Fig. 2.5 Telluric absorption and spectroscopy: the spectrum of the star γ Cas. The spectrum was obtained using the 3.6 m Canada–France–Hawaii telescope at the summit of Mauna Kea, Hawaii, USA (4 200 m), with a Fourier transform interferometer (see Sect. 8.3.4), in the near infrared atmospheric transmission window, with a resolution of $\Delta\sigma = 0.5\,\mathrm{cm}^{-1}$ (Chalabaev A., Maillard J.-P., Ap. J. **294**, 640, 1984). Atmospheric absorption bands are indicated, together with photometric windows I, J, and K (see Sect. 3.3). The star has both a continuum and emission lines (mainly H recombination lines). The *inset* shows the 3–4 Paschen α line ($5331.6\,\mathrm{cm}^{-1}$) extracted from a heavily absorbed part of the spectrum: the spectrum of γ Cas was divided by that of a reference star (α Leo) to eliminate atmospheric bands. As α Leo (spectral type B7) also has hydrogen lines, the absolute value of Pα is not significant. Observation of Pα would be impossible at lower altitude. (With the kind permission of the Astrophysical Journal)

Telluric Bands

In astronomical spectroscopy (see Chap. 8), the need often arises to distinguish a spectral line from a nearby atmospheric absorption band.[3] Such a situation is illustrated in Fig. 2.5. A precise knowledge of the *profile* of the atmospheric absorption band is then required.[4]

---

[3] A detailed review of atmospheric bands in the radiofrequency range can be found in *Methods of Experimental Physics*, Vol. 12B. An inventory of telluric bands in the visible and near-infrared can be found in an atlas of the solar spectrum, where they arise as absorption bands.

[4] In millimetric astronomy, a very accurate measurement of the amount of water along the line of sight, combined with a careful model of the bands, can be used to almost completely eliminate them from observed spectra.

The profile of a molecular absorption band (rotational or vibrational–rotational) is generally a *Lorentz profile*, characterising lines in which the damping is dominated by collisions (pressure broadening). In the impact approximation (see AF, Sect. 2.20.2), the absorption cross-section thus has a frequency dependence $\phi(\nu, \nu_0)$, with the integral over all frequencies normalised to unity,

$$\phi(\nu, \nu_0) = \frac{\Delta\nu_L/2\pi}{(\nu - \nu_0)^2 + (\Delta\nu_L/2)^2},$$

where ν_0 is the central frequency of the transition, and $\Delta\nu_L$ is its total width. $\Delta\nu_L$ is related to the mean time τ between collisions according to

$$\Delta\nu_L = (\pi\tau)^{-1}.$$

This profile, illustrated in Fig. 2.6, gives a good representation of the cores of atmospheric bands, but generally underestimates the wings. The increase in the average attenuation with frequency is due to the cumulative effect of the wings of many weak H_2O bands. The attenuation is measured in dB km^{-1} and can be converted to optical depth by the formula $\tau = 0.23 \times$ attenuation [dB km^{-1}].

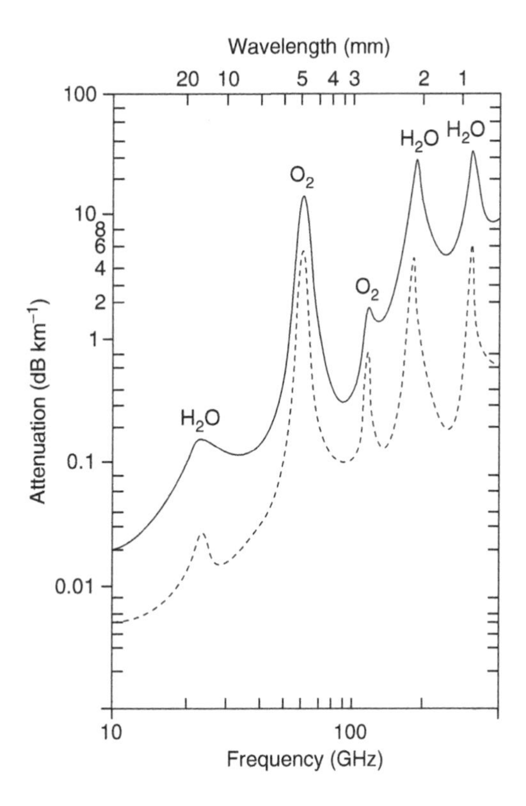

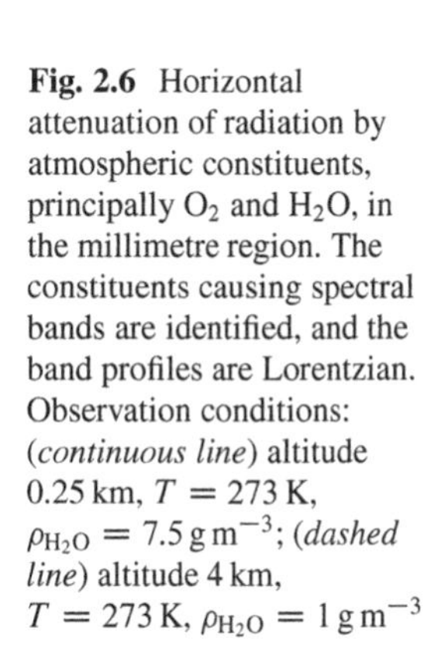

Fig. 2.6 Horizontal attenuation of radiation by atmospheric constituents, principally O_2 and H_2O, in the millimetre region. The constituents causing spectral bands are identified, and the band profiles are Lorentzian. Observation conditions: (*continuous line*) altitude 0.25 km, $T = 273$ K, $\rho_{H_2O} = 7.5$ g m^{-3}; (*dashed line*) altitude 4 km, $T = 273$ K, $\rho_{H_2O} = 1$ g m^{-3}

Ionospheric Plasma

The ionised, and hence conducting, layers of the upper atmosphere have a refractive index n related to the electron density N_e [cm^{-3}] by the expression

$$n^2 = 1 - \frac{\omega_p^2}{\omega^2} = 1 - \left(\frac{\lambda}{\lambda_p}\right)^2,$$

where the plasma frequency $\nu_p = 2\pi/\lambda_p$ is given by

$$\nu_p\,[\mathrm{Hz}] = \frac{\omega_p}{2\pi} = \left(\frac{N_e e^2}{4\pi^2 \varepsilon_0 m}\right)^{1/2} = 8.97 \times 10^3\ N_e^{1/2}.$$

For example, the F-layer ($N_e = 2 \times 10^6$ cm^{-3}) causes total reflection at $\lambda = 23.5$ m ($\nu \approx 12$ MHz), the wavelength for which $n = 0$.

The ionosphere is thus generally transparent to both centimetre and millimetre wavelengths, a fact which explains the rapid development of ground-based radioastronomy.

2.3 Atmospheric Emission

The Earth's atmosphere emits photons, both by fluorescence and by thermal emission. The ability to discriminate between this emission and that of the astronomical sources under study determines one of the main limitations for ground-based observation in the visible, infrared, and millimetre regions. It is necessary to study, not only the intensity and wavelength dependence of these emissions, but also their possible fluctuation in space and time. Apart from these photons, which arise through conversion of incident solar light radiation, the atmosphere can also generate photons as a result of other incident particles, e.g., electrons. This case is discussed briefly in this section, and in more detail in the next on γ radiation.

2.3.1 *Fluorescent Emission*

The recombination of electrons with ions, which have been produced by daytime reactions of photochemical dissociation, leads to the emission of photons. This is called *fluorescent emission*, because the probabilities of de-excitation are small, and emission may occur up to several hours after excitation. Fluorescent emission comprises both a continuum and emission lines. It is also known as *airglow* (Fig. 2.7) and can only be detected at night, being swamped by scattering (see Sect. 2.4) during the day.

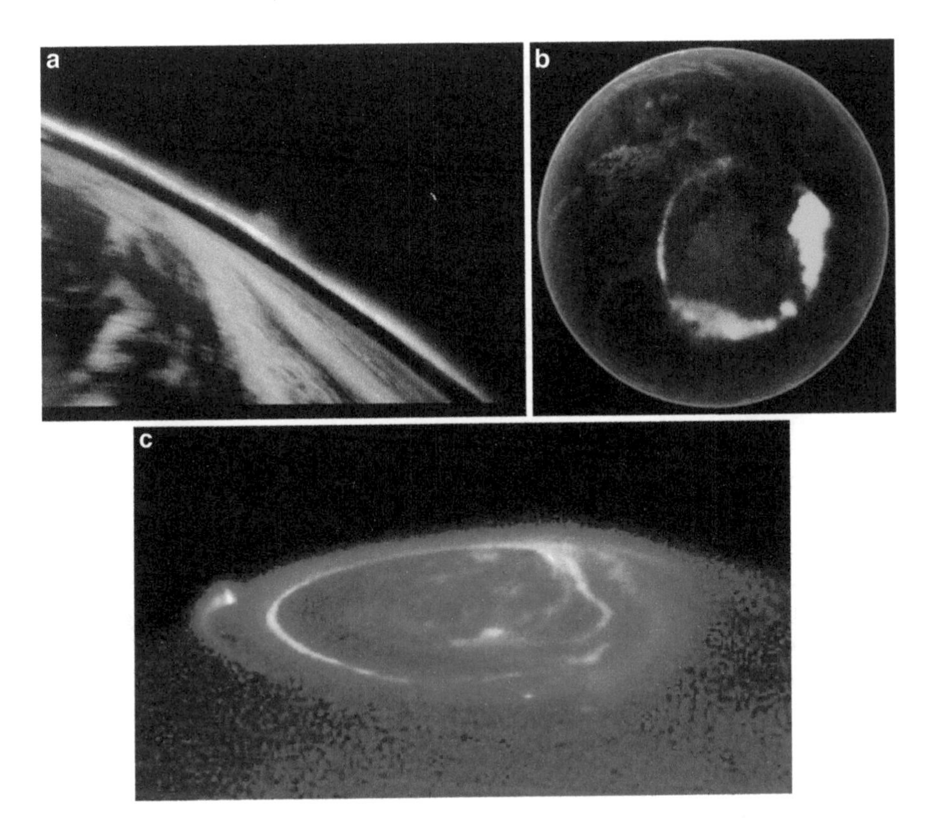

Fig. 2.7 (**a**) The Earth's atmosphere photographed tangentially from the Space Shuttle (mission STS-039, May 1991). One clearly sees the green airglow of the high atmospheric oxygen at altitude 80–120 km. The tangential observation increases the optical depth, hence the observed brightness (NASA). (**b**) The auroral ring above Australia, photographed by the NASA Image mission (2000–2006). (**c**) A polar aurora on the north pole of the planet Jupiter, photographed in the ultraviolet by the Hubble telescope (NASA/ESA, Clark J, University of Michigan). One sees the footprints of three of Jupiter's satellites (*from left to right*): Io, Ganymede, and Europa, creating currents in Jupiter's magnetosphere

Most fluorescence occurs at altitudes of around 100 km or above, where the density is low enough for spontaneous radiative decays, with spontaneous transition probability $A > 10^2\,\mathrm{s}^{-1}$, to occur before collisional de-excitation.

The main sources of emission are the atoms and radicals O I, Na I, O_2, OH, and H. In the spectroscopic notation used by astronomers, O I is the neutral oxygen atom, O II is once-ionised oxygen, and so on. Emission intensity is measured in rayleighs [R], the unit being defined by

$$1 \text{ rayleigh} = \frac{10^6}{4\pi} \text{photons cm}^{-2}\,\mathrm{s}^{-1}\,\mathrm{sr}^{-1} = \frac{1.58 \times 10^{-7}}{\lambda[\mathrm{nm}]} \mathrm{W\,m}^{-2}\,\mathrm{sr}^{-1}.$$

Table 2.2 Atmospheric emission. R resonant scattering, C chemical reaction, I ionic reaction. Å denotes the old *angstrom* unit, 1 Å = 0.1 nm. (After B.M. MacCormac, 1971)

λ [nm]	Emitter (state)	Altitude [km]	Intensity [rayleigh]	Process
102.5	H Lyβ	200	10	R
121.6	H Lyα	10^2–10^5	2 000	R
260–380	$O_2 A^3\Sigma_u^+$	90	600	C
500–650	NO_2	90	1 RÅ^{-1}	C
557.7	O^1S	90–300	250	C, I
589.3	Na^2P	90	20–150	C
630	O^1D	300	10–500	I
761.9	O_2	80	6 000	C

Table 2.2 summarises the main atmospheric fluorescences in the visible, and in the near infrared and ultraviolet. Note that, with the exception of auroral zones (magnetic latitude greater than 70°), in which ionisation is caused by injected electrons, the sky's emission varies little with latitude. Note also the presence of hydrogen emission lines caused by multiple resonant scattering of sunlight in the highest layers of the atmosphere (the hydrogen *geocorona*). The fluorescence of the radical OH is important in the near infrared (Fig. 2.8).

The Hubble Space Telescope, placed in low orbit, can easily measure the sky background. These measurements reveal (Caulet A. et al, Astron. Astrophys. Suppl. Ser. **108**, 1-8, 1994):

- Geocoronal emission at the wavelength of the Lyα (121.6 nm) line, with intensity from 3 to 20 k rayleigh, depending on the angle between the line of sight and the Sun.
- Emission by residual atomic oxygen O I, only measurable during the day, with intensity around 5 k rayleigh.
- A continuum of zodiacal scattering, in agreement with models of scattering by the zodiacal atmosphere (see Sect. 2.9.2).

Figure 2.9 gives the visible and ultraviolet magnitudes of the sky background, for observations from the ground and from space near the Earth. These magnitudes are given for a reference solid angle equal to one square arc second. They measure the flux produced by one square arc second of the sky background. Note the intensity of the geocoronal hydrogen Lyα (121.6 nm) line ($\sim 7\times10^{-3}$ photon s^{-1} cm^{-2} arcsec^{-2}). For wavelengths greater than 400 nm, the following correspondence is observed: a magnitude $m = 27$ arcsec^{-2} corresponds to a monochromatic flux of 10^{-7} photon s^{-1} cm^{-2} nm^{-1} arcsec^{-2}.

The ready excitation of atmospheric components, so harmful to astronomical observation, is quite essential for the detailed study of the atmosphere in terms of local composition, temperature, and physical chemistry. The radar-like technique Lidar (LIght Detection And Ranging) involves excitation of atmospheric constituents, generally by tuneable laser, and the study of their de-excitation by measurement of re-emitted light. These techniques inspired a way of creating artificial stars, crucial for adaptive optics (see Sect. 6.3).

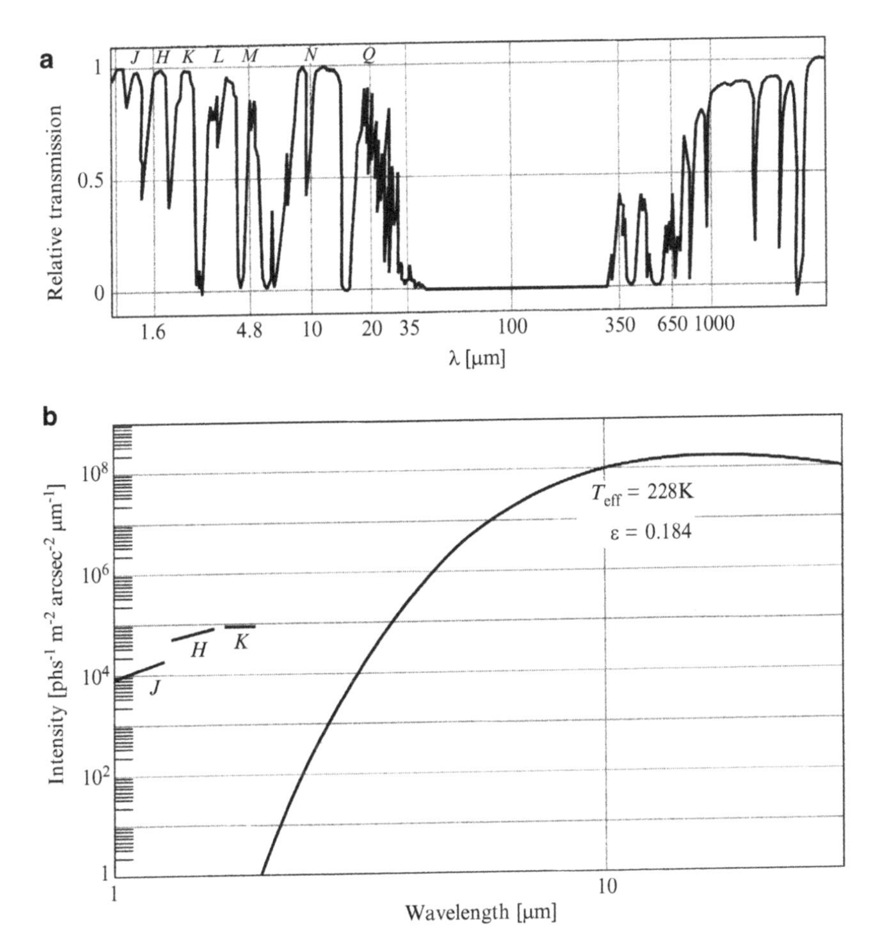

Fig. 2.8 (**a**) Atmospheric transmission (at altitude $z_0 = 4\,200$ m) across the whole infrared and submillimetre range. Spectral bands (see Sect. 3.3) are indicated with the usual notation. The relatively good submillimetre transparency corresponds to a quantity of precipitable water less than 1 mm. (**b**) Sky background brightness in the near infrared, at the altitude of Mauna Kea (4 200 m). The blackbody temperature (indicated at the *top right*) which best represents the emission is $T = 228$ K, with an emissivity $\varepsilon = 0.184$ [after Traub and Stier, 1976]. Upper limits for emission by the radical OH in spectral bands J, H, and K are indicated on the *left* (after McLean I., 1993)

Discrimination of Weak Astronomical Sources

The mean angular diameter of a stellar image, observed through the turbulent atmosphere of the Earth, is typically of the order of one second of arc (see Sect. 6.2.2), and consequently, it is usual to measure the sky background in magnitudes (the logarithmic units defined in Sect. 3.3) per square arc second.

We now calculate the magnitude of a square of sky background, with side one second of arc, which would correspond to fluorescence of intensity 1 RÅ^{-1} at 550 nm, using the fact that magnitude 0 corresponds to a monochromatic flux of $3.92 \times 10^{-8}\,\mathrm{W\,m^{-2}\,\mu m^{-1}}$:

$$\frac{1.58 \times 10^{-7}}{4\pi \times 550} \times (2.35 \times 10^{-11}\,\mathrm{sr}) \times 10^4 = 0.537 \times 10^{-17}\,\mathrm{W\,m^{-2}\,\mu m^{-1}\,arcsec^{-2}}.$$

This gives $m_V(\mathrm{sky}) = 24.6\,\mathrm{arcsec}^{-2}$. Even using sky subtraction techniques between two neighbouring points, the space and/or time fluctuations of this emission would make it difficult to extract the signal of a galaxy with magnitude significantly less than $m_V(\mathrm{sky})$ from a photographic plate or an electronically digitised image. A contrast less than 1% (that is, five magnitudes) is thus difficult to detect.

In addition, the development of imaging techniques at the diffraction limit of large telescopes (*adaptive optics*, see Sect. 6.3) requires great care to be taken over the grain of the background at high spatial frequencies. Without this attention, information at these frequencies could be lost.

The limit for observing very faint objects from the ground is thus set by the intrinsic emission of the night sky. There is a considerable gain in observing from space, as shown in Fig. 2.9, but there nevertheless remains a radiation background caused by the Solar System environment (see Sect. 2.9).

If the source has an intrinsic angular dimension much less than one second of arc, and if it is possible on the ground to reduce the effects of atmospheric turbulence, that is, reduce the solid angle of sky simultaneously observed, the sky background contribution will be greatly diminished. The contrast, and therefore the detection, will then be improved. This is the achievement of *adaptive optics* (see Sect. 6.3).

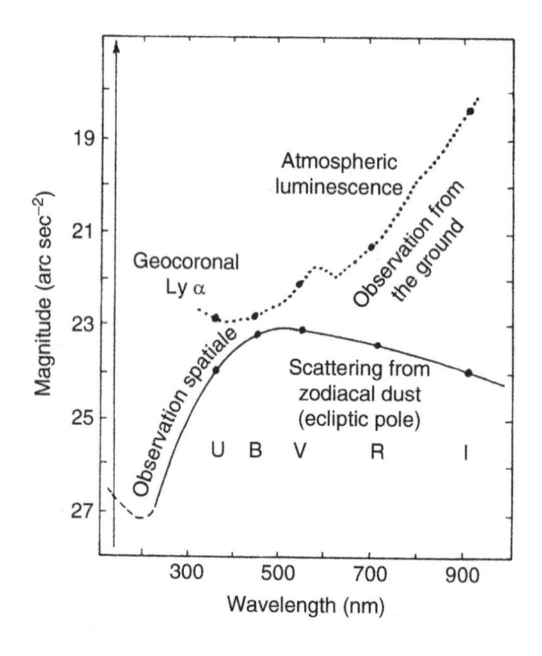

Fig. 2.9 Visible and ultraviolet magnitudes of the sky background per square arc second, for observations from the ground and from space near the Earth. Letters denote the spectral bands of the photometric system (see Sect. 3.3), corresponding to the points (•). (After Courtès G., personal communication; Smith H.E., Burbridge E.M., Ap. J. **210**, 629, 1979; Leinert C., Sp. Sci. Rev. **18**, 281, 1975; Machetto F. et al, ESA-SP **1028**, 1980, European Space Agency)

2.3.2 *Thermal Emission*

The Earth's atmosphere can be considered as a gas in local thermodynamic equilibrium (LTE) up to an altitude of 40 to 60 km, beyond which collisions occur too infrequently to ensure the thermal population of all energy levels. A full radiative transfer calculation gives the specific intensity received at the ground for any given wavelength. A simple approximation is possible when the atmosphere is optically shallow, that is, when the optical depth τ_λ along the zenith line is much less than unity, corresponding to an atmosphere which is transparent at the wavelength considered. In this case the intensity received at altitude z and at zenith distance θ is

$$I_\lambda(z) = \tau_\lambda B_\lambda(\overline{T}) \frac{1}{\cos\theta},$$

where $B_\lambda(\overline{T})$ is the Planck function at the mean temperature $\overline{T}$ of the atmosphere.

The wavelengths at which the two conditions $\tau_\lambda \ll 1$ and $B_\lambda(\overline{T})$ non-negligible are satisfied, correspond to the near infrared (1 to 20 μm) and millimetre (0.5–2 mm) atmospheric windows.

Table 2.3 gives the infrared magnitude (quantity defined in Sect. 3.3) of the sky per square arc second, in the zenith direction, taking $\overline{T} = 250$ K for the mean temperature in the emissive region of the atmosphere. These are, of course, mean values, and scatter about these values could be significant for some sites.

Detectable astronomical sources can be several orders of magnitude fainter than this sky emission. The problem of discriminating faint sources is similar to that encountered above for fluorescence.

The first part of Fig. 2.8 shows graphically the atmospheric transmission in the infrared and millimetre bands. The second part of the same figure shows a graph of atmospheric emission in the near infrared, from one of the best sites in the world for ground-based observation (the summit of Mauna Kea, Hawaii, $z_0 = 4\,200$ m). On the first graph are shown the observational photometric windows defined in Sect. 3.3. On the second can be seen the extremely rapid growth of atmospheric emission on the Wien side of the Planck function $B_\lambda(T)$, leading to very high sky background emission at 10 μm (see also Table 2.3). At shorter wavelengths, thermal emission is overtaken by fluorescent emission of the radical OH. The spectral

Table 2.3 Mean thermal emission of the atmosphere. 1 jansky = 10^{-26} W m^{-2} Hz^{-1}

Spectral band (see Sect. 3.3)	L	M	N	Q
Mean wavelength [μm]	3.4	5.0	10.2	21.0
Mean optical depth τ	0.15	0.3	0.08	0.3
Magnitude [arcsec^{-2}]	8.1	2.0	−2.1	−5.8
Monochromatic intensity [Jy arcsec^{-2}]	0.16	22.5	250	2 100

resolution of the second graph in the figure is not sufficient to resolve the discrete line structure of this emission, which severely limits detection sensitivity by ground-based observation between 1.6 and 2 μm.

It it is in fact possible to design a comb-like spectral filtering device which, applied to the highly dispersed spectrum, cuts out these unwanted atmospheric bands and makes it possible to search for faint galaxies, for example. In the mid-1990s, this idea was applied to several telescopes, in particular the Japanese 8.2 m Subaru telescope in Hawaii, where it is now operational (see Sect. 5.2).

It is essential to make careful measurements of the precipitable water vapour when seeking new sites for infrared and millimetre observatories (see Sect. 2.8). To this end, a *spectral hygrometer* is used. This is a differential photometer, which compares the emission received at one wavelength (around 13 μm), where the atmosphere is entirely opaque and thus radiates practically like a black body ($\overline{T} = 250$ K), with the emission received at a wavelength (1.8 or 6 μm) dominated by an H_2O emission line, and whose optical depth is close to unity, in order to be as sensitive as possible to the value of h_{H_2O}. An example would be $\lambda = 6\,\mu$m, for which the absorption coefficient is $\kappa(H_2O) = 10^2\,\text{cm}^2\,\text{g}^{-1}$, whence $\tau(H_2O) = 0.1$ when $h_{H_2O} = 0.1$ mm, a typical value for a site of exceptional quality such as the Antarctic plateau.

This thermal emission fluctuates in time because of the turbulent motions of the atmosphere. The temporal power spectrum for these fluctuations is generally inversely proportional to the frequency.[5] The emission also fluctuates with the line of sight, as the flux integration is carried out over differing air masses. The result is a granularity of the thermal sky background emission which is a source of noise in some observations (see Sect. 9.4).

2.3.3 *Differential Measurement Techniques*

The intense emission of the atmosphere and the sensitivity of detectors make it essential to use *differential measurement* to eliminate sky background radiation, whether it be of thermal or fluorescent origin.

This applies to the detection of faint sources in the visible (magnitude $m_V > 15$), just as in the infrared, submillimetre, and millimetre ranges. It becomes imperative, even for a bright source, near $\lambda = 10\,\mu$m, where the maximum for atmospheric thermal emission is located. Even for a small pixel, for example, several seconds of arc across, the intensity received from the atmosphere can be several times greater than that of the signal received from the source.

The same is true in the millimetre range, where the exceptional sensitivity of detectors allows detection of atmospheric emission quite far from its maximum. Consider, on Fig. 2.6, the wavelength $\lambda = 2.6$ mm of a well-studied transition of the ^{12}CO molecule. The atmospheric transmission at a good quality site ($z_0 \sim 2\,000$ m),

[5] A quantitative analysis can be found in Kaüfl et al., Exp. Astron. **2**, 115, 1991.

is around 76% at this wavelength, that is, $\tau(2.6\,\text{mm}) = 0.27$. The atmospheric transmission therefore corresponds to an antenna temperature (see Sect. 7.5) of $0.27 \times 273 = 75\,\text{K}$, or to a spectral flux of 10 Jy ($1\,\text{Jy} = 10^{-26}\,\text{W}\,\text{m}^{-2}\,\text{Hz}^{-1}$), whereas the astronomical source under study may be much less intense, by three or four orders of magnitude.

To distinguish the source from the sky background in each of these cases, an *offsetting* or *on–off technique* is used, which involves pointing the telescope successively at the source and then at a neighbouring part of the sky, supposed to be free of sources. A neighbourhood of several minutes of arc is adequate, and the hypothesis that atmospheric emission remains spatially uniform over such a region seems to be borne out. Taking the difference between the two measurements, the signal of the source alone can be reconstituted (*subtraction of sky background*). However, new techniques in spectroscopy, using fibre optics (see Sect. 8.3), detect light, very locally, over several hundred pixels spread across a field of several minutes of arc. Subtraction of the sky background can then reveal spatial fluctuation residues, representing an additional *spatial noise*. We shall discuss this in more detail in Chap. 9 (see Sect. 9.4), which deals with the precautions needed to carry out this type of observation.

A second effect creates difficulties for this method. Atmospheric emission generally fluctuates in time (movement of invisible clouds of water vapour, variable excitation of fluorescence during the night, ionospheric winds, and so on), and the frequencies are rather high ($f \sim 0.1$–$10\,\text{Hz}$). The differential measurement thus compares two emissions which differ randomly, and the result of the subtraction will be affected by a random noise, referred to as *sky noise*.

This noise is weaker as the modulation between the two fields increases. To use this fact, the secondary mirror of a Cassegrain or Coudé telescope (see Chap. 5) is given a sinusoidal, or better, a square-wave vibration which displaces the source image on the diaphragm determining the field (*sky chopping*). Meanwhile the thermal emission produced by the telescope itself, which also crosses the field diaphragm and reaches the detector, is kept absolutely constant. Radiotelescopes are generally too large, and so have too great an inertia, to allow this rapid modulation ($> 1\,\text{Hz}$) of the pointing or of the secondary mirror. It is thus usual to obtain modulation by applying the on–off technique to the whole instrument, although naturally this is less fast.

To be more precise, the beam is displaced on the primary mirror during modulation by the secondary mirror, since the *field diaphragm* is generally chosen to be the secondary mirror itself. Any variation, even minimal, in the temperature or emissivity of the primary mirror then produces a further modulation, called the *offset signal*. These signals which cause very significant perturbation, are discussed in Chap. 9. It is this effect which renders infrared and millimetre measurements so difficult, and which disappears whenever it is possible, in space, to cool the telescope itself.

The principle of differential modulation has been described here in the context of dealing with atmospheric background emission, but it applies equally well in any situation where the problem is one of distinguishing a weak and spatially restricted source from a more or less uniform and bright background. Thus the detection of infrared galactic sources requires elimination of the zodiacal background (see Sect. 2.9.2). The same is true in the search for fluctuations in the diffuse cosmological background, which are of the order of 10^6 times less than the mean radiative background and require, this time in the submillimetre range, a subtraction of galactic interstellar emission. Figure 2.10 shows the grain of this galactic background quantitatively, by its spatial power spectrum.

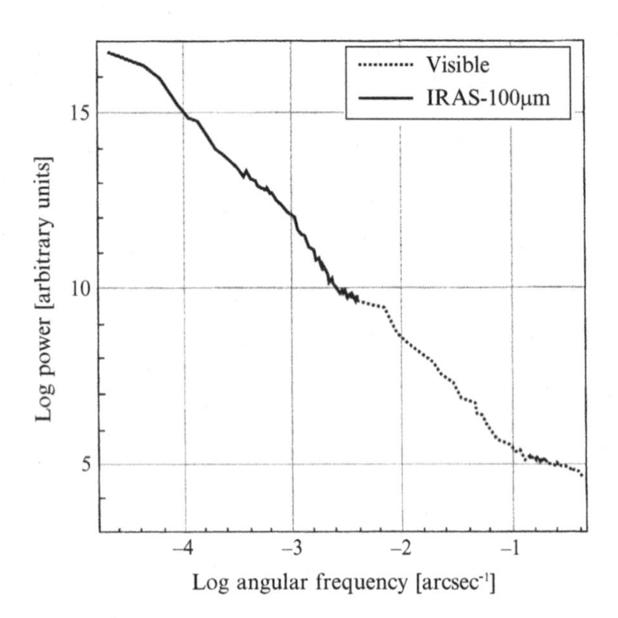

Fig. 2.10 Power spectrum of sky brightness as observed from an orbit around the Earth, in the direction of the North Galactic Pole, at $\lambda = 100\,\mu\text{m}$. *Continuous curve* : low frequency part sampled by the satellite IRAS (1983) with resolution 2 arcmin. *Dotted curve* : high frequency part deduced from CCD measurements in the visible. (After Low F.J., Cutri R.M., Infrared Phys. **35**, 29, 1994)

2.4 Scattering of Radiation

Atmospheric scattering is caused by the molecules which make up the air and by aerosols suspended in it. The vertical distribution of the former is directly related to density and thus decreases with altitude, whilst that of the latter is more capricious, depending on winds, climate, type of ground, volcanic activity, industrial pollution, and many other factors. The scattering of light by the brighter celestial bodies — the Sun and the Moon — gives rise in the daytime to the blue colour of the sky, and at night, to a veiling of faint astronomical sources. In the same way as atmospheric emission, scattering makes it difficult to discriminate faint sources. The highly chromatic nature of scattering only disturbs observation in the visible and very near infrared, for the day and night alike.

The molecular scattering in the visible and near infrared is *Rayleigh scattering*. The cross-section for Rayleigh scattering, integrated over all directions, is given by

$$\sigma_{\mathrm{R}}(\lambda) = \frac{8\pi^3}{3}\frac{(n^2-1)^2}{N^2\lambda^4},$$

where n is the refractive index and N the number of molecules per unit volume. The refractive index of air at $\lambda = 0.5\,\mu\text{m}$ can be approximated by Gladstone's law, which gives

$$n - 1 \simeq 80 \times 10^{-6}\frac{P\,[\text{mb}]}{T\,[\text{K}]}.$$

Rayleigh scattering is not isotropic and $\sigma_R = \sigma_R(\theta, \lambda)$ is a function of the angle θ between the directions of the incident and scattered radiation [see *Astrophysical Quantities*, 3rd edn., Athlone, London 1973 (AQ), Chap. 5]. For an incident intensity I, the intensity j scattered into solid angle $\mathrm{d}\omega$ is given by

$$j = \sigma_R \frac{3}{4}(1 + \cos^2\theta)\frac{\mathrm{d}\omega}{4\pi} I .$$

Even in the absence of any aerosols, the sky brightness due to Rayleigh scattering cannot be neglected. At an altitude of 2 000 m, at an angular distance $\theta = 90°$ from the Sun, and at the wavelength of red light $\lambda = 700$ nm, the ratio of sky brightness to that of the Sun's disk is about 10^{-7}.

Aerosol scattering is governed by a different law, because the scattering particles are bigger than molecules.

The total effective cross-section of a sphere of radius a is given by Mie's theory as

$$\sigma = \pi a^2 (Q_{\text{scattering}} + Q_{\text{absorption}}) .$$

The general theory of scattering by small particles can be found in the classic text by van de Hulst H.C., *Light Scattering by Small Particles* (Wiley, New York, 1957). Some useful results are also included in AF, Sect. 1.41.

If $a \gg \lambda$, $Q_s = Q_a = 1$, the scattered power is equal to the absorbed power, and the effective cross-section is twice the geometrical cross-section. If $a > \lambda$, Q_s and Q_a have a complicated λ-dependence, but for dielectric spheres (water droplets or dust grains such as silicates), the relation $Q_s \propto \lambda^{-1}$ holds, i.e., the scattered intensity varies as λ^{-1}.

The above description of the scattering of sunlight applies equally to the scattering of moonlight, whence infrared observation from Earth-based observatories is favoured on nights close to the full Moon. The scattering of urban lighting is also a problem (see Sect. 2.8).

Daylight Observations from the Ground

The thermal emission of the atmosphere grows exponentially (Wien part of the Planck curve) as the wavelength increases from the visible to the near infrared. Indeed, there is a wavelength beyond which this emission exceeds daytime scattering emissions, and hence in this range the brightness of the sky is largely independent of the day–night cycle. This is shown in Fig. 2.11. For telescopes with absolute pointing (i.e., to an accuracy of several seconds of arc, and without reference star), observations become feasible during the day. This could significantly increase the time available for use at large ground-based optical telescopes. Unfortunately, there are problems associated with the thermal control of instruments which limit this possibility. Of course, in the millimetre range, round-the-clock observation has become commonplace.

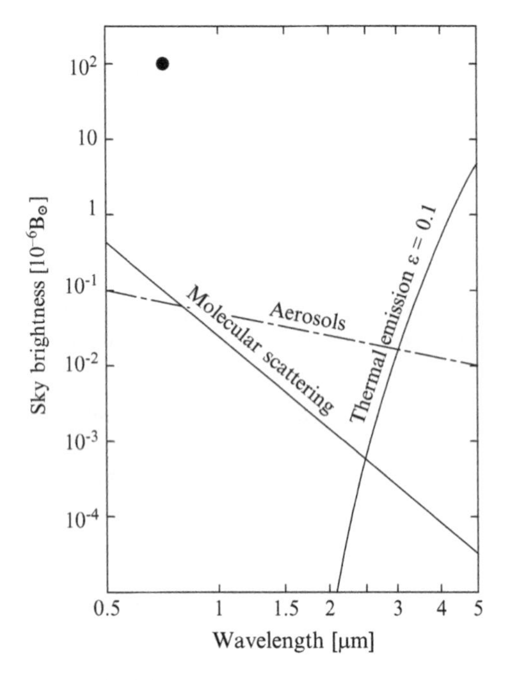

Fig. 2.11 Contributions to the sky brightness in the visible and near infrared. Molecular scattering is given for the altitude $z = 2\,000$ m, at 90° from the Sun. The wavelength dependence is λ^{-4}, and does not include correction for variation of refractive index $n(\lambda)$ with λ. Thermal emission is also shown, assuming uniform mean emissivity of 0.1, the exact value depending in particular on the quantity of precipitable water. The intersection of the two curves shows, in the absence of aerosols, the wavelength beyond which observation can be carried out during the night or the day regardless. The *dashed line* shows the possible additional contribution due to aerosols, varying as λ^{-1}, which moves the above wavelength threshold to the right. For comparison, (•) marks the sky brightness measured at 0.5 arcmin from the Sun's limb at an astronomical site (Kitt Peak National Observatory, Arizona. Source P. Léna, unpublished)

Solar Eclipses

Visible emissions from the solar corona immediately above the Sun's limb ($1.03\ R_\odot$ from the centre of the disk, or 30 arcsec from the limb) are around 10^6 times less bright than the photospheric emission. The coronal emission is thus greater than the Rayleigh scattered intensity, but generally less than the light scattered by aerosols (*solar aureole*), which is generally quite considerable for scattering angles very close to zero (see Fig. 2.11).

Even from a high altitude site, observation of the white corona (continuous or K corona) is therefore very difficult, except during an eclipse of the Sun (see Fig. 2.12), for the intensity lies between 10^{-6} and 10^{-9} times that of the photosphere, and within just 1 $R_\odot$ of the Sun's limb. On the other hand, for wavelengths at which chromospheric (Hα line) or coronal (Fe XIV line at $\lambda = 530.3$ nm) emission is significant and quasi-monochromatic, atmospheric scattering can be practically eliminated by using a narrow filter. A ground-based coronagraph (see Sect. 6.6),

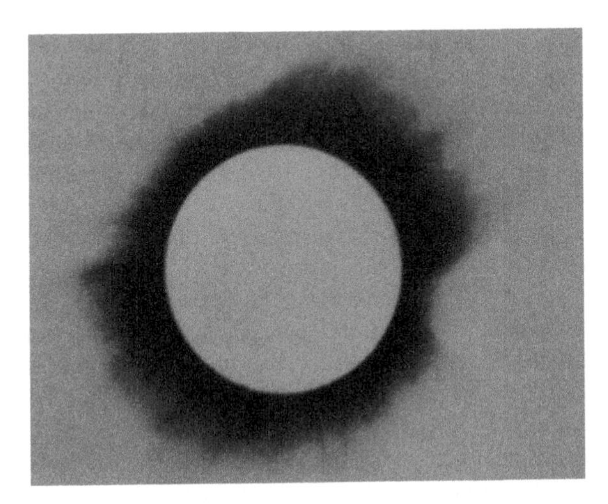

Fig. 2.12 The eclipse observation which confirmed general relativity. This photograph was obtained by the British astronomer Arthur Eddington during the total solar eclipse of 29 May 1919. By comparing the relative position of stars close to the solar limb with their positions on a photograph taken several months before, the deflection of light while grazing the Sun was measured and shown to be in agreement with the prediction made by Einstein on the basis of the curvature of space due to the Sun's gravitational field. Picture from the original article Dyson F.W., Eddington A.S., Davidson C., A determination of the deflection of light by the Sun's gravitational field, from observations made at the total eclipse of May 29, 1919, Phil. Trans. R. Soc. Lond. A **220**, 291–333, 1920

fitted with a filter, eliminates part of the instrumental and atmospheric scattering, which have values around 10^{-5} times the photospheric emission for a good site. This allows observation of the monochromatic corona out of eclipse. In space, only instrumental scattering remains and it is easier to observe the corona. Photospheric emission at X-ray wavelengths is negligible, and the X-ray corona can therefore be observed from space without the use of a coronagraph. Eclipses, although romantic events in the life of an astronomer, have had their importance diminished by space observation.

2.5 Atmospheric Refraction and Dispersion

The refractive index of air $n(\lambda)$ depends on the vacuum wavelength λ_0. In dry air, between 0.2 and 4 μm, the following expression holds at pressure $P = 1$ atm and temperature $T = 15°$C:

$$[n(\lambda) - 1] \times 10^6 = 64.328 + \frac{29\,498.1}{146 - 1/\lambda_0^2} + \frac{255.4}{41 - 1/\lambda_0^2},$$

where λ_0 is measured in μm.

At radiofrequencies (centimetre and beyond), refraction is hardly chromatic at all. Atmospheric refraction is manifested by a flattening of the solar disk at the horizon. Its dispersive character produces the *green flash* phenomenon at sunrise and sunset, and also the sometimes coloured appearance of stellar scintillation. More important for observation is the fact that the zenith distance θ' of a star is deviated from its true value θ. Refraction is a function of the thickness of air traversed, thus, for an atmosphere stratified in parallel planes, it is also a function of the quantity $m = 1/\cos\theta$, called the *air mass*. This effect is weak, but not negligible: $\theta' - \theta \sim -2''$ between 0.3 and 0.8 μm, for $\theta = 45°$. In addition, there is chromatic dispersion, whence the coloured spreading in images of stars observed far from the zenith. These effects of differential refraction and dispersion cannot be ignored when angular resolution is significantly greater than 1 arcsec, which is the case in high resolution imaging, and more specifically in adaptive optics (see Sect. 6.3) and optical interferometry (see Sect. 6.4). A compensating optical device (an *atmospheric dispersion corrector*) must then be used.

The analogous effects of ionospheric refraction and dispersion are observed at radiofrequencies, since the refractive index of the ionospheric plasma is highly chromatic (see Sect. 2.2).

2.6 Turbulence Structure of the Earth Atmosphere

A static and homogeneous description of the atmosphere is clearly inadequate. It is a fundamentally variable medium, whose fluctuations have wide-ranging effects on astronomical observation, but cannot be described by a single model of those fluctuations. Its spatial fluctuations have wavelengths ranging from several thousand kilometres (depressions and anticyclones) down to millimetres (viscous dissipation); and likewise, its temporal fluctuations have time scales ranging from the secular, for certain climatic changes, to the millisecond, for image deformations.

In addition, these spatial and temporal variations are clearly not stationary. Large-scale spatial fluctuations ($\gg$1 km) can lead to systematic effects, for example, errors in position determinations in *astrometry* resulting from corrections for atmospheric refraction (see Chap. 4). The present discussion will be limited to a description of effects on a small spatial scale ($\lesssim$1 km), and a small time scale ($\lesssim$1 s), these being typical scales at which atmospheric turbulence phenomena can directly affect astronomical observation.

The *stationarity* assumption, often inescapable in any reasonably simple statistical treatment, does not generally correspond to the reality in the context of observation. It is nevertheless possible, under this assumption, to derive a statistical description of atmospheric fluctuations, and to determine the various moments (mean, standard deviation) of those quantities which influence the propagation of electromagnetic waves. This will be dealt with in detail in Sect. 6.2, in the context of image formation.

2.6.1 Turbulence in the Lower and Middle Atmosphere

The layers under consideration here are the troposphere ($z \lesssim 12\,\mathrm{km}$) and possibly the lower regions of the stratosphere ($12 \lesssim z \lesssim 20\,\mathrm{km}$). We study here the way turbulence develops in these layers in the presence of winds, giving a rather simplified model, but one which satisfies our requirements.

A Simple Model of Turbulence

Turbulence develops in a fluid when the *Reynolds number*, a dimensionless quantity characterising the flow, exceeds a critical value. This number is defined as

$$\mathrm{Re} = \frac{VL}{\nu},$$

where V is the flow velocity, ν the kinematic viscosity of the fluid, and L a characteristic length (e.g., diameter of the pipe in which the flow takes place, or width of an obstacle normal to the flow).

In air, $\nu = 1.5 \times 10^{-5}\,\mathrm{m^2\,s^{-1}}$. Taking $L = 15\,\mathrm{m}$ and $V = 1\,\mathrm{ms^{-1}}$, we obtain $\mathrm{Re} = 10^6$, and this is far greater than the critical value $\mathrm{Re} \approx 2\,000$ which corresponds to the transition from laminar to turbulent flow. For such high values of Re, turbulence is highly developed; the kinetic energy of large scale movements ($\sim L$) is gradually transferred to smaller scales of movement, with isotropisation, down to a scale l at which the energy is dissipated by viscous friction. The local velocity of a turbulent fluid is a random variable $V(\boldsymbol{r}, t)$ (see Appendix B), depending on the point $\boldsymbol{r}$ and the time t. At any instant $t = t_0$, $V(\boldsymbol{r}, t_0)$ can be decomposed into spatial harmonics of the wave vector $\boldsymbol{\kappa}$, giving a further random variable $\vartheta(\boldsymbol{\kappa})$:

$$V(\boldsymbol{r}) = \iiint \vartheta(\boldsymbol{\kappa}) \mathrm{e}^{2\mathrm{i}\pi\boldsymbol{\kappa}\cdot\boldsymbol{r}}\,\mathrm{d}\boldsymbol{\kappa}.$$

The mean kinetic energy (in the sense of an ensemble average) $\mathrm{d}E(\kappa)$ in the interval between κ and $\kappa + \mathrm{d}\kappa$, where $\kappa = |\boldsymbol{\kappa}|$, is proportional to $\langle|\vartheta(\boldsymbol{\kappa})|^2\rangle$, which is the power spectrum of the velocity $V(\boldsymbol{r})$. Straightforward dimensional analysis leads to the formula

$$V \propto (\varepsilon_0 R)^{1/3}$$

for the fluid velocity, where ε_0 is the rate of production (or dissipation) of turbulent energy, and R the scale under consideration ($R = 1/\kappa$). We thus obtain

$$\mathrm{d}E(\kappa) \propto \kappa^{-2/3}\,\mathrm{d}\kappa.$$

The dimensional argument mentioned above is based upon the following observation: at scales R which are large compared with the scale at which viscous dissipation takes place, the turbulent velocity V should be independent of viscosity. Indeed, it should depend only on the energy production rate ε [$\mathrm{J\,s^{-1}\,kg^{-1}}$], the scale R, and the mass density ρ of the fluid. The only combination of these quantities having the dimensions of a velocity is the one given above. See Landau L., Lifschitz E., *Fluid Mechanics* (Pergamon, Oxford).

By integration, we obtain the spectrum of the kinetic energy, known as the *one-dimensional Kolmogorov spectrum*,[6]

$$E(\kappa)d \propto \kappa^{-5/3},$$

provided κ is in the so-called inertial range

$$L_0^{-1} \lesssim \kappa \lesssim l_0^{-1},$$

where l_0 and L_0 are called the *internal* and the *external scale* of the turbulence, respectively. The assumed *ergodicity* (see Appendix B) of fully developed turbulence allows us to identify the ensemble and time averages

$$\left\langle |\vartheta(\boldsymbol{\kappa})|^2 \right\rangle = \left\langle |\vartheta(\boldsymbol{\kappa}, t)|^2 \right\rangle_t .$$

Turbulence described by a $\kappa^{-5/3}$ one-dimensional power spectrum is said to be *homogeneous*. Outside the inertial range, the spectrum is governed by generation and dissipation processes. Near ground level, the value of the external scale L_0 varies between several metres and several hundred metres.

When turbulence occurs in a layer with a temperature gradient differing from the adiabatic one, it mixes air of different temperatures at the same altitude, and hence produces temperature fluctuations.

It can be shown that the equations governing velocity and temperature fluctuations are analogous, and that the one-dimensional spatial spectral density of the temperature fluctuations is

$$\Phi_\theta(\kappa) \propto \kappa^{-5/3},$$

where

$$\Phi_\theta(\kappa) = 4\pi\kappa^2 \left| \iiint \Theta(\boldsymbol{r})\, \mathrm{e}^{-2\mathrm{i}\pi \boldsymbol{r}\cdot\boldsymbol{\kappa}}\, \mathrm{d}\boldsymbol{r} \right|^2,$$

and $\Theta(\boldsymbol{r})$ is the temperature fluctuation about the mean $\langle T(\boldsymbol{r})\rangle$

$$\Theta(\boldsymbol{r}) = T(\boldsymbol{r}) - \langle T(\boldsymbol{r})\rangle.$$

More generally, Obukhov A.M. (Izv. Nauk. SSSR. Ser. Geograf. Geofiz. **13**, 58, 1949) has shown that the concentration of an impurity which is passive (i.e., does not affect the dynamics of

[6]Named after the hugely productive Soviet mathematician Andreï Kolmogorov (1903–1987). Apart from the first detailed study of turbulence, he also worked on signal processing, the subject of Chap. 9.

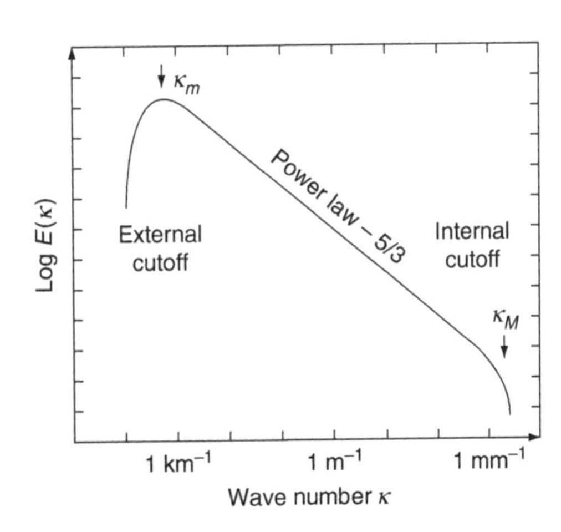

Fig. 2.13 One-dimensional power spectrum $E(\kappa)$ of the velocity fluctuations in a turbulent fluid, where the turbulence is isotropic and fully developed between the two scales L_0 and l_0 (turbulence obeying Kolmogorov's law in this interval). The corresponding wave numbers are $\kappa_m = 1/L_0$ and $\kappa_M = 1/l_0$. The ordinate is $\log E(\kappa)$. A variation in intensity of the turbulence (or of the energy injected at the scale L_0) results in a vertical shift of the curve

the fluid) and additive (i.e., does not react chemically) is also governed by the Kolmogorov law.

The spectral density of the temperature fluctuations also has low and high frequency cutoffs, just like the energy spectrum (Fig. 2.13).

It thus follows that the three-dimensional spectrum is

$$\Phi_T(\boldsymbol{\kappa}) \propto \kappa^{-11/3}.$$

The covariance of the temperature fluctuations is defined by

$$B_T(\boldsymbol{\rho}) = \langle \Theta(\boldsymbol{r}) \Theta(\boldsymbol{r} + \boldsymbol{\rho}) \rangle,$$

where $\langle \quad \rangle$ denotes a mean over the space variable $\boldsymbol{r}$. $B_T(\boldsymbol{\rho})$ is the Fourier transform of $\Phi_T(\boldsymbol{\kappa})$ (Wiener's theorem, see Appendix B):

$$B_T(\boldsymbol{\rho}) = \iiint \Phi_T(\boldsymbol{\kappa}) \mathrm{e}^{-2\mathrm{i}\pi \boldsymbol{\kappa} \cdot \boldsymbol{\rho}} \, \mathrm{d}\boldsymbol{\kappa}.$$

This integral diverges if $\Phi_T(\boldsymbol{\kappa})$ follows the power law down to zero, but remains finite with the cutoff at $\kappa_m = 1/L_0$.

The *structure function* of the random variable $\Theta(\boldsymbol{r})$ is defined as

$$D_T(\boldsymbol{\rho}) = \langle |\Theta(\boldsymbol{r} + \boldsymbol{\rho}) - \Theta(\boldsymbol{r})|^2 \rangle,$$

and remains defined for all values of $\boldsymbol{\rho}$, which indeed motivates the definition. It can then be shown, once again by dimensional considerations, that

$$D_T(\boldsymbol{\rho}) = C_T^2 \rho^{2/3},$$

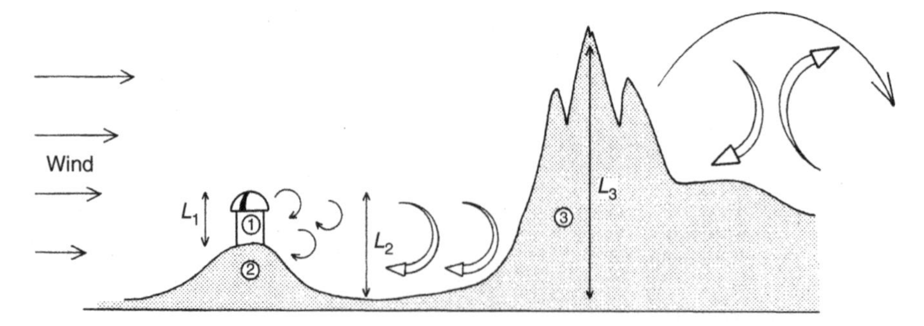

Fig. 2.14 Schematic representation of the generation of turbulence in the atmosphere by different obstacles. The amplitude of the temperature fluctuations depends on the amplitude of the turbulence and on the deviation of the actual temperature gradient from the adiabatic gradient. The scales L_1, L_2, L_3 are characteristic of the external scales of turbulence caused by wind around the obstacles 1, 2, and 3. Clearly, the choice of an astronomical observation site must take into account the prevailing winds and the relief upwind, as whirlwinds and eddies in turbulence damp down very slowly

and the three-dimensional temperature fluctuation spectrum is

$$\Phi_{\mathrm{T}}(\kappa) = \frac{\Gamma(8/3)\sin(\pi/3)}{4\pi^2} C_{\mathrm{T}}^2 \kappa^{-11/3} = 0.033\, C_{\mathrm{T}}^2 \kappa^{-11/3}.$$

C_{T}^2 is called the *structure constant* of the temperature fluctuations for homogeneous turbulence. The numerical value of C_{T}^2 characterises the intensity of the turbulence.

Turbulence in the atmosphere can be generated on different scales. Near ground level, the planetary boundary layer has thickness of the order one kilometre, and turbulence is produced by the flow of winds over surface (orographic) irregularities (see Fig. 2.14). Figure 2.15 shows a measurement of the vertical structure of turbulence in the first twenty kilometres above ground level. The structure constant $C_{\mathrm{T}}^2(z)$ is determined locally by measuring the correlation in scintillation of the two components of a double star, using a technique called SCIDAR (SCIntillation Detection And Ranging).

The external scale L_0 is still not very well known. It varies enormously with the site and the conditions of excitation of the turbulence (wind), and its values range from several metres to several hundred, or thousand, metres. A good knowledge of L_0 is particularly important for optical interferometry (see Sect. 6.2.3), determining as it does the amplitude of the differential phase perturbation (*piston effect*) between two widely separated pupils.

At any altitude, the shearing motions of winds can produce a turbulent interface between layers in laminar flow. In the tropopause, this mechanism can lead to large values of C_{T}^2, because of the high velocity (100–200 km h^{-1}) of the winds associated with the general circulation of the atmosphere (the jet stream). In addition, this general circulation can interact with large surface features (mountains) and cause gravity waves, which are likely to break in the non-linear regime of turbulent flow.

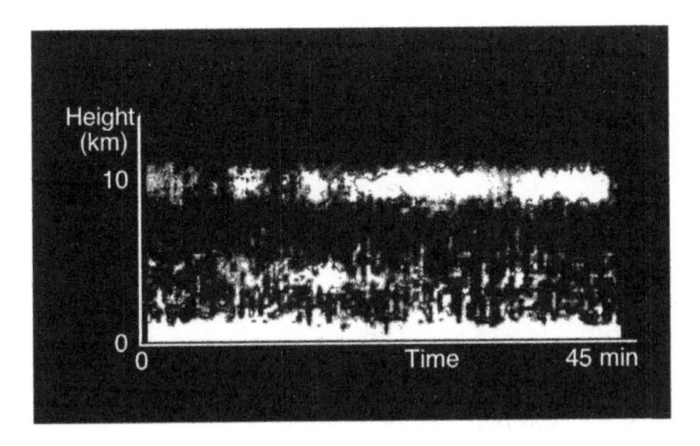

Fig. 2.15 Optical sounding of atmospheric turbulence. The *ordinate* gives the altitude and the abscissa the time (total duration of measurement was 45 min). The level of intensity gives the value of $C_T^2(z,t)$. The vertical resolution is ± 800 m, and the time resolution 20 s. Three turbulent layers can be seen: the *planetary boundary layer* near ground level, a layer at 4.8 km, and the layer at the tropopause (10–11 km). The sudden time variations show the intermittent nature of the turbulence. (Measurements carried out at the Observatoire de Haute-Provence, 1981, by Azouit M., Vernin J., J. Atm. Sci. **37**, 1550)

The criterion for the onset of turbulence caused by wind shear is given by the *Richardson number*

$$\mathrm{Ri} = \frac{g}{T} \frac{\partial T/\partial z}{(\partial V/\partial z)^2},$$

where $V(z)$ is the wind velocity at altitude z. When $\mathrm{Ri} < 0.25$, the flow generally becomes turbulent.[7] This is a commonly occurring situation in the generation of terrestrial turbulence, when there is density and/or temperature stratification.

Strictly speaking, the quantity which should appear in the above expression is T_P, the *potential temperature*, and not T. The potential temperature is the temperature that a volume element of dry air would have if it were transformed adiabatically to a pressure of 1 bar.

It is thus possible, from a set of vertical soundings of the atmosphere (sounding balloons), to establish the distribution $C_T^2(z)$. The structure function $D_T(\boldsymbol{\rho})$ of temperature variations is measured at various altitudes using rapidly-responding thermal probes with spatial separation $\boldsymbol{\rho}$. This distribution can also be measured from the ground using optical methods (Fig. 2.15), or indirectly from its effect on astronomical images, by determining the Fried parameter r_0 and its time development (see Sect. 6.2).

Knowledge of local turbulence is an essential parameter in the choice of an observing site.

[7]Woods J.D., Radio Sc. **4**, 1289, 1969, discusses the transition between laminar and turbulent regimes in atmospheric flow.

Temperature Fluctuations and the Refractive Index of Air

Fluctuations in temperature and the concentration of H_2O cause fluctuations in the refractive index of air. This can be written

$$n = n_0(T, C_{H_2O}) + ik(C_{H_2O}),$$

where the imaginary part k corresponds to absorption by H_2O and is important in the infrared and millimetre regions. (The concentrations of other constituents are ignored, their mixing ratios varying only by small amounts locally.) Fluctuations $\Delta n = n_0 - \langle n_0 \rangle$ of the real part can be written

$$\Delta n = \frac{\partial n_0}{\partial T}\Theta + \frac{\partial n_0}{\partial C}c.$$

For low values of the concentration C_{H_2O}, such as characterise astronomical sites, we obtain

$$\langle \Delta n^2 \rangle = \left(\frac{\partial n_0}{\partial T}\right)^2 \langle \Theta^2 \rangle.$$

As the fluctuations are isobaric, Gladstone's relation gives

$$\frac{\partial n_0}{\partial T} = \frac{80 \times 10^{-6}}{T^2} P,$$

for P in millibars (1 millibar [mb] = 100 pascal [Pa]) and T in kelvin [K]. The structure function for the refractive index can be written

$$D_n(\rho) = C_n^2 \rho^{2/3},$$

with a structure constant C_n related to C_T by

$$C_n = \frac{80 \times 10^{-6} P \text{ [mb]}}{T^2 \text{ [K]}} C_T.$$

Any fluctuation in the concentration of the absorber will lead to fluctuation in the imaginary part of the refractive index, and hence to fluctuation in the thermal absorption and emission of the volume element under consideration (*sky noise*, see Sects. 2.3.3 and 9.4).

As a first approximation, when studying the propagation of a wave, it is sufficient to know the integral of fluctuations along the line of sight. Hence the atmospheric turbulence of a given layer is characterised by the product $C_T^2 \Delta h$ for the temperature, or $C_n^2 \Delta h$ for the refractive index, where Δh is the thickness of that layer. Table 2.4 gives some typical values of the latter product, which is the only relevant parameter in image formation (see Sect. 6.2), for thin layers (several hundred metres) situated at various altitudes.

Table 2.4 Typical intensities of atmospheric turbulence

Altitude of layer [km]	3	6	10
$C_n^2 \Delta h$ [cm$^{1/3}$]	4×10^{-13}	13×10^{-13}	7×10^{-13}

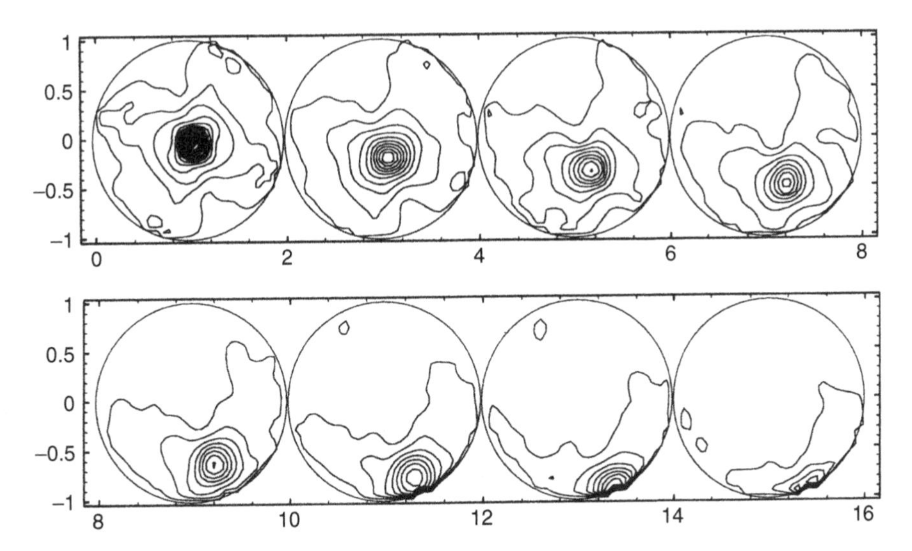

Fig. 2.16 Motion of a 'frozen' phase screen. The sequence of pictures shows the time development of the cross-correlation function (contours) of the phase of the wave emitted by a star (quasi-point source), reaching the circular pupil of a telescope (diameter $D = 3.6$ m) after crossing the atmosphere. The cross-correlation is calculated between time 0 and successive instants at intervals of 40 ms (see Appendix B). The uniform displacement of the cross-correlation peak can be clearly seen, pushed across by the wind, whose speed and direction can then be determined ($14\,\mathrm{m\,s^{-1}}$). 3.6 m ESO telescope, La Silla, Chile. (Gendron E., 1994)

Time Dependence of Turbulence

At a given point $\boldsymbol{r}$, the temperature is a random function $T(\boldsymbol{r}, t)$ of time. In the stationary case, it becomes possible to define a temporal power spectrum, characterising the time frequency content of the phenomenon (see Appendix B).

A simple model is made by considering the turbulence as 'frozen' (the *Taylor hypothesis*), with the above spatial power spectrum $E(\kappa)$, and assuming that a uniform wind is translating the air mass with velocity V (see Fig. 2.16). The temporal spectrum is then found by letting

$$\rho = V\kappa .$$

The temporal cutoff frequency is thus $f_c = V/l_0$, which gives, for $l_0 = 10$ mm and $V = 10\,\mathrm{m\,s^{-1}}$, the value $f_c = 10^3$ Hz. Indeed, this value is in good agreement with the order of magnitude of the highest frequencies observed in image deformation.

At a given site, the turbulence occurring in the various atmospheric layers, if indeed there are several, is generally a superposition of 'frozen' turbulence and of a turbulence which is more local, in the sense that it does not result from translation by a horizontal wind, but rather from vertical convection, for example near the tube or the dome of the telescope (*boiling*).

The physical origin of the Taylor hypothesis is simple: the time scales involved in development of turbulence are much longer than the time taken for a turbulent volume element, displaced by wind, to pass across the aperture of a telescope. It is therefore more correct to speak of a 'frozen phase screen', for a short time scale, than of 'frozen turbulence'.

2.6.2 *Ionospheric Turbulence*

In the weakly ionised plasma which makes up the ionosphere, the electron density N_e fluctuates. The expression for the refractive index $n(\nu)$ is

$$n(\nu) = \left(1 - \frac{\nu_p^2}{\nu^2}\right)^{1/2} .$$

Observation reveals that the relative electron density fluctuations are small.[8] It is found that

$$\langle \Delta N_e^2 \rangle^{1/2} \simeq 10^{-3} \langle N_e \rangle .$$

The correlation function for the refractive index is well represented by a Gaussian distribution

$$\frac{\langle n(\boldsymbol{\rho}) n(\boldsymbol{r} + \boldsymbol{\rho}) \rangle}{\langle n(\boldsymbol{\rho})^2 \rangle} \approx \exp\left[-\frac{r^2}{2a^2(\nu)}\right] ,$$

where $a(\nu)$ is a typical correlation length. For $N_e = 10^5\,\mathrm{cm}^{-3}$ ($\nu_p = 45\,\mathrm{MHz}$), it is found that a takes values between 0.1 and 1 km. The power spectrum, which is the Fourier transform of the autocorrelation function, is then also Gaussian.

2.7 The Atmosphere as Radiation Converter

2.7.1 *Ground-Based Gamma-Ray Astronomy*

The detection of photons with energies of several hundred GeV takes us into the regime of very high energy γ ray astronomy. Here an atmospheric emission that is at

[8] The book *Methods of Experimental Physics*, Vol. 12A, contains a detailed discussion of fluctuations in the ionospheric refractive index by T. Hagfon.

first sight undesirable is used to detect the γ radiation from sources of astronomical interest. The atmosphere can be considered here as a huge, indirect detector (in fact, an electromagnetic calorimeter) of γ radiation. The following could thus have been discussed in Chap. 7.

When high-energy γ rays of astronomical origins enter the atmosphere, they produce e^+e^- which emit other high-energy photons as they slow down. These photons in turn produce other pairs and the process repeats itself. In the upper atmosphere, the particles involved in these cascade reactions are relativistic, and in fact move faster than the speed of light for this medium, whence they emit Cherenkov radiation (see Sect. 7.6). This very short emission of blue light, lasting only a few nanoseconds, can be detected. The signal is nevertheless very weak and must be focused by very big mirrors to be recorded by photomultiplier tubes placed in their focal plane. Observation conditions are the same as for observatories in the visible region, i.e., the sky must be very clear and without light pollution. The short duration of the Cherenkov signal is used to eliminate photon noise due to night sky emission, whence the main source of noise is in fact from cosmic radiation. Indeed, high-energy cosmic charged particles produce air showers, analogous to those initiated by γ-rays photons. Only a careful analysis of events can distinguish one effect from the other.

2.7.2 *Air Showers and Cosmic Rays*

When a proton or charged heavy nucleus of cosmic origin reaches the Earth, it will collide with a nucleus in the upper atmosphere, thereby producing a large number of secondary particles. The latter, which share the energy of the incident particle, will in turn collide with other atmospheric nuclei, creating an even greater number of high-energy particles. This process is repeated right down to ground level, where billions of particles of many different kinds are observed, covering an area of several tens of square kilometers. The first ground-based or balloon-borne cosmic ray observatories sought high altitudes in order to be closer to the primary source of the shower (the French cosmic ray laboratory which operated between 1940 and 1950 was situated at 3842 m on the Aiguille du Midi in the Mont Blanc massif, France). We shall examine the many uses of this phenomenon in Chap. 7.

2.8 Terrestrial Observing Sites

The high cost of a modern observatory makes it essential, for ground-based observation, to choose the best possible site, whatever logistic difficulties it may involve. It is quite probable that there are very few exceptional sites, taking into account the criteria mentioned in the following, which refer, firstly, to the visible, infrared and millimetre ranges, and then to radio waves (centimetre and above).[9]

2.8.1 Visible, Infrared, and Millimetre Observations

The visible and infrared wavelength range is $\lambda \lesssim 30\,\mu\text{m}$, while the millimetre wavelength range is $\lambda \gtrsim 0.5\,\text{mm}$. Criteria common to the choice of these sites are: absence of cloud, photometric quality, infrared and millimetre transparency, and image quality.

Cloud Cover

Tropical and desert regions are clearly the best with regard to satisfying this criterion. In such regions, diurnal convection, caused by large increases in surface temperature, can lead to cloud formation, especially around mountain peaks. The most favourable situation is one in which temperature inversion occurs (Fig. 2.17), stabilising those layers closest to ground level and preventing the cloud layer from rising above the inversion layer. Such a configuration exists at the volcanic peaks of Teide and La Palma (Canary Islands), and Mauna Kea (Hawaii), and also along the coastal mountain ranges of Chile and Namibia. Satellite surveys of the Earth provide information accurate to resolutions of around ten metres (Spot satellites).

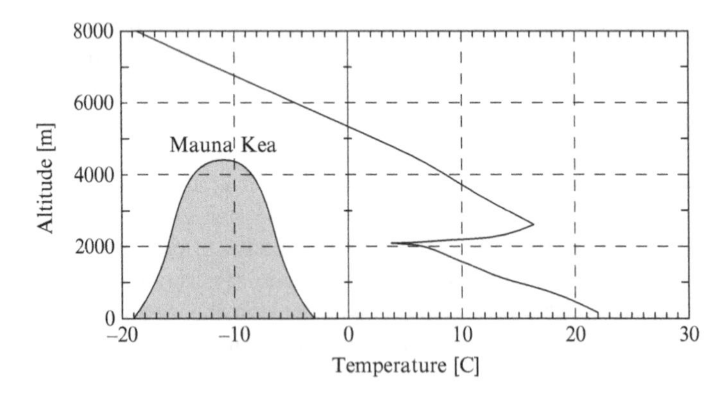

Fig. 2.17 Inversion layer above the Pacific Ocean, near the island of Hawaii (latitude 30° N). This is a *subsidence inversion*, caused by reheating of cold, descending air layers; the latter movement is itself caused by the general circulation of the atmosphere observed in *Hadley cells*. (Sounding balloon measurements, kindly communicated by P. Bely and the Hilo Weather Bureau, Hawaii, USA)

[9] An excellent review of the problems involved in selecting sites for optical astronomy can be found in *Site Testing for Future Large Telescopes*, European Southern Astronomy Workshop, J.P. Swings, Ed. Garching, 1983. See also the bibliography at the end of the book.

Regions with the least cloud are located in two bands on either side of the equator (10° to 35° N and 0–10° S to 35–40° S), but fluctuate over different longitudes.[10]

Photometric Quality

The photometric quality of a site refers to the stability of atmospheric transparency in the visible. The necessary photometric quality is considered to be reached in the visible when there are six consecutive hours of clear sky, for this allows sufficient time to apply extinction corrections (see Sect. 2.2).

Fluctuations in transmission are then of the order of 2%, or about 0.02 magnitudes. The above criterion is inadequate for infrared observation, affected by the passage of invisible clouds of water vapour, which mean that considerable fluctuations in IR transmission can coexist with an apparently clear sky.

Infrared and Millimetre Transparency

It was shown above that atmospheric absorption, thermal emission, and the spatial and temporal fluctuation (grain and sky noise, respectively) of the latter are all harmful phenomena caused by the presence of water vapour. For this reason, the choice of a site requires minimisation of the height of precipitable water, a criterion which once again favours polar sites and dry tropical sites, with seasonally anticyclonic weather conditions, and high altitude sites, given that the scale height of water is of the order of 2 to 3 km in the lower troposphere (Table 2.5). For example,

Table 2.5 Column of precipitable water at various sites. Values of h are mean values and may be subject to large seasonal variations. (After Chandrasekar T., Sahu K.C., Desai J.N., Infrared Phys. **23**, 119, 1983)

Site	Country	Altitude [m]	Mean column of precipitable water [mm]
South Pole	Antarctic	3 000	<1.0 (summer)
Jungfraujoch	Switzerland	3 570	2.8
Kardung La	India	5 200	1.5
Kitt Peak	Arizona	2 130	7.1
La Silla	Chile	2 440	3.9
Mauna Kea	Hawaii	4 200	2.2
Mount Lemmon	Arizona	2 600	4.9
Mount Palomar	California	1 706	6.0
Cerro Paranal	Chile	2 660	2.3
Chajnantor	Chile	5 100	<0.5
Tenerife	Canaries	3 600	3.8
Zelenchuskaya	Caucasus	2 070	5.7

[10] See Miller D.B., *Global of Cloud Cover*, U.S. Dept. of Commerce, 1971.

the site Cerro Chajnantor, in the Andes at the frontier between Chile, Argentina, and Bolivia, was chosen at the end of the 1990s to set up the Atacama Large Millimeter Array (ALMA) (see Sect. 6.5.1).

Image Quality

Variations in temperature, and hence in the refractive index of the air, perturb the phase of electromagnetic wavefronts (see Sect. 6.2). This in turn affects the quality of images and measurements at visible and millimetre wavelengths.

To first order, these effects on astronomical images are characterised by several parameters:

- The integral of the turbulence along the line of sight

$$\int_{z_0}^{\infty} C_{\mathrm{T}}^2(z)\,\mathrm{d}z \simeq \sum_i C_{\mathrm{T}_i}^2\,\Delta h_i\,,$$

  where the summation is over i distinct turbulent layers of thickness Δh_i.
- The evolution time of the turbulence, which is directly related to wind speed V_i in the various layers.
- The external scale L_i of the turbulence in each of the layers.

A more careful analysis of the effects of turbulence on images shows that the same turbulence but occurring at different altitudes will have different effects on an image.

In addition, turbulence is clearly not stationary over the periods of time considered (days, months, or even years), and so the histogram of its intensity over time must also be taken into consideration. Thus a site of average quality which benefits from periods of exceptional quality, such as the Pic du Midi in the French Pyrenees, may be preferred to a site with more uniform quality.

Local effects due to surface irregularities, as shown in Fig. 2.14, are often the dominant ones in the generation of turbulence, and so general rules cannot be given. Each potential astronomical site must undergo an in-depth study.

2.8.2 Centimetre and Metre Wave Radioastronomy

Radiofrequency interference is the main cause of perturbation in this range, and its avoidance has guided the choice of sites such as the Nançay centimetre telescope (France), the Very Large Array (VLA) in New Mexico (USA), and the Giant Meterwave Radio Telescope (GMRT) in Pune (India).

Other criteria enter into the choice of a good site: the latitude, with a view to covering as much as possible of the two celestial hemispheres; the horizontal surface area available for setting up interferometers (the Very Large Array occupies an area about 35 km in diameter); and the accessibility. This last criterion will no doubt become less and less relevant, given the possibility of remote control of observatories and transmission of their observational data. Completely automated space observatories have clearly demonstrated the feasibility of this procedure, which will be a characteristic feature of observatories in the twenty-first century.

2.8.3 Very High Energy Gamma-Ray Astronomy

The detection of very weak blue Cherenkov radiation from the upper atmosphere imposes the same conditions as for observatories observing in the visible and near infrared: clear skies, almost total absence of precipitation, and absence of light pollution. To satisfy these conditions, combined with the need to view the galactic center which can only be seen from the Southern Hemisphere, the site for the four 12 metre mirrors of the High Energy Stereoscopic System (HESS) was chosen in Namibia at the beginning of the 2000s (see Sect. 7.6).

2.8.4 Very High Energy Cosmic Radiation

The optimal altitude for direct detection of the particles producing extensive air showers, themselves created by very high energy cosmic rays, lies between 1 000 and 1 500 m. Sites must be flat over several thousand square kilometers so that radio links between detectors are not blocked by orographic features. In addition, the detection of near ultraviolet radiation ($\approx$ 300 nm) produced by fluorescence of atmospheric nitrogen, itself excited by passing particles, involves similar constraints to those affecting observatories in the visible region of the spectrum.

The international Pierre Auger observatory,[11] located in the flat deserts of the Pampa Amarilla in the west of Argentina, uses two detection techniques: observation of Cherenkov radiation produced locally in 1 600 water tanks, and observation of fluorescence from atmospheric nitrogen.

2.8.5 Man-Made Pollution and Interference

Human activity generates many difficulties for the running of a modern observatory, with its extremely sensitive detectors. To name but a few: light pollution in the visible, radiofrequency interference, heat sources such as nuclear power stations, which modify microclimates, vibrations, industrial aerosols, and the risks of an over-exploitation of space in the vicinity of the Earth.

---

[11]The French physicist Pierre Auger (1889–1993) observed cosmic rays at the Jungfraujoch observatory (Switzerland).

Light pollution results mainly from scattering of light from street lighting, by atmospheric aerosols and molecules.

The following empirical expression can be used to estimate light pollution:

$$\log I = 3 - 2.5 \log R + \log P,$$

where P is the population of a town in units of 10^5 inhabitants, R is the distance in kilometres, and I is the ratio of artificial intensity to natural intensity (see Sect. 2.3.1) at a zenith distance of 45°. A town of 100 000 inhabitants at 40 km thereby increases the brightness of the night sky by 10%. For a source of luminosity L lumens at a distance R, the expression becomes

$$\log I = -4.7 - 2.5 \log R + \log L.$$

Aeroplanes also perturb photographic plates with their light signals, as do artificial satellites with scattering of sunlight. Such objects often turn up on astronomical images!

These preoccupations are clearly analysed in Cayrel R. et al. (eds.), *Guidelines for Minimizing Urban Sky Glow near Astronomical Observatories*, IAU, 1980; in *Rapport sur la protection des observatoires astronomiques et géophysiques*, Académie des Sciences, Paris, 1984; and also in publications of the *International Dark Sky Association*, NOAO, Tucson, Arizona.

Radiofrequency pollution comes from various sources: fixed or mobile emitters, radiotelephones, radars, high voltage transmission lines, industrial ovens, and many others. Specific spectral bands essential for radioastronomy are protected as part of the general problem of allocating frequencies, and the International Astronomical Union has established a threshold of $2 \times 10^{-6}\,\mathrm{W\,m^{-2}}$ as the maximum tolerable level.

The question of man-made *vibrations* has become relevant with the advent of optical interferometry and the detection of gravitational waves, which both require a high degree of stability for the instruments, of the order of 10 nm rms or better. Up until now, tests carried out in good sites have always shown that natural micro-seismicity dominates over distant perturbations. It will nevertheless be essential to regulate nearby sources, in particular road traffic.

The control of *aerosols* and, more generally, the rejection of industrial effluents into the atmosphere is a general preoccupation in environmental protection. However, industrial activity and, in particular, mining, which may develop near to astronomical sites, has to be watched with great care.

For a while it was feared that *commercial exploitation of space* might lead to a proliferation of low orbit light sources (advertising, lighting of ground areas by large scale solar mirrors), and also of radio-reflecting screens, and stations producing energy transported by microwave beams. All these possibilities would represent a threat to Earth-based astronomical observation, but regulation, monitored by the International Astronomical Union (UAI), has been introduced to control this.

2.8.6 The Antarctic

The ice cap covering the Antarctic continent offers possibilities for astronomy which have hardly been investigated yet. The plateau, at a mean altitude of

3 000 m and reaching a maximum height of 5 140 m, includes a wide range of different sites which are already equipped for intensive geophysical study under international collaboration (the South Pole, and nearby peaks on which bases have been established: Dome A, Dome C).

The temperature is very low on this plateau, the tropopause lying practically tangent to ground level. The atmosphere is extremely dry, with a quantity of precipitable water which can be less than 0.1 mm. Transmission in the infrared, submillimetre, and millimetre regions is certainly higher than anywhere else on the Earth's surface, and the emissivity correspondingly weak, both factors which contribute to increasing sensitivity of observation. It is likely that turbulence is much reduced by the absence of both low altitude convection and the high altitude jet stream, the atmosphere being stratified with a weak vertical temperature gradient. This fact also implies vastly improved image quality. Indeed, the systematic analyses[12] of image quality, now carried out both in summer and winter alike, have shown that there is a weakly turbulent layer confined to within roughly 30 m of the surface (Fig. 2.18).

Furthermore, astronomical objects are observable at zenith distances which remain constant over long periods of time. This is invaluable for making uninterrupted time sequences, when studying solar or stellar oscillations, determining stellar oscillation periods, or observing temporal micro-variations, with sensitivity greater than one thousandth of a magnitude.

Antarctic sites manifest some similarities with space or Moon-based sites, providing a transition from more conventional terrestrial sites. The establishment of permanent, high capacity astronomical observatories seems a likely possibility during the twenty-first century, bearing in mind that some instruments (e.g., a 1.7 m millimetre antenna) are already operating at the South Pole, set up by the United States in 1995, while many projects are under assessment for the French–Italian Concordia Station at Dome C, under the impetus of the International Polar Year (2007–2008).

2.9 Observation from Space

Space travel, from its beginnings in 1959, has revolutionised astronomical observation. The equipment put into orbit, usually unmanned, has shown a steady increase in both size and complexity, and its lifetime in orbit around the Sun, the Earth, or other planets has also increased, reaching more than a decade in the case of the Voyager probes and the satellite IUE (International Ultraviolet Explorer).

In the present book we have chosen to leave aside the possibilities of *in situ* observations and measurements in the Solar System, exemplified in the many exploratory missions to the planets and their satellites. These missions would require a whole book for themselves, as already stressed in Chap. 1. We shall thus restrict

[12]This is a fast-moving area, but the many investigations into the quality of Antarctic sites can be followed at http://arena.unice.fr.

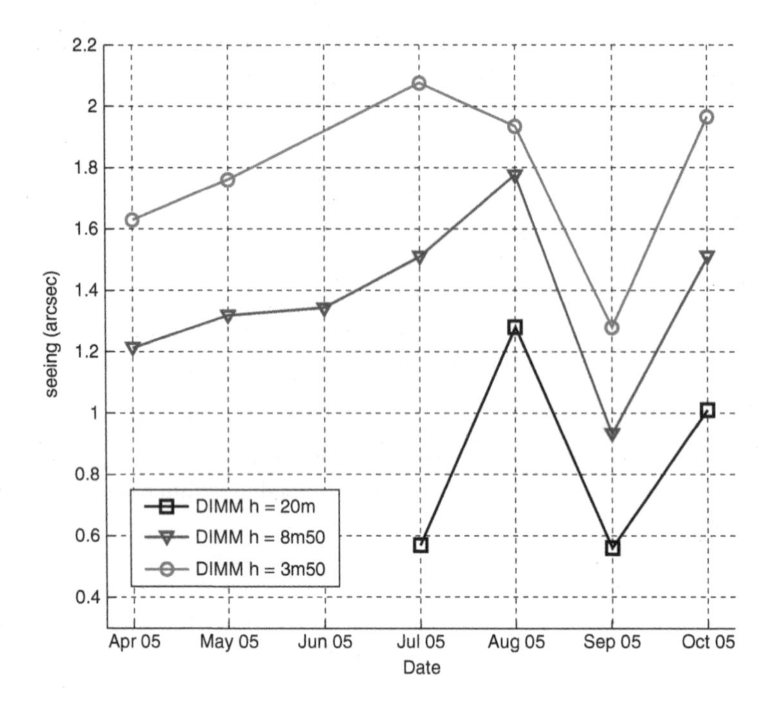

Fig. 2.18 Image quality at Dome C on the Antarctic plateau. The average size of a stellar image (seeing due to turbulence) is measured at different times of the year and at three different altitudes above the ground (3.5 m, 8.5 m, and 20 m), with an instrument called a *differential image motion monitor* (DIMM), used in site survey campaigns. These average values do not show the daily variability, which may occasionally reveal an exceptional seeing. From Aristidi, E., Fossat, E., Agabi, A., Mékarnia, D., Jeanneaux, F., Bondoux, E., Challita, Z., Ziad, A., Vernin, J., Trinquet, H., Dome C site testing, Astron. Astrophys. **499**, 955 (2009)

the discussion here to space observatories designed to observe the remote universe. A space observatory is not fundamentally different in principle from an observatory on the surface of the Earth. It contains telescopes, spectrographs, detectors, data processing equipment, and so on, the principles of which are dealt with in the following chapters. It is almost invariably unmanned, and therefore does not require any human intervention beyond remote control, although possibly in one of its more advanced forms, such as artificial intelligence or techniques involving virtual reality.

It is clearly not possible to cover all the details of space technology in this book. It is a fascinating subject, which includes many different aspects: the *launchers*, which determine the orbits and the mass of equipment; the *energy supply*, which determines manoeuverability and data transmission capacity; the various *protection systems*, which fend off particles, micrometeorites, and the like, and the *cryogenic systems* which guarantee whatever lifetime is required for the mission; and the *quality control* and *reliability studies*, which test the system as a whole. Some of these points are discussed briefly in the following chapters, in particular in Sect. 9.2, which deals with models for space observatories.

Traditionally, observations from atmospheric platforms, such as *aircraft* (10–20 km), *stratospheric balloons* (20–40 km), and *rockets* (up to 300 km), have been included under the denomination of space observation. Figure 2.4 shows the spectral ranges relevant to each of these platforms.

2.9.1 The Advantages of Observation from Space

The discussion earlier in this chapter has exposed three main causes of perturbation for astronomical observation: *absorption of radiation*, *turbulence*, and *interfering emissions*. The first of these considerably restricts access to the electromagnetic spectrum; the second affects image quality (resolution); and the third has a double effect, creating a uniform background from which faint sources must be extracted, but also, through *spatial fluctuations* (grain) and *temporal fluctuations* (sky noise), introducing a source of noise which reduces the sensitivity of observations.

Although atmospheric absorption practically disappears when an observatory is placed in low orbit ($z > 500$ km), some interference remains. Indeed, the transition is continuous between the upper atmosphere, the solar wind, with which it reacts (the magnetosphere), and the zodiacal dust cloud, which scatters the light from the Sun and emits its own thermal radiation. Moreover, the flux of particles coming from the Sun or diffusing through the Galaxy (cosmic rays) can interfere with detectors on board a space observatory, or even the materials they are made from. In the same way as the atmospheric signal, albeit invaluable for the study of the atmosphere, perturbs observation from the Earth, so the astronomical signals of nearby sources (the zodiacal nebula, the Sun, or even the Galaxy) are invaluable for the study of those sources, but perturb the observation of more distant and often fainter objects.

Some of these effects can be overcome by suitable choice of orbit; the problems caused by particles trapped in the terrestrial magnetosphere are to a large extent avoided by using low equatorial orbits (below 500 km) or very distant orbits (60 000 km and beyond). Other effects, such as zodiacal emission, are unavoidable as long as the orbit of the observatory is close to the Earth.

2.9.2 Sources of Perturbation

Although the conditions of observation from space are free from all those perturbing atmospheric phenomena which have been examined in this chapter, other perturbations come into play when an observatory has been put into space. These will be examined in the present section.

The Zodiacal Nebula

This is a distribution of dust grains in orbit around the Sun, in the neighbourhood of a plane of symmetry very close to the ecliptic (inclination $\sim 3°$). The thermal emission from this zodiacal cloud is considerable, because of the temperature of the dust grains, heated by the Sun (~ 300 K at 1 AU). The spatial distribution of the dust grains, and also the distribution of their sizes, has been determined by studying the way they scatter sunlight (see Fig. 2.19).

The number of grains with radius between a and $a + da$, at a distance r from the Sun (measured in astronomical units), and at ecliptic latitude β, is given by

$$n(r, \beta, a) = N_0(a) \frac{a^{-k}}{r\,[\text{AU}]} \exp(-2.6|\sin\beta|^{1.3}),$$

where the value of k determines one of the three grain populations involved, as shown in Table 2.6.

This early model [Frazier E.N., *Infrared Radiation of the Zodiacal Light*, SPIE **124**, 139, 1977] uses too high a value for the grain albedo (ratio of scattered to received energy), as shown by Mauser M.G., Ap. J. **278**, L19, 1984.

The monochromatic intensity received at the Earth can be expressed as

$$I(\lambda, \varepsilon, \beta) = I_S(\lambda, \varepsilon, \beta) + I_E(\lambda, \varepsilon, \beta),$$

where ε and β are the ecliptic coordinates of the line of sight (solar elongation ε and latitude β), and I_S, I_E are the contributions from scattered sunlight and intrinsic thermal emission, respectively.

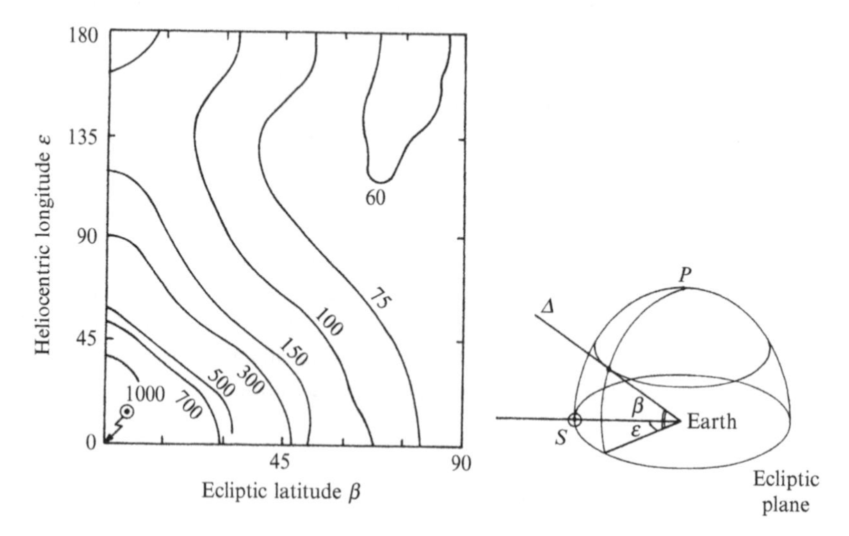

Fig. 2.19 Scattered zodiacal light. Isophotes of the scattered intensity measured in the direction Δ, characterised by its azimuth ε (measured from the Sun), and its ecliptic latitude β. P is the ecliptic north pole. The intensity contours are specified in S_{10} units, that is, in numbers of stars of magnitude $m_v = 10$ per square degree (see AQ, Sect. 73). For $S_{10} = 1$, the brightness at $\lambda = 540$ nm is $1.26 \times 10^{-8}\,\text{W m}^{-2}\,\mu\text{m}^{-1}\,\text{sr}^{-1}$, or $4.3 \times 10^{-16}\overline{B}_{\odot}$, where $\overline{B}_{\odot}$ is the average brightness of the Sun at that wavelength. (Levasseur-Regourd A.C., Dumont R., Astr. Ap. **84**, 277, 1980)

Table 2.6 Size of interplanetary dust grains

a [µm]	k	N_0 [cm^{-3}]
0.008–0.16	2.7	10^{-12}
0.16–0.29	2.0	1.1×10^{-14}
0.29–340	4.33	1.9×10^{-17}

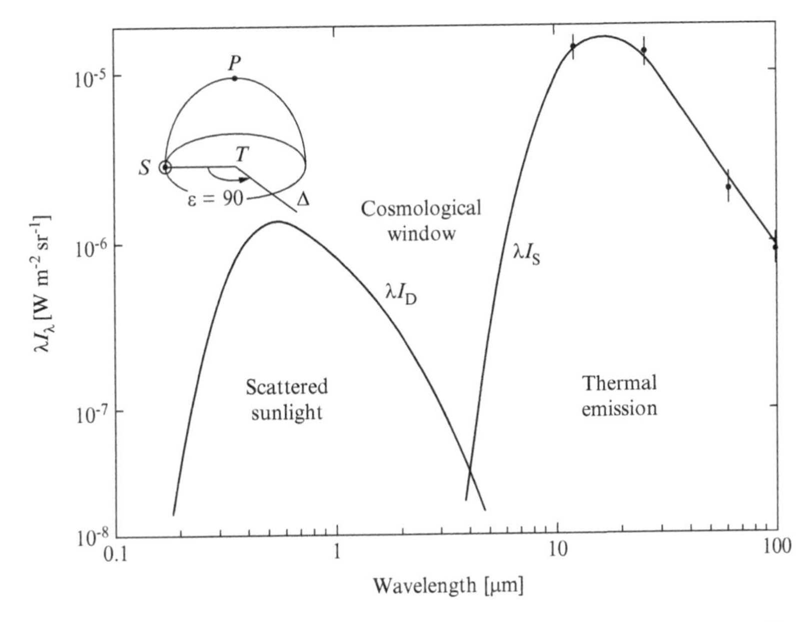

Fig. 2.20 Spectrum of zodiacal light in the visible and the whole of the infrared. The *ordinate* is the reduced brightness $\lambda I(\lambda)$. (The advantage of using this quantity is explained in Sect. 3.1.) The *inset* shows the line of sight Δ. The measurement points for I_S are from the satellite IRAS. (After Hauser M.G. et al., Ap. J. **278**, L19, 1984)

The scattered intensity I_S manifests exactly the colour of the Sun's photosphere, which has colour index $(B - V)_\odot = 0.65$, to be compared with $(B - V)_S = 0.64$. The spatial dependence of I_S is shown in Fig. 2.19 (see also AQ, Sect. 84) and the spectrum is given in Fig. 2.20.

Figure 2.20 gives the radiated intensity of the zodiacal nebula in the whole of the infrared region. This intensity depends very much on the position of observation, which corresponds, in the figure, to a choice of orbit close to the Earth and a line of sight lying in the ecliptic plane, at 90° to the direction of the Sun. The emission is fitted to measurement points obtained by the satellite IRAS (InfraRed Astronomical Satellite, 1983). It can be well represented by the curve corresponding to grey grains of constant emissivity 3.8×10^{-7}, with a mean temperature of $T = 235$ K; but even better by the curve for grains of λ-dependent emissivity (in fact, a function of λ^{-1}, see Sect. 2.4), with temperature 164 K. Note the deep minimum between the scattered and thermal components, near 3.5 µm. This minimum in the background emission due to the interplanetary 'atmosphere', is sometimes called the *cosmological window*, permitting, as it does, observation of distant, and therefore faint, objects. At wavelengths above 100 μm, great care must be taken to extract the contribution of the zodiacal nebula when measuring the weak signals of the cosmological background: the WMAP and Planck missions (the latter was launched in 2008) will thus refine our knowledge of this emission (see Sect. 7.4).

The observations made possible by this window may be carried out from the ground or from higher altitudes. The background zodiacal emission, which decreases as the wavelength increases from 1 to 3 μm, makes way for atmospheric emission, or indeed emission from the instrument itself, should the latter not have been cooled. Hence, the point of minimum emission is shifted anywhere between $\lambda = 1.6\,\mu$m (for the non-cooled Hubble telescope, emissivity 0.2) and $\lambda = 2\,\mu$m (for a ground-based telescope optimised in the infrared, emissivity 0.03), but attaining its lowest value at 2.5 μm (as envisaged for a 6 m telescope project, at temperature 200 K, emissivity 0.03, to be carried by stratospheric dirigible at altitude 12 km over the Antarctic).

High-Energy Particles and Photons

Diffuse Cosmic Background

In the X and γ ray regions, this background (see Fig. 2.21) consists mainly of a superposition of emissions with different redshifts, originating in active galactic

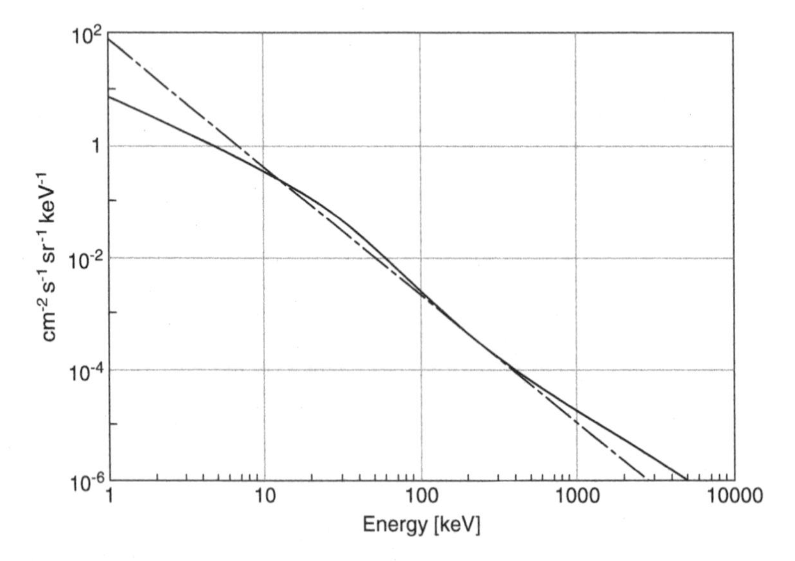

Fig. 2.21 Observed spectrum of diffuse electromagnetic cosmic background between 3 keV and 10 MeV (X and γ rays). The *straight line* represents the power law $E^{-2.3}$, which gives a good fit to the observations. (After Gruber D.E. et al., Ap. J. **520**, 124 1999)

nuclei (*quasars, Seyfert galaxies*). Visibility of a specific object in the field of view is reduced, because fluctuations in this cosmic background act as a source of noise when measuring the flux from the object. Between 10 keV and 1 MeV, the spectrum is governed approximately by the power law $I = 87.4 \times 10^{-2.3}\,\mathrm{cm}^{-2}\,\mathrm{s}^{-1}\,\mathrm{sr}^{-1}\,\mathrm{keV}^{-1}$. This source of noise, which dominates in the X region ($E < 50\,\mathrm{keV}$), becomes negligible at higher energies ($E > 500\,\mathrm{keV}$), where another background signal takes over, namely the one produced by interactions of charged particles with the detector and its environment. Collimators restricting the instrument's field of view are used to minimise background noise from this source.

Solar Wind

The solar wind is hydrogen plasma ejected from the Sun, which travels at high speeds along the field lines of the *heliosphere*. It varies with solar activity and its intensity determines the size of the heliosphere. In active solar periods, *solar flares* send huge quantities of particles into interplanetary space (see Fig. 6.37). The flux of particles can then increase by a factor of 1 000 for almost one day. Such events greatly perturb the running of observatories in orbit. Indeed, their measurements often become unusable, and the instruments, or even the spacecraft, may suffer considerable damage (e.g., electronic failure). It is of the utmost importance to assess the radiation impinging on each part of an experiment, and this throughout the mission, in order to select components of the appropriate quality (particularly, resistance to irradiation).

Radiation Belts

The trajectories of charged particles in the solar wind, electrons and protons, are modified by the lines of force of the Earth's magnetic field, and some particles are trapped in what are referred to as radiation belts (*van Allen belts*).[13] The proton belt extends roughly between altitudes 1 000 and 15 000 km. The electron belts extend as far as 50 000 km, and at high geomagnetic latitudes have 'horns' which descend to low altitude. There are two belts, or two maxima, in the distribution of electrons, the main one being at about 20 000 km, with a secondary maximum at about 3 000 km (Fig. 2.22). Moreover, the motion of the geomagnetic axis relative to the axis of the Earth's rotation tends to bring the belts closer together in the austral region of the Atlantic ocean, and this creates the so-called *South Atlantic Anomaly* (Fig. 2.23).

[13]Named after US physicist and astronomer James van Allen (1914–2006), who discovered the radiation (in fact, particle) belts which carry his name during the first space flights (the Explorer mission in 1958).

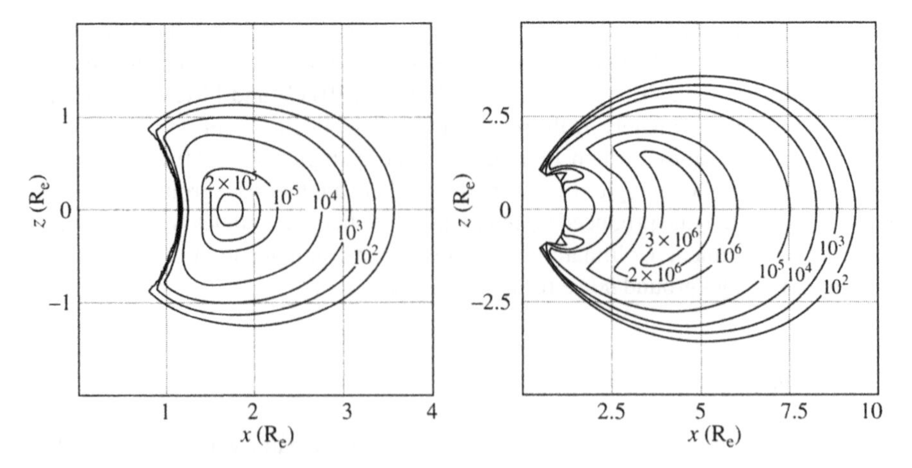

Fig. 2.22 Radiation belts of the Earth. Iso-intensity curves represent the mean flux, integrated over all directions, of electrons (*right*) and protons (*left*); z is the axis of the Earth's magnetic poles and x lies in the plane perpendicular to it, with units equal to R_e, the Earth's radius. Flux units are the number of particles $cm^{-3}\,s^{-1}$ above an energy threshold indicated in MeV. [After Daly E.J., Evaluation of the space radiation environment for ESA projects, ESA Journal **88** (12), 229, 1988]

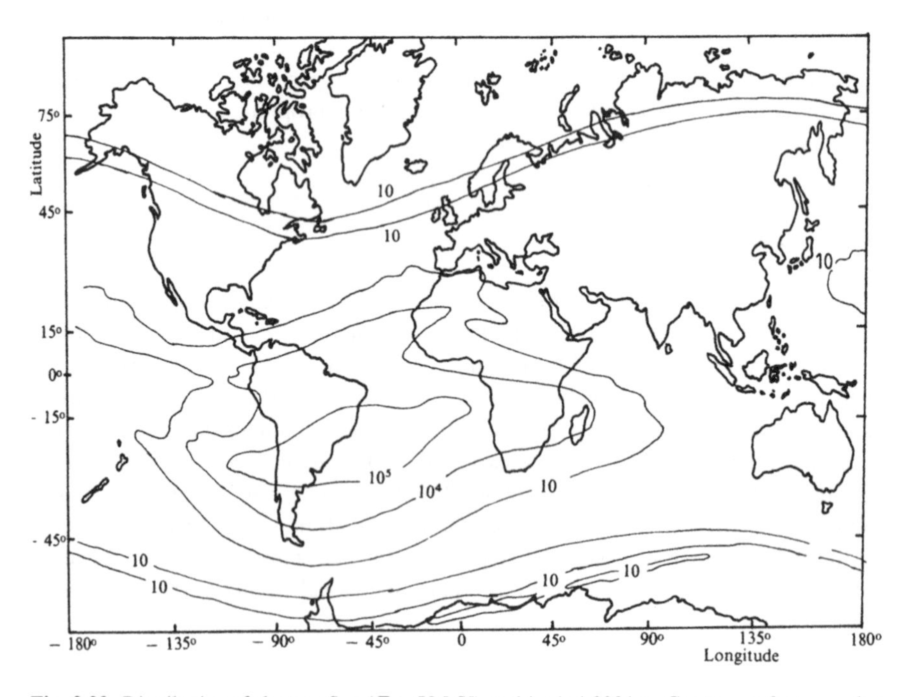

Fig. 2.23 Distribution of electron flux ($E > 5$ MeV) at altitude 1 000 km. Contours refer to number of electrons $cm^{-2}\,s^{-1}$. The flux is essentially directed along the lines of force of the Earth's magnetic field. Note the *South Atlantic Anomaly*, and the concentrations in the auroral zones. The data shown is only quantitatively correct for the epoch 1968–1970. (After Stassinopoulos E.G., NASA SP-3054, 1970. With the kind permission of E.G. Stassinopoulos)

Cosmic Rays

The protons, electrons, and nuclei comprising cosmic rays enter the Solar System and interact with the heliosphere, which opposes their penetration. The extent of the heliosphere, made by the solar wind, increases as the Sun is active, and hence the penetration of cosmic rays is governed by solar activity; the flux of cosmic rays in the neighbourhood of the Earth is maximum when solar activity is minimum, and conversely (Fig. 2.24). This phenomenon is called *solar modulation*. The Earth's magnetosphere also opposes the penetration of cosmic rays.

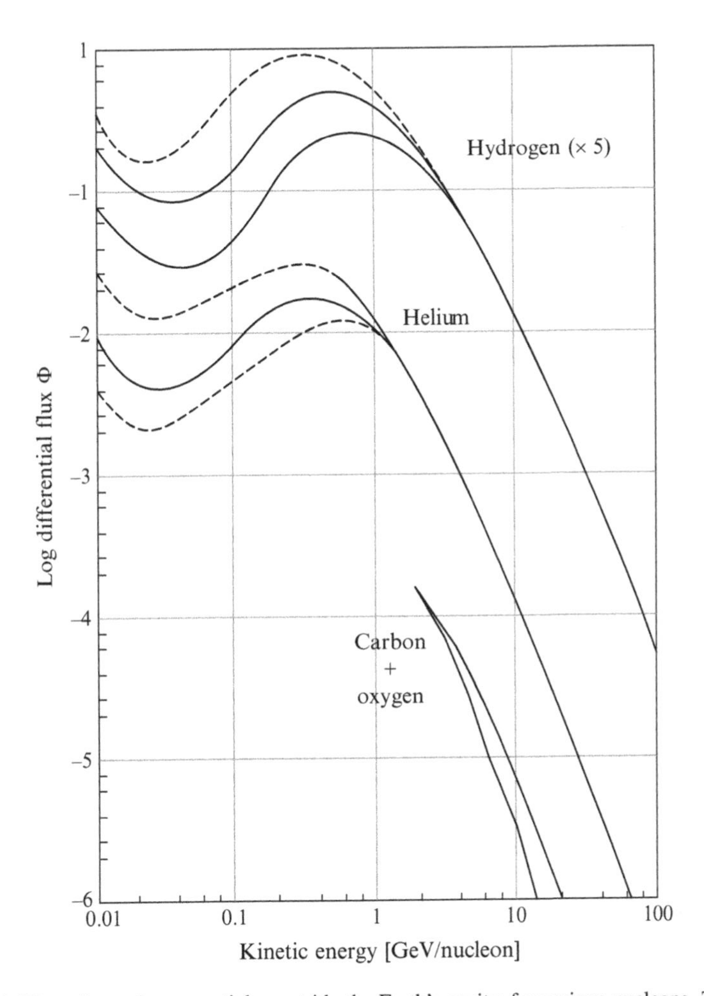

Fig. 2.24 Flux of cosmic ray particles outside the Earth's cavity, for various nucleons. The *ordinate* Φ is a differential energy distribution (number of nucleons considered $m^{-2}\,sr^{-1}\,s^{-1}\,MeV^{-1}$), and the *abscissa* is the nucleon energy. Modulation due to solar activity is visible at low energies: *dashed curves* indicate maximum and minimum levels of the solar cycle (*upper* and *lower curves*, respectively); *continuous curves* refer to mean solar activity. (After Webber W.R., Lezniak J.A., Astrophys. Sp. Sci. **30**, 361, 1974)

Background from Interaction with Surrounding Matter

Orbiting observatories, and all measurements they carry out, are affected by the arrival of these particles. Protons from the Sun or in cosmic rays induce *spallation* in all matter constituting or surrounding experiments. This matter thus becomes radioactive, emitting neutrons or γ rays, either immediately (instantaneous de-excitation), or after a lapse of time determined by the lifetime of whatever radioactive isotopes have been created. Hence, not only do charged particles interfere directly with γ-ray detectors, simulating the detection of high energy photons, but also the detectors are blinded by γ ray emission from de-excitation of surrounding matter.

This emission has a highly complex spectrum, containing many de-excitation lines superposed upon a continuous emission. This emission contributes to the background and limits the sensitivity of the experiment. As its spectrum has a slope less than the spectrum of diffuse emission, the former greatly exceeds the latter in the γ ray region. The contribution varies with the particle environment of the satellite, which is due to belts and cosmic rays, and therefore depends on the orbit, the position of the satellite on this orbit, and solar activity.

2.9.3 Choice of Orbits

With the advent of space travel, it has become possible to launch *probes* to make measurements, and to explore *in situ*, or in the vicinity of, objects in the Solar System; and also to launch *observatories* to measure electromagnetic radiation, for the main part, but possibly gravitational radiation in the future, which has been emitted from distant objects outside the Solar System, thus complementing in this case the role of ground-based observatories.

The choice of orbit for a probe depends at the outset on the object under study, e.g., the Moon, Mars, Saturn, cometary nucleus, and so on, the problem being one of organising an encounter, with the help of calculations in the realm of celestial mechanics. The aims of the observation dictate the choice of location for the observatory. But other factors must nevertheless be taken into account: the power and availability of launchers; the cost of the mission; and the location of receiving stations on the Earth or of relay satellites in direct view of the space observatory, given that rates of data production are often such that it cannot be stored even temporarily aboard, but must be continuously transmitted. A compromise is often necessary between the ideal location and the one eventually opted for.

In *low equatorial orbits* (300 to 500 km), which are used by the space shuttle and the International Space Station (ISS), communication is easy and repairs are possible. On the other hand, their lifetime is reduced by residual atmospheric friction, the Earth blocks 2π sr of the field of view, and changes between night and day take place very quickly, leading to breaks in visibility of the studied source about once every hour. Such an orbit was chosen for the Hubble telescope because it could be easily accessed by the space shuttle.

In *high circular orbit* (6 000 to 100 000 km), pointing is easier, observation periods are long, blocking of the field of view by the Earth is reduced, and interference from scattering, or radiofrequency and thermal emissions, due to the Earth is weaker. On the other hand, the energy required for the launch and for communication is greater, and the costs are therefore higher. Such an orbit was chosen for the satellite Hipparcos (1989), but was never attained, owing to failure of a boost motor.

A compromise can be made by choosing a highly *elliptical orbit*, which requires less power for the launch, and passes close to the Earth to transmit its data, although this involves storing data on board; the satellite nevertheless spends most of its time far from the Earth and its associated interference emissions. The distance of the ISO satellite (Infrared Space Observatory, 1995), placed in an elliptical orbit, thus varied between 2 000 and 70 000 km.

The best orbits for γ ray astronomy are either very distant, avoiding the radiation belts, or else close circular equatorial orbits, avoiding the South Atlantic Anomaly and protected from cosmic rays by the magnetosphere. The latter offer by far the most suitable conditions for observation, but, unfortunately, are rather inaccessible from most of the larger launch pads (Cape Canaveral in Florida and Baikonour in Central Asia), and represent no interest whatsoever from the economic point of view (for telecommunications or remote sensing), or from a military point of view (observation of the Earth's surface). They are thus left unexploited by the space agencies. Indeed, there are some problems associated with them. Their occultation by the Earth limits observation periods and complicates manoeuvres. Radio links with the satellite are usually very short, unless a large number of relay satellites are available (a system of relay satellites called Transmission Data Relay Satellite System, TDRSS). In general, a mass memory must be installed aboard the satellite. For all these reasons, distant circular orbits (>60 000 km), or eccentric orbits (apogee around 200 000 km), are considered to represent the best compromise; although there is more noise, observation and transmission conditions are less restrictive.

Finally, the *Lagrange points*[14] should be mentioned. These are rather special locations in space at which the gravitational fields of the Earth and the Sun combine to give a local minimum of gravitational potential, thus favouring a stable orbit there. Given two bodies, there are five points at which a small object can remain in stable or unstable equilibrium while co-orbiting with the Earth (E) around the Sun: three are located on the line SE, L_1 (between S and E) and L_2 at 1.5×10^6 km symmetrically with respect to E, L_3 opposite E; L_4 and L_5 sit at the vertices of two symmetrical equilateral triangles with base SE. If a satellite is placed at one of these points, it can maintain a fixed position with respect to the Earth and the Sun, which is a great help for communications, but also for controlling thermal conditions. The European solar observatory SOHO (1995) occupies the point L_1, while L_2 is or will be occupied by the European missions Planck (2008), Herschel (2008), and Gaia (2011), then NASA's JWST (2013), successor to the Hubble Space Telescope.

As yet, no observatory operating beyond the confines of the Solar System, and therefore outside the interference emissions of the Sun or the zodiacal nebula, has been seriously envisaged, for reasons of cost and owing to the problem of fast transmission to Earth.

2.10 The Moon as an Astronomical Site

The Moon was explored for the last time by the astronauts of the mission Apollo 17 in 1972, and completely surveyed by the unmanned probe Clementine in 1994. The density of the lunar atmosphere is a factor of 10^{14} less than that of the Earth at

---

[14]Named after the French mathematician Joseph-Louis Lagrange (1736–1813), originally from the Piemonte, author of the *Treatise on Analytical Mechanics*, which uses differential calculus. He proved the existence of the stable points in the Solar System which carry his name.

its surface, and the Moon can be considered to be practically without atmosphere. It thus offers observation conditions very similar to those encountered in interplanetary space, which give access to the whole of the electromagnetic spectrum.

We briefly discuss the main features of the Moon's surface that are relevant to setting up astronomical observatories:

- The lunar day lasts 27.321 terrestrial days. Such a long night allows long integration periods on a single source. In the polar craters, and in particular in the deep crater discovered at the south pole, the Sun only grazes the horizon, offering simultaneously a source of energy and a permanent state of half-night.
- The lunar surface is stable, seismic activity being less by a factor of 10^8 than on Earth. This could provide an almost infinitely stable foundation for telescopes. The stability would be invaluable for long baseline interferometry, which could extend over hectometres or kilometres, in the seas or in the depths of craters, and could observe at wavelengths anywhere from the ultraviolet to the millimetre range.
- The absolute instantaneous position of the Moon relative to stellar reference frames (see Chap. 4) is known to a very high degree of accuracy, far greater than the accuracy to which the position of an orbiting observatory can be known. The absolute position of the Moon, and its position relative to the Earth, are known to within a few millimetres, which is quite remarkable.
- The ground temperature varies widely between day and night, from 90 to 400 K except perhaps at high latitudes. This is a major disadvantage as regards thermal protection of instrumentation, but the very low temperatures reached by the ground provide a potential cryogenic source which would be invaluable for the cooling of telescopes, in the infrared and submillimetre regions, and for the cooling of detectors.
- The weak gravity on the Moon (0.16 g) makes it possible to build large structures which are both rigid and light. It is sufficiently strong, however, to bring surface dust back down to the ground, whereas any debris in space tends to co-orbit with its source. There nevertheless remains some uncertainty concerning the effects of a surface electric field on the dust; the surface of the dust grains is charged by the photoelectric effect of solar UV and by the space charge associated with it.
- The face of the Moon which remains permanently hidden from the Earth is entirely free of man-made radiofrequency interference, a fact which would strongly favour the setting up of radiotelescopes. Moreover, there is no ionosphere, and this would open the way to very low frequency radioastronomy, between 30 MHz and 100 kHz, which means wavelengths between 10 m and 30 km, a range quite inaccessible on the Earth, unless conditions are exceptional. Combination with a radiotelescope carried by satellite would then provide very long baseline observations, even at these frequencies. This part of the spectrum has hardly been explored as yet.
- Apart from the cost of setting up a lunar base, either manned or operated by robots, the main disadvantages are: the continual bombardment of the lunar surface by the solar wind and cosmic rays; the intense solar radiation in the EUV and X-ray regions; and the incessant impacts of micrometeorites (100 microcraters larger than 0.05 mm across are created per square metre per year).

Even though the construction of an astronomical base on the Moon is not for the immediate future, it is important to bear the idea in mind, because in some ways a lunar site could compete with orbital sites.

Problems

2.1. Using the fractional content of water vapour in air (Fig. 2.3), calculate the total quantity of precipitable water above the altitudes 4 km (Hawaii-sized mountain) and 12 km (airborne observatory).

2.2. The absorption cross-section of molecular oxygen is roughly $\kappa = 10^{-1}\,\mathrm{cm^2\,g^{-1}}$. Calculate the horizontal optical depth of the atmosphere over a length $l = 1$ km at altitude $z = 4$ km, in the molecular oxygen absorption band at wavelength $\lambda = 4.8$ mm ($\nu = 62.5$ GHz). Compare the result with the values given in Fig. 2.6. Calculate the integrated optical depth for the whole atmosphere.

Answer. The mass of O_2 along a horizontal line of length l, and per unit normal area, is

$$\frac{l\,\rho_0(z)\,M(O_2)}{5M_0},$$

where $\rho_0(z)/M_0$ is the molecular density at altitude z, and $M(O_2)$ is the molecular mass of O_2. At $z = 4$ km, $M_0 = 29$ g and

$$\rho_0(z) = e^{-z/H} M_0 \frac{P}{RT_0} = 774\,\mathrm{g\,m^{-3}},$$

which gives $\tau = 10^3 \times 774 \times 32 \times 10^{-1} \times 10^{-4}/29 \times 5 = 1.7$. This should be compared with the value given in Fig. 2.6; the attenuation in db km^{-1} is approximately 5, which implies $\tau = 5 \times 0.23 = 1.15$.

Vertically, the mass of O_2 per unit area normal to the vertical line is

$$\frac{M(O_2)}{5M_0} \int \rho_0(z)\,dz,$$

which implies that $\tau = 22.6$ from ground level, and $\tau = 13.6$ from $z = 4$ km. The atmosphere is thus optically thick at this wavelength.

2.3. Find the damping coefficient γ for the pure rotational transition of the H_2O molecule at $\lambda = 2.1$ mm, assuming that the band profile is Lorentzian (Fig. 2.6). Compare the value obtained with the collision frequency at the given pressure.

2.4. From the variation of horizontal attenuation of a band of O_2 as a function of altitude (Fig. 2.6), calculate the scale height of this component in the troposphere. Calculate the scale height of H_2O for the concentrations used in Fig. 2.6.

Answer. The z dependence of the horizontal attenuation A is

$$A(O_2) \propto \rho_{O_2}(z) \propto e^{-z/H},$$

which implies $H = (z_2 - z_2)/\log(A_1/A_2)$. Then, from Fig. 2.6, $H = 8$ km for O_2 and $H = 4.5$ km for H_2O.

2.5. What is the wavelength limit for near ultraviolet observations from a site at an altitude of $z = 2.86\,\text{km}$ (Pic du Midi observatory, France), or $z = 4.2\,\text{km}$ (Mauna Kea, Hawaii)?

2.6. Calculate the collision frequency of atmospheric constituents at an altitude of $z = 100\,\text{km}$, and compare it with the frequency of spontaneous radiative decay of neutral oxygen O I by the transition ^1S to ^1D, with wavelength $\lambda = 557.5\,\text{nm}$, for which $A_{\text{rad}} = 1.34\,\text{s}^{-1}$. Adopt a collision cross-section of $\sigma \approx 10^{-20}\,\text{m}^2$.

Answer. For $z = 100\,\text{km}$, $T = 210\,\text{K}$ and $P = 10^{-3}\,\text{mb}$, and the typical speed v of a molecule is given by

$$\frac{3}{2}kT = \frac{M_0\langle v^2\rangle}{2\text{N}} \quad \text{implying} \quad v = \sqrt{\frac{3kT\text{N}}{M_0}},$$

where N is Avogadro's number. The typical molecular density is $n = P\text{N}/\text{R}T$, and the typical time between two collisions

$$t = A_{\text{col}}^{-1} \approx \frac{1}{n\sigma v} = 0.8\,\text{s},$$

implying $A_{\text{col}} = 1.2\,\text{s}^{-1} \gg A_{\text{rad}}$. At this altitude, the collisional de-excitation rate has decreased sufficiently for spontaneous radiative de-excitation to become significant.

2.7. At an astronomical site of average quality, the seeing disc of a star has an angular size of 2″. At an exceptional site, and at the right moment, an optimised

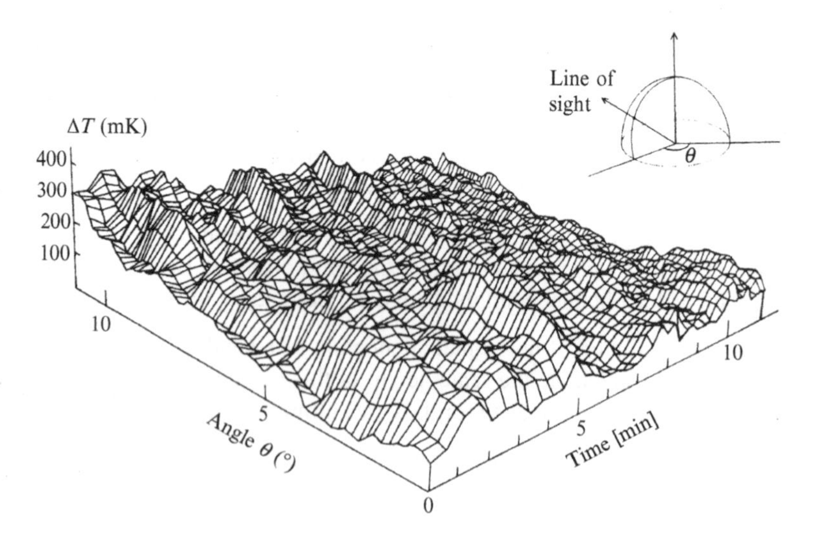

Fig. 2.25 Space and time fluctuations of the atmosphere at submillimetric wavelengths (Mauna Kea, Hawaii, 4 200 m)

telescope can give an image ten times smaller. What is the gain in contrast obtained for the detection of a quasar (angular size $\ll 1''$), or for the detection of a galaxy (angular size $> 1''$)?

Answer. The energy of an unresolved source (in this case, a quasar) is diluted in the marks it generates across the image plane. The gain in spatial resolution thus leads directly to a gain in contrast. On the other hand, for a resolved source, there is no effect whatsoever, the flux received per pixel being the same. Apart from improving image resolution, the techniques of *adaptive optics* can thus increase the capacity to detect weak unresolved sources.

2.8. Submillimetre Observations on Hawaii. The space and time fluctuations $I_\sigma(\theta, t)$ of atmospheric emission are measured at Mauna Kea, altitude $z = 4.2$ km, $\sigma = 11\,\text{cm}^{-1}$, where θ is the azimuth of the line of site and t the time. Rayleigh–Jeans emission is assumed, with brightness temperature $\langle T \rangle = 100$ K. Figure 2.25 shows the fluctuation ΔT (mK) (ordinate) for this brightness temperature. From the figure, estimate the mean emissivity ε of the atmosphere, and also the standard deviations σ_θ and σ_t, in space and time, respectively, of the received emission. Given that the average wind speed is $v = 35\,\text{km}\,\text{hr}^{-1}$ during these measurements, calculate the average size of the atmospheric inhomogeneities causing the observed fluctuations. (After Pajot F., Thesis, 1983.)

Chapter 3
Radiation and Photometry

This chapter deals with several fundamental properties of electromagnetic radiation.

We first define the basic notions of *radiometry*, namely, the physical quantities associated with the energy transported by electromagnetic radiation. Those photometric quantities and units related to lighting techniques, photography, and so on, are all adapted to the sensitivities of the human eye. This means that they refer to visible wavelengths, and thus are not suitable for astronomical observations, which cover the whole electromagnetic spectrum and must therefore be independent of the properties of the human eye.

We then consider the *coherence properties of radiation*, which are related to the space and time distributions of the energy transported. These properties are particularly relevant in the study of image formation (see Chap. 6), and the study of spectra (see Chap. 8).

Blackbody radiation plays a central, although not exclusive, role in the emission of astrophysical objects, and also in the limitations on observation. Several relevant properties will be discussed. A detailed treatment of radiative transfer in astrophysical objects, or the physical mechanisms producing the various emissions, will not be given here.

Absolute calibration aims to provide a way of converting measurements into the absolute energy values emitted at one or another wavelength by the objects under study, be they stars, interstellar clouds, molecules or dust grains, pulsars, galactic nuclei, accreting X-ray sources, cosmological background, or any other. These values determine the overall balance of mass or energy in the Universe, and its evolutionary time scales, even revealing, from time to time, the presence of previously unknown physical phenomena. Hence their great importance in astrophysics and physical cosmology.

The difficulty involved in constructing *calibration sources* in the laboratory, which remain accurate over a wide range of wavelength regions, should not be underestimated. Comparison of these sources with astronomical sources is also a difficult task, when the Earth's atmosphere, or the observation conditions of unmanned telescopes in space, are taken into account. It is not unusual, under these conditions, for certain physical quantities to be revised by quite large amounts,

P. Léna et al., *Observational Astrophysics*, Astronomy and Astrophysics Library,
DOI 10.1007/978-3-642-21815-6_3, © Springer-Verlag Berlin Heidelberg 2012

with significant consequences for any previously elaborated models. As a result, astrophysics has sometimes earned itself the undeserved reputation of being a science in which only orders of magnitude are possible, rather than precise values. This situation is rather a sign of the great experimental difficulty involved in observation.

3.1 Radiometry

Originally, the term *photometry* was used by opticians, and referred to measurement of the various energy quantities associated with *visual* sensations. It was thus closely linked to the spectral properties of a rather specific detector, the human eye. *Radiometry*, in the same classical sense, deals more generally with the energy transported by electromagnetic radiation, at whatever wavelength.

Astronomy has developed its own terminology, although more or less in agreement with the existing official definitions, as it passed through stages of visual, then photographic, and then photoelectric observation. And astrophysics, which deals with *radiative transfer* from a theoretical point of view, has also adopted a language not always perfectly coherent with existing terms. We shall try to clarify these points.

The term *photometry*, as it is used in astronomy, covers the quite general meaning of radiometry. We use it in this way, noting that most of the quantities to be defined refer to some specified wavelength or frequency, and that, strictly speaking, the correct term would be *spectrophotometry*, which is indeed often used. Photometry is very closely linked to the general ideas of *spectroscopy*, dealt with in Chap. 8.

Definitions

Etendue (Throughput)

This *geometrical* notion defines the limits of a set of rays carrying energy (Fig. 3.1). Considering a pencil of rays emitted by the surface element $\mathrm{d}S_1$, and received by the element $\mathrm{d}S_2$ at a distance r, the *étendue* or *throughput* is defined as

$$\frac{1}{r^2}\,\mathrm{d}S_1\,\cos\theta_1\,\mathrm{d}S_2\,\cos\theta_2 \;=\; \mathrm{d}\sigma_1\,\mathrm{d}\Omega\,,$$

where $\mathrm{d}\sigma_1$ is the projection of $\mathrm{d}S_1$ normal to the direction of propagation $\boldsymbol{k}$, and $\mathrm{d}\Omega$ is the solid angle subtended by the pencil at $\mathrm{d}S_1$.

It can be shown that conservation of energy by the propagation implies that the product of the étendue and the square of the refractive index is constant. In vacuum, $n = 1$ and therefore the étendue is a conserved quantity.

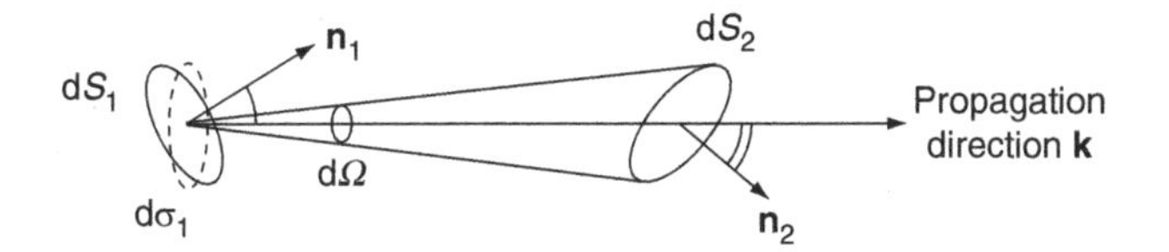

Fig. 3.1 Beam étendue

Transported Power

The power transported by radiation of wave vector $\boldsymbol{k}$, across surface element $\mathrm{d}\sigma$ normal to $\boldsymbol{k}$ at a point M, in the frequency range between ν and $\nu + \mathrm{d}\nu$, and in the solid angle $\mathrm{d}\Omega$, is written

$$\mathrm{d}P_\nu = I_\nu(\boldsymbol{k}, M)\,\mathrm{d}\sigma\,\mathrm{d}\Omega\,\mathrm{d}\nu.$$

I_ν is called the *specific intensity* at M in the direction $\boldsymbol{k}$ and is measured in W m^{-2} sr^{-1} Hz^{-1}. The term is used in the theory of radiative transfer, and is also the usual terminology in astronomy, although it differs from the one adopted internationally in standard radiometry, where *intensity* refers to the total emitted power (in watts), integrated over the whole surface of the emitter.

It will be shown in Chap. 7 that a distinction must be made between detectors which measure intensity and detectors which measure electric field $\boldsymbol{E}_\nu$. The quantities I_ν and $\boldsymbol{E}_\nu$ are related by the Poynting vector, whose average flux across a surface is the power crossing that surface. This gives

$$I_\nu = \frac{1}{2}\varepsilon_0\, c\,\langle E_\nu^2\rangle.$$

When the quantity I_ν refers to the power emitted by a surface, it is also called the *specific* or *monochromatic intensity*. The integral of this intensity over all space directions, divided by 4π, is called the *mean monochromatic intensity* [W m^{-2} Hz^{-1}].

Monochromatic Flux

The power received per unit area normal to the propagation direction $\boldsymbol{k}$ is

$$\frac{\mathrm{d}P_\nu}{\mathrm{d}S_2\cos\theta_2} = \mathscr{E}_\nu\,\mathrm{d}\nu,$$

where $\mathscr{E}_\nu$ is called the *monochromatic flux* produced by those sources illuminating the receiving surface (the observation site), and has units W m^{-2} Hz^{-1}. In radiometry, it is also called the *monochromatic irradiance*. It follows that

$$\mathscr{E}_\nu = \int_{\text{source}} I_\nu \cos\theta_1 \, d\Omega,$$

noting that the total monochromatic flux does not depend on the angular size of the source.

Although this quantity is called a *flux*, and this use is in perfect accord with the terminology of radiative transfer, it does not agree with the official use of the term in radiometry, where the flux is the total power transported by the radiation, and is thus measured in watts.

The *flux unit* (FU), which is often used in astrophysics, is actually a unit of monochromatic flux, also called the jansky (Jy), in honour of Karl Jansky who was the first to make radioastronomical observations:

$$1 \text{ jansky} = 10^{-26} \, \text{W m}^{-2} \, \text{Hz}^{-1}.$$

Magnitudes are logarithmic units of relative monochromatic flux, and their frequent use in astronomy justifies a detailed presentation (see Sect. 3.3). The term *monochromatic flux* should be adhered to in astrophysical discussions.

Other Units

Units measured per unit frequency interval, as above, can be replaced by units measured per interval of photon energy ($h\nu$), per wavelength interval (λ), or per wave number interval ($\sigma = \lambda^{-1}$, measured in cm^{-1}). Similarly, power can be measured in watts, but also in photon number per unit time, for photons of given frequency. Table 3.1 summarises the various units and the terminology encountered in astrophysical literature.

There is not always a satisfactory correspondence between the names of quantities used in astronomy and their official counterparts in radiometry. *Power* in the former becomes flux (in watts) in the latter, *mean intensity* becomes emittance or intensity (in W m^{-2}), and *flux* becomes irradiance (in W m^{-2}). The terms *monochromatic* and/or *spectral*, and the dimension Hz^{-1}, can be added quite naturally to whichever term is used.

Table 3.1 Units used in astronomy for the measurement of radiation

Quantity	Symbol	S.I. unit	Other units
Etendue		$\text{m}^2 \text{ sr}$	
Specific intensity	I_ν	$\text{W m}^{-2} \text{ sr}^{-1} \text{ Hz}^{-1}$	
	I_λ	$\text{W m}^{-2} \text{ sr}^{-1} \text{ m}^{-1}$	$\text{W m}^{-2} \text{ sr}^{-1} \, \mu\text{m}^{-1}$, etc.
	I_σ		$\text{W m}^{-2} \text{ sr}^{-1} \text{ cm}^{-1}$
	N_ν	$\text{quanta m}^{-2} \text{ sr}^{-1} \text{ Hz}^{-1}$	$\text{quanta m}^{-2} \text{ sr}^{-1} \text{ eV}^{-1}$, etc.
Monochromatic flux	$\mathscr{E}_\nu$	$\text{W m}^{-2} \text{ Hz}^{-1}$	jansky
	$\mathscr{E}_\lambda$	$\text{W m}^{-2} \, \mu\text{m}^{-1}$	$\text{quanta m}^{-2} \, \mu\text{m}^{-1}$, etc.

Reduced Brightness

The quantity

$$\lambda I_\lambda = \nu I_\nu \qquad [\mathrm{W\,m^{-2}\,sr^{-1}}]$$

is independent of the units of wavelength or frequency. Figure 3.2 shows how useful it can be in cases when spectra extend over broad wavelength ranges. Clearly, for a graph giving the quantity λI_λ (or νI_ν) as a function of $\log \lambda$ (or $\log \nu$, respectively), the powers emitted in various spectral intervals can be compared directly by comparing the areas under the relevant parts of the curve. This is a consequence of the equation

$$\int_{\Delta\nu} \nu I(\nu)\,\mathrm{d}\ln\nu = \int_{\Delta\nu} I(\nu)\,\mathrm{d}\nu.$$

A further illustration is given in Fig. 2.20.

Luminosity

The total power in watts emitted by an astrophysical source is called its *luminosity*, and is given by the integral

$$L = \int_S \int_{4\pi} \int_\nu \mathrm{d}P_\nu = \int_\nu L_\nu\,\mathrm{d}\nu,$$

where S denotes a surface bounding the source and L_ν is called the *monochromatic luminosity* $[\mathrm{W\,Hz^{-1}}]$.

Polarisation

Several factors make this an important parameter. Firstly, it may reveal properties of the source (magnetic field, aligned dust grains, anisotropic scattering, etc.). Secondly, the detector may only be sensitive to one component of polarisation (e.g., at radiofrequencies). And thirdly, the telescope may alter the polarisation (optics, waveguides).

Consider a plane wave described by

$$E_x = a_1 \cos(2\pi\nu t - \boldsymbol{k}\cdot\boldsymbol{r} + \phi_1),$$

$$E_y = a_2 \cos(2\pi\nu t - \boldsymbol{k}\cdot\boldsymbol{r} + \phi_2),$$

where a_1 and a_2 are the amplitudes in the x and y directions, respectively, ν the frequency, $\boldsymbol{k}$ the wave vector, and ϕ_1 and ϕ_2 the phases. There are four quantities to be determined, namely a_1, a_2, ϕ_1, and ϕ_2. Putting $\phi = \phi_2 - \phi_1$, the polarisation of

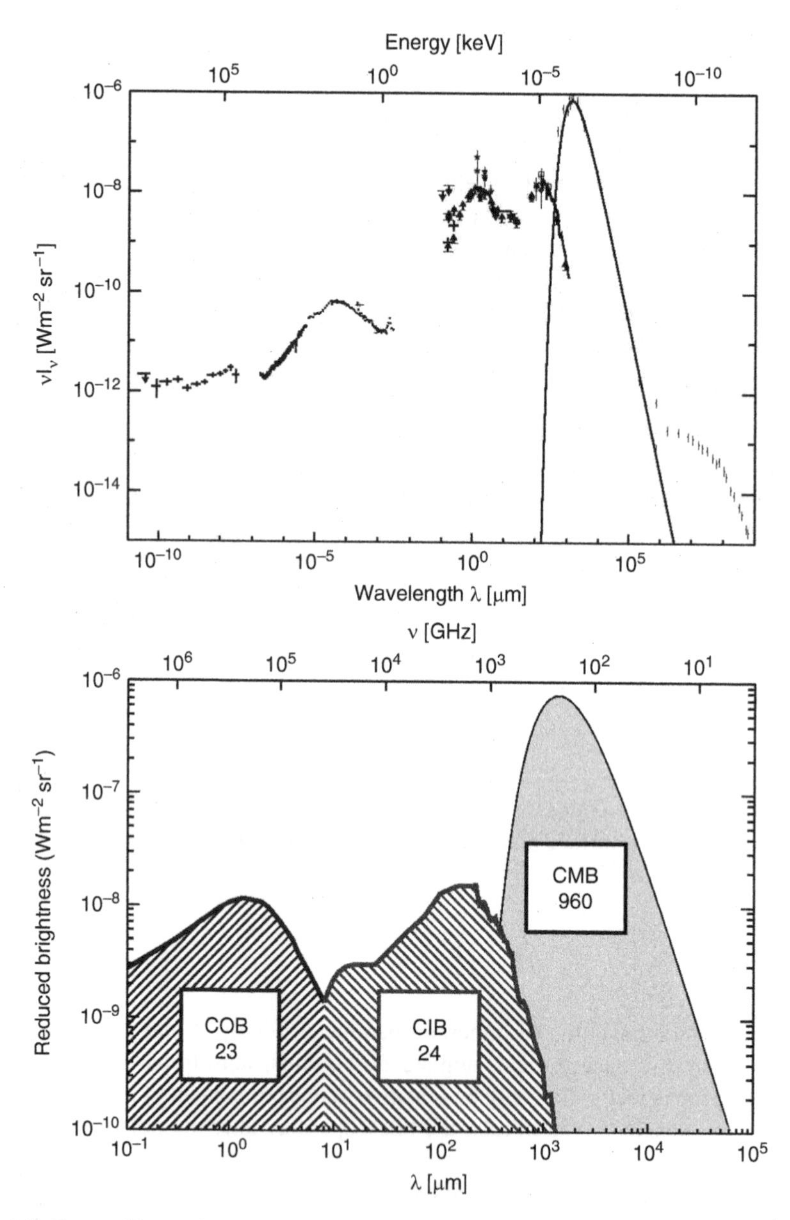

Fig. 3.2 *Upper*: Observed sky background radiation, plotting reduced brightness νI_ν against wavelength (*lower scale*) or against photon energy (*upper scale*). From left to right are the extragalactic ultraviolet, visible, and infrared contributions, the cosmological blackbody contribution, and finally the radiofrequency radiation in the Galaxy. The interplanetary component (zodiacal emission) has been subtracted. *Lower*: Smoothed extract from the above spectrum, showing the cosmic optical background (COB), cosmic infrared background (CIB), and cosmic microwave background (CMB). Numerical values indicate the total brightness of each component in $\mathrm{nW\,m^{-2}\,sr^{-1}}$. Note that the infrared background (formation and evolution of galaxies) is comparable with the optical background (stars), and that the latter is a small fraction of the microwave background. (Figures with the courtesy of Hervé Dole, private communication, 2007)

an arbitrary wave is conveniently specified by the *Stokes parameters*

$$\begin{aligned} I &= a_1^2 + a_2^2, \\ Q &= a_1^2 - a_2^2 = I \cos 2\chi \ \cos 2\psi, \\ U &= 2a_1 a_2 \cos\phi = I \cos 2\chi \ \sin 2\psi, \\ V &= 2a_1 a_2 \sin\phi = I \sin 2\chi, \end{aligned}$$

which are related by

$$I^2 = Q^2 + U^2 + V^2.$$

The degree of polarisation Π of a general wave is defined as

$$\Pi = \frac{(Q^2 + U^2 + V^2)^{1/2}}{I},$$

where $I,\ Q,\ U,\ V$ are time averages. For the plane wave (above), we obtain $\Pi = 1$.

3.2 Aspects of Radiation

3.2.1 Blackbody Radiation

Although no astronomical source emits exactly as a blackbody, many are sufficiently close to this case to allow their emitted radiation to be approximately characterised by a single temperature. Stellar spectra and the cosmological background provide good examples. Moreover, as telescopes are never at absolute zero temperature, they radiate, as do their surroundings. And detectors are immersed in this radiation, a fact which partly determines their performance.

The specific intensity of a blackbody is given by the Planck law

$$B_\nu(T) = \frac{2h\nu^3}{c^2}\left[\exp\left(\frac{h\nu}{kT}\right) - 1\right]^{-1},$$

with units W m^{-2} sr^{-1} Hz^{-1}. At high frequencies, this is approximated by the Wien distribution

$$B_\nu(T) \sim \frac{2h\nu^3}{c^2}\exp\left(-\frac{h\nu}{kT}\right), \qquad h\nu \gg kT.$$

At low frequencies, it is approximated by the Rayleigh–Jeans law

$$B_\nu(T) \sim \frac{2\nu^2}{c^2}kT = \frac{2kT}{\lambda^2}, \qquad h\nu \ll kT.$$

The total power radiated per unit surface of the blackbody is

$$\int_{\Omega} \int_{\nu} B_\nu(T)\, \mathrm{d}\nu\, \mathrm{d}\Omega = \sigma T^4 \quad [\mathrm{W\,m^{-2}}],$$

where $\sigma = 5.670 \times 10^{-8}\ \mathrm{W\,m^{-2}\,K^{-4}}$ is the Stefan–Boltzmann constant. Blackbody radiation obeys Lambert's law

$$\mathrm{d}P_\nu = B_\nu\, \mathrm{d}S\, \cos\theta\, \mathrm{d}\Omega\, \mathrm{d}\nu = B_\nu\, \mathrm{d}\sigma\, \mathrm{d}\Omega\, \mathrm{d}\nu,$$

which says that intensity is independent of direction of observation θ. The fact that the brightness of the Sun, or indeed a star, varies between the centre of the disk and the limb demonstrates that it is not strictly radiating as a blackbody.

The specific intensity is a maximum for a given temperature when

$$\frac{c}{\nu_\mathrm{m}} T = 5.0996 \times 10^{-3}\,\mathrm{m\,K} \quad \text{for} \quad \frac{\mathrm{d}B_\nu}{\mathrm{d}\nu} = 0,$$

$$\lambda_\mathrm{m} T = 2.898 \times 10^{-3}\,\mathrm{m\,K} \quad \text{for} \quad \frac{\mathrm{d}B_\lambda}{\mathrm{d}\lambda} = 0.$$

The intensity given by the Planck formula is only a mean intensity. We will see in Sect. 7.2, in the context of noise phenomena, that this mean value is subject to thermodynamic fluctuations.

3.2.2 *Coherence*

Coherence can be understood intuitively by considering two extreme types of radiation. Total coherence is manifested by a uni-directional monochromatic wave (frequency ν, wave vector $\boldsymbol{k}$). Indeed, it is possible to define the relative phase of this wave at two arbitrarily separated points in space and time. On the other hand, blackbody radiation exhibits the minimal coherence; at two separate points in space, or at two different instants of time, the phase relation of the field is arbitrary, unless the spatial and temporal separations between these two points are very small.

Intermediate cases exist: radiation which is more or less pure spectrally, and non-thermal radiation. The idea of coherence can be defined either in terms of classical electromagnetic wave propagation, or from the photon point of view.

The idea of spatial coherence is crucial in the treatment of image formation (see Chap. 6), whereas temporal coherence comes into the treatment of spectral analysis (see Chap. 8).

Field Coherence and Mutual Degree of Coherence

We consider a complex field $\mathbf{V}(t)$, which will be taken in the most general case as an ergodic and stationary random process, with power spectrum $\mathbf{S}(\nu)$. It is assumed that $\langle \mathbf{V}(t) \rangle = 0$, so that the autocorrelation and autocovariance are equal (see Appendix B for definitions and properties). We consider a well-defined polarisation of the field.

Let $\mathbf{V}(\boldsymbol{r}_1, t)$ and $\mathbf{V}(\boldsymbol{r}_2, t)$ be the fields measured at any two points in space, denoted $\mathbf{V}_1(t)$ and $\mathbf{V}_2(t)$ to simplify. Then the *complex degree of mutual coherence* of these fields is the quantity

$$\gamma_{12}(\tau) = \frac{\Gamma_{12}(\tau)}{[\Gamma_{11}(0)\Gamma_{22}(0)]^{1/2}},$$

where

$$\Gamma_{12}(\tau) = \langle \mathbf{V}_1(t)\mathbf{V}_2^*(t+\tau) \rangle$$

is the cross-correlation of $\mathbf{V}_1$ and $\mathbf{V}_2$. The quantity $\Gamma_{11}(0) = \langle \mathbf{V}_1(t)\mathbf{V}_1^*(t) \rangle$ represents the mean intensity of the field at point 1. The ensemble averages $\langle \ \rangle$ can be taken over time, given the ergodicity assumption (see Appendix B).

The quantity $\gamma_{12}(\tau)$ describes the *spatial coherence*, since it correlates the fields at two distinct points of space, and also the *temporal coherence*, since it compares them at two different instants of time, separated by τ.

Quasi-Monochromatic Radiation

Radiation whose spectral density is confined to a neighbourhood of some frequency ν_0 is referred to as *quasi-monochromatic radiation*. For example, for $\Delta\nu \ll \nu_0$,

$$S(\nu) = \exp\left[-\frac{(\nu - \nu_0)^2}{\Delta\nu^2}\right].$$

The autocorrelation of the field $\mathbf{V}(t)$ is then given by

$$R(\tau) = \exp\left(-\frac{\tau^2}{\tau_c^2}\right),$$

with the following important relation holding between the *spectral width* and the *temporal width* (Fig. 3.3):

$$\tau_c \, \Delta\nu \simeq 1.$$

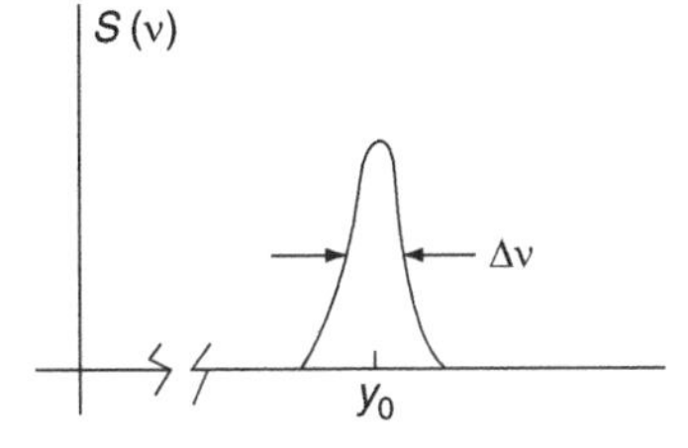

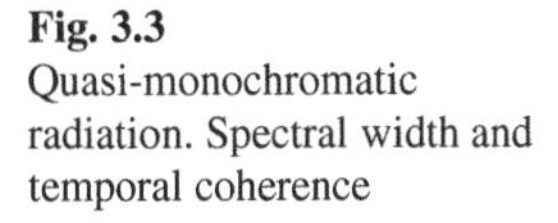

Fig. 3.3 Quasi-monochromatic radiation. Spectral width and temporal coherence

The *coherence length*, measured along the wave vector $\boldsymbol{k}$, is $c\tau_c$. It is the length over which the field retains the memory of its phase. Further

$$l_c = c\tau_c = \frac{\lambda_0^2}{\Delta\lambda}, \qquad \frac{\Delta\lambda}{\lambda_0} = \frac{\Delta\nu}{\nu_0},$$

and the coherence length is the distance beyond which the waves λ and $\lambda + \Delta\lambda$ are out of step by λ.

When the length l is such that $l \ll c\tau_c$, it follows that

$$\gamma_{12}(\tau) \sim \gamma_{12}(0)e^{-2i\pi\nu_0\tau},$$

and the coherence is determined by $\gamma_{12}(0)$. Consequently, $\gamma_{12}(0)$ is often referred to as the degree of coherence, in the case of quasi-monochromatic radiation.

Interference Measurements of Coherence

Consider a simple Young slit experiment (Fig. 3.4), in which two diffracted beams interfere. The *visibility* of the observed interference fringes is defined to be

$$\mathscr{V} = \frac{I_{max} - I_{min}}{I_{max} + I_{min}},$$

where I_{max} and I_{min} are the maximum and minimum intensities observed. It is not difficult to show that

$$|\gamma_{12}(\tau)| = \mathscr{V}.$$

In the same way, a Michelson interferometer (Fig. 3.5) measures temporal coherence.

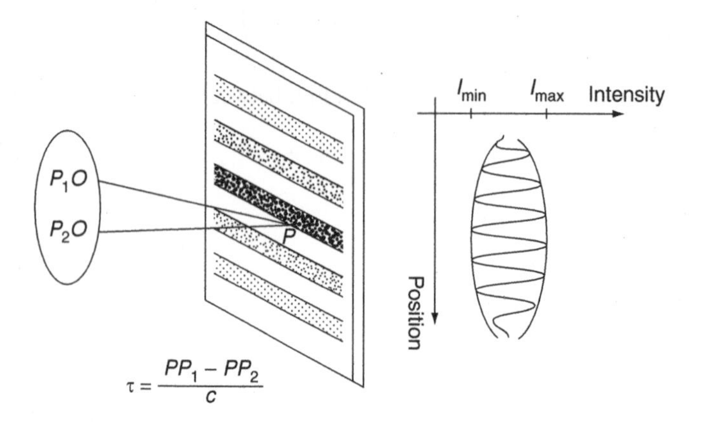

Fig. 3.4 Measurement of spatio-temporal coherence by interference

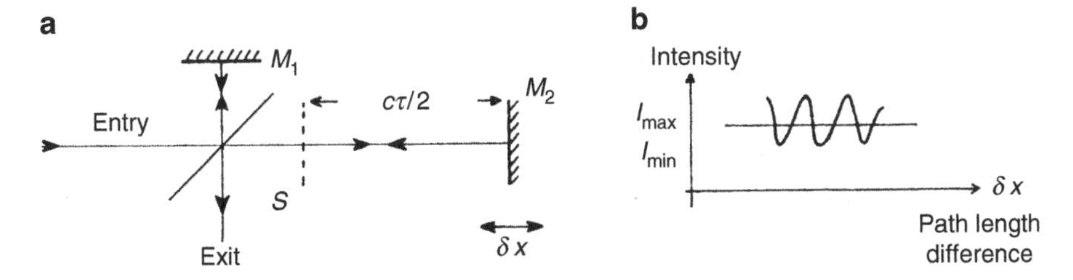

Fig. 3.5 Measurement of temporal coherence by interference. (**a**) Schematic representation of a Michelson interferometer. (**b**) Measurement of output intensity

Coherence of Blackbody Radiation

Quasi-monochromatic radiation is a particular case of a Gaussian process, where $R(\tau)$ is itself Gaussian. We recall that there exist infinitely many stationary Gaussian processes, determined by their mean and their autocorrelation $R(\tau)$ (see Appendix B).

As the radiation of a blackbody is a stationary Gaussian process, its coherence is completely defined by the quantity $R(\tau)$. This temporal coherence can be deduced by Fourier transform from the Planck blackbody spectrum $S(\nu)$.

Higher Order Moments

In the case of an arbitrary stationary process, the *higher order moments* must be specified:

$$\Gamma(x_1, x_2, \ldots, x_{2n}) = \langle \mathbf{V}(x_1)\mathbf{V}(x_2)\ldots\mathbf{V}^*(x_{n+1})\ldots\mathbf{V}^*(x_{2n})\rangle,$$

where x_i denotes a point $(\boldsymbol{r}_i, t)$. The *complex degree of coherence of order n* is defined as

$$\gamma_n(x_1, \ldots, x_{2n}) = \frac{\Gamma(x_1, \ldots, x_{2n})}{\left[\prod_i \Gamma(x_i, x_i)\right]^{1/2}}.$$

It is straightforward to check that a strictly monochromatic wave gives $\gamma_n = 1$ for any n.

Coherence and Photon Statistics

Coherence has been presented here from the wave point of view, as the quantity

$$\langle \mathbf{V}(t)\mathbf{V}^*(t+\tau)\rangle.$$

We now consider how this relates to the particle point of view.

Strictly monochromatic radiation $\exp(2\pi i\nu_0 t)$ has infinite coherence time. The wave point of view has nothing to say about fluctuations. In contrast, photons can be described as classical particles, and for any τ, the number of photons detected in time τ obeys a Poisson distribution. If $\overline{n}$ is the mean number of photons detected per unit time, the variance of n is

$$\langle \Delta n^2 \rangle = \overline{n}\tau.$$

Quasi-monochromatic radiation (for example, an optically broad spectral line, formed at local thermodynamic equilibrium LTE, and of Gaussian profile), has a finite coherence time $\tau_c \sim 1/\Delta\nu$. If $\tau \gg \tau_c$, the fluctuation in the number of photons detected in time τ is given by calculation using Bose–Einstein statistics

$$\langle \Delta n^2 \rangle = \overline{n}\tau(1+\delta), \qquad \delta = (e^{h\nu/kT} - 1)^{-1},$$

and contains both particle and wave terms.

In contrast, for thermal radiation (Gaussian), if $\tau < \tau_c$, the photons no longer obey stationary Poisson statistics. Indeed, the photons show a tendency to group together (*bunching*). The effect becomes significant when $\delta \sim 1$ (Fig. 3.6).

Finally, for non-thermal radiation, this phenomenon becomes more significant as the quantity δ, known as the *degeneracy factor*, increases.

The study of coherence, that is, the correlation of photons over short time intervals ($<\tau_c$), can be used to detect deviations from LTE in spectral line formation. Progress in the development of quantum detectors may lead to interesting applications of this idea in astrophysics.

For a general introduction to the effects of coherence, and a bibliography, see Harvey A., *Coherent Light*. See also the interesting discussion in Dravins D. et al., Appl. Opt., 1995. A detailed analysis of thermodynamic fluctuations can be found in Landau L., Lifshitz E.M., *Statistical Physics*. See also Harvey A., *Coherent Light*, Chap. 1.

3.3 Magnitudes

The magnitude system[1] has its origins in the classification of stars according to their visual brightness, introduced by the Greek astronomers. In the nineteenth century, this classification was formulated (by Pogson, 1856) in the form of a logarithmic law, which corresponded roughly to the visual sense.

Magnitudes are *relative measures* of the monochromatic flux of a source. When referring to stars, which are effectively point-like for observational purposes, the

[1] A more detailed discussion of these classic issues can be found, for example, in *Stars and Stellar Systems*, Vol. II, Astronomical Techniques. A more recent and very complete reference can be found at the website www.astro.virginia.edu/class/oconnell/astr511/lec14-f03.pdf (University of Virginia, USA).

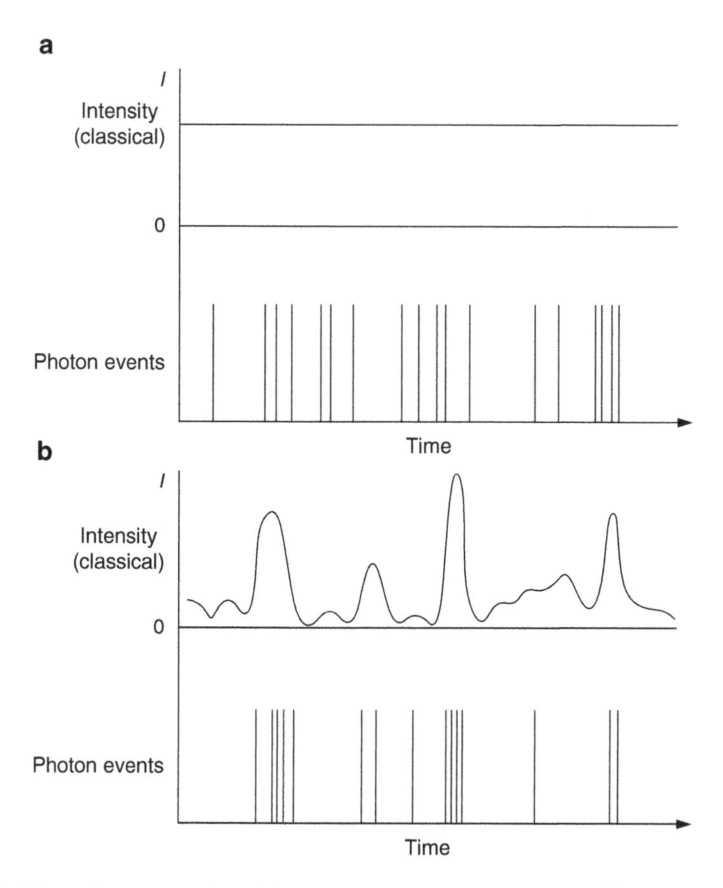

Fig. 3.6 Schematic representation of the semi-classical model of radiation. (**a**) Constant classical intensity leading to photoelectric events with Poisson time distribution. (**b**) Classical intensity of a thermal source, with fluctuations leading to a Poisson process combined with a Bose–Einstein distribution, and bunching. (Extract from Dainty J.C., in *Image Science*)

apparent size of the object is irrelevant; but in the case of extended sources, such as galaxies or the sky background, the extent must be taken into account, and a further unit, the *magnitude per square second of arc*, is introduced, so that the solid angle of observation is equal to 1 arcsec × 1 arcsec.

If $e(\lambda)$ is the monochromatic flux of a source, measured outside the Earth's atmosphere, the magnitude at wavelength λ_0 is defined as

$$m_{\lambda_0} = -2.5 \log \frac{e(\lambda_0)}{e_0} = -2.5 \log e(\lambda_0) + q_{\lambda_0}, \tag{3.1}$$

where the constant q_{λ_0} defines magnitude zero.

Table 3.2 Standard photometry. 1 Jy = 10^{-26} W m^{-2} Hz^{-1}. It is convenient to note that the magnitude $m_v = 0$ corresponds to a monochromatic flux of 10^{11} photons m^{-2} s^{-1} μm^{-1}

Name	λ_0 [μm]	$\Delta\lambda_0$ [μm]	e_0 [W m^{-2} μm^{-1}]	e_0 [Jy]	
U	0.36	0.068	4.35×10^{-8}	1 880	Ultraviolet
B	0.44	0.098	7.20×10^{-8}	4 650	Blue
V	0.55	0.089	3.92×10^{-8}	3 950	Visible
R	0.70	0.22	1.76×10^{-8}	2 870	Red
I	0.90	0.24	8.3×10^{-9}	2 240	Infrared
J	1.25	0.30	3.4×10^{-9}	1 770	Infrared
H	1.65	0.35	7×10^{-10}	636	Infrared
K	2.20	0.40	3.9×10^{-10}	629	Infrared
L	3.40	0.55	8.1×10^{-11}	312	Infrared
M	5.0	0.3	2.2×10^{-11}	183	Infrared
N	10.2	5	1.23×10^{-12}	43	Infrared
Q	21.0	8	6.8×10^{-14}	10	Infrared

In practice, any measurement necessarily includes some finite band of wavelengths, possibly characterised by a transmission filter $t_0(\lambda)$, and this explains why there exist several *magnitude systems*, depending on the choice of spectral bands. The magnitude m_{λ_0} of a source producing a monochromatic flux $e(\lambda)$ is then

$$m_{\lambda_0} = -2.5 \log \int_0^\infty t_0(\lambda) e(\lambda)\,\mathrm{d}\lambda + 2.5 \log \int_0^\infty t_0(\lambda)\,\mathrm{d}\lambda + q_{\lambda_0}.$$

Table 3.2 gives the reference intensities q_{λ_0} and the central wavelengths λ_0 for the *photometric magnitude system*, which makes the best use possible of the atmospheric transmission windows in the near UV, the visible, and the near and mid-infrared. The use of magnitude systems to measure fluxes is indeed limited to these three spectral ranges, for historical reasons.

The functions $t_0(\lambda)$, centred on λ_0, can be found in AQ. See also Low F., in *Methods of Experimental Physics*, Vol. 12A. *Photographic visual magnitudes* differ from those given in the table; the associated functions $t_0(\lambda)$ are chosen to represent the sensitivity of the eye or photographic plates as closely as possible.

The photometry of a source, sometimes obtained through the Earth atmosphere, sometimes from space, uses filters optimised for these different conditions (chromatic sensitivity of the detector, atmospheric absorption bands to be eliminated, etc.). The various magnitude systems have sought to optimise the choice of bands, while new observing conditions, e.g., space, can lead to new systems being devised, which must then be matched up with the previous ones via a form of absolute calibration — the star Vega as primary standard, with which a whole series of secondary standards can subsequently be associated. The proliferation of magnitude systems up until the year 2000 can also lead to a certain confusion, and difficulties in going from one system to another.

In 2007, a more intrinsic definition of magnitudes was gradually introduced, based on (3.1):

- **Space Telescope Magnitude (STMAG).** If the flux unit is e_λ, hence given per wavelength interval with units W m^{-2} μm^{-1}, the STMAG magnitude (used for observations by the Hubble Space Telescope) is defined by

$$m_\lambda = -2.5 \log_{10} e_\lambda - 18.6.$$

- **AB Magnitude.** If the flux unit is e_ν, hence given per frequency interval with units W m^{-2} Hz^{-1}, the AB magnitude is defined by

$$m_\nu = -2.5 \log_{10} e_\nu - 56.1.$$

  This magnitude system is the one used by the Sloan Digital Sky Survey, for example (see Sect. 10.2).

The advantage with these systems is that they allow immediate conversion, whatever the wavelength or frequency, from the magnitude to a monochromatic flux without atmosphere, which is a perfectly well-defined physical quantity.

Bolometric Magnitude

The apparent bolometric magnitude is a measure of the integral of the monochromatic flux over all wavelengths:

$$m_{\mathrm{bol}} = -2.5 \log \frac{\int_0^\infty e(\lambda)\,\mathrm{d}\lambda}{e_\mathrm{b}}, \qquad e_\mathrm{b} = 2.52 \times 10^{-8}\,\mathrm{W\,m^{-2}}.$$

It thus gives the luminosity, if the source radiates isotropically, if its distance D is known (in parsecs), and if there is no absorption between the source and the observer:

$$m_{\mathrm{bol}} = -0.25 + 5 \log D \; - 2.5 \log \frac{L}{L_\odot},$$

where $L_\odot = 3.827 \times 10^{26}$ W is the luminosity of the Sun.

Absolute Magnitude

Apparent magnitude is a measure of the flux received at the Earth. Comparison between sources is made possible by introducing *absolute magnitude*, which is the apparent magnitude the source would have at a distance of 10 pc (where 1 pc $= 3.086 \times 10^{16}$ m $= 3.261$ light years), correction being made for any interstellar absorption effects (*reddening*, denoted by A). The relation between apparent magnitude m and absolute magnitude M of an object at distance D is

$$M = m + 5 - 5 \log D - A.$$

Colour Indices

Quantities like U–B and V–K, and more generally, differences $m_{\lambda_2} - m_{\lambda_1}$, are called *colour indices*. They measure the ratio of fluxes at two different wavelengths.

The quantities q_{λ_0} are chosen so that all the colour indices beyond the V band of a star with spectral type A0 (dwarf) are zero. Explicitly, for a high temperature blackbody, whose spectrum thus obeys the Rayleigh–Jeans law ($B_\lambda \propto \lambda^{-4}$) in the UV, the visible, and the infrared, the colour indices are

$$\mathrm{B} - \mathrm{V} = -0.46, \qquad \mathrm{U} - \mathrm{B} = -1.33,$$

$$\mathrm{V} - \mathrm{R}, \quad \mathrm{V} - \mathrm{I}, \quad \ldots, \quad \mathrm{V} - \mathrm{N} = 0.0.$$

By comparing colour indices, the fluxes at two wavelengths can be related using this reference spectrum, and the spectral slope or *colour* of the emitted radiation can be calculated (Fig. 3.7). The colour index is extremely useful for classifying large numbers of sources or stars, observed in *sky surveys*, into different types of objects.

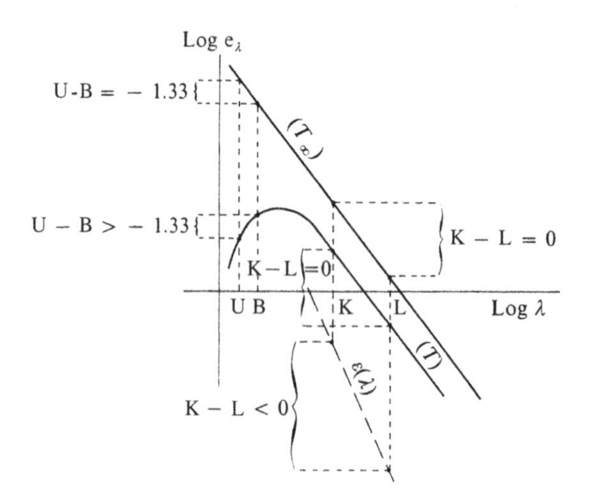

Fig. 3.7 Colour indices. The figure shows the fluxes e_λ of three sources: a greybody source, whose spectrum has the same slope as a blackbody of infinite temperature, a blackbody source of temperature T, and a source of temperature T and variable emissivity $\mathscr{E}(\lambda)$. The colour indices U–B and K–L are also shown

In the Hertzsprung–Russell diagram, the colour index B−V, which defines a colour temperature for the star, is plotted as a function of the absolute magnitude M_V, which itself is directly related to the luminosity L.

3.4 Photometry Through the Atmosphere

Even at wavelengths for which the atmosphere is almost transparent (see Sect. 2.2), there is nonetheless a modification of radiation measured at ground level. An *extinction correction* must be applied. This is straightforward, provided the atmosphere can be modelled as a set of parallel-plane absorbing layers, with absorption coefficient $k_\nu(z)$. If the line of sight makes an angle α to the zenith, and the incident intensity is $I_{\nu 0}$, the transmitted intensity can be calculated from

$$I_\nu(\alpha) = I_{\nu 0} \exp\left[-\sec\alpha \int_0^\infty k_\nu(z)\,\mathrm{d}z\right],$$

where the quantity $\sec\alpha = 1/\cos\alpha$ is called the *air mass*. If the atmosphere is stable in time, measurements made at different zenith angles and plotted on a graph of $\log I_\nu(\alpha)$ against $\sec\alpha$ will lie on a straight line *Bouguer's line*. Extrapolating this line to zero air mass, determines the intensity outside the atmosphere. The assumptions of temporal stability and horizontal stratification often fail, and *precision photometry* at ground level becomes a difficult exercise. This is particularly true in the infrared, affected by the considerable lack of homogeneity in the space and time distribution of water vapour, its principal absorber. Clearly, precision photometry is improved by measurements made in space, and indeed, these measurements are the only ones possible at wavelengths which are totally absorbed by the Earth's atmosphere. With great care, a photometric accuracy of one thousandth of a magnitude can be achieved.

Measurement conditions in space are excellent, since variations in atmospheric extinction are eliminated, and also because objects can be followed over longer periods of time. Space missions observing stars, their variability, and especially their oscillations (*asteroseismology*), such as the Corot mission of the French space agency CNES in 2006 (see Sect. 9.4) or NASA's Kepler mission, can achieve a photometric accuracy of 2×10^{-7} on a star of magnitude $m_V = 8$, or of 5×10^{-6} on a star of magnitude $m_V = 15$, using a 1 m telescope and an integration time of around one month.

In order to calculate an emitted intensity, correction for *interstellar extinction* is required. This is an important correction, demanding an astrophysical interpretation of the measurements (for example, direction of the line of sight relative to the galactic plane), and goes beyond the present discussion.

3.5 Calibration and Intensity Standards

Determining the *absolute values* of astrophysical quantities in terms of the basic units of power, time, frequency, and so on, is a fundamental and difficult problem for observational astronomy: it is fundamental because the consistency of the physical models depends on these values; and difficult because astronomical instruments are complex machines, observing in conditions which are often hard to reproduce, whereas, in the laboratory, reference standards are permanently available and highly controllable.

Calibration can be divided into several categories:

- *Energy calibrations*, in which the aim is typically to measure a specific intensity $B_\nu(\theta)$, as a function of the angular position θ of the source, and at a given frequency ν. The techniques of energy calibration vary enormously with wavelength, and will be reviewed below for each of the main spectral regions.
- *Spectral calibrations*, in which the aim is to establish the absolute frequency of observed spectral lines. This will be discussed briefly, in the context of each of the main spectral regions, since techniques vary once again.
- *Angular calibrations*, in which the aim is to find the absolute angular positions of a set of reference sources, and then determine the angular position of a given source relative to these. This is *astrometry*, dealt with in Chap. 4.
- *Time calibration*, or *chronometry*, in which the aim is to time the variations of a source (for example, a pulsar).

3.5.1 *Radiofrequencies*

We take this to mean the wavelength range $\lambda \gtrsim 0.5$ mm. The calibration standard is the *blackbody*, whose intensity is given by the Rayleigh–Jeans approximation, which is normally valid for the temperatures under consideration:

$$B_\nu \approx \frac{2\nu^2}{c^2} kT.$$

We will see later (see Sect. 7.5) that radiofrequency detectors are limited to a *coherence étendue* $A\Omega = \lambda^2$, and only detect one polarisation of the field.

Method One

Consider a receiving antenna totally immersed in blackbody radiation at temperature T. Assuming its response to be a box function $\Pi(\nu/\Delta\nu)$, the power detected in the frequency interval $\Delta\nu$ is

$$P = \frac{1}{2} \int_{4\pi} \int_0^{\infty} B_\nu(\theta) \, d\nu \, d\omega = kT\Delta\nu.$$

Measuring $\Delta\nu$ and T now calibrates the receiver output voltage. For this method, the Earth's surface can be used as the blackbody source ($T \approx 300$ K, emissivity ≈ 1), the radiotelescope being pointed at the ground to carry out calibration.

Method Two

The receiver is disconnected from the antenna and linked to a *noise source*; a resistance R at temperature T generates a randomly fluctuating signal. The available electric power is given by the average value of the square of the instantaneous voltage across R caused by the thermal noise (Sect. 7.2.2):

$$P = \frac{\langle V \rangle^2}{R} = \frac{4kTR\Delta\nu}{R} = 4kT\Delta\nu.$$

If the resistance is matched to the entry impedance Z of the receiver (i.e., $R = Z$), the received power is $2kT\Delta\nu$. A measurement of the transmission of the antenna, taking into account the imperfect reflectivity of the telescope surface, and waveguide losses between focus and detector, is required to complete calibration. Systematic error can be reduced by carrying out a differential measurement using sources at different temperatures, one of which is cryogenic.

Method Three

The voltage $V(t)$ across a resistance fluctuates randomly, and with white spectral density if $h\nu < kT$ (see Sect. 7.2.2). Any other voltage with the same features can be used for calibration, the result being all the more precise as the signal is stronger. A *noise generator* can provide a random voltage $V(t)$ with spectral density roughly constant over the frequencies used, and such that $\langle V(t) \rangle = 0$. This generator could be a *noise diode*, in which the passage of a current produces a randomly fluctuating voltage, or a *discharge tube*, in which a discharge in a gas produces a high electron temperature and, consequently, a random electric field. Such sources do not have a well-defined thermodynamic temperature but, in so far as they are stable in time, they do provide convenient secondary standards, once calibrated themselves relative to some reference blackbody.

Spectral Calibration

The availability of a signal in the form of an electrical oscillation in a waveguide, a cavity, or a circuit, allows direct comparison with the clocks or oscillators providing frequency standards; and these are directly linked to the measurement of time (see Chap. 4), and hence to the frequency unit, the hertz. These frequency calibrations are therefore extremely accurate.

3.5.2 Submillimetre, Infrared, and Visible

Absolute Photometric Calibration

Blackbody radiation is the main standard in this domain. The source temperature must be higher as the wavelength is shorter, for the intensity decreases exponentially at wavelengths below that corresponding to the maximum of the Planck function, and this would make the source too weak and introduce inaccuracies. The maximum of $B_\lambda(t)$ occurs for $\lambda = \lambda_m$, where

$$\lambda_m T = 2\,898\ \mu\mathrm{m\,K}.$$

In a calibration setup, the power received by the detector is

$$P = \iiint \mathscr{E}(\lambda)\, B(\lambda, T)\, t(\boldsymbol{\theta}, \boldsymbol{r}, \lambda)\, \mathrm{d}\lambda\, \mathrm{d}\boldsymbol{\theta},$$

where $\mathscr{E}(\lambda)$ is the *emissivity* of the reference source, $t(\boldsymbol{\theta}, \boldsymbol{r}, \lambda)$ the transparency and/or spectral transmission of the instrument, defined by its diaphragms, its beam geometry, its filters, and the spectral reflectivity of the mirrors. $\boldsymbol{r}$ is a variable defining a point of the surface S of the pupil, and $\boldsymbol{\theta}$ the direction of the radiation. Each of these quantities must be determined to great accuracy, using spectroscopic techniques (Chap. 8).

Temperature

A precise determination can be made using standard methods of *pyrometry*, for example, thermocouples or optical pyrometry. Relay calibration by lamps with ribbon filaments can be used.

Emissivity

The quantity $\mathscr{E}(\lambda)$ depends on the geometry of the source, and the type of material constituting the filament or the emitting cavity. The exact determination of $\mathscr{E}(\lambda)$ remains the main source of uncertainty in absolute infrared calibrations, whereas uncertainty over temperature measurements dominates in calibration at visible wavelengths.

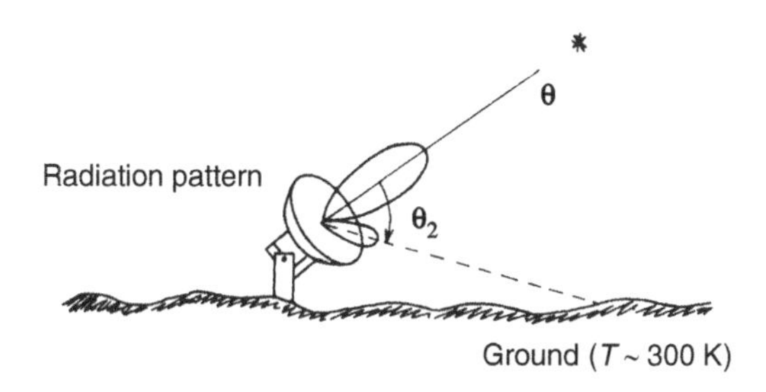

Fig. 3.8 Sidelobes of a radiotelescope. The source is in the direction $\boldsymbol{\theta}$, but a non-zero power is nonetheless received from the direction $\boldsymbol{\theta} - \boldsymbol{\theta}_2$ (thermal emission from the ground)

Beam Geometry

This comes in through the term $t(\boldsymbol{\theta}, \boldsymbol{r}, \lambda)$. At short wavelengths ($\lambda \lesssim 5$ μm), for which the sizes of optical elements are large in comparison, diffraction barely enters the definition of t, and it can be considered as resulting wholly from geometric considerations, such as the sizes of diaphragms, distances, and so on. This is no longer the case in the mid-infrared and submillimetre regions, where diffraction produces *sidelobes*. The detector then receives power from other directions than the line of sight (Fig. 3.8). The *radiation pattern* is the polar representation of the relative sensitivity of the detector as a function of $\boldsymbol{\theta}$.

This phenomenon is particularly awkward for very low temperature detectors, when the observed source is very faint, whereas the surroundings of the detector, for example, the mirror of the telescope, emit significant power because of their higher temperature, thus making a non-negligible contribution to the signal. Absolute calibration methods for spaceborne submillimetre telescopes, e.g., the WMAP or Planck missions (see Sect. 7.4.7), have come a long way since 1990, stimulated by the difficulties involved in accurate measurement of the cosmological blackbody radiation.

Apodisation methods can reduce the diffraction lobes and bring the effective étendue closer to the geometric value. (See, for example, Born and Wolf, *Principles of Optics*, and also the discussion on diffraction by pupils in Sect. 6.6 and Problem 8.1.)

Synchrotron radiation sources are becoming increasingly available and could soon be used in the submillimetre to visible regions instead of blackbodies. They are already the classic calibration source at higher energies, as described in Sect. 3.5.3.

Relative Calibration

Absolute calibration implies the use of a standard source, something which is not always easily available during observation. Secondary standards, such as stars or reference objects, are then used. These are assumed to be constant in time and are more accessible during an observing sequence.

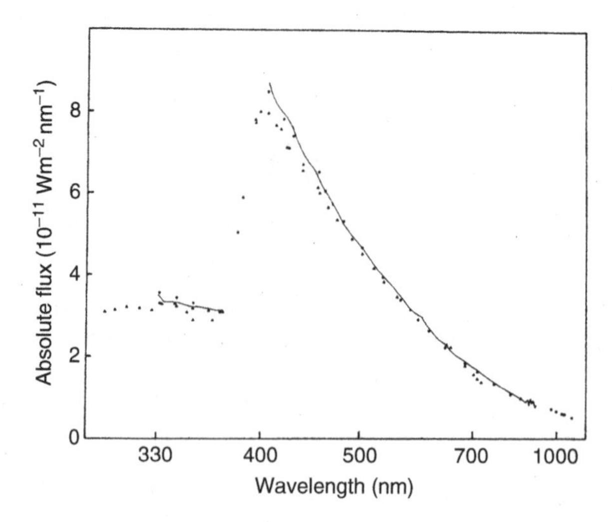

Fig. 3.9 Absolute flux of the star Vega, in the visible and near infrared regions, after various authors (Tug et al., Oke and Schild, Kharitonov, Glushneval)

Visible

This is the region $0.3 \leq \lambda \leq 0.7\ \mu\mathrm{m}$. The reference object is the star Vega (α Lyrae), whose absolute spectrum is shown in Fig. 3.9. Note the non-negligible fluctuations caused by uncertainties of measurement. Using the functions $t_0(\lambda)$ which come into the definitions of magnitudes, the spectrum gives the magnitudes of Vega, and thus by comparison (flux ratios), the magnitudes of other stars. The Sun is rarely used as a secondary standard, its brightness being too great for the dynamic range of most detectors.

Absolute stellar photometry can reach an accuracy of 0.001 magnitudes in the visible.

Near Infrared

This region ($\lambda \lesssim 30\ \mu\mathrm{m}$) is observable from ground level. A series of stars has been chosen as secondary standards, and their fluxes have been determined in the following way:

- It is assumed that the absolute solar flux has been precisely measured in the range 1 to 25 μm. As an example, Fig. 3.10 shows the determination of the specific intensity of the Sun.[2] This measurement results from a comparison of the solar

[2]The Sun is considered as a standard in the visible, but its spectral intensity fluctuates, in particular at radio and X-ray wavelengths. Even at visible wavelengths, at the time of writing (2007), more

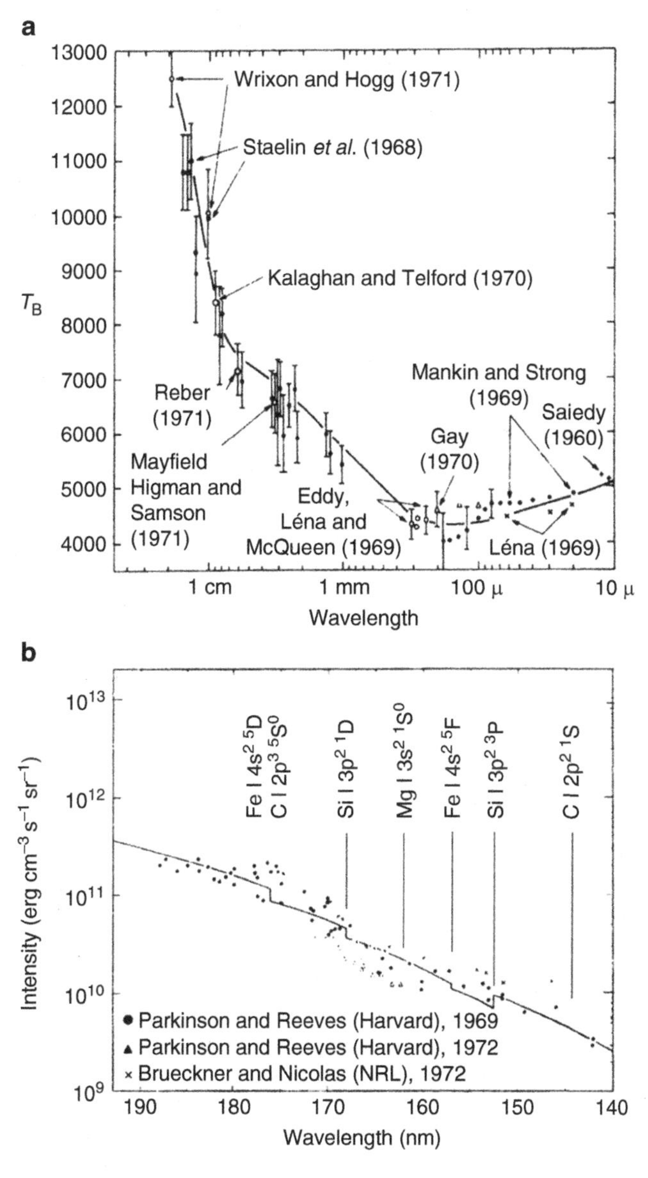

Fig. 3.10 Observational determination of (**a**) solar brightness (expressed in terms of brightness temperature T_B) in the far infrared, and (**b**) monochromatic flux from the centre of the disk (quiet region) in the ultraviolet. The *continuous curve* shows the prediction of solar atmosphere models. Note the scatter of the measurements even for an object as intense as the Sun. (After Vernazza J. et al., Ap. J. **184**, 605, 1973. With the kind permission of the Astrophysical Journal)

flux with the standards discussed above. Having fixed the constants q_{λ_0} of the magnitude scale, the colour indices of the Sun can be calculated.

- A series of reference stars is chosen, having a similar *spectral type* to the Sun (G0 to G4), and whose infrared spectra are assumed to be strictly proportional to that of the Sun.

  This assumption implies another, namely that these stars are not surrounded by dust clouds, whose thermal emission in the near infrared would add to that of the photosphere of the star. The existence of a zodiacal nebula around stars, which have been assumed to be 'naked' up until recently, has been demonstrated in the case of Vega (α Lyr), and indeed a certain number of other stars (e.g., β Pictoris, Polaris, Fomalhaut, etc.).

  The magnitude V of these stars is measured, for example, by comparison with α Lyr, and their magnitudes R, ..., Q can be deduced.
- This intermediate series of stars is then compared with bright stars, in order to establish secondary standards with a good signal-to-noise ratio. These standards should be well placed in the sky, at small zenith distances, and accessible at any time. Their spectral type is arbitrary and their magnitudes R, ..., Q are determined by comparison with the previous series.

Far Infrared

This is the region 30 μm $\lesssim \lambda \lesssim$ 0.5 mm, which is observed from above the atmosphere. There are no standard spectral bands for normalising measurements, and so the literature gives values of the monochromatic fluxes $B(\lambda)$ of sources, or else the flux in jansky when the source is limited in spatial extent or unresolved by the observation.

Observations made from aircraft or balloons use the intrinsic radiation of the planets as a secondary standard. Planetary radiation is assumed identical to that of a blackbody with a brightness temperature $T(\lambda)$ in the spectral region considered, and $T(\lambda)$ is found by independent comparison with a standard blackbody.

The brightness temperature of a planet is a function of wavelength. The approximate values given in Table 3.3 are valid to within 10% for the far infrared, and are improved with each space exploration mission.

Table 3.3 Brightness temperatures of the planets, for $50 \lesssim \lambda \lesssim 500$ μm

	$T_B(\lambda)$
Venus	255 K
Mars	235 K
Jupiter	145 K
Saturn	85 K

---

and more accurate measurements are required to analyse the possible impact of solar fluctuations on climate change.

The sensitivity of a spaceborne telescope is greatly superior, and the planets are too bright, or sometimes too extended in space, to be used as standard sources. Calibration proceeds in several steps. First, an absolute internal standard is provided on board, by a blackbody whose emissivity has been calibrated before the launch, and whose temperature is maintained at an accurately known value. Such is the case for the instruments on board the satellite ISO (1995), and also for the Spitzer observatory (2004), named after the astrophysicist Lyman Spitzer and dedicated to the near and far infrared, with a cooled telescope of diameter 0.85 m (see Sect. 5.2.3).

The next step is the observation of asteroids. The dimensions of these objects are known, as are their brightness temperatures and their emissivities. Indeed, these quantities are relatively easy to model, depending on the mineralogical composition of the asteroid and its distance from the Sun. Magnitudes and fluxes can thus be determined at any wavelength.

During the mission it is also necessary, in order to better establish such a calibration, to use a standard star which is simultaneously observable from the ground, and hence well calibrated. The star α Tau has been used in this way as a 'photometric relay' at $\lambda = 12\ \mu$m for calibration of the satellite IRAS (1983). In the range 2–42 μm, the Spitzer mission (2004) uses a set of standard stars whose synthetic spectra (at a resolution of 1 200) have been calculated on the basis of everything we now know about these stars.

Spectral Calibration

In the visible, this calibration is carried out using spectral lamps (thorium, for example) which emit lines at very precisely known wavelengths. Certain spectroscopic techniques (see Chap. 8) provide an absolute measurement of wavelengths, either by comparison with lasers (Fourier spectrometers), or by comparison with the atomic emission line of a gas (resonance cells, see Sect. 8.5). In the near infrared, the atmospheric emission lines of the radical OH provide a good standard for ground-based observations. Beyond, the instrument must be pre-calibrated using a laser standard, for example, with many lines (see Chap. 8).

3.5.3 Ultraviolet and X Rays

Use of blackbodies becomes more difficult in this region ($0.1 \lesssim \lambda \lesssim 300$ nm), where very high temperatures are required to obtain an acceptable brightness. Inaccuracy in temperature measurements limits the feasibility of absolute calibration using *thermal sources* in these spectral regions. On the other hand, the availability of *synchrotron radiation sources*, which are non-thermal, provides an accurate and reliable method of absolute calibration for UV and X-ray instrumentation. In either case, calibration is made easier by the absence of interfering radiation from the environment, which is so harmful in the infrared region.

Thermal Sources

These are considered in two wavelength regions.

The Region 100–300 nm

It is possible to produce very hot plasmas ($T \approx 14\,000$ K) whose temperature is nevertheless accurately known (∼2%). On experimentally accessible length scales l (1–10 cm in Ar), this plasma is only optically thick in the *resonance lines* of the gas atoms, viz., the Lyα line (121.5 nm) of hydrogen, N II lines (108.5 nm), and various lines of C I, Kr I, and so on.

In these lines, the emitted intensity

$$I_\lambda = B_\lambda(T)\,[1 - \exp(-\kappa_\lambda l)]$$

is very close to that of a blackbody $B_\lambda(T)$, if the optical depth $\kappa_\lambda l$ is larger than unity. Nevertheless, with $\Delta T/T \sim 2 \times 10^{-2}$, we still have $\Delta B/B = 0.2$, which is far from ideal.

The Region 0.1–100 nm

Optically thick sources can no longer be created and the idea of using blackbody radiation to calibrate instruments must be abandoned. Instead, the *thermal free–free emission* of a hot and optically thin plasma can be used. The power radiated per unit volume of plasma (*volume emissivity*) is

$$\mathscr{E}_\nu = 6.36 \times 10^{-47}\, N_\mathrm{i}\, N_\mathrm{e}\, Z^2\, g(\nu, T) \frac{\exp(-h\nu/kT)}{(kT)^{1/2}} \quad \mathrm{W\,m^{-3}\,sr^{-1}\,Hz^{-1}},$$

where N_i and N_e [$\mathrm{cm^{-3}}$] are the ion and electron number densities, respectively, Z the ionic charge number, and $g(\nu, T)$ the Gaunt factor. (The radiation emitted by a hot plasma is given in AF, Sect. 1.30, and also in AQ, Sect. 43, where a table of Gaunt factors can be found.)

This plasma is created in a *pulsed discharge tube*, where the temperature reaches $T = 1.5 \times 10^6$ K, and the electron density $N_\mathrm{e} = 2 \times 10^{17}\,\mathrm{cm^{-3}}$. Initial ionisation is achieved by applying a radiofrequency field which preheats the discharge. Heating then results from compression of the matter by an intense magnetic field (0.5 T). Electromagnetic emission only lasts for a few microseconds.

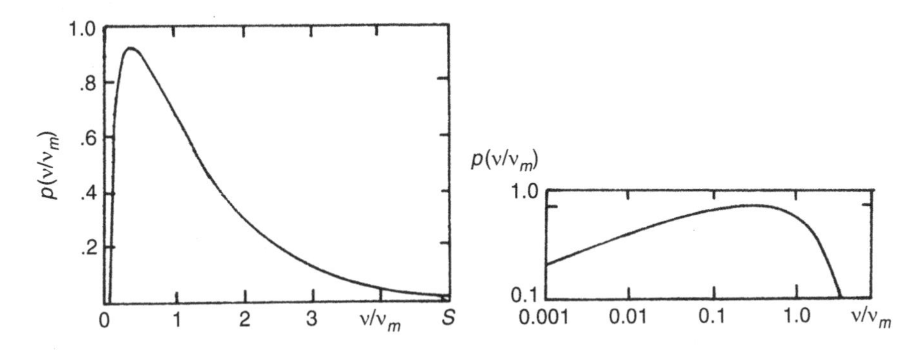

Fig. 3.11 Spectral distribution of synchrotron radiation from an electron in a magnetic field. (Lang K.R., *Astrophysical Formulae*)

Non-Thermal Sources: Synchrotron Radiation

Synchrotron radiation is emitted by relativistic electrons accelerated in a magnetic field, and its easily quantifiable and reproducible character makes it a useful calibration standard.

The radiation emitted by an electron of energy $\mathscr{E}$, orbiting in a magnetic field B, is concentrated around a frequency ν_m given by[3]

$$\nu_m = \frac{eB}{2\pi m_e}\left(\frac{\mathscr{E}}{m_e c^2}\right)^2,$$

where m_e is the mass of the electron. Figure 3.11 shows how the spectrum is distributed around ν_m. The radiated power is given by

$$dP(\nu) = \frac{16e^3 Bc}{4\pi\varepsilon_0 m_e c^2} p(\nu/\nu_m)\, d\nu.$$

The radiation is directional, in a narrow cone with axis along the tangent to the electron trajectory, and opening angle 2θ given by

$$\theta(\text{rad}) \simeq \frac{m_e c^2}{\mathscr{E}} = \frac{0.511}{\mathscr{E}[\text{MeV}]}.$$

Although synchrotron radiation is particularly suitable in the UV and X-ray regions, it also provides an excellent calibration source in the visible and near infrared, in view of its precise directionality.

[3]Synchrotron radiation is described in *Astrophysical Formulae*, Sect. 1.25.

Since the mid-1990s, many synchrotron radiation sources have become available, e.g., the European Synchrotron Research Facility (ESRF) (Grenoble, France), DORIS (Hamburg, Germany), Brookhaven (New York, USA), Tsukuba (Japan), Daresbury (Great Britain), and Argonne (Chicago, USA). To give some idea of the values of the parameters, the ESRF provides a beam with $\theta = 0.1$ mrad and intensity reaching between 10^{14} and 10^{18} photons s^{-1} mm^{-2} $mrad^{-2}$ spectral interval^{-1}. The spectral interval at an energy E is of the order of $\Delta E/E \approx 10^{-3}$. It is also possible to produce γ radiation from synchrotron electrons, using the inverse Compton effect on a laser beam. γ radiation with well-defined spectral properties is obtained, in the range 1–1.5 GeV.

3.5.4 Gamma-Ray Radiation

In this region, detectors always function in *counting mode*, and with two possible objectives: an energy calibration of each photon (*spectrometry*), and some measurement of the efficiency of detection (*photometry*).

At low energies (< 2 MeV), the reference standards are radioactive samples with known emission (for example, ^{60}Co, which emits lines at 1.17 and 1.33 MeV), and whose activity has been calibrated.

In the case of photometry, there are two problems worth noting with radioactive calibration sources. To begin with, the activity of these radioactive sources is not generally known to better than a few percent. Furthermore, for sources with a short half-life, less than 10 yr, the activity of the source must be regularly updated. Indeed, when the half-life is very short, say less than a week, the decay of the source must be taken into account during its use.

At high energies, particle accelerators are used. Once again, the accuracy obtained in photometry is barely better than a few percent.

Calibration of instruments can also be carried out in flight (on balloons, or in space), for the response of an instrument may evolve in time, particularly after launching. Photometric calibration can be made by observing some stable reference object of known spectrum (the Crab Nebula, for example); and spectral calibration can be carried out by observing emission lines from radioactive de-excitation, either of sources intentionally placed aboard, or of the material of the instrument itself, activated by radiation received in orbit.

3.5.5 Some Examples of Spectrophotometry

Determining the specific intensity of a source with spatially limited extent, in various wavelength ranges, requires very diverse instrumental techniques. Accuracy of calibration itself varies with wavelength. Putting together all the observations produces a global spectrum of the object, and this is the only way to penetrate the underlying physics of the processes involved. A few examples are given here.

The Sun

It would be easy to imagine that the absolute spectrophotometry of such a close and bright object would be perfectly established over the whole electromagnetic spectrum. This is not the case, as is shown in Figs. 3.10 and 3.12. Apart from the obvious difficulty in obtaining measurements of λE_λ which cover eighteen powers of ten, the definition of a uniform flux over time is made impossible in certain spectral domains, by the variability of the Sun in those domains. The same remark applies equally to most stars, or other astrophysical objects, and consequently, it is rare for some improvement in photometric accuracy not to reveal a variability in the luminosity of objects which were previously supposed to be stable.

Extragalactic Radiofrequency Sources

Determining the spectrum of an object across a wide range of frequencies is essential in order to understand the physical origins of the radiation which it emits; for example, to establish whether it is thermal or non-thermal. The difficulty here comes from the fact that the sources are distant and weak, requiring long integration times, and thus opening the way to all kinds of interference from the atmosphere, or from internal fluctuations of the instrument, which can modify any initial calibration. Figure 3.13 gives several examples of absolute photometry of extragalactic sources, from which either thermal or non-thermal spectra can be attributed to them.

Cosmological Background Radiation

The great difficulty involved in measuring this radiation provides a remarkable illustration of the accuracy which can be achieved in absolute spectrophotometry, and also of the fundamental consequences that these measurements can have for astrophysics. The satellite COBE (COsmic Background Explorer, 1989) carried an absolute spectrophotometry instrument (FIRAS), comprising a Michelson interferometer, followed by bolometric detectors, with the aim of determining the spectrum of the cosmological background in the spectral range 0.1 mm $< \lambda <$ 10 mm. Absolute calibration was carried out using an internal thermal radiation source, approximating a blackbody in this range by an emissivity better than 0.9999.

After subtracting the intrinsic emission of the Galaxy, also detected by the instrument and suitably modelled, the spectrum of the cosmic background can be calculated. A best fit over 0.5 mm $< \lambda <$ 5 mm is obtained with a blackbody of temperature $T = 2.726 \pm 0.010$ K. Across this range, the greatest deviation from the Planck spectrum is only 0.03 %, a quite exceptional accuracy, achieved in

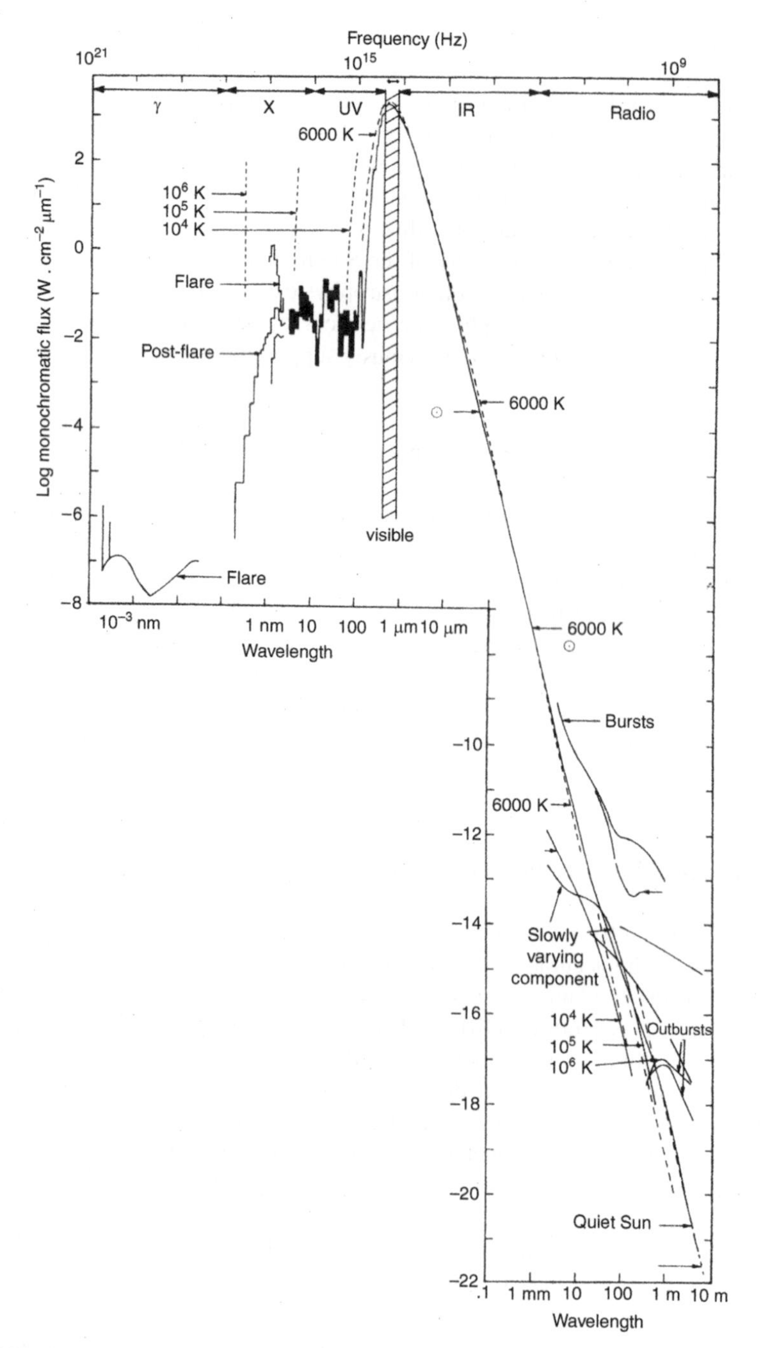

Fig. 3.12 Composite view of the integrated radiation from the solar disc. The monochromatic flux outside the Earth's atmosphere is plotted as a function of wavelength. The temperature of various blackbodies is given. The symbol ⊙ denotes the Sun. (After White O.R., *The Physical Output of the Sun*, Boulder, 1976. With the kind permission of Boulder Press, CO)

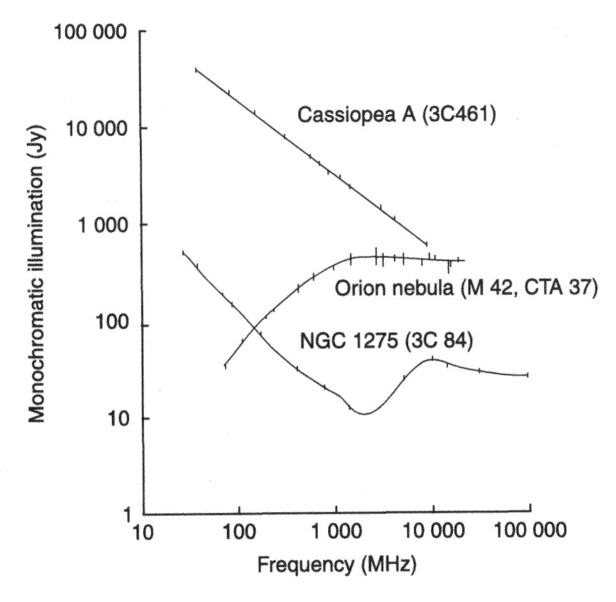

Fig. 3.13 Photometry in the radio region: the radio source Cassiopeia A (synchrotron radiation); the thermal bremsstrahlung source of the Orion Nebula (H II region); the galactic nucleus NGC 1275, which has a partly thermal and partly synchrotron spectrum at low frequency. (After Lang K.R., *Astrophysical Formulae*. The figure collects the results of several authors.) Note the model indicated by *continuous curves*, measurement points, and error bars. (With the kind permission of the publisher)

extremely difficult experimental conditions.[4] In 2007, not only was the accuracy of these measurements improved, but a detailed map of the spatial fluctuations of this radiation was established by NASA's WMAP mission (2000). This in turn will be improved by the European Planck mission (2008), discussed in Sect. 7.4.7.

3.6 Calibration of Angular Dimensions

With the advent of very high angular resolutions, which can range from the millisecond of arc to the microsecond of arc (5×10^{-12} rad) thanks to optical and radio interferometers (see Sect. 6.4), it has become necessary to be able to specify an *unresolved source* on this same scale. This is then called a point source, with the understanding that it is a point source for the given instrument. Indeed, when such an instrument measures the visibility of the source at a given spatial frequency, the

[4]A complete set of references concerning the extraordinary COBE mission is given by Mather J.C., Infrared measurements from the Cosmic Background Explorer, Infrared Phys. Technol. **35**, 331–336, 1994.

observation mixes instrumental effects, discussed in detail in Chap. 6, and spatial properties of the object, e.g., if it is a star, its angular diameter and its centre–limb darkening.

To distinguish the two parameters, we must be able to observe, with the same instrument and under like conditions, a source that can be considered as known spatially, i.e., in practice, as unresolved or pointlike (having very small dimensions compared with the resolution of the instrument). A perfect solution would be to use the disk of a distant star, observed at visible wavelengths (radiation from the photosphere), but its magnitude will generally be too faint to be observable. If for instance we were to choose a star of the same type as the Sun, and if we required its angular diameter to be less than 100 μarcsec, it would have to be at a distance of 10 kpc, and its magnitude (V) would then be around 19, much too faint for the sensitivity of the interferometer.

The solution to this problem in 2007 is to use less remote stars, hence partially resolved, but for which the center–limb darkening, and more generally the photospheric brightness distribution, can be considered as sufficiently well modelled to allow calibration of the interferometer. In the near-infrared, the problem is complicated by the possible, but not necessarily known, presence of circumstellar material which emits radiation and may be modifying the observed spatial structure.

Problems

3.1. Colour Indices. Table 3.4 gives the colour indices of main sequence stars. Find the *colour temperature* of the various spectral types (that is, the temperature of the blackbody with the most similar spectrum in a given wavelength interval). For simplicity, represent each band by its central wavelength.

Answer. A blackbody at temperature T has colour index

$$\mathrm{V}-\mathrm{L}\ (T) = \mathrm{constant}_{\mathrm{VL}} - 2.5 \log \frac{B(\lambda_{\mathrm{V}}, T)}{B(\lambda_{\mathrm{L}}, T)},$$

where λ_{V} and λ_{L} are the representative wavelengths of the corresponding bands (to simplify). The constant is determined by the requirement that V–L= 0 for a

Table 3.4 Colour indices of main sequence stars

Spectral type	V–R	V–I	V–J	V–K	V–L	V–N
AO	0.00	0.00	0.00	0.00	0.00	0.00
FO	0.30	0.47	0.55	0.74	0.80	0.00
GO	0.52	0.93	1.02	1.35	1.50	1.40
KO	0.74	1.40	1.50	2.0	2.50	
MO	1.10	2.20	2.30	3.5	4.30	
M5		2.80			6.40	

blackbody of infinite temperature. The flux of such a blackbody varies as λ^{-2} so

$$\text{constant}_{\text{VL}} = 2.5 \times 2 \times \log(\lambda_{\text{L}}/\lambda_{\text{V}}) = 3.96.$$

From Table 3.4, the colour temperature T_{MO} of an MO star satisfies

$$3.96 - 2.5 \log \frac{B(\lambda_{\text{V}}, T_{\text{MO}})}{B(\lambda_{\text{L}}, T_{\text{MO}})} = 4.30,$$

which, in turn, implies that $T_{\text{MO}} = 4{,}100$ K.

3.2. A telescope cooled to a very low temperature is used to measure the cosmological background at $\lambda = 1$ mm. Calculate the power received in the coherence étendue λ^2 per frequency interval $\Delta\nu$. Calculate also the power received from the Earth, assuming its temperature to be 300 K, observed at 90° from the principal line of sight (the telescope is in orbit and pointing towards the local zenith). Comparing these two values, deduce the angular rejection needed to make a measurement of the cosmological background, accurate to within 1%. (The configuration of the diffraction lobes of a pupil can be found in Chap. 6.) Take $D = 10$ cm for the diameter of the telescope.

3.3. A thermal source (H II region of ionised hydrogen) is optically thick at all wavelengths, and its thermal radiation (bremsstrahlung) is such that $I_\nu =$ constant (see AF, Sect. 1.30). Given that it radiates 10 mJy at radiofrequencies, calculate its magnitude m_{v} at visible wavelengths.

Answer. Firstly, the radiation from the source satisfies

$$I_\nu = 10\,\text{mJy} = 10^{-28}\,\text{W}\,\text{m}^{-2}\,\text{Hz}^{-1},$$

for all wavelengths, and in particular, in the visible. Secondly, in the visible

$$m_{\text{v}} = -2.5 \log\left(I_\lambda / 3.92 \times 10^{-8}\,\text{W}\,\text{m}^{-2}\,\mu\text{m}^{-1}\right).$$

The link between these two relations is made using

$$I_\nu \Delta\nu = I_\lambda \Delta\lambda \quad \text{where} \quad \Delta\nu = (c/\lambda^2)\Delta\lambda.$$

Thus, $I_\lambda = 9.92 \times 10^{-14}\,\text{W}\,\text{m}^{-2}\,\mu\text{m}^{-1}$, and the visible magnitude of the source is $m_{\text{v}} = 14$.

Chapter 4
Space–Time Reference Frames

An essential part of astrophysical observation and data analysis consists in measuring the positions of astronomical objects and in dating events. Obtaining the photometric magnitudes of the emitted radiation is not enough. The source must be identified in catalogues or data bases, and the position must be determined to great accuracy, since spatial resolutions of the order of several tenths of a millisecond of arc can now be attained in images, and this trend is certain to continue. Concerning the absolution positioning accuracies of celestial objects, over the past 20 years or so, these have gone from a few hundredths of a second of arc before the Hipparcos satellite was launched to about ten microseconds of arc with the Gaia mission. In terms of sensitivity, detectors can now measure the characteristics of phenomena whose time of variation is less than one millisecond. And further, extremely accurate time scales, taking into account the effects of general relativity, are required for space travel and exploration of specific sites in the Solar System.

This chapter treats the recent progress in astronomical reference systems, first for space and then for time, even though the primary source for time scales is now linked to a phenomenon quite distinct from the Earth's movement. The notions involved also have important applications for the everyday life of Earthlings like ourselves, e.g., with the advent of GPS or the synchronisation of clocks in transport and telecommunications. The objective is to provide a basis for the space and time coordinates needed in practical calculations that do not aim for accuracies better than a few $0.01''$ for absolute positions. To compare this requirement with past and current performance in astrometry, we need only examine the way accuracies in position and parallax measurements have evolved over time, as shown in Fig. 4.1.

The Hipparchos–Ptolemy catalogue[1] as transmitted in the Almagest can be used to calculate true star positions relative to the equator or the ecliptic of that epoch

[1] Hipparchos was the greatest astronomer of Ancient Greece, living in the second century BC. He discovered the precession of the equinoxes and compiled the first star catalogue. Ptolemy was an astronomer and mathematician who lived in Alexandria in the second century AD. He recorded the astronomical knowledge of his day in the Almagest, which served as a reference for the geocentric theories of planetary motions for over 15 centuries.

P. Léna et al., *Observational Astrophysics*, Astronomy and Astrophysics Library,
DOI 10.1007/978-3-642-21815-6_4,

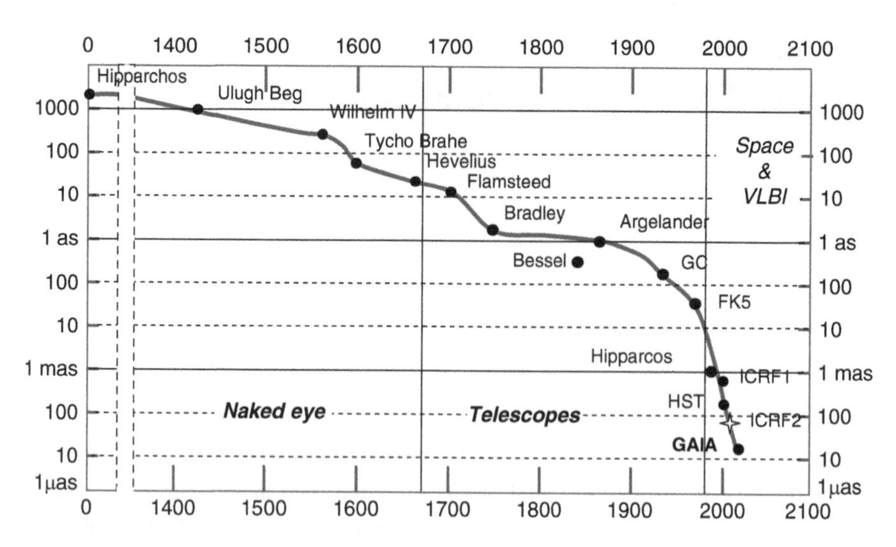

Fig. 4.1 Improving astrometric accuracy in the observations of stars and planets at different epochs

and thereby determine the statistical quality of the measurements. The result is an accuracy of the order of 0.3–0.5 degrees. Figure 4.1 shows a very slow improvement in the accuracy of astronomical pointing. There is a very slight inflexion with the advent of optical instruments toward the end of the seventeenth century. The refracting telescope became available with Galileo[2] in 1610, but its use as a sighting instrument came rather later, coinciding with the foundation of the observatories in Paris and Greenwich. Refracting telescopes were above all used to observe faint objects, but that step forward did not necessarily improve direction measurements. Later, progress in mechanical construction and clock-making proved to be just as important in this respect as developments in optics.

However, the most recent part of the curve is highly instructive and illustrates the significant progress made in astrometry (measurement of positions, proper motions, and parallaxes of heavenly bodies) by virtue of space techniques. While it took about 20 centuries to improve astrometric accuracy by 4 to 5 orders of magnitude, it subsequently took a mere 20 years with Hipparcos, Gaia, and SIM to achieve a similar leap forward. In the end, the DIVA and FAME projects in Germany and the USA, respectively, put forward just after the success of the Hipparcos mission, were never funded.

Modelling requirements depend significantly on the level of accuracy one hopes to achieve, and the degree of difficulty grows rapidly with this accuracy. It is thus essential to adopt the right tools for each situation. It is pointless to use the

[2]G. Galileo (1564–1642) was an Italian physicicst and astronomer. With Newton, he founded modern mechanics and was the first to use a telescope for scientific observation of the heavens. He proved himself to be a talented and stubborn defender of the ideas of Copernicus.

Table 4.1 Angular units in position astronomy with an illustration by apparent diameter of a 1 euro coin. The last column indicates the epoch at which the given accuracy was achieved

Angle	Radians	1 euro coin at distance	Epoch
1 degree	$\sim 10^{-2}$	1 m	Hipparchos
1′	$\sim 2 \times 10^{-4}$	100 m	Tycho Brahe
1″	$\sim 10^{-5}$	4 km	1900
1 mas	$\sim 10^{-8}$	4 000 km	Hipparcos
10 μas	$\sim 10^{-10}$	4×10^5 km (on the Moon)	Gaia
1 μas	$\sim 10^{-11}$	1 hair at 4 000 km	SIM

relativistic formulation of aberration for astrometry to arcsec accuracy, but it is essential to do so for accuracies of millisecond (mas) order. Typical angular values for astronomical positions are given in Table 4.1, with an estimate of the epoch at which this accuracy was (or will be) attained.

4.1 Spatial Reference Systems

4.1.1 Definitions of Spatial Frames

The determination of coordinates for astronomical events is closely connected to the gathering of information carried by photons. Fixing the position of celestial objects, such as planets, stars, galaxies, quasars, and so on, is an essential step towards identifying and cataloguing them, so that they can be found for further observation, or used for statistical studies on the distribution of matter in the Universe. But far more than being a mere list, this information about the positions and motions of objects is as important scientifically as information about their specific physical properties.

Throughout this chapter, space will be referred to an origin O and three fundamental directions Ox_1, Ox_2, Ox_3 defined by unit orthogonal vectors $\boldsymbol{e}_1, \boldsymbol{e}_2, \boldsymbol{e}_3$. The triad will be taken to be right-handed, that is $\boldsymbol{e}_3 = \boldsymbol{e}_1 \times \boldsymbol{e}_2$ (vector product). A point M is fixed either by its Cartesian coordinates (x_1, x_2, x_3), or by its spherical coordinates (r, θ, ϕ) with

$$\begin{aligned} x_1 &= r \cos\phi \cos\theta, \\ x_2 &= r \sin\phi \cos\theta, \\ x_3 &= r \sin\theta, \end{aligned}$$

where $r > 0$, $0 \leq \phi < 2\pi$, $-\pi/2 \leq \theta \leq \pi/2$. We shall call ϕ the longitude and θ the latitude (see Fig. 4.2), but it should be noted that the definition of latitude as the angle above the equator is common usage in astronomy, whereas physicists prefer to use the colatitude measured from the polar direction. This should be borne in mind when using tangent frames that are not oriented in the same way.

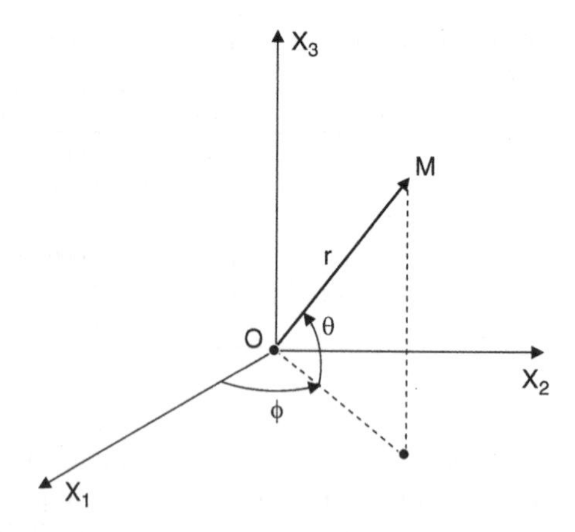

Fig. 4.2 Reference triad and spherical coordinates, with astronomers' conventions

The definition and the practical application of such a frame require a choice of fundamental plane normal to $\boldsymbol{e}_3$, and some practical realisation of these directions in the sky.

The Cartesian coordinates (x_1, x_2, x_3) of the point M can be calculated using the above formulas, given its spherical coordinates (r, θ, ϕ). Care should be taken over the inverse transformation. Indeed, determination of an angle in $[0, 2\pi]$ requires knowledge of both its sine and its cosine, or at least one of the trigonometric lines and the sign of the other. Many disagreements and errors in automated calculations can be traced to the neglect of this rule. But, bearing this in mind, the following formulas allow transformation from the normalised Cartesian components of $\boldsymbol{OM}$ to its spherical coordinates:

$$\theta = \arcsin x_3,$$

$$\phi = \mathrm{angle}(x_1, x_2) = \arctan 2, (x_2, x_1),$$

where the function $\alpha = \mathrm{angle}(x, y) \iff x = \cos\alpha$ and $y = \sin\alpha$.

In order to avoid tests in programs and to use only the function $\arctan x$, which is available in all programming languages, the longitude can be determined more directly by

$$\tan\frac{\phi}{2} = \frac{\sin\phi}{1 + \cos\phi} = \frac{x_2}{r\cos\theta + x_1},$$

and

$$\phi = 2\arctan\left(\frac{x_2}{r\cos\theta + x_1}\right),$$

which automatically gives an angle ϕ between $-\pi$ and $+\pi$, without any ambiguity as to the quadrant. The angle θ lies between $-\pi/2$ and $+\pi/2$ and is thus completely determined by its sine. These formulas are easily adapted when the position vector is not unit length.

4.1.2 Astronomical Reference Frames

Characteristics of the Reference Frames

In order to study the positions and motions of celestial bodies (planets, satellites, stars, galaxies, etc.), spatial reference frames must be defined and constructed practically. A major part of astronomical activity in the past was devoted to such constructions, and it is only in the last 150 years that the physical study of these bodies has taken an important place in astronomical research. The triads used in astronomy can be distinguished from one another by the features listed in Table 4.2.

The vertical and the celestial pole correspond to ideal physical definitions which we try to realise materially as best as we can from observations, whereas poles by convention are just defined by a set of numbers which allow us to attach them to physical systems. For example, the plane of the solar equator used to fix the positions of sunspots is inclined by definition at an angle of 7.2 degrees to the ecliptic of 1850. The origin in the equatorial plane is determined by the ascending node of the solar equator on the ecliptic at noon on 1/1/1854. The meridian origin makes one sidereal revolution every 25.38 mean solar days. Since the ecliptic plane of 1850 can be linked to the equatorial plane at any date by the theory of the Earth's motion, the heliographic frame can be defined at any moment by a few numerical constants. The accuracy of the definition is compatible with the accuracy to which the system can be observed: there is no point trying to locate a sunspot, with its uncertain contours, to a greater accuracy than a few seconds of arc.

The Horizontal Frame

This is the most natural reference frame for recording the positions of celestial bodies and their apparent motion during the diurnal cycle. We consider a triad with origin at the position of the observer. At each point and at any moment, there exists a natural and easily determined direction, namely the local vertical specified by

Table 4.2 Standard reference frames for position astronomy

Origin of coordinates	
Place of observation	Topocentric frame
Centre of the Earth	Geocentric frame
Centre of the Sun	Heliocentric frame
A centre of mass	Barycentric frame
Choice of fundamental directions	
The local vertical	Horizontal system
The celestial pole	Equatorial system
The normal to the ecliptic	Ecliptic system
A pole by convention	Galactic system, heliographic system

the weight (resultant of the gravitational attraction and the centrifugal force due to the Earth's rotation). The plane normal to the vertical is the horizon, or horizontal plane. The point on the celestial sphere associated with the ascendant vertical is the zenith, and the opposite point is the nadir. The plane containing the local vertical and the direction of the celestial pole (this plane may well not contain the axis of rotation of the Earth, but we shall ignore this slight difference in what follows) is the local meridian plane, and its intersection with the horizon defines the origin for longitude in the horizontal frame. The intersection of the local meridian plane with the celestial sphere is called the *astronomical meridian*. This is indeed an astronomical notion and not a geographical one, for the latter would be fixed relative to terrestrial points of given longitude on Earth. The exact shape of the Earth and the complexity of its gravitational field do not enter directly into the definition. Only the notion of local vertical is important, and it can be materialised by means of a plumb line, or by the surface of a liquid at rest, whose plane will be normal to the vertical.

The two spherical angles determining a direction relative to this triad are the azimuthal angle A and the height h. The azimuthal angle is measured in the horizontal plane using various conventions of orientation, depending on whether the application is in astronomy, geodesy, sea navigation, or space geodesy. There are four possible conventions, depending on the origin (north or south), and the orientation (left- or right-handed). Following common usage in astronomy, we take the south as origin. Astronomers used to measure azimuthal angles in the retrograde direction (so that angles increase for a typical body in its daily motion and for an inhabitant of the northern hemisphere). Today, with the advent of automatic computation and matrix formulas, a right-handed or direct triad orientation is much preferred, so that calculations can be made using rotation operators without the need to be over-concerned with the orientation of the angles. Hence, in the following formulas, azimuthal angles increase from 0 to 90° toward the east. Note, however, that it is not uncommon to find slightly different formulas in works on spherical astronomy, especially in older editions.

In the horizontal system (Fig. 4.3), the unit vector in the direction $\boldsymbol{OM}$ takes the form

$$\boldsymbol{U}(A,h) = \begin{bmatrix} x_1 \\ x_2 \\ x_3 \end{bmatrix} = \begin{bmatrix} \cos A \cos h \\ \sin A \cos h \\ \sin h \end{bmatrix} .$$

Note. The fundamental direction of the horizontal system is the vertical of the point in question. This is the direction of a plumb line, viz., the resultant of the Earth's Newtonian attraction, the inertial force, at the place of observation, due to the Earth's rotation relative to an inertial system, attractions of the Moon and the Sun (tidal forces), and, to a lesser extent, the attraction of the planets. As the intensity and direction of tidal forces vary in time, so therefore does the local vertical relative to tangible reference points (walls, telegraph poles, rocks, buildings, and so on). The local astronomical triad is thus slightly variable in relation to any terrestrial frame, with an amplitude of the order of 0.015″, and monthly and annual periods.

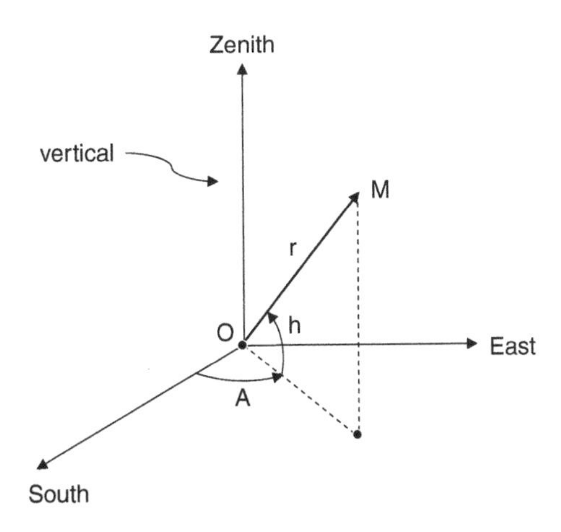

Fig. 4.3 Definition of the horizontal reference system

The Hour Frame

As a result of diurnal motion, the coordinates of a star in the horizontal frame are variable, and subject to rather complicated laws. This renders the frame impractical for the cataloguing of celestial objects. Such a system is just adapted to the instantaneous description of the positions of objects in the sky for an observer at the Earth's surface. The situation is improved by choosing the celestial equator as the fundamental plane, and thereby placing $\boldsymbol{e}_3$ in the direction of the celestial pole. In this plane, the origin for longitudes is given by the intersection of the celestial equator with the local celestial meridian (see Fig. 4.4). There are two such intersections and the standard choice has been the direction in which the Sun culminates, i.e., the south for inhabitants of the Mediterranean region or China. The associated spherical coordinates are respectively:

- The declination δ for the angle of latitude, that is, the height of the star above the celestial equator. This takes values between $-\pi/2$ and $+\pi/2$ and, for a given star, varies slightly in time due to the precession of the celestial equator and the proper motion of the star itself. This precession, i.e., the systematic motion of the celestial pole about the pole of the ecliptic, produces an annual variation in the declinations of stars equal to at most $20''$ per year, whereas the true motion of the star in space leads to proper motions of 10 mas/year, except in exceptional cases. Precession is a motion of the frame relative to sources and not a true motion of the sources.
- The hour angle H is the dihedral angle between the local meridian and the plane containing the directions of both the star and the celestial pole. For historical reasons connected with techniques for determining time, the convention fixing the orientation of H is $\mathrm{d}H/\mathrm{d}t$, which gives H a left-handed orientation. Great care is therefore needed when applying rotation matrices in the hour frame. In contrast to the orientation of the azimuthal angle, the convention for orientating the hour angle involves no exceptions and we shall stick with it, despite the resulting mathematical drawback.

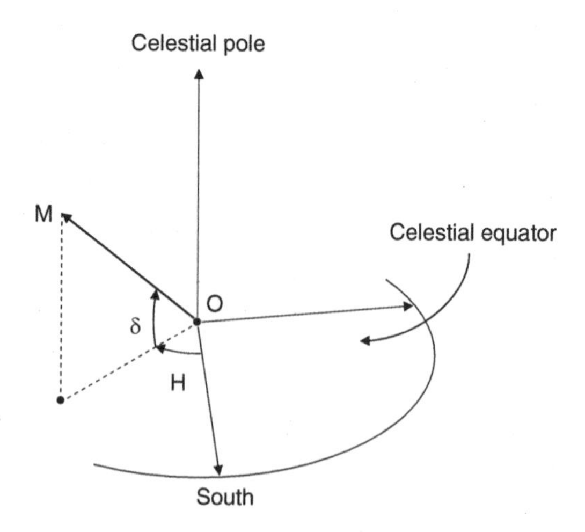

Fig. 4.4 The hour system of coordinates

The hour angle is traditionally expressed in units of time (hours, minutes, seconds), although it would be more appropriate to use units of angle (decimal degrees, radians) to carry out calculations:

$$\boldsymbol{U}(H,\delta) = \begin{bmatrix} x_1 \\ x_2 \\ x_3 \end{bmatrix} = \begin{bmatrix} \cos H \cos\delta \\ -\sin H \cos\delta \\ \sin\delta \end{bmatrix}.$$

The Equatorial Frame

The hour frame is still not perfectly adapted to a description of the sky which is independent of the observer, since it refers to the observer's local meridian. This limitation is overcome by removing the large component of motion due to the Earth's rotation (a period of 24 hours, to be compared with the small components due to nutation, period 18 years, and precession, period 26 000 years). The celestial pole still defines the direction $\boldsymbol{e}_3$, and declination is therefore the angle of latitude (see Fig. 4.5). In the equatorial plane, a non-rotating point is chosen as longitudinal origin, namely the intersection between the mean orbital plane of the Earth and the instantaneous celestial equator. This point is traditionally known as the vernal equinox and denoted γ. The longitude in this system is called right ascension and is measured anti-clockwise as viewed from the celestial pole, so that the unit vector in the direction $\boldsymbol{OM}$ becomes

$$\boldsymbol{U}(\alpha,\delta) = \begin{bmatrix} x_1 \\ x_2 \\ x_3 \end{bmatrix} = \begin{bmatrix} \cos\alpha \cos\delta \\ \sin\alpha \cos\delta \\ \sin\delta \end{bmatrix}.$$

Star, galaxy, and radio-source catalogues use this system, specifying precisely the date corresponding to the choice of equator and the epoch at which the positions of

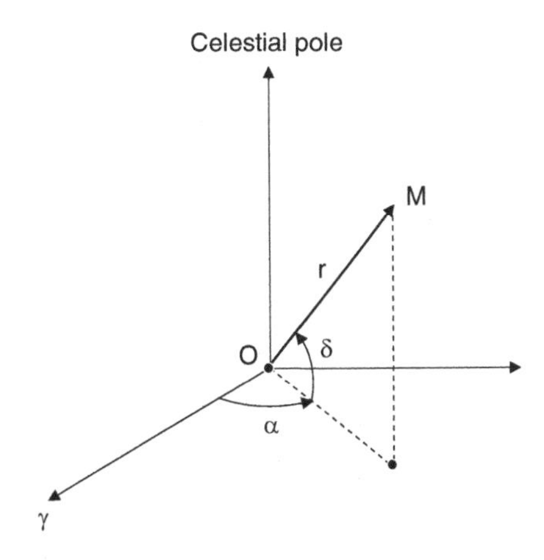

Fig. 4.5 The equatorial reference system

the objects in the catalogue are given. There are, in fact, two main causes of variation in the equatorial coordinates of stars:

- On the one hand, the change of frame which, through precession and nutation, modifies the position of the celestial equator relative to the stars and thus leads to variation of their coordinates, even if they have no real motion with respect to an inertial frame. The annual changes in coordinate values due to precession are of the order of 50″.
- On the other hand, the stars are not fixed in space but rotate with the Galaxy, and each has its own proper displacement in time, leading to a change of equatorial coordinates, even in a fixed frame. This is manifested, in the sky, by a relative displacement of the stars with respect to each other, and by a slow evolution in the shapes of the constellations. The proper motions are generally very small, only 100 stars having an annual displacement greater than 0.1″, and the most rapid amongst these, Barnard's star,[3] a proper motion of 10″ per year.

A stellar catalogue referred to the equatorial system is supplied with an equinox, i.e., a date corresponding to the orientation of the chosen equatorial systems, and an epoch, i.e., a date at which star positions are specified. In general the equinox is a neat conventional date (see Sect. 4.3.6), whereas the epoch of the catalogue is close to the median date of the observations used to compile it. Since the introduction of the ICRS reference system, the equatorial system is no longer formally attached to the celestial equator (although it is, by construction, very close to the celestial equator of 1 January 2000 at 12 h) and must be considered as fixed in

[3]The American astronomer E.E. Barnard (1857–1923) was above all a talented observer, either visually or with photographic plates. He discovered Amalthea, the first satellite of Jupiter to be detected since Galileo, and noticed the exceptional proper motion of the star that now carries his name.

the future. Its relationship with the equator at the given date is very complex for high accuracy observations (typically better than the millisecond of arc), but for the kind of accuracies discussed in this introduction, the classical precession and nutation transformations are largely sufficient.

The Ecliptic Frame

The fundamental plane is the orbital plane of the Earth about the Sun, with origin at the vernal equinox (see Fig. 4.6). The whole problem here is to give a precise definition of the ecliptic and its relation to the equatorial system. If the Solar System contained only the Earth and the Sun, the orbital motion of the Earth would take place in a fixed plane which could be used as a definition and would be accessible to observation. The instantaneous orbital plane, containing the vector Sun–Earth (Sun–centre-of-mass of the Earth–Moon system would be a better definition) and the velocity vector, is not fixed because of planetary perturbations. It oscillates with amplitude $1''$ of arc about a mean plane, on a time scale of several years. In other words, the ecliptic latitude of the Sun is not zero, but can reach $1''$. A very slow evolution of amplitude $\pm\, 1.3$ degrees on a time scale of 100 000 years is superposed on this periodic motion, and this is one of the factors contributing to climatic oscillations. Over short periods, the ecliptic is defined by the mean plane with the rapid oscillations removed, and this rotates by about $47''$ per century relative to an inertial frame.

The mean ecliptic at date T is inclined to the reference ecliptic at epoch J2000 (see Sect. 4.3.6) at an angle

$$\pi = 47.0029''T - 0.033''T^2,$$

where T is measured in Julian centuries of 36 525 days.

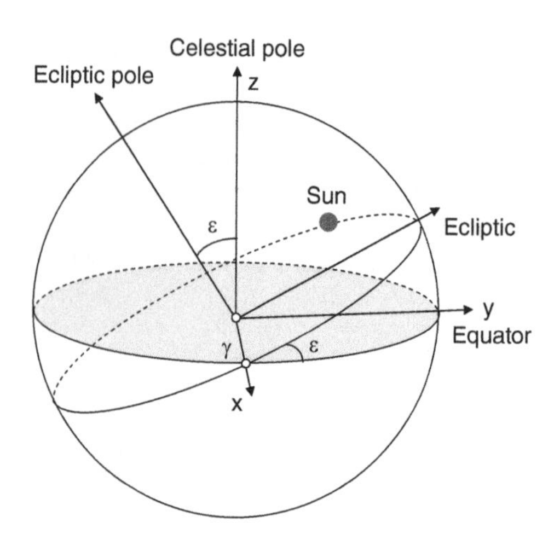

Fig. 4.6 Position of the ecliptic in relation to the celestial equator

The inclination ε_0 of the ecliptic to the celestial equator (called the *obliquity of the ecliptic*) at the epoch J2000 is one of the constants in the system of astronomical constants. It is given by

$$\varepsilon_0 = 23°26'21.406'' \approx 23.439\,279°,$$

and its value at another epoch T is given by

$$\varepsilon(T) = \varepsilon_0 + aT = 23.439\,279° - 46.815''T + O(T^2),$$

where T is measured in Julian centuries from J2000. (Note here that the rate of change is given in seconds of arc, not degrees.) This secular decrease, due to the movement of the ecliptic, has been exposed through obliquity measurements made from the time of Hipparchos in the second century BC. We know that Eratosthenes obtained $\varepsilon = 23°50'$ in the second century BC, i.e., a difference of just $7'$ from the value given by the above formula.

Many periodic oscillations are superposed on this regular variation of the obliquity, the largest resulting from the nutation of the Earth's axis, with a period of 18.6 years and an amplitude of $9.20''$. However, this is a motion of the celestial equator relative to the ecliptic, and not a true motion of the ecliptic relative to the stars.

The Galactic Frame

A system of galactic coordinates is also required for studies in which the galactic plane enters as the natural plane of symmetry, e.g., when studying the spatial distribution of different populations of stars, and for kinematic or dynamic studies of the Galaxy. The galactic plane is thus used as the fundamental reference plane. Since no precise physical definition exists for the latter, a plane is chosen by convention, close to the plane perpendicular to the rotation vector of the Galaxy, which is determined by observation of the hydrogen 21 cm line. Although equivalence with some physical definition is not essential in defining this frame (there is no accurate access to the system), it is of the utmost importance that astronomers should agree on a common definition. In order to orient the galactic triad with respect to the equatorial triad (see p. 134), three parameters are required, corresponding to the three degrees of freedom of a solid rotating about a fixed point, e.g., the three Euler angles shown in Fig. 4.7.

With sufficient accuracy for calculation purposes, we can use

$$\psi = 282.86\,\text{deg},\ \theta = 62.87\,\text{deg},\ \phi = 327.06\,\text{deg},$$

which gives for the coordinates of the galactic pole (direction of z_2) and the galactic centre (direction of y_1) in J2000

$$\alpha_\text{p} = 12\text{h}\,51\text{m}\,26\text{s} = 192.859\,\text{deg},\ \delta_\text{p} = 27°07'41'' = 27.128\,\text{deg},$$

$$\alpha_\text{c} = 17\text{h}\,45\text{m}\,37\text{s} = 266.405\,\text{deg},\ \delta_\text{c} = -28°56'10'' = -28.936\,\text{deg}.$$

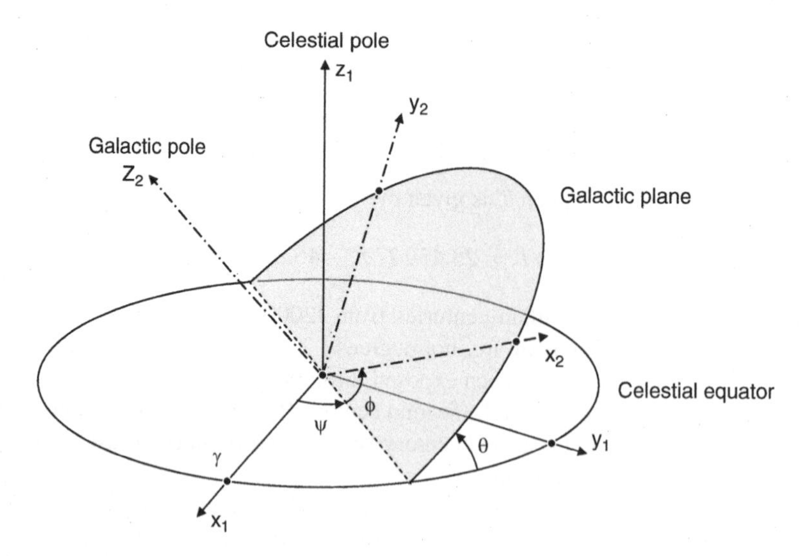

Fig. 4.7 The galactic coordinate system

Note that the usual definition through the coordinates of the pole and the centre makes use of four parameters instead of three. The four coordinates are not independent since the galactic centre lies in the plane normal to the direction defined by the galactic pole.

4.1.3 Change of Frame

In every case, changing from one celestial frame to another means finding the rotation matrix which transforms the vector basis of one frame into that of the other. The vectorial approach is far better adapted to modern calculation techniques than the traditional method based on trigonometric relations in spherical triangles, and naturally it leads to exactly the same analytical formulas. However, as far as computation time is concerned, linear transformations are much more efficient for numerical computations than evaluating trigonometric lines occurring in the full formulas, and the programming aspect is more straightforward, too. While it should not be forgotten that the trigonometric calculations are embodied by the rotation operators, the latter become tools and their calculation can be optimised once and for all.

Given the vector $\boldsymbol{OM}$, let x_1, x_2, x_3 be its Cartesian components in frame R, and x_1', x_2', x_3' its components in frame R'. We transform from R to R' by rotating the triad through an angle ϕ about a direction $\boldsymbol{I}$. The transformation is parametrised by a 3×3 orthogonal matrix R.

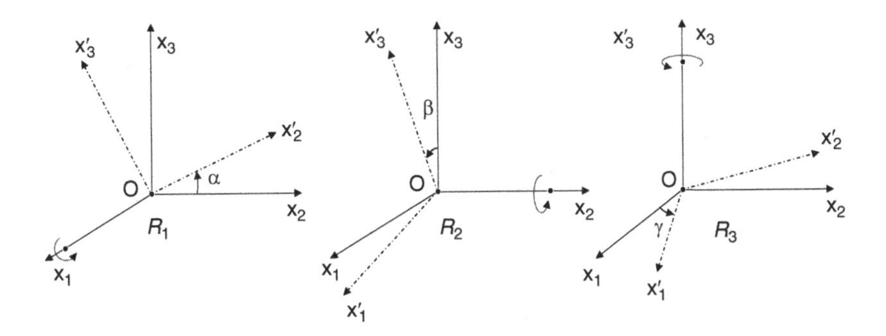

Fig. 4.8 Elementary rotations about coordinate axes

Thus,

$$\begin{bmatrix} x'_1 \\ x'_2 \\ x'_3 \end{bmatrix} = [\mathsf{R}] \begin{bmatrix} x_1 \\ x_2 \\ x_3 \end{bmatrix}.$$

In the elementary case of rotations about coordinate axes, which are the most important in practice, the matrices R have the following form (see Fig. 4.8):

$$\mathsf{R}_1(\alpha) = \begin{bmatrix} 1 & 0 & 0 \\ 0 & \cos\alpha & \sin\alpha \\ 0 & -\sin\alpha & \cos\alpha \end{bmatrix}, \quad \mathsf{R}_2(\beta) = \begin{bmatrix} \cos\beta & 0 & -\sin\beta \\ 0 & 1 & 0 \\ \sin\beta & 0 & \cos\beta \end{bmatrix},$$

$$\mathsf{R}_3(\gamma) = \begin{bmatrix} \cos\gamma & \sin\gamma & 0 \\ -\sin\gamma & \cos\gamma & 0 \\ 0 & 0 & 1 \end{bmatrix}.$$

Note. The above rotation matrices correspond to passive rotations. It is the axes that rotate and not the source. When applied to a vector, the components of the vector relative to the first system are transformed to the components of the same vector relative to the second system.

Transformation from Horizontal to Hour Frame $(A, h) \longleftrightarrow (H, \delta)$

The two triads have a common vector $\boldsymbol{e}_2$, hence the transformation is just a rotation about axis Ox_2. The angle between the vectors $\boldsymbol{e}_3$ of each system is equal to the co-latitude at the point, that is, $\pi/2 - \phi$, where ϕ is the latitude of the observer. The transformation between the unit vectors is thus

$$\boldsymbol{U}(A, h) = \mathsf{R}_2(\pi/2 - \phi)\, \boldsymbol{U}(H, \delta).$$

So, applying the rotation matrix to the Cartesian components, we obtain the transformation of the spherical coordinates directly:

$$\boxed{\begin{aligned}&\cos h \cos A = -\cos\phi \sin\delta + \sin\phi \cos\delta \cos H,\\&\cos h \sin A = -\cos\delta \sin H,\\&\sin h = \sin\phi \sin\delta + \cos\phi \cos\delta \cos H.\end{aligned}}$$

The last equation in this group, giving the height of a body as a function of time via H, is the fundamental equation of diurnal motion, used to calculate risings and settings, the height at culmination, the duration of a sunset, and so on.

Likewise for the inverse transformation,

$$\boldsymbol{U}(H,\delta) = \mathsf{R}_2(-\pi/2 + \phi)\,\boldsymbol{U}(A,h),$$

which gives

$$\boxed{\begin{aligned}&\cos \delta \cos H = \cos\phi \sin h + \sin\phi \cos h \cos A,\\&\cos\delta \sin H = -\cos h \sin A,\\&\sin\delta = \sin\phi \sin h - \cos\phi \cos h \cos A.\end{aligned}}$$

Note. The trigonometric formulas are given here for reference. It is always preferable to work with Cartesian components and the rotation matrices for numerical computations, only bringing in angles as intermediate quantities between input and output. Nevertheless, qualitative and analytic studies are more easily handled using the angular expressions.

Transformation from Hour to Equatorial Frame $(H, \delta) \longleftrightarrow (\alpha, \delta)$

These two systems, described in Sect. 4.1.2, share the axis $\boldsymbol{e}_3$ which corresponds to the orientation of the Earth in space, and only differ through their rotation about this axis. This rotation positions the observer's astronomical meridian relative to the vernal equinox. Denote the dihedral angle between this meridian and the plane $\alpha = 0$ passing through the vernal equinox by α_z. It is also the right ascension of the observer's zenith at the instant of observation. It varies with time, increasing by 360 degrees for each sidereal rotation of the Earth, that is, every 86 164.10 s. With the help of Fig. 4.9, we observe that this angle is numerically equal to the hour angle of the vernal equinox, otherwise known as the *local sidereal time*. This terminology, which has its origins in the meridian observations practised in all observatories since the seventeenth century, should be avoided. The quantity in question is an angle, not a time, and relating its definition to an hour angle orientates it in the retrograde direction. Moreover, time measurement today has no connection with the rotation of the Earth. Use of the right ascension of the zenith instead of sidereal

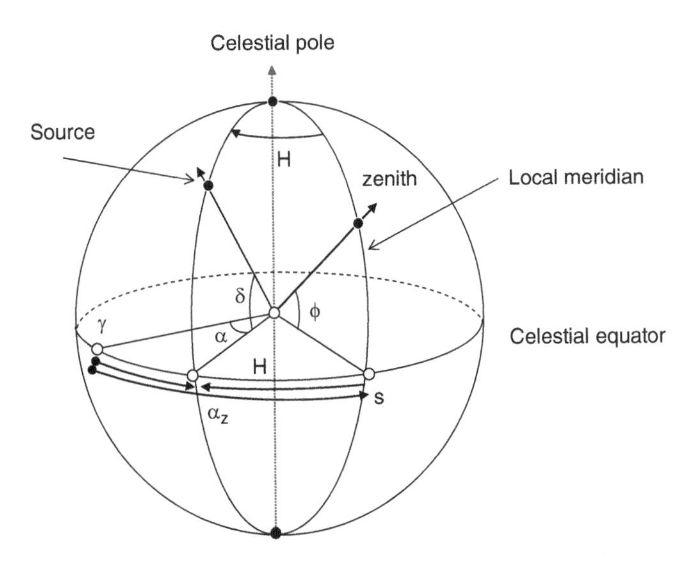

Fig. 4.9 Relation between hour and equatorial systems

time confirms this separation, avoiding confusion and returning the question to the realm of geometry. When the rotational state of the Earth is specified by the hour angle of the vernal equinox, i.e., the sidereal time, we are thinking of the skies as in motion around the Earth, since the hour angle increases with time. It is therefore a description of the apparent motion as it would be perceived by an Earth-based observer. Using the right ascension of the local zenith, it is the Earth that rotates relative to the stars, and this *sidereal angle* is also an increasing function of time. Its interpretation conforms better to true motions.

In fact, the transformations between hour and equatorial coordinates are very simple, reducing to

$$\boldsymbol{U}(H,\delta) = \mathsf{R}_3\,(\alpha_z)\ \boldsymbol{U}(\alpha,\delta),$$

which gives for the spherical coordinates

$$\boxed{H = \alpha_z - \alpha, \quad \delta = \delta.}$$

In order to carry out this transformation in practice, we need to know α_z. Since this angle describes the Earth's rotation relative to the celestial sphere, we must have recourse either to observation, or to a theory of the Earth's rotation accurate enough for the purpose in hand. At the introductory level supposed in the present book, it is sufficient to consider the Earth as rotating uniformly about its axis once every sidereal day, with angular speed

$$\omega = \frac{360}{86\,164.10} \times 3\,600 \sim 15.041\,07\,\mathrm{deg\,h^{-1}}.$$

The deviation from uniformity amounts to about 0.03″ on a time scale of several years. Over periods of several centuries, the secular slowing of the Earth's rotation has to be taken into account. The latter effect, although very slight regarding the speed of rotation, is nevertheless cumulative on the phase, increasing as the square of the time. In practice, the precise orientation of the Earth in space can only be obtained to a very high level of accuracy by regular observation using space geodesy techniques and very long baseline interferometry. The measured orientation and angular velocity parameters are available online, almost in real time, and always accompanied by predicted values for the coming months.

The right ascension at Greenwich is given in the astronomical almanacs in the form of a table, for each day at 0 hours (0 h), and from this the value of α_z at the same instant of time can be calculated at any longitude λ (measured positively towards the east) by

$$\alpha_z(0\,\mathrm{h}) = \alpha_G(0\,\mathrm{h}) + \lambda\,,$$

and at any time t during the day by

$$\alpha_z(t) = \alpha_z(0\,\mathrm{h}) + \omega\, t\,.$$

To an accuracy of about 0.005 degrees, $\alpha_G(0\,\mathrm{h})$ can be calculated directly, without reference to the almanacs, using the formula

$$\alpha_G(0\,\mathrm{h}) = 100.4606 + 0.9856473\, d\,,$$

where d is the number of days between J2000 (JD 2 451 545, where JD is discussed later) and the day under consideration at 0 h UT. Since the reference epoch J2000 begins at 12 h, in the above expression, d takes values ± 0.5, ± 1.5, and so on, rather than continuous or integer values. This is a stroboscopic expression, in which we remove multiples of 360 degrees corresponding to integer numbers of rotations of the Earth about its axis. Since TU is very close to mean solar time, the numerical coefficient in the formula for $\alpha_G(0\,\mathrm{h})$ is nothing else but the mean displacement of the (tropical) Sun in right ascension per day, namely $360/365.2422 = 0.985647$ deg. For example, for the sidereal angle of Greenwich on 01/01/2007 at 0 h UT, we have JD $=$ 2 454 101.5, $d = 2\,556.5$, and $\alpha_G = 100.267\,9°$. The value given by a high accuracy astronomical almanac is 6 h41 m4.55 s $=$ 100.268 96°. The discrepancy comes from the nutation effect on longitude, of the order of 0.001 deg, or 0.2 s.

Transformation from Equatorial to Ecliptic Frame $(\alpha, \delta) \longleftrightarrow (l, b)$

This is another trivial case, being simply a rotation through angle ε about the axis $\boldsymbol{e}_1$ which is common to the two frames:

$$\boldsymbol{U}(l, b) = \mathsf{R}_1\,(\varepsilon)\; \boldsymbol{U}(\alpha, \delta),$$

$$\boxed{\begin{aligned}\cos b\,\cos l &= \cos\alpha\,\cos\delta,\\ \cos b\,\sin l &= \sin\varepsilon\,\sin\delta + \cos\varepsilon\,\cos\delta\,\sin\alpha,\\ \sin b &= \cos\varepsilon\,\sin\delta - \sin\varepsilon\,\cos\delta\,\sin\alpha,\end{aligned}}$$

and for the inverse transformation

$$U(\alpha,\delta) = \mathsf{R}_1(-\varepsilon)\,U(l,b),$$

$$\boxed{\begin{aligned}\cos\delta\,\cos\alpha &= \cos l\,\cos b,\\ \cos\delta\,\sin\alpha &= -\sin\varepsilon\,\sin b + \cos\varepsilon\,\cos b\,\sin l,\\ \sin\delta &= \cos\varepsilon\,\sin b + \sin\varepsilon\,\cos b\,\sin l.\end{aligned}}$$

It is interesting to specialise the second group for the Sun, i.e., when $b = 0$:

$$\begin{aligned}\cos\delta\cos\alpha &= \cos l,\\ \cos\delta\sin\alpha &= \sin l\cos\varepsilon,\\ \sin\delta &= \sin l\sin\varepsilon,\end{aligned}$$

which gives the equatorial coordinates of the Sun as a function of its ecliptic longitude l. The longitudinal motion is very easy to calculate from Kepler's equations, with an accuracy of $\sim 10''$, so it is a straightforward matter to write a little autonomous program a few lines long that gives a low accuracy ephemeris for the Sun. Eliminating the declination between the first two equations, this yields

$$\tan\alpha = \cos\varepsilon\,\tan l,$$

whence it is possible to express the right ascension as a function of longitude using a rapidly converging expansion, viz.,

$$\alpha = l - \tan^2\frac{\varepsilon}{2}\,\sin 2l + \frac{1}{2}\tan^4\frac{\varepsilon}{2}\,\sin 4l - \frac{1}{3}\tan^6\frac{\varepsilon}{2}\,\sin 6l + \cdots,$$

or numerically in degrees:

$$\alpha = l - 2.4658\,\sin 2l + 0.0531\,\sin 4l - 0.0015\,\sin 6l + \cdots.$$

The quantity $\alpha - l$ is usually called the *reduction at the equator*. It is important in the definition of the mean solar time, or when reading a sundial.

Transformation from Equatorial to Galactic Frame $(\alpha,\delta) \longleftrightarrow (\lambda,\beta)$

This last case is the most general, requiring a complete rotation matrix, composed of three elementary rotations through each of the three Euler angles ψ, θ, ϕ, whose numerical values were given in Sect. 4.1.2:

$$U(\lambda, \beta) = \mathsf{R}_3(\phi)\ \mathsf{R}_1(\theta)\ \mathsf{R}_3(\psi)\ U(\alpha, \delta).$$

Calculation of the product of the three above matrices is straightforward but tedious. Numerically, the Cartesian coordinates (x'_1, x'_2, x'_3) of $\boldsymbol{OM}$ in the galactic frame can be calculated from its Cartesian coordinates (x_1, x_2, x_3) in the equatorial frame by

$$\begin{bmatrix} x'_1 \\ x'_2 \\ x'_3 \end{bmatrix} = \begin{bmatrix} -0.05488 & -0.87344 & -0.48384 \\ 0.49411 & -0.44483 & 0.74698 \\ -0.86767 & -0.19808 & 0.45598 \end{bmatrix} \begin{bmatrix} x_1 \\ x_2 \\ x_3 \end{bmatrix}.$$

For example, the Crab Nebula has equatorial coordinates $\alpha = 5\text{h}\,31\text{m} \approx 82.75\,\text{deg}$, $\delta = 21°59' \approx 21.98\,\text{deg}$. The vector $\boldsymbol{OM}$ thus has the Cartesian coordinate representation $(0.11702, 0.91990, 0.37428)$. Applying the above matrix, its Cartesian coordinates in the galactic frame are $(-0.99100, -0.07186, -0.11303)$ which gives angular coordinates $\lambda = 184.14\,\text{deg}$, $\beta = -6.49\,\text{deg}$, implying a location very close to the galactic plane.

4.2 Practical Realisation of Spatial Frames

4.2.1 Celestial Reference Systems

From the moment astronomers wished to record their observations in the form of catalogues or charts, either to be able to describe the sky, or just to remember them, they had to choose a reference system by defining a fundamental plane and an origin for longitudes. They then defined a frame, that is, a system of coordinates, and, finally, some practical realisation of those coordinates in space, in order to be able to determine them by measurement. It is no easy task, while making the first observations of some new body, or following the path of an asteroid, to establish the coordinates of such an object lost in the heavens. An idea which comes immediately to mind would be to measure its position relative to known neighbouring objects and to determine its coordinates from the coordinates of these nearby stars, either by direct micrometric measurement, or by means of photographic plates or CCD images. But, as in every system of standards, there then arises the problem of fixing the coordinates of those reference objects themselves by some other method, and indeed with great care, given their prior position in the order of things.

The aim of these reference catalogues, the counterpart of standards in physical metrology, is to place markers of known position across the sky, not only at the precise date when the catalogue is made, but also at other times both before and after that date, which involves determining the motion of these markers. The transition from a theoretical reference system to its practical realisation in the sky amounts to establishing by convention a system of reference as close as possible to the ideal one, as it was conceived by its inventors. In practice, the difference between the two may be obscured, but, on a conceptual level, it remains fundamental. The same distinction could be made between the first definition of the metre as being a certain fraction of the terrestrial meridian, and its realisation in the form of a standard metre stored in the archives. The latter was indeed the standard, an accessible material realisation, and not the metre by definition, a situation to be contrasted with the case of the kilogram even today.

Astronomical reference frames, unlike milestones along a main road, are more than just kinematic reference points. It is required of these frames that the laws of dynamics should hold in them without recourse to any other forces than gravity and, in certain cases, well-understood dissipative forces. This implies that the frames should be inertial, and therefore non-rotating. Once again, the problems posed by this ideal go outside the realm of the present text, and we shall merely assume the absence of any rotation relative to the general distribution of extragalactic matter beyond a distance of 100 Mpc.

4.2.2 Fundamental Catalogues

By definition, these catalogues, which contain only a small number of stars, represent the inertial system by convention. In the case of the equatorial system, the position and proper motion of the stars, i.e., their right ascension and declination at different epochs, must be determined. To this end, the declinations present no difficulty in principle, using the right meridian instrumentation. With a good model of the Earth's precession, values measured at different dates can be related to a unique reference equator. Concerning right ascension, however, a major difficulty arises because the vernal equinox is not accessible by observation of the stars. Only the difference between the right ascensions of two stars can be easily obtained by means of a clock and a meridian circle, which constitutes only a relative observation and not a fundamental one. This difficulty was to be expected, for the definition of the vernal equinox refers to the annual motion of the Earth and not to the stars. The solution to the problem lies in simultaneous observation of both the Sun and the stars, with a view to linking the right ascension of each star to that of the Sun, and thus to the vernal equinox.

In this sense, these fundamental catalogues are realisations of a dynamical reference system. The point is crucial, demonstrating the great difference between the celestial reference systems, based on fundamental catalogues, and the extragalactic system which has replaced them as the conventional system. Nevertheless, once a fundamental catalogue is available, it is the right ascensions and declinations taken as a whole which define in practical terms the equator and the equinox. Only further observations of the Sun (or the planets) and the stars allow the relation between the definition of the reference system and its practical realisation to be checked, and only then can any correction to the equinox be published, if necessary, and applied to the stars.

Table 4.3 The main star catalogues

Catalogue	Number of stars	Mean epoch	Accuracy of 1990 positions	Accuracy of proper motions
FK4	1 500	1949	0.1″	
FK5	1 500	1949	0.05″	0.0008″/year
FK5 extended	3 000		0.08″	0.002″/year
Hipparcos	120 000	1991.25	0.001″	0.001″/year
Tycho-2	2 000 000	1991.25	0.03″	0.003″/year
PPM	380 000		0.3″	0.006″/year
SAO	250 000		1.5″	
GSC	20 000 000		1.5″	
UCAC	50 000 000	2001	0.02–0.07″	

The first fundamental system was devised by F.W. Bessel[4] who, in the nineteenth century, used J. Bradley's[5] observations to link 14 bright stars directly to the Sun. A few years later the system was extended to 36 stars and thereafter remained the basis for determining all right ascensions until the beginning of the twentieth century. The number of fundamental stars increased gradually up until the publication of the FK4 (*Vierter Fundamental Katalog*) in 1963, with its 1 535 stars. Then came a further revision, the FK5, in 1986, with the same stars but a new determination of the equinox and the constant of precession, and a correction to the proper motions.

Between the two catalogues, right ascensions differ systematically according to the relation

$$\alpha_{\mathrm{FK5}} = \alpha_{\mathrm{FK4}} + 1.163'' + 1.275''T,$$

where T is the time in Julian centuries measured from J2000 (see Sect. 4.3.6). Consequently, the proper motions in right ascension given in the two systems differ by 1.275″ per century.

By 1990, the accuracy was 0.05″ on the positions, and 0.001″ per year on the proper motions. Comparison with recent observations and, in particular, with the Hipparcos catalogue also reveal systematic errors of amplitude around 0.1″. A further 3 000 stars, measured to an accuracy of 0.08″ and 0.002″ per year, had been added.

All other high-quality determinations of stellar position are based on extensions of the fundamental catalogue, either through relative meridian observations, which are so much easier than absolute observations, or through photographic compilations of relatively low accuracy ($\approx$ 1″). General catalogues of position and/or proper motion include between 10^4 and 10^7 stars, extending the fundamental catalogues in terms of both spatial density and magnitude, to varying degrees of accuracy. A short list of these catalogues and their features is given in Table 4.3.

[4]The German mathematician and astronomer F.W. Bessel (1784–1846) founded the German school of observational astronomy. He made the first uncontested measurement of stellar parallax in 1838.

[5]The British astronomer J. Bradley (1693–1762) discovered and explained stellar aberration and later nutation. He succeeded E. Halley as Astronomer Royal at the Greenwich Observatory.

4.2.3 *The Extragalactic System*

A Kinematic Reference System

Up until now we have been trying to give the inertial reference system a practical realisation in terms of sources in perpetual movement relative to each other and only indirectly related to the theoretical frame (in this case, the equator and the equinox). But there exists a much more geometrical alternative. If the stars had no relative motion, one relative to the other, they would constitute a set of luminous points on the sky, globally unchanging, which could be used to mark positions. The purely kinematic aspect of the reference frame would be satisfied by this solution.

Now, there would be nothing to guarantee that the whole system were not rotating relative to some inertial system. But if this were the case, it would be possible to demonstrate the fact, by comparing the motion of the planets relative to this framework with their theoretically predicted motion deduced from Newton's law in the absence of any inertial forces. By applying this global rotational correction, i.e., attributing a suitable proper motion to each star to ensure that the resulting system were inertial, one would then be able to materialise a reference frame with the fundamental features. Unfortunately, the stars in the Galaxy (the only stars we can actually see individually) have relative motions, and the constellations are not unchanging patterns.

Such a reference system, determined by the relative positions of sources, is described as a *geometric system* or *kinematic system*, given the absence of any dynamical criteria in its definition.

For the straightforward task of locating objects not in the original catalogue, the fact that the frame is not inertial presents no great difficulty, and a system as arbitrary as this might be sufficient to build up a catalogue of positions. But from the moment dynamical studies are envisaged (rotation of the Earth, Solar System, rotational movements of the Galaxy, and so on), this problem must be faced. The most satisfying way of fulfilling the two requirements (no relative motion and no global rotation) is to use extragalactic quasi-stellar objects (QSO) which are virtually fixed relative to each other. The idea is not new, having been proposed by both Herschel[6] and Laplace[7] over 200 years ago. The fact that extragalactic objects constitute a non-rotating sphere cannot be demonstrated in itself, but rests on a series of assumptions which are confirmed by observation. The most distant galaxies have an isotropic distribution and the radial components of their velocities increase systematically with their distances. The 3 K background,

---

[6]The British physicist and astronomer J. Herschel (1792–1871) compiled catalogues of double stars and nebulas in the southern hemisphere. He was an influential teacher, a pioneer of photography, and a major figure in nineteenth century British science.

[7]The French mathematician and physicist P.S. Laplace (1749–1827) contributed to astronomy through his monumental *Traité de Mécanique Céleste* and *Exposition du Système du Monde*. His *Théorie Analytique des Probabilités* is another major work.

uniform to a very high degree, supports the idea that the Universe as a whole has no rotation.

The physical significance of this declaration of faith nevertheless leaves the way open to several interpretations. This choice must be viewed in the light of Mach's principle,[8] which attributes inertia to the general distribution of mass in the Universe, but is independent of relativity theories. Experimentally, comparison of the most accurate measurements of a dynamical system, namely those obtained by laser telemetry of the Moon, with those of a kinematic system, based on radio interferometry observations, has not revealed any rotation between the two systems. Moreover, if we suppose that the tangential speeds of quasars are statistically comparable to their recession speeds, then their proper motion would be

$$\alpha = H < 0.02\,\mathrm{mas/year},$$

where H is Hubble's constant[9] and $1\,\mathrm{mas} = 0.001''$. It is reasonable to suppose that the actual speeds are much less than this value and that, up to a few μas ($= 10^{-6}$ seconds of arc), quasars show no systematic movement. At the 1991 General Assembly of the International Astronomical Union, it was recommended that the future conventional inertial reference system should from that time comprise a set of distant radio sources referred to the barycenter of the Solar System. The system specified in this way is called the International Celestial Reference System (ICRS). The basic principle is quite general and says nothing about the list of sources or observational techniques to be used to observe them. A practical realisation of this system will involve physical construction of the reference frame materialised by a highly accurate list of extragalactic sources accompanied by their coordinates. The sources must therefore be selected and observed in a consistent way, so that the angular distances between pairs of sources on the sky are in perfect agreement with a calculation based on their coordinates. This realisation is called the International Celestial Reference Frame (ICRF).

The difference between the words 'system' and 'frame' is fundamental: a system is above all a set of principles and prescriptions specifying the concepts to be used to construct the frame, i.e., the practical realisation of the system. These principles must be sufficient to specify the origin and orientation of the reference triad at any given epoch. There is only one system (here the ICRS), which can have different realisations, even if, at a given epoch, only one of these can be qualified as reference frame (here the ICRF). The current primary reference frame is based on very long baseline interferometry (VLBI) observations, but in the future, optical observations by the Gaia mission will provide a new realisation.

[8]The Austrian physicist and epistemologist E. Mach (1838–1916) contributed to the theory of waves, the propagation of sound, and thermodynamics. He was particularly concerned by foundational issues in physics, and his thoughts regarding the origin of inertia and mass influenced Einstein.

[9]The American astronomer E.P. Hubble (1889–1953) made two of the greatest scientific discoveries of the twentieth century: the extragalactic nature of the nebulas and the expansion of the Universe.

Observational Methods

In order to construct a kinematic frame, a reference plane and origin for longitudes must once again be defined. There is no a priori natural definition since the system is geometric. It is the coordinates themselves of the objects which both define the frame and realise it physically. The declinations of two objects are sufficient to construct the pole (with a discrete choice between two possibilities), and the longitude of any particular object defines the origin of longitude. For reasons of continuity, one ensures that the frame thereby constructed remains very close to the equatorial frame (see Sect. 4.1.2), previously materialised by FK5, and the origin of longitudes is chosen as close as possible to the dynamical equinox of J2000. We then speak of declinations and right ascensions, even though the latter owe nothing to the definiton of the equinox, but rather to a conventional origin in the fundamental plane.

The radio source observations are made by very long baseline interferometry (VLBI) to an accuracy better than one millisecond of arc (see Fig. 4.10). In contrast with the case of connected interferometers, correlation of signals received in each radiotelescope is delayed, bearing in mind that the very high stability of the local oscillators and the possibility of a posteriori synchronisation of the signals allows for digital rather than analogical phasing of the two signals (see Sect. 6.5.1). The observable quantity is not the phase of the fringes but rather the delay between arrivals at the two antennas, together with its derivative. For a wavelength λ of several centimetres and a base line length B of several thousand kilometres, a resolution of the order of

$$\lambda / B \approx 1 \,\text{mas}$$

can be achieved. If $\boldsymbol{r}_1$ and $\boldsymbol{r}_2$ are the radius vectors from the centre of the Earth to the two stations, the base line is $\boldsymbol{B} = \boldsymbol{r}_2 - \boldsymbol{r}_1$. Calling the unit vector in the direction of the source $\boldsymbol{s}$, we obtain the fundamental relation of a VLBI observation:

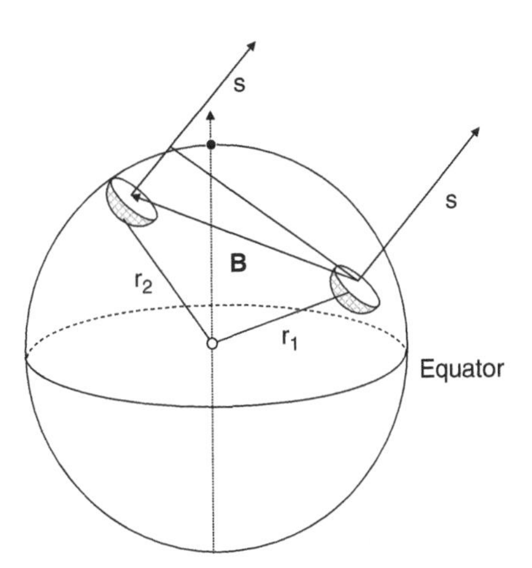

Fig. 4.10 Geometry of VLBI observations

$$\tau = \frac{\boldsymbol{B} \cdot \boldsymbol{s}}{c},$$

where τ is the delay between wavefront arrivals and c is the speed of light in vacuum. Letting δ and H be the declination and hour angle, respectively, of the source, and d and h the corresponding quantities for the unit vector of the base line $\boldsymbol{B}$, the delay is given by

$$\tau = \frac{B}{c} \left[\sin d \ \sin \delta + \cos d \cos \delta \ \cos(H - h) \right].$$

The observation thus reveals the position of the base line relative to the celestial reference system, as well as the direction of the source. For base lines known in the terrestrial frame, we can determine parameters describing the orientation of the Earth and also the coordinates of the sources in the celestial reference system. Conversely, observation of known sources can be used to follow the rotation of the Earth and locate the stations on the various continents. The implication is thus that we have a global treatment of the observations made by a network of radiotelescopes.

At present, the VLBI astrometry program comprises about twenty antennas, most of these being located in the northern hemisphere. Analysis of the data has created a fundamental catalogue, the observations relating to the whole of the sky, object by object, without any need to link those objects progressively to each other. The set of arcs joining the sources into pairs constitutes the reference system, although, in practice, because of the choice of origin for right ascensions, the actual coordinates of the sources constitute the system.

Properties of the ICRF

The first ICRF solution was officially published in 1998 and formally adopted as reference frame replacing the FK5 on 1 January 1998, following an IAU resolution during the general assembly of August 1997. In order to keep the orientation as close as possible to that defined by the FK5, the optical coordinates of the radiosource 3C273 have been determined, by lunar occultation, in the dynamical reference system. The right ascension (α =12h 29m 6.6997s in J2000) of 3C273 was thus fixed initially in the treatment of data, imposing a constraint on the origin of the kinematic reference system. In the final version of the reference system, this origin was actually set by imposing mean difference of zero on a set of 23 sources. This version of the ICRF (known as ICRF1) was followed by two extensions, with slightly more sources. The last extension of the ICRF1 released in 2004 (denoted ICRF1-Ext.2) contained 717 sources (compared with 608 in the initial version of 1998), including 212 defining sources, 294 candidate sources, and 211 other sources.

A new version was adopted by the IAU in August 2009 (the ICRF2) as the current official primary realisation of the celestial reference frame. It contains

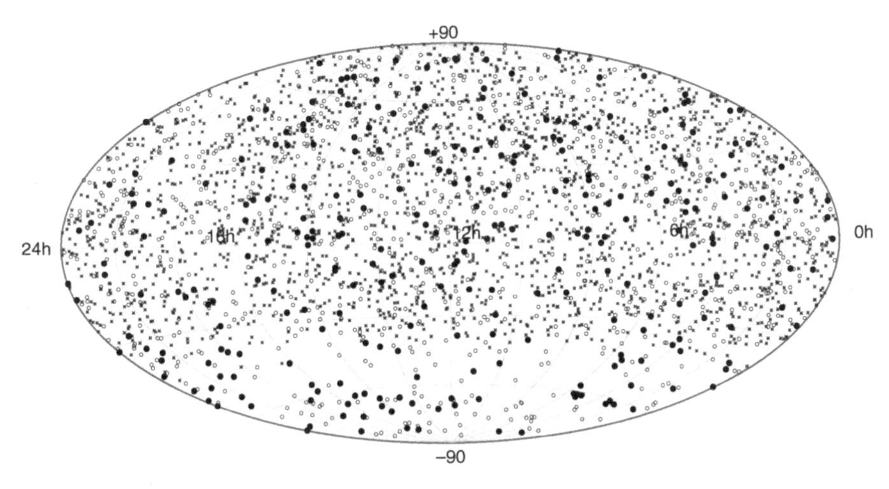

Fig. 4.11 Distribution of ICRF2 sources in equatorial coordinates. *Black circles*: the 295 defining sources. *White circles*: the 922 additional VLBI sources. *Stars*: the remaining 2197 sources, from the calibration sessions.

precise positions of 3414 compact radio astronomical sources, with a noise floor of 40 μas and a stability in its orientation of 10 μas. The ICRF2 axes rest upon a set of 295 new defining sources selected on the basis of positional stability and the absence of extensive intrinsic source structure. It is complemented by another set of 922 sources observed in VLBI, with good distribution over the sky and a larger set of 2197 sources used as calibrators for the radio astronomical community. They have also been included in the ICRF2, given their high astrometric quality. The distributions of the three categories of sources are shown in Fig. 4.11.

The ICRF pole (and hence also the reference equator) do not coincide exactly with the celestial pole (and celestial equator) at the epoch J2000, but every effort has been made to ensure that it is as close as possible. The discrepancy is -17 ± 0.2 mas in the direction $\alpha = 0$ deg and -5.1 ± 0.2 mas in the direction $\alpha = 90$ deg. For calculations of medium accuracy (> 100 mas), it can be taken to coincide with the celestial pole at J2000. The origin in the ICRF equator is shifted by $+55.4$ mas relative to the inertial equinox (intersection of the ecliptic with the ICRF equator), and is also located at 22.9 mas from the origin of the FK5 right ascension.

4.2.4 The Hipparcos Frame

Various extensions of the extragalactic system are necessary, in the first place because of the small number of reference sources ($\sim$200), but also because stellar astronomy has a strong preference for optical and infrared wavelengths. The optical counterparts of the ICRF sources are all faint in the visible and difficult to observe. Since the density is very low (one source corresponds on average to

∼70 square degrees), direct use of these sources would be impossible, even in wide-field astronomy. As far as optical sources are concerned, only the Hipparcos catalogue can link them accurately to the extragalactic system. Before explaining how the Hipparcos system can be attached to the radiosource system, we describe some of the technical details of this space mission, and present the main properties of the results.

The Hipparcos Mission

Hipparcos is an acronym for HIgh Precision PARallax COllecting Satellite. The Hipparcos satellite was launched by Ariane, from the Kourou launch pad, on 8 August 1989. Owing to failure of the apogee boost motor, it proved impossible to place the satellite on the circular geostationary orbit of radius 42 000 km for which the mission had been planned. A few weeks later the mission had been adapted to the new environment, and the satellite began operation, working correctly until 25 August 1993 when the mission controllers in the ESA command centre at Darmstadt decided to put it out of action after a series of problems with the guidance system and on-board computer.

Aim of the Mission

The objective of the mission was to carry out a very high precision survey of the celestial sphere, involving observation of around 120 000 carefully selected stars spread evenly across the sky, with an accuracy for position measurements of $0.002''$. Apart from positions, the parallax of the stars on the programme was also to be obtained to a high degree of accuracy. Indeed, because of the importance of this parameter in relation to theories of stellar evolution, and in understanding the history of the Universe, these results were considered by many astronomers to be the most interesting. The main features of the results of the Hipparcos mission are summarised in Table 4.4, together with those of the complementary mission Tycho, which included a greater number of stars but with measurements to a lesser degree of accuracy.

The Hipparcos satellite was equipped with a modestly sized telescope, 30 cm in diameter, which was nevertheless sufficient to observe stars down to magnitude

Table 4.4 Main features of the Hipparcos mission

	Main mission Hipparcos	Complementary mission Tycho
Number of stars	120 000	1 000 000
Maximum magnitude	12.4	11.0
Accuracy of positions	$0.001''$	$0.02''$
Accuracy of speeds	$0.001''$ per year	–
Accuracy of parallax	$0.001''$	–
Accuracy of magnitude	0.001 mag	0.01 mag

12.5. Stars were observed simultaneously in two directions at an angle of 58 degrees to each other, and their images formed in the focal plane of the instrument on an analysing grating composed of 2 688 slits of period 1.208″, the transparent part comprising 39%. The field of view was 0.9° × 0.9° and, on average, two stars were visible in each of the two directions. The detector located behind the grating analysed the intensity fluctuations recorded when the stellar images were displaced across the grating, these displacements being the result of an extremely regular rotational movement of 167 degrees per hour imposed on the satellite.

Hipparcos was a scanning satellite whose mission had been completely planned and optimised before the launch. In order to ensure that the survey would cover the whole sky, observing all 120 000 stars on the programme and each of those at thirty different epochs, the exploration of the celestial sphere had to follow a well-defined rule. Owing to its rotation, the stellar images moved across the analysis grating, thereby modulating the light received by the detector placed just behind the grating.

By analysing the signal received when a star crossed the grating, the position of the centre of the image could be determined within a slit of the grating, at any precise moment (at least, in the case of a non-multiple star). This was an entirely local measurement within the reference system of the instrument. It was already very accurate, to around 0.015″ for a magnitude 9 star.

The signal was then expanded in two harmonics

$$I(t) = I + B + I\left[M\cos(\omega t + \phi) + N\cos(2\omega t + \psi)\right],$$

where I is the intensity in Hz and B is the non-modulated part of the signal, including, among other things, the stellar background noise. M and N are the modulation parameters with typical values, for a non-multiple star, of 0.7 and 0.25, respectively. The phases ϕ and ψ of the two harmonics characterised the position of the image on the grating at the moment under consideration. The positions on the grating were determined, every 2.13 s, for all stars within the field of view of the instrument. The determination was repeated for each star during the 19 s it took for its image to cross the grating. The two amplitudes were used for the photometric catalogue and variable star survey.

As a consequence of the way the satellite scanned the sky, all the information collected over a period of 10 h (approximately one orbit) relates to stars very close to a single great circle of the celestial sphere. As a result, the abscissæ of around 1 500 stars were determined on the circle. Repeating the operation for each orbit of 10.7 h, over a period of 37 months, produced 2 500 circles of 1 500 stars, and hence an average of (2 500×1 500)/120 000=31 observation epochs per star.

After processing, this delivered an astrometric solution giving the position, parallax, and two components of the proper motion of each star on the programme. About 20 000 objects were detected as multiple systems and further processing was carried out to resolve the components whenever possible. The final accuracy for a star of magnitude 8.5 of the Hipparcos catalogue is 0.001″ for position and parallax and $0.001''$ per year for proper motions, and about twice these values for stars of magnitude 11. The catalogue was published by the ESA in 1996, with all its appendices relating to multiple systems and photometry. Fifteen years on, Hipparcos remains the ultimate reference among fundamental catalogues in astronomy and has been used, with its complement Tycho covering some 2 million stars, as a reference to construct fainter and denser catalogues.

In particular, the CCD Astrograph Catalog of the US Naval Observatory in Washington (UCAC), complete up to magnitude $R = 16$, gives the astrometric positions in the Hipparcos system of more than 50 million stars with accuracies ranging from 20 to 70 mas depending on the magnitude. It is without doubt the most important recent contribution to astrometry since Hipparcos and the first CCD survey in this domain. It was achieved in a few years using an astrograph placed for three years in the southern hemisphere, then two years in the northern hemisphere to complete the programme based on observations made between 1998 and 2004. The latest version, UCAC-3, has been released in 2009 and a new version (UCAC-4) is in its final stage for a release in 2012.

Relating to the Kinematic Reference System

As a consequence of the observation method, the solution is invariant under rotation, and the Hipparcos system would be rotating freely relative to the inertial system if this were not avoided by attaching it to the inertial system. This can be visualised by considering the Hipparcos catalogue as being made up of a very large number of arcs, of perfectly determined dimensions, joining all 120 000 stars on the programme. Clearly, there will be only one way to position the stars relative to each other whilst satisfying all the constraints imposed by these arcs, and this will leave just the degrees of freedom for the rotation of a solid. As the stars have proper motion, the rotation mentioned will be variable in time, and this implies a total of six parameters required to fix the Hipparcos sphere relative to the extragalactic system of reference.

At any moment t, consider a source common to the two systems, with unit vector $\boldsymbol{X}(t)$ in the kinematic system and $\boldsymbol{X}'(t)$ in the Hipparcos system. Then there exists some rotation $\mathsf{R}(t)$ such that

$$\boldsymbol{X}' = [\mathsf{R}]\,\boldsymbol{X}\,.$$

This equation holds for all t and, when linearized in this variable, breaks into two subsystems for values of t near the reference epoch t_0:

$$\begin{aligned}
\boldsymbol{X}(t) &= \boldsymbol{X}_0 + \boldsymbol{V}(t - t_0),\\
\boldsymbol{X}'(t) &= \boldsymbol{X}'_0 + \boldsymbol{V}'(t - t_0),\\
\mathsf{R} &= \mathsf{R}_0 + \mathsf{S}(t - t_0),
\end{aligned}$$

whence, by identification,

$$\begin{aligned} \boldsymbol{X'}_0 &= [\mathsf{R}_0]\,\boldsymbol{X}_0, \\ \boldsymbol{V'} &= [\mathsf{R}_0]\,\boldsymbol{V} + [\mathsf{S}]\,\boldsymbol{X}_0. \end{aligned}$$

As the rotation R is through only a few tens of mas, it can be decomposed in terms of three infinitesimal rotations (α, β, γ) about the coordinate axes:

$$\mathsf{R}_0 = \begin{bmatrix} 1 & \gamma & -\beta \\ -\gamma & 1 & \alpha \\ \beta & -\alpha & 1 \end{bmatrix}, \qquad \mathsf{S} = \begin{bmatrix} 0 & \dot{\gamma} & -\dot{\beta} \\ -\dot{\gamma} & 0 & \dot{\alpha} \\ \dot{\beta} & -\dot{\alpha} & 0 \end{bmatrix}.$$

Knowing the values of $\boldsymbol{X'}_0$, $\boldsymbol{V'}$ and $\boldsymbol{X}_0$, $\boldsymbol{V}$, the values of the six rotation parameters can be found by a method of least squares, so that the Hipparcos solution can finally be expressed in the extragalactic frame.

In the case of Hipparcos, the objects common to the two systems were not the quasars of the kinematic system, too faint to be observed by Hipparcos, but rather a dozen or so compact radio stars that were sufficiently bright at both optical and radio wavelengths. A specific VLBI observation programme, over a period of nearly ten years, has made it possible to fix these objects in the kinematic system, and to measure their proper motion in that system. Moreover, these stars were high priority candidates of the Hipparcos programme, and were observed as often as possible. Once the residual rotation had been calculated, it only had to be applied to the whole solution to produce the final catalogue. This optical realisation of the celestial reference system was adapted by the IAU in 2000 and called the Hipparcos Celestial Reference Frame (HCRF). It contains all the non-multiple stars of the Hipparcos catalogue.

4.2.5 The Near Future: The Gaia Mission

Introduction

Hipparcos was related to the ICRF but was not itself able to produce an independent realisation since it could not observe sources fainter than $V \simeq 13$, i.e., it was only able to observe a single ICRF source. Following the success of this mission, many similar projects were put forward. The simple fact of replacing a photoelectric detector with total efficiency of the order of 0.004 by a CCD led to a gain of several magnitudes and considerably increased the size of the catalogue. In every case, the missions, proposed by American, German, or European scientists in the ESA, opted for the Hipparcos two-field observation principle with a scanning satellite. This was necessary in order to achieve absolute astrometry, even if the priorities for these missions were astrophysical.

In the end, only the Gaia mission stood out, and was finally selected by the European Space Agency for launch in 2013. The scientific objectives are very broad in the areas of stellar and galactic physics, fundamental physics, and detection of exoplanets, and in addition to astrometry, a complete photometric study over about thirty spectral bands and spectral measurements for objects with $V < 16$. Considering only the astrometry, Gaia should carry out a full survey of the sky up to magnitude 20 with an astrometric accuracy of 25 μas at $V = 15$. The number of objects surveyed will be of the order of 10^9 over the 5 years of the mission. Concerning the astrometric aspects of the mission, there is no great difference with the principles underlying Hipparcos. However, the following points should be noted:

- There is no observation programme based on a previously established star list, as was done for Hipparcos. Gaia has its own autonomous system for detecting sources passing through the field of the instrument, and every detected source is observed and recorded on CCDs.
- There are two large telescopes ($1.45 \times 0.50\,\mathrm{m}^2$) and the beams are combined before focusing on a single focal plane.
- Detection and measurements occur on about 110 CCDs (4.5 k × 2 k) with a pixel 10 μm long in the scanning direction.
- The angle between the two fields is 106.5° as compared with 58° for Hipparcos. In practice, this difference is of no particular importance, provided that the angle is large enough and bears no simple ratio with 360.
- Gaia will be placed at the L2 point of the Sun–Earth system, at a distance of 1.5 million km from Earth in the Sun–Earth direction.

Realisation of the Reference System

In contrast with Hipparcos, Gaia will be able to observe the extragalactic sources of the ICRF and many other quasars. The typical density of quasars brighter than $V = 20$ in the visible is of the order of 20 sources per square degree. That gives about 800 000 sources across the whole sky, of which half are well outside the galactic plane. This is roughly the number of sources that Gaia should detect and measure. By restricting to sources brighter than $V \sim 18$, that gives about 30 000 extragalactic sources that can be used to build up a realisation of the reference system.

Most of these sources are not identified as quasars today, and a recognition system will combine photometric and astrometric properties (there is no significant parallax for a quasar and its proper motion should be very small or non-existent). It will be possible to produce a sample without stellar contamination. The astrometric solution will be constrained to have no overall rotation (ICRS paradigm) and the origin of the system will be fixed on the basis of the present ICRF solution, still with the aim of ensuring continuity with metrological references.

If the set of quasars has an overall rotation $\boldsymbol{\omega}$ in the initial Gaia solution (this will occur because the observing system is invariant under rotation, exactly as for Hipparcos), this will be reflected in a systematic proper motion of the sources given by

$$\mu_\alpha \cos\delta = \omega_x \sin\delta\cos\alpha + \omega_y \sin\delta\sin\alpha - \omega_z \cos\delta,$$
$$\mu_\delta = -\omega_x \sin\alpha + \omega_y \cos\alpha.$$

The observed proper motions can therefore be adjusted to recover this rotation and eliminate it from the solution.

The system provided by Gaia will constitute a new primary realisation of an ICRF satisfying the specifications of the ICRS and directly accessible in the visible for optical and near infrared astronomy. It is estimated that the residual rotation relative to the diffuse cosmic background will not exceed 0.3 μas/yr, whereas for Hipparcos the uncertainty is 0.25 mas/yr. Note that, when eliminating the overall rotation, one must simultaneously determine the acceleration of the Solar System as a whole relative to distant sources, and this will reveal the rotation of the Galaxy.

4.3 Temporal Reference Systems

4.3.1 Time Scales

Constructing a temporal reference system, that is, a time scale together with a basic unit, involves the same steps as for spatial frames: to begin with, the idea of quantifying which, applied to time, has provoked much discussion between scientists, philosophers, and psychologists; then the definition based on the laws of physics and observable quantities; and, finally, the practical realisation of one or several concrete systems, subject to revision and improvement. It is paradoxical that, although the first step, establishing the very concept itself, remains in an unsatisfactory state — What is time?[10] — the measurement of time and frequencies should have become, in a few decades, the most precise in the whole of physics and continues to improve rapidly with the advent of cold atomic clocks on the one hand and optical standards on the other (Table 4.5).

Since 1967, the definition of time has been in the hands of physicists rather than astronomers, thus ending a tradition many centuries long. However, one of the two major astronomical almanacs, published by the *Institut de Mécanique Céleste et de Calcul des Ephémérides* (IMCCE) under the auspices of the *Bureau des Longitudes* in Paris, still has the same name that it has carried since its foundation: *La Connaissance des Temps*. This should serve to remind us that, even though astronomers are no longer the keepers of the unit of time, the definition and practical

[10]This age-old question has been raised by many thinkers. Saint Augustin sums it up perfectly: "So what is time? So long as no-one asks me, I know the answer. But if I must reply, I cannot say." *Confessions*, Book 11, Chap. 14, 17.

Table 4.5 The evolution of time scales

Epoch	Physical phenomenon	Definition of the second	Scale of time
Before 1960	Rotation of the Earth	1/86 400 of the mean solar day	Universal Time (UT)
1960–1967	Orbital motion of the Earth	1/31 556 925.974 7 of the 1900.0 tropical year	Ephemeris Time (ET)
1967	Transition between two atomic levels	9 192 631 770 transition periods of Cs 133	Atomic time scale
1971			Atomic time scale of BIH (then BIPM) becomes International Atomic Time (TAI)

realisations of this quantity are still of the utmost importance to them. There are at least two reasons:

- In every step of its development, celestial mechanics has used the uniform time of Newton, or one of its close relatives imposed by a choice of coordinates in the relativistic theory of gravitation, as argument for its theories; and it is indeed the same argument which is found in the tables of the almanacs. Given that the latter will be used either to compare observation with calculation, or to prepare for further observation, this argument must be defined as fittingly as possible, and its relations with commonly-used time scales determined.
- The SI second and the international atomic time scale were only introduced in 1967. The first caesium clocks date from the 1950s. Before these times, it is not possible to date events relative to the new scales and one must resort to some sort of astronomical time scale. The earliest observations were expressed in solar time, whence the need to keep some trace of this time scale, and its relation to the time of the almanacs, which is, so it would seem, a good extrapolation of international atomic time into the past.

Until the beginning of the twentieth century, the determination of time was entirely based on the assumption that the Earth's rotation rate about its axis was constant. The second was by definition a perfectly specified fraction of the mean solar day. However, observations of the Moon and planets, together with studies of past eclipses, threw doubt on the principle of invariability of the Earth's rotational speed. In the 1930s, seasonal variations were found and also apparently unpredictable periods of acceleration and deceleration, reaching several thousandths of a second per day.

As a consequence, astronomers gradually abandoned the Earth's rotation as fundamental standard, switching to the motion of the Earth around the Sun and hence redefining the second, although maintaining continuity with the old definition as far as possible. The new scale of time, ephemeris time, had all the qualities of uniformity required, but it was less accessible, due to the apparent slowness of the Sun's motion and the difficulty of the measurements involved. Very soon, it was realised physically by the motion of the Moon, theoretical constants being adjusted to ephemeris time. Unfortunately, the theory for the Moon is much more complicated than for the Sun, including dissipative terms which we are still unable to model correctly.

By the middle of the 1950s, the first laboratory frequency standards had appeared, based on very high frequency atomic transitions which, because of their stability and accessibility, were soon adopted in favour of astronomical phenomena as the more suitable for the construction of a uniform time scale. Finally, in 1976, with the guidance of the International Astronomical Union, several scales of time were defined, taking into account relativistic effects on clocks due to position and motion of the observer in the neighbourhood of the Earth or the centre of mass of the Solar System. The relation between these scales and atomic time is still the subject of much discussion, although it is generally agreed that they are necessary. The history of the successive definitions of the second is summarised in Table 4.5.

At this point we define some standard terms used when speaking about time scales.

Apparent Solar Time and Mean Solar Time

The apparent solar time at a given place and a given instant, is the hour angle of the Sun at that place and time. When compared to the uniform time of the mechanical model, it manifests periodic variations included in the so-called equation of time. The latter contains the equation of centre describing the effects of eccentricity of the terrestrial orbit, the reduction at the equator due to the inclination of the ecliptic to the equator, and variations in sidereal time due to precession and nutation. During the year, its amplitude never increases beyond 20 min.

Mean solar time is the apparent solar time with adjustment made for all the periodic terms affecting the motion of the Sun in right ascension. It would be a linear function of the uniform time of mechanics, were it not for variations in the Earth's rotation.

Let t be ideal Newtonian time and T and T_{true} its realisations by mean solar time and true solar time, respectively. By definition, $T_{\mathrm{true}} = H_{\odot}$ with a scale factor to transfer angles to units of time. We then have

$$H_{\odot} = \alpha_z - \alpha_{\odot},$$

and for the average values with periodic effects removed,

$$\langle H_{\odot} \rangle = \alpha_z - \langle \alpha_{\odot} \rangle,$$

or again,

$$H_{\odot} - \langle H_{\odot} \rangle = \langle \alpha_{\odot} \rangle - \alpha_{\odot}.$$

So the differences appearing in the true solar time are exactly those featuring in the motion of the Sun in right ascension. Let

$$E = \alpha_{\odot} - \langle \alpha_{\odot} \rangle = \langle H_{\odot} \rangle - H_{\odot}$$

denote all the periodic terms in the right ascension of the Sun. This expression has been called the *equation of time* for more than two centuries, a name that aptly expresses its origin (see Fig. 4.12).

From the properties of the elliptical motion and the reduction at the equator discussed above, we find

$$E = 2e \sin M - \tan^2 \frac{\varepsilon}{2} \sin 2(\varpi + M) + O(e^2, \varepsilon^4),$$

where $e = 0.0167$ is the eccentricity of the Earth's orbit, $M = nt$ the mean anomaly throughout the orbit, and $\varpi = 282°$ the longitude of the periastron of the orbit. Finally, $n = 0.9856°$/day is the mean motion of the Sun. The equation of time thus includes an annual term and a term of period six months. Taking into account higher order terms for the amplitudes, we have in minutes

$$E = 7.7 \sin M - 9.8 \sin(204 + 2M).$$

With the sign convention adopted for E (there is no commonly accepted convention and the sign depends on the source and national preferences), if $E > 0$, the true Sun is ahead of the mean Sun in right ascension, whence true noon (when the Sun passes

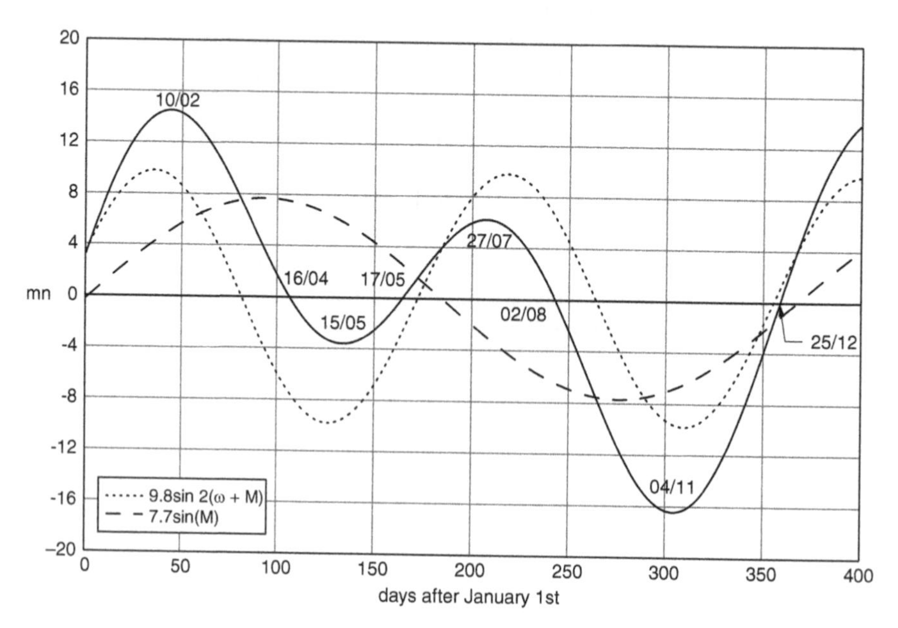

Fig. 4.12 Values of the equation of time and its two main components during the year

the local meridian, i.e., noon for a sundial) occurs later than 12 h mean time for this location. Since our official time is close to the mean time of a reference meridian, there is still one correction to be made to the longitude in order to read official time from a sundial.

Civil Time and Universal Time (UT)

The civil time in a certain place is the mean solar time plus twelve hours which are added so that the day begins at midnight and not at midday. Universal time is the civil time of the international meridian. The civil time in a given place should not be confused with the official time in the corresponding country. This official time is fixed by the administration of the country and, in general, differs from universal time by a whole number of hours (except in a few cases where there may be a difference of half an hour).

4.3.2 Atomic Time

Frequency Standards

During the second world war, the development of radars led to considerable improvements in the technology of microwave circuits and high-frequency cavities. These cavities could be controlled sufficiently accurately to correspond to atomic and molecular transition frequencies in the centimetre range. In 1948, the U.S. Bureau of Standards was able to use a cavity adjusted to the transition frequency of ammonia. The ammonia molecules were excited and returned to the ground state by emission of a photon; the intensity of the observed signal increased as the cavity frequency approached the transition frequency of the ammonia molecule.

It thus became possible to tune the cavity to the ammonia transition frequency by a feedback loop, thereby creating a frequency regulator: if the radiofrequency of excitation of the cavity differed somewhat from the absorption frequency of the ammonia molecule, the output signal would weaken. The excitation frequency of the cavity was then altered until it approached the transition frequency of 23 870 MHz.

It turns out that the transition resonance of ammonia molecules is extremely narrow compared with those of classical oscillators, and therefore, it is only for an excitation radiofrequency almost exactly equal to the ammonia transition frequency that any significant number of molecules will undergo stimulated transition. This is one of the main features contributing to the success of atomic frequency standards.

The Caesium Clock

The same principles were applied to atoms, and in particular, to Cs 133. This atom contains 55 protons, 78 neutrons and, of course, 55 electrons. The outer electron is not paired and has zero orbital angular momentum. The ground state of Cs 133 has a two-state hyperfine structure coming from the interaction of the spin of the outer electron ($S = 1/2$) with the angular momentum ($J = 7/2$) of the nucleus. The angular momentum addition rule is straightforward in this case, leaving only two possibilities: if the angular momenta are parallel, the total angular momentum is $F = 4$, and if they are anti-parallel, it has value $F = 3$. A photon of frequency $\nu_0 \sim 9$ GHz is emitted when spontaneous transition from $F = 4 \rightarrow F = 3$ occurs.

In 1950, H. Lyons and his colleagues at the National Bureau of Standards measured the frequency of the $F = 4 \rightarrow F = 3$ transition of ^{133}Cs (in terms of the SI second as it was then defined) to a relative accuracy of 10^{-7}. This was still not precise enough to challenge the astronomical standards. Then, in June 1955, L. Essen and J.V.L. Parry of the National Physical Laboratory in the UK produced the first caesium frequency standard worthy of the name, reaching an accuracy of order 10^{-10}. From 1955 to 1958, the National Physics Laboratory and the Naval Observatory in Washington carried out a joint experiment to determine the relation between the transition frequency of caesium and the second of ephemeris time. They found a value of 9 192 631 770$\pm$20 Hz (in ephemeris seconds), the error originating rather from the quality of the determination of ephemeris time than from inability to read the frequency. From then on, it became possible to compare two frequencies with an accuracy of 10^{-12}. The moment had come to consider replacing celestial standards in favour of some system based on laboratory clocks.

At the thirteenth General Conference on Weights and Measures (*Conférence Générale des Poids et Mesures* CGPM), a new definition of the second was adopted, which became the SI unit in 1967:

> One second is the duration of 9 192 631 770 periods of the radiation corresponding to the transition between the two hyperfine levels of the ground state of the ^{133}Cs atom.

The practical realisation of the international second by means of an atomic standard turned the latter into a frequency standard, whereas previously it had only been a frequency generator. It became a time standard through the counting of periods. Figure 4.13 shows how an atomic standard works. A source, heated to around 100°C, produces a beam of caesium atoms about equally distributed between the $F = 4$ and $F = 3$ states. An intense inhomogeneous magnetic field removes atoms in the $F = 3$ state, so that only those in the $F = 4$ state enter the cavity. A hyperfrequency field at frequency ν very close to ν_0 causes stimulated emission, returning atoms to their $F = 3$ ground state and emitting photons at frequency $\nu = \nu_0$. At the other end of the cavity, a second magnetic sorting once again separates those atoms remaining in the $F = 4$ state from those having undergone a transition to the $F = 3$ state. After ionisation, the current obtained is used to control the cavity frequency, via a quartz oscillator and a feedback mechanism.

Specialised laboratories construct caesium frequency standards with a view to improving the realisation of the second, that is, aiming to establish ever greater accuracy. These standards, realised today by means of cooled caesium and rubidium atomic fountains, are accurate to within 10^{-16}. Standards of this type are developed at the Paris Observatory and at the NIST laboratories in Boulder. Furthermore,

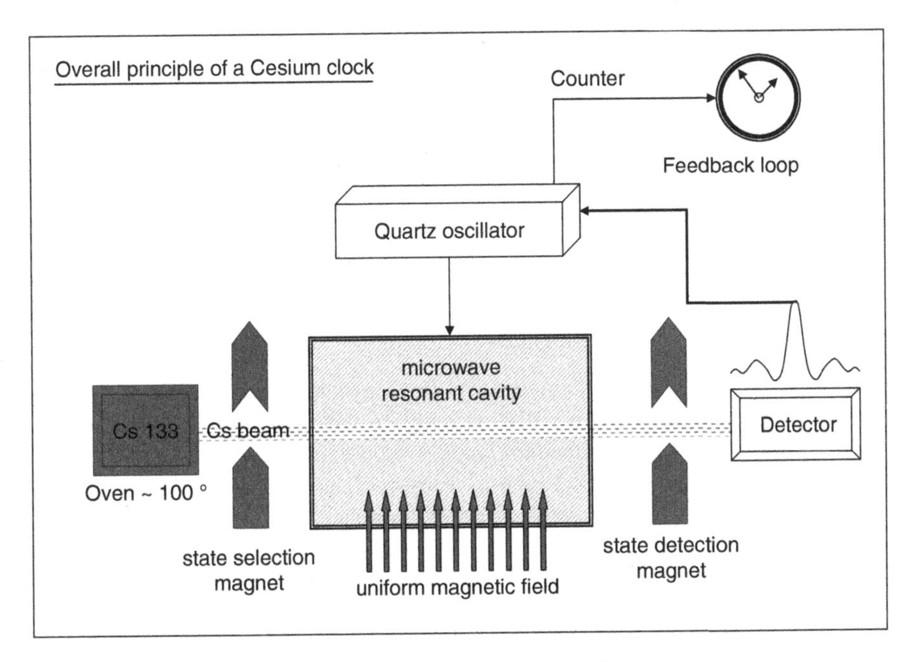

Fig. 4.13 Operating principle of the caesium frequency standard

slightly less accurate caesium jet clocks available from manufacturers are nevertheless accurate to between 1 and 2×10^{-12}, which is quite remarkable.

Definition of International Atomic Time (Temps Atomique International TAI)

The frequency standard can be used cumulatively to construct a continuous scale of time, insofar as the clock or clocks which realise this scale operate continuously. A definition was proposed in 1970 by the Consultative Committee for the Definition of the Second (*Comité Consultatif pour la Définition de la Seconde* CCDS) and approved in the same year by the International Committee for Weights and Measures (*Comité Internationale des Poids et Mesures* CIPM). Finally, in October 1971, the 14th General Conference on Weights and Measures (*Conférence Générale des Poids et Mesures* CGPM) defined the new temporal reference system, international atomic time:

> International atomic time is the temporal reference coordinate established by the *Bureau International de l'Heure* on the basis of the indications of atomic clocks operating in various establishments in conformity with the definition of the second, unit of time of the *Système International d'Unités*.

The *Bureau International de l'Heure* applied this definition and constructed a time scale cumulatively from SI seconds. Unlike the dynamical scales based on arbitrary division of the scale associated with some flow of events, the TAI is an integrated time scale.

There were at least two possible ways of achieving this international atomic time scale:

- Select one particular clock, a very high quality caesium standard clock, in permanent operation, whose readings would be, by definition, the practical realisation of the TAI.
- Establish the TAI scale by coordinating, in one centre, the readings of a wide-ranging network of clocks in several laboratories. The TAI is then calculated from this data using a well-defined algorithm.

The first solution, although simple (TAI fixed by one clock, in one place), was obviously open to the risk of a discontinuity, in the event of operating problems or even total failure of the standard clock, which could never guarantee anything like the immutability of the celestial movements. The second solution, which was finally the one adopted, almost completely eliminates this risk, and furthermore, the large set of readings provides for a statistical improvement. The readings of the 260 participating clocks are weighted according to the quality of each clock and submitted to an algorithm.

Today, production of the TAI is handled by the International Bureau of Weights and Measures (*Bureau International des Poids et Mesures* BIPM), following the transfer, in 1985, of the time section of the *Bureau International de l'Heure* (BIH) from the Paris Observatory to the BIPM. (In 1987, the BIH was split into two organisations: the International Earth Rotation Service (IERS), which is still at the Paris Observatory; and the Time Section of the BIPM in Sèvres, near Paris, which deals with construction of the TAI scale and maintenance of the SI second.)

Given the TAI scale, we might be tempted to speak of a TAI second which, if we were not careful, could differ from the second of the *Système International d'Unités* (SI second). In order to deal with this potential weak point, the frequency of the TAI is determined from primary frequency standards at the National Research Council (Canada) and at the *Physikalisch-Technische Bundesanstalt* (Germany). To give an example, the relative value of the frequency of the TAI was reduced by 1×10^{-12} on 1 January 1977. With these precautions, the stability of the TAI lies somewhere between 1 and 5×10^{-14} over periods of one month to several years. The TAI second is coherent with the SI second at the level of 2×10^{-14}.

4.3.3 Coordinated Universal Time (CUT or UTC)

As the atomic time scale is quite independent of the celestial movements, it follows that the mean solar day is not exactly 86 400 TAI seconds. The length of the mean solar day was equal to approximately 86 400 SI seconds in 1820. It was shorter before, and since then has been increasing all the time to reach 86 400.0025 SI seconds around 1980. These ≈ 2.5 ms per day accumulate during the year to give a typical annual shift of approximately 1 s between a uniform scale and a scale constructed on the basis of the Earth's rotation. It is this second that is adjusted from time to time to maintain the relationship with the Earth's rotation. Even if the duration of the second in the definition had been adjusted, agreement at one particular instant would soon be lost as a result of irregularities in, and the secular slowing of, the Earth's rotation. It was indeed this very same phenomenon, and in particular its unpredictability, which led to the abandoning of the Earth's motion as a source for a uniform time.

However, when the transition was made from astronomical standards of time to atomic standards, it was considered useful not to completely forego the close connection between time and the Earth's orientation in space. For this reason, a scale of time called coordinated universal time was created (officially denoted UTC). This is a hybrid, possessing the uniformity of atomic time, at least, piecewise, but undergoing one second discontinuities whenever necessary in order to keep the Earth's rotation in phase with laboratory clocks. In brief, UTC is an approximation to mean solar time, but read on a much better time-keeper than the Earth's rotation (see Fig. 4.14).

Having converted the Earth's rotation into a time scale known as UT1, the following relations hold after 1 January 1972:

$$\mathrm{TAI} - \mathrm{UTC} = n\ \mathrm{s} \quad (n \in \mathbb{Z}),$$

$$|\mathrm{UT1} - \mathrm{UTC}| \leq 0.9\,\mathrm{s}.$$

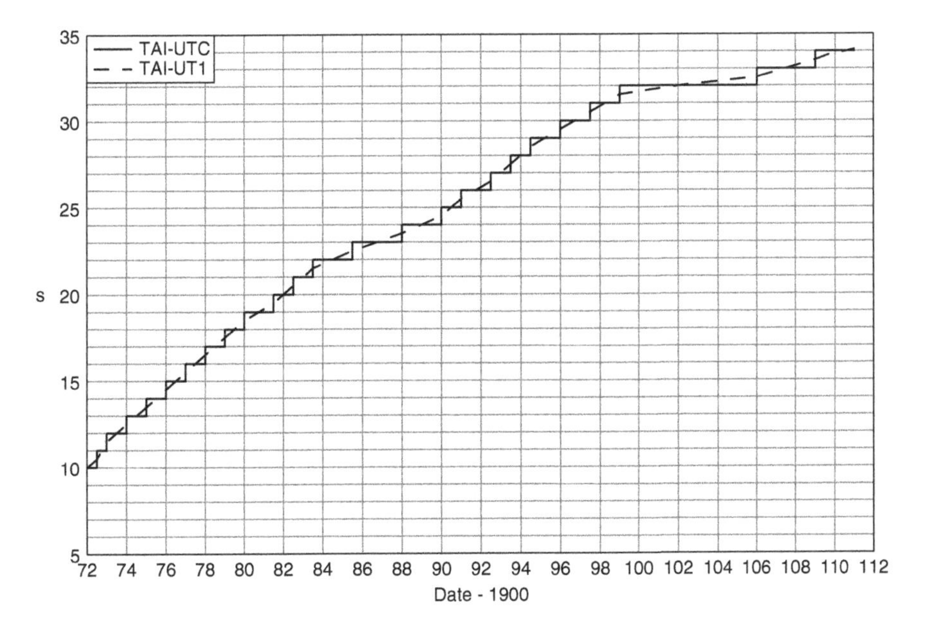

Fig. 4.14 Difference between TAI and UTC from 1972 to 2011

Depending on how urgent it is, the leap second is placed at the end of December or June, or at the end of any month, and announced at least eight weeks beforehand. The exact relation between UT1 and UTC can only be known after the event, when measurements of the Earth's rotation, using space geodesy, laser–Moon telemetry, and very long baseline interferometry, have been published. The question here is no longer one of time metrology, but rather a study of the Earth's orientation in space, a prerequisite for the preparation and processing of these observations since the time scale available in an observatory is usually UTC. The decision to introduce a leap second is taken by the central office of the International Earth Rotation Service (IERS).

4.3.4 GPS Time

Although this time scale does not enjoy a fully official existence in the world of metrology, it is nevertheless of considerable practical importance by virtue of its accessibility. Each satellite of the GPS constellation (Global Positioning System) carries one caesium and one rubidium atomic clock. The constellation comprises 28 operational satellites of the second generation. The network distributes a continuous time scale (the GPS time scale) which follows UTC to within 1 μs, but without the leap seconds. It is the computational algorithm or software of the receiver that ensures the correct UTC display. GPS time coincides with UTC at the origin of its definition at 0 h on 6 January 1980. At this instant, TAI was ahead of UTC by 19 s, and hence ahead of T_{GPS}, as can be seen in Fig. 4.14. This value remains constant due to the servocontrol on TAI through a continuous UTC. Hence,

$$\mathrm{TAI} - T_{\mathrm{GPS}} = 19\,\mathrm{s}.$$

In 2011, the deviation from UTC reached 15 s (see Fig. 4.14), with

$$T_{\mathrm{GPS}} - \mathrm{UTC} = 15\,\mathrm{s\ in\ 2011}.$$

Galileo time should follow the same convention and deliver a continuous time that is rigidly linked to TAI. For the Russian system GLONASS, the leap seconds are included in the primary scale of the system, which is therefore tied to UTC (in fact, UTC + 3 h, i.e., Moscow time).

The Future of UTC

The future of UTC as a widely distributed and accessible time scale for civilian needs is under discussion by the relevant international bodies, including astronomers and geophysicists, and telecommunications and navigation organisations. The main problem with this time scale is the presence of discontinuities due to unpredictable leap seconds, requiring the tables to be updated in computer programs. Furthermore, some automatic systems are unable to integrate such leaps, and there are many problems relating to the dating of events that arrive during the leap second, with a non-negligible risk of ambiguity in the coding of the exact date.

Various solutions are under examination by working groups, including:

- maintenance of the current system, with an increasing number of leap seconds in the future owing to the lengthening of the day,
- suppression of the leap seconds which would allow the UTC scale to drift away from UT1,
- greater tolerance, leading to bigger but less frequent leaps, possibly combined with regular and predetermined leaps,
- a shift to another time scale which could be TAI or another scale linked to the GPS or Galileo.

But the most spectacular possibility arising during these discussions (with absolutely no chance of being realised) would be to change the definition of the second to maintain the approximate agreement of 86 400 seconds per mean solar day.

The secular relationship between civil time and the apparent motion of the Sun is a feature that few would be ready to abandon, and the groups supporting this point of view have certainly made themselves heard during the discussions. However, extrapolating into the future, what we know about the secular slowing down of the Earth's rotation indicates that it would take over 2 000 years before the discrepancy between solar time and a uniform time without leap seconds reached one hour. It would then be easy to reset the clocks if that were judged necessary by means of a leap hour. This would not then be totally unlike the Gregorian reform of the calendar which removed 10 days between 4 and 15 October 1582.

At the time of writing (2011), the result of these discussions is not yet known, but it seems that the most likely outcome will be that leap seconds will be abandoned, and that the time distributed will be a scale strictly related to TAI, with a constant shift of a few tens of seconds. Another possibility is that the GPS or Galileo time scales will win through for purely practically reasons, with a constant relationship to TAI (which is already the case for all intents and purposes). After the transition from the astronomical definition of the second to the physical one in 1967, a new page in the measurement of time and its relationship with the celestial world will then be turned.

4.3.5 Dynamical Time Scales

Ephemeris Time

Newton's law and its application to the dynamics of the Solar System assume the existence of a uniform and absolute time at the very outset of the physical theory. Motion only being uniform relative to some particular time scale, it is clear that we cannot define uniform motion without first being in possession of some time scale which itself bears no relation to motion. The alternative adopted by Newton, and thereafter by the astronomical community, as a way of realising a dynamical time scale recognises uniform motion a priori by the absence of force, or a complete modelling of the forces in an inertial system. Thus, the choice of a theory of motion, and of the numerical constants required to completely define initial conditions, leads to an ephemeris, that is, a position vector $\boldsymbol{r}(t)$. Observation of the body gives the position $\boldsymbol{OM}$ and the value of the parameter t is found by solving the equation in t, $\boldsymbol{OM} = \boldsymbol{r}(t)$. The definition of the unit is then some fraction of a well-defined interval, such as the day or the year. This is quite the opposite of what happens for atomic time, where the scale is constructed by cumulating units, by integration, in some sense, whilst the unit of the dynamical scale is obtained by derivation.

Solar time and its close relative universal time are dynamical times based on a very simple model of the Earth's rotational movement: $\omega =$ constant, which gives for the phase, the observable quantity, $\phi = \omega t$. This equation is so simple to invert that the nature of its terms (the angle as configuration parameter and time as independent variable) has become somewhat obscured, rather as happens in thermodynamics with temperature and energy, or in nuclear physics with the mass and energy of particles.

More important is ephemeris time (ET) and its modern derivatives, *temps dynamique terrestre* (TDT), *temps terrestre* (TT), *temps dynamique barycentrique* (TDB), and *temps-coordonées barycentrique* (TCB). Ephemeris time is realised practically by adopting a motion of the Sun by convention, such as the motion of the mean geometric longitude of the Sun, given by

$$\langle L \rangle = 279°41'48.04'' + 129\,602\,768.13''\, t_{\mathrm{E}} + 1.089''\, t_{\mathrm{E}}^2 ,$$

where t_{E} is measured in Julian centuries of 36 525 ephemeris days. The above expression was obtained using the best available theories and observations, with a time scale coming from the Earth's rotation. Relative to the latter scale, the numbers appearing in the longitude expression are only known to a certain degree of accuracy. But as the source for a new time scale, the mean longitude is exact *by definition*, despite the experimental origin of its coefficients. Time $t = 0$ is defined as 0.5 January 1900 (31 December 1899 at 12 h), and the unit of time is a fraction of the tropical year at the same date, deduced from the value of the coefficient of t_{E}. The observable event corresponding to $t_{\mathrm{E}} = 0$ is $\langle L \rangle = 279°41'48.04'' =$ 279.696 678 deg and can be reconstructed from later measurements using the theory of the Sun.

As mentioned earlier, the practical realisation of this scale proved difficult and it was rapidly superseded when atomic time arrived on the scene. However, it remains an important scale for two reasons:

- Comparison with the TAI scale since 1955 has revealed no systematic difference between the two scales. This agreement is quite remarkable, relating a scale of time based on the laws of quantum mechanics, to another depending only on the laws of gravity. The discrepancy between the two scales is constant and given experimentally by

$$\mathrm{ET} \approx \mathrm{TAI} + 32.184\,\mathrm{s}.$$

- Astronomical observations pre-dating the TAI (1955) are still much in use and can be dated with ET. Therefore, ET can be considered as an extrapolation into the past of the present scale of uniform time.

Coordinated Time Scales

In 1976 and 1991, the International Astronomical Union made decisions leading to the introduction of other time scales based on the TAI but maintaining continuity with ET. Up until 1991, these scales went by the unfortunate names terrestrial dynamical time (*temps dynamique terrestre* TDT) and centre-of-mass dynamical time (*temps dynamique barycentrique* TDB), whereas their definition linked them to TAI without any reference to a dynamical ephemeris. In 1991, the time scales TT (*temps terrestre* or terrestrial time) and TCB (*temps-coordonnées barycentriques* or centre-of-mass coordinated times) were introduced. The following relations hold to a very high degree of accuracy:

$$\mathrm{TT} = \mathrm{TDT} = \mathrm{TAI} + 32.184\,\mathrm{s},$$

$$\mathrm{TDB} = \mathrm{TDT} + \mathrm{P},$$

where P is a series of about 500 periodic terms of amplitude greater than 0.1 ns describing relativistic differences between the running of a clock at the centre of mass of the Solar System and another in orbital motion on the orbit of the Earth. The main term has period one year and amplitude 1.656 ms, whilst the second has amplitude 22 μs. Note that all these terms are essential in order to handle, without loss of accuracy relative to the measurements, the timing of millisecond pulsars.

These new definitions are justified by the requirements of a metrology of space and time conceived within the framework of relativistic gravitational theories, but also by the exceptional stability of atomic clocks and the availability of cold atom standards with stabilities on the ground of 10^{-16}.

4.3.6 Dates and Epochs. Dealing with Long Periods

The system of astronomical constants fixed by the International Astronomical Union introduces a fundamental epoch to which all other events are related, using multiples or submultiples of a time unit.

Concerning time, the unit is the SI second, defined from the transition of Cs 133 and explained in Sect. 4.3.2. One day is 86 400 s, one Julian year is 365.25 d, and one Julian century is 36 525 d. These derived units are thus rigidly fixed relative to the SI second and depend in no way on the motions of celestial bodies, even though obviously chosen with reference to the solar day and the year of the seasons. Recall that the basic unit was, previously, the mean solar day divided into 86 400 s, and later, some fraction of the tropical year 1900. The year was then the tropical year of 365.2422 d, which was the year of the seasons, bearing no relation to the Julian year.

Starting from these relatively short durations, ways have to be found to mark off the epochs over longer periods, thus making the calendar we are familiar with in everyday life to follow the years. However, this calendar, even though based on a very precise algorithm relating a calendar date with a precise day, does not allow for simple calculation of the time interval between two events such as 15.00 h on 12 February 1853 and 8.00 h on 15 January 2007. For the observation of many events such as eclipses, planetary motions, variable stars, supernovæ, and so on, not to mention the exploitation of archives from previous centuries, the astronomer requires a continuous reckoning of time which simplifies the evaluation of time intervals in the decimal system.

The solution consists in counting days on a counter which has sufficiently many digits to cover several thousand years. The method was suggested in 1606 by J. Scaliger who devised the term Julian period in memory of his father, Julius, and not through any connection with the Julian calendar. At each instant, the Julian date of an event is the number of days (including the decimal part) since midday on 1 January −4712 (4713 BC), the beginning of the Julian period. This origin results from Scaliger's desire to return to the instant of time when the three fundamental cycles (the 15 year Roman indiction, the 19 year Meton cycle, and the 28 year Julian cycle) were simultaneously equal to 1. In this system for counting days, 1.5 January 2000 = JD 2 451 545 and 1 January 2007 at 0 h = JD 2 454 101.5. For recent epochs, the modified Julian day is commonly used, MJD = JD − 2 400 000.5, for which the origin is 0 h on 17 November 1858. It should be remembered that the Julian day begins at 12 h and not 0 h. The fraction 0.5 in the definition of the MJD is there precisely so that modified Julian days begin at 0 h. We can check that 15.00 h on 12 February 1853 is JD 2 397 897.125 and that 8.00 h on 15 January 2007 is JD 2 454 115.833, so that the time interval between the two epochs is 56 217.708 d.

Even though the definition is clear, it assumes that we know how to count the days after a reference date, and also that the definition of the day throughout this period is unchanging. In practice, it is more convenient to work with a reference epoch called J2000, which corresponds to 1 January 2000 at 12.00 h TDB (where TDB $\approx$ TAI + 32.184 s) and whose Julian date is 2 451 545.0 TDB. Then every event is given by a Julian Epoch (JE) expressed by its Julian date (JED) as follows:

$$\mathrm{JE} = 2000.0 + \frac{\mathrm{JED} - 2\,451\,545.0}{365.25}.$$

This date is written Jaaaa.aa, where aaaa.aa denotes the year in decimals. For example, 1 January 2007 at 0.00 h = JD 2 454 101.5 gets transformed to Julian epoch J2006.999315. Conversely, J2007.0 gives JD 2 454 101.75, i.e., 1 January 2007 at 6 h in the morning. An ordinary calendar date is transformed to its Julian

Table 4.6 Correspondence between epochs, Julian dates, and calendar dates

Besselian epoch	Julian epoch	Julian date	Calendar date
B1900.0	J1900.000857	2 415 020.3135	31/12/1899 19 h 31 m 26 s
B1950.0	J1949.999789	2 433 282.4235	31/12/1949 22 h 09 m 50 s
B2000	J1999.998722	2 451 544.4334	31/12/1999 22 h 24 m 06 s
B1950.000210	J1950.0	2 433 282.5	1/01/1950 0 h
B2000.001278	J2000	2 451 545.0	1/01/2000 12 h

date, or vice versa, by fairly simple algorithms available in basic astronomical works and computer libraries. In order to extend the validity of these algorithms to dates before the Gregorian reform, we must introduce the discontinuity in the calendar for the year of 1582 and take into account the change in the definition of leap years. The extension of the Gregorian calendar to a date prior to the reform is called the proleptic Gregorian calendar. In general, all historical dates are given relative to the calendar that was valid at the given date, so it is important to make clear mention if a date refers to the proleptic Gregorian calendar. Furthermore, since the date when the Gregorian calendar was adopted depends on the country (December 1582 in France), there are some uncertainties over dates coming from England, the United Provinces, and Germany up until the eighteenth century, and dates are often given in both forms.

The reader may encounter other epochs in the literature, particularly in slightly older sources. These are reviewed in Table 4.6 to an accuracy of ± 2 s. The epoch Bxxxx refers to a year equal to the tropical year of 365.242 198 781 d, used in the astronomical constants until 1976, in particular for the theory of precession. For example, 0.00 h on 1 January 1995 = JD2 449 718.5 corresponds to the Besselian epoch B1995.00048.

Part II
Data Collection

Chapter 5
Telescopes and Images

The telescope is so often considered as the astronomical tool *par excellence* that it would be easy to forget that it is just one of the components of an observing system, and that it would serve no purpose without spectrometers or detectors. The aim of this chapter is first of all to examine the double function of the telescope: it must collect, then form an image of the source of that radiation. Naturally, these two aspects are very closely connected. It is to improve the performance of both that astronomers have always striven, and still strive, to build telescopes of ever increasing size, freeing them if necessary from the constraints of the Earth atmosphere by placing them in space. The close relationship between these functions means that the telescope *stricto sensu* is considered as just one element of an integrated system, whence the design of the telescope is intimately bound to the design of focal instruments such as cameras, spectrometers, and so on.

Image formation is such a central issue when using a telescope that we shall treat it here from the relatively simple viewpoint of geometrical optics, then discuss the limitations of this picture. This elementary approach will suffice to present the main features of instruments (telescopes) comprising a single primary mirror. Even though most of the ideas are quite general and presented, according to the basic principle of this book, independently of the wavelength under consideration, actual realisations of instruments are in fact highly dependent on the wavelength. We shall attempt to give as complete a presentation as possible of this rapidly evolving area, covering the full extent of the electromagnetic spectrum currently accessible to astronomical observation.

The study of the spatial coherence of radiation will be reserved for Chap. 6, along with the resulting theory of diffraction and image formation in the presence of diffraction. Several major applications will be described: the ultimate angular resolution of telescopes, aperture synthesis or coherent combination of several telescopes to form an array, and image degradation by the Earth atmosphere and ways of overcoming it.

Gravitational optics has recently become a promising branch of astrophysics. Although it does not involve the construction of optical systems in the classical sense of the term, the idea of a *gravitational telescope* rests upon principles which,

P. Léna et al., *Observational Astrophysics*, Astronomy and Astrophysics Library,
DOI 10.1007/978-3-642-21815-6_5,

at least in the simple cases, are very close to those of geometrical optics. This rapidly moving area will thus be included along with discussion of the more conventional aspects of geometrical optics.

5.1 Image and Object in Astronomy

The telescope is the instrument which produces an 'image of the sky' in order to identify and measure the energy received from space. But what exactly do we mean by this word 'image'? Indeed, despite the three-dimensional structure of space, the terrestrial observer can only ever perceive a two-dimensional projection of it on the celestial 'sphere'. The received intensity is the integral of the locally emitted energy, integrated along the *line of sight*. Unfolding the measured information, in order to reconstruct the local conditions at each point of the line of sight, involves the delicate operation of inverting the integral, an operation included within the more general category of *inverse problems*.

Throughout this chapter, the 2-component vector quantity $\boldsymbol{\theta}$ denotes a point of the celestial sphere, which is to say, a direction characterised by two angles. Different reference frames can be used (horizontal, equatorial, galactic, etc.), and these are discussed in Chap. 8. It serves no purpose, in the present context, to specify the particular frame chosen to represent $\boldsymbol{\theta}$.

As seen by the observer, each angular direction $\boldsymbol{\theta}$ in the sky is associated with a monochromatic flux (defined in Chap. 2) of frequency ν received by the observer:

$$I_\nu(\boldsymbol{\theta}) \quad \left[\mathrm{W\,m^{-2}\,sr^{-1}\,Hz^{-1}}\right] ,$$

which can also be expressed in terms of a number of photons using

$$N_\nu(\boldsymbol{\theta}) = \frac{I_\nu(\boldsymbol{\theta})}{h\nu} \quad \left[\mathrm{photons\ s^{-1}\,m^{-2}\,sr^{-1}\,Hz^{-1}}\right] .$$

The distribution $I_\nu(\boldsymbol{\theta})$ is the monochromatic image or map of the observed region of the celestial sphere. The ν-dependence of this map describes numerous physical properties of the source: opacities responsible for absorption or emission by the source, correlation between phenomena of differing energies, excitation of atoms or molecules in the source, magnetic fields, and many others. The flux $I_\nu(\boldsymbol{\theta})$ may also depend on time (variable source) and the detected polarisation (polarised sources).

The observation process degrades $I_\nu(\boldsymbol{\theta})$ in at least four ways:

1. Through imperfections in the realisation of optical surfaces, which leads to geometrical aberrations, and the finite size of telescopes, antennas, and light gatherers used to collect the incident flux, which leads to diffraction.
2. Through the finite number of photons incident during the measurement time, which introduces measurement noise, and leads to imperfect restitution of $I_\nu(\boldsymbol{\theta})$, relative to a measurement of infinite signal-to-noise ratio.

3. In the case of ground-based observations, through the consequences of crossing the Earth's heterogeneous and turbulent atmosphere.
4. Through the existence of a finite spectral resolution $\delta\nu_0$ and a finite angular resolution $\delta\theta_0$, thereby mixing up radiation coming from different directions and having different frequencies.

In the case where $I_\nu(\boldsymbol{\theta}) = \delta(\boldsymbol{\theta})$, an intensity distribution characterising a point source, the domain $|\boldsymbol{\theta}| < \delta\theta_0$ in which the intensity is actually received determines the *angular resolution* $\delta\theta_0$ of the instrument. This notion of resolution, which is so important in analysing the capacities of an instrument, can also be defined when $I_\nu(\boldsymbol{\theta})$ is no longer studied in the space of values of $\boldsymbol{\theta}$, but rather in the Fourier conjugate space $\boldsymbol{w}$. Such is the case for the large group of instruments known as *telescope arrays* or *interferometers* (see Chap. 6).

5.1.1 The Telescope and Geometrical Optics

Geometrical optics provides the first approximation in telescope design.[1] However many mirrors or lenses are used, the maximum energy collecting capacity is always determined by the aperture of the *primary mirror* or *lens*, the first optical element encountered by light from the source. This first element is generally followed by other optical elements (secondary or tertiary) which modify the position, the scale, or the quality of the primary image. At the end of the optical path is the focal plane of the telescope, where the final image is formed. This is the image which will be analysed by spectrometers, polarimeters, and other analysis systems, and/or light detection systems.

The aim of this optical configuration is to obtain, for a source located at infinity and for a given aperture, an image that is as anastigmatic as possible, over as wide a field as possible. The system is *perfectly anastigmatic* when all rays from a direction $\boldsymbol{\theta}$ converge exactly to a single point $A(\boldsymbol{\theta})$ in the focal plane of the telescope. If a system is approximately anastigmatic, the image is spread out, rays converging to some neighbourhood of $A(\boldsymbol{\theta})$ whose dimension determines the quality of the image at A. The *field of the telescope* is the region of the focal plane in which the image is sufficiently bright and sufficiently anastigmatic to fulfil the requirements of the instrument. It is clear that the larger the field, the better will be the use of the available photons and the observation time, with the proviso that the analysis and detection systems be capable of fully exploiting all points of this field.

The perfect anastigmatic surface for producing the image of a point at infinity on its axis, is the paraboloid of revolution. It is natural, therefore, that this should be the surface the most commonly used as primary mirror in telescopes. However, this

[1]Telescope design is a huge subject. We thus refer the reader to the excellent and extremely comprehensive recent books Wilson, R. *Reflecting Telescope Optics*, Springer (1996) and Bely, P. *The Design and Construction of Large Optical Telescopes*, Springer (2003).

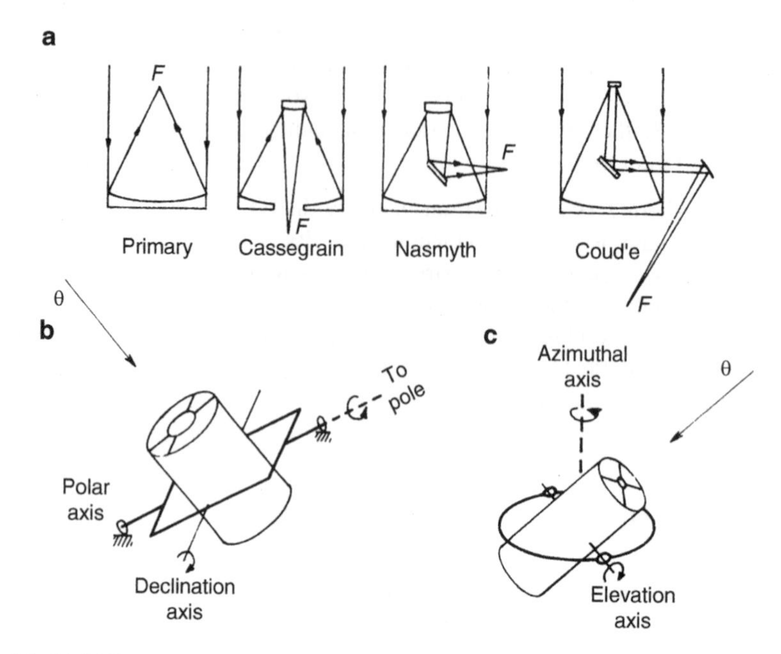

Fig. 5.1 (**a**) Different focal configurations. Pointing of (**b**) equatorial and (**c**) altazimuth ground-based telescopes

surface used in isolation is not anastigmatic for any points off its axis, and therefore has no field, so it is equally natural that optical systems should be designed which incorporate more than one mirror. By increasing the number of degrees of freedom of the optical combination, these composite systems offer the best compromise possible between the approximately anastigmatic quality of the image, and the field, bearing in mind the requirements of the user. Nevertheless, it is much more expensive to polish a paraboloid than to polish a sphere, so spherical primary mirrors are sometimes used, with the unavoidable extra cost of then correcting for their aberrations.

Optical Configuration of Telescopes

Figure 5.1 shows some of the most common configurations, used for both ground-based and spaceborne telescopes, and in very varied wavelength domains, whose design specifications will be described later. The figure shows the single mirror configuration (image at the prime focus), and configurations using two or more components (Cassegrain, Nasmyth, and Coudé).

The *off-axis configuration* is not shown in Fig. 5.1. In this setup, an aspherical primary mirror forms an image outside the incident beam, a method which avoids any obstruction by secondary or tertiary elements, but which is more complex and more costly to construct. We shall see interesting applications of this in Chap. 6.

The *focal ratio*, *aperture ratio*, or *numerical aperture* f/D, where f is the focal length (or equivalent focal length in the case of a combination of several mirrors) and D the diameter of the pupil, is an important parameter in the design and use of a telescope. The prime focus configuration gives a small focal ratio (1.2–2.5), the Cassegrain and Nasmyth configurations give intermediate focal ratios (7–30), and the Coudé configuration gives the largest focal ratios (>50).

The focal ratio of a single spherical mirror is directly related to its radius of curvature R, since $f = R/2$. The construction of mirrors of comparable quality and large curvature, possessing values of f/D only just greater than unity, leads to technical difficulties in polishing very open mirrors, but is nevertheless the result sought after. Indeed, for equal values of D, smaller values of f mean less cumbersome lengths, less couple on guiding equipment, reduced mechanical inertia, and a more favourable design generally.

The equivalent focal length, obtained after crossing or being reflected by all the optical surfaces in the system, also governs the *scale* of the image in the final focal plane. This scale, measured in millimetres per second of arc, is directly proportional to the value of f:

$$\text{focal scale} = 4.85 \times 10^{-6} f \quad \text{mm arcsec}^{-1} ,$$

where f is in millimetres.

Optimal sampling of images aims to match the detector pixel size a to the size of the element of angular resolution $\delta\theta$:

$$f\delta\theta \approx a .$$

Detectors with small pixels (e.g., photographic plates, or CCD with $a < 10\,\mu\text{m}$) can thus easily be placed at the prime focus. Even in this case, increasing diameters and the constant decrease in CCD pixel size often lead to *oversampling*. When the detector system requires or tolerates narrower beams, or accepts the use of an optical annex to modify the magnification of the image (*focal reducer*), the detector can be inserted at the Cassegrain or Nasmyth focus. This is the case for photometers, echelle spectrographs, or even Fourier transform spectrometers.

Optical Aberrations

The compromise between image quality and field means that the image of a geometric point on the source is not a point. This phenomenon is called *optical aberration*. There are several methods for determining the aberrations of an optical system. In the light ray approximation (geometrical optics), computer ray tracing methods assume some shape for the optical surface (spherical, parabolic, or other aspherical surfaces, such as ellipsoids, hyperboloids or higher order surfaces), and then numerically simulate illumination from a point source S in a given direction $\boldsymbol{\theta}$. The distribution of the impact points of the rays in the focal plane represents the energy distribution in the astigmatic image of S, and can be assessed with respect to the allowed tolerance.[2]

[2] We could mention Zemax, for example, a commercial software for modelling optical systems.

A detailed analysis of aberration, or the methods developed by opticians to overcome it, will not be given here. These are classical topics which can be found in Born and Wolf, *Principles of Optics*, and also in *Lunettes et Télescopes* by Danjon and Couder (Blanchard, Paris, 1983) (see the bibliography). A few remarks will suffice to illustrate the problem. Considering a set of rays coming from a point A of the object and reaching the focal plane after passing through the telescope optics, they will be distributed about some point A' which would be the ideal image in the absence of any aberration. This scatter of a given ray around A' is basically a function of two variables. The first is the angular distance α between A and the optical axis; the larger the angle α, which ranges over the field of the instrument, the greater will be the scatter. The second variable is the distance r between the point of impact of the ray on the pupil and the centre of the pupil: the larger this distance is, the greater will be the scatter about A'. The parameter α_{max} thus determines the field of the instrument, whilst the maximal value of r in keeping with the required image quality at A' will determine the maximal diameter D.

The term *field* refers to the range of values of α such that the image quality there is acceptable. The *fully illuminated field* refers to that region over which the source A can be displaced without the intensity of the image A' being reduced by obstructions due to intermediate diaphragms of the optical system. Beyond the fully illuminated field is a region known as the *vignetted field*, across which the illumination of the focal plane progressively decreases.

When a telescope is being designed, the problem is approached from the other end: first, D and α_{max} are fixed, and then the optical arrangement (the shape and number of reflecting surfaces, etc.) is sought which best maintains the quality of the images A' within the angle α_{max}.

The five *primary* or *Seidel aberrations* are:

- *Spherical aberration*, which only depends on the value r, that is, on the aperture D. In other words, it only depends on the focal ratio f/D of the focal length to the diameter, and is smaller as this ratio is greater, thus favouring narrow optical systems.
- *Coma*, which depends only on α and occurs even for very small values of D, if α is non-zero.
- *Astigmatism* and *field curvature*, which involve the values of both f/D and α.
- *Distortion*, which destroys geometrical similarity between object and image, whilst nevertheless maintaining a one-to-one correspondence between points of the object and points of the image.

Although useful, a study based upon geometrical optics is not sufficient to completely determine the image quality of an optical system, neglecting as it does the phenomenon of diffraction, caused mainly by the finite dimension of the pupils.

It should be added that the anastigmatic character of perfect optical surfaces is not the only factor bearing on image quality. Diffraction comes in again, since mirrors and lenses can never be perfectly polished, and this introduces a further cause of image degradation. This is observed when the size of irregularities on the optical surface becomes comparable with the wavelength of the light in the transverse direction, i.e., along the mirror surface, or in the perpendicular direction. This effect can be measured by quantifying the residual phase perturbations produced on the reflected or refracted wave. Suffice it to say in the present context that polishing defects smaller than the wavelength by a factor of 20 or more have a negligible effect on image quality. Such demanding criteria are commonplace in astronomy, so that only the five main types of aberration mentioned above need be taken into consideration.

One particular case is worth mentioning, however. A telescope, for example, at radiofrequencies, may have been optimised for observation at a certain wavelength λ_0, maybe in the decimetre or centimetre range. A later use at much shorter wavelengths λ, maybe in the centimetre or millimetre range, will generally produce less good images, flaws in the optical path or phase with root mean square deviation σ_ϕ, induced by reflecting surfaces, notably on the primary mirror, no longer satisfying the condition $\sigma_\phi < \lambda/20$.

Constructing an instrument always involves a compromise between image quality (approximately anastigmatic) and the field within which this quality will be preserved. The right compromise can be found by taking the following steps: firstly, choosing the geometry of the surfaces (spherical, thus less costly, or aspherical); secondly, choosing the combination of individual components, a larger number offering more degrees of freedom and better optimisation, although at the risk of greater light loss, only partially compensated for by optimal treatment of the surfaces (high reflectivity of mirrors, anti-reflection dioptrics).

These considerations are illustrated in Fig. 5.2, which shows *ray tracing* by numerical simulation, a method universally used in studying the properties of optical systems. This is the result for a parabolic primary mirror with opening $f/2$, slightly deformed in a controlled manner (*active optics*), associated with an arrangement of three lenses (*triplet corrector*) close to its prime focus.

The description and analysis of optical aberrations, and likewise, the consequences of inadequately polished optical surfaces, provide a particularly convenient way of introducing a presentation of the local phase errors produced on a wavefront. This analysis cannot, of course, come from the theory of geometrical optics. The formalism of *Zernike polynomials* is invaluable in the study of general perturbations on wavefronts. In the present case, these perturbations are caused by imperfect or astigmatic optical surfaces, but they may well also be produced when a wave crosses some non-homogeneous medium, and indeed, it is in this context that they will be discussed (see Sect. 6.2).

The Main Features of a Telescope and Its Optics

The principal parameters characterising a telescope and its optics are as follows:

1. The *focal arrangement*, illustrated in Fig. 5.1. Note that a single instrument can have several focal points; transformation from one to another simply requires insertion or removal of a mirror. The great advantage of such an arrangement is that focal instrumentation can be located in several different places at any one time, each place having its own focal ratio, or other characteristics.

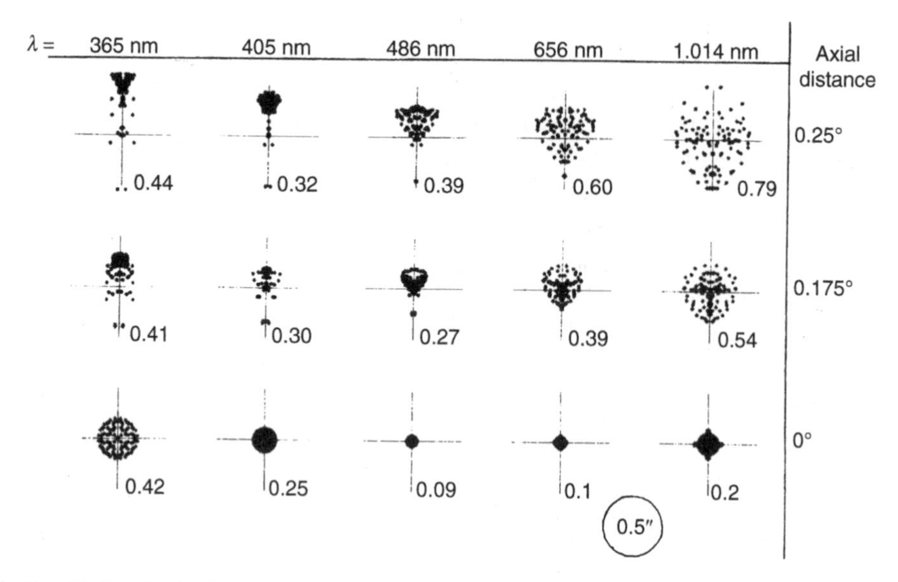

Fig. 5.2 Ray tracing by computer. A point source at infinity, placed at various angular distances from the optical axis, illuminates the pupil of a telescope. Each elementary point represented here in the focal plane corresponds to the impact of a ray coming from the same point source but passing through a different point of the pupil. As the calculations also take into account the effects of diffraction, which is a chromatic phenomenon, the final result depends on the wavelength, given on the *horizontal axis*. The number next to each image gives the radius in arcsec of the disk containing 80% of the light energy. The *scale* is given by the *circle* at the bottom which has radius 0.5″. The linear scale in the focal plane is 292 mm degree^{-1}. (Simulation by Enard D., Feasibility Study for the 8 m European VLT. European Southern Observatory)

2. The *scale* of the image in each focal plane, generally given in mm per arcsec. This scale has to be adapted to the choice of detector pixel size, unless the instrument placed at the focus itself modifies the scale through a further optical transformation.
3. The *focal ratio* of the beam, given by the ratio f/D between the equivalent focal length and the primary mirror diameter.
4. The *field*, in the two senses of the term. Firstly, the area of the focal plane in which the image quality remains acceptable according to whatever criteria have been predefined; and secondly, the area over which the illumination remains uniform (fully illuminated field), or partial (vignetted field).
5. The *image quality*, which can be quantified in various ways: using the modulation transfer function (MTF, defined in Sect. 6.1.4) at each point; observing the angular diameter of the circle in which a certain fraction (50% or 80%) of the energy from a conjugate point source is concentrated around its ideal image; or, finally, by ray tracing (Fig. 5.2). The image quality that is finally achieved is governed by optical design, quality of optical surfaces, diffraction, and also external factors, such as the presence of a medium upstream whose refractive index is not homogeneous (Earth atmosphere).

6. The *curvature of the focal plane*, which is not necessarily a plane surface, but may indeed produce the best anastigmatic approximation by being curved. This is the case of the *Schmidt telescope*, whose field can thereby attain several degrees, an unusually large value.

Further specifications could be added to this list, depending on the wavelength region under observation. We return to this in Sect. 5.2, which deals with specific realisations of telescopes.

5.1.2 Gravitational Optics

In empty Euclidean space, light rays are described by the geometrical approximation and follow straight lines. When the refractive index $n(r)$ depends on the space variable, the trajectories are curved, as illustrated by the simple example of atmospheric refraction (see Sect. 2.5).

In general relativity the geometry of space is curved by the distribution of matter, and the trajectories of light rays, or *geodesics*, are curves which always traverse the local minimal distance.[3] Such effects are present in the Universe on different scales, but especially in the vicinity of a black hole, or on the very large scale of clusters of galaxies. A given mass distribution thus produces convergence or divergence effects in light beams, in a way exactly analogous to the effects of variable refractive index in refracting media. The generic term for these effects is *gravitational optics*. By extension, a *gravitational telescope* refers to any mass distribution producing this kind of image.

Without going too deeply into the problem, we may consider the simple case of a light ray passing in the vicinity of some mass M. Recall first Fermat's principle, in differential form:

$$\delta \int n(r)\, \mathrm{d}\xi = 0 \,,$$

where $\mathrm{d}\xi$ is the length element along the light path. The equation of the trajectory in a medium with variable index follows from Lagrange's equation,

$$\frac{\mathrm{d}}{\mathrm{d}\xi}\left[n(r) \frac{\mathrm{d}\boldsymbol{x}}{\mathrm{d}\xi} \right] = \nabla n(r) \,,$$

where x is the space coordinate.

In a gravitational field, the condition $\mathrm{d}s^2 = 0$ on the generalised distance s is satisfied by the propagation of the light along its trajectory. In a simple metric of

---

[3]What distinguishes the geodesics of a curved space from those in a Euclidean space is the fact that the projection of a geodesic is not a geodesic in the projected space.

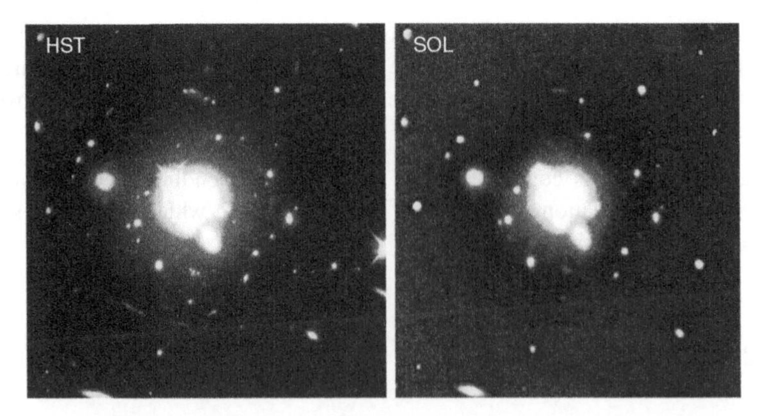

Fig. 5.3 Gravitational lens effect due to the cluster of galaxies MS 0440+0203, redshift $z = 0.19$. The approximately concentric arclets are 'gravitational images', created by the cluster of galaxies located beyond it. *Left*: Image obtained by the Hubble telescope, with a six hour integration time (six orbits), and resolution $0.1''$ (camera WFPC2), sufficient to resolve the thickness of the arclets. *Right*: Image obtained from ground level by the 3.6 m CFH telescope (Hawaii), with angular resolution $0.5''$ and limited by seeing. (Document kindly provided by Hammer F. et al., 1995)

general relativity, the space and time differentials, $\mathrm{d}r$ and $\mathrm{d}t$, respectively, at distance r from the mass M, are related by the equation

$$c\frac{\mathrm{d}t}{\mathrm{d}r} = 1 + \frac{GM}{r} = \frac{c}{v(r)} ,$$

which thus gives a value $v(r)$ for the apparent speed, in perfect analogy with the effect of a refractive index. This value can be put into Lagrange's equation, which, upon integration, yields the light paths.

It can thus be understood how a sufficiently large mass distribution may act as a lens, forming an image for the Earth-based observer of some very distant object (Fig. 5.3).

5.2 Telescopes

In this section, covering the whole electromagnetic spectrum, we shall examine the specific features of telescopes designed for each given wavelength range. The concept[4] was of course first developed for the visible region (by Galileo and Newton in the seventeenth century, Foucault in the nineteenth century, George Hale, Aden Meinel, Raymond Wilson, and others in the twentieth century), then extended to the

[4] A good reference here is King H.C., *The History of the Telescope*, Dover, 1977.

radio region after 1945. And finally after 1960, it was extended to cover the whole spectrum, observed from the Earth's surface but also from space.

The discovery or prediction of other information carriers (gravitational waves and neutrinos) has opened the way to speculation about the possibility of building systems, detecting these forms of energy, which fulfil the same role as the classical telescope. This would be an abuse of language at the present rudimentary stage of technical development, in which the tasks of collecting energy, detecting sources, and localising sources in space cannot yet be approached separately. These questions will be approached in Chap. 7, in the context of detection, which remains the main obstacle to observing such signals.

Returning to optical instruments, the quest for maximal collecting area is the common preoccupation in all spectral regions. The limitations, apart from cost, are usually of technological origin. Three basic points are worth remembering:

- The departure of actual optical surfaces from the ideal must remain small compared to the relevant wavelength, in order to obtain the best anastigmatic approximation.
- When the angular resolution is determined in the most favourable case by diffraction (see Chap. 6), it varies as λ/D, where D is the dimension of the primary mirror (or first optical collecting surface). In a global strategy, astronomers of different countries aim in the long term to obtain a uniform angular resolution on the sky, e.g., 0.1 arcsec, in order to be able to compare the properties and morphology of a given object on the same scale and at various wavelengths. However, actually making these different instruments involves a very wide range of sizes and technologies, and the aim of a uniform resolution is almost never achieved.
- The nature of the detection process, which takes place at the focal point of the telescope, determines to a certain extent the design of the telescope.

Because of all these wavelength dependencies, telescope design will be discussed here primarily in terms of the main spectral regions, but distinguishing Earth-based from space observation when necessary. The design of telescope arrays, which significantly improve angular resolutions, will be discussed in Chap 6.

5.2.1 Radiotelescopes

Telescopes designed to receive radiofrequency radiation (0.5 mm $\lesssim \lambda \lesssim$ 1 km) are called *radiotelescopes* or *radio dishes*. However, the boundary between radio wavelengths (from kilometres to submillimetre) and far infrared wavelengths is not well defined, being somewhere around 0.5 mm. Since it does not correspond to any fundamental property, this transition arises rather because of a change in the type of detector from heterodyne detectors to another kind used at shorter wavelengths, i.e., bolometers, then photoconductors, as will be explained in Chap. 7. The transition from one type to the other requires certain modifications in the design and use of the resulting telescopes, so it is logical enough to describe here some of the key features of instruments operating in the radio region.

These radiotelescopes comprise a very big primary mirror (10–100 m). It is usually parabolic and the surface accuracy is determined by the operating wavelength. The surface is made from a grid for long wavelengths and continuous panels for shorter wavelengths, with increasing accuracy (λ/20 environ) for decreasing λ. The growing importance of the submillimetre region has led to the construction of better quality mirrors, and even the boundary between radiotelescopes and optical instruments is growing fuzzy. The 15 m JCMT radiotelescope (James Clerk Maxwell Telescope) is thus described as the biggest optical telescope in the world (in 2007). The receivers, which also mix strictly radiofrequency detection techniques with optical techniques at the shortest wavelengths (see Sect. 7.4), are placed either directly at the primary focus or at a secondary focus in a Cassegrain setup.

On Earth, the telescope is generally mounted on an altazimuth mount and follows the source in its diurnal motion. At these wavelengths, since sky emission does not vary between day and night, these instruments can operate at all times, whether the sky is cloudy or not. Since the mirrors on these telescopes are robust, they can be built outdoors without a protective dome, which reduces the cost.

More than 20 instruments of this kind with diameters greater than 15 m are currently planned or operating around the world (Fig. 5.4). A rather special radiotelescope is the Arecibo radiotelescope in Puerto Rico, which has a primary mirror 300 m in diameter with a fixed vertical axis. This is a *zenith telescope* in which the image forms at the focus, but the focus moves as time goes by, so that the receiver must follow the source. The primary mirror of the decimeter wavelength radiotelescope in Nançay (France), dating from 1961–1964, is rectangular (300×35 m), vertical, but also fixed. It is preceded by a moving plane mirror, which allows a given source to be monitored for longer, and followed by a moving carriage house carrying the receiver.

At radiofrequencies, there are very few multipixel receivers, like photographic plates or CCDs. Obtaining an image of a source with only one pixel available usually means 2D scanning of the source (*raster* or TV-type scanning), with each pointing direction sampling one pixel of the image. However, at the time of writing (2007), multipixel receivers are being developed, although their capabilities remain modest as yet, both for millimeter wave incoherent imaging (with 12×12 bolometer arrays), and for heterodyne detection with a focal plane paved with an array of 5×5 receiver horns (see Sect. 7.5).

Note also that propagative perturbations in the Earth atmosphere (turbulence in the troposphere or ionosphere), which alters the phase of the waves, only affects observation at wavelengths less than the centimetre.

The angular resolution of radiotelescopes is generally diffraction limited, i.e., at the value λ/D, where D is the diameter of the primary mirror (see Sect. 6.1). With $D = 50$ m, for example, this leads to rather mediocre resolutions, namely 4 arcsec at $\lambda = 1$ mm to more than one degree at $\lambda = 1$ m. To improve these values, coupled arrays of radiotelescopes have been constructed since the 1950s, as discussed in Sect. 6.5.1.

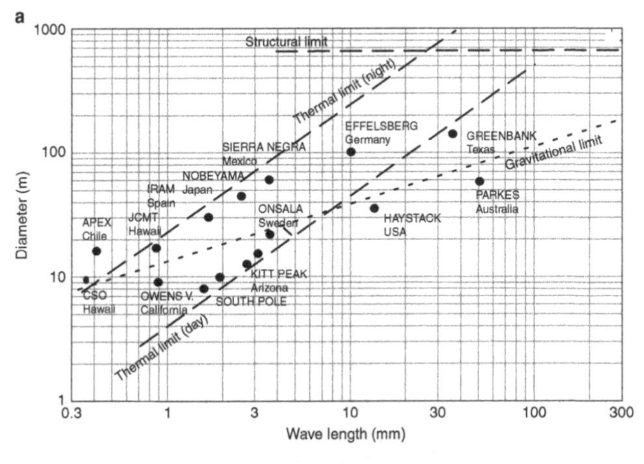

b

Fig. 5.4 (**a**) Large steerable single-dish radiotelescopes around the world. Existing instruments are plotted against the shortest wavelengths for which they will operate at normal efficiency. Limits imposed by considerations of structural rigidity, and thermal or gravitational distortion, are indicated. These perturbations limit the accuracy with which the desired parabolic shape can be achieved. (Updated version of Blum E.J., Adv. Electron. Electr. Phys. **56**, 97, 1981. With the kind permission of the editor.) (**b**) The Effelsberg radiotelescope (Max Planck Institute, Bonn). The usable part of the mirror is 100 m in diameter

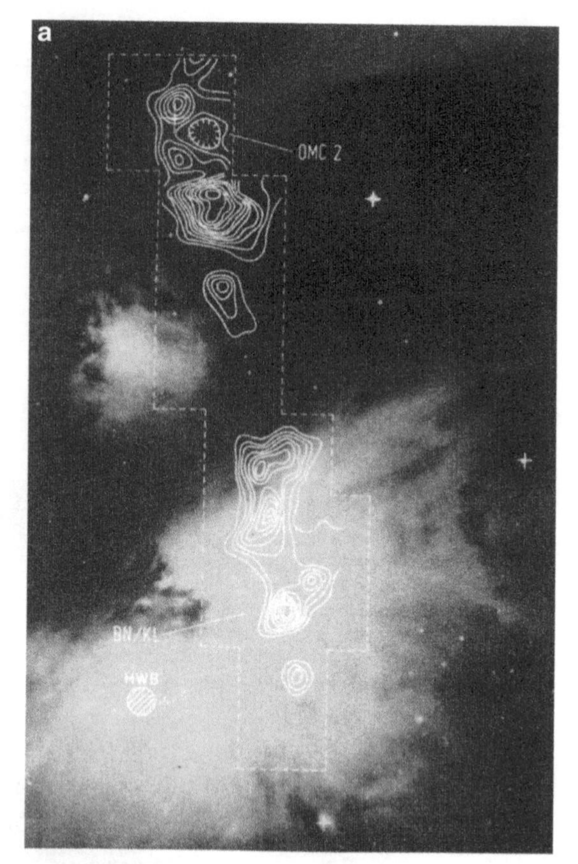

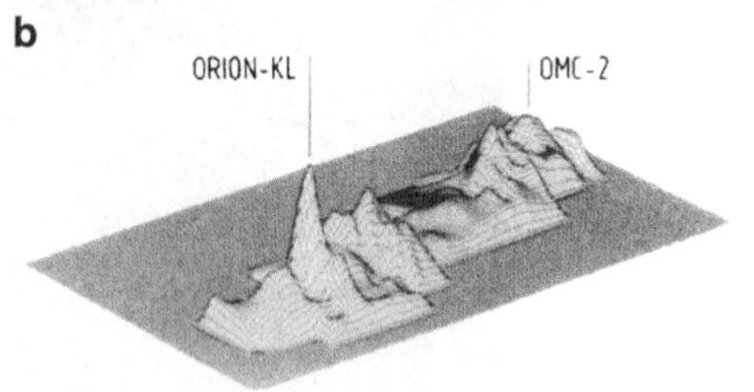

Fig. 5.5 Map of one of the regions of intense star formation in Orion, including the Becklin–Neugbauer (BN) object. The map was made in the ammonia (NH_3) band ($\lambda = 1.3$ cm) with the Effelsberg radiotelescope (Germany). The 4″ lobe is shown (half width beam HWB). (**a**) Constant intensity contours superposed on a photograph. (**b**) 3D intensity plot. See also the infrared image of the same region in Chap. 7, Fig. 7.35. (After Batrla W. et al., Astr. Ap. **128**, 279, 1983. With the kind permission of Astronomy and Astrophysics)

The Effelsberg Radiotelescope (Germany)

The largest steerable dish in existence, this 100 m diameter instrument has a sufficient surface quality (using metal plates) to operate at wavelengths $\lambda \gtrsim 7$ mm. At shorter wavelengths, the energy concentration in the diffracted image becomes inadequate (Fig. 5.4b).

The 30 m Millimetre Radiotelescope at Pico Veleta (Spain)

This is a steel structure supporting aluminium panels machined to a nominal precision of 35 μm rms, and which can produce images limited by diffraction to wavelengths in the range 0.87 to 4 mm.

5.2.2 Ground-Based Optical Telescopes: Visible and Near Infrared

Telescopes operating in the visible region are the oldest imaging instruments used in astronomy, apart from the human eye, of course. Diameters grew in stages, depending on technological progress made with materials. With the gradual disappearance of the *refracting telescope*, which effectively ceased to play any important role during the twentieth century, instruments with 5 to 6 m mirrors began to appear in the first half of that century, followed by instruments up to 10 m in the second half. By 2020–2030, we shall no doubt see instruments reaching 30–50 m in diameter. The extension to the near and mid-infrared ($\lambda < 30\ \mu$m, approximately) was made in the 1970s, as sensitive detectors became available, and sites were discovered which could provide good atmospheric transmission windows. Tables 5.1 and 5.2 list all visible and infrared optical telescopes on the Earth's surface with diameter greater than 3.5 m at the time of writing (2007), specifying their location.[5]

It is interesting to invoke several recent developments profoundly affecting some concepts which appeared to have reached their limits. Indeed, the technological difficulties, degradation of images by atmospheric turbulence, and the advent of space travel, seemed for a while at the end of the 1970s to make the further construction of large ground-based telescopes, with diameters greater than 5–6 m, somewhat improbable. But this point of view rapidly became outdated, for a number of reasons:

- The *technological limitations* on the construction of very large mirrors have been largely overcome by the use of new materials (e.g., ceramics) and the possibilities of active control leading to a substantial decrease in weight, by building mirrors with segmented surfaces.
- Methods *correcting for the effects of atmospheric turbulence* appeared at the end of the 1980s (adaptive optics), preceded by numerical methods correcting images *a posteriori* (speckle interferometry) developed in the 1970s. Overcoming the effects of turbulence (see Sect. 6.2) meant that full advantage could be taken of improved angular resolution, proportional to the diameter D of the telescope.

---

[5] At the end of the book, the reader will find a table of websites for these instruments.

Table 5.1 Large ground-based optical telescopes in the northern hemisphere

Latitude	Altitude [m]	Site	Country	Diameter [m]	Remarks	Date
47°N	2 070	Zelenchuskaya (Caucasus)	Russia	6	Altazimuth mount	1972
42°N	2 500	Xing Long	China	4	LAMOST siderostat	2007
37°N	2 160	Calar Alto (Spain)	Germany and Spain	3.5		1981
34°N	1 706	Palomar (California)	USA	5	First VLT	1948
32°N	2 130	Kitt Peak, Arizona	USA	3.8	Mayall	1974
	3 266	Mt. Graham, Arizona	USA, Italy, Germany	2 × 8.2	LBT	2007
30°N	2 076	Mt. Locke (Texas)	USA	9.2	Hobby–Eberly fixed elevation	1997
28°N	2 370	La Palma (Canaries)	UK	4.2	WHT (Herschel)	1984
			Spain	10.4	GranTeCan	2008
19°N	4 200	Mauna Kea (Hawaii)	UK	3.8	Infrared UKIRT	1979
			Canada, France, Hawaii	3.6	CFHT	1974
			USA (CalTech)	2 × 10	Keck I and II	1994
			Japan	8.4	Subaru	1999
			USA (NSF)	8.0	Gemini N	1999

Table 5.2 Large ground-based optical telescopes in the southern hemisphere

Latitude	Altitude [m]	Site	Country	Diameter [m]	Remarks	Date
23°S	2 650	C. Paranal (Chile)	Europe	4 × 8.2	VLT	1998
				4	VISTA	2008
29°S	2 280	Las Campanas (Chile)	USA	2 × 6	Magellan	2002
29°S	2 430	La Silla (Chile)	Europe (ESO)	3.6		1977
				3.5	NTT	1989
30°S	2 700	C. Tololo (Chile)	USA	4	Blanco	1974
	2 738	C. Pachon (Chile)	USA	8.1	Gemini S	2001
			Brazil and USA	4.1	SOAR	2005
32°S	1 500	Sutherland (South Africa)	South Africa and others	11	SALT	2005
34°S	1 165	Siding Springs (Australia)	Australia and UK	3.9	AAT	1974

- In *high resolution spectrography* (Chap. 8), sensitivity depends critically on the diameter D, whilst the luminosity contribution from the sky background is a nuisance but remains acceptable. The need to observe from space is thus less acute than in the case of low resolution photometric or spectral measurements.
- The *cost of space missions* is still at least one order of magnitude greater than the cost of building a comparable instrument at ground level. Following the beautiful results of the Hubble telescope, it has now become clear that ground-based and spaceborne observation are complementary and essential.
- The design of specialised telescopes, e.g., for spectroscopy, results in the development of less costly instruments, but without the flexibility of a conventional telescope. This is the case for the Large Sky Area Multi-Object Fiber Spectroscopic Telescope (LAMOST) at Xing Long in China (2007), which has a fixed primary mirror like the decimetre wavelength radiotelescope in Nançay, France, or the South African Large Telescope (SALT), which has fixed elevation (2005).

Construction of Very Large Mirrors

The strategy adopted for constructing very large mirrors (with diameters greater than 2–3 m) is now to make them lighter, i.e., thinner. Whatever technical solution is used, the primary becomes too light to ensure its own rigidity. Deformation of the mirror under its own weight would lead to significant and unacceptable aberration, and worse, it would vary with the pointing direction.

Two main technological choices have come to the fore:

- Thin single-piece mirrors, e.g., the European 8.2 m VLT, or thick but lightened mirrors, e.g., the 8.2 m Large Binocular Telescope in Arizona.
- Segmented mirrors, i.e., made up of small optical surfaces (around 1 m), arranged side by side and polished in such a way as to reconstitute the desired paraboloid, e.g., the 10 m Keck Telescope in Hawaii.

Beyond diameters of 8–10 m, for the generation of extremely large telescopes (ELT) of the years 2005–2040, only the segmented solution is considered.

In both cases, the deformation is accepted but corrected in real time by means of mechanical control from behind the monolithic mirror or the segments making up the mirror. These corrections can in part be programmed as a function of temperature, and also the altitude of the mirror, for gravitational effects, but they are mainly deduced from real time analysis of the image of a star, as given by the mirror. This is the principle of *active optics* (Fig. 5.6). Benefiting from all the resources of real time control provided by computer science, it underpins the development of mirrors for all the large Earth-based optical telescopes both now and in the future, until new, more effective technologies are devised for lightening these structures (perhaps membrane mirrors).

Ground-based telescopes were traditionally mounted on *equatorial mounts*, but these have now been practically abandoned, being both cumbersome and heavy, and

Fig. 5.6 An example of active optics. The New Technology Telescope (NTT) at the European Southern Observatory, opened in 1989, has a thin, actively supported primary mirror of diameter 3.5 m. Shown here is the model used to finalise the dorsal assemblage of jacks which control the shape. These act in such a way as to correct for any mechanical (e.g., gravitational, wind) or thermal deformations of frequency less than 1 Hz. (Document R. Wilson, European Southern Observatory)

hence expensive. Today they have been replaced by the *altazimuth mount*, which uses two axes to compensate for diurnal motion and is computer driven. A pointing accuracy of the order of 0.1 arcsec rms is achieved. The disadvantage with this arrangement is that it creates a *rotation of the field* which must be compensated for by a *field derotator*.

Features Specific to the Infrared

At ground level, telescopes can be used indifferently in the visible or the infrared. Note that some large instruments have been built specially for the infrared, an example being the 3.8 m UK telescope at Mauna Kea (Hawaii), known as UKIRT. Two factors distinguish the optimal design of an infrared instrument, and the same factors are relevant in the submillimetre region:

- A *chopping secondary mirror*, usually vibrating at a frequency less than 50 Hz, makes it possible to alternate very quickly between source and sky background as imaged at the detector, and thus to make a differential measurement. This technique, well adapted to single pixel detectors, has developed hand in hand with more sophisticated background subtraction methods, as described in Sects. 2.3.3 and 9.4. It is nevertheless true to say that the chopping frequency should be as high as possible owing to the rapid time evolution of atmospheric emissions.
- *Low thermal pollution*. The optics are designed so that the detector receives as little radiation as possible from sources other than the optical surfaces themselves. This aim is achieved by using suitable arrangements of diaphragms

and baffles. In addition, the optical surfaces are treated to reduce emissivity, and a good telescope should not have an emissivity greater than a few percent, taking into account all its surfaces. These considerations apply equally to submillimetre and millimetre telescopes, for the same reason, namely, to avoid drowning out the astronomical signal by the thermal background emitted from the optical system itself.

The New Generation of Optical Telescopes

We shall describe several examples out of the dozen or so new generation telescopes going into operation around the year 2000, and one project for the next generation.

The two *Keck telescopes*, built in 1992, have diameter 10 m, the primary mirrors each consisting of 36 independent and actively supported hexagonal segments of 1.8 m, separated by 3 mm spaces. The focal ratio f/D is equal to 1.75. Several focal points are available: primary, Cassegrain, and Nasmyth. The two telescopes are $B = 75$ m apart at the summit of Mauna Kea (Hawaii). They can be coupled coherently (interferometry), and the resulting diameter will lead to great sensitivity, and a large angular resolution ($\lambda/B = 2.75 \times 10^{-3}$ arcsec at $\lambda = 1$ μm).

The European *Very Large Telescope*, or VLT, went into operation in 1998. It consists of four independent telescopes with thin single-piece primary mirrors of effective diameter 8.2 m. Each is mounted on active supports, $f/D = 1.8$ m, and Nasmyth, Coudé, and Cassegrain focal points are available. The four telescopes (Fig. 5.7) are located at the corners of a trapezoid at Cerro Paranal (Chile), and this arrangement allows either an independent use or a coherent combination

Fig. 5.7 The European Very Large Telescope (VLT), Cerro Paranal, (2 635 m, Chile). The VLT consists of four 8.2 m telescopes, and also a complementary network of mobile interferometric telescopes (foreground). Channels allow transportation of light for recombination at a common focal point. (Drawing kindly supplied by the European Southern Observatory)

(interferometry). In the latter case, the resolution is determined by the largest possible separation, $B = 120$ m. Four smaller movable telescopes (1.8 m) complete the interferometric system, providing bases of variable length and orientation up to 200 m long.

The *Large Binocular Telescope* (LBT or Magellan) is based on a hybrid technology, since it uses thick but lightened single-piece active mirrors of 8.4 m forming a binocular pair on the same altazimuth mounting. As in the original MultiMirror Telescope (Mount Hopkins, Arizona), which was its predecessor, built in 1976 and including six mirrors on the same mounting, the beams of light can be either coherently or incoherently superposed. The LBT was set up on Mount Graham (Arizona) and saw its first light in 2005.

The *Thirty Meter Telescope* (TMT) project results from a joint effort by private and public institutions in Canada and the United States. At the time of writing (2006), this planned 30 m telescope is the most advanced of a new generation that will no doubt become a reality in the first half of the twenty-first century. Equipped with a primary mirror of diameter $D = 30$ m comprising 492 individual 1.4 m hexagonal mirrors 4.5 cm thick, open at $f/1$, it will be set up at the Mauna Kea site in Hawaii. As for all instruments in this category, the construction is only justified if the ultimate angular resolution λ/D can be achieved, and this will require an extraordinarily high-performance adaptive optics system (see Sect. 6.3).

Europe (European Southern Observatory ESO) has also been studying the idea of a comparable project, namely the European Extremely Large Telescope (E-ELT) of diameter 42 m, and this was approved in 2006.

5.2.3 Space Telescopes, from Ultraviolet to Submillimetre

By placing a telescope in space, all the obstacles to astronomical observation caused by the day–night cycle and the Earth atmosphere can be entirely avoided, while covering a very broad region of the electromagnetic spectrum from the ultraviolet to the submillimetre.

Space travel offers possibilities ranging from the use of aeroplanes and stratospheric balloons, to space shuttles, low-orbiting space stations (altitude 400–500 km), satellites orbiting the Earth at various distances and sometimes stabilised at Lagrange points, probes investigating objects in the Solar System (planets, moons, comets, asteroids, the Sun, etc.), and instruments on or around the Moon.

Onboard telescopes observe deep space, beyond the confines of the Solar System. Apart from the possibilities of *in situ* investigation, more specific instruments can also provide images of objects, e.g., mapping the surface of Mars from an instrument in Martian orbit. The quest for large diameters, usually justified by the increased sensitivity required of instruments, is limited by technological and economic issues, just as it is for ground-based instruments. These telescopes are remote controlled and highly reliable both for control of acquisition and pointing, for example, and for reception, processing, and transmission of measurement data. For deep space observation, operational and planned missions are genuine space observatories, specifically designed for a given spectral region.

Ultraviolet

In the ultraviolet, the spectral limits are rather clear. The atmosphere blocks wavelengths below 350 nm (or 320 nm at high altitude sites), and this has made space observation unavoidable, using balloons and rockets in the 1960s and 1970s, and orbiting observatories thereafter. UV space telescopes are distinguished from visible telescopes only by the quality of their optical surfaces (flaws $< \lambda/20$, approximately), and the need for efficient reflecting surfaces (MgF_2 layers). The production of reflecting layers by means of multiple coatings gives acceptable reflectivity (10 to 20%) down to near X ray wavelengths (10 nm), and thus avoids the constraints imposed in the latter spectral region by the need to use grazing incidence devices.

The Orbiting Astronomical Observatory (OAO) satellites provided the first mapping and photometry of the UV sky in the 1970s. They were followed, in 1978, by the remarkably fertile International Ultraviolet Explorer (IUE) (see Fig. 8.9). This instrument remained active for almost 20 years (1978–1996), equipped with a spectrophotometer in the region 115–320 nm.

In the near ultraviolet region, the main project of the 1990s and 2000s has been the Hubble Space Telescope, launched in 1989 and expected to remain in operation up until 2010. With a diameter of 2.4 m and angular resolution $0.1''$ (after a few initial difficulties), it covers all wavelengths between 110 and 700 nm. In 2002, during a visit by NASA astronauts, it was equipped with a new instrument, the Advanced Camera for Surveys (ACS), able to provide images and spectra from the ultraviolet to the very near infrared (115–1 100 nm).

Among more specialised missions, we should mention:

- The Far Ultraviolet Spectroscopic Explorer (FUSE) run by NASA and other countries, launched in 1999 but affected by a serious failure in 2006, and covering the region 90.5–118.7 nm.
- The Galaxy Evolution Explorer (GALEX), a 35 cm telescope launched by NASA in 2003, and covering the region 135–280 nm.

The extraordinary results harvested by the Hubble telescope stimulated the preparation of a successor, namely the James Webb Space Telescope (JWST).[6] This observatory will be placed at Lagrange point L2 and carry a 6.5 m telescope equipped with a segmented mirror deployed in space (see Fig. 5.8). It will cover the region 0.6–27 μm, so is designed rather as a near infrared telescope than an ultraviolet telescope. However, it will be observing the very remote Universe, where spectral shifts due to expansion ($z \gtrsim 1$) will bring radiation emitted in the ultraviolet into the visible and near infrared range. Launch is planned for 2013.

---

[6] James Webb ran NASA at the time of the Apollo lunar exploration programme, from 1961 to 1968.

Fig. 5.8 Artist's impression of the future 6.5 m James Webb Space Telescope, to be placed at Lagrange point L2. The figure shows the alignment of the telescope with the Earth and Sun, and a screen protecting the optical system. Credit NASA

Infrared and Submillimetre

The infrared refers to 1–100 μm and the submillimetre to 0.1–1 mm. Although the near and mid-infrared (up to about 30 μm) can in part be observed from the Earth's surface, space observation is inevitable beyond that. And even below, certain spectral regions absorbed by the Earth atmosphere (e.g., between 5 and 9 μm) contain emissions of considerable importance (e.g., the spectra emitted by aggregates of aromatic hydrocarbons). Beyond 30 μm, some useful transmission windows exist at ground level in several exceptional sites, such as the Antarctic, but only space observation can provide systematic and precise data. Nevertheless, the concentration of absorbing molecules (mainly H_20) at low altitudes means that aeroplanes and balloons are less costly for observation than a deep space instrument.

During 1980–2010, a considerable wealth of discoveries has been made by infrared observation in this vast spectral region: emission by cold objects in galaxies, but also the intense radiation associated with star formation, the spectra of primordial galaxies shifted toward the infrared by cosmic expansion, the 2.7 K cosmic background radiation, and many others. This success justifies a major effort to prepare space observatories covering the whole of this spectral region. But one of the main difficulties with these instruments comes from the need to cool the telescopes, essential for improving sensitivity and reducing the associated background signal and thermal noise (see Chap. 9). In an *aircraft* or *balloon*, this cooling remains modest (−30 to −50°C) given the presence of residual atmosphere. It will be pushed as far as possible in satellites.

Aircraft. An example here is the telescope of the Kuiper Airborne Observatory (KAO), named after the Dutch astronomer G. Kuiper (1906–1973), which had a diameter of 0.9 m, and was carried on an aeroplane (USA). It operated in the

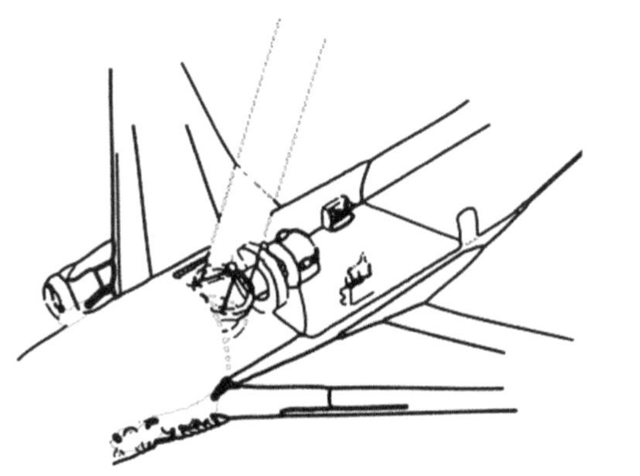

Fig. 5.9 The 2.5 m submillimeter telescope of the NASA's Stratospheric Observatory for Infrared Astronomy SOFIA on its Boeing 747-SP aircraft. *Insert*: The 747-SP flying with the telescope open, during its final tests over California in August 2010 (NASA). Drawing kindly provided by *Max Planck Institut für Extraterrestrische Physik, Garching*. Photo NASA

stratosphere in a very similar way to a ground-based observatory. Its successor aboard a Boeing 747SP (collaboration between NASA and Germany, flown in 2008) is the Stratospheric Observatory for Infrared Astronomy (SOFIA) shown in Fig. 5.9, equipped with 2.5 m mirror and flown at an altitude of 14 km.

Balloons. Many instruments have been carried aboard stratospheric balloons since the beginning of the 1960s, some reaching diameters of around 1 m. Among these missions, one of the most remarkable is BOOMERANG, run by NASA in partnership with many other countries since 1998 and designed to study the cosmological background radiation (imaging and polarisation). The cooled telescope has a 1.3 m primary mirror and observes wavelengths[7] in the range 1–2.1 mm from a stratospheric balloon flying at an altitude of 42 km above the Antarctic continent.

Satellites. On board a satellite, passive cooling by radiation into space leads to temperatures around 80 or 100 K. The use of cryogenic fluids or closed circuit refrigerators allows for cooling of optical parts and detectors down to 10 or 20 K, sometimes even much lower (0.1 K for dilution refrigerators), and this totally removes the instrumental background.

The InfraRed Astronomy Satellite (IRAS), launched in 1983, included a 0.6 m telescope completely cooled by liquid helium. The Infrared Space Observatory (ISO), which operated from 1995 to 1998, had the same diameter and was cooled in

[7] As we have already pointed out, the boundary between the infrared and submillimetre region on the one hand and the millimetre wave radio region on the other is rather poorly defined. The transition region, of considerable cosmological and astrophysical importance, lying between 0.5 and 2 mm, can be attached to one side or the other depending on the detection techniques brought into service (see Chap. 7). Here, the extreme measurement conditions tend to attach it to the first region.

the same way. The optics of the COsmic Background Explorer (COBE), launched in 1992, were also cooled, during the first part of the mission, by liquid helium, whose gradual evaporation limited the period of operation.

Let us briefly describe three major missions over the period 2000–2020, the first two (Spitzer and Herschel) dedicated to a wide range of galactic and extragalactic observations and the third (Planck) specifically designed for detailed observation of the cosmological background:

- The Spitzer observatory[8] (a mission originally called SIRTF) includes a cooled 0.85 m telescope operating in the region 3–180 μm. It was launched by NASA in 2003 with an expected lifetime of 5 years. It is more sensitive than all previous missions (IRAS, ISO).
- The Herschel observatory[9] (a mission originally called the Far InfraRed Space Telescope FIRST) includes a cooled 3.5 m telescope designed to carry out galactic and extragalactic observations in the range 60–670 μm, with a lifetime of 3 to 4 years (ESA, launch in 2008). It will be placed at Lagrange point L2. Its focal instruments combine detection techniques specific to the infrared (photoconductors) with others for the radio region (heterodyne, see Sect. 7.5).
- The Planck observatory[10] is mainly designed to study the cosmological background radiation, but its extreme sensitivity means it should be able to address many issues of galactic and extragalactic astronomy. Equipped with a telescope cooled to 20 K and detectors operating at 0.1 K, covering frequencies 30–860 GHz in fields of 30 to 5 arcmin, this observatory, also placed at Lagrange point L2, should operate for at least 21 months (launch 2008).

Infrared and submillimetre instrumentation on the surface of the Moon would benefit from the very low temperatures (about 100 K at the bottom of permanently shaded craters in the polar regions), a natural cryogenic source for cooling the optics. This option is still not seriously envisaged (see Sect. 2.10).

[8]The very talented US astrophysicist Lyman Spitzer (1914–1997), professor at Princeton, pioneered investigation of the interstellar medium and was the first, in the 1950s, to suggest building a large ultraviolet telescope that could be placed in space. This initiative led to the launch of NASA's Hubble Space Telescope.

[9]Sir William Herschel (1728–1832) was a British astronomer, born in Germany. He was the first to show, around 1800, the existence of infrared radiation beyond the visible spectrum. Extraordinarily talented and productive, he discovered the planet Uranus and built many telescopes.

[10]Max Planck (1858–1947) was one of the great physicists of the end of the nineteenth century and the beginning of the twentieth. In his famous paper of 14 December 1900, he introduced the idea of energy quanta and the constant h which now carries his name, considering them essential to explaining the blackbody radiation spectrum, as measured by Wien and Stefan, on the basis of the fundamental principle of thermodynamic irreversibility. The universal Planck constant h, one of the most fundamental constants of microscopic physics, enters into the formulas for two key cosmological quantities: the Planck time $t_\mathrm{P} = \sqrt{\hbar G/c^5}$ and the Planck length $l_\mathrm{P} = ct_\mathrm{P}$.

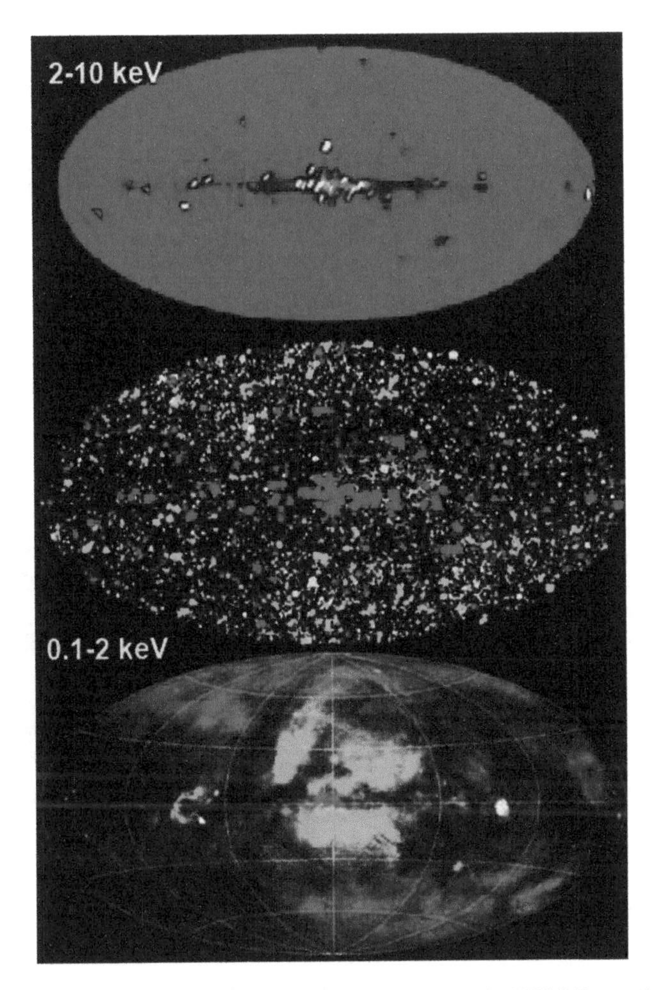

Fig. 5.10 X-ray sky in galactic coordinates. *Upper*: X-ray survey by HEAO-I, covering 2–20 keV (Boldt, 1987). *Middle*: Survey by ROSAT. *Dots* denote the brightest of the 100 000 point X-ray sources detected (Voges et al., 2000). *Lower*: ROSAT survey of diffuse X-ray emission. Energies are colour coded, from red (0.1–0.5 keV) to blue (0.9–2.4 keV) (Snowden et al., 1997). (Figure kindly communicated by G. Hasinger, Garching)

5.2.4 X-Ray Telescopes

This concerns photons with energies in the range (0.1–10 keV). Since the discovery in 1962 of the first X-ray source outside the Solar System, X-ray astronomy has progressed enormously, observations being made almost exclusively by instruments in orbit. In 1978, the launch of the Einstein satellite (HEAO-II) resulted in a large quantity of data and, notably, X-ray images (Fig. 7.27). Figure 5.10 gives some idea of the progress that has been made.

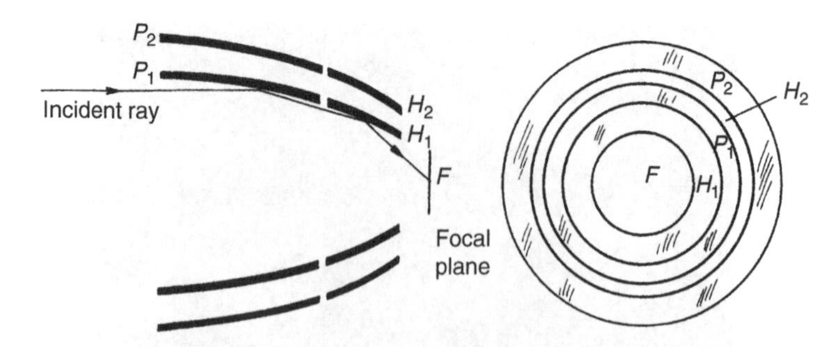

Fig. 5.11 Side and front views of a Wolter X-ray telescope. P and H denote parabolic and hyperbolic surfaces of revolution, whose common axis points to the source. See also Fig. 5.12

Fig. 5.12 Aperture of the 7 convergent X-ray telescopes planned for the E-ROSITA mission (launch 2015). Each telescope, with entry diameter 16 cm, consists of 27 nested mirrors. (Source Max Planck Institute für Extraterrestrische Physik)

Grazing Incidence Telescopes

As metallic surfaces absorb X-rays at all but very high incidence angles, the *Wolter telescope* uses grazing incidence ($\approx 89°$) (Figs. 5.11 and 5.12). Two confocal segments of a paraboloid P and a hyperboloid H form, after two reflections, an image of a source at infinity at their common focus.

To increase the effective area presented to the incident radiation, several mirrors are nested together. The system is perfectly anastigmatic on the axis, has no spherical aberration, and satisfies *Abbe's condition* (aplanetism), which means that an image formed from paraxial rays has no coma.

In the Einstein telescope, the collecting area is 300 cm^2 for energies less than 2 keV, and decreases to 50 cm^2 at 3.5 keV, since only those parts of the mirror at the greatest incidence angles play any role at higher energies. The image quality, greatly inferior to the one which would be imposed by diffraction, is limited by defects in the crystalline structure of the metallic mirror surfaces. These defects disturb the periodicity of the crystalline lattice and cause phase errors at the pupil, and hence a deviation from the ideal image in the focal plane (see Sect. 6.2).

Table 5.3 Image quality of X-ray telescopes. The values given in arcsec are the half power beam widths of the point spread functions. The *asterisk* indicates that image qualities are given by the radius (arcsec) enclosing 50% of the energy. See the table of space missions at the end of the book to identify the satellites and their instruments, and also planned missions at the time of writing (2007)

Observatory (satellite)	Energy [keV]	Distance from optical axis [arcmin]			
		0′	5′	10′	20′
Einstein	0.28	8″	10″	25″	–
	3	20″	25″	40″	–
Exosat*	–	18″	–	–	40″
Rosat (HRI)*	1	3″	3″	7″	26″
XMM	$\lesssim 2.5$	20″	–	–	–
Chandra	–	<1″	2″	5″	20″

By the mid-1990s, the energy limit had progressed from 1 to around 100 keV, by use of mirrors with multiple coatings, which increase reflectivity (Rosat, XMM, and AXAF missions).

Adaptive optics techniques can be transferred from the infrared and visible to X-ray imaging, in this case to correct for phase defects in the mirrors themselves. By the mid-1990s, there has not yet been any application in astronomy, but correcting mirrors are used in the synchrotron at the European Synchrotron Research Facility (Grenoble, France) in order to improve beam focusing.

Table 5.3 gives the image qualities of several X-ray telescopes, measured by the half-power diameter of the image, in seconds of arc.

Application of aperture synthesis methods to the X-ray region is possible, in principle, but will not be justified until diffraction becomes the limiting factor to resolution in X-ray telescopes, and further progress is required.

The availability of powerful and geometrically well-defined X-ray sources in the laboratory (synchrotron radiation at the ESRF in Grenoble, France, see Chap. 3) has greatly encouraged the development of X-ray optical techniques since 1990: interference mirrors, transmitting *Fresnel lenses*, reflecting *Bragg–Fresnel lenses*, and so on. This amounts to a generalised transposition to the nanometre scale of totally classical concepts developed for visible optics, or infrared optics, at the micrometre scale.

5.2.5 Gamma-Ray Telescopes

Telescopes in the γ ray region (≥ 10 keV) differ from those at other wavelengths in that it becomes difficult to use mirrors in this energy range, and this for two reasons. To begin with, since the grazing incidence angle, already used in X-ray mirrors, is inversely proportional to the photon energy, the focal length, which is of the order of 10 m at 20 keV, would be of order 500 m at 1 MeV. In addition, the size of tolerable surface defects on the mirrors would also be inversely proportional to the energy.

The first of these two difficulties will soon be solved by flying two satellites in formation, one carrying the mirror and the other the detector. The distance between the two satellites, i.e., the focal length, can then be arbitrarily large. This is the plan for the French–Italian mission SIMBOL-X which, with two satellites a distance 20 m apart, is expected to operate up to almost 100 keV with multilayer mirrors.[11] At the time of writing (2007), developments are underway to try to reduce the second difficulty, with the hope of making mirrors operating up to a few hundred keV. So long as such mirrors are not available, images are formed in the γ ray region using other principles.

In the region dominated by the *photoelectric effect* (see Figs. 7.18 and 7.20), *coded masks* are generally used. However, in energy ranges dominated by *Compton scattering* or *pair creation*, the dynamics of each of these phenomena can be used to estimate the direction of each incident photon and form images. The pair effect gives information about the direction of each photon. Compton scattering generally only allows an event to be localised within a circle drawn on the celestial sphere. In both cases, the telescope and the detector become one. This is why the operation of these rather special telescopes will be postponed until Chap 7.

Here we present the coded mask telescope. Coupled with a photoelectric absorption detector, which gives no information about the direction of the incident photon, it forms images just as other telescopes do. However, mirrors are used in other regions not only to form images, but also to concentrate radiation, and it is this aspect which is the most difficult to replace in the γ ray region. Below we shall also discuss other γ ray observation setups, such as Bragg lenses or detector arrays carried in space and specially designed for the observation of γ ray bursts.

One particular feature of γ ray observation is the following: the telescope serves mainly to determine as accurately as possible the position in the sky of a generally unresolved source. This identification is essential when we seek the physical nature of the source, or its association with an object like a star or galaxy observed at a different wavelength. It is less common to obtain maps of resolved objects, and more difficult too, given the necessary image processing.

Coded Masks

The principle of the *stenope* or *camera obscura* is an old one: a hole in a mask allows one to form an image on a screen. Making the hole smaller increases the resolution. Since diffraction becomes negligible in the X and γ ray regions, there is no risk of it degrading the resolution and the hole can be arbitrarily small. However, the smaller the hole, the lower the sensitivity. A mask containing a large number of small holes provides a way around these contradictory constraints. The small size of the holes guarantees a good resolution and their large number a high sensitivity. The image is then a linear superposition of the images given by each hole. The idea is to

[11]This mission is described at www-dapnia.cea.fr/Phocea/.

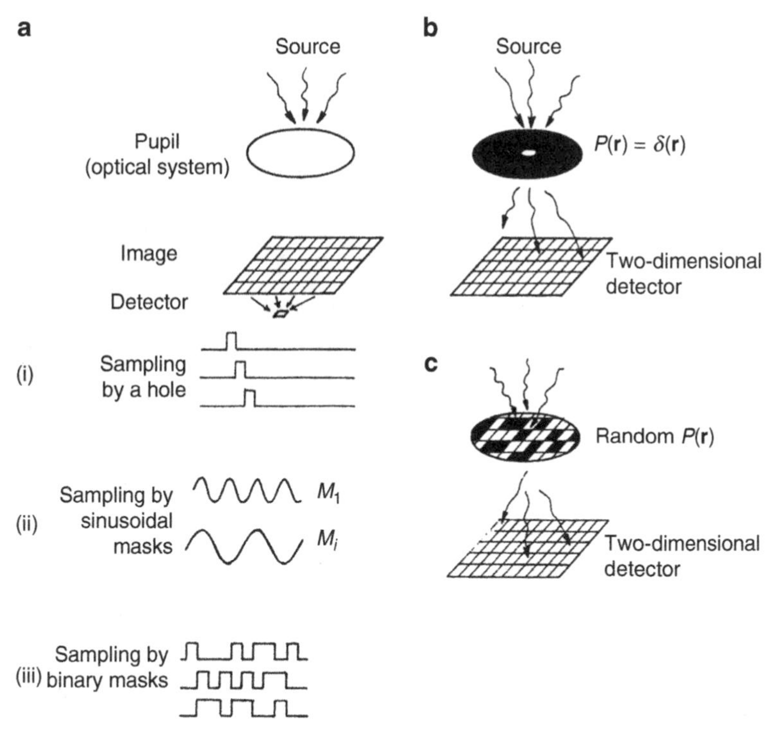

Fig. 5.13 Different coding methods. (**a**) An optical system forms an image of an object. This image is decomposed into $p \times q$ pixels and sampled by a succession of masks, the energy being focused on a single detector: (*i*) sampling by a single hole, positioned successively on each pixel; (*ii*) sampling by a succession of masks M_i with sinusoidal transmission; (*iii*) sampling by binary masks with random distribution of holes. (**b**) Stenope. A two-dimensional detector receives the pseudo-image of the source. (**c**) Random transmission pupil with two-dimensional detector

distribute the holes on the mask, which constitutes the aperture, in such a way that the object can be unambiguously reconstructed from measurements of the intensity distribution in the image. The solution to this problem is provided by the *coded mask*, whose surface is described by a variable transparency function (Fig. 5.13).

Let P be the mask pattern used to modulate the flux of an extended source S. Then the distribution I observed on the detector can be written as a convolution

$$I = P \star S + B ,$$

where B represents the background at the detector. An estimate S' of the source S can be obtained by applying a decoding function D:

$$S' = D \star P \star S + D \star B .$$

The quality of reconstruction thus depends on the choice of P and D (see Sect. 9.6). In the absence of noise, there must exist a one-to-one correspondence between S and

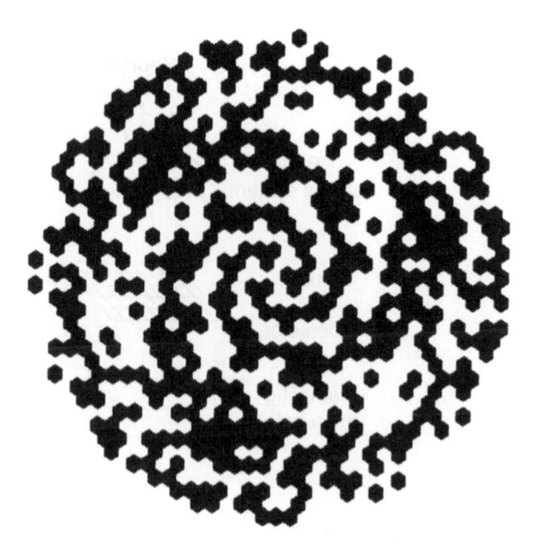

Fig. 5.14 Example of a coded mask based on the HURA principle (Hexagonal Uniformly Redundant Array)

S', which amounts to requiring that $D \star P = \delta$. Furthermore, the background effect should preferably be uniform in the deconvoluted image. Mask patterns P which satisfy these two requirements are said to be *optimal*. Several different types of optimal pattern are shown in Figs. 5.14 and 5.15.

It can be shown that the efficiency of a coded mask is optimal when the mask transmits around 50%, but this optimum is not critical, for any transparency between 30 and 70% gives excellent results. The sensitivity of a coded mask telescope also depends on the ratio of the size s_e of the elements of the mask to the spatial resolution r_s of the detector, i.e., the capacity of the detector to sample the image of the mask. In practice, a ratio s_e/r_s greater than 2–3 ensures an acceptable imaging efficiency. The angular resolution of a coded mask telescope is directly related to the size s_e of the mask elements and the distance d between the mask and the detector:

$$\theta = \arctan \frac{s_e}{d} .$$

For a fixed distance between mask and detector, and if we impose the imaging efficiency, this sensitivity is thus directly dependent on the spatial resolution of the detector.

More often than not, the basic pattern of the mask covers the same area as the detector, and the mask is composed of a partial repetition of this pattern around a complete central pattern. The field of the telescope depends on the extent of the repetition. Sensitivity is not uniform across the whole field of view, but is directly related to the coded fraction, that is, the fraction of the detector illuminated by a basic mask pattern (even incomplete). In order to characterise the field of view of a coded mask telescope, three limits can be identified: the fully-coded limit, the semi-coded limit, and the zero-coded limit. If there is no repetition of the basic pattern,

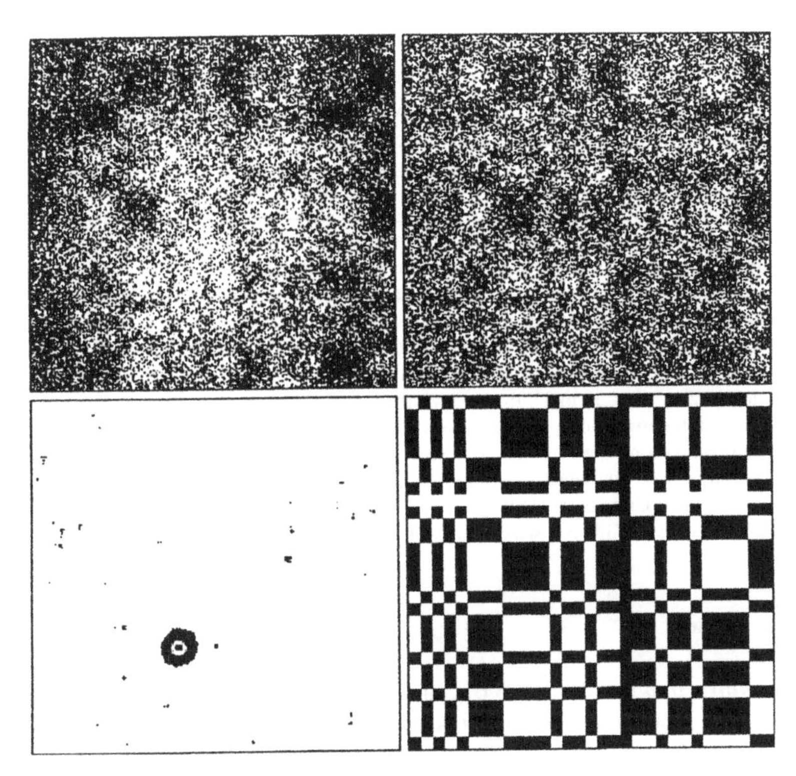

Fig. 5.15 Image restitution by coded mask. *Top left*: Image recorded by the γ ray telescope of the Sigma instrument (Granat mission 1989–1997). *Top right*: Image corrected for pixel uniformity defects. *Bottom right*: Pattern of the Sigma coded mask. *Bottom left*: Deconvolution of the image top right reveals the Crab Nebula as viewed between 75 and 150 keV

that is, if the mask has the same size as the detector, the field fully coded is zero, and the field at half-sensitivity will be equal to the solid angle subtended by the mask at the detector.

In the French telescope Sigma, on board the Russian satellite Granat (1989–1997), the mask was of the Uniformly Redundant Array (URA) type, composed of 49×53 tungsten elements, each measuring $0.94 \times 0.94 \times 1.5\ \text{cm}^3$. It was placed at 2.5 m from a crystal scintillator (NaI) γ-ray camera and had a theoretical separating power of the order of 13′. The γ-ray camera consisted of a NaI disk with diameter 57 cm and thickness 1.2 cm, and 61 hexagonal photomultiplier tubes (see Sect. 7.6). The location of an interaction was determined by the relative amplitudes of the photomultiplier impulses, to an accuracy of within a few millimetres. The principle of the γ-ray camera came from applications in medicine. For orbiting instruments, the accuracy in locating events is reduced by the presence of charged particles. Indeed, the signal corresponding to a proton is such that it disturbs measurements for more than 100 μs, a lapse of time comparable with the average time between the arrivals of two protons. Consequently, measurements are always perturbed.

As observed above, masks are designed to give a uniform noise level in the deconvoluted image. This design assumes a uniform noise level at the detector, which is never actually the case. In addition, the pixels of an image never all have the same sensitive area. These two sources of non-homogeneity in the background must therefore be corrected for before deconvoluting the image. Figure 5.15 illustrates the results of such a treatment on data from Sigma.

The imaging telescope IBIS carried aboard the European satellite Integral (launched in 2002), is equipped with a MURA (Modified URA) type coded mask comprising 53×53 tungsten elements, each measuring $1.12 \times 1.12 \times 1.6$ cm^3. Placed at 3.2 m from two γ ray cameras that work together to ensure efficient detection of photons with energies in the range from 15 keV to 10 MeV, this mask has a separating power of 12 arcmin. ISGRI, the first of the two cameras, is a new generation instrument comprising 16 384 CdTe semiconductor crystals of dimensions $4 \times 4 \times 2$ mm^3, each detecting photons in the range from 15 keV to 1 MeV with a resolution of the order of 7% at 100 keV. The second layer comprises 4 096 scintillating crystals of caesium iodide (CsI) of dimensions $9.2 \times 9.2 \times 30$ mm^3, from which the signal is collected by photodiodes. In this way, a passing proton perturbs only one pixel of ISGRI or PICSIT, and the average time between the passage of two protons is greater than one second. Note that such a large number of measurement channels can only be envisaged in association with dedicated integrated electronics (Application Specific Integrated Circuits, or ASIC) and data compression (see Sect. 9.1.3). The processing of images produced by ISGRI is discussed in Sect. 9.3.4.

Bragg Lenses

Bragg diffraction (see Sect. 8.3.6) can in principle be applied in the γ ray region. Using diffraction in the bulk of a crystal lattice, this avoids the problems relating to the surface state of a mirror. Furthermore, it requires a shorter focal length than with a mirror ($\approx$20 m at 500 keV). The focusing it produces should in principle offer excellent sensitivity.

A Bragg lens can be made from small crystals ($\approx$2 cm) placed on concentric circles and oriented in such a way that the crystal planes diffract the photons towards a focal point. Several serious problems are nevertheless encountered. Two of these are intrinsic. The Bragg diffraction condition implies a very narrow field of view ($\approx$35 arcsec at $E = 500$ keV) and a very narrow spectral dynamic range ($\approx$6 keV at $E = 500$ keV). These limitations can be reduced either by using crystals in which the atoms are not perfectly aligned on the lattice, or by using various types of crystal or various crystal planes for diffraction. Unfortunately, the resulting improvements in the field and spectral dynamics are obtained at the expense of the sensitivity. Moreover, the orientation of the mirrors must be adjusted and maintained with very great accuracy, of the order of a few arcsec, a real challenge for a spaceborne experiment which is likely to undergo significant vibration during launch.

Gamma-Ray Burst Telescopes

A γ ray burst is an intense flash of γ rays, with a luminosity that generally exceeds all other sources in the sky. Some of these are known to be caused by supernova explosions. Any γ ray telescope can, of course, detect a burst produced within its field of view, these bursts being extremely bright given the sensitivity of instruments. However, even after five years of operation, the Sigma telescope operating in the 1990s did not detect a single such event in the partially coded field of view of its mask. This is a consequence of the rarity of these events, for the main part, since they occur at the rate of about once per day over the whole sky. Quite clearly, the main requirement to be made of a γ ray burst telescope is that it should survey the whole sky at once.

The BAT instrument (Burst Alert Telescope), carried aboard the SWIFT satellite launched in 2004 and dedicated to γ ray burst observation, comes close to meeting this objective. It is a coded mask telescope, similar to SIGMA or IBIS, except that it has a field of view of 2 steradians, whence it can detect and accurately locate a hundred or so bursts each year.

Before the advent of these high performance imaging telescopes, other techniques were used to locate bursts. Although they did not use an imaging method in the strict sense of the term, these telescopes are presented here because they do not locate a source by measuring the direction of each photon, as happens in the Compton, pair creation, or Cherenkov telescopes discussed in Sect. 7.6.

Satellite Networks

A network of satellites equipped with detectors can constitute a γ ray burst telescope, provided that each can date events to within 1 millisecond. Knowing the position of each satellite at the moment its detector is triggered, the position of the source (the *burster*) can be found. The accuracy of this method, known as *triangulation*, depends on the number of satellites used, their distances, and their relative orientations. This is a relatively old method, but has provided, since 1995, the most accurate results.

Weighting Telescope

This is an extremely simple system, consisting only of detectors, which are oriented to cover the whole sky. It is based on the idea that a burst, when it occurs, is the brightest source in the sky, which means that the counting rate of any detector during the burst depends only upon its orientation relative to the source (the burster). Schematically, the direction of the burster is estimated by weighting the various lines of sight of the detectors according to their counting rates. The BATSE telescope, on board the Compton Space Observatory (1990–2000), was the first telescope of this type.

Chapter 6
Diffraction and Image Formation

In the last chapter, we discussed telescopes as optical devices able to form images in a way that could be explained by geometrical optics, at least to first order in the light intensity distribution. The wave nature of electromagnetic radiation produces diffraction effects that modify this distribution and introduce a fundamental limitation on the angular resolution of telescopes. Since astronomers always want to obtain images containing more and more detail, it is essential to come to grips with these effects, whose amplitude is directly related to the wavelength of the radiation. Using the notion of coherence discussed in Chap. 3 (see Sect. 3.2), we begin by examining the process by which images are formed in the presence of diffraction, and translate the results in terms of *spatial frequency filtering*.

There are several reasons for improving the angular resolution of observations. The most obvious is to be able to see the details of an object, such as the surface of a planet, solar granulation, circumstellar disks, stellar diameters, etc. The second and absolutely essential reason is to avoid *confusion*. Galileo's refracting telescope allowed him to resolve the milky appearance of our own galaxy, the Milky Way, into stars. Aperture synthesis at radiofrequencies allowed the individual identification of radiogalaxies back in the 1950s. The sky background radiation at X-ray wavelengths was discovered by the first observations of the X-ray sky in 1962.[1] But it was not until the Chandra space observatory went into operation some 40 years later that this background was resolved into individual sources, viz., active galactic nuclei. Today, one of the key objectives of the Planck space mission is to discern the individual primordial galaxies at submillimetre wavelengths in the extragalactic background radiation (see Sect. 7.4.10).

Having noted that the property of coherence is closely related to the image formation process and image quality, it is natural to examine the loss of coherence occurring when light passes through inhomogeneous media, and in particular the

[1]The Italian physicist and astronomer Riccardo Giacconi, born in 1931, who later adopted US nationality, launched the first sounding rockets which discovered this radiation. For this work, he shared the 2002 Nobel Prize for Physics with Raymond Davis Jr. and Masatoshi Koshiba.

P. Léna et al., *Observational Astrophysics*, Astronomy and Astrophysics Library,
DOI 10.1007/978-3-642-21815-6_6,

Earth atmosphere, together with the harmful consequences of such losses for astronomical images (seeing). Here we shall discuss a convenient general formalism for treating such coherence losses.

The recent discovery, in the 1980s, of ways of limiting these effects by *adaptive optics* is well worth describing, since it has already contributed to many discoveries, and especially to the present and future development of very large Earth-based optical telescopes.

It was with this tool that twentieth century astronomers, in the tradition of ideas due to Fizeau[2] and Michelson,[3] were able to increase the angular resolution of their instruments almost at will by building *telescope arrays*, often called *interferometers*, able to reconstitute images, by means of *aperture synthesis*, with very high angular resolution, several orders of magnitude higher than the value λ/D given by a telescope of diameter D, which as we have seen could not exceed several tens or hundreds of metres.

As we have done for telescopes, we shall discuss the way this method is implemented for each region of the electromagnetic spectrum, from radiofrequencies to X-ray radiation. However, since γ-ray imaging methods are closely related to those used for detection and spectral analysis, we shall discuss γ-ray imaging in Chap. 7 on detectors.

The last section of this chapter deals with *coronagraphy* which, with the help of *apodisation* methods, provides a way of observing, under good conditions, images with a very broad dynamic range, i.e., exhibiting a significant range of fluxes (from 10^3 to 10^{10}). While it was traditionally used to study the solar corona, coronagraphy has made a comeback as a way of detecting extrasolar planets, which are faint objects and difficult to observe directly, being swamped by the radiation from the star they orbit so close to.

All questions relating to digital image *processing* will be postponed to the discussion of signal processing in Chap. 9. This is an area that has witnessed a considerable development with progress in computing.

6.1 Diffraction by an Arbitrary Aperture

Here we consider the most general case of diffraction of a light wave by an arbitrary aperture. After presenting the basic mathematical tool, which is simple but powerful and provides a way of treating the whole range of physical cases we are likely

[2]Armand Hippolyte Fizeau (1819–1896) was one of the great French physicists of the nineteenth century who demonstrated the power of the wave model of light. Apart from his paper in 1868, which serves as a foundation for astronomical interferometry, he is also famous for measuring the speed of light, and for discovering the spectral shift effect so often attributed only to Doppler.

[3]Albert Abraham Michelson was an American physicist (1852–1931) and brilliant experimentalist who devised the interferometer now named after him which is used in spectroscopy (see Sect. 8.3).

to encounter, we discuss the concept of coherent radiation which is essential to understanding image formation.

6.1.1 The Zernike Theorem

We now reconsider quasi-monochromatic radiation, introduced in Chap. 3. We have already shown that it can be regarded as coherent over a coherence length l_c along the wave vector $\boldsymbol{k}$. We now seek a rough answer to the following question: if this radiation illuminates an extended surface, and with various orientations of the wave vector, over what area and in which directions can it nevertheless be regarded as coherent? Figure 6.1 shows that a spatial displacement a, normal to $\boldsymbol{k}$ and associated with an angular displacement θ from $\boldsymbol{k}$, is acceptable provided that

$$a\theta \sim \lambda \, .$$

The *associated étendue* or *throughput*, which is the product of the beam area and its solid angle, is then

$$\omega \simeq a^2\theta^2 \simeq \lambda^2 \, .$$

The quantity λ^2 is called the *étendue of coherence* of quasi-monochromatic radiation of wavelength λ. The remainder of this section will be devoted to a more rigorous demonstration of this result.

Consider a source of area A_S, bounded by a closed curve and illuminating a screen (Fig. 6.2). Each point of the source is assumed to emit quasi-monochromatic radiation. Two points (atoms) of the source, separated by a distance much less than the wavelength λ, are assumed mutually incoherent. The aim is to determine, at two points 1 and 2 of the screen, the quantity measuring the correlation of the electric fields

$$\langle \mathbf{V}_1(t)\mathbf{V}_2^*(t) \rangle \, ,$$

where $\mathbf{V}_i(t) = \mathbf{V}(\boldsymbol{r}_i, t)$ denotes the electric field at the point $\boldsymbol{r}_i$, and the asterisk denotes complex conjugation.

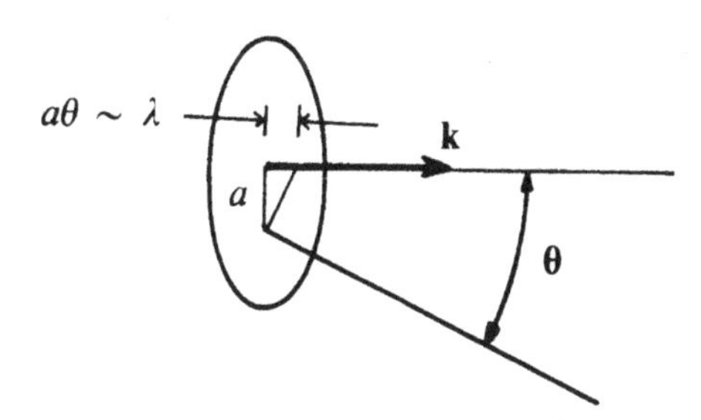

Fig. 6.1 Coherence in a beam of finite throughput

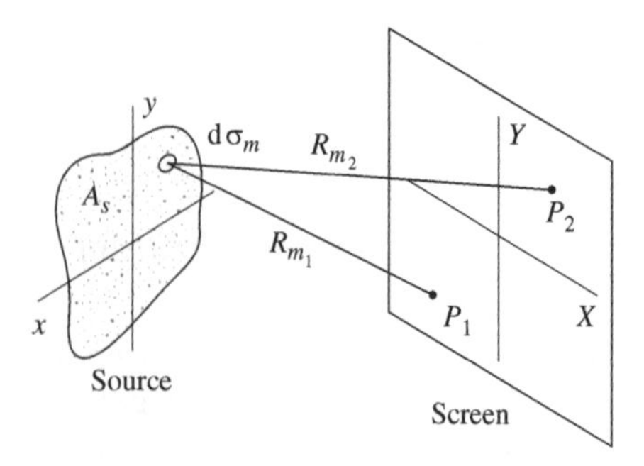

Fig. 6.2 Coherence of the field on a screen illuminated by a source

Decomposing A_S into elements $d\sigma_m$ of dimensions much less than λ, at distances R_{m_1} and R_{m_2} from points 1 and 2, it follows that

$$\begin{aligned}\langle \mathbf{V}_1(t)\mathbf{V}_2^*(t)\rangle &= \Big\langle \sum_m \mathbf{V}_{m_1}(t) \sum_j \mathbf{V}_{j_2}^*(t)\Big\rangle \\ &= \sum_m \langle \mathbf{V}_{m_1}(t)\mathbf{V}_{m_2}^*(t)\rangle + \sum_m \sum_{j\neq m} \Big\langle \mathbf{V}_{m_1}(t)\mathbf{V}_{j_2}^*(t)\Big\rangle ,\end{aligned}$$

where $\mathbf{V}_{m_1}$ represents the contribution of the element $d\sigma_m$ to $\mathbf{V}_1$. The second term has zero mean.

The complex field $\mathbf{V}_m(t)$ is given by

$$\mathbf{V}_m(t) = A_m\left(t - \frac{R_m}{c}\right)\frac{\exp\left[-2\pi i\nu_0(t - R_m/c)\right]}{R_m} ,$$

where $|A_m|$ is the amplitude and $\arg A_m$ the phase of the emission from $d\sigma_m$. By homogeneity, A_m is a complex vector. However, to simplify the notation, the vector notation is understood in the term A_m.

It now follows that

$$\langle \mathbf{V}_{m_1}(t)\mathbf{V}_{m_2}^*(t)\rangle = \left\langle A_{m_1}(t)A_{m_2}^*\left(t - \frac{R_{m_2} - R_{m_1}}{c}\right)\right\rangle \frac{\exp\left[2\pi i\nu_0(R_{m_1} - R_{m_2})/c\right]}{R_{m_1}R_{m_2}} ,$$

and, taking the stationarity of $\mathbf{V}$ into account,

$$\left\langle A_{m_1}(t)A_{m_2}^*\left(t - \frac{R_{m_2} - R_{m_1}}{c}\right)\right\rangle = \left\langle A_{m_1}\left(t - \frac{R_{m_1}}{c}\right)A_{m_2}^*\left(t - \frac{R_{m_2}}{c}\right)\right\rangle .$$

If $(R_{m_1} - R_{m_2})/c \ll \tau_c$ (temporal coherence), then

$$\langle \mathbf{V}_1(t)\mathbf{V}_2^*(t)\rangle = \sum_m \langle A_{m_1}(t)A_{m_2}^*(t)\rangle \frac{\exp\left[2\pi i\nu_0(R_{m_1} - R_{m_2})/c\right]}{R_{m_1}R_{m_2}} .$$

Characterising the intensity from $\mathrm{d}\sigma_m$ by

$$I(\boldsymbol{r}_m)\,\mathrm{d}\sigma_m = \langle A_m(t)A_m^*(t)\rangle ,$$

we now obtain

$$\langle \mathbf{V}_1(t)\mathbf{V}_2^*(t)\rangle = \int_{A_S} I(\boldsymbol{r})\frac{\exp\left[\mathrm{i}k(R_1-R_2)\right]}{R_1R_2}\,\mathrm{d}\boldsymbol{r} ,$$

where $k = 2\pi\nu_0/c$ and $\boldsymbol{r}$ denotes an arbitrary point of the source S, at distances R_1 and R_2 from the points 1 and 2. Finally, the complex degree of coherence can be written

$$\gamma_{12}(0) = \frac{1}{(\langle|V_1|^2\rangle\langle|V_2|^2\rangle)^{1/2}}\int_{\text{source}} I(\boldsymbol{r})\frac{\exp\left[\mathrm{i}k(R_1-R_2)\right]}{R_1R_2}\,\mathrm{d}\boldsymbol{r} .$$

This is the *Zernike–Van Cittert theorem.*[4]

Special Case: Large Distance from Source to Screen

The result of the theorem greatly simplifies in this case. Using the notation $\boldsymbol{r}(x,y)$, $P_1(X_1,Y_1)$, and $P_2(X_2,Y_2)$ in Fig. 6.2, and retaining only first order terms, with $R_1 \approx R_2 \approx R$,

$$R_1 - R_2 = \frac{(X_1^2+Y_1^2)-(X_2^2+Y_2^2)}{2R} \pm \frac{(X_1-X_2)x+(Y_1-Y_2)y}{R}$$

and

$$\gamma_{12}(0) = \exp\mathrm{i}k\left[(X_1^2+Y_1^2)-(X_2^2+Y_2^2)\right]^{1/2}$$
$$\times\frac{\iint I(x,y)\exp\mathrm{i}k\left[\dfrac{(X_1-X_2)x}{R}+\dfrac{(Y_1-Y_2)y}{R}\right]\mathrm{d}x\,\mathrm{d}y}{\iint I(x,y)\,\mathrm{d}x\,\mathrm{d}y} .$$

Introducing angular variables, assumed small,

$$\frac{x}{R} = \alpha , \qquad \frac{y}{R} = \beta , \qquad \boldsymbol{\theta} = (\alpha,\beta) ,$$

---

[4]The Dutch physicist Frederik Zernike (1888–1966) won the Nobel Prize for Physics in 1953 for his invention of the phase contrast microscope. The Dutch mathematician Van Cittert also discovered this theorem in 1931.

which describe the source as seen from the screen, and the reduced conjugate variables of the Fourier transform which bring in the wavelength, the simplified expression of the theorem becomes

$$|\gamma_{12}(0)| = \left| \frac{\iint I(\boldsymbol{\theta}) \exp\left\{-2\pi \mathrm{i} \left[\frac{(X_2 - X_1)\alpha}{\lambda} + \frac{(Y_2 - Y_1)\beta}{\lambda} \right]\right\} \mathrm{d}\boldsymbol{\theta}}{\iint I(\boldsymbol{\theta})\, \mathrm{d}\boldsymbol{\theta}} \right| .$$

In the present case, the Zernike–Van Cittert theorem can be stated as follows:

Theorem. *If the linear dimensions of a quasi-monochromatic radiation source and the distance between two points of the screen are both small compared with the distance between the source and the screen, the modulus of the complex degree of coherence is equal to the modulus of the spatial Fourier transform of the source intensity, normalised by the total intensity of the source.*

Special Case: Circular Source

Consider a circular source of radius r_0 and uniform intensity, illuminating a screen. We calculate the complex degree of coherence $\gamma_{12}(0)$. Let P_1 be the centre of the screen, P_2 a point at distance ρ from the centre, and J_1 the Bessel function of the first kind. Then it follows that

$$I(\boldsymbol{\theta}) = \Pi\left(\frac{r}{2r_0}\right) = \Pi\left(\frac{\theta}{2\theta_0}\right) , \qquad \theta_0 = \frac{r_0}{R} ,$$

$$|\gamma_{12}(0)| = |\gamma(\rho, 0)| = \left| \frac{J_1(2\pi\theta_0\rho/\lambda)}{\pi\theta_0\rho/\lambda} \right| = \frac{2|J_1(u)|}{u} .$$

Figure 6.3 shows the modulus $|\gamma|$ of the degree of coherence between the centre of the screen and an arbitrary point of the screen at distance ρ. The phase difference is unknown, but is irrelevant provided $\tau \ll \tau_c$. The coherence, which takes the value of one at the origin, by definition, decreases in an oscillatory fashion with the distance separating the two points.

6.1.2 Coherence Etendue

In the case of a point source at infinity, $r_0/R \to 0$, $\gamma \to 1$ over the whole screen. Hence, in this limit, the screen is illuminated by a plane quasi-monochromatic wave.

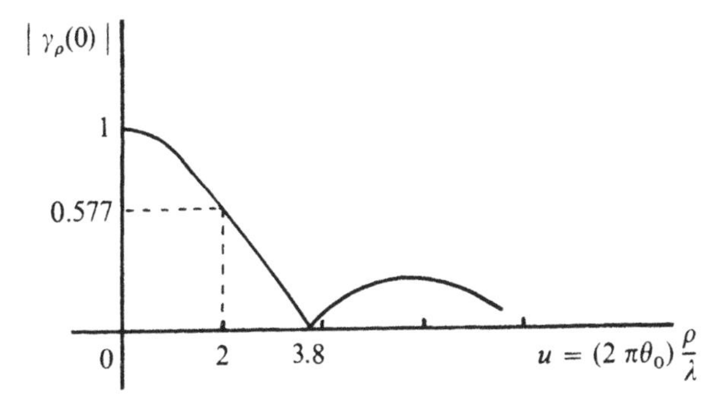

Fig. 6.3 Coherence of the field produced by a circular source of angular radius θ_0

If the source has a finite angular size of radius θ_0, it subtends a solid angle Ω at a point of the screen, and some area $S = \pi\rho^2$ of the screen corresponds to a *beam étendue* $\mathcal{E}$ given by

$$\mathcal{E} = S\Omega = \pi\rho^2\pi\theta_0^2 = \frac{\lambda^2}{4}u^2 .$$

Choosing $u = 2$ arbitrarily, so that

$$\left|\gamma\left(\rho = \frac{\lambda}{\pi\theta_0}\right)\right| = \frac{2J_1(2)}{2} = 0.577 ,$$

it can be deduced that the coherence on the screen remains significant ($|\gamma| > 0.577$) in the étendue

$$\mathcal{E} = \lambda^2 .$$

This is a quantitative demonstration of the rather intuitive argument given earlier. The fundamental result is:

The coherence étendue (or throughput) of quasi-monochromatic radiation is λ^2.

Coherence of the Radiation Received from a Star

Consider a red giant star, of radius $r_0 = 1.5 \times 10^{11}$ m, at a distance of 10 pc (1 pc = 3×10^{16} m). For this object, $\theta_0 = 5 \times 10^{-7}$ rad = 0.1 arcsec. If the star is observed at $\lambda = 0.5$ μm, the value of the coherence radius ρ, on a screen normal to the rays, on Earth, is

$$\rho = \frac{\lambda}{\pi\theta_0} = 32 \text{ cm} .$$

In the infrared, at $\lambda = 25$ μm, the radius ρ is increased fifty-fold, an observation which will be put to good use later on.

The received radiation is thus coherent, in the sense described above, within a circle of radius $\rho(\lambda)$. Clearly, this is no longer exact if the radiation is compared at two points within the circle at times differing by Δt, such that $\Delta t \gg \tau_c = 1/\Delta\nu$.

The surface of the star plays the part of the plane source. The fact that the star is spherical simply modifies the phase of the elements $d\sigma_m$, and this has no effect on the result.

Reception of Radiofrequencies

A radiofrequency receiver superposes the fields emitted by different parts of the source (Ω) and received at different points (S). This superposition can only be carried out in a coherent way, producing constructive interference and hence a significant field amplitude, if $S\Omega \lesssim \lambda^2$.

6.1.3 *Diffraction at Infinity*

Having received the wave at the surface of the screen, we must now examine the consequences of the fact that only a finite part of the wavefront can be analysed in the receiving system. This is indeed what happens at the telescope pupil when the incident wave (generally, plane) encounters the primary mirror, i.e., the mirror truncates the wave when it reflects it.

Fraunhofer Diffraction

Consider a purely monochromatic wave, emitted at P_0, and received at P_1, after diffraction by a screen $\mathscr{A}$ (Fig. 6.4). A calculation very similar to the previous one

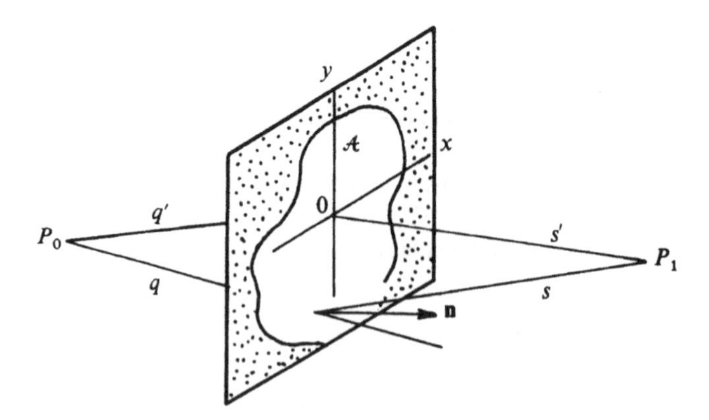

Fig. 6.4 Fraunhofer diffraction

leads to the *Kirchoff–Fresnel relation* or *integral*,[5] giving the complex amplitude of the field at P_1

$$\mathbf{V}_1(t) = \frac{\mathrm{i}}{2\lambda}\mathrm{e}^{2\pi\mathrm{i}\nu t}\iint A(\boldsymbol{r})\mathrm{e}^{\mathrm{i}\phi(\boldsymbol{r})}\frac{\exp\left[\mathrm{i}k(q+s)\right]}{qs}\left[\cos(\boldsymbol{n},\boldsymbol{q})-\cos(\boldsymbol{n},\boldsymbol{s})\right]\mathrm{d}\boldsymbol{r}\ ,$$

where the notation is specified in Fig. 6.4, with $A(\boldsymbol{r})$ the field amplitude at the screen, which can be taken as constant, and $\phi(\boldsymbol{r})$ represents any phase change created by the screen (referred to as a phase mask).[6] In general, $\phi = 0$.

Diffraction observed at a finite distance from the diffracting object is called *Fresnel diffraction*,[7] and diffraction observed at an infinite distance is called *Fraunhofer diffraction*.[8]

We now consider *diffraction at infinity*, where the linear dimensions of the screen are small compared with (q, s). Introducing angles $\boldsymbol{\theta}_0 = (x_0/p,\ y_0/q)$, $\boldsymbol{\theta}_1 = (x_1/s,\ y_1/s)$, and putting $\boldsymbol{r} = (x, y)$, the amplitude of the diffracted field in direction $\boldsymbol{\theta}$ can be expressed as

$$\mathbf{V}_1(\boldsymbol{\theta}_1) = C\iint G\left(\frac{\boldsymbol{r}}{\lambda}\right)A\left(\frac{\boldsymbol{r}}{\lambda}\right)\exp\left[-2\pi\mathrm{i}\,(\boldsymbol{\theta}_1-\boldsymbol{\theta}_0)\cdot\frac{\boldsymbol{r}}{\lambda}\right]\frac{\mathrm{d}\boldsymbol{r}}{\lambda^2}\ ,$$

where C is a constant. $\boldsymbol{r}/\lambda$ and $\boldsymbol{\theta}$ appear as conjugate Fourier variables and $G(\boldsymbol{r}/\lambda)$ denotes a *pupil function*, satisfying

$$G(\boldsymbol{r}) = \begin{cases} 1 \text{ inside } \mathscr{A}\ , \\ 0 \text{ outside } \mathscr{A}\ . \end{cases}$$

The generality of the notion of pupil function makes it invaluable. If the pupil introduced a phase difference $\phi(\boldsymbol{r})$ at the point $\boldsymbol{r}$, this could be represented by putting $G(\boldsymbol{r}) = \exp(\mathrm{i}\phi(\boldsymbol{r}))$, rather than $G = 1$, on $\mathscr{A}$. This property is used for apodisation of a pupil (see Sect. 6.6).

In order to determine C, we use the fact that the energy E coming from P_0 and passing through the area $\mathscr{A}$ of the pupil is conserved across the diffraction pattern, so that

$$\iint |\mathbf{V}_1(\boldsymbol{\theta}_1)|^2\,\mathrm{d}\boldsymbol{\theta}_1 = E\ .$$

Given the above Fourier transform, Parseval's theorem can be written

$$\iint |\mathbf{V}_1(\boldsymbol{\theta}_1)|^2\,\mathrm{d}\boldsymbol{\theta}_1 = |C|^2\iint\left|G\left(\frac{\boldsymbol{r}}{\lambda}\right)\right|^2\frac{\mathrm{d}\boldsymbol{r}}{\lambda^2}\ ,$$

[5]The German physicist Gustav Kirchhoff (1824–1887) founded spectral analysis in astronomy.

[6]Detailed demonstrations of the results given here can be found in Born M., Wolf E., *Principles of Optics*, Pergamon (1980), and also Françon M., *Optique*, and Hecht E., *Optique*, Pearson (2005).

[7]The French physicist Augustin Fresnel (1788–1827) was one of the founders of wave optics.

[8]The German optician Joseph-Franz Fraunhofer (1727–1826) built the first spectroscope. With this instrument he was able to study the Sun's spectrum, where he discovered the absorption lines that carry his name.

and this gives

$$E = |C|^2 \frac{\mathscr{A}}{\lambda^2} .$$

C is determined and the final result is

$$\mathbf{V}_1(\boldsymbol{\theta}_1, t) = \lambda \left(\frac{E}{\mathscr{A}} \right)^{1/2} \iint_{\text{screen}} G\left(\frac{\boldsymbol{r}}{\lambda} \right) \exp\left[-2\pi \mathrm{i} (\boldsymbol{\theta}_1 - \boldsymbol{\theta}_0) \cdot \frac{\boldsymbol{r}}{\lambda} \right] \frac{\mathrm{d}\boldsymbol{r}}{\lambda^2} .$$

This equation is the fundamental result for diffraction at infinity, and forms the basis of the following discussion. It can be stated as:

Theorem. *When a screen is illuminated by a source at infinity, the amplitude of the field diffracted in any direction is the Fourier transform of the pupil function characterising the screen, the conjugate variables being the angular direction and the reduced coordinates* $\mathbf{r}/\lambda$ *on the screen.*

Note. If a lens with no aberrations is placed behind the screen, all the rays of the same $\boldsymbol{\theta}_1$ converge at the same point of the focal plane of the lens, and the Fourier transform is moved back to a finite distance in this focal plane. It is clearly more convenient, and more common, to observe or measure images in a plane situated at a finite distance.

Relationship Between Image and Object

Consider an object at infinity made up of a set of point sources P_0, emitting quasi-monochromatic radiation, and illuminating a screen (Fig. 6.5). A contour C determines the region of the screen, called the *pupil*, letting the wave through. Each point of the image is characterised by a direction $\boldsymbol{\theta}$, or, if a lens of focal length f and without aberration forms this image on its focal plane, by a position $\boldsymbol{R} = f\boldsymbol{\theta}$ in this image plane. In elementary geometrical optics, there is a one-to-one correspondence between the points of the object and those of the image, and hence the name of the latter.

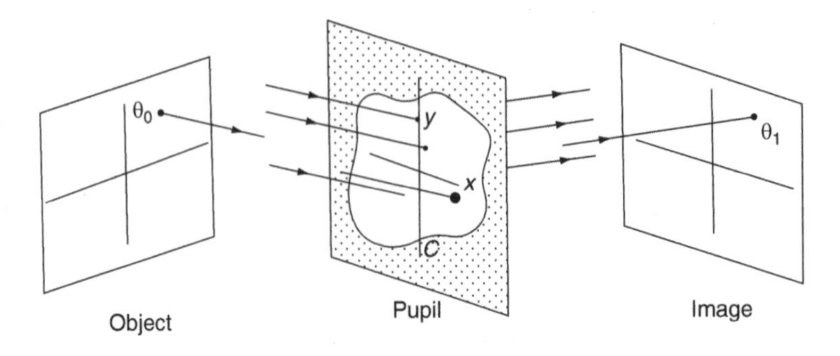

Fig. 6.5 Relationship between image and object

Denote by $K(\boldsymbol{\theta}_0\,;\,\boldsymbol{\theta}_1)$ the transmission of the system, that is, the complex amplitude per unit solid angle round $\boldsymbol{\theta}_1$, obtained in the image for unit amplitude and zero phase at the point $\boldsymbol{\theta}_0$ of the object. Denote by V the complex field amplitude, dropping the time dependence to simplify notation.

Coherent Sources

Consider first a point source at $\boldsymbol{\theta}'_0$, so that

$$V(\boldsymbol{\theta}_1) = K(\boldsymbol{\theta}'_0\,;\,\boldsymbol{\theta}_1)\,, \qquad V(\boldsymbol{\theta}_0) = \delta(\boldsymbol{\theta}_0 - \boldsymbol{\theta}'_0)\,.$$

Then $K(\boldsymbol{\theta}'_0\,;\,\boldsymbol{\theta}_1)$ is the response function of the system, and the image brightness is

$$|V(\boldsymbol{\theta}_1)|^2 = |K|^2\,.$$

If the optical system between object and image planes is perfect (without aberration), then

$$K(\boldsymbol{\theta}'_0\,;\,\boldsymbol{\theta}_1) = K(\boldsymbol{\theta}_1 - \boldsymbol{\theta}'_0)\,,$$

and the image of an off-axis source can be found by translation of the image of a source on the axis (*isoplanicity*). Even in an imperfect system which has been reasonably well corrected for geometrical aberrations, this condition is usually satisfied in some small neighbourhood of a point (*domain of isoplanicity*), to which we can limit ourselves.

The image of an extended object is obtained by linear superposition

$$V(\boldsymbol{\theta}_1) = \iint_{\text{object}} V_0(\boldsymbol{\theta}_0) K(\boldsymbol{\theta}_1 - \boldsymbol{\theta}_0)\,\mathrm{d}\boldsymbol{\theta}_0\,,$$

where V_0 denotes the complex amplitude emitted by the point $\boldsymbol{\theta}_0$ of the object. Then the expression obtained for the diffraction at infinity gives

$$V(\boldsymbol{\theta}_1) = \iint V_0(\boldsymbol{\theta}_0) K(\boldsymbol{\theta}_1 - \boldsymbol{\theta}_0)\,\mathrm{d}\boldsymbol{\theta}_0\,,$$

$$K(\boldsymbol{\theta}) = \iint G(\boldsymbol{r}) \exp\left(-2\pi\mathrm{i}\frac{\boldsymbol{r}}{\lambda}\cdot\theta\right) \frac{\mathrm{d}\boldsymbol{r}}{\lambda^2}\,,$$

where $G(\boldsymbol{r})$ is the pupil function.

This is a convolution equation, which can be conveniently expressed in the Fourier space, where the variable is the spatial frequency $\boldsymbol{w}$. Setting

$$\tilde{V}_0(\boldsymbol{w}) = \iint_{\text{object}} V_0(\boldsymbol{\theta}_0) \exp(-2\pi\mathrm{i}\boldsymbol{\theta}_0\cdot\boldsymbol{w})\,\mathrm{d}\boldsymbol{\theta}_0\,,$$

and similarly

$$\tilde{V}(\boldsymbol{w}) = \mathrm{FT}\,[V(\boldsymbol{\theta}_1)] \;, \qquad \tilde{K} = \mathrm{FT}\,[K] \;,$$

the convolution equation becomes

$$\tilde{V} = \tilde{V}_0 \tilde{K} = \tilde{V}_0 G \;.$$

The complex quantity $\tilde{K}$ is the modulation transfer function (MTF) of the amplitude for coherent illumination.

The variable $\boldsymbol{w}$ is dimensionless and conjugate to the angular variable $\boldsymbol{\theta}$ (rad). It can thus be expressed in rad^{-1}, and considered as a spatial frequency. The amplitude of $\tilde{V}_0(\boldsymbol{w})$ expresses the effect of the component of frequency $\boldsymbol{w}$ in the object, whilst the phase of $\tilde{V}_0(\boldsymbol{w})$ expresses the translated position of this component in the image.

The relation simply asserts that the complex amplitude of a Fourier component of the image can be found from that of the object by multiplication by the filter $\tilde{K}$, which thus acts as a *spatial filter*. Note that K is itself the Fourier transform of the pupil function, expressed in reduced coordinates $\boldsymbol{r}/\lambda$. These results are summarised in the example shown in Fig. 6.6.

Incoherent Sources

This is, of course, the more common case, since extended sources of radiation in astrophysics, whether they be thermal or non-thermal, do not manifest large scale spatial coherence.

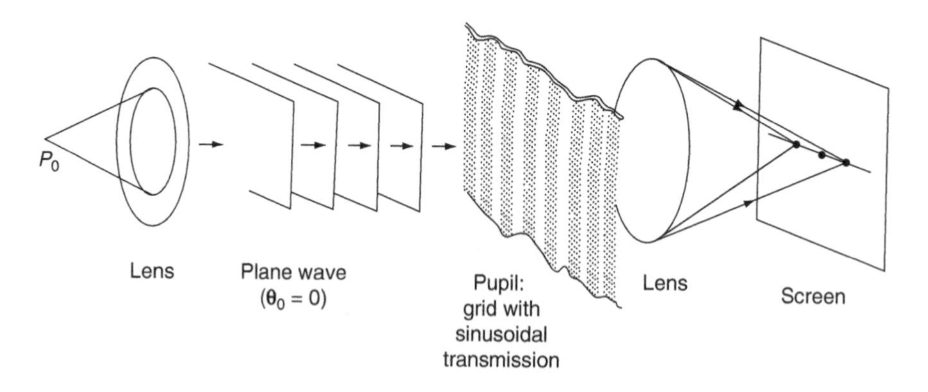

$$G(\mathbf{r}) = 1 + \sin\left(2\pi \frac{x}{a}\right)$$

Fig. 6.6 Spatial frequencies of a grid. The grid, with sinusoidal transmission $G(\boldsymbol{r})$, is illuminated by a plane monochromatic wave. Its diffraction pattern at infinity is brought to the focal plane of a lens. It has three Dirac spikes, the autocorrelation function of the Fourier transform of $G(\boldsymbol{r})$, convoluted by the point source response associated with the finite size of the lens

The method is essentially the same, except that intensities, rather than amplitudes, are added in this case. If there is isoplanicity, it follows that

$$I(\boldsymbol{\theta}_1) = \iint_{\text{object}} I_0(\boldsymbol{\theta}_0)|K(\boldsymbol{\theta}_1 - \boldsymbol{\theta}_0)|^2 \, \mathrm{d}\boldsymbol{\theta}_0 \, .$$

With the same notations for the Fourier transforms (Appendix A), we obtain

$$\tilde{I}(\boldsymbol{w}) = \tilde{I}_0(\boldsymbol{w})\tilde{H}(\boldsymbol{w}) \, ,$$

where $\tilde{I}(\boldsymbol{w})$ and $\tilde{I}_0(\boldsymbol{w})$ are the Fourier transforms of the image brightness and of the object intensity, respectively. Denoting convolution by $\star$,

$$\tilde{H}(\boldsymbol{w}) = \mathrm{FT}\left[|K|^2\right] = \mathrm{FT}\left[KK^*\right] = G(\lambda\boldsymbol{w}) \star G^*(-\lambda\boldsymbol{w}) \, .$$

Using the result obtained here, the image brightness can be found from the object and pupil structures, when diffraction is the only phenomenon to be taken into account.

6.1.4 *Spatial Filtering by a Pupil*

The image intensity manifests a spectrum of spatial frequencies, resulting from that of the object by a transformation referred to as *linear filtering*. This filter only depends on the pupil function $G(\boldsymbol{r}/\lambda)$ (which is complex, in general), expressed as a function of the reduced space variable $\boldsymbol{w} = \boldsymbol{r}/\lambda$.

Modulation Transfer Function

We have shown that diffraction by a finite sized pupil amounts to a spatial filtering of the object.

In the special (and common) case of a centrally symmetric pupil, the autoconvolution is just the autocorrelation, and we obtain

$$\tilde{H}(\boldsymbol{w}) = \iint_{\text{pupil plane}} G(\lambda\boldsymbol{w} + \boldsymbol{r})G^*(\boldsymbol{r}) \frac{\mathrm{d}\boldsymbol{r}}{\lambda^2} \, .$$

Normalising to the area of the pupil, in the same units, we obtain

$$\tilde{T}(\boldsymbol{w}) = \frac{\tilde{H}(\boldsymbol{w})}{\displaystyle\iint G(\boldsymbol{r})G^*(\boldsymbol{r}) \, \mathrm{d}\boldsymbol{r}/\lambda^2} \, .$$

The function $\tilde{T}(\boldsymbol{w})$ is called the *intensity modulation transfer function* of the system. It is often simply denoted MTF, the context removing any risk of confusion with the amplitude MTF.

The spatial frequency plane, over which ranges $\boldsymbol{w} = (u, v)$, is often referred to as the (u, v) plane. The function $|K|^2 = H(\boldsymbol{\theta})$ is called the *point source response* of the system, or the *point spread function* (PSF). Clearly, $H(\boldsymbol{\theta})$ depends on the shape of the pupil. There may be sidelobes, of greater or lesser importance, reducing the energy concentration in the central regions of the image. A frequently used approximation, when $H(\boldsymbol{\theta})$ is sufficiently compact and has circular symmetry, is to characterise it by a *half power beam width* (HPBW), i.e., the angular size $\Delta\theta$ within which half of the beam energy resides. This gives an idea of the order of magnitude of the angular resolution.

Example. Referring to the arrangement shown in Fig. 6.6, if the modulation of the grid G is limited to a pupil of finite width a (a slit pupil), it follows that

$$G(\boldsymbol{r}) = \Pi\left(\frac{x}{a}\right), \quad G \star G^* = \Lambda\left(\frac{x}{a}\right), \quad \boldsymbol{r} = (x, y),$$

$$\tilde{T}(\boldsymbol{w}) = \Lambda\left(\lambda\frac{u}{a}\right), \quad \boldsymbol{w} = (u, v),$$

$$H(\boldsymbol{\theta}) = \mathrm{sinc}^2\left(\frac{a}{\lambda}\theta_x\right).$$

Image Sampling

An important consequence of the filtering theorem just established is the following: since all physical pupils have a finite size in the pupil plane, the function T must have *bounded support*. In every direction of the $\boldsymbol{w}$ plane, there is a *cutoff frequency* $w_{\mathrm{c}} = (u_{\mathrm{c}}^2 + v_{\mathrm{c}}^2)^{1/2}$. The pupil acts as a *low-pass filter* on the spatial frequencies of the object. All other physical dimensions being equal, the spatial frequency cutoff w_{c} is lower as the wavelength is higher.

Since the FT of $I(\boldsymbol{\theta})$ has bounded support, $I(\boldsymbol{\theta})$ can be completely determined by a discrete sampling of the $\boldsymbol{\theta}$ plane. If w_{c} is the largest spatial frequency contained in $I(\boldsymbol{\theta})$, Shannon's theorem (Sect. 9.1) shows that I can be sampled with a rate at least equal to $\Delta\boldsymbol{\theta} = 1/2w_{\mathrm{c}}$. This does not mean that the support of $I(\boldsymbol{\theta})$ is bounded. Nevertheless, in practice, if $I_0(\boldsymbol{\theta})$ has bounded support, then so does $I(\boldsymbol{\theta})$, and a finite number of discrete sampling points will suffice.

Note. Even if $I(\boldsymbol{w})$ contains no information about the object for $w > w_{\mathrm{c}}$, it can nevertheless include noise at frequencies greater than w_{c}. There is no reason why the grain of a photographic plate should have the same cutoff frequency as the optical system which produced the image on the plate.

Circular Pupils

Circular pupils play such an important role in astronomy that they deserve description. Let r_0 be the radius of the pupil, which is used for quasi-monochromatic light of wavelength λ. Then

$$G(\boldsymbol{r}) = \Pi\left(\frac{r}{2r_0}\right) .$$

The pupil is symmetric, hence the MTF is real, the autoconvolution is just the autocorrelation, and

$$G(\boldsymbol{r}) \star G(\boldsymbol{r}) = \pi r_0^2 \left[\frac{2}{\pi} \arccos\left(\frac{r}{2r_0}\right) - \frac{r}{r_0}\left(1 - \frac{r^2}{4r_0^2}\right)^{1/2}\right] .$$

The MTF is

$$\tilde{T}(w) = \frac{2}{\pi}\left[\arccos\left(\frac{\lambda w}{2r_0}\right) - \frac{\lambda w}{r_0}\left(1 - \frac{\lambda^2 w^2}{4r_0^2}\right)^{1/2}\right] ,$$

and is shown in Fig. 6.7. Note the cutoff frequency $w_c = 2r_0/\lambda$, the circular symmetry, and the gradual attenuation with increasing frequency.

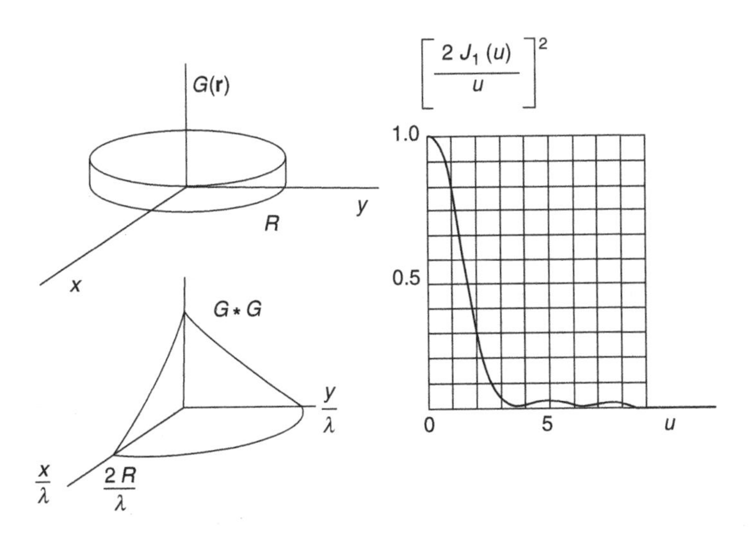

Fig. 6.7 Modulation transfer function of a circular pupil (Airy). The reduced variable $u = 2\pi r_0 \theta/\lambda$ has been used with $R = r_0$

The *point source response* is the image intensity distribution resulting when the pupil is illuminated by a point source. If J_1 denotes the Bessel function of the first kind,

$$I_0(\boldsymbol{\theta}) = \delta(\boldsymbol{\theta}) , \qquad I_1(\boldsymbol{\theta}) = \left[\frac{2J_1(2\pi r_0\theta/\lambda)}{2\pi r_0\theta/\lambda} \right]^2 .$$

This function is often called the *Airy function*[9] and has a ring-like structure. A telescope of diameter 6 m, used at $\lambda = 0.5\ \mu\mathrm{m}$, has cutoff $w_c = 60\ \mathrm{arcsec}^{-1}$, whereas a radiotelescope of diameter 100 m, used at $\lambda = 18$ cm (OH radical line), will cut off spatial frequencies beyond the much lower frequency of $w_c = 3 \times 10^{-3}\ \mathrm{arcsec}^{-1}$.

Most telescopes include a central secondary mirror of diameter D' which partially obstructs the pupil of diameter D. The pupil function is, in this case,

$$G(\boldsymbol{r}) = \begin{cases} 0 \text{ for } & r < \dfrac{D'}{2} \text{ and } r > \dfrac{D}{2} , \\[2ex] 1 \text{ for } & r \in \left[\dfrac{D'}{2}, \dfrac{D}{2} \right] . \end{cases}$$

It is then straightforward to calculate the new expressions for the MTF and the point source response.

The Rayleigh Resolution Criterion

The image of two identical point sources with separation θ is the incoherent superposition of two identical Airy functions. The limit at which the two sources are separated or resolved is arbitrarily fixed by the condition

$$\theta > \theta_0 = 0.61 \frac{\lambda}{r_0} ,$$

the value for which the maximum of one Airy function coincides with the first zero of the other. This angular value is often used to characterise the *spatial resolving power* or discriminating power of a pupil. This so-called Rayleigh criterion[10] is more approximate than the one given by the MTF.

In certain cases, it is possible to resolve two point sources closer than θ_0. Indeed, knowing beforehand that two objects are unresolved, for example, the two components of a double star, and that the measurement is made with an excellent

---

[9] Sir George Biddell Airy (1801–1892) was a British astronomer and physicist.

[10] John Strutt (Lord Rayleigh, 1842–1919) was awarded the Nobel Prize for Physics in 1904 for the discovery of argon.

signal-to-noise ratio ($\gtrsim$100), then the profile $I_1(\boldsymbol{\theta})$ will differ measurably from the profile of a single source, even if the separation of the two components is less than θ_0.

This remark illustrates the importance of knowing a priori some of the properties of an object, when reconstructing that object from an image (see Sect. 9.6).

Apodisation

The intensity distribution in the point source response can be modified by influencing the pupil function $G(\boldsymbol{r})$, which directly determines it. If $G(\boldsymbol{r})$ is replaced, in the region where it equals unity, by a complex transmission $\exp \mathrm{i}\phi(\boldsymbol{r})$, which does not affect the amplitude but simply introduces a position-dependent phase change (a *phase mask*), a judicious choice of the function $\phi(\boldsymbol{r})$ can significantly reduce the secondary lobes of the PSF (the 'feet' of the diffraction). But this is at the expense of broadening the central lobe, energy being conserved. This operation is referred to as *apodisation* (see Problem 6.1), discussed further in Sect. 6.6.

An Overview of Pupil Diffraction

The various Fourier correspondences between quantities describing image formation are summarised in Table 6.1.

Lunar Occultations. There is one particularly interesting case of diffraction which leads to a high angular resolution and is not related to the size of the telescope. This occurs if an astronomical source is occulted by the lunar limb, when the proper motion of the Moon causes it to pass in front of the source. If it is a point source, the monochromatic wave, diffracted at infinity, produces fringes (screen edge fringes), observable as a variation in the intensity $I_0(t)$ received over time by a telescope. If the source is extended, $I(t)$ results from the convolution of $I_0(t)$ with the angular profile of the source (integrated in the direction parallel to the edge), in the direction normal to the lunar limb (the time dependence being obtained from the speed at which the Moon scans

Table 6.1 Quantities related to image formation. Corresponding entries in *left-* and *right-hand columns* are Fourier pairs

Source $O(\boldsymbol{\theta})$	Spatial spectrum $\tilde{O}(\boldsymbol{w})$
Image $I(\boldsymbol{\theta})$	Spatial spectrum $\tilde{I}(\boldsymbol{w})$
Point spread function PSF($\boldsymbol{\theta}$)	Modulation transfer function MTF($\boldsymbol{w}$)
$I(\boldsymbol{\theta}) = O(\boldsymbol{\theta}) \star \mathrm{PSF}(\boldsymbol{\theta})$	$\tilde{I}(\boldsymbol{w}) = \tilde{O}(\boldsymbol{w}) \cdot \mathrm{MTF}(\boldsymbol{w})$
	Pupil function $G(\boldsymbol{r})$ (spatial units)
	Pupil $G(\lambda\boldsymbol{w})$ (spatial frequency units)
	Autoconvolution $G(\lambda\boldsymbol{w}) \star G(-\lambda\boldsymbol{w})$ (pupil not centrally symmetric)
	Autocorrelation $G(\lambda\boldsymbol{w}) \otimes G(\lambda\boldsymbol{w})$ (pupil centrally symmetric)
PSF($\boldsymbol{\theta}$)	FTM($\boldsymbol{w}$) normalised autoconvolution

across). The method can be used to reconstruct the angular profile in one dimension, and with high resolution. The latter is determined by the Fresnel scale of the screen edge diffraction, i.e., $(\lambda/2D_{\text{Earth-Moon}})^{1/2}$, ranging from a few arcsec at radiofrequencies to a few mas (10^{-3} arcsec) in the near infrared. Before the advent of astronomical interferometry, lunar occultations were used to determine the size of radiosources and stellar envelopes in the infrared, e.g., the galactic object IRC+10216.

Disconnected Pupils and Bandpass Frequency Filtering

In the last section, we described a formalism for determining the MTF and the intensity distribution in the image given by an optical system with entrance pupil of arbitrary shape. These general results were applied to the special case of a circular pupil, for which we may thus determine the limiting angular resolution.

We now examine another case of major importance in astronomy, viz., the case where the pupil comprises *two* circular apertures, often called *sub-pupils*, of diameter d, placed at a center-to-center distance D from one another, where $D > d$, so that the pupil is disconnected.

The MTF of this pupil (Fig. 6.8) is given simply by the quantity $G(\boldsymbol{r}) \otimes G(\boldsymbol{r})$, after normalisation. This function $\tilde{T}(\boldsymbol{w})$ behaves as a spatial filter, which allows through frequencies close to zero and also a band of frequencies, narrow if $D \gg d$, centred on the frequency D/λ, of maximum width $2d/\lambda$. The point source response is the Fourier transform of $G \otimes G$, that is, a sinusoidal modulation of the intensity,

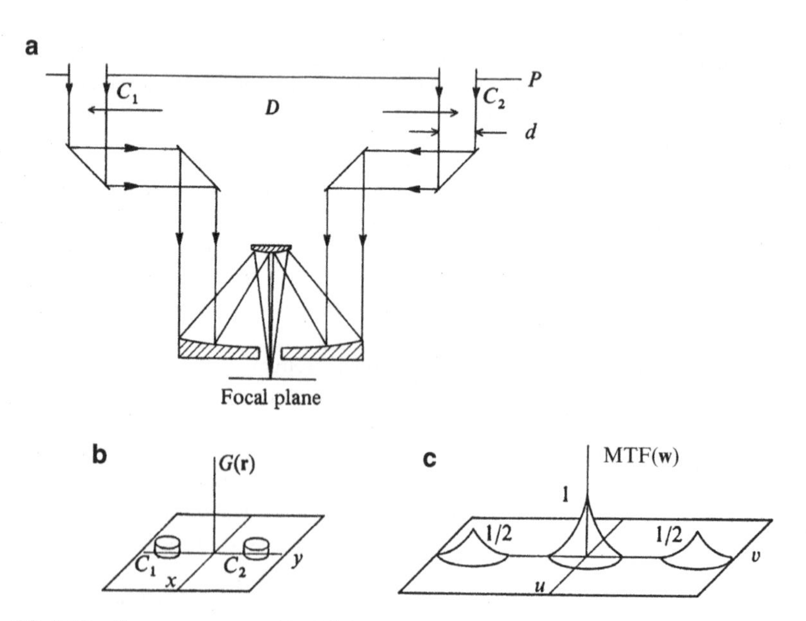

Fig. 6.8 (**a**) Pupil arrangement used by Michelson and Pease on the 2.5 m Mount Wilson telescope (California). (**b**) The pupil function $G(\boldsymbol{r})$. (**c**) $\tilde{T}(\boldsymbol{w}) = G \otimes G$, MTF for this pupil

having the Airy function of an individual pupil as envelope (see Fig. 6.8c). In other words, the image of a point source placed at infinity through an optical system with this entrance pupil will be a series of narrow fringes which modulate the intensity of a more spread out Airy function.

If the object is now given by the intensity distribution $I(\boldsymbol{\theta})$ with spatial spectrum $\widetilde{I}_0(\boldsymbol{w})$, the spatial spectrum of the image is simply given by the filtering operation

$$\widetilde{I}(\boldsymbol{w}) = \widetilde{T}(\boldsymbol{w})\widetilde{I}_0(\boldsymbol{w}) \ .$$

The Fourier transform of $\widetilde{I}(\boldsymbol{w})$ is the intensity distribution $I(\boldsymbol{\theta})$ in the final image.

Note that the observed quantity $I(\boldsymbol{\theta})$ describes an intensity modulation (fringes), with *visibility* $\widetilde{I}_0(\boldsymbol{w})/\widetilde{I}_0(\boldsymbol{0})$. Note also that this visibility $\mathscr{V}(\boldsymbol{w})$ is a complex quantity. The modulation factor of the fringes is given by the modulus $|\mathscr{V}|$ of this complex quantity and is equal to unity for a point source. We may thus say that this pupil, acting as a filter, measures the spatial coherence of the incident wave between two points separated by a displacement $\boldsymbol{D}$.

Aperture Synthesis

For each value of the vector $\boldsymbol{D}$ separating the apertures (distance and direction), the pupil behaves as a *bandpass spatial filter* which samples the $\boldsymbol{w}(u, v)$ plane of spatial frequencies. It is no longer the image which is sampled point by point, but rather its Fourier transform, with each value of $\boldsymbol{D}$ leading to a value of $\tilde{I}(\boldsymbol{w} = \boldsymbol{D}/\lambda)$. In contrast to the case of a circular pupil, which transmits all spatial frequencies with varying degrees of attenuation and thereby produces an image that 'looks like' the object, this no longer happens in the present case, where the spatial filtering due to the pupil is much more radical.

The usual notions of angular resolution and field in the image space can be transposed to the frequency space. The highest frequencies are $|\boldsymbol{w}| = D_{\text{max}}/\lambda$, where D_{max} denotes the maximal separation of the sub-pupils, and the frequency width of a frequency information element is of the order of $\delta w = 2d/\lambda$.

Supposing that all values of $\boldsymbol{w}$ between 0 and D_{max}/λ have been sampled, the image can be calculated by

$$I(\boldsymbol{\theta}) = \text{FT}\left[\tilde{I}(\boldsymbol{w})\right] ,$$

$\tilde{I}(\boldsymbol{w})$ only being known on the bounded support $|\boldsymbol{w}| < D_{\text{max}}/\lambda$. All the information in the image that would have been provided by an aperture of diameter D_{max} has thus been reconstructed, and this reconstruction of a large pupil from pairs of smaller pupils is referred to as *aperture synthesis*. We shall discuss the many applications of this later.

Note that, just as it is sufficient to sample an image at a finite number of points when its spatial frequency content is bounded (*Shannon's theorem*, see Sect. 9.1.3), so it is sufficient in this case to sample $\tilde{I}(\boldsymbol{w})$ at a discrete set of frequencies $\boldsymbol{w}$.

Instead of combining the radiation in an image plane common to the two telescopes, it is also possible to coherently superpose the two pupils, a method referred to as *pupil plane interferometry*. From the point of view of information content, the two methods are strictly equivalent. All information entering the pupil of each telescope is contained in the image, and vice versa. Choosing between one or the other way of analysing coherence is rather a question of experimental convenience, detector layout, and finding the best way to handle image degradation due to atmospheric turbulence (see Sect. 6.4).

6.2 The Earth Atmosphere and Coherence Losses

The study of image formation reveals the importance of spatial coherence in this analysis. What happens to the image of a point source at infinity (the simplest case) when the coherence of the incident plane wave is modified by phase perturbations? We shall address this question in the present section, accounting for perturbing phenomena produced by crossing the Earth's atmosphere. As discussed in Chap. 2, the atmosphere has fluctuating refractive index, due either to non-uniformities in the temperature or the electron density, depending on the wavelength.

Let us examine these effects empirically in the particular case of visible and near infrared wavelengths. There are several different effects (Fig. 6.9). Firstly, *scintillation*, the variation in brightness observed by the eye, or a very small diameter pupil, which corresponds to the spreading or concentration of the wavefront energy. Note immediately that scintillation is only observed for stars (almost point sources), and not for bright planets, which are extended sources (of the order of 10 arcsec at the most). Secondly, *agitation* of the image in the focal plane of the instrument, which corresponds to a temporal variation in the angle of the mean tangent plane to the wavefront, an angle which determines the position of the centre of the image. And thirdly, *smearing* of the image, which leads to image sizes much greater (of the

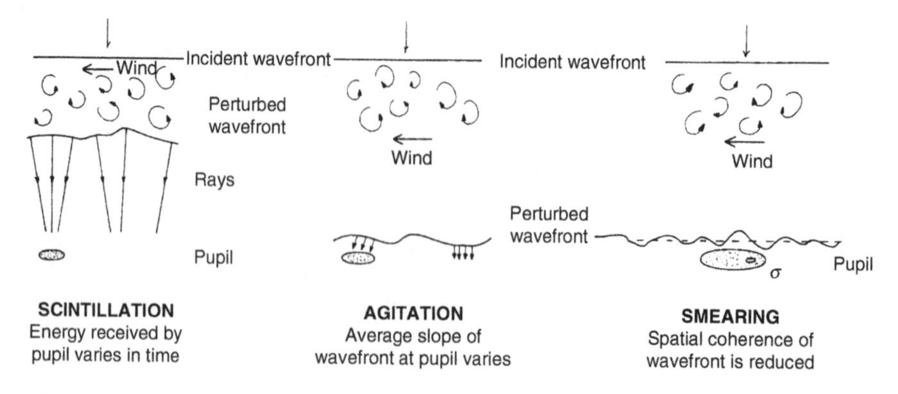

Fig. 6.9 The main effects of atmospheric turbulence on images. Shown schematically are the incident plane wave, the turbulent atmospheric region, the light rays normal to the perturbed wavefront, and the pupil (*shaded*). Also shown is the reduced coherence area σ

order of 1 arcsec) than would result from diffraction alone, caused by loss of spatial coherence at the entrance pupil. It is this smearing effect that is largely responsible for the loss of angular resolution.

The formalism required to interpret and quantify these familiar observational effects will be developed in the following. This same formalism is also used in adaptive optics to correct for such effects.

A complete analysis is given in Roddier F., *The Effects of Atmospheric Turbulence in Optical Astronomy* in Prog. Opti. **XIX**, 281, 1981. This reference gives details and justification for the following calculations. See also Woolf N.J., ARAA **20**, 367, 1982 and Roddier F. (Ed.), *Adaptive Optics in Astronomy*, Cambridge University Press, 1999.

6.2.1 Perturbations of the Wavefront

We consider the behaviour of the coherence of a monochromatic wave subjected to the effect of a medium with random refractive index $n(\boldsymbol{r},t)$ through which it propagates. The process is treated as stationary and ergodic (see Appendix B), although this may well be unrealistic. Denote by τ_c the characteristic correlation time of the refractive index, which would be of the order of a few milliseconds in the troposphere. An observation made over a short time relative to τ_c 'freezes' the turbulence, and two observations separated by a much greater time than τ_c are regarded as independent. Note that the period of the electromagnetic waves considered, i.e., 10^{-9} to 10^{-15} s, is totally negligible relative to τ_c.

Consider a plane wave $\psi_\infty = 1$ impinging on a turbulent atmospheric layer of thickness Δh, sufficiently small for diffraction on the turbulent elements to be negligible across Δh (Fig. 6.10). The wave is quasi-monochromatic. In the geometrical optics approximation, the wave leaving the layer is given by

$$\psi_h(\boldsymbol{x}) = \exp \mathrm{i}\phi_h(\boldsymbol{x}) ,$$

$$\phi_h(\boldsymbol{x}) = k \int_h^{h+\Delta h} n(\boldsymbol{x}, h)\, \mathrm{d}h , \qquad k = \frac{2\pi}{\lambda} .$$

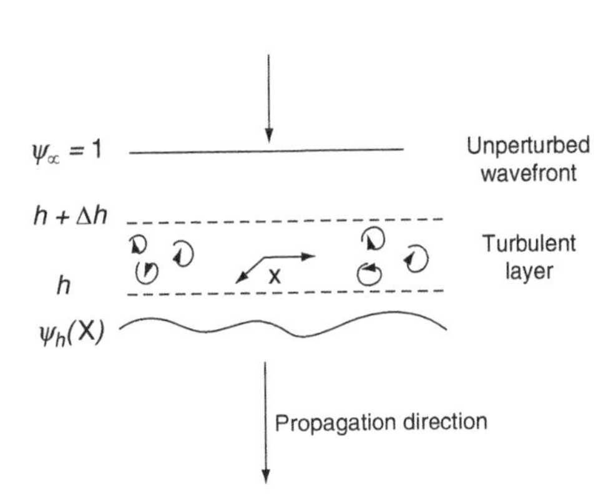

Fig. 6.10 Perturbation of a wavefront by a turbulent layer of thickness Δh

The phase ϕ_h of the wave as it leaves the layer is a random variable, whose spatial statistics are determined by the properties of the random variable $n(\boldsymbol{x},h)$. We assume the phenomenon is ergodic (see Appendix B) whence we may use the same notation for time and space averages. The moments of orders 1 (phase excursions greater than 2π) and 2 are written

$$\langle \psi_h(\boldsymbol{x}) \rangle = 0 \,,$$

$$B_h(\boldsymbol{x}) = \langle \psi_h(\boldsymbol{x}+\xi)\psi_h^*(\xi) \rangle = \langle \exp \mathrm{i}\,[\phi_h(\boldsymbol{x}+\xi) - \phi_h(\xi)] \rangle \,,$$

where the averages are taken over space. The output phase $\phi_h(\boldsymbol{x})$ can be regarded as resulting from a large number of perturbations, which are phase independent once $\Delta h \gg L$, where L is the *external scale* of the turbulence. The distribution of ϕ_h is thus Gaussian (by the central limit theorem), with zero mean.

If u is a real Gaussian random variable with zero mean, it is easily shown that

$$\langle \exp(\mathrm{i}u) \rangle = \langle \cos u \rangle = \exp\left(-\frac{1}{2}\langle u^2 \rangle\right) \,,$$

and consequently

$$B_h(\boldsymbol{x}) = \exp\left[-\frac{1}{2}\langle |\phi_h(\boldsymbol{x}+\xi) - \phi_h(\xi)|^2 \rangle\right] = \exp\left(-\frac{1}{2}D_\phi\right) \,.$$

The *structure function* for the phase has been introduced:

$$D_\phi(\boldsymbol{x}) = \langle |\phi_h(\boldsymbol{x}+\xi) - \phi_h(\xi)|^2 \rangle \,.$$

This can also be written

$$D_\phi(\boldsymbol{x}) = 2\langle \phi_h^2(\xi) \rangle - 2\langle \phi_h(\xi)\phi_h(\boldsymbol{x}+\xi) \rangle \,.$$

The expression for the phase as a function of n then gives

$$\langle \phi_h(\xi)\phi_h(\boldsymbol{x}+\xi) \rangle = k^2 \Delta h \int_0^h \mathrm{d}z \int \langle n(\boldsymbol{x}+\xi, z+\zeta) n(\xi,\zeta) \rangle \, \mathrm{d}\xi \, \mathrm{d}\zeta \,.$$

Using the expression (see Sect. 2.6) for the structure function of the refractive index in a medium of homogeneous and isotropic turbulence,

$$D_n(\boldsymbol{x},z) = \left\langle |n(\boldsymbol{x}+\xi, z+\zeta) - n(\xi,\zeta)|^2 \right\rangle = C_n^2 (|\boldsymbol{x}|^2 + z^2)^{1/3} \,,$$

it follows, after integration, that

$$D_\phi(x) = 2.91\, k^2\, C_n^2\, \Delta h\, x^{5/3} \,.$$

Recall that C_n^2 is the *structure constant* for the refractive index.

The spatial correlation function for the complex amplitude, on leaving the turbulent layer, is therefore

$$\langle \psi_h(\boldsymbol{x}+\xi)\psi_h^*(\xi)\rangle = \exp\left(-1.45\,k^2\,C_n^2\,\Delta h\,x^{5/3}\right) .$$

Since the turbulent layers may be a long way from the entrance pupil of the instrument, we must determine the effect of propagation on this function. The wave field at zero altitude ($h = 0$) results from *Fresnel diffraction* (i.e., long range) of the wave on leaving the layer. It can be shown that the correlation function $B_z(\boldsymbol{x})$ is invariant under Fresnel diffraction (Problem 6.15), and hence

$$B_0(\boldsymbol{x}) = B_h(\boldsymbol{x}) = \exp\left(-\frac{1}{2}D_\phi\right) .$$

Some important conclusions can be drawn from these results.

- A short distance behind the layer, only the phase is perturbed. The effects of this phase perturbation are *smearing* and *image motion*. Further away, both phase and amplitude are perturbed, and *scintillation* then occurs. Fluctuations in the amplitude $|\psi_0(\boldsymbol{x})|$ can often be neglected when the turbulence is not very great (and we shall quantify this later).
- On leaving the layer, the perturbed wavefront displays a correlation function with a complex amplitude, which gives the random phase distribution. This function $B_0(\boldsymbol{x})$ has an isotropic profile in the plane which is 'near Gaussian', so a *correlation length* x_c can be defined by the relation

$$\frac{\langle \psi_h(x+x_c)\psi_h^*(x)\rangle}{\langle |\psi_h(0)|^2\rangle} \sim \frac{1}{e} ,$$

so that

$$x_c \sim (1.45\,k^2\,C_n^2\,\Delta h)^{-3/5} .$$

Inserting typical values 10^{-12} cm$^{1/3}$ for $C_n^2\,\Delta h$, at $\lambda = 0.5$ μm, the correlation length turns out to be $x_c = 9.6$ cm. This implies a significant reduction in the spatial coherence of the incident wave. Recall the earlier calculation (see Sect. 6.1.2) which showed that, for a star of apparent diameter 0.1 arcsec, there is a wavefront whose coherence radius reaches 32 cm at this wavelength 0.5 μm.
- Note the highly chromatic nature of the coherence, x_c varying as $\lambda^{6/5}$. The coherence length grows more quickly than λ, favouring the infrared over the visible.
- If turbulence extends across the whole depth of the atmosphere, the above formulas can be shown to remain valid, replacing $C_n^2\,\Delta h$ by the expression

$$\int_0^\infty C_n^2(z)\,dz .$$

The value of this expression is found from the vertical distribution of the turbulence (see Sect. 2.6.1). When this distribution is known, e.g., by means of sounding balloons or SODAR, the effect and hence the relative importance of the different layers can be identified for a given site.

- If the effects described above are transposed to radiofrequencies, e.g., $\lambda = 1$ cm, the resulting value of x_c reaches several hundred metres and therefore does not perturb image formation, even for the largest radiotelescope diameters. However, phase fluctuations can arise when combining the signals from two widely separated telescopes in interferometry (see Sect. 6.5.1).

6.2.2 The Perturbed Image

In the following discussion, we confine ourselves to phase perturbations, neglecting the amplitude perturbations which lead to scintillation.[11] An incident plane wave $\psi_\infty = 1$ reaches the pupil perturbed, in the form $\psi_0(\boldsymbol{r}, t)$. In the absence of perturbation, the image would be limited by diffraction, and would be deduced from the pupil function $G(\boldsymbol{r})$ defined above (see Sect. 6.1). As the incident wave is plane, the point source response of the pupil would be obtained, e.g., an Airy function in the case of a circular pupil.

In the perturbed case, it is sufficient, in order to calculate the image, to replace the pupil function by a new instantaneous pupil function, which is now a random complex function of time:

$$G(\boldsymbol{r})\psi_0(\boldsymbol{r}, t) .$$

The point source response is then an intensity distribution given by the Fourier transform of the autocorrelation of this pupil function. The image of a point source is therefore a *random intensity distribution* in the focal plane, whose characteristic evolution time is given by the atmospheric coherence time τ_c.

Long-Exposure Images

When the measurement time (or integration time, in the case of a photographic plate or photoelectric detector) is much larger than τ_c, the observed intensity at each point of the image is simply the mean value of the instantaneous intensity:

$$I(\boldsymbol{\theta}) = \langle I_0(\boldsymbol{\theta}) \star T(\boldsymbol{\theta}, t) \rangle .$$

[11] We understand here that only almost pointlike sources will scintillate. According to the phase screen approximation, each point of an extended source produces its own diffraction pattern, with intensity variations. The patterns produced by the different points after passing through different non-uniformities will tend to balance out to yield an almost constant intensity over time.

The mean MTF $\langle \tilde{T} \rangle$ is written as the normalised mean autocorrelation function of the pupil

$$\langle \tilde{T}(\boldsymbol{w}) \rangle = \frac{\iint G(\boldsymbol{s})G(\boldsymbol{s}-\boldsymbol{w})\langle \psi(\boldsymbol{s},t)\psi^*(\boldsymbol{s}-\boldsymbol{w},t)\rangle \, \mathrm{d}\boldsymbol{s}}{\iint G^2(\boldsymbol{s})\mathrm{d}\boldsymbol{s}} ,$$

where $\boldsymbol{w} = \boldsymbol{r}/\lambda$. The second order moment of the complex amplitude of the wavefront is given directly by the mean value of the MTF. Limiting discussion to the simple case of a circular pupil, of diameter $D \gg x_\mathrm{c}$, the mean MTF, denoted here by $B_0(\boldsymbol{w})$, becomes

$$\langle \tilde{T}(\boldsymbol{w}) \rangle \approx B_0(\boldsymbol{w}) = \exp\left[-1.45\, k^2\, C_n^2\, \Delta h\, (\lambda w)^{5/3}\right] .$$

Figure 6.11 illustrates the *spatial filtering* effect of averaged turbulence. The image has lost its high frequencies. The point source response is the FT of $\langle \tilde{T} \rangle$, and even though an analytic expression has not been given for the image profile, it can be seen that its angular dimension now has order λ/x_c rather than λ/D. In other words, the image is smeared by the loss of spatial coherence over the pupil plane. Furthermore, the time averaging effect has removed the agitation from the image, at least as long as the ergodicity hypothesis is justified.

We conclude immediately that increasing the diameter D of the instrument does not give better resolution than an instrument whose diameter is of the order of x_c, which means a few centimetres at visible wavelengths! The figure also shows the appearance of a characteristic frequency r_0/λ, which we now discuss.

The Fried Parameter $r_0(\lambda)$

In order to be able to compare with greater precision the diffraction-limited Airy image, described by the intensity $I_\mathrm{A}(\boldsymbol{\theta})$, with the long-exposure image $\langle I \rangle$, we

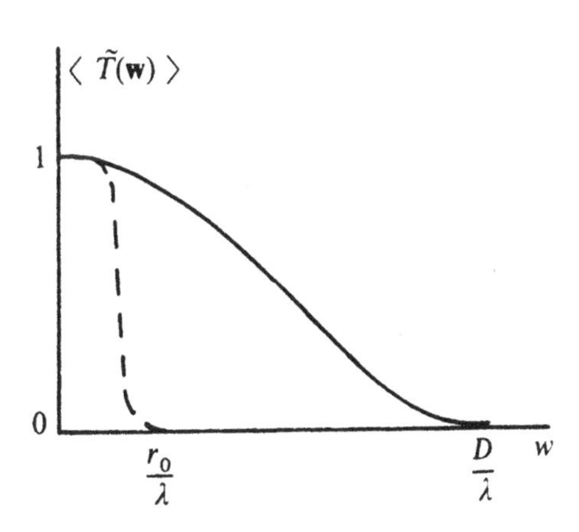

Fig. 6.11 MTF of the perturbed pupil. *Continuous curve*: ideal MTF. *Dashed curve*: long exposure

calculate the diameter r_0 of a diffraction-limited circular pupil which would give an image of the same resolution as the image degraded by passing through the atmosphere. Up to now we have related this quantity intuitively to the width at half-maximum of the spreading function of the image. A better and more quantitive definition of the resolution $\mathscr{R}$ is conventionally taken to be

$$\mathscr{R} = \int \langle \tilde{T}(\boldsymbol{w}) \rangle \, \mathrm{d}\boldsymbol{w} \simeq \int B_0(\boldsymbol{w}) \, \mathrm{d}\boldsymbol{w} = \int \mathrm{FT}\,[I_{\mathrm{A}}(\boldsymbol{\theta})] \, \mathrm{d}\boldsymbol{w} \, .$$

It can be shown that the quantity r_0, also called the *Fried parameter*[12] or the linear dimension of the *coherence area* (spatial coherence being understood), is given by the expression

$$r_0(\lambda) = \left[0.423\, k^2 \int_0^\infty C_n^2(z) \, \mathrm{d}z \right]^{-3/5} = 0.185\, \lambda^{6/5} \left[\int_0^\infty C_n^2(z) \, \mathrm{d}z \right]^{-3/5} .$$

This quantity can be calculated directly from the turbulence and is sufficient to determine the perturbed image, a spot whose diameter is of order λ / r_0. The angle $\Delta\theta = \lambda / r_0$ is often called the *seeing angle*, or simply the *seeing*. It is thus the size of star images on a photographic plate or CCD when a long exposure is made.

Note a key point here, namely that r_0 is highly chromatic, varying as $\lambda^{6/5}$. For example, $r_0(0.5\ \mu\mathrm{m}) = 10$ cm implies $r_0(20\ \mu\mathrm{m}) = 8.4$ m. Hence, a telescope like one of the 8.2 m telescopes at the VLT, while highly perturbed in the visible, is almost completely unperturbed in the mid-infrared. Consequently, for a given diameter, a telescope will give images closer to the diffraction limit in the infrared than in the visible.

Short-Exposure Images

Consider now an *instantaneous image*, the intensity distribution observed over an integration time which is short compared with τ_{c}. The atmosphere can thus be considered as 'frozen' here.

The turbulent medium produces a complex MTF, with a random phase whose mean is zero for frequencies above r_0/λ. Only the mean value of the modulus of the MTF may be non-zero. Hence,

$$\langle |\tilde{T}(\boldsymbol{w})|^2 \rangle = \frac{1}{A^2} \iint \langle \psi^*(\boldsymbol{s}) \psi(\boldsymbol{s} - \boldsymbol{w}) \psi^*(\boldsymbol{s}' - \boldsymbol{w}) \psi(\boldsymbol{s}') \rangle$$
$$\times G(\boldsymbol{s}) G(\boldsymbol{s} - \boldsymbol{w}) G(\boldsymbol{s}') G(\boldsymbol{s}' - \boldsymbol{w}) \, \mathrm{d}\boldsymbol{s} \, \mathrm{d}\boldsymbol{s}' \, ,$$

---

[12] The American optician D.L. Fried developed the theory of light propagation in random media in the 1960s.

where $s = r/\lambda$ is the running variable in the pupil plane, and A the area of the pupil in units of r/λ. The fourth order moment of the complex amplitude ψ appears. If $|w| \gg r_0/\lambda$, then $\psi(s)$ and $\psi(s-w)$ can be regarded as decorrelated, in which case the expression in $\langle\ \rangle$ under the integral sign becomes

$$\langle\ \rangle = \langle \psi^*(s)\psi(s')\rangle\langle\psi(s-w)\psi^*(s'-w)\rangle \simeq B_0^2(s-s') .$$

This leads to

$$\begin{aligned}\langle|\tilde{T}(w)|^2\rangle &\simeq \frac{1}{A^2}\iint B_0^2(s-s')G(s)G(s-w)G(s')G(s'-w)\,\mathrm{d}s\,\mathrm{d}s' \\ &\simeq \frac{1}{A^2}\int G(s)G(s-w)\mathrm{d}s\int B_0^2(u)G(s-u)G(s-u-w)\mathrm{d}u .\end{aligned}$$

Moreover, if the condition $D \gg r_0(\lambda)$ holds, then

$$\int B_0^2(u)G(s-u)G(s-u-w)\,\mathrm{d}u \sim \int B_0^2(u)\,\mathrm{d}u = \sigma B_0^2(\mathbf{0}) = \sigma ,$$

where σ is the Fried coherence area, $\sigma = 0.342(r_0/\lambda)^2$.

The remarkable expression

$$\langle|\tilde{T}(w)|^2\rangle \simeq \frac{\sigma}{A}\frac{\int G(s)G(s-w)\,\mathrm{d}s}{A} = \frac{\sigma}{A}\tilde{T}_0(w)$$

is illustrated in Fig. 6.12. $\tilde{T}_0(w)$ is the unperturbed pupil transfer function. High spatial frequencies, between the long-exposure cutoff frequency r_0/λ and the cutoff frequency D/λ of the unperturbed pupil, are transmitted, but attenuated in a constant ratio. Note that the square $\langle|\tilde{T}(w)|^2\rangle$ is proportional to $\tilde{T}_0$, and not to its square $\tilde{T}_0^2$.

What then is the instantaneous image $I(\theta, t)$ of a quasi-point source, such as an unresolved star? The star is unresolved if the wave in the absence of turbulence

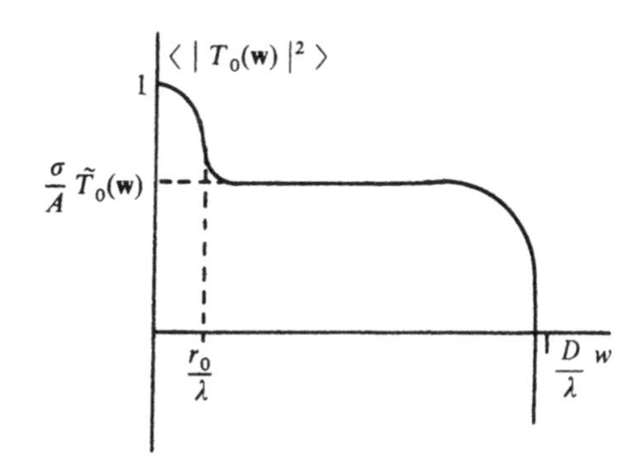

Fig. 6.12 MTF of the perturbed pupil for short exposure. Note the attenuated but non-zero transmission at high frequencies. The *ordinate scale* is logarithmic

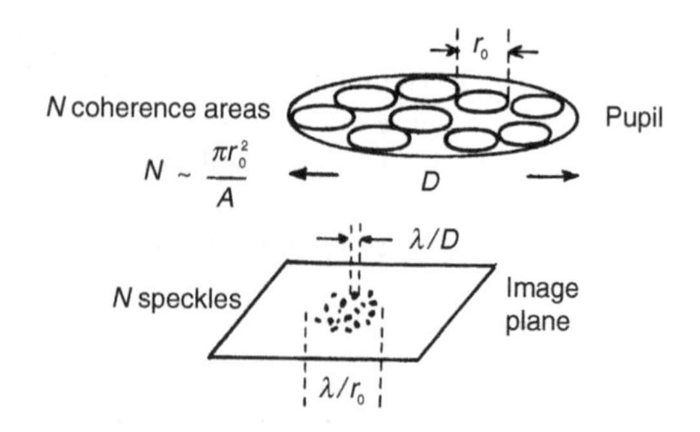

Fig. 6.13 Speckles. Note the distribution of coherence over the pupil, and the distribution of intensity over the image, as well as the sizes of the various patterns

is coherent on a scale greatly exceeding that of the pupil. $I(\boldsymbol{\theta}, t)$ is then the FT of an instantaneous instance of the random quantity $\tilde{T}(\boldsymbol{w})$. It is a random intensity distribution in the focal plane (Fig. 6.13, see also Fig. 6.28c). The presence of bright and dark *speckles* reveals the existence of high spatial frequencies. From one image to another, these speckles move randomly, and their superposition over a period of time produces the mean long-exposure image described earlier. This image could also be interpreted as the pattern of the (Fraunhofer) diffraction at infinity of the random phase perturbations produced at the pupil by crossing the turbulent medium.

Laser Speckles. Scattering of a laser beam, which is, of course, coherent, by an irregular surface, gives the same granular appearance when viewed at infinity, because of the random interference of the scattered waves.

In 1970, while observing the image of Betelgeuse (α Ori) with a large telescope, Antoine Labeyrie[13] noted the presence of similar structures and deduced that the high frequency information was being degraded, but was not completely eliminated from the image, whence it might be possible to reconstitute it. This fundamental observation was the starting point for *speckle interferometry*, followed by a whole series of parallel methods, e.g., *shift and add* (see Sect. 9.6). It subsequently contributed (as of the 1980s) to the development of adaptive optics, to be discussed in the next section.

Isoplanatic Patch

Consider the instantaneous images of the two components 1 and 2 of a binary star, with angular separation α. Each image will exhibit speckles, each corresponding to a wavefront perturbed by slightly different atmospheric layers, depending on the

[13]The French astronomer Antoine Labeyrie, born in 1944, pioneered many developments in high angular resolution visible and infrared observation over the last decades of the twentieth century. The 'bunch of grapes' structure of star images had already been noted in the 1940s by the French astronomer Jean Rösch (1915–1999) at the Pic-du-Midi Observatory (France), but without drawing any particular conclusions about its significance.

angle α. Indeed, the degree of similarity of the two simultaneous instantaneous MTFs $\widetilde{T}_1(\boldsymbol{w})$ and $\widetilde{T}_2(\boldsymbol{w})$ will depend on the value of α. Quantitatively, their average correlation $C_T(\alpha) \propto \langle T_1 \otimes T_2 \rangle$ over time will give a measure of this similarity. For example, the width at half-maximum of $C_T(\alpha)$ defines the *isoplanatic angle* α_0 of the atmosphere in the given state. By a similar calculation to the one leading to $r_0(\lambda)$, it can be shown that

$$\alpha_0(\lambda) = 0.314 \frac{r_0(\lambda) \cos \gamma}{\bar{h}} ,$$

where the angle γ is the zenith distance of the viewing direction and $\bar{h}$ a weighted altitude for the vertical distribution of turbulence. In the simple case where there is only one thin turbulent layer at altitude h, we deduce immediately that, in the vertical direction, $\alpha_0(\lambda) = r_0(\lambda)/h$.

Like r_0, the isoplanatic patch defined by α_0 is highly chromatic. All things being equal, we understand intuitively that this field will depend on the altitude distribution of the turbulent layers. If the turbulence is limited to layers close to ground level, as happens on Antarctic Dome C, for example, α_0 will be large. Typical values at 0.5 μm are $\alpha_0 \approx$ 5–10 arcsec. These considerations are very important in adaptive optics. Indeed, when images are processed after observation, we shall show in Sect. 9.6 how much the processing differs when the effect of the instrument is a simple convolution of the object in the whole observed field (isoplanicity) and when this is not the case (outside this region).

The impact of a turbulent layer or region on propagation, and hence on an image, is not only relevant for propagation through the atmosphere. Radioastronomers have observed *interplanetary* and *interstellar scintillation* affecting the intensity received from a source by radiotelescopes. As for the ionosphere, this phenomenon is caused by electron density fluctuations in the interplanetary or interstellar plasma, the intensity of the fluctuation depending on the angular size of the source. It was through such measurements back in 1950 that the first upper limit to the size of the Cygnus A radiosource, discovered slightly earlier, could be obtained.

Speckle Interferometry

Speckle interferometry[14] is a method for restituting the unperturbed image, or some of its characteristics. This method played an important role in the years 1975–1990, but it has now been largely superseded by adaptive optics, to which it made a significant contribution. However, it may make a comeback in ground-based mid-infrared observation by large telescopes.

Let $I_0(\boldsymbol{\theta})$ be an arbitrary source whose image we wish to determine. The image is given by the convolution

$$I(\boldsymbol{\theta}) = I_0(\boldsymbol{\theta}) \star T(\boldsymbol{\theta}) ,$$

[14]See Labeyrie A., Stellar interferometry methods, ARAA **16**, 77, 1978.

and hence

$$\langle|\tilde{I}(\boldsymbol{w})|^2\rangle = \left|\tilde{I}_0(\boldsymbol{w})\right|^2 \left\langle\left|\tilde{T}(\boldsymbol{w})\right|^2\right\rangle .$$

If stationarity conditions are satisfied, it suffices to evaluate $\langle|\tilde{T}(\boldsymbol{w})|^2\rangle$ by observing a point source (in practice, an object that is not resolved by the given telescope) in some arbitrary direction, but at the same r_0, in order to be able to write

$$|\tilde{I}_0(\boldsymbol{w})| = \left[\frac{\langle|\tilde{I}(\boldsymbol{w})|^2\rangle_{\text{observed}}}{\langle|\tilde{T}(\boldsymbol{w})|^2\rangle_{\text{observed}}}\right]^{1/2} .$$

This gives the modulus of the spatial spectrum up to the cutoff frequency D/λ of the pupil, assuming only that $\langle|\tilde{T}(\boldsymbol{w})|^2\rangle$ can be found with a good signal-to-noise ratio. By the Wiener–Khinchine theorem (see Appendix A), this is equivalent to finding the average autocorrelation function of the image.

In the visible, the resolution at the diffraction limit has been reached by this method. Examples are the separation of binary stars, or the determination of stellar diameters. At $\lambda = 0.5\,\mu$m, the ultimate, i.e., diffraction-limited, resolution of a 5 m telescope attains $D/\lambda = 20$ mas (1 mas $= 10^{-3}$ arcsec). In the infrared, the increase in r_0 with λ implies that the attenuation factor $\pi r_0^2/A$ improves, thus increasing the signal-to-noise ratio, where A is the area of the collecting pupil.

Note that this method of restoring just the modulus $|\tilde{I}(\boldsymbol{w})|$ is not perfect. It does not restore $I(\boldsymbol{\theta})$, for that would require knowledge of both the amplitude and the phase of $\tilde{I}(\boldsymbol{w})$, unless the object had circular symmetry. This is not relevant, of course, for the separation of a pair of sources. It can be shown that the phase information in arg $\tilde{I}(\boldsymbol{w})$ is also present in the image and can be restored by the appropriate methods.[15]

Differential speckle interferometry, or simply, differential interferometry, treats sources whose morphology is wavelength dependent, and indeed this is usually the case for astronomical sources. But here we consider the case in which the quantities $O(\boldsymbol{\theta}, \lambda_1)$ and $O(\boldsymbol{\theta}, \lambda_2)$ differ greatly for two neighbouring wavelengths λ_1 and λ_2. This is the situation, for example, where a star is observed at wavelength λ_1, in the continuous photospheric spectrum, and λ_2 is a chromospheric emission line. The two spectra $I(\boldsymbol{w}, \lambda_1, t)$ and $I(\boldsymbol{w}, \lambda_2, t)$ differ greatly in their wavelength dependence, but are totally correlated temporally if measured simultaneously. They can therefore be compared without loss of information in $\boldsymbol{w}$, and without loss of angular resolution in the final image. The resolution λ/D of the telescope can be reached, or even surpassed (*super-resolution*), if a priori information about the object $O(\boldsymbol{\theta}, \lambda)$ is available and the signal-to-noise ratio in spectral measurements is good enough.

6.2.3 *Effect of the Atmosphere on Interferometry*

In an interferometry configuration (see Sect. 6.4), and if the wave is not perturbed, both the phase and the amplitude of the complex quantity $\tilde{I}(\boldsymbol{w} = \boldsymbol{D}/\lambda)$ can be measured for different values of $\boldsymbol{w}$, so that an image of the object can be reconstructed.

[15] See, for example, Alloin D. & Mariotti J.M. (Eds.), Diffraction-limited imaging with large telescopes, Kluwer, 1989.

For a telescope array, and in particular for a pair of sub-pupils 1 and 2, the problem is to examine how the atmosphere affects the random *phase difference* $\delta\phi_{1,2}(t) = \phi_2(t) - \phi_1(t)$ it produces between the centers of these sub-pupils. Clearly, in a stationary atmosphere, the mean value $\langle \delta\phi_{1,2}(t) \rangle$ will be zero. However, the phase excursion will be characterised by the variance (or standard deviation):

$$\langle | \delta\phi_{1,2}(t)^2 | \rangle = \sigma_\phi^2 \, .$$

For Kolmogorov turbulence, this standard deviation σ_ϕ increases with the separation of the sub-pupils. It can be shown by the same method as above that it is then given by

$$\sigma_\phi = 2.62 \left[\frac{D_{1,2}}{r_0(\lambda)} \right]^{5/6} \, .$$

This standard deviation saturates at a maximal value when this distance $D_{1,2}$ exceeds the external scale $\mathscr{L}_0$ of the turbulence. This saturation provides a simple way of measuring $\mathscr{L}_0$ at a given site. Later we shall see the extent to which this effect, which randomly translates the interference fringes obtained between the sub-pupils, perturbs ground-based astronomical observation, at both radiofrequencies and optical frequencies. The so-called *phase closure* methods provide a way of avoiding to some extent the harmful consequences of this *phase noise*.

Its quantitative effect on the complex MTF is a random phase rotation. The instantaneous MTF becomes

$$\widetilde{T}(\boldsymbol{w}) \exp\left[\mathrm{i}\Delta\varphi_{1,2}(D,t)\right] \, , \qquad \Delta\varphi_{1,2}(D,t) \ \text{random phase difference} \, .$$

At radiofrequencies, phase fluctuations are mainly introduced into the wavefront by ionospheric irregularities (Fig. 6.14). At millimetre and infrared wavelengths on the one hand, and visible wavelengths on the other, refractive index fluctuations are

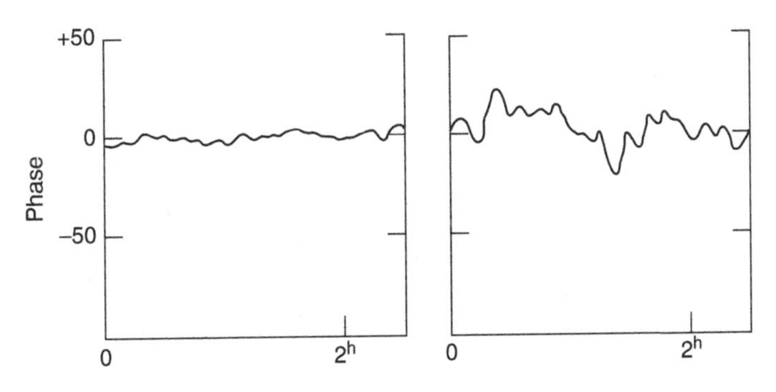

Fig. 6.14 Phase evolution (in degrees) due to ionospheric fluctuations, at $\lambda = 6$ cm. Note that the standard deviation is small compared with 180°

dominated by those of water vapour or the air, as already mentioned. However, water vapour is not without influence at radiofrequencies, particularly at short wavelengths (cm and mm).

Many methods have been developed to restore the phase of the signal, degraded by the atmosphere but essential for good image formation (see Sect. 9.6).

6.3 Adaptive Optics

The methods of speckle interferometry show that information concerning the high spatial frequencies ($r_0/\lambda \leq |\boldsymbol{w}| \leq D/\lambda$) of the source is not completely absent from the image, despite the spatial filtering effect of the atmosphere. Under certain conditions, it is possible to restore spatial frequencies, filtered by the atmosphere, in real time. This is the aim of *adaptive optics*.[16]

The principle is very simple. Consider the deformed wavefront

$$\psi(\boldsymbol{r}, t) = \psi_0 \mathrm{e}^{\mathrm{i}\phi(\boldsymbol{r}, t)} ,$$

given here for a point source at infinity, without modulation of the amplitude ψ_0 (no *scintillation* in moderate turbulence), $\phi(\boldsymbol{r}, t)$ denoting the phase perturbation introduced by the atmosphere. Naturally, it is assumed that this astronomical source is temporally incoherent. Using a mechanical or electro-optical device, a phase conjugation $\phi'(\boldsymbol{r}, t)$ is introduced to give a wave

$$\psi'(\boldsymbol{r}, t) = \psi(\boldsymbol{r}, t)\mathrm{e}^{-\mathrm{i}\phi'(\boldsymbol{r}, t)} .$$

If it were possible to make the exact correction $\phi' = -\phi, \forall\, \boldsymbol{r},\, t$, the device would be perfect and the image perfectly restored to what would have been observed without the signal having crossed the atmosphere. However, various instrumental limitations mean that a *phase error* remains:

$$\delta\phi(\boldsymbol{r}, t) = \phi(\boldsymbol{r}, t) - \phi'(\boldsymbol{r}, t) .$$

The result is a corrected image, but exhibiting some deviation from the ideal PSF, which would only be limited by diffraction at the pupil.

The problems encountered in adaptive optics vary widely, from measurement of the perturbed wavefront $\psi(\boldsymbol{r}, t)$, to choice of the device producing the phase

[16]The website www.ctio.noao.edu/atokovin/tutorial/ maintained by A. Tokovinin contains an excellent and detailed analysis of the ideas presented somewhat succinctly here, translated in several languages.

correction ϕ', assessment of the final image, choice of reference source, and determination of the ultimate potential of the method. We shall consider each of these in turn.

By endowing the eye with an optical system that could vary its focal length (accommodation), nature achieved the first optical device that could be described as adaptive, while maintaining the quality of the image on the retina. In 1953, the American astronomer Harold Babcock suggested correcting for deformations in solar images by analysing the wavefront and introducing phase corrections in real time. However, he never constructed such a device. In 1957, the Soviet physicist V.P. Linnick made a similar proposal, apparently independently. He had the idea of a segmented correcting mirror, and even the idea of an artificial reference source carried by an aircraft (see below). These suggestions were then left by the wayside for some time, partly because it seemed like an impossible dream, and partly because of the difficulty in carrying out the calculations sufficiently quickly in real time. Around 1970, it was taken up again secretly by the US military, but it was in Europe, stimulated by the recent commitment to the Very Large Telescope, that the first astronomical image — of a binary star — was published with corrections made by adaptive optics in 1990.

Between 1990 and 2006, adaptive optics moved forward quickly, and now equips practically all the major Earth-based telescopes, in particular for their near-infrared observations. Indeed, it has become a key technique for future 30–50 m telescopes. In addition, many new ideas are emerging, as we shall see below.

6.3.1 *Wavefront Measurement*

This measurement constitutes the first stage of adaptive optics. For convenience, the wavefront is expanded in terms of a proper basis of vectors of the pupil (usually the primary mirror of the telescope):

$$\phi(\boldsymbol{r}, t) = \sum_{i=1}^{\infty} a_i(t) Q_i(\boldsymbol{r}) ,$$

where Q_i are the basis vectors (*spatial modes*). There are infinitely many such orthonormal bases. One of the most commonly used is the set of *Zernike polynomials* Z_i, which describe the classical optical aberrations in terms of phase and not in terms of rays, as was done in Sect. 5.1.

The Zernike polynomials are defined by:

$$Z_{n,m,\text{even}} = \sqrt{n+1}\, R_n^m(r)\sqrt{2}\cos(m\theta) , \qquad m \neq 0 ,$$

$$Z_{n,m,\text{odd}} = \sqrt{n+1}\, R_n^m(r)\sqrt{2}\sin(m\theta) , \qquad m \neq 0 ,$$

$$Z_{0,n} = \sqrt{n+1}\, R_n^0(r) , \qquad m = 0 ,$$

Table 6.2 The first few Zernike polynomials and corresponding optical aberrations. By convention, the double index n, m of the Zernike polynomials is replaced here by a single sequential index i

Radial degree n	Azimuthal frequency m			
	$m = 0$	$m = 1$	$m = 2$	$m = 3$
0	$Z_1 = 1$ Piston			
1		$Z_2 = 2r\cos\alpha$ $Z_3 = 2r\sin\alpha$ Tip–tilt		
2	$Z_4 = \sqrt{3}$ $(2r^2 - 1)$ Defocus		$Z_5 = \sqrt{6}r^2 \sin 2\alpha$ $Z_6 = \sqrt{6}r^2 \cos 2\alpha$ Astigmatism	
3		$Z_7 = \sqrt{8}(3r^3 - 2r)$ $\sin\alpha$ $Z_8 = \sqrt{8}(3r^3 - 2r)$ $\cos\alpha$ Coma		$Z_9 = \sqrt{8}r^3$ $\sin 3\alpha$ $Z_{10} = \sqrt{8}r^3$ $\cos 3\alpha$
4	$Z_{11} = \sqrt{5}$ $(6r^4 - 6r^2 + 1)$ Spherical aberration		$Z_{12} = \sqrt{10}$ $(4r^4 - 3r^2)\cos 2\alpha$ $Z_{13} = \sqrt{10}$ $(4r^4 - 3r^2)\sin 2\alpha$	

where

$$R_n^m(r) = \sum_{s=0}^{(n-m)/2} \frac{(-1)^s (n-s)!}{s!\,[(n+m)/2 - s]!\,[(n-m)/2 - s]!} r^{n-2s} ,$$

and m is the azimuthal frequency, n the radial degree.

Table 6.2 gives the classification, the formula and the equivalent aberrations in classical optics for the first few Zernike polynomials. By convention, the indices m, n are replaced by a single sequential index. This basis is orthogonal on a circular pupil:

$$\frac{1}{\pi} \iint G(\boldsymbol{r}) Z_i Z_j \, \mathrm{d}\boldsymbol{r} = \delta_{ij} \qquad \text{with} \quad G(\boldsymbol{r}) = \Pi\left(\frac{r}{2}\right) .$$

Note, however, that this orthogonality relation does not strictly hold for a telescope pupil which is partially blocked by a central obstruction, such as a secondary mirror, as is often the case.

Now let $\phi(\boldsymbol{r}, t) = \phi(R\rho, \theta, t)$, where R is the pupil radius. The coefficients a_i of the expansion of ϕ in this basis are given by

$$a_i(t) = \iint G(\boldsymbol{r}) \phi(\rho, \theta, t) Z_i(\rho, \theta) \, \mathrm{d}\boldsymbol{r} .$$

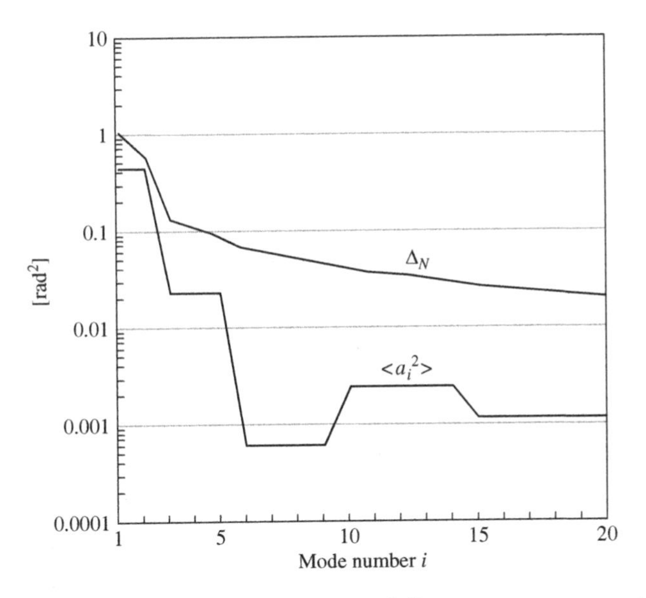

Fig. 6.15 Variance (in rad^2) of the Zernike coefficients $\langle a_i^2 \rangle$, due to Kolmogorov turbulence with infinite external scale L_0, where i denotes the number of the polynomial. Also shown is the residual phase error (*Noll residual*), after perfect correction of the first N modes. The values of $\langle a_i^2 \rangle$ and Δ_N are expressed in units of $[D/r_0(\lambda)]^{5/3}$, where D is the pupil diameter, and r_0 the coherence scale of the atmosphere, at the wavelength λ at which the phase is being studied. Recall that, to first order, the phase differences result from differences in optical path, and these are achromatic, i.e., not wavelength dependent, in the Earth atmosphere

In the presence of atmospheric turbulence, the a_i are random variables in time, and ergodic, if the turbulence is stationary. The basis Z_i is not quite adequate for projecting the effects of Kolmogorov turbulence. Indeed, the coefficients a_i are not statistically independent, as can be seen by calculating the covariances $\langle a_i a_j \rangle \neq 0$ for certain pairs of values of i, j. It may therefore be necessary to choose another basis (the *Karhunen–Loeve polynomials*, which preserve the non-correlation of the coefficients).

In fully developed (Kolmogorov) turbulence, the variance of each coefficient can be calculated exactly, together with the variance residue Δ_N, if it is assumed that the phase conjugation has been achieved perfectly for the values $i \leq N$:

$$\Delta_N = \langle |\phi^2| \rangle - \sum_{i=2}^{N} \langle a_i^2 \rangle = \sum_{i=N+1}^{\infty} \langle a_i^2 \rangle \,.$$

Figure 6.15 gives the variance and the residues for the first few values of N. It can be used to calculate the phase error remaining after correction. For large N (>10 or 20), there is an asymptotic expression

$$\Delta_N = 0.2944\, N^{-\sqrt{3}/2} \left(\frac{D}{r_0} \right)^{5/3} .$$

Note, in Fig. 6.15, that the first few modes contain a large part of the total variance. If they are imperfectly corrected, then the higher order corrections become superfluous. This is indeed one of the problems to be faced in the design of an adaptive system.

Angular properties are important. Is it possible to use the wavefront coming from a source situated in a certain direction to correct the image of another source, observed in a different direction? This can also be characterised. If two sources are observed, in directions $\boldsymbol{\theta}_1$ and $\boldsymbol{\theta}_2$, the relevant quantity is the angular decorrelation (*anisoplanicity*)

$$\left\langle a_i(\boldsymbol{\theta}_1, t)\, a_j(\boldsymbol{\theta}_2, t)\right\rangle .$$

The correlation decreases more rapidly as N is large and the wavelength λ is short. At $\lambda = 1\ \mu\text{m}$, the isoplanatic patch of the first ten polynomials is of the order of 10 arcsec for average turbulence.

Measurement of the perturbed phase requires *wavefront analysers*, applying quality control principles used when polishing optical elements, with the only difference being that here the analyser must necessarily operate in temporally incoherent light. Such analysers include *shearing interferometers*, and, more commonly, analysers segmenting the pupil into sub-pupils (*Shack–Hartmann*), which measure the phase gradient $\nabla\phi$, curvature analysers, which measure the Laplacian $\Delta\phi$, and pyramid analysers with broad dynamic range.

Whatever analysis technique is chosen, it suffices to sample the phase discretely on the wavefront, at N regularly located points on the pupil. The bigger the value of N, the smaller will be the residual phase errors. In practice, if these measurements of $\varphi(\mathbf{r}, t)$ are made at a specific wavelength λ_0, a sampling rate of order $r_0(\lambda_0)$, i.e., $N = (D/r_0)^2$, is chosen. For example, for a 10 m telescope operating at $\lambda_0 = 1\ \mu\text{m}$ with $r_0 = 50$ cm, a suitable number of measurement points would be $N = 20^2 = 400$.

Note also that, up to a factor of $2\pi/\lambda$, $\varphi(\boldsymbol{r}, t)$ is achromatic. A favourable consequence of this is that the value of λ_0 used for the measurement can be chosen from considerations of experimental convenience (see below) and the correction can thus be applied at a different wavelength.

Tip–Tilt Correction. Figure 6.15 shows that a significant fraction of the phase error comes from the first two terms ($n = m = 1$, see Table 6.2), which correspond to the image centroid in the image plane. This is called a *tip–tilt fluctuation*. The image could thus be significantly improved by correcting for this fluctuation alone. Indeed, this was done very early on in observations of the Sun, in particular by H. Babcock in 1953. More recently, a posteriori processing techniques for short exposure images (see above and Sect. 9.6) have used *image refocusing*, which superposes short exposure images after refocusing each of them. This yields images equivalent to a long exposure, but with much better angular resolution.

Tip–tilt can be corrected in real time, using a simple mirror that tilts around two axes and thereby hold the image centroid fixed. The tilt angles are continually reassessed by a tip–tilt sensor. This yields an effective and cheap adaptive optics system, although still somewhat rudimentary. This kind of device, useful whenever $r_0 \leq D$, has even been commercialised to equip amateur telescopes and significantly improve image quality.

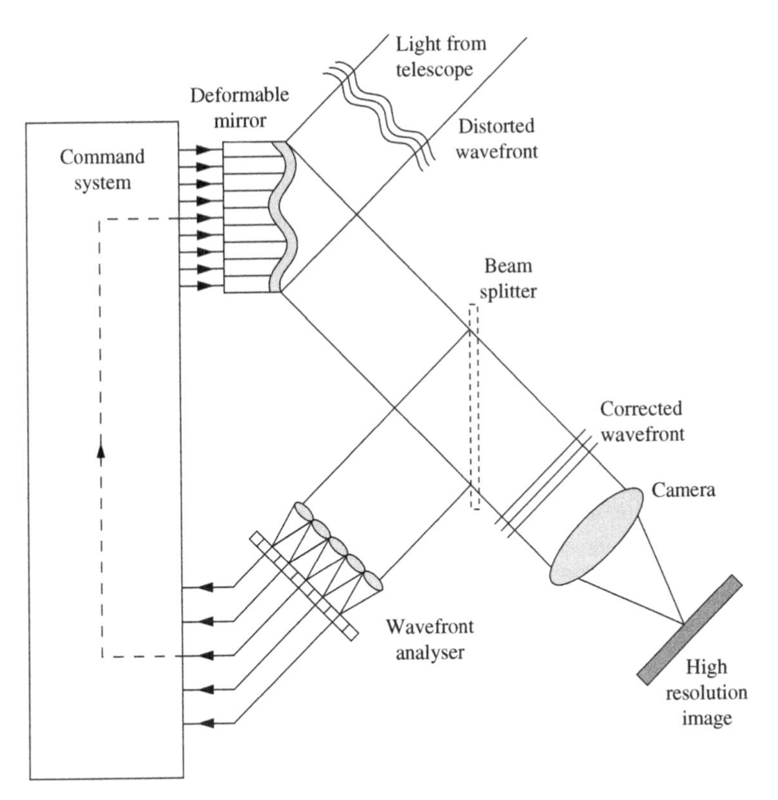

Fig. 6.16 Basic setup in adaptive optics. A computer is placed between analysis and control. The *dotted line* represents a beam splitter (possibly dichroic), dividing the light between wavefront sensor (λ_0) and camera measuring the corrected image (λ). The wavefront sensor does not necessarily operate at the same wavelength as the camera forming the corrected image

6.3.2 Phase Correction Devices

The standard setup is shown in Fig. 6.16. A thin mirror, whose surface can be deformed by mechanical actuators, reflects the wave to be corrected. Ideally, this would be the primary mirror, but it has too much inertia and it is not possible to make corrections quickly enough. The possibility, discovered recently (2006), of making the correction on the secondary mirror of the telescope is particularly attractive (Fig. 6.17). Whichever mirror is chosen, by forming an image of the pupil on it, the phase correction $\phi'(\boldsymbol{r}, t)$ is calculated by computer from measurements made by the wavefront sensor and introduced by means of a servoloop. Various mirror arrangements can be used, e.g., mirrors with actuators, *bimorphic* mirrors modifying curvature locally, and microelectromechanical systems (MEMS). The aim is generally to reduce the residual phase error below $\lambda/4$ (Rayleigh criterion), although some applications such as EXAO (see below and Sect. 6.6) may require a

Fig. 6.17 The first adaptive secondary mirror, with $N = 336$ actuators, on the 6.5 m MMT telescope (Mt. Hopkins, Arizona). This elegant solution, obtained with a very thin (1.8 mm) and flexible mirror, reduces the optical complexity of the adaptive optics. Minimising the number of optical surfaces, it also reduces the intrinsic thermal emissivity of the instrument

lower value. In terms of the Zernike polynomials, this constraint fixes the highest degree that needs to be corrected.

In the particular case of a Shack–Hartmann sensor with N sub-pupils, the mirror must also have N degrees of freedom and the correction is then acquired up to the polynomial of degree $2N$. We thus obtain the expression

$$N(\lambda) \approx \left[\frac{D}{r_0(\lambda)}\right]^2 ,$$

which is the number of degrees of freedom here. Note that the $\lambda^{6/5}$ dependence of $r_0(\lambda)$ leads to a very rapid decrease in N with increasing λ. Corrections are therefore much easier in the infrared.

A second constraint imposed on the deformable mirror is its passband. Indeed, it is inserted in a servocontrol loop that must respond perfectly to temporal phase variations, which themselves have the characteristic time $\tau_c(\lambda_0)$, fixed by the atmosphere and also varying as $\lambda^{6/5}$. Since a servocontrol system must always have a much broader passband (greater by a factor of 10) than the phenomenon to be corrected, i.e., around $10/\tau_c$, the deformable mirror must respond to frequencies of several kHz.

6.3.3 *The Final Image*

The quality of the final image can only be completely assessed by examining its profile $I(\boldsymbol{\theta})$, in order to compare it with the profile of the ideal image. Since the correction is never perfect, the profile of the instantaneous image exhibits

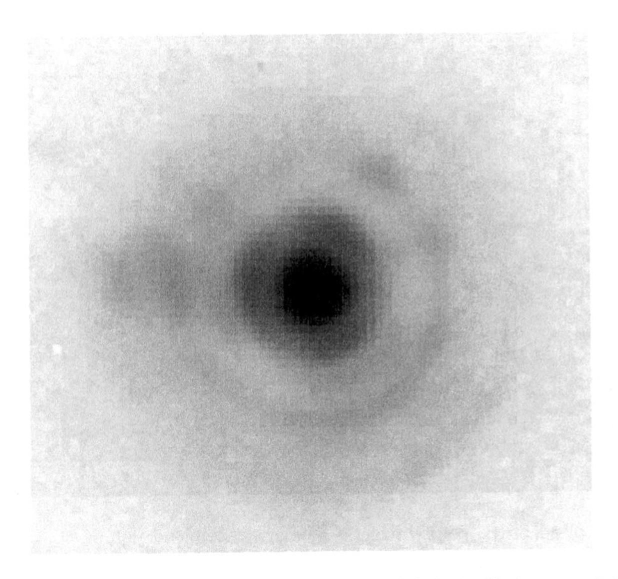

Fig. 6.18 Image of the unresolved star HIC69495 ($m_V = 9.9$) in the K photometric band (2.2 μm) with a 100 s exposure time, displayed on a logarithmic scale. This (PSF) was obtained using the multi-application curvature adaptive optics (MACAO) system that has equipped each of the 8.2 m telescopes of the VLT since 2003. The mirror has deformable curvature and includes $N = 60$ actuators. The Strehl ratio is $\mathscr{S} = 0.55$ and the image FWHM reaches 0.06 arcsec. This remarkable image clearly shows the three first Airy diffraction rings. Document European Southern Observatory

random fluctuations (residual speckles), whereas these are averaged out in the long exposure image. When the correction is reasonably good, the long exposure profile generally exhibits a core very close to the Airy profile (the diffraction limit), surrounded by a halo due to the non-corrected phase residues $\delta\phi$ (Fig. 6.18). As this residue increases, the proportion of energy in the core is attenuated in favour of the halo, until the classic uncorrected long-exposure image in the presence of seeing is reached (see above). Hence the terms *partial* and *total correction*. This correction is quantified by the *Strehl ratio* $\mathscr{S}$. The Strehl ratio of the image of a point source is defined as the ratio of the maximal intensity of the corrected image to the maximal intensity of an image limited only by diffraction at the wavelength considered. It is given by

$$\mathscr{S} = \frac{\displaystyle\int \tilde{S}(\boldsymbol{w})\, \mathrm{d}^2\boldsymbol{w}}{\displaystyle\int \tilde{T}_{\mathrm{diff}}(\boldsymbol{w})\, \mathrm{d}^2\boldsymbol{w}} ,$$

and varies from 1, for total correction, to less than 10^{-3}, for a highly speckled image.

Figure 6.19 shows the result of an adaptive correction carried out on the image of a double star.

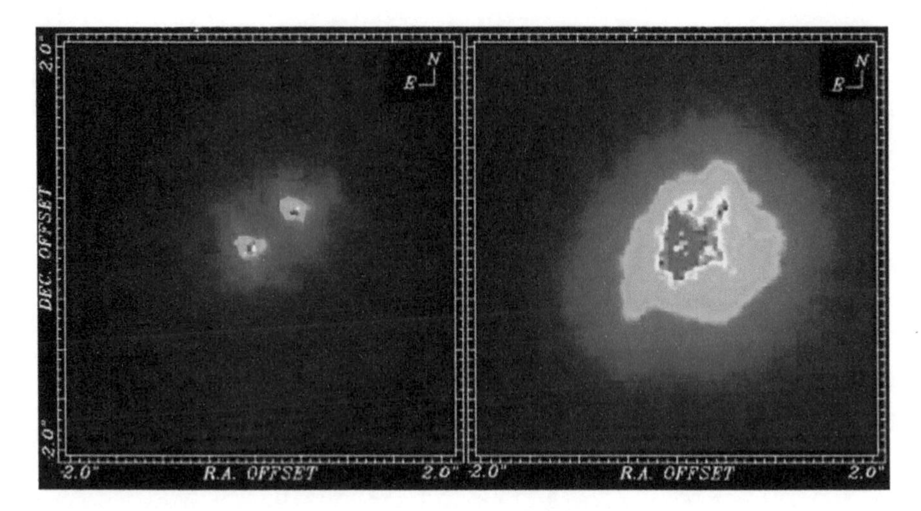

Fig. 6.19 Birth of adaptive optics. Image of a binary system (magnitude $m_v = 13.1$, separation $0.59''$) using a 3.6 m telescope at wavelength 1.65 μm (European Southern Observatory, La Silla, Chile). *Right*: Image affected by atmospheric turbulence (seeing $1.7''$). *Left*: Corrected image, exposure time 10 s, final resolution $0.12''$ (FWHM), 1 pixel = 50 mas. Adaptive system Come-On Plus, 1993. (Picture by Beuzit J.L., Paris Observatory and ESO)

6.3.4 Sensitivity and Reference Sources

The operation of the servoloop in an adaptive optics system depends on the quality of information received from the wavefront sensor and transformed into commands to the degrees of freedom of the active mirror. The signal-to-noise ratio of this command is thus critical. Naturally, it depends on the magnitude of the source under observation and the sensitivity of the detector equipping the analyser. There will clearly be a limiting magnitude beyond which it is no longer possible to close the loop. This is the Achilles heel of adaptive optics, limiting its application to relatively bright objects. It led to the development of a *laser guide star* or artificial reference source. The wavelengths λ_0 and λ for analysis and measurement of the image, respectively, can differ, provided that the condition $\lambda_0 \leq \lambda$ is satisfied, in order to correctly measure high spatial frequencies.

The ideal detector, associated with the wavefront sensor, must have a quantum efficiency η as close as possible to unity and a response time compatible with the atmospheric turbulence, characterised by correlation times which vary from $\tau_c \approx 1$ to 100 ms, depending on the value of λ_0, and the cause of the turbulence (e.g., wind, convection). Avalanche diodes ($\eta \approx 0.8$), EB-CCDs in counting mode ($\eta \approx 0.08$), or rapid low-noise CCDs, which are difficult to make ($\eta \approx 0.6$), can be used (see Sect. 7.4.6 for more about these detectors).

When specifying an adaptive optics system, the first step is to fix the number of modes one would like to correct at a given wavelength and under given turbulence

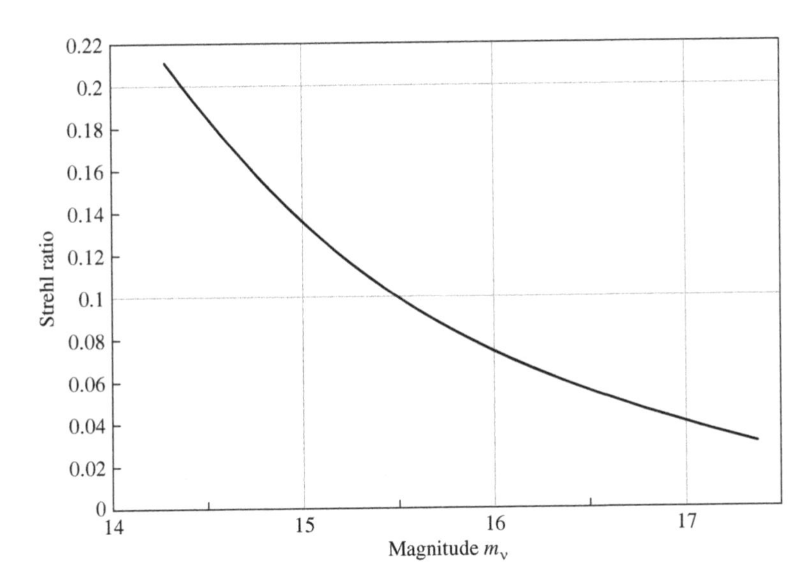

Fig. 6.20 Sensitivity of adaptive optics. *Ordinate*: Strehl ratio $\mathscr{S}$ for an image corrected at $\lambda = 2.2\ \mu\mathrm{m}$ (spectral band K), in average turbulence conditions. *Abscissa*: Magnitude m_V of the source used by the wavefront analyser (hence analysis wavelength $\lambda_0 = 0.55\ \mu\mathrm{m}$). We assume here that the sensor is equipped with a detector with high quantum efficiency (CCD, $\eta = 0.6$) and a very low readout noise ($2e^-$ rms). From Gendron E., doctoral thesis, 1995

conditions. This determines the number N of actuators required for the deformable mirror and the number of sub-pupils of the analyser (almost the same). Once the properties of the analyser are established, it is straightforward to calculate the Strehl ratio obtained for a source of given magnitude m_{λ_0}. The fainter the source, the more difficult it is to correct the high modes (or spatial frequencies), since the signal-to-noise ratio in the measurement of the variance of these modes tends to decrease. Under reasonable hypotheses, Fig. 6.20 shows the correction prospects as a function of the magnitude of the observed object. It can be seen that a high quality correction requires a magnitude m_V less than 13–14. In fact, taking sensor performances into account, the spectral band R is often chosen for the analyser, rather than V.

Note here that the diameter of the telescope is irrelevant. Indeed, the light intensity available in each sub-pupil of the analyser is the determining factor, and that only depends on the value of the Fried parameter $r_0(\lambda)$.

Choice of Reference

All the above reasoning is based on the initial assumption that the observed source is a point, or at least, not resolved at the diffraction limit of the telescope. Can adaptive optics be used to correct the image of an extended and resolved source S ? The answer is affirmative, but two cases need to be considered:

- In the immediate vicinity of the source S, there is another source R, called the reference, which is a point source bright enough and close enough to S to ensure that the correlation between the waves $\psi_S(\boldsymbol{r},t)$ and $\psi_R(\boldsymbol{r},t)$ is good: R is in the isoplanatic patch of S, whose value was specified above, i.e., several tens of arcsec under suitable observing conditions. The wavefront analyser calculates the correction at λ_0 by observing R, and the camera forms the image at λ of the chosen object S.
- When there is no such nearby source, the only solution would be to use S itself as reference. But is that possible when S is resolved? Although this may seem paradoxical, the answer is affirmative, provided that the angular size of S is significantly smaller than the size of the isoplanatic patch. Indeed, S may then be treated as comprising an ensemble of point sources, detected by the analyser and all suffering the same phase perturbation. The measurement made by the analyser does indeed then determine this phase perturbation globally for S. In practice, it is the field accepted by each of the sub-pupils of the analyser (generally 2–3 arcsec) which fixes the maximal acceptable size of S for the adaptive optics system to work properly.

How could adaptive optics be applied to observation of the Sun, which is a very large object? The idea here is to use very small details on the solar surface as reference, provided that they contrast sufficiently with their neighbourhood, e.g., granules, faculae, or spots. The results are spectacular.[17]

Artificial Reference: Laser Guide Star

Given the isoplanatic patch for a given wavelength λ_0, together with the density of stars of adequate magnitude to reach a given value of the Strehl ratio $\mathscr{S}$, it is easy to show that the *sky coverage* obtained by an adaptive optics system is low, of the order of a few percent at best. This nevertheless gives access to many valuable observations of stars and galactic objects, although only a small number of extragalactic objects, which are rather faint and rarely come with a bright star that could serve as reference, especially in the vicinity of the galactic pole. It was thus necessary to create an *artificial reference star* that could be set up at will not far from the object S whose high resolution image is to be obtained.

This idea, first proposed by V.P. Linnik in 1957, then shelved for thirty years, was picked up again by R. Foy and A. Labeyrie (France) in 1987, in the following form. In the Earth atmosphere, there is a layer at an altitude of about 80 km containing a high density of sodium (Na) atoms. These atoms are deposited there,

[17] Some beautiful examples can be found at the website of the National Solar Observatories (NSO) in the US, at the address www.nso.edu.

and regularly refreshed, by meteoritic particles bombarding the Earth. If a laser beam of wavlength $\lambda = 589.3$ nm is shone from the ground up to this layer, it locally excites the sodium resonance transition, which has a particularly high absorption cross-section. The spontaneous emission of atoms returning to their ground state can then be observed from the ground, and provides an almost point source of monochromatic light which is easily picked out from the atmospheric light background by spectral filtering. This source, which is thus like an *artificial star*, can then be used as reference for a wavefront analyser, provided that the laser power is high enough to give it a sufficient magnitude. Aligned with the telescope, it can be positioned with great accuracy in the immediate vicinity of the observed source. The first truly astronomical laser guide star was set up on the 3.5 m Calar Alto telescope (*Max Planck Institut für Astronomie*, Spain) in 1996 (the US military having pursued their own investigations in the meantime).

Although the idea of an artificial *laser guide star* (LGS) is an attractive one, several problems arise in comparison with a *natural guide star* (NGS):

- The reemitted light must come back through the same turbulent layers as the light emitted by the source S, situated effectively at infinity. But since the sodium layer is at an altitude of 80–90 km, this condition is not entirely satisfied, giving rise to a geometric *cone effect* which reduces the quality of the correction. This can be partially corrected by using several laser spots in the field.
- The laser light goes up through the atmospheric layers and the reemitted light comes back down through the same layers, whence the tip–tilt effects on the wavefront are suffered twice in opposite directions and therefore cancel. While these effects dominate the image perturbation, they escape measurement here and are thus left uncorrected. The first method for dealing with this, which is still employed, is to use a natural star to measure those, taking advantage of the large isoplanatic patch associated with tip–tilt (several arcmin), which improves the chances of finding a star there of adequate magnitude. Other, more subtle methods are under investigation, e.g., using photons that are backscattered at different wavelengths in combination with the very slightly chromatic character of the tip–tilt affecting their images.
- Finally, given the remoteness of the sodium layer, the power required of the laser is very high (typically around 10 W continuous wave), and the laser must be carried by the telescope, and accurately aligned parallel to the telescope axis. This is why another kind of artificial star known as a *Rayleigh star* has been under investigation. These are formed by Rayleigh backscattering from atmospheric molecules in the lower atmosphere. The resulting correction is then only valid for turbulence occurring very close to the telescope aperture.

Despite these limitations, many instruments at the time of writing (2006) are equipped with a laser and adaptive optics, as exemplified by the Keck Telescope (Hawaii) in 2004, the European VLT in 2005 (see Fig 6.21), Gemini South (Chile) in 2007, and others.

Fig. 6.21 At the Paranal site in Chile, the Yepun telescope of the VLT is equipped with a laser producing an artificial guide star, used as a reference for NAOS adaptive optics. Image kindly provided by Stefan Seip (astromeeting.de) and insert by T. Ott, ESO

6.3.5 New Concepts

The above discussion highlights the principles and difficulties of adaptive optics. Since such systems will be absolutely essential for telescopes with diameters in the range 30–50 m, currently planned for ground-based observation, every attempt is being made to improve them in any way possible. The emergence of new potential sites with very different turbulence characteristics, such as the Antarctic (see Sect. 2.8) where turbulence is almost exclusively restricted to the immediate vicinity of ground level, considerably changes the conditions for setting up an adaptive optics system.

Multi-Conjugate Adaptive Optics (MCAO)

Multi-conjugate adaptive optics (MCAO) is a version of adaptive optics exploiting the fact that the phase error $\varphi(\boldsymbol{r}, t)$ is the sum total of effects produced in the different atmospheric layers encountered, whence the adaptive optics correction will be improved if the effect of each layer can be corrected by a different deformable mirror, whose surface is optically conjugate to the position of the layer (since these layers are not at infinity, their images thus form at different positions within the telescope). Figure 6.22 shows the general principle and Figure 6.23 a detailed MCAO simulation. In practice, it suffices to correct two or three layers separately by creating the same number of artificial stars (because of the geometric effect) to significantly improve the final correction.

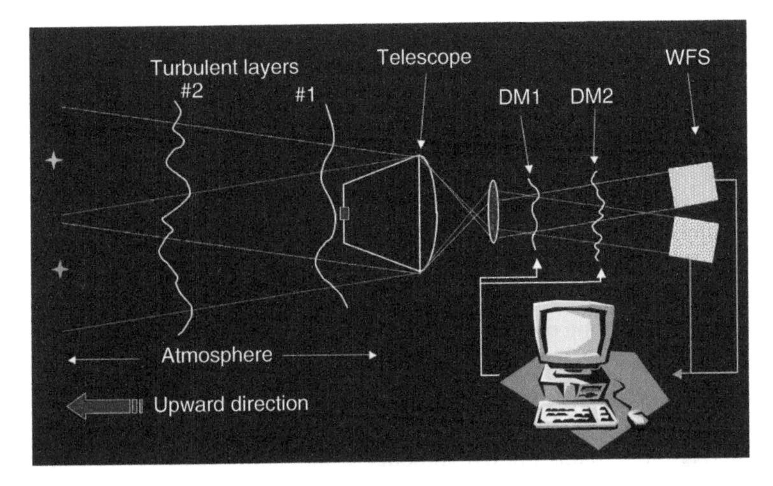

Fig. 6.22 General setup for multi-conjugate adaptive optics (MCAO). The two deformable mirrors DM1 and DM2 are optically conjugate to the two atmospheric layers 1 and 2 to be corrected for — usually the layer close to ground level and another higher up, at 8–10 km — while the wavefronts from the two spatially distinct laser guide stars are analysed by two wavefront sensors (WFS). The calculator uses these measurements to deduce the corrections to be applied to the two mirrors. Diagram kindly provided by F. Rigaut, 2006

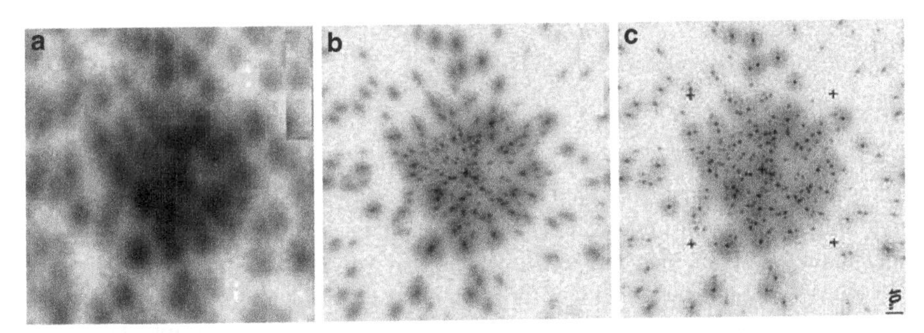

Fig. 6.23 Simulation of multi-conjugate adaptive optics operating in spectral band H ($\lambda = 1.65\,\mu\text{m}$) and designed for the Gemini South telescope (Chile). The field is 150 arcsec (2.5 arcmin). (**a**) Uncorrected image of star field. (**b**) Image corrected by conventional adaptive optics. Note the significant anisoplanicity of the correction, with lengthened images as one moves off the axis. (**c**) Image corrected by multi-conjugate adaptive optics. The quality of the correction is restored practically throughout the field. Document kindly provided by F. Rigaut, 2006

Ground Layer Adaptive Optics (GLAO)

Ground layer adaptive optics (GLAO) is a version specifically designed to correct for the effects of turbulence close to ground level, noting that in certain sites this largely dominates image degradation. In this case, the isoplanatic patch can be much larger, since to a first approximation it is trivially given by

$$\alpha_{\text{iso}}(\lambda) = \frac{r_0(\lambda)}{H},$$

where H is the altitude of the dominant layer. The correction is clearly more limited here (smaller value of $\mathscr{S}$), but the corrected field can easily reach several arcmin. This version of adaptive optics is highly promising for use in the Antarctic (Dome C), where most turbulence occurs very close to ground level, within about 50 m at most.

Multi-Object Adaptive Optics (MOAO)

Multi-object adaptive optics (MOAO) is a technique designed for the study and high-resolution imaging of galaxies, in regions where there are many galaxies in the field of view. It is inspired by the technique of multi-object spectroscopy described in Chap. 8. A corrective microsystem is installed by means of a moving arm on the image of each galaxy.[18] Each corrector contains only a small number of actuators, so the correction remains limited, although it may nevertheless represent an order of magnitude in resolution compared with the image obtained without adaptive optics, enough to be able to carry out a spectrographic analysis of different regions of the galaxy, e.g., to determine its rotation. The perturbed phase is measured simultaneously on several stars in the field in order to improve the signal-to-noise ratio, with the correction calculator adapting the latter to each point of the field to be corrected.

Extreme Adaptive Optics (EXAO)

When the correction gives a Strehl ratio $\mathscr{S}$ greater than about 0.6, the core of the image and also the first rings are practically identical to those of a diffraction-limited image. However, the uncorrected residual phase $\delta\varphi$ produces a halo around the core, with random intensity fluctuations in the instantaneous images. These average out in the long exposure image to produce a uniform halo, with an intensity that is never less than about one thousandth of the peak intensity and which makes it impossible to detect a faint object close to the star, e.g., an exoplanet. Coronagraphy techniques (see Sect. 6.6) aim at such detections, but operate better if the halo is reduced.

To reduce it, the number of corrected Zernike modes must be increased, and hence likewise the number N of sub-pupils and correction actuators. This is what is provided by extreme adaptive optics systems. For example, in the near infrared (spectral bands J, H, and K) and for a 10 m telescope, using a mirror with 4 000 actuators (MEMS), it seems possible, over a very small field (about one arcsec), to obtain a value of $\mathscr{S}$ greater than 0.9, and to reduce the halo intensity of a bright star ($m_R \approx 6$–7) to 10^{-7} at a distance of a few times λ/D from the central peak.

Adaptive optics has considerably improved the possibilities for observation using large ground-based visible and near infrared telescopes. Some have even claimed,

---

[18]The Falcon project, developed at the Paris Observatory and going into operation on the VLT in 2007, can be used to observe 15 galaxies at the same time, with spectral resolution $\mathscr{R} \geq 600$. See www.onera.fr/dota/publications.php.

perhaps with a slight exaggeration, that it represents the greatest revolution in Earth-based optical observation since Galileo. In any case, the growing diameters of telescopes, already over 10 m, can now be accompanied by a growth in angular resolution. Apart from the gain in resolution, the possibility of reducing the size of the PSF has many consequences. To name but a few:

- Improved contrast in images of unresolved objects in the presence of a background, and hence a gain in signal-to-noise ratio.
- Coronagraphy.
- Reduced size of spectrometer slits for the same spectral resolution.
- Possibility of injecting the image directly into single-mode optical fibres for interferometric applications.
- Reduced confusion in rich fields, e.g., globular clusters.

The often highly accurate knowledge of the point source response (PSF) provided by adaptive optics allows elaborate data processing of corrected images such as deconvolution, which can improve the quality. This point will be discussed further in Sect. 9.6.

To end this section, which deals with a rapidly evolving field, let us mention one of the most outstanding results obtained to date: the observation of stellar orbits very close to the center of the Galaxy, leading to a determination of the mass of the black hole which is very likely located there (Fig. 6.24).

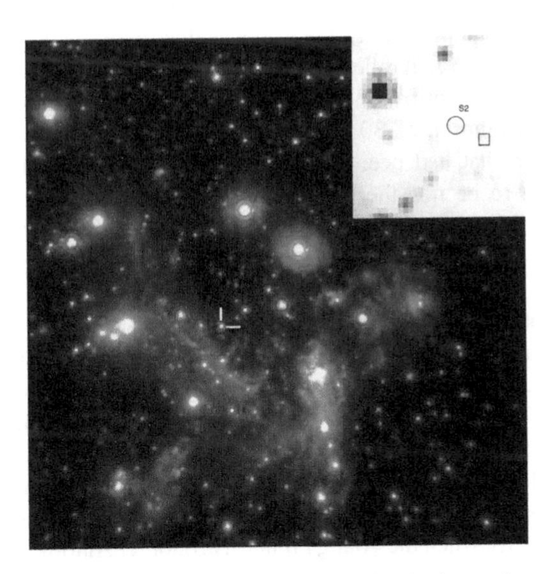

Fig. 6.24 Observation of the galactic centre using NAOS adaptive optics on the Very Large Telescope. Image obtained at $\lambda = 3.5\ \mu\text{m}$ (spectral band L′), in a field of 21.9×21.9 arcsec corrected by adaptive optics. The intensity scale is logarithmic. *Insert*: Magnification of the central 2×2 arcsec region around the star S2 (○) marked by two dashes in the main image and close to the central radiosource SgrA* (□). The mass of the central black hole has been estimated by monitoring the motion of this star, knowing the distance from the galactic center to the Sun. Document kindly provided by Y. Clénet, in collaboration with R. Genzel, 2004

6.4 Astronomical Interferometry

We saw earlier that two sub-pupils 1 and 2 sampling the wavefront from the source at two points a distance D apart works like a bandpass spatial filter, whence it is possible to measure the Fourier components of the source. By varying D, the spatial spectrum of the source is then explored point by point, whereupon its image can be reconstructed in principle, with a spatial resolution that is then only limited by the maximal value of D.

It was the insuperable limitation imposed by the diameter of the primary mirror of the telescope, for technological (optical and mechanical) reasons, which led to the development of astronomical interferometry. It all began with a remarkable analysis by Hippolyte Fizeau in 1868, followed by the first attempts to measure star diameters by Edouard Stephan in 1874. The latter only obtained an upper limit because his telescope was too small. It was Albert Michelson who first measured the diameters of the the Galilean moons of Jupiter (1891). Then 20 years later at Mt. Wilson (California), Francis Pease succeeded in measuring the diameters of several stars, including Betelgeuse (α Ori) (Fig. 6.8). Only the extraordinary talents of Michelson seemed able to overcome the extreme experimental difficulties (due to vibrations) of maintaining the coherence of the light in two optical paths. But after this success, optical interferometry stopped in its tracks for half a century (apart from one particular form, namely homodyne interferometry, discussed below).

During World War II, radar operators noticed that the radar echos reflected from an aircraft varied in amplitude. The reason was that there was interference between the direct signal and a signal reflected by the mirror surface. This phenomenon, well known in optics (Lloyd mirrors), resulted in a maritime interferometer in Australia (1946): with a resolution of 30 arcmin, observation of the Sun showed that the variable emission that had been noticed came from a region that was too small for the emission to be non-thermal.[19] This was the beginning of *radio interferometry*. And it was at radiofrequencies that it continued. In 1946, in Cambridge (UK), Martin Ryle[20] was working at decimetre wavelengths, where it was much easier from the experimental point of view to maintain coherence. A series of radio interferometers was used to catalogue the positions of newly discovered radiosources (catalogues 1C–4C). The wavelength decreased gradually from the metre to the centimetre, sensitivity improved, baselines extended to culminate in the *Five Kilometer Radiotelescope*, and aperture synthesis came gradually into being.

The technique developed very quickly and the harvest of results was rich, resulting in a large family of radiotelescope arrays. Between 1921, when the Mt. Wilson telescope was successfully used in interferometric mode by Michelson and Pease, and 1975, when the young French astronomer Antoine Labeyrie obtained

[19]This remarkable observation is mentioned by Moran, Ann. Rev. Astron. Astroph., 2001. Much here is borrowed from this excellent reference.

[20]The British astronomer Martin Ryle (1918–1984) won the Nobel Prize for Physics with Anthony Hewish in 1974 for his work on radiofrequency interferometry.

his first fringes from the star Vega, optical interferometry was never completely abandoned. The main problem was to measure star diameters, where almost no progress had been made since the measurements by Pease. The methods of phase closure, optical telescope arrays, pupil masking (already used by Edouard Stephan in 1874) were discussed (D. Currie). In 1970, at the Pulkovo observatory (Crimea, USSR), E.S. Kulagin published a measurement of the orbit of the star Capella made with a Michelson type interferometer, using a 6 m baseline. He thus preceded Labeyrie by a few years. In parallel, the homodyne method was put into practice by Robert Hanbury-Brown and Robert Twiss at Jodrell Bank (UK) in 1956, then in Australia where they measured the diameters of Sirius and a few other bright stars. Finally, in 1972, Jean Gay at the Paris Observatory successfully applied the radiofrequency heterodyne technique in the mid-infrared (10.6 μm). This approach had also been developed by Charles Townes[21] and his students at Berkeley (United States).

This work renewed interest in astronomical interferometry in the optical (visible and infrared) region, leading to considerable progress in angular resolution, and the construction of several arrays by the beginning of the twenty-first century, as we shall describe below.

In the present section, we shall discuss the various forms of astronomical interferometry on the basis of the principles given above. These different forms depend to some extent on the wavelength, but we shall always be able to identify a certain number of common functions: sub-pupils collecting incident radiation, transfer of the amplitude and phase data received by each sub-pupil to a recombination device, correction where necessary for effects due to the different optical paths of the two signals, and detection of the recombined signal, usually in the form of interference fringes, by a suitable detector for the given wavelength.

6.4.1 *Obtaining an Interferometer Signal*

There are several ways to recombine the radiation from the sub-pupils to obtain and measure either the interference pattern or any other quantity containing the same information (Fig. 6.25). The choice of one method rather than another depends on the means applied to detect the radiation (Chap. 7).

Direct (Fizeau) Interferometry

The spatial frequency filter that constitutes an interferometer is exactly realised by small apertures (diameter d) forming a *pupil mask* that covers the primary mirror of a telescope. An image detector like a camera, if there is such a thing at the given

[21]The American physicist and astronomer Charles Townes, born in 1915, received the Nobel Prize for Physics in 1988 for his work on maser pumping, the maser being a precursor of the laser.

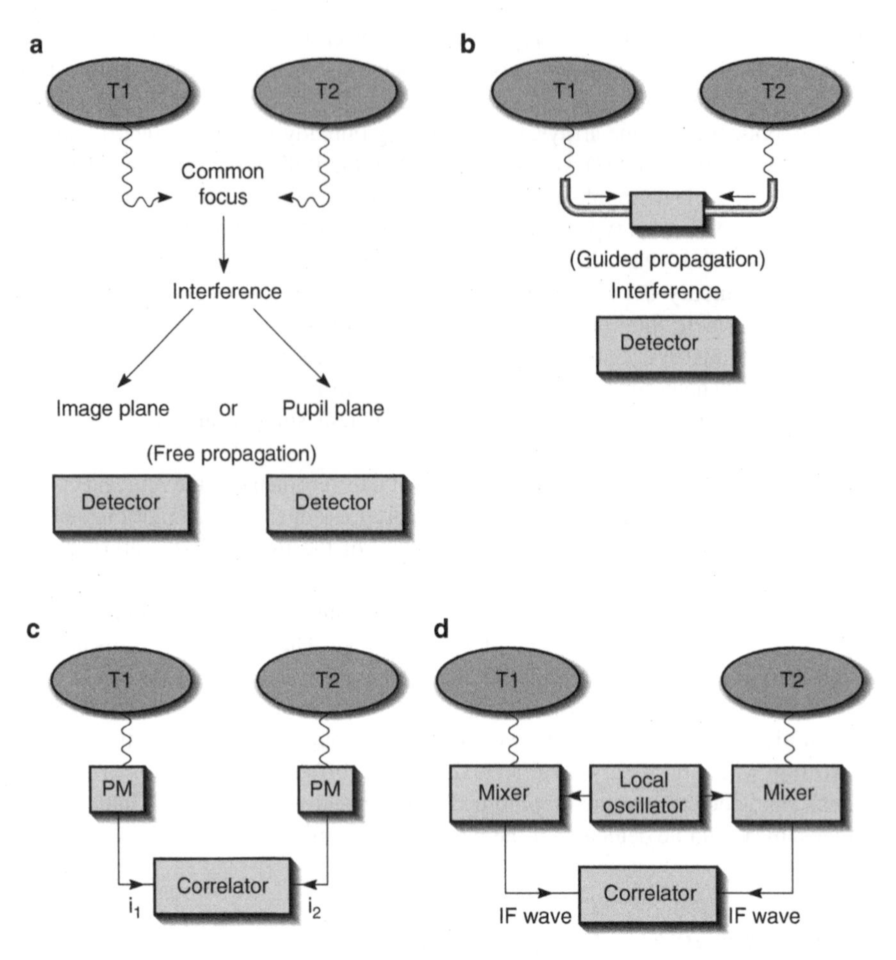

Fig. 6.25 Different kinds of interferometric combination. T_1 and T_2 denote two widely separated telescopes. (**a**) Direct interferometry. The waves travel to the common focus, where an interference pattern (image plane or pupil plane) forms on the detector (Fizeau, Michelson, or densified pupil configurations). (**b**) Direct interferometry by guided optics, using optical fibres. (**c**) Homodyne interferometry. Two photomultipliers in photon counting mode deliver currents i_1 and i_2, analysed by a correlator. (**d**) Heterodyne interferometry (radiofrequencies and mid-infrared). Frequencies of incident waves are changed and a correlator analyses the intermediate frequency (IF) signals

wavelength, placed at the focus, measures the resulting fringes. This is the *Fizeau configuration*. It has the advantage of having a wide field, that of the telescope itself, since the optical paths between each aperture and the focal point remain equal for all points of the field. This can also be achieved with two separate but rigidly joined telescopes, as with the Large Binocular Telescope (LBT) discussed in Sect. 5.2.2, or the planned European space observatory Darwin. Since the two telescopes and their common base are rigidly fixed, so is their pointing and the optical paths from the source to the common focus are identical.

Direct (Michelson) Interferometry

This is the most commonly used configuration in optical (visible and infrared) interferometers, e.g., the Very Large Telescope Interferometer. A disconnected and independent pair of telescopes of diameter D, constituting two sub-pupils, arranged on a baseline B, collect light which is then transferred by optical train and combined at a common focus. Here the complex coherence of the radiation is measured (Fig. 6.26), after eliminating the optical path difference by a delay line. Since the telescopes are independent, the effect of the Earth's daily rotation must be corrected for using delay lines, equalising the lengths of the optical paths between the sub-pupils and the common focus.

Like the last method, this one requires an image detector (camera) to analyse the fringes after recombination. The difference here is in the way the beams are transferred from the sub-pupils to the common focus. Before recombination, an exit pupil must be reconstituted that is homothetic to the entrance pupil. The interferometric field of view is generally small (a few arcsec).

Also known as direct interferometry, this setup measures the complex visibility directly, with a spectral band that can be quite broad ($\Delta\lambda/\lambda \leq 0.1$), i.e., with reduced temporal coherence, or again, with few fringes observable near the zero path length difference (PLD), and therefore high sensitivity. This works best in

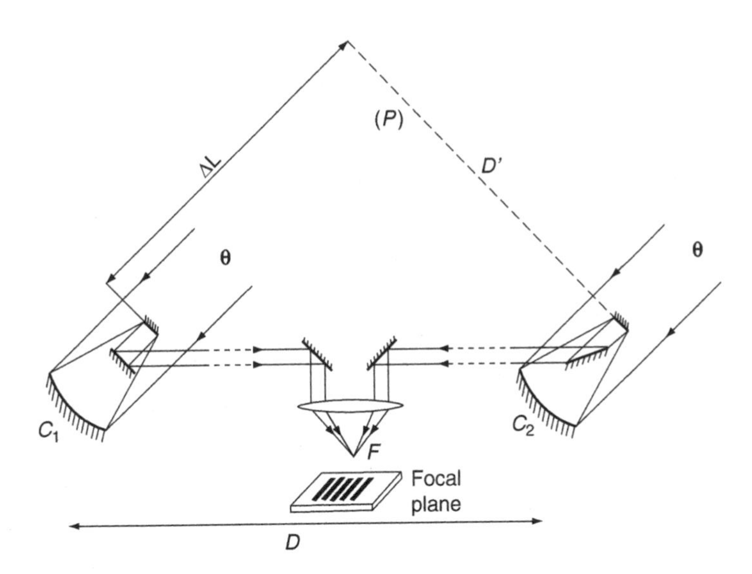

Fig. 6.26 Optical (visible or infrared) interferometry from the ground (Michelson method). The two telescopes point in the direction $\boldsymbol{\theta}$. The afocal beams are guided by mirrors or optical fibres, then recombined in a common focal plane F, where various combination schemes can be used (image plane, pupil plane). The complex visibility $\mathcal{V}$ is measured. A continuously variable delay line compensates for the daily rotation of the Earth (variation of the zenith distance θ): $FC_2 - FC_1 = \Delta L(t)$. The random residual differential phase due to the atmosphere applies to the visibility $\mathcal{V}$

the near infrared spectral region (1–12 μm), because the properties of atmospheric phase effects are generally more favourable: the seeing size r_0, the characteristic time τ_c, and the isoplanatic patch α_{iso}, all three of which vary as $\lambda^{6/5}$, whereas the constraints imposed by mechanical and optical accuracy vary as λ. This explains why the method is more difficult to use in the visible, and hence still limited in this region.

Heterodyne Interferometry

Here the electric fields in the direction $\boldsymbol{\theta}$ are detected locally behind sub-pupils 1 and 2 (Fig. 6.25d). The amplitude and phase are measured by interfering each with a field produced by a heterodyne local oscillator (see Sect. 7.5.2). The intermediate frequency (IF) signals are easily carried by cable to a common point, or even recorded and recombined later, as in very long baseline interferometry (VLBI). Since the problem is to measure the coherence between the two parts of the incident wavefront sampled by sub-pupils 1 and 2, we only need to calculate the *correlation product* of these fields in order to determine the complex coherence, which is simply $I(\boldsymbol{w})$ normalised:

$$|\gamma_{12}(\tau)| = \left|\left\langle E_0 E_1(t) \mathrm{e}^{2\pi \mathrm{i}(\nu-\nu_0)t} \otimes E_0 E_2(t+\tau) \mathrm{e}^{-2\pi \mathrm{i}(\nu-\nu_0)(t+\tau)}\right\rangle_t\right| .$$

In contrast with direct combination of the beams, heterodyne combination works first by a change of frequency, which conserves the phase of the signal, at the focus of each telescope, then by recombination of the intermediate frequency (IF) signals in a correlator. Although the aim is the same, i.e., to measure the level of spatial coherence $\gamma_{12}(\tau)$, there is an important difference between the two methods. In the first case, the usable bandwidth is broad, typically $\Delta\lambda/\lambda \sim 10^{-2}$ at $\lambda = 0.5$ μm for $r_0 = 10$ cm. In the second case, the bandwidth is limited by the *mixer*, with value up to a few GHz. Going from radiofrequencies towards the infrared and the visible, the relative bandwidth, and hence the detectable energy, decreases, as can be seen from Table 6.3. The sensitivity is thus limited, and all the more so as the wavelength gets shorter.

As for radiofrequencies, the gain in *temporal coherence* with decreasing relative bandwidth makes it easier to detect the fringes and measure their visibility, even in the presence of large phase differences due to optical path differences caused

Table 6.3 Spectral width in interferometry for four different wavelengths. The table gives values of relative spectral width $\Delta\lambda/\lambda = \Delta\nu/\nu$

		1 mm	100 μm	10 μm	1 μm
Heterodyne interferometry $\Delta\nu = 300$ MHz		10^{-3}	10^{-4}	10^{-5}	10^{-6}
Direct interferometry	Ground	–	–	10^{-1}–10^{-2}	10^{-2}–10^{-3}
	Space			$\lesssim 1$ if desired	

by mechanical or atmospheric perturbations. This explains the successful use of the heterodyne method in the mid-infrared, especially as CO_2 lasers can provide a stable and powerful local oscillator at 10.6 μm.

The *heterodyning technique* is mainly applied in radiofrequency interferometry (from submillimetre to metre wave) and in the mid-infrared.[22] It has not been extended to shorter wavelengths.

Photon Correlation Measurement

Here the photocurrent is measured directly using sensitive photoelectric detectors placed at the focus of sub-pupils 1 and 2. The time correlation of the two currents can then be measured, transferring them by cable to the correlator (Fig. 6.25c). Zernike's theorem shows that a measurement of this correlation gives the coherence $|\gamma_{12}(\tau)|$ between 1 and 2 directly, and this is indeed the desired quantity. In the 1960s, this method was successfully applied by Hanbury-Brown[23] in Australia.[24] The future of this so-called *homodyne interferometry* is relatively limited, being applicable only to measuring the diameters of bright stars, and requiring telescopes with diameters of several metres, while the two apertures can of course be widely separated. Note, however, that the method does not require any special image quality at the focus of the sub-pupils, and does not therefore require adaptive optics. It could one day become relevant again, with the advent of giant telescopes on the one hand, and with the availability of very high bandpass quantum photodetectors (GHz, or even THz).

In practice, then, there are four different interferometric setups, the first two being rather similar. The usual terminology tends to contrast these methods, referred to as interferometric (because they produce fringes rather like Young's fringes), with conventional imaging as it occurs at the focus of a classical mirror. It has been shown above that there is, in fact, no difference whatsoever, and we might just as well say that any image is the interference pattern formed by the whole set of waves diffracted at the pupil, whether that pupil be connected or not. The only difference is in the frequency filtering imposed by the given pupil: low pass, it provides an image that closely resembles the object; bandpass, the image is made up of interference fringes and no longer resembles the object.

[22]See, for example, Townes C.H., in *Optical and Infrared Telescopes for the 1990s*, A. Hewitt (Ed.), Kitt Peak National Observatory, 1980.

[23]The British physicist and astronomer Robert Hanbury-Brown (1916–2002) invented intensity interferometry with the Indian-born British mathematician Richard Twiss (1920–2005), developing their idea in Australia.

[24]A very complete description is given in Hanbury-Brown R., The intensity interferometer, ARAA **6**, 13, 1968.

6.4.2 *Light Transfer*

Since an interferometer has two sub-pupils at least several tens or hundreds of metres apart, this raises the problem of transferring light or information without loss of coherence (see Fig. 6.25). As it happens, there are several ways to transfer light, with or without loss of coherence. We shall review these briefly, then consider the use of optical fibres (see Fig. 6.25b), which recent technological progress has extended to all kinds of optical arrangements.

Diversity of Light Transfer Techniques

Depending on the wavelength and the optical constraints imposed by the given instrumentation, several transfer methods are possible:

- At optical (visible or infrared) wavelengths, conventional setups always involve mirrors or lenses, whatever the wavelength being studied. Transfer should ideally be made in vacuum, so that there is no phase perturbation. This is clearly achieved in space, but would be extremely expensive in ground-based interferometers with baselines of a hundred metres or more. These generally transfer through air, care being taken to avoid thermal instability in light ducts.
- When heterodyne combination is possible (radiofrequencies or infrared), the IF signal downstream of the frequency change device is easily transported by cables, in which the risks of phase perturbation are very small.
- Upstream of the frequency change, *waveguides* provide a way of transferring an electromagnetic signal from one point to another while preserving the phase. This is generally a hollow tube, with rectangular or circular cross-section, and inner walls made from some reflecting metal. It offers an alternative to free propagation in the air. As well as being made to gradually change direction, the radiation can be mixed with other radiation, or coupled with a detection system. Waveguides are commonly used at radiofrequencies. The transverse dimension of a waveguide is comparable with the wavelength of the radiation it can transport, and the propagation of the radiation is described by Maxwell's equations, taking into account the boundary conditions imposed by the reflecting walls. At the time of writing (2007), technological progress with miniaturisation is such that hollow waveguides can now be made for submillimetre and even mid-infrared applications.
- The recent development of *optical fibres* preserving the coherence of the wave (single-mode fibres) and behaving like hollow waveguides means that signals can be transported in the visible and infrared regions without loss of phase information from the sub-pupils to the common focus, where the complex coherence can be measured. At the time of writing (2007), this mode of transport is under experimental investigation ('OHANA project, see below), but may replace the conventional mode of transfer using mirrors.

Optical Fibres

The considerable recent development of these components for telecommunications has allowed many astronomical applications, in interferometry and spectroscopy, in the visible and near infrared regions ($\lambda \lesssim 10\ \mu\text{m}$), which justifies the discussion here.

An optical fibre consists of two concentric cylinders, the core and the cladding, made from materials of different refractive index (*step-index fibres*), or of a single cylinder in which the refractive index varies continuously from the axis outwards (*graded-index fibres*). When the dimension a of the fibre is large compared with the wavelength of radiation being carried, the geometrical approximation is suitable. On the other hand, when the transverse dimension and the wavelength are of the same order, diffraction must be taken into account, and this case resembles what happens in ordinary waveguides.

The idea of propagation mode is useful. During propagation in the fibre or the waveguide, boundary conditions require a stationary transverse wave (perpendicular to the axis), the moving wave being directed along the axis. The *number of modes* of propagation is, for given dimensions and given wavelength, the number of different stationary configurations allowed by the fibre. Thus, for $a \gg \lambda$, the system is said to be multimode, and the geometrical approximation applies. When $a \approx \lambda$, the single-mode condition can be satisfied.

A single-mode fibre or waveguide has the property of only accepting and propagating one coherence étendue, viz. λ^2, and therefore behaves as a *spatial filter*.

It is useful to define several parameters associated with optical fibres:

- The *numerical aperture* $\sin\theta$ of a step-index fibre, between core and cladding (of index n_{c} and n_{cl}, respectively), is given by

$$\sin\theta = \left(n_{\text{c}}^2 - n_{\text{cl}}^2\right)^{1/2} ,$$

  and the angle θ defines the half-angle at the apex of a cone with axis along the fibre and within which radiation is accepted by the fibre (*acceptance angle* of the fibre).
- The *attenuation* of a waveguide or a fibre is measured in decibel m^{-1} (or decibel km^{-1}). It depends on properties of the materials used and describes absorption and reflection losses. These losses can be less than 0.1 dB km^{-1}. Fibres which are transparent in the infrared have been developed (fluoride glasses, chalcogenides). Rayleigh scattering in the material of the fibre inevitably limits transmission in the blue and precludes it completely in the ultraviolet.
- The *cutoff wavelength* of a single-mode fibre is the highest wavelength that can be propagated for a given dimension a.
- The *coupling efficiency* (as a percentage) measures losses due to mismatching of impedance between free propagation in the air and guided propagation in the fibre. It takes values in the range 40 to 75%.
- The behaviour with respect to *polarisation*: certain fibres have axial symmetry and transmit all polarisations or natural light equally. If this symmetry is broken (elliptical core, torsion in the fibre), certain polarisations are preferred.

- There is *dispersion* in optical fibres, since the propagation speed usually depends on the wavelength. This dispersion is a serious drawback when using single-mode fibres to transfer light in the two arms of an interferometer, because it is then essential to have equal lengths of fibre whenever the aim is to observe a source across a relatively broad spectral band. In the future, it should be possible to make fibres without dispersion from *photonic crystals*.

Single-mode waveguides and optical fibres, which conserve the phase, are invaluable for direct coupling of telescopes used in aperture synthesis (see the discussion of the 'OHANA interferometer below), but also for coupling a single coherence étendue (a single mode) between focal plane and detector, which is the case for radiofrequency imaging and also infrared adaptive imaging at the diffraction limit.

At optical wavelengths, multimode fibres are used in the straightforward transportation of light energy between focal plane and instrumentation such as spectrograph and detectors (see Sect. 8.3). However, the transfer rates obtained by telecommunications fibres are already sufficient (2006) to use the fibres to transport the intermediate frequency (IF) signal over very great distances between radiotelescopes operating in very long baseline arrays, making real-time VLBI possible (see below).

6.4.3 Temporal Coherence

Up to now we have been assuming monochromatic radiation, and hence maximal temporal coherence. A fixed optical path difference between beams 1 and 2 is in this case of no importance when measuring the visibility $\mathcal{V}$. However, a real observation requires a finite spectral passband $\Delta\lambda$, implying a reduced temporal coherence and hence a finite number of fringes near the zero optical path length difference (PLD). One must then ensure that path length differences are equal for the radiation following paths 1 and 2, up to a few λ in the case of a broad spectral band, the number of observable fringes being given approximately by $\lambda/\Delta\lambda$. This is all the more relevant given that the Earth's rotation continually modifies this path length difference (Fig. 6.26). Any ground-based interferometer, whatever the wavelength it operates at, must include a compensation system, i.e., a *delay line*, and in some cases *fringe tracking*, except in the very special and obviously convenient case where the two sub-pupils are placed on the same mount, as in the LBT telescope.

6.4.4 Loss of Spatial Coherence

In the ideal picture of an interferometer with two sub-pupils presented above, a measurement of the amplitude and phase of the complex visibility of the

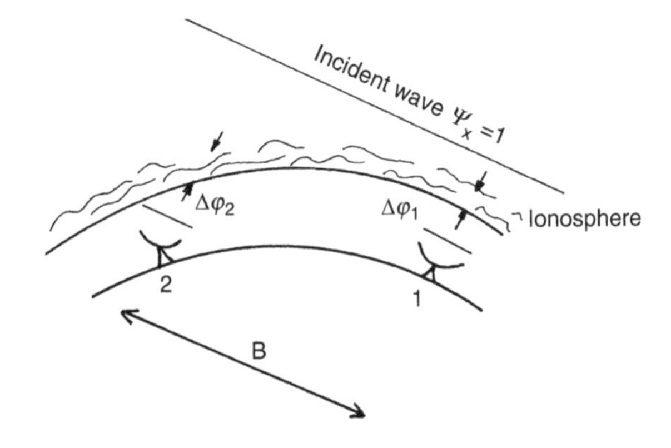

Fig. 6.27 Phase perturbations in aperture synthesis at radiofrequencies. The phase perturbation due to the ionosphere is $\Delta\phi_2 - \Delta\phi_1$

fringes at the common focus provides the desired quantity $\widetilde{I}(\boldsymbol{w})$, the normalised Fourier component of the spatial spectrum of the object. However, there are many limitations to this ideal case in a real astronomical interferometer.

Atmospheric Phase Noise

We have already seen that the complex visibility $\mathcal{V}$ of the fringes may be affected by a perturbing phase noise due to random fluctuations of the optical path in the Earth atmosphere, along the distinct paths followed by the two beams upstream (Fig. 6.27) or downstream of the sub-pupils. Furthermore, this phase noise is chromatic and hence reduces the *temporal* coherence of the signal. These effects, predominant at optical wavelengths, disappear in space, and thus constitute one of the main arguments for building space-based interferometers.

Instrumental Decoherence

Other sources of perturbation may remain, such as vibrations along the optical path, asymmetric polarisation effects, and so on. They have the same result, producing phase noise and loss of coherence, i.e., a reduction in the amplitude, either instantaneous or averaged over time, of the fringe visibility $\mathcal{V}$.

Non-Coherent Sub-Pupil

Up to now it has been implicitly assumed that the radiated electric fields collected at sub-pupils 1 and 2 are those of the wave emitted by the source, and hence that a measurement of these fields would exactly reflect the degree of coherence of the source. This is clearly true in space. However, for a ground-based interferometer, it is no longer true unless the size d of the sub-pupil remains less than the Fried length $r_0(\lambda)$. In the latter case, the coherence of the incident wavefront is maintained over the whole surface of the sub-pupils. This situation is found in practically all radiotelescopes, which are not affected by atmospheric seeing, whatever their diameter. However, very wide aperture antennas (more than 10 m) operating at wavelengths of millimetre order may suffer from seeing, e.g., the IRAM interferometric array in France, and the ALMA array under construction in Chile (see below).

The problem is much more serious at visible and infrared wavelengths. The first generation of optical interferometers (1970–1995) usually operated with telescopes (sub-pupils) smaller than r_0, or of the same order (around ten centimetres), which guaranteed the coherence of the incident wave on the sub-pupil. In a much bigger instrument, the images given by each sub-pupil are affected by speckles, and their coherent superposition produces an interference pattern from which it is very difficult to extract a measurement of $\widetilde{I}(\boldsymbol{w})$. The most one could do is to determine its modulus, all phase information having been destroyed by the loss of coherence. Adaptive optics provides a way round this difficulty, so that telescopes of very large diameter can then operate in interferometric mode.

To sum up, the effect of atmospheric turbulence will depend on the relative size D of the sub-pupil and the Fried parameter $r_0(\lambda)$ (see Sect 6.2):

- If $D < r_0(\lambda)$, the fringe system (Fig. 6.28) appears where the images given by the two sub-pupils are superposed. If there is an elementary adaptive optics device to correct for agitation (tip–tilt stabilisation), the two images remain superposed and the fringe system is observable. However, the differential phase noise between the two sub-pupils will shift the whole pattern randomly as time goes by.
- If $D \gg r_0(\lambda)$, the size of the image spot will be $\lambda / r_0(\lambda)$ and the fringes will be perturbed by speckles (see Sect. 6.2). It then becomes very difficult to analyse the image and determine $\mathcal{V}$.

When the observed source is bright, this means that there may be enough photons to be able to measure the modulus of the visibility with an acceptable signal-to-noise ratio in a time shorter than the coherence time τ_c, whence the atmospheric non-uniformities will not have had time to modify the instantaneous phase between the two beams. On the other hand, if the object is too faint to satisfy this condition, the signal-to-noise ratio on a measurement of duration τ will be less than unity. In this case, one must take the *spectral density* of the intensity distribution in each exposure, where the fringes are invisible, i.e., $S_i(f)$, then take the average $\langle S_i(f) \rangle$ over all exposures and deduce the modulus of the spectral density of the object at this frequency (Fig. 6.29).

All these factors make it difficult to use ground-based optical interferometry and thus require very careful observing protocols. In addition, the ever present phase noise, whether due to ionospheric perturbations at radiofrequencies or tropospheric perturbations in the infrared and visible, means that only $|\mathcal{V}|$ can be measured. When the phase is not known, the image of the source cannot then be reconstructed

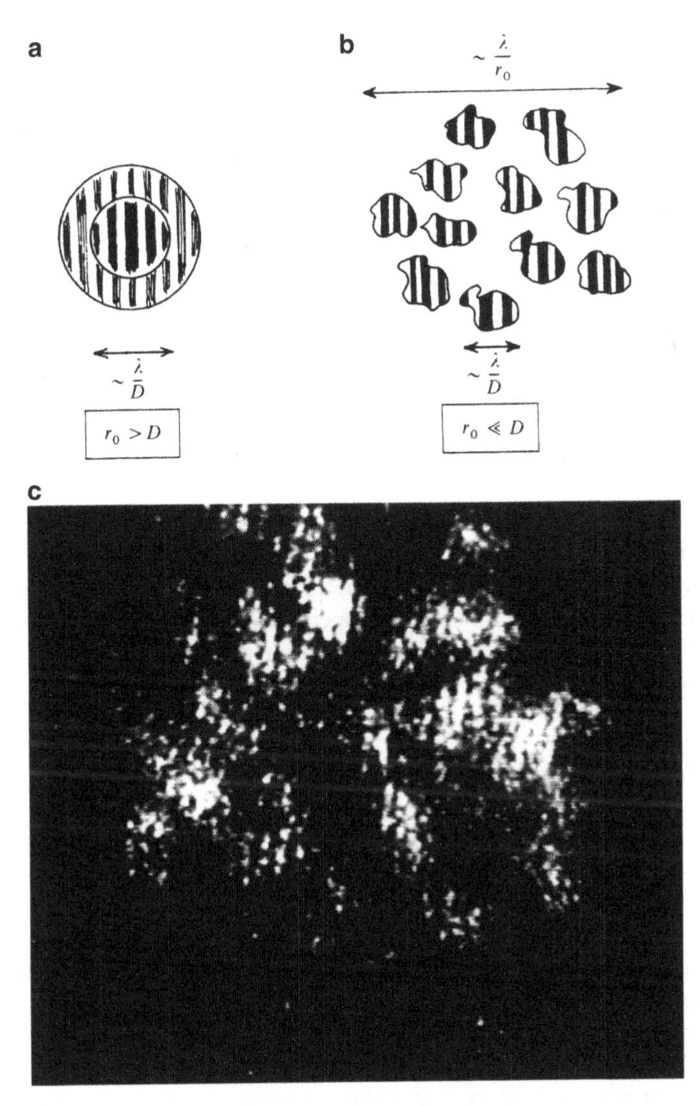

Fig. 6.28 Views of the focal plane of an interferometer. (**a**) Perfect coherence of the wavefronts at the two apertures. Interference is produced in the Airy image given by a single pupil. (**b**) Coherence of the wavefronts at each pupil, limited by atmospheric turbulence (Fried parameter r_0). Fringes are present in each speckle, with random phase. (**c**) Interference pattern obtained from two mirrors of the Multi-Mirror Telescope, with centre-to-centre distance 4.6 m. $\lambda = 600$ nm, exposure $1/60$ s. Note the presence of speckles containing fringes. (Photograph courtesy of Hege E.K., Beckers J.)

from a set of values of $\boldsymbol{w}$, except in the special case where one can use information known by other means, e.g., one may know that an object is centrally symmetric, whereupon the phases $\phi(\boldsymbol{w})$ must be zero. This limitation has led to the development of *phase closure* methods (see Sect. 6.4.6).

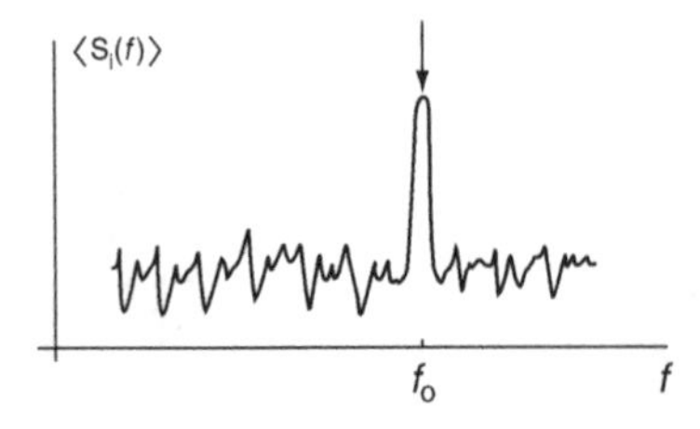

Fig. 6.29 Mean spectral density of a large number of instantaneous images, in a direction normal to the fringes of spatial frequency f_0. The spike shows that fringes are present, and the ratio $2\langle S(f_0)\rangle / S(0)$ gives the modulus of their visibility at this frequency

6.4.5 Calibrating the Instrumental MTF

The instrumental modulation transfer function (MTF) and its consequences in the image, i.e., the point spread function (PSF), are relevant to interferometers and arrays just as they are relevant to telescopes. However, the fact that it is a bandpass system (just one, or several in aperture synthesis) implies certain specific characteristics that we shall examine here. In interferometry, it is often simpler to refer to the modulation transfer function than to consider the PSF.

Two Sub-Pupils

In order to reconstruct the spectrum $\widetilde{I}(\boldsymbol{w})$ of a source S, one must know the MTF of the instrument used to observe it. The ideal way to determine this is to observe a point source S_0 called a reference source in the sky, and to measure the amplitude and phase of the complex quantity $\widetilde{I}_0(\boldsymbol{w})$, i.e., simply the complex visibility $\mathcal{V}_0$ for this reference source. Let us suppose for a moment that there are no atmospheric phase perturbations, in which case the reference S_0 can be distinct from S. But then the angular separation between S and S_0 must be accurately known in order to correct for the phase introduced by this separation when determining $\mathcal{V}_o$.

This idea was first applied in radioastronomy interferometry in the 1960s (at decimetre wavelengths), localising S in the sky to within a few arcsec, and reconstructing its image $I(\boldsymbol{\theta})$ by aperture synthesis. Quasars provide radiosources that can be treated as unresolved by the instrument, hence pointlike.

At optical wavelengths and on the ground, one can no longer assume that there are no atmospheric effects. The fringes continually move around under the effects of phase noise. How can one then determine the (complex) instrumental MTF? This is only now possible for its amplitude. By observing the fringe system of a point source S_0, one generally obtains a value of $|\mathcal{V}_0|$ less than unity, which must be measured as accurately as possible, and which calibrates coherence losses within the instrument. This source S_0 can be in a different direction, although preferably close to the direction of S. A more important point is to choose S_0 with similar magnitude and spectral distribution (colour) to S, in order not to suffer the consequences of any non-linearities in the system.

Finite Angular Size of Reference Star. The angular resolutions achieved by optical interferometers are now less than the mas (10^{-3} arcsec). There are no stars with apparent size smaller than this value and with sufficient magnitude to give a measurable signal. For example, a star like the Sun at 10 pc would subtend an angle of 1 mas. One must therefore accept a calibration using a resolved star, but with a diameter α_0 that is assumed known, by virtue of the spectral type and assuming there is no stellar envelope. The visibility at frequency $\boldsymbol{w}(u, v)$ is simply given by the Fourier transform of the uniform brightness distribution of the stellar disk (see Sect. 6.1):

$$\mathscr{V}_0(u) = \frac{2J_1(\pi\alpha_0 u)}{\pi\alpha_0 u} .$$

With the achievable measurement ranges, of order 10^3 or better, the uniform disk assumption gives an incorrect value $\mathscr{V}_0$, because there is generally a darkening of the star from the centre out toward the limb (*center-to-limb variation* or CLV), which modifies the above expression. One must then make a model of the star which makes a quantitative prediction of the CLV effect, and replace this expression by a hopefully better calculated expression (see Sect. 3.6).

This is a serious limitation. It can give rise to errors in the determination of the instrumental factor $\mathscr{V}_0$, and hence in the visibility $\mathscr{V}$ of the source S. This problem also occurs for a space-based interferometer. However, at radiofrequencies, the angular size of the central emission of quasars or active galactic nuclei (AGN) is less than 0.1 mas, whereas the decimetre wave resolution of a baseline equal to the diameter of the Earth is ‘only’ 2 mas.

Multiple Sub-Pupils

Return once again to the simple case where there are no atmospheric effects, but consider now a pupil with many sub-pupils covering a certain number of spatial frequencies at a given instant of time. This is the case in radiofrequency aperture synthesis, in an array of telescopes, with either direct coupling or delayed coupling (VLBI). The transfer function (MTF) is the autocorrelation function of this pupil and behaves as a spatial filter transmitting a certain number of frequencies apart from the frequency 0 and its immediate vicinity (which corresponds to the total received power). The array can be *dense* or *dilute*, depending on whether the MTF covers a region of the (u, v) plane almost entirely or only rather sparsely. In the first case, the image of a point source (PSF) is rather close to an Airy function, and in the second, it will be rather different. The second case is referred to as a *dirty beam*.

It is easy to see that the image of a somewhat complicated source, convolution of the object by this PSF, will be very hard or even impossible to interpret, because too much information will be lacking about the unmeasured spatial frequencies. These gaps in the spatial filter create unwanted artifacts in the raw image, such as secondary lobes, feet, ghost images, and so on, and can even distort it to the point that it becomes unintelligible. Such an image is sometimes called a *dirty map*. The PSF and especially the MTF can, as above, be determined by observing an unresolved source.

Information Restitution: The CLEAN Algorithm

The CLEAN algorithm[25] was designed by Jan Högbom in Holland in 1974 to process images from the Westerbork array, for which the PSF (dirty beam) was too dispersed. Extra information is added to the rather poor information extracted from the measurement, in order to restore an object that is compatible not only with the measurement, but also with an a priori physical idea of the object. It is difficult to be certain that the solution found in this way is necessarily unique. One accepts merely that it is highly probable, given the constraints one introduces regarding the object.[26] At the time of writing (2007), these methods have been significantly refined, and will be discussed further in Sect. 9.6.

With the CLEAN algorithm, the constraint is expressed by assuming that the object comprises a series of point sources. This is appropriate for a field filled with unresolved stars or radiogalaxies, for example. Other methods for supplying 'probable information' were devised, some of them very powerful. These were applied not only to interferometry, but also to other kinds of signal processing, in spectroscopy, for example. For this reason, the discussion of these issues will be postponed until Chap. 9.

Suppose the intensity $I_0(\boldsymbol{\theta})$ of the object consists of a background which is zero almost everywhere, upon which are superposed several localised sources, whose arbitrary intensity profile is to be determined. Let $\tilde{T}(\boldsymbol{w})$ be the instrumental profile (MTF), assumed to have been determined independently by observing a point source, and $\tilde{I}(\boldsymbol{w})$ the observation. Define

$$t_1(\boldsymbol{w}) = g_1\,\tilde{T}(\boldsymbol{w})\,\mathrm{e}^{2\mathrm{i}\pi\boldsymbol{w}\cdot\boldsymbol{\alpha}_1}$$

and choose $\boldsymbol{\alpha}_1$, which is the position of source 1, to minimise the quantity

$$r_1(\boldsymbol{w}) = \tilde{I}(\boldsymbol{w}) - t_1(\boldsymbol{w})$$

by least squares. The weighting factor g_1 is taken less than or equal to unity. This operation is equivalent to removing the brightest source, assumed pointlike to a first approximation, from the observed area. The procedure is repeated on the residue $r_1(\boldsymbol{w})$ with $t_2(\boldsymbol{w})$, then $t_3,\ldots,t_i$, using as many components i as there are discernible sources in the observed area. It can be shown that $r_i(\boldsymbol{w})$ tends to zero, and this implies that the approximation

$$\hat{\tilde{I}}(\boldsymbol{w}) = \sum_i g_i\,\mathrm{e}^{2\mathrm{i}\pi\boldsymbol{w}\cdot\boldsymbol{\alpha}_i}$$

[25] See in particular Bracewell R., Computer image processing, ARAA **17**, 113, 1979, and Pearson T., Readhead A., Image formation by self-calibration in astronomy, ARAA **22**, 130, 1984.

[26] Schwarz U.J., Astron. Astrophys. **65**, 345, 1978.

tends to the quantity $\tilde{I}(\boldsymbol{w})$, which was to be determined. Fourier transforming, a better approximation of the desired image $\tilde{I}(\boldsymbol{\theta})$ is recovered. The method is illustrated in Figs. 6.30 and 6.31. CLEAN is a method which takes account of the boundedness of the image, but not its positivity, the latter being another piece of a priori information that other methods put to use. CLEAN is useful for images consisting of well-separated sources, without a significant continuous background.

6.4.6 Phase Closure

The effects of coherence losses due to the atmosphere became significant in radio interferometry as soon as it began to move toward short wavelengths, and later these effects became immediately critical in optical interferometry. To deal with them, Roger Jennison introduced the idea of *phase closure* in his thesis (1951). The idea was then rediscovered in 1974 by A.E.E. Rogers, and since then, the method has been widely used at radiofrequencies, and was applied for the first time in optical interferometry by John Baldwin (Cambridge, UK) in 1998. It is interesting

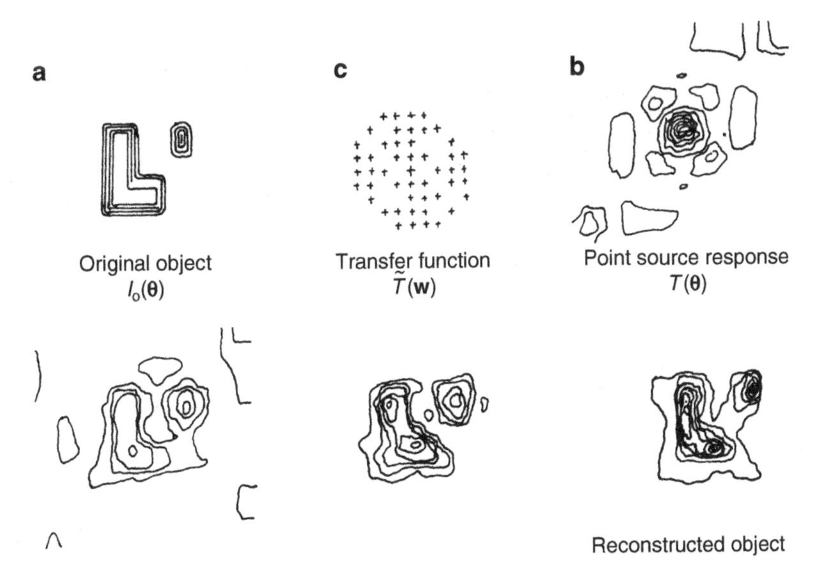

Fig. 6.30 Image reconstruction (numerical simulation). The transfer function $\tilde{T}(\boldsymbol{w})$ has gaps in its spatial frequency coverage, and the point source response (PSF) is not 'clean'. (**a**) Reconstruction by inverse FT of $\tilde{I}_0(\boldsymbol{w})\,\tilde{T}(\boldsymbol{w})$. (**b**) Reconstruction using CLEAN. (**c**) Reconstruction using the maximum entropy method (see Sect. 9.5). Compare (**a**), (**b**), and (**c**) with the original object $I(\boldsymbol{\theta})$. From Rogers A.E., *Methods of Experimental Physics*, Vol 12C. With the kind permission of the editor

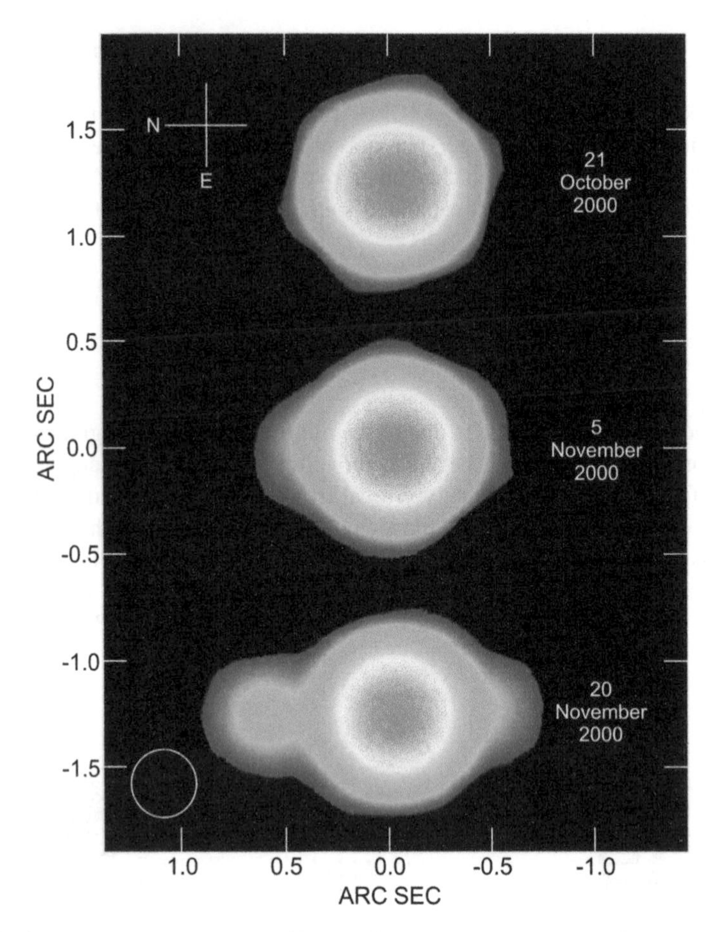

Fig. 6.31 Image processing using the CLEAN algorithm. The radiosource CygX-3, discovered in 1967 with the first rocket-borne X-ray observations of the sky, is a binary system including a Wolf–Rayet star and a compact variable object (microquasar). Observed with the VLA array at frequency 5 GHz, rapid bipolar ejection from the microquasar became clearly visible for the first time (speed $0.5c$). The CLEAN process is adapted here to the case of a rapidly varying object, with a final resolution of $0.3''$ and a logarithmic intensity scale. (J. Marti et al., Astron. Astrophys. **375**, 476, 2001. With the kind permision of NRAO-AUI)

to note that the idea of phase closure, under the name of *structural invariance*, was already used in X-ray crystallography back in the 1950s to determine the electron distribution in solids (pointed out by K. Kellermann & J. Moran).

Consider to begin with an array comprising three non-redundant sub-pupils, e.g., arranged at the corners of a triangle. With no atmosphere and a perfect instrument, the phase of the fringes given by each pair of sub-pupils is $2\pi\lambda(\boldsymbol{D}_{ij}\cdot\boldsymbol{s})$, where $\boldsymbol{s}$ is the unit vector in the direction of the source. Since $\sum \boldsymbol{D}_{ij} = 0$, the sum of the three phases is also equal to zero. If an atmospheric or instrumental perturbation should affect one of the sub-pupils, the sum remains zero, the perturbation adding to one baseline and subtracting from the other. If all the sub-pupils are affected, the result is the same, and the sum of the *measured* phases is equal to the sum of the *true* phases from the source. Clearly, the object must be bright enough to be able to measure the fringes (and hence their phase), with a high enough signal-to-noise ratio and over a lapse of time during which the perturbation remains stable, hence shorter than the coherence time τ_c of the atmosphere. This is called *snapshot mode*.

In the case of a non-redundant array of N telescopes, a total of $N(N-1)/2$ frequencies is measured, but only $(N-1)(N-2)/2$ independent phase closure relations are obtained, which is thus not enough to reconstruct the N unknown (because perturbed) phases of the object at the measured frequencies. However, if N is large, the fraction of phase information actually recovered, viz., $(N-2)/N$, tends to 100%. Since the spectrum of the source S is known in amplitude and almost completely known in phase, high quality images of S can be reconstructed, without a priori assumptions about S. If this process is combined with the CLEAN procedure, i.e., by including a priori information, the reconstructed image can be further improved. This combination is then referred to as a *hybrid mapping*. An extension of the method has become known as *self-calibration*.

All of these procedures have their limitations, residues of artifacts due to a poorly corrected PSF. At the time of writing (2006), a residual artifact less than 0.1% is achieved with large radio arrays (VLA, VLBA). The magnificent images of extragalactic objects produced by these arrays are thus obtained through a large collecting area and long exposures (access to faint objects), very long baselines allowing very high angular resolutions, and the high quality of the final PSF thanks to phase closure and numerical image restitution procedures. It is not yet possible in 2007 to boast equivalent performances for interferometry at optical wavelengths, but there is little doubt that progress will be fast given that the problems are very similar.

Pupil Masking. In 1874, Edouard Stephan in Marseille (France) built upon Fizeau's ideas for reducing the effects of atmospheric perturbations by arranging two small apertures at variable distances apart on a diameter of the mirror of his telescope pointing toward a star. The apertures were only a few centimetres in diameter, hence smaller than the Fried parameter r_0. The apertures produced fringes agitated by a random translatory motion (phase noise). He estimated the visibility of the fringes and concluded the observed stars to have diameters smaller than the diffraction limit (0.16 arcsec) of the telescope he was using.

Now suppose there are, not two, but N apertures, all smaller than r_0, on the surface of a large primary mirror (or an image of it, reformed downstream). At optical wavelengths, the image of a star is poorly visible, formed from N systems of agitated fringes. This is a fortiori the image of a resolved object. However, if the apertures are non-redundant, Fourier analysis of the instantaneous image (snapshot) will perfectly separate each of the frequencies $\boldsymbol{w}_i$ of the image, and provide its amplitude and phase. The situation is strictly the same as what we have just described at radiofrequencies, and phase closure methods along with information restitution techniques like CLEAN and others can be applied in exactly the same way, to obtain from the source an image of resolution $\lambda/D_{\text{mirror}}$, limited only by diffraction (Fig. 6.32). In some cases, this approach can compete with adaptive optics, since the instrumental setup is very simple here (comprising only a mask), whereafter all processing is done numerically.

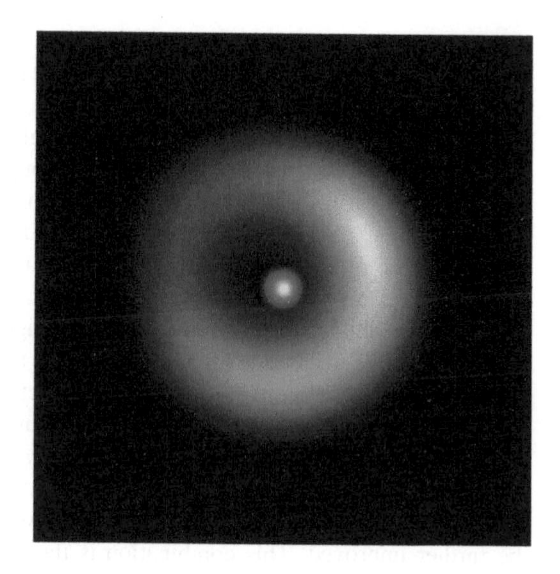

Fig. 6.32 Image of the LkHα 101 circumstellar envelope (accretion), obtained by pupil masking and phase closure on the 10 m Keck telescope (Hawaii) at $\lambda = 2.2\ \mu$m. The image has been processed using a certain amount of a priori information in order to reveal the central star Herbig Ae-Be (see Sect 9.6). With the kind permission of P. Tuthill. See Tuthill, P. et al., Ap. J. **577**, 826, 2002

6.5 Astronomical Interferometers

In this section, we shall describe current and future realisations of interferometer arrays in various wavelength ranges, from radiofrequencies, through visible and infrared, to certain projects for X-ray observation. This kind of tool is developing fast, aiming for angular resolutions less than the mas (10^{-3} arcsec), or even the μas (10^{-6} arcsec). The latter corresponds to a resolution of 1 000 AU at a distance of 1 Gpc, and requires a baseline of 2 million kilometers at 1 cm, or 200 m at $\lambda = 1$ nm! We shall also describe some of the observations made using these techniques.

6.5.1 Radiotelescope Arrays

Although Stephan and Michelson made their pioneering experiments in the visible prior to 1921, the technique of astronomical interferometry only really developed with the advent of radioastronomy, in the 1950s. This is because the accuracies required of the instrumentation are obviously less stringent than at optical wavelengths, and above all it is easy to transfer the Hertzian antenna signals by cable whilst preserving phase information, so that they can subsequently be recombined. Finally, the frequency bandwidth used is relatively narrow ($\sim 10^6$ to 10^8 Hz) compared with optical bandwidths ($\sim 10^{14}$ Hz). The temporal coherence length is therefore large, and this means that it is easier to obtain an interference signal from two coherent signals originating at widely separated antennas.

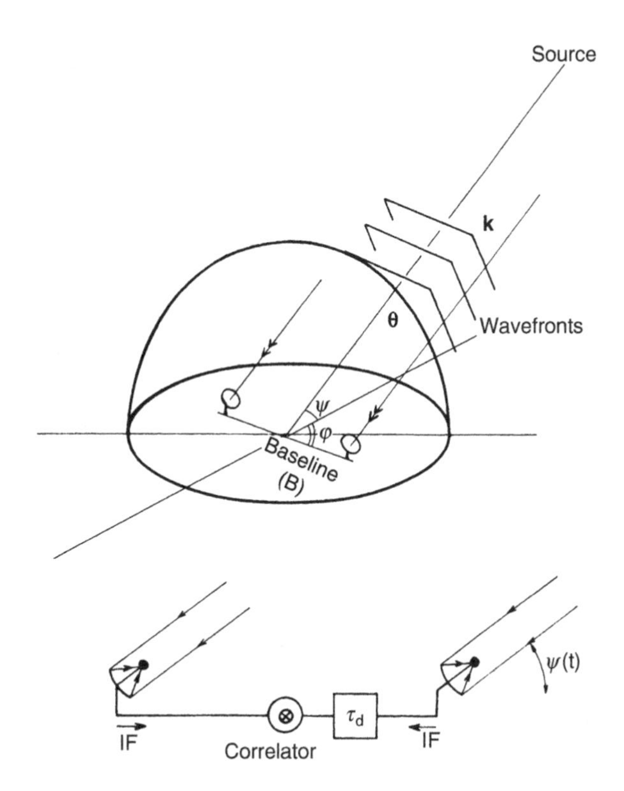

Fig. 6.33 Radio interferometry. A pair of antennas with baseline B point in the direction $\boldsymbol{\theta}(t)$ of the source. The angles $\psi(t)$ and $\phi(t)$ defining $\boldsymbol{\theta}(t)$ are functions of time due to the Earth's rotation. The phase difference between the two antennas can easily be calculated. An additional phase difference $\Delta\phi$ can be introduced electrically (delay line τ_d) and the signals mixed in a correlator which provides the fringe signal. The temporal coherence of the signals is sufficient (the pass band $\Delta\nu$ is small compared with the signal frequency), so there will be many fringes around the zero path length difference

Ground-based radio interferometers[27] cover a broad spectrum, from decametre wave ($\lambda = 30$ m corresponds to 10 MHz) to millimetre wave (1 mm corresponds to 300 GHz). Even though the underlying principles of construction and use are largely the same, the breadth of the region justifies presenting these instruments in subfamilies. An important place is given to *very long baseline interferometry* (VLBI), which is the only one to appeal to observation from space.

Obtaining the Interferometric Signal (Radiofrequencies)

A widely separated pair of telescopes (antennas), comprising the two sub-pupils 1 and 2 and forming a *baseline* of length $\boldsymbol{B}_{12}$, or simply $\boldsymbol{B}$, are both pointing in the direction $\boldsymbol{\theta}$ (Fig. 6.33). The electric fields received at their foci, which may be transported locally by a waveguide, first undergo a change of frequency (intermediate frequency or IF) at 1 and 2, whereupon these two IF signals are carried to a *correlator* which measures their coherence $\gamma_{12}(\tau)$, or equivalently, the visibility

[27]K.I. Kellermann, J.M. Moran, The development of high resolution imaging in radioastronomy, ARAA **39**, 457–509 (2001).

of the fringes. In order to restore the temporal coherence between signals 1 and 2, a delay line can introduce a suitable phase shift $\delta\varphi = 2\pi\nu\tau_d$. The time delay τ is a function of the length and orientation of the baseline $\boldsymbol{B}$, and also the diurnal rotation of the Earth. The spatial frequency analysed is determined by the value of the baseline $\boldsymbol{B}$ projected onto a plane normal to the incident wave vector. The sighting angle $\boldsymbol{\theta}(t)$, with components $\psi(t)$ and $\varphi(t)$ (Fig. 6.33), induces a time shift which is given as a function of time by

$$\tau = \tau_d + \frac{B}{\lambda c}\left[\cos\varphi(t)\cos\psi(t)\right] .$$

The spatial frequency analysed, with components u and v, is determined by the projected baseline, and given by

$$|\boldsymbol{w}| = (u^2 + v^2)^{1/2} = \frac{B}{\lambda}\left[\sin^2\varphi(t) + \sin^2\psi(t)\cos^2\varphi(t)\right]^{1/2} .$$

It lies in the direction of the baseline projected onto the plane of the wavefront, i.e., onto the sky. The Earth's rotation thus gives, for a fixed value of B, a partial sampling of the spatial frequency plane (u, v).

Measuring the phase of the fringes completely fixes the complex Fourier component. These phase measurements can lead to extremely precise results. Over a baseline $B = 1$ km, at $\lambda = 1$ cm, a determination of the phase to within 10 degrees, of a source known to be effectively pointlike will fix the angular position of that source in the coordinate system of the base (terrestrial coordinates) to an accuracy of

$$\delta\theta = \frac{10}{360}\frac{\lambda}{B} = 0.06 \text{ arcsec} .$$

Radioastronomical interferometry is thus a powerful tool in *astrometry*.

Note that such a phase measurement is never absolute and that a reference network of point sources across the sky (usually quasars, see Sect. 4.2) is therefore necessary in order to establish reference phases, as explained earlier.

Aperture Synthesis

Heterodyne detection of radiofrequencies has one remarkable property that does not occur for visible and infrared interferometers. Suppose that, instead of a pair of sub-pupils, we have some number N greater than 2, forming many baselines $\boldsymbol{B}_{i,j}$ between them. The IF signal of an arbitrary antenna i can then be distributed over several correlators, to form all possible correlations (i, j), making a total of $N(N-1)/2$ correlations. Thanks to the local oscillator, this division of the signal can be achieved *without reducing the signal-to-noise ratio*, as explained in Chap. 7 on detectors (see in particular Sect. 7.5).

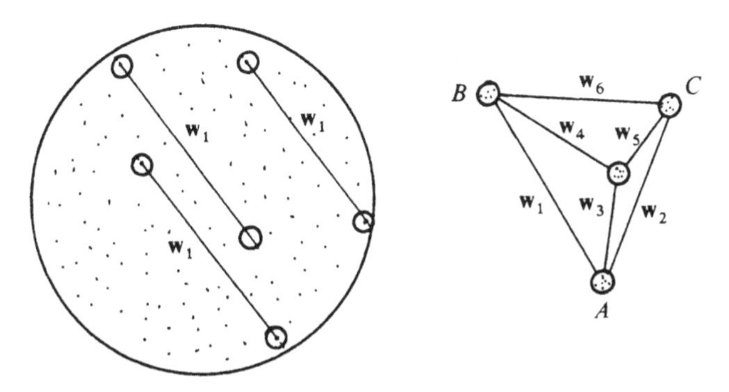

Fig. 6.34 Redundancy of a pupil. *Left*: A redundant pupil. Several pairs of elements sample the same spatial frequency $\boldsymbol{w}_1$. *Right*: A non-redundant pupil. Only the pair AB samples $\boldsymbol{w}_1$, the other pairs each sampling different frequencies

In practice, a large number of antennas are used (an *array*), paired in such a way that, varying $\boldsymbol{B}$ for each pair, coverage of the $\boldsymbol{w}$ plane can be optimised. If each pair corresponds to a single value of $\boldsymbol{w}$ without any overlapping, the pupil is said to be *non-redundant*, in contrast to the highly *redundant* pupil formed by a standard circular aperture (Fig. 6.34). A dedicated computer combines the signals from all the antennas, calculating all the possible correlation functions, after first introducing the dephasing $\tau_{i,j}(t)$ required to centre the correlations on a given instant, which amounts to referring all the measurements to a common wave plane. This is the operation known as *aperture synthesis*.

Since the spatial frequencies filtered by a pair of telescopes is determined by the *projection* of the baseline $\boldsymbol{B}$ on the sky, the Earth's rotation varies this quantity as time goes by, thereby increasing the number of spatial frequencies covered in the plane (u, v). This combination of aperature synthesis and the Earth's rotation is sometimes called *aperture supersynthesis*.

The image $I(\boldsymbol{\theta})$ of the source is then obtained by a numerical Fourier transformation of $\widetilde{I}(\boldsymbol{w})$. This operation requires sampling of the MTF of the array (see above), and also image processing, to be discussed later (see Sect. 9.6). The extraordinary progress in computer technology since 1980, combined with powerful processing algorithms based on information theory, have made it possible to produce excellent images even though the arrays do not sample the whole range of frequencies up to some cutoff value, in the way that a single circular pupil would.

Arrays for Metre to Centimetre Observation

The Very Large Array (VLA)

After the arrays in Cambridge (UK) and Westerbork (Netherlands), which played an important role in the development of radio interferometry over the decades

Fig. 6.35 Aerial view of the antennas of the VLA (Very Large Array) aperture synthesis array. The antennas are arranged in the shape of a Y, with the northern branch on the right, the south-eastern branch on the left, and the south-western branch at the top. Each branch includes 24 possible antenna positions at which the 27 mobile antennas can be placed. (NRAO photograph, Associated Universities, with the kind permission of the National Science Foundation)

1960–1980 and remain active today, the VLA of the National Radio Astronomy Observatory (NRAO) in New Mexico (USA) has become the most powerful tool of its kind in the world today at centimetre wavelengths, since it went into operation in 1980. It comprises 27 mobile antennas (Fig. 6.35) with separations up to 30 km, and operates from $\lambda = 4$ m down to $\lambda \approx 7$ mm, achieving a resolution of 20 mas at this wavelength (Fig. 6.36). It provides more than 1 000 images ($1\,024 \times 4\,096$ pixels) per day of observation and has 512 spectral channels. As commented by the radioastronomer Jim Moran at the Center for Astrophysics in Harvard in 2001:

> Since its completion in 1980, the VLA has been used by over 2 000 scientists from hundreds of laboratories and universities around the world for more than 10 000 individual observing programs. It exceeds its design specifications by orders of magnitude in nearly all parameters and remains, today, by far the most powerful radiotelescope in the world, although much of the instrumentation dates from the 1970s.

The Giant Meter Radio Telescope (GMRT)

This instrument, which went into operation in 2001, is located near Pune in Maharashstra (India). The array comprises 30 fixed antennas, each of diameter 45 m, of which 14 are within 1 km^2 from the centre, surrounded by 16 others, arranged on the branches of a Y with baselines up to 25 km. With this flexible configuration, a total of $N(N-1)/2 = 435$ different baselines are available, covering a wavelength region from 6 m to 21 cm, and reaching a maximal resolution of 2 arcsec.

This instrument was built by India alone and is today the best metre wave array in the world. Apart from its scientific production, it is being used to develop techniques for a future instrument which could be the subject of a very broad international collaboration for the 2020s, namely the Square Kilometric Array (SKA). This would operate in the centimetre and decimetre wave range and could have a collecting area as great as 1 km^2. An interesting idea is under investigation here. Instead of orienting each antenna of the array individually to point in the direction of the source, the direction $\boldsymbol{\theta}$ of the maximum point source response of the array (peak of the PSF) is fixed by modifying the phase of the signal received by each antenna, which can thus be flat, horizontal, and fixed, giving what is called a *phased array*. This technique

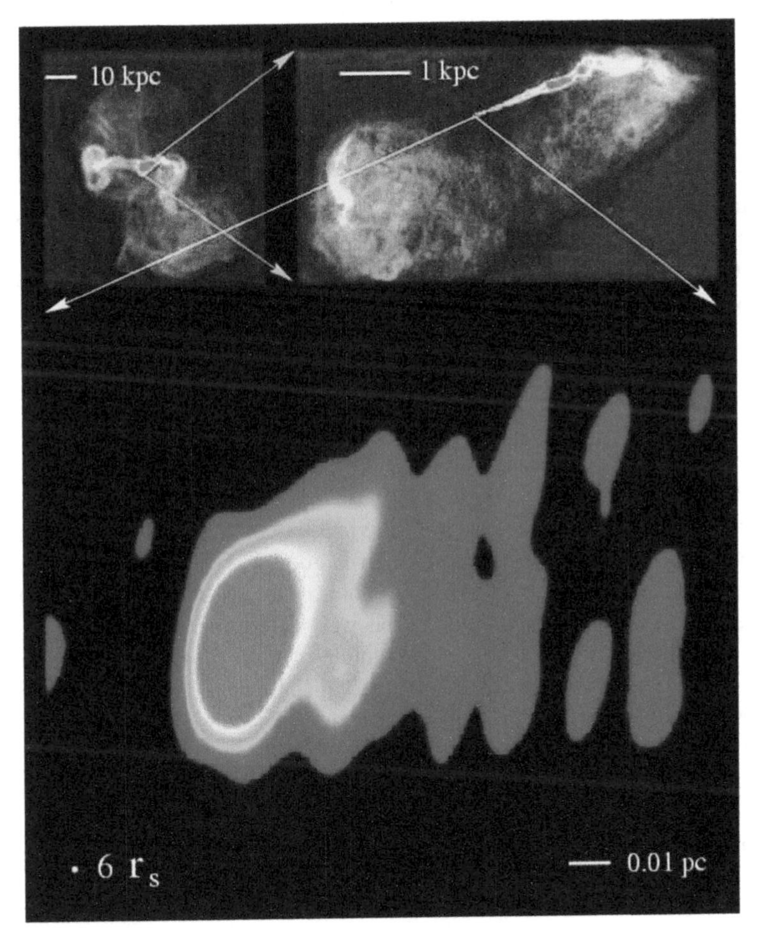

Fig. 6.36 The galaxy M 87 and its jet, about 15 Mpc from the Sun, observed at radiofrequencies. *Upper*: VLA observations, *left* at $\lambda = 90$ cm, *right* at $\lambda = 20$ cm. *Lower*: Observations by the VLBA interferometric array at $\lambda = 7$ mm. Scales are specified on each image. On the lower image, r_s is the Schwarzschild radius of the central black hole, of mass $M = 3\times 10^9 M_\odot$. Image by Junor, W., Biretta, J., Livio, M., courtesy of NRAO/AUI/NSF

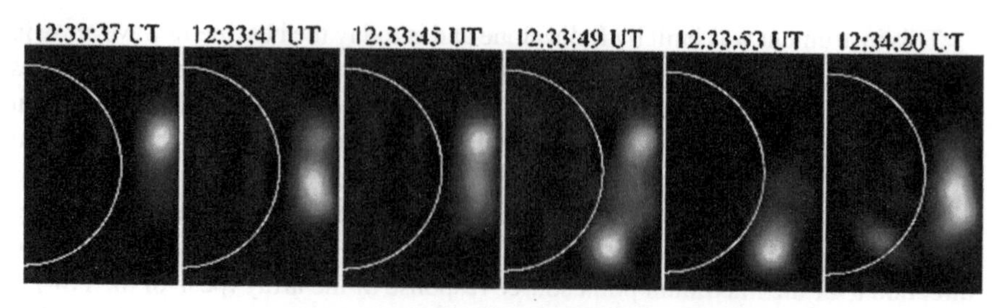

Fig. 6.37 A sequence of images of the Sun taken by the Nançay radioheliograph at $\lambda = 0.92$ m during a solar flare. The emission is due to electron beams propagating in the magnetic structures of the corona at speeds of the order of $c/3$, and lasts only a few seconds. The *white circle* marks the edge of the visible disk of the Sun. Image kindly provided by L. Klein and published by Maia, D. et al., Solar Phys. **181**, 121–132 (1998)

is highly conventional in the construction of certain radars, and spatial scanning is done without rotating the antenna mechanically, but rather by modulating the phase of the signal emitted or received by each of the elements making it up.

Another project known as the Low Frequency Array (LOFAR) is also under study at the time of writing (2007) to build an instrument comprising several thousand antennas distributed over more than 100 km of baseline. These would be simple and relatively inexpensive, because they would operate at a long wavelength $\lambda = 30$ m.

Solar Surveillance by Radio Interferometry

The radioheliograph at Nançay (France) is a metre wave instrument, consisting of two 10 m antennas and sixteen 3 m antennas, and operates at 169 MHz ($\lambda = 1.7$ m). There are 32 possible combinations, giving baselines of lengths varying from 100 to 3 200 m. The east–west resolution is good ($\sim 1.2'$) near the solar equator, and mediocre in the direction of the poles. The solar longitude of an event (a prominence) on the Sun can thus be well determined. Because of the great sensitivity of the detectors, the rapid evolution of *solar flares* can be followed with a time constant of less than one second, and this allows for correlations to be made between radio emissions and X-ray or EUV emissions observed in the corona from space (Fig. 6.37).

Millimetre Arrays

The first millimetre interferometers were those at Hat Creek (California), Owens Valley (California), and Nobeyama (Japan). They were followed later by another at the Plateau de Bure in France in the 1980s. Millimetre wave radio interferometry encounters specific difficulties due to atmospheric absorption by H_2O, which increases at these wavelengths, and the correlative effects of atmospheric turbulence

Table 6.4 The main millimetre and submillimetre arrays. N number of antennas, D diameter of antennas. Updated in 2007

Organisation and partners	Year	Site	Telescopes $N \times D$	Baseline [km]	λ_{min} [mm]
Japan	1993	Nobeyama	6×10 m	≤ 0.6	1–3
CARMA CalTech and others	2005	BIMA + OVRO (California)	9×6.1 m	≤ 2	1.0
		Owens Valley	6×10.4 m		
IRAM CNRS/MPG/Spain	1985	Plateau de Bure (France)	6×15 m	≤ 1	1.0
ALMA Europe/USA	2011	Chajnantor 5 100 m (Chile)	50×12 m	14	0.35
Smithsonian, Acad. Sinica SubMillimeter Array (SMA)	2005	Mauna Kea 4 080 m	8×6 m	≤ 0.5	0.6

which begin to make themselves felt. On the other hand, for the same length of baseline $\boldsymbol{B}$, the resolution increases considerably as compared with centimetre wave observation. The millimetre region is also very rich in astrophysical information, particularly with regard to molecular spectroscopy, not only in the galactic interstellar medium but also in very remote quasars and galaxies. The possible combination of very high angular resolution and high sensitivity inspired an ambitious project at the beginning of the 2000s, with worldwide participation (Europe, USA, and Japan), namely the Atacama Large Millimeter Array (ALMA). The first images should become available around 2010.

Table 6.4 shows the very fast development of millimetre interferometry, which now extends to higher frequencies by using very high altitude sites like Mauna Kea in Hawaii and Cerro Chajnantor in Chile, where atmospheric transmission allows observation down to 650 μm (460 GHz).

The Plateau de Bure Interferometer

This instrument has 6 mobile 15 m antennas on an east–west baseline 1 000 m long (Fig. 6.38). The resolution obtained at $\lambda = 3$ mm is of the order of 2 arcsec. The site is chosen for its atmospheric properties (altitude and hygrometry). However, variations in the refractive index of the air due to the humidity lead to phase fluctuations in the wavefront. These fluctuations limit the operating wavelength to 3 mm in the summer, while observation at shorter wavelengths ($\lambda \gtrsim 1$ mm) is possible in the winter.

The Combined Array for Research in Millimeter Astronomy (CARMA)

This array in California began observations in 2005. On a baseline of about a kilometre, it brings together two older arrays, BIMA and OVRO, set up by different

Fig. 6.38 The radio interferometer at the Plateau de Bure (France), operating at millimetre wavelengths

US institutions, whence the name Combined Array for Research in Millimeter Astronomy (CARMA). It includes 23 telescopes of different sizes, from 3.5 to 10.4 m, now connected to form the highest performing millimetre interferometer in the world, at least until ALMA goes into operation. It illustrates the remarkable flexibility of interferometric combination now achieved at radiofrequencies.

The Atacama Large Millimeter Array (ALMA)

The Atacama Large Millimeter Array (ALMA) is a joint project involving Europe (ESO), the United States (National Science Foundation), and Japan, to build an interferometer covering the Chajnantor plateau, which is about 10 km across and situated at an altitude of 5 100 m and latitude 20°S in the Atacama desert in the Chilean Andes. The array will include about 50 mobile antennas of diameter 12 m, providing a total collecting area of 5 000 m^2 and operating down to a wavelength of 350 μm (950 GHz). The sensitivity and image quality of the array and the prevailing atmospheric stability at this altitude will allow a very wide-ranging scientific programme, from exoplanets in the Galaxy to the study of galaxies and quasars in the early Universe (see Fig. 6.39).

Very Long Baseline Interferometry (VLBI)

To reach the resolutions required by the sizes of radiosources (mas) at radiofrequencies, the only solution is to extend the baseline B to the dimensions of continents. But this means solving the problem of recording IF signals locally and correlating the phases of the incident fields on each aperture, since cable transfer of the signals is no longer feasible. This particular coupling technique is called very long baseline interferometry (VLBI) and has been evolving now since 1967. At the time of writing (2007), extraordinary levels of resolution and flexibility have become possible.

The IF signal is recorded locally with a clock reading (atomic clock or hydrogen maser), the two clocks having been synchronised beforehand. These recordings,

Fig. 6.39 Artist's view of the ALMA interferometric array under construction in 2008 at the Chanjnantor site in Chile, at an altitude of 5 100 m. The array is represented here at maximal deployment. *Insert*: Prototype 15 m antenna. Document courtesy of the ESO

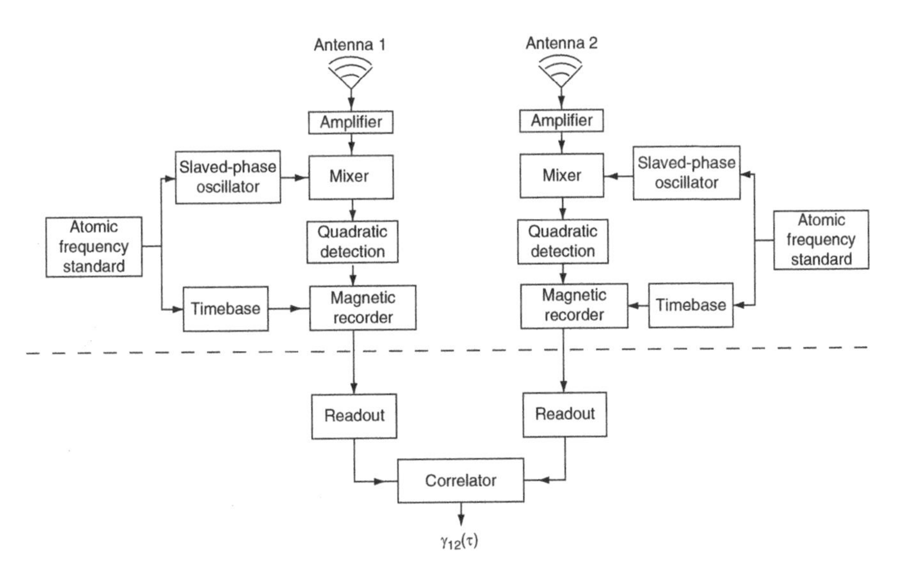

Fig. 6.40 Schematic view of a conventional VLBI arrangement. From Moran J.M., *Methods of Experimental Physics*, Vol. 12C

originally made on magnetic tape, but today on disk (2006), are then conveyed to a common point some time after the observation and 'replayed' in such a way that they interfere in a correlator and produce the interferometric signal (Fig. 6.40). Soon it may be possible to digitise the signal directly and send it by Internet, at least for frequencies below a few GHz.

Many corrections must be made to ensure that the correlation function effectively measures the coherence of the field at the two points considered. These must deal with effects related to the Earth's rotation, phase drifts of the clocks, frequency drifts of the local oscillators, and others. The stability of the clocks (better than 10^{-14}) is such that the ultimate limitation for long exposures on intercontinental baselines is imposed by random phase fluctuations due to the Earth's atmosphere.

The primary clock (see Sect. 4.3) is either a broadcast time signal from the GPS (Global Positioning System) satellite network, accurate to within 0.1 μs after correction for propagation irregularities introduced by the ionosphere, or a caesium clock reading, with a drift of around 0.25 μs per month. The local oscillator is a hydrogen maser, whose phase is locked to the clock, and whose relative phase stability is about 10^{-14} over a period of up to 10^4 s.

The signal is converted to an intermediate frequency (IF) signal with bandwidth $\Delta\nu \sim 0.1$ MHz, implying a timebase stability of the order of $1/\Delta\nu \sim 10^{-5}$ s.

VLBI measurements make use of the more or less random locations of the large radiotelescopes around the world. The $\boldsymbol{w}$ plane is thus only partially covered, even though improved by the Earth's rotation (Fig. 6.41). VLBI resolution at centimetre wavelengths has long held the absolute record ($\gtrsim 10^{-3}$ arcsec) by comparison with

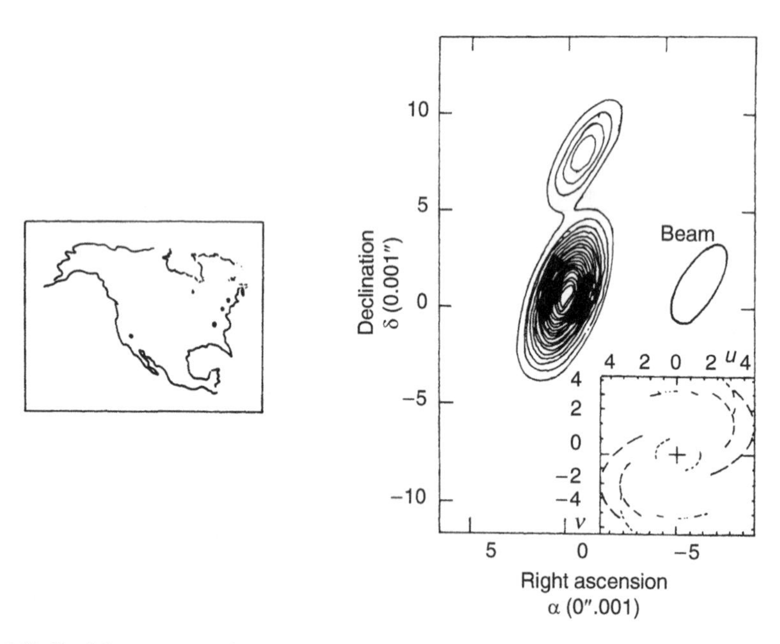

Fig. 6.41 Partial aperture synthesis by VLBI. *Left*: Location of North American antennas used. *Right*: Map of H_2O maser emission ($\lambda = 1.35$ cm, $\nu = 22.2$ GHz) from the young galactic source GL 2591, an extremely dense cloud situated at 2 kpc from the Sun. Note the high angular resolution and the beam profile. *Inset*: Spatial frequency coverage of the (u, v) plane, resulting from the pairing of antennas and the Earth's rotation (u, v have units $D/10^7\lambda$). The baselines have lengths varying from 228–3 929 km. After Walker et al., Ap. J. **226**, 95, 1978. With the kind permission of the Astrophysical Journal

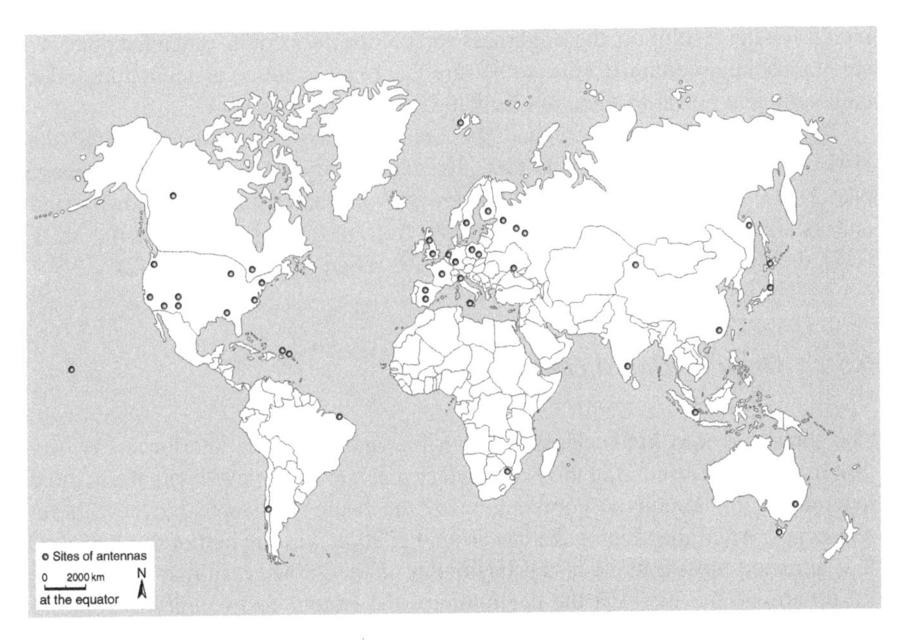

Fig. 6.42 Worldwide network of radiotelescopes which can be combined in VLBI observations (2007). Regional networks are organised in Europe with the European VLBI Network (EVN), and in the USA with the Very Long Baseline Array (VLBA). Most of the instruments featured here are radiotelescopes, but some are specific to terrestrial geodesy, using VLBI to study continental drift and the deformations of the land masses over time. New instruments are frequently added to this network, e.g., the Antarctic

other wavelengths, and it is only recently, since the 1990s, that aperture synthesis in the visible has begun to attain similar angular resolutions.

At the end of the decade 1980–1990, VLBI had been extended to millimetre wavelengths, thus putting large new antennas (30 to 50 m) to good use. In addition, the stability of frequency standards is improving (superconducting cavity oscillators, whose frequency width is thus reduced), the frequency range of recording equipment is increasing (>100 MHz), and so, too, is the number of spectral channels (up to 1 024, see Sect. 8.4). Resolutions better than 10^{-4} arcsec have been achieved in the millimetre region.

There is a very interesting application of VLBI to continental geodesy. Before the development of the Global Positioning System (GPS) in the 1990s, VLBI was able to detect the slow movements of continental drift over distances of the order of the size of the Earth.

The extension of VLBI to space observation is an interesting prospect. A telescope in a more or less elliptical orbit around the Earth, coupled to ground-based telescopes, would sweep out a large part of the spatial frequency plane during its orbit, and thus lead to measurements of high spatial frequencies, with good coverage of the (u, v) plane. The onboard reference phase would be provided by a ground-based clock, of relative stability around 10^{-14}, whose signal would be transmitted to the satellite and then returned to the ground by two-way telemetry. Before arbitrarily increasing the resolution, the brightness and compactness of the intended objects of study must be ascertained, so as to be sure that one resolution element does indeed emit sufficient power to stand out from the instrument noise.

The VLBI satellite now called the Space Observatory Project (VSOP then HALCA, Japan) was launched in 1997. Its maximum distance from Earth or apogee is at 21 400 km. Its 10 m antenna operates at $\lambda = 6$ and 18 cm (OH radical band), and its signals can be combined with data from terrestrial arrays like the VLBA (US), the European VLBI Network, or the Australian network (Fig. 6.42).

6.5.2 *Ground-Based Optical Arrays*

The three possible kinds of optical (visible and infrared) interferometry were described above, discussing the performance that can be achieved with each: direct, heterodyne, or homodyne combination of the beams delivered by the different telescopes. After abandoning the last around 1980, it was the first of these methods that attracted most interest at the beginning of the 2000s, resulting in the arrays set up around the world at the beginning of the twenty-first century. Figure 6.43

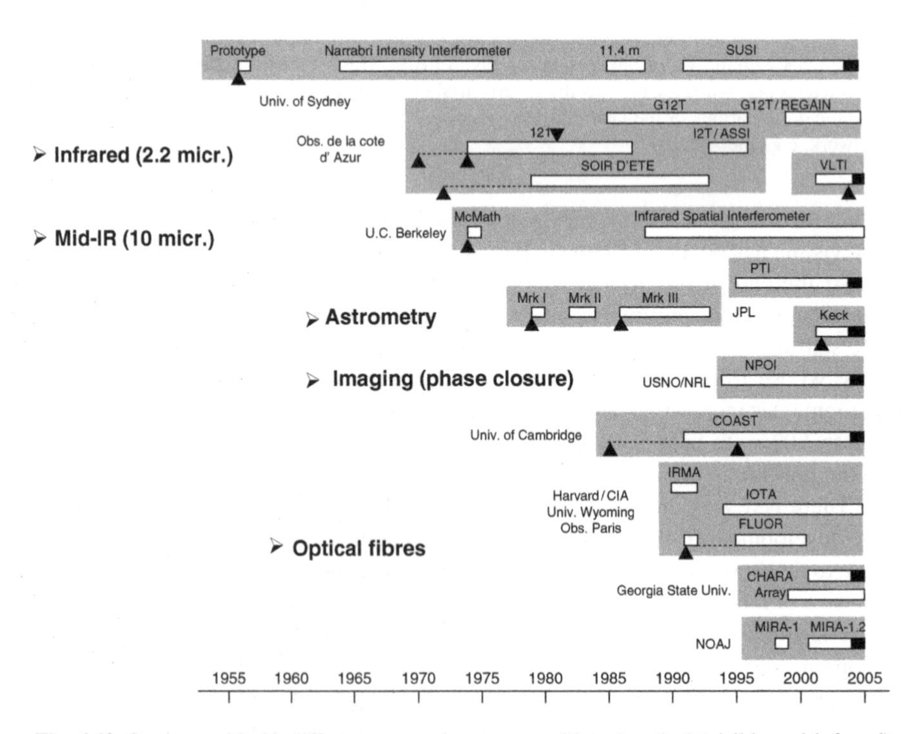

Fig. 6.43 Progress with the different approaches to ground-based optical (visible and infrared) interferometry between 1974 and 2006. Systems remaining in activity after this last date are extended with a ■. Diagram courtesy of Peter Lawson and completed by the authors

shows how these developments have gradually led to the arrays and networks we find today. However, the homodyne and heterodyne methods may reappear at some point in the future in other configurations, e.g., giant ground-based telescopes, space arrays, wavelengths in the far infrared, and so on.

Here we examine the arrays based on two of the above methods, viz., heterodyne, then direct, and look at some of the results.

Optical Interferometry: Infrared Heterodyne Method

The Infrared Spatial Interferometer (ISI) at the University of California, built on Mount Wilson in the 1980s, includes two mobile, 1.65 m telescopes, with baselines of 4 to 70 m. Its special feature is that it uses the heterodyne technique at a wavelength of 10.6 μm, a CO_2 laser providing the local oscillator signal, and a mixer producing an IF signal in the usual way, with maximal frequency band 0.2 to 2.8 GHz. The sensitivity of the instrument is thus reduced by the narrowness of this spectral band. Nevertheless, the transfer of IF signals by cable is more convenient, using techniques tried and tested at radiofrequencies, and there exist several hundred stars in dust clouds which are bright enough to provide interference fringes with satisfactory signal-to-noise ratio, so the instrument is indeed a high quality tool (Fig. 6.44). On the other hand, high angular and spectral resolutions can be combined here, an advantage so far unique to this type of instrument, which can achieve a spectral resolution of 5×10^2. It has thus been possible to study the formation of ammonia NH_3 or silane SiH_4 molecules in evolved stellar atmospheres (like IRC+10216).[28]

Ground-based heterodyne interferometry is unlikely to be a technique of the future for the reasons already stated. But in space, Table 6.3 shows that there is a crossover with the performance of direct interferometry in the far infrared and submillimetre regions, in particular when studying sources emitting in a spectral band. The future will no doubt tell whether this method can be successfully applied in space for these wavelength regions, which are still the least explored in the electromagnetic spectrum.

Optical Interferometry: Michelson Method

As soon as optical interferometry came back on the scene in 1975, the idea of telescope arrays which had proved so effective at radiofrequencies became a major objective (Fig. 6.45). The family of optical interferometry arrays has moved forward very quickly since the 1980s, and the main technical problems have been resolved

[28] Hale, D.D. et al., The Berkeley infrared spatial interferometer: A heterodyne stellar interferometer for the mid-infrared, Ap. J. **537**, 998–1012 (2000).

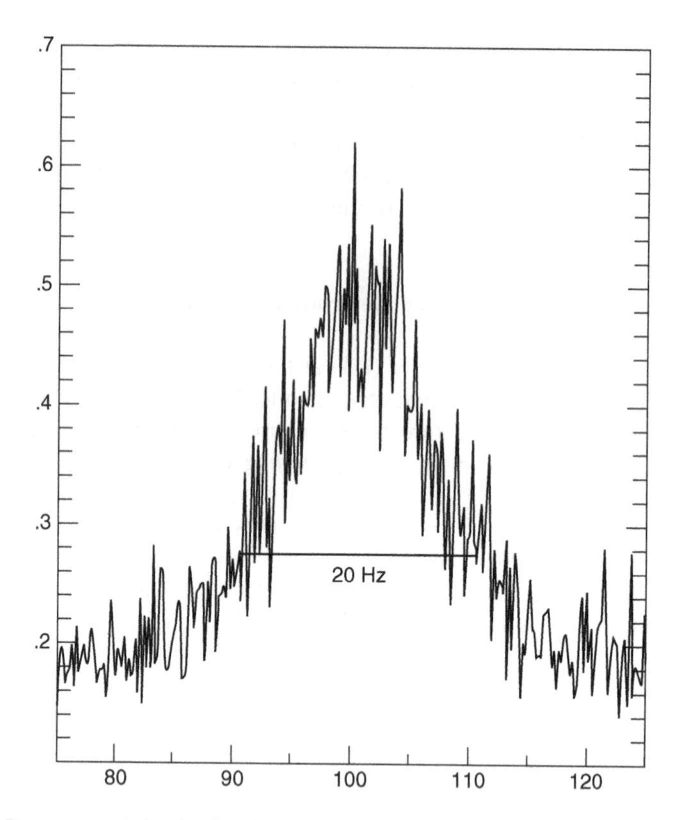

Fig. 6.44 Power spectral density (in arbitrary units) of the interferometer signal (fringes) obtained using the heterodyne interferometer ISI at $\lambda = 10.6\ \mu$m on the star α Orionis (Betelgeuse) in 1992. The frequency of the fringes, close to 100 Hz, is determined by the Earth's rotation, the position of the source in the sky, and the baseline vector $\boldsymbol{B}$. During the observation, the atmosphere causes a large and time-variable differential phase between the two optical paths, producing the 'accordion', or frequency modulation, effect on the fringes. This significantly widens the peak of the spectral density (by around 20 Hz) and correlatively reduces its amplitude. Total observation time 4.5 min. After Townes C.H. et al., Infrared Phys. **35**, 1994

(Fig. 6.43 and Table 6.5). The advent of adaptive optics, which could make very large apertures (up to 10 m) spatially coherent, resulted in the use of large telescopes like VLT and Keck in interferometric mode.

The Very Large Telescope Interferometer (VLTI)

The optical interferometer associated with the European Very Large Telescope went into operation in 2001, and at the time of writing (2007) is the most flexible and powerful instrument of its kind. High up on the Cerro Paranal in Chile, the four fixed 8.2 m telescopes are supplemented by an array of four mobile 1.8 m telescopes that can be moved to any of 30 different positions (Fig. 6.46). Each

Fig. 6.45 Artist's impression of an array of 1.5 m telescopes used for visible and infrared aperture synthesis. Note the rails allowing variation of $\boldsymbol{B}$, and also the common focus where the beams are recombined. The spherical mounting of the telescopes is novel. (Labeyrie A., La Recherche **76**, 421, 1976. With the kind permission of *La Recherche*)

telescope is equipped with adaptive optics. In the interferometric focal plane, the beams from two to at most four telescopes (in 2007), of either diameter, can be made to interfere after being brought by means of delay lines to a path length difference close to zero. Fringe detectors are positioned at the common focus. The first generation of instruments measures fringe visibility in the range 8–13 μm (MIDI instrument, with spectral resolution up to 230) and 1–2.2 μm (AMBER instrument, spectral resolution up to 300). In addition, the system can operate in phase closure in order to measure the amplitude and phase of complex visibilities when simultaneously accepting the beams from three telescopes: 254 baselines are possible, extending up to 200 m, and able to supply up to 3 025 closure relations. A *fringe tracker* can compensate for random path length differences produced by atmospheric turbulence, provided one uses a bright enough star to ensure that the signal-to-noise ratio is sufficient during the atmospheric coherence time τ_c (similar magnitude limits to those for adaptive optics).

After 2009, a new generation of instruments should allow phase closure with four telescopes, high enough astrometric accuracy (around 100 μas) for detecting the reflex motion of stars with massive exoplanets, and eventually the possibility of imaging the region close to the central black hole of the Galaxy with a desired astrometric accuracy of a few tens of μas.

The Keck Interferometer (KI)

Since 2001, the two 10 m telescopes located at the top of Mauna Kea (Hawaii), which are about 80 m apart, can be coherently coupled. This interferometer

Table 6.5 Optical interferometric networks. In the uses column, $|\mathcal{V}|$ spectral amplitudes, I imaging, S spectroscopy, A astrometry, and N nulling. Data updated as of 2007

Name	Site	Number and diameter of telescopes	Baselines [m]	Region [μm]	Uses
SUSI	Narrabri (Australia)	5 × 0.14	5–640	0.40–0.75	I, A
GI2T (Regain)	Calern (France)	2 × 1.5 m	12–65	0.57–0.70	$|\mathcal{V}|$, S
VLT interferometer	Cerro Paranal (Chile)	4 × 8.2 m 4 × 1.8 m	20–200	1–12	I, S, A, N
PTI	Mt. Palomar (California)	3 × 0.4 m	110	2–2.4	A
Keck interferometer	Mauna Kea (Hawaii)	2 × 10 m	85	2.2–10	$|\mathcal{V}|$, N
NPOI	Flagstaff (Arizona)	6 × 0.5 m	2–437	0.45–0.85	A
COAST	Cambridge (UK)	5 × 0.65 m			I, S
IOTA	Mt. Hopkins (Arizona)	3 × 0.45 m	5–38	1–2.4	I, S
CHARA	Mt. Wilson (California)	6 × 1.0 m	34–331	0.55–2.4	I, S
MIRA 1 and 2	Mitaka (Japan)	2 × 0.3 m	30	0.8	$|\mathcal{V}|$
MROI (2008)	Magdalena Mt. (New Mexico)	6 × 1.4 m	7.5–350	0.6–2.4	I, A, S
LBT (Fizeau)	Mt. Graham (Arizona)	2 × 6.5 m	22.8	1–20	I, S
ISI (heterodyne)	Mt. Wilson (California)	3 × 1.65 m	4–70	10.6	I, S

Fig. 6.46 Three of the mobile 1.8 m telescopes of the VLTI, which is the interferometric configuration of the Paranal VLT. Original photo with the courtesy of Pierre Kervella

was specifically designed for observing exoplanets, and since 2005 has been equipped with a *nulling fringe mode* (see Sect. 6.6), which requires the use of high performance adaptive optics.

Optical Arrays and Densified Pupils. Consider a source S_1 placed on the optical axis of two or more telescopes forming an interferometer. A fringe system is obtained at the common focus. Will a second source S_2 situated at an angular distance α from S_1 then be observable? For this to be the case, several conditions must be fulfilled. First of all, it must remain in the field of each telescope, which is usually quite wide. Secondly, the light beam from S_2 must be able to propagate to the common focus, in particular through the delay lines, without being diaphragmed. This condition determines a maximum value α_{m}, which depends on the dimensions of the transfer optics. In the VLT interferometer, $\alpha \approx 2$ arcsec, which is a very small value, resulting from the hectometric optical paths to be travelled between telescope and common focus. One final condition concerns the configuration of the exit pupils, which must be *homothetic* to the configuration of the entrance pupils in order to ensure the fringe contrast for all $\alpha \leq \alpha_{\mathrm{m}}$. The latter condition, long considered intangible, is even referred to as the *golden rule* of an interferometer.

Now consider the common focal plane when an interferometer comprises several dilute entrance pupils, i.e., the diameter D of the telescopes is very small compared with their separations B, and observe a point source while satisfying the above homothety condition. The diffraction pattern of this telescope array leads to a significant spread of the PSF in the focal plane, making it very difficult to implement any coronagraph technique for eliminating the signal from this source in order to observe any very faint companions in its gravitational sphere. Suppose then that the exit pupil is rearranged, e.g., so that the exit pupils are now in contact with one another. The homothety condition is no longer satisfied, and the field α_{m} will be significantly reduced. On the other hand, the exit pupil will be more like that of a single telescope, and the PSF will have a bright central peak surrounded by secondary lobes. It will thus be easy to mask. It is straightforward to show that the reduction of the field is proportional to the rearrangement factor measuring how much the pupils are brought together as compared with the homothetic situation.

This remarkable idea of *pupil densification*, put forward by Antoine Labeyrie in 2000, led to the notion of *hypertelescope*. This could include, especially in space, several tens of apertures of metre dimensions, possibly several kilometers apart, but nevertheless delivering, in a tiny field, a highly concentrated interference pattern that would be easier to detect and cancel out if necessary.[29]

Optical Interferometry: Fizeau Method

This refers to an arrangement in which delay lines serve no purpose since the equality of the optical path lengths between the different pupils and the common focal plane is satisfied by construction. Here is an example of just such a configuration.

The Large Binocular Telescope (LBT)

This instrument on Mt. Graham (Arizona) has been gradually set in operation since the end of 2005. It consists of two telescopes with diameter $D = 8.4$ m, fitted on the same altazimuth mount with their optical axes $B = 14.4$ m apart. The

[29] Pedretti, E., Labeyrie, A., Arnold, L., Thureau, N., Lardière, O., Boccaletti, A., Riaud, P., First images on the sky from a hypertelescope, Astron. Astrophys. Suppl. **147**, 385, 2000.

instrument thus looks like a conventional telescope, except that the entrance pupil is disconnected, comprising two distinct apertures. The highest observable spatial frequency is therefore $(B + D)/\lambda$. The baseline of the two telescopes is always normal to the incident direction, so no delay line is required to compensate — except one of very small amplitude to adjust the zero PLD — and the interferometric field therefore has the same limitations as the normal imaging field of a telescope, i.e., several arcmin, much higher than an interferometer in the Michelson configuration. The mirror train is reduced to the strict minimum, i.e., primary, secondary, tertiary, then a return mirror on each beam is enough to supply the common focus. Alternatively, the two telescopes can be used with two independent instruments, but constrained to observe the same source, or at least the same region of the sky.

Roger Angel[30] put forward the interesting idea known as 20+20, in which two independent telescopes of diameter 20 m each (for example) can move around a circular rail of diameter B_M. Positioning the two telescopes at the ends of a chord of this circle of length $B \leq B_M$, arranging for the common focal plane to pass through the middle of this chord, and moving the two telescopes in such a way that the chord is always normal to the azimuth of the line of sight, the result is an interferometer with adjustable baseline B, operating in Fizeau mode, hence without delay line. As in all interferometric setups using large pupils ($D \gg r_0$), the latter must be equipped with adaptive optics to restore coherence of the incident waves.

Fibre Interferometry: Optical Hawaiian Array for Nanoradian Astronomy ('OHANA)

In fact, the acronym 'OHANA suggested by the French astronomer Jean-Marie Mariotti (1955–1998) has a double meaning, since it is also the Hawaiian word for 'family'.

It is easy to understand how the need to transport the beams from each telescope to the common focus using a set of mirrors makes it practically impossible to achieve very long baselines, i.e., of hectometric order or more, on the ground. This stumbling block is avoided by extracting the light from the focus of each telescope and transferring it to the common focus by a *single-mode optical fibre*, which propagates the phase of the wave while preserving its temporal coherence (see Sect. 6.4.2). By choosing fibres of equal length from each telescope to the common focus, one can be sure that the optical paths are equal, and of course without atmospheric effects. It remains to obtain a zero path length difference using a variable delay line, as in a Michelson setup. Demonstrated in 2006 in the near infrared between the two Keck telescopes on Mauna Kea (Hawaii),[31] this promising idea would make it possible,

[30] The British optician and astronomer Roger Angel, born in 1941, is professor at the Stewart Observatory, Tucson, where he had the idea of the LBT.

[31] Perrin, G. et al., Interferometric coupling of the Keck telescopes with single-mode fibers, Science **311**, 194, 2006.

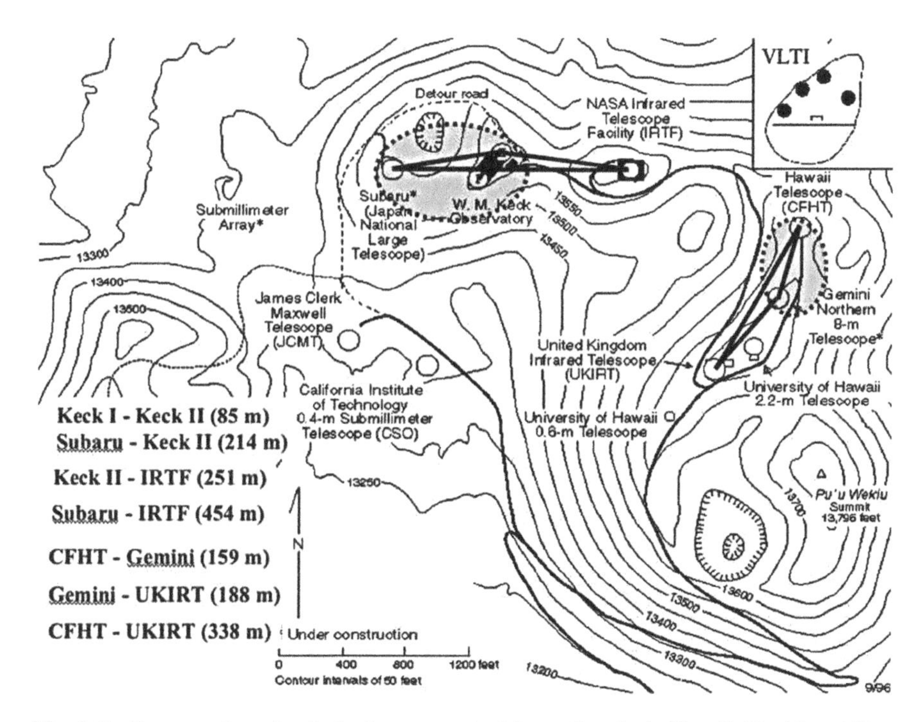

Fig. 6.47 Concentration of optical telescopes at the Mauna Kea site in Hawaii. The idea of fibre optic coupling can be used to connect two or more of these telescopes interferometrically, directly extracting light from Coudé or Cassegrain focal points. The possible interferometric baselines are indicated *on the left*, while the *insert* on the *upper right* compares the arrangement of the four 8.2 m telescopes of the VLT. *Dotted circles* show the first two configurations implemented in the first decade after 2000. From Mariotti J.M., 1995 and Perrin, G., 2006. Map due to the Institute for Astronomy, Honolulu. Altitudes are given in feet, where 1 foot = 0.305 m

by coupling all the large telescopes located on this peak (Fig. 6.47), to obtain images of the cores of active galactic nuclei (AGN) like NGC1068 with resolutions as high as 200 μas.

Some Results from Optical Interferometry

Since the 2000s, optical interferomety has become a rapidly expanding technique in astronomy thanks to results that could be obtained in no other way. To see an example of this, consider Fig. 6.48, which shows the astrometric performance obtained in the 1970s with the first optical interferometers. The aim there was to measure stellar diameters, and the diameters of the orbits in close binary star systems from which the star masses could be deduced.

Figure 6.49 shows the accuracy that could be achieved in 2006 in the measurement of a visibility, and the way its wavelength dependence could be used to accurately specify the radial intensity distribution of a stellar atmosphere. In the

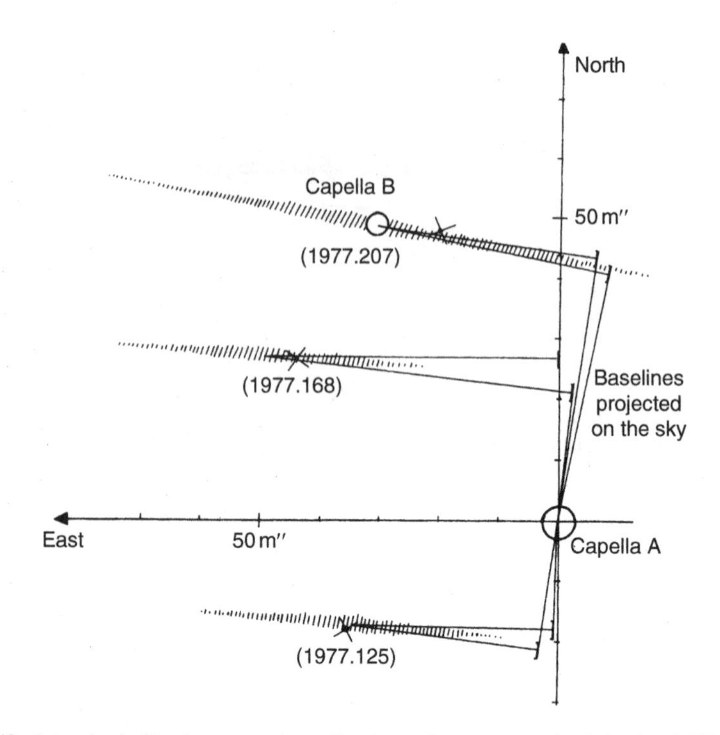

Fig. 6.48 Astrophysical and astrometric applications of aperture synthesis in the visible. Observations of the double system Capella (α Aurigae) A and B. *Circles* mark the two component stars, resolved and shown with their apparent diameters. X marks positions predicted for B on its orbit (Finsen). *Shading* indicates measured positions of B at different epochs (projections of baselines on the sky are shown at the same Julian dates, which are marked). Note the excellent resolution in the direction of the baseline, and the poor resolution in the perpendicular direction. Measurements obtained at the Plateau de Calern, France, with two 25 cm telescopes. From Blazit A. et al., Ap. J. **217**, L55, 1977. With the kind permission of the Astrophysical Journal

example chosen here, the interferometric observation of Cepheid variables, crucial for determining distances in the Universe, provides a way of refining the accuracy of this fundamental length scale.

6.5.3 Space-Based Optical Interferometry

The difficulties caused by the Earth's atmosphere, like those due to the Earth's rotation (the need for continuously variable delay lines), have been sufficiently emphasised to show why space looks such a good place to set up optical interferometers, observing from the visible to the far infrared. But one should not underestimate the difficulties of quite another order raised by this undertaking: the need for telescopes flying in formation as soon as baselines exceed about ten metres and a supporting beam can no longer be used; variation of baselines to sample spatial

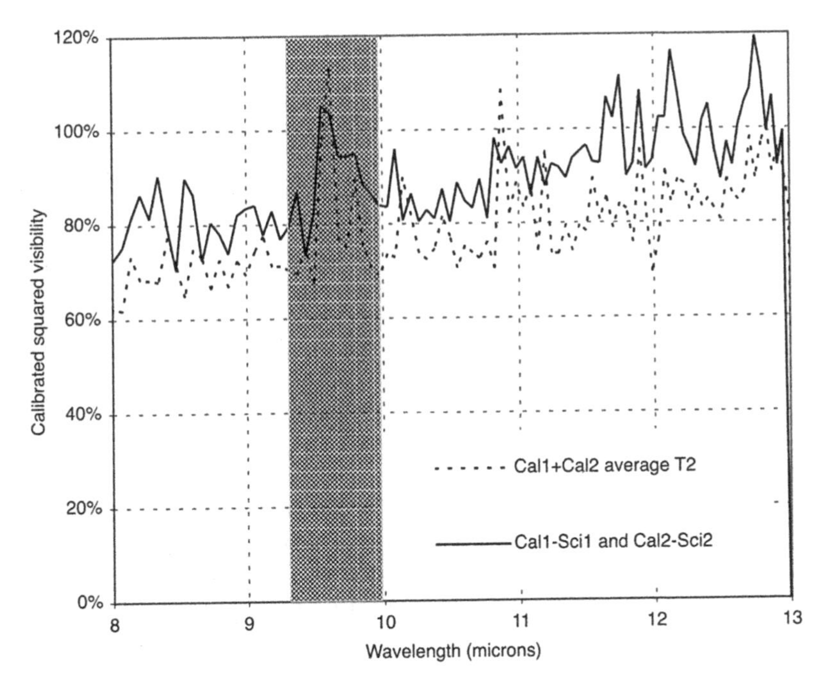

Fig. 6.49 First demonstration of the presence of interstellar matter around the Cepheid star l Carinae, using optical interferometry. The measured visibility (squared modulus) $\mathcal{V}(\lambda)^2$ is plotted versus λ, in the atmospheric window around $\lambda = 10\ \mu$m. The VLTI works here with two 8.2 m telescopes, on a baseline $D = 40$ m, which sets the spatial frequency D/λ at each λ. At these wavelengths, the accuracy on $\mathcal{V}(\lambda)^2$ is limited by atmospheric and thermal noise. The star, with a photospheric radius of 2.70 mas, is practically unresolved, while the measured $\mathcal{V}(\lambda)^2$ shows the presence of a resolved envelope of 8 ± 3 milliarcsec radius. Such envelopes are due to the periodic ejection of matter by the pulsating star. The two curves (*continuous* and *dashed*) correspond to two different interferometric calibrations, using an unresolved star without envelope. Document Kervella, P. et al., Astron. Astroph. **448**, 623 (2006)

frequencies; and holding optical paths to within a fraction of a wavelength. This is why, despite many proposals and projects, the spaceborne optical interferometers will not appear before 2010–2020. Two of these missions, currently under study, will be discussed here.

The Space Interferometry Mission (SIM) and NASA's PlanetQuest

This astrometric mission aims to study extrasolar planets and at the same time to make exceptionally accurate measurements of the parallax of stars in the Galaxy.[32]

[32] See the detailed description and updates on the Space Interferometry Mission (SIM) at the website of the Jet Propulsion Laboratory (JPL) in California: planetquest.jpl.nasa.gov/SIM/.

A 9-m beam, which can be oriented in space, will carry a series of 30 cm telescopes that can be combined to form several interferometers. The common focal plane will be equipped with a CCD detector on which the fringes will form, in the spectral region 0.4–0.9 μm. The instrument, placed in orbit around the Sun, will gradually move away from the Earth during its 5 year mission.

The key feature of the Space Interferometry Mission (SIM) is its high angular resolution. The diffraction limited resolution λ/D is 20 mas. However, two factors allow the system to improve considerably on this limit. The first is instrumental stability since, in the absence of any atmosphere, the fringes are stable to better than $\lambda/100$ for several minutes, allowing long time integrations, and hence extreme sensitivity (limiting magnitude $m_{\mathrm{V}} = 20$). The second factor is that it can operate in differential mode, comparing the positions of two nearby objects by the observed fringe shifts they each produce. This is how SIM can achieve resolutions in the range 1–4 μas. This mission could be launched in the decade from 2010.

In order to detect extrasolar planets, SIM would not observe the planet directly directly, since it would be too faint, but rather would measure the angular distance between the star S thought to hold the planet and a distant reference star S_{R}. Over time, the angle between S and S_{R} will vary for several reasons, i.e., proper motion of the stars and different parallaxes, but it will also vary if S carries a planet, orbiting like S around the centre of mass of the system. With a sensitivity of a few μas, it will be possible to detect planets with comparable or even smaller masses than the Earth, after eliminating the two other effects, which can be determined independently.

To measure the distance of a star in the Galaxy, SIM will measure its parallax during the year, by simultaneous and oft-repeated observation of a reference star belonging to a reference grid distributed across the whole sky. With an ultimate resolution of 4 μas, equal to the parallax of an object at a distance of 250 kpc, SIM will be able to reach stars almost throughout the whole Galaxy with a very high level of accuracy.

With these two goals, SIM will complement the Corot and Kepler missions in search of exoplanets and also the GAIA mission for astrometry (see Sect. 4.2).

ESA's Darwin Mission

First proposed in 1993, just before the discovery of the first extrasolar planet in 1995, this mission[33] exploits the technique of *nulling interferometry* This will be discussed in detail later (Sect. 6.6), in the context of coronagraphy techniques. For the moment, we shall just describe the array of telescopes, each with diameter 3–4 m, placed in a plane perpendicular to the line of sight as shown in Fig. 6.50. At the common focus, a dark interference fringe is formed at the position of the star,

[33] The complexity and cost of the mission must be balanced against its extraordinary aim, namely the detection of life on extrasolar planets. This is why, in 2007, it was once again proposed to the European Space Agency (ESA).

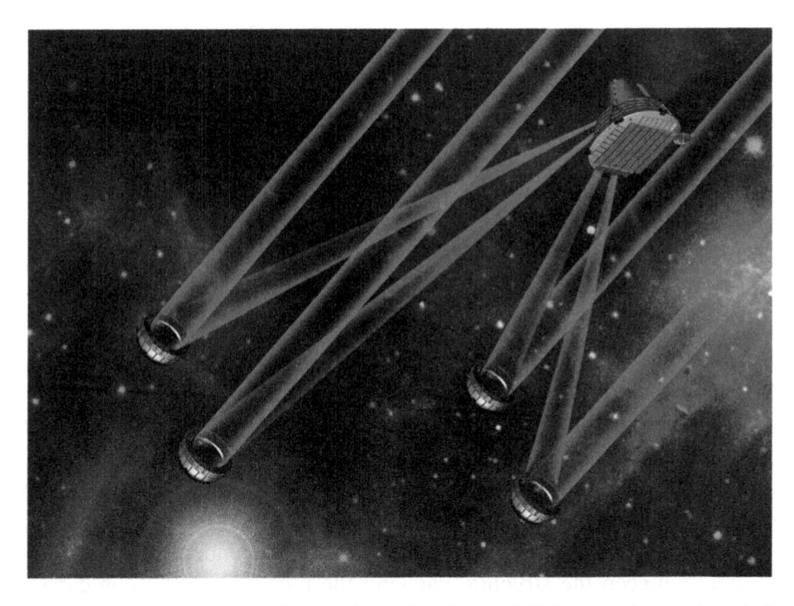

Fig. 6.50 Artist's view of a possible configuration for the ESA's Darwin mission. A fleet of four independent satellites, each carrying a telescope and flying in formation, constitutes an interferometer. The interferometric signal is produced and processed by a fifth spacecraft, at which the four beams from the others are made to converge. It also provides the communications link with the Earth and the necessary energy via its solar panels. The whole system is placed at Lagrange point L2. Since its design in 1990, the configuration has been continually improved, in particular, by reducing the number of telescopes and hence the cost. (European Space Agency, 2007)

while the interference remains constructive at the suspected position of the planet, leading to a maximum amount of light, which must then be distinguished from the noise background, or the exozodiacal light, if there is any, due to dust present in the planetary system (as in our own Solar System). The instrument will operate in the wavelength range 6.5–18 μm, in order to maximise the contrast between the star and the planet on the one hand, and in order to fall within the spectral region containing the spectral signature of ozone O_3 on the other. A low spectral resolution ($R = 20$) suffices to reveal the presence of this molecule, which in turn indicates the presence of oxygen. Placed at Lagrange point L2, it is relatively easy to protect it against radiation from the Earth, the Sun, and the Moon, since, viewed from L2, these three bodies remain more or less aligned all the time.

Interferometry and X-Ray Radiation

It is already difficult at visible wavelengths to maintain the equality of optical paths at a small fraction of the wavelength. So how could one go about doing this at wavelengths a thousand times shorter, in the nanometre range, in order to resolve the X-ray emission produced in the immediate neighbourhood of a black hole? This is the challenge faced by the MAXIM space mission,[34] under study at NASA. The environment of the black hole in the galaxy M87 (see Fig. 6.36), with estimated mass $M = 10^8 M_\odot$, is a bright X-ray source. The black hole has a Schwarzschild radius of 10^{11} m, which would require a resolution of 0.05 μas at the distance of 18 Mpc to resolve it.

Young fringes have already been obtained with X rays in the laboratory,[35] a natural consequence of their wave nature. In space, considering X rays of 1 keV ($\lambda \approx 1$ nm), an angular resolution of 200 μas could be achieved with a baseline of $B = 1$ m, and telescopes with collecting areas of 100 cm^2 would provide a comparable sensitivity to the Einstein and Rosat missions. But the difficulty here lies in the detection of the fringes. On the one hand, their separation must be increased by interfering the beams (two or more, depending on the number of pupils used) over a very small angle. To obtain an fringe spacing of 100 μm with X rays of wavelength 1 nm, the necessary amplification is 10^5, which thus requires an angle of 10^{-5} rad. The distance between the structure carrying the mirrors and the focal plane is therefore $10^5 B$, or at least 100 km. As in the Darwin or LISA missions, the problem is to achieve highly accurate navigation of independent satellites (*formation flight*), here between the structure carrying the mirrors and the one carrying the detectors. At the same time, sufficient spectral selectivity is needed so that it is not necessary to obtain zero PLD in order to find the fringes: a spectral resolution of 10^2 to 10^3 is required to achieve this, leading to fringes that could be observed over several millimetres.

In a second stage, the baseline could be extended by further separating the satellites carrying each of the collecting mirrors. These ambitious missions involve complex technological development and are unlikely to see the light of day before 2020–2030.

6.6 High Dynamic Range Imaging (HDRI)

When imaging certain astronomical objects, there are some special cases in which the contrast in brightness between different regions of the field of view is extremely high, varying by a factor of as much as 10^{10}. This is the case, for example, with the solar corona in the vicinity of the photosphere, circumstellar disks or exoplanets not far from a star, and active galactic nuclei. The generic term *coronagraphy* refers to all those methods aiming to obtain the image of a faint field around a very bright source by effectively masking the latter. These methods are also called *high contrast imaging* or *high dynamic range imaging* (HDRI).

---

[34]The site maxim.gsfc.nasa.gov/docs/pathfinder/pathfinder.html describes the various aspects of the MAXIM mission.

[35]See casa.colorado.edu/~wcash/interf/cuxi/FringeWriteup.html.

6.6.1 *Coronagraphy and Apodisation*

In the case of an almost point source such as a star or active galactic nucleus, surrounded by structure of some kind separated by only a very small angle, e.g., disks or planets, high angular resolution is required, and adaptive optics and interferometry become essential. The dynamic range can be considerable, even as high as 10^9–10^{10}.

Coronagraphy comes from the Latin word *corona*, meaning crown. Indeed, the technique was invented by Bernard Lyot,[36] when attempting to image the solar corona in the years 1940–1950, by masking the solar disk, a million times brighter than the corona, in order to reveal the corona's flamboyant activity.

Following the work by Lyot, who had the extremely powerful idea of putting a mask, not only in the focal plane, but also in the plane conjugate to the pupil, the only notable innovation over several decades was conceived by Jacquinot.[37] He had the idea of reducing the amplitude of the feet of the diffraction pattern of the bright source by modulating the transmission by the pupil. This is the basic idea of *apodisation*, already mentioned briefly in Sect. 6.1.

It was only at the beginning of the 2000s, following the discovery of the first extrasolar planets, that this technique suddenly came back into the fore. The aim of forming the image of such planets in orbit around their star became a powerful driving force, resulting in an explosion of new concepts. It is important to realise just how difficult this is. The problem is to capture and identify, at a typical angular distance of 0.1 arcsec from a star, the photons from a companion that is 10^{10} times (25 magnitudes) less bright, these values corresponding to the case of a system like the Sun–Earth system at a distance of 10 pc from the observer (a frequently used case study).

The General Problem

In almost all imagined solutions, the rule is to mask or otherwise extinguish the bright source, because the smallest leak of its photons would be disastrous. In addition, the wings of the diffraction pattern, which extend a long way, are sources of photon noise. Using the Fourier formalism, we first express the problem in complete generality, considering the optical system shown in Fig. 6.51 (lower). The entrance pupil modulates the amplitude of the incident wave, a phase and/or amplitude mask is placed in the focal plane on the axis of the bright source, and an image is once again formed on a pupil diaphragm, in order to block any residual diffracted light.

---

[36]The French astronomer Bernard Lyot (1897–1952) used images taken at the Pic-du-Midi observatory to produce a magnificent film entitled *Flammes du Soleil*, which showed the public the coronal activity of the Sun.

[37]Pierre Jacquinot (1910–2002) was a French optician whose work led to the Fourier transform spectrometer (see Sect. 8.3).

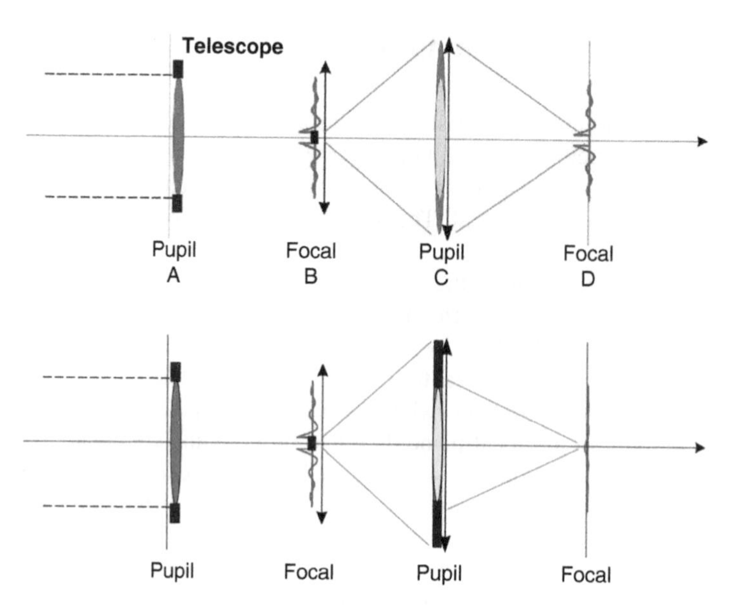

Fig. 6.51 Configuration of the Lyot coronograph. Light goes from left to right. *Top*: Without a stop in pupil C, the wings of the PSF appear in the final image plane D because of the diffracted light (edge rays). *Bottom*: The introduction of a stop in the reimaged pupil C blocks this contribution. The lens placed in the first focal plane B reimages the pupil, while the second lens in C provides the final focal plane image D. Source C. Aime & R. Soummer

Let $u(x, y)$ be the amplitude in the pupil plane (1 in the pupil and 0 outside, if there is no modulation). Then the amplitude in the focal plane is $v(x, y) = \tilde{u}(x, y)$, where $\tilde{u}$ is the Fourier transform of u. Applying the focal mask amounts to writing

$$v(x, y) = \tilde{u}(x, y)\,[1 - M(x, y)] \;,$$

where M is a complex quantity whose modulus (≤ 1) and argument describe the attenuation and phase shift introduced by the mask, respectively. In particular, M equals unity where the mask is opaque and $M = 1 - \exp i\pi = 2$ where the mask introduces a phase shift of π. In the conjugate pupil plane, the amplitude then becomes

$$w = \tilde{v} = \left(u - u \star \tilde{M}\right) \times D \;,$$

where D is the diaphragm function. The amplitude w thus results from the difference between the direct wave and the diffracted wave, a difference to which the pupil diaphragm is then applied.

The general problem raised by high dynamic range imaging is to make w zero or minimal within the diaphragm. If we consider the space of free parameters represented by the modulation of the amplitude of the wave transmitted by the pupil, the modulations of the amplitude and phase of the wave transmitted by the focal plane, the amplitude of the wave transmitted by the pupil diaphragm and the shape of the latter, it is easier to understand how so many different solutions could have been put forward. However, we may consider two main families of solutions based on this scheme: those playing on the focal mask, referred to under the heading of *coronagraphy*, and those seeking to reconfigure the PSF, either on a bounded support, or by processing the residual faint regions in the final field, which come under the heading of *apodisation*.

Finally, the general heading of *interferometric techniques* will cover all other solutions, using a more sophisticated optical setup than the one shown in Fig. 6.52 and aiming explicitly to produce destructive interference to extinguish the bright source.

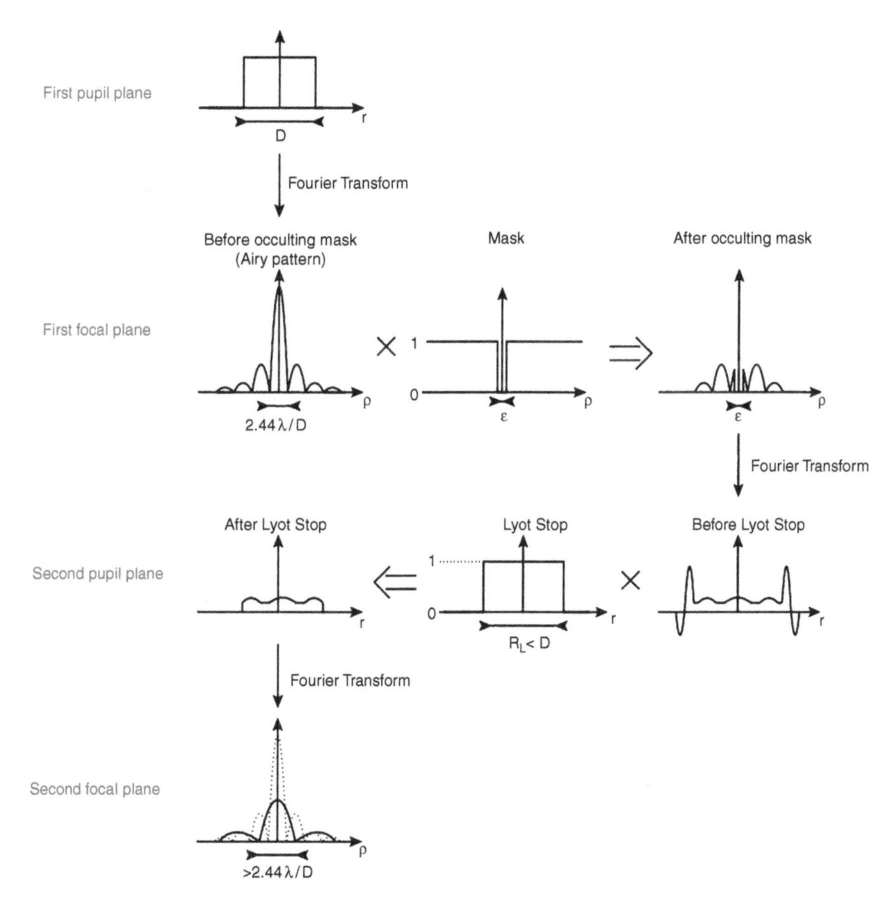

Fig. 6.52 Example of the progression of light in a coronagraphic system. This basic Lyot scheme offers many possible variations and improvements, discussed in the text. *From top to bottom*: Uniform wave amplitude on the entrance pupil. Effect of an occulting mask in the first focal plane. Effect of a Lyot stop in the second pupil plane. Final image in the second focal plane. From R. Galicher

Coronagraphy

The problem faced by Lyot when viewing the solar corona was to considerably reduce the light from the Sun's disk, diffracted and scattered by all the optical defects.

The Lyot Coronagraph

The first and most natural idea is to place an opaque mask in the focal plane, slightly larger than the image of the solar photosphere. What Lyot then noted was that the diffraction by the mask itself, but also by all the tiny defects in the optical system, such as dust or phase defects, was the source of a significant diffuse background.

Indeed, consider an opaque particle of diameter d in the focal plane. It diffracts a wave coming from a given direction into a solid angle of characteristic half-angle $\theta = \lambda/d$. For example, if d is of the order of the size of the diffraction spot, i.e., $f\lambda/D$ (f being the focal length of the telescope and D its diameter), then $\theta = D/f$, i.e., the same angle as the geometrical beam. The convolution of the two solid angles (geometrical and diffracted) means that a significant part of the light ends up outside the geometrical beam (of the order of 75% in our example). If an opaque mask containing a circular hole slightly smaller than the geometric pupil is placed in the plane conjugate to the pupil plane, then most of this diffracted light gets blocked, while light from another point of the field, not affected by our particle, will follow the path predicted by geometrical optics and will go through almost unaffected. Figure 6.51 shows this configuration.

Let us examine this result using the Fourier formalism introduced above. In the expression

$$w = \mathrm{FT}(v) = \left(u - u \star \tilde{M}\right) \times D ,$$

D and M are box or top-hat functions, equal to unity within the radius of the mask or diaphragm and zero outside. $\tilde{M}$ is a cardinal sine function which gets broader as the mask gets smaller. The left-hand side of Fig. 6.52 shows the evolution of the wave amplitude in successive planes, and in particular the different components of the amplitude in the pupil plane C: the direct wave (Fig. 6.52e), the wave diffracted by the mask (Fig. 6.52f), and their difference before (Fig. 6.52g) and after (Fig. 6.52h) inserting the mask. We see that the maximum intensity (modulus squared of the amplitude) is concentrated at the radius of the image of the pupil in the form of a bright ring. This is precisely what caught Lyot's attention. We thus understand why the pupil diaphragm must be slightly smaller than the geometric image of the pupil if the diffracted light is to be effectively blocked.

The Roddier & Roddier Coronagraph[38]

François and Claude Roddier had the idea of replacing the amplitude focal mask by a phase mask, i.e., a transparent mask producing a path length difference as a

[38] François Roddier, born in 1936, and Claude Roddier, born in 1938, are French astronomers and opticians

function of its thickness. If this difference is equal to $\lambda/2$, it then introduces a phase shift of π, in other words a change in the sign of the amplitude.

The relation established for the Lyot mask becomes

$$v(x, y) = \tilde{u}(x, y)\,[1 - 2M(x, y)] \ ,$$

and in the conjugate pupil plane,

$$w = \mathrm{FT}\,(v) = \left(u - 2u \star \tilde{M}\right) \times D \ .$$

If the size of the phase mask is chosen judiciously (radius $0.53\lambda/D$), the level of extinction will be high and the amplitude of w can be further reduced with respect to the Lyot solution. The increased efficiency of this solution can be interpreted by considering that the incident wave has been divided into two parts, with approximately equal integrated amplitudes, one undergoing a phase shift of π, whereupon the destructive interference of these two waves produces the desired extinction effect.

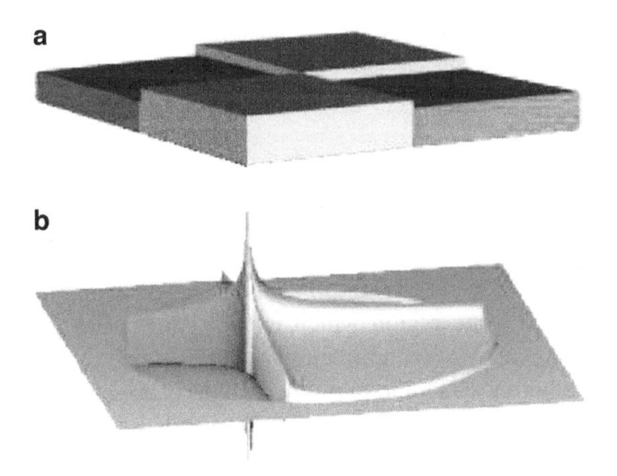

Fig. 6.53 Four-quadrant phase mask coronagraph. In the focal plane, the image of the star to be extinguished is placed at the centre of symmetry of the mask. (**a**) The phase mask (schematic). (**b**) Function to be integrated to obtain the amplitude in the pupil. It can be shown that the integral is zero, however far off centre the disk may be

Rouan Coronagraph[39]

This solution is also based on the phase mask idea, producing destructive interference once again. The mask, which affects the whole focal field, is formed by juxtaposing four quadrants, two of which, in diagonally opposite positions, produce a phase shift of π (Fig. 6.53a). It can then be shown that a source perfectly centered at the meeting point of the four quadrants will be totally extinguished, i.e., $w = 0$ uniformly in the pupil, the diffracted light being rejected entirely outside the pupil.

The full pupil condition, in particular without central obstruction of the telescope, must nevertheless be satisfied for this to work. This is one of the desired cases where $w = 0$. Indeed, the term $u \star \tilde{M}$ reduces here to the convolution product of the pupil function and $1/xy$, which is simply the integral of the function $1/xy$ over a region corresponding to an off-centre disk, as shown in Fig. 6.53b. Exploiting the power of a point with respect to a circle, there is an elegant proof due to Jean Gay (2002) that this integral must be strictly zero, however far off centre the disk may be.

Several coronagraphs based on this idea have already been installed in telescopes: two devices operating at 2.2 μm on the NACO adaptive optics system of the VLT, and three others in the mid-infrared instrument (MIRI) camera, one of the four focal instruments of the James Webb Space Telescope (JWST) developed by NASA in collaboration with Europe. The latter instrument (expected launch in around 2014) is designed for direct detection of giant extrasolar planets in the 11–16 μm range. The instrument known as spectropolarimetric high-contrast exoplanet research or SPHERE, expected to equip the VLT in 2012, is designed for an analogous purpose but at shorter wavelengths, and uses similar components.

Notch Filter Mask

This solution also uses a focal mask, but here it is the transmission that is modulated by what is known as a *notch filter mask*, whose radial transmission varies as $1 - \sin r/r$. It can then be shown that $\tilde{M}$ is a box function and that its convolution with u is flat over a large region which covers part of the pupil. The quantity $(u - u \star \tilde{M})$ is thus zero over the main part of the pupil, as required, except for a ring-shaped region on which the pupil diaphragm is applied. Note that this mask can also be designed to modulate in just one dimension.

There are several limitations to the use of focal masks:

- To arrange for u and $u \star \tilde{M}$ to be as similar as possible (the condition for cancelling w), $\tilde{M}$ must be as close as possible to a δ function, i.e., it must have a very narrow support. This in turn means that the focal mask, its Fourier conjugate, must have a very broad support, which is barely compatible with the fact that the aim is to form an image of the field as close as possible to the central source. The four-quadrant phase mask is an exception here, and for this reason is rather large, since it extends over the whole field, while conserving good 'transmission' close to the axis.
- The formalism used here assumes that the wavefront is perfectly plane, which is never the case in practice due to optical defects and atmospheric perturbations. In particular, it is absolutely

[39]Daniel Rouan is a French astronomer born in 1950

essential to use adaptive optics for ground-based observation, because the PSF of the bright source must be quasi-coherent and must be maintained very accurately on the axis of the focal mask. This is therefore the main limitation of coronagraphs in use at the time of writing (2007), including those soon to be used in space. Ways of reducing the consequences of this limitation will be discussed below.

- In phase mask systems, the phase shift introduced by the mask is not independent of the wavelength, since it is generally achieved using a slab of thickness e and refractive index n ($\phi = 2\pi en/\lambda$). This means that it will only be effective in a narrow spectral band centered on the wavelength $\lambda = 2en$.

Various solutions have been suggested to make the π phase shift *achromatic* in the latter case. Undoubtedly one of the most promising, proposed by Mawet and coworkers, is to use a zero order grating (ZOG). Such components are obtained using nanotechniques, etching grooves of width smaller than the wavelength (so-called sub-lambda grooves) on a substrate. It can be shown that these gratings have birefringent properties with a different refractive index between the ordinary and extraordinary axes which can be made linear in λ. This amounts to having an achromatic phase shift between two perpendicular polarisations. The idea then is to etch grooves in one direction on two opposite quadrants, and perpendicular grooves on the other two quadrants. Whatever the polarisation state of the incident light, the desired π phase shift is indeed introduced.

Following the same line of thinking, a very elegant solution is the *achromatic groove phase mask* (AGPM). This picks up the ZOG idea except that the sub-lambda grooves have circular symmetry. The effect of the differential path length difference between two perpendicular polarisations is indeed obtained, as with the four-quadrant ZOG, but the effect of the boundary between the quadrants disappears, which is a non-negligible advantage. Figure 6.54 illustrates the solutions based on the ZOG.

Apodisation

In this family of techniques, the idea is to play rather on the function u in the equation $w = \mathrm{FT}(v) = (u - \epsilon u \star \tilde{M})$. There are two aims:

- To ensure that the feet of the PSF of the bright source do not extend too far, or at least decrease more abruptly than in a classical Airy disk (in which the intensity decreases on average as θ^{-3}).
- To round off the profile u of the pupil so that it matches as well as possible the profile of the diffracted wave, i.e., the convolution product $u \star \tilde{M}$. Indeed, just as we have seen that a narrow focal mask implies $\tilde{M}$ with wide support, so the convolution product is described by a function closer to a bell shape than a top hat.

Non-Circular Pupil

A first solution is to consider particular shapes of aperture so that, in one or more selected directions, the profile of the PSF decreases much faster than the Airy disk. This is the case with a square pupil: indeed, in the directions corresponding to the diagonals, the brightness profile of the PSF varies as θ^4, whereas it only decreases as θ^2 in any other direction. Another solution, put forward by Kasdine and Spergel, is to consider apertures of Gaussian shape. Figure 6.55 shows such a pupil along with the corresponding PSF and the brightness profile in the optimal direction. Other

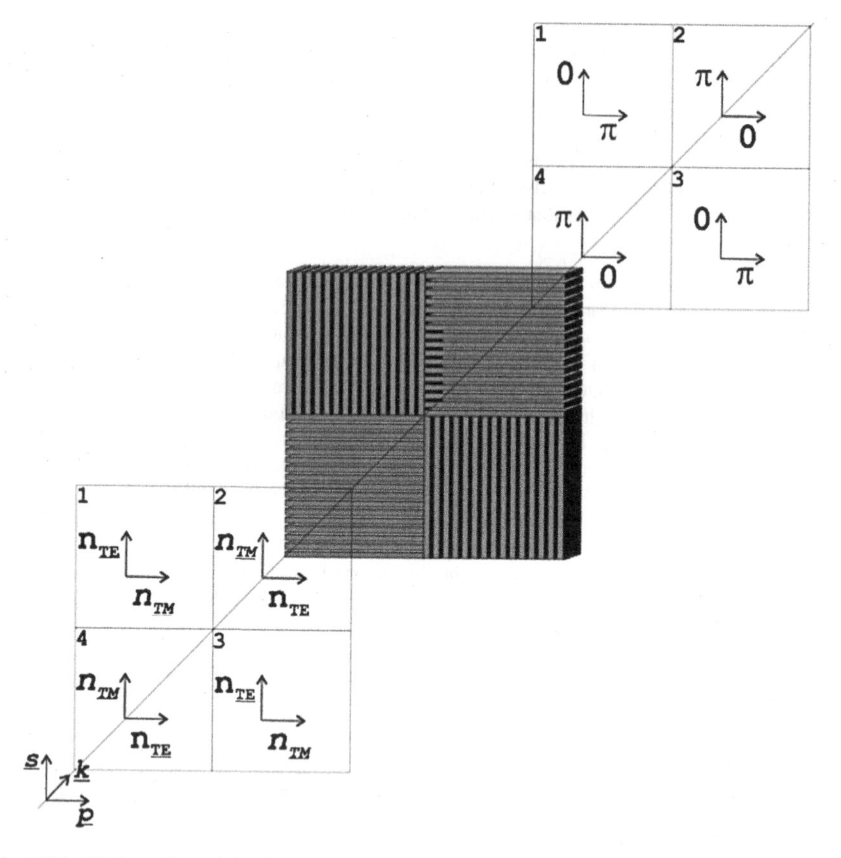

Fig. 6.54 ZOG version of the four-quadrant phase mask coronagraph. The quadrants are etched with grooves with spacing $p < \lambda$ using nanotechniques. The shape birefringence introduced in this way produces the desired π phase shift between quadrants, as illustrated by the two configurations of the electric field components. Source: D. Mawet

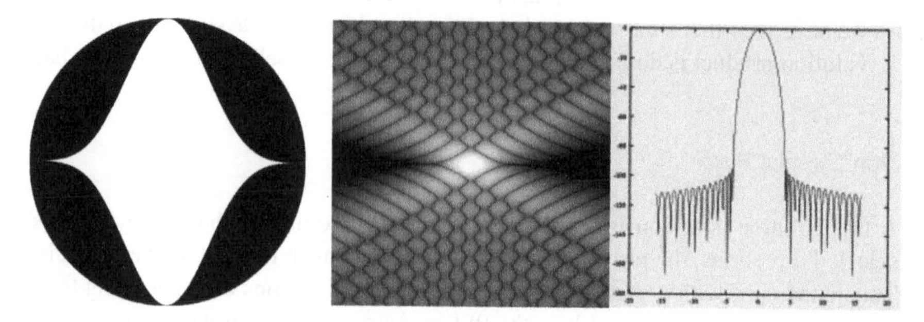

Fig. 6.55 The Kasdine–Spergel coronagraph. Left: the pupil shape; center: the point spread function; right: a cut along the x-direction revealing the very low level of residual light on each side of the main core

combinations in which several different Gaussian profiles are juxtaposed are also possible.

For all these solutions, the high contrast part of the field, i.e., the part in which faint structures might be sought, is rather limited. In particular, it must be possible to rotate the telescope about its optical axis in order to explore the whole environment of the bright source.

Variable Transmission Pupil Mask

If M is an indicator function (taking values 1 or 0), Aime and Soummer have shown that the equation $w = (u - \epsilon u \star \tilde{M}) \times D = 0$ can be construed as an eigenvalue problem with solution the pair $[u = \Phi, \Lambda]$, where the eigenvalue, if it exists, is $\Lambda = 1/\epsilon$. Such solutions exist for the Roddier–Roddier coronagraph ($\epsilon = 2$) for a circular or rectangular aperture. The function Φ is then a particular mathematical function, viz., a linear or circular prolate spheroidal function, respectively. In the case of a Lyot coronagraph ($\epsilon = 1$), there is no exact solution, but approximate solutions can be defined using these same functions, solutions which become more efficient as the mask gets larger. What is remarkable in this case is the fact that the amplitude of the wave diffracted by the mask is proportional to the amplitude in the entrance pupil. The process can thus be reiterated by introducing a series of focal mask/pupil diaphragm pairs, each introducing the same reduction factor for the stellar light. The extinction then increases exponentially with the number of stages. This idea is under investigation for the SPHERE instrument to equip the VLT, as mentioned earlier.

The difficulties in making masks where the transmission varies according to a precise law may be overcome by using high-energy beam sensitive (HEBS) materials. These can become opaque in a stable manner under the effects of high energy electron bombardment. A narrow electron beam can be computer controlled to scan a HEBS film deposited on a glass slab in order to obtain a mask with very precise transmission profile.

In this same family of apodising pupil masks, a very elegant solution is *phase induced amplitude apodization* (PIAA), suggested by Olivier Guyon. To get round the fact that the resolution and equivalent transmission are degraded in conventional amplitude modulation (since a pupil mask necessarily absorbs light), Guyon suggested playing on the shape of the optical surfaces to concentrate or dilute the light rays in the exit pupil of an afocal system, as shown in Fig. 6.56. The apodisation effect on the PSF is the same as for a pupil mask but without the drawbacks. The collecting area of the primary mirror of the telescope is fully exploited.

Optical surfaces capable of producing the desired brightness distributions are highly aspherical, and since the optical arrangement no longer respects the Gauss conditions, the useful field becomes very small. However, this is not a serious problem, since the structures sought, e.g., exoplanets, are always very close in angular terms to the optical axis of the bright source. Moreover, by using an identical but reversed optical system after passing through a coronagraphic mask — a simple Lyot mask is perfectly suitable, since the resulting PSF has bounded support — the isoplanicity properties of the field can be reestablished. The main difficulty with this solution is in producing

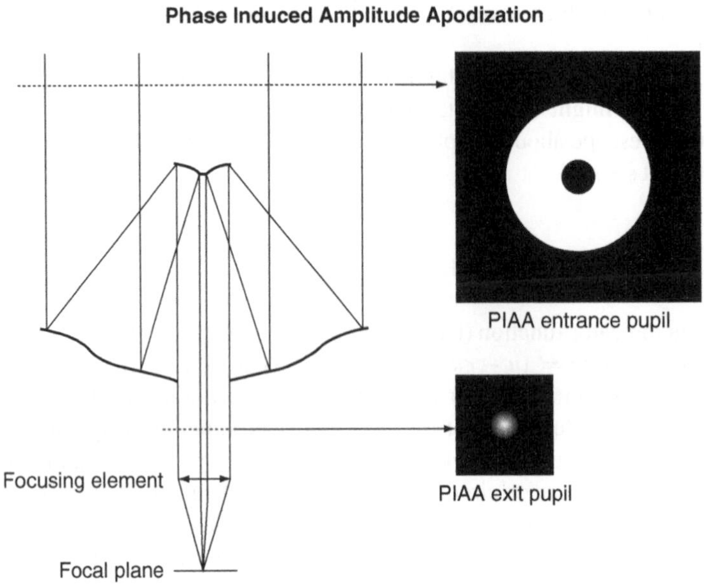

Fig. 6.56 The Guyon PIAA coronagraph. The pupil is reshaped thanks to aspheric optics. Source: O. Guyon

specific aspherical surfaces, whose radius of curvature must vary by several orders of magnitude along a diameter.

Interferometric Solutions

Although inspired by Lyot's idea, devices using phase masks already exploit the principle of destructive interference of the wave coming from the bright source. In this last part, we discuss methods aiming explicitly to produce such destructive interference, but without resorting to the focal mask/pupil diaphragm scheme originally proposed by Lyot.

Achromatic Interferential Coronagraph (AIC)

It was in fact just such a method, proposed by the French astronomers Jean Gay and Yves Rabbia, that triggered new interest in coronagraphy at the end of the 1990s. Achromatic interferential coronagraphy (AIC) exploits a Michelson interferometer to make the wave from the bright source interfere destructively with itself. The novelty here is to reverse the wave in one arm of the interferometer by passing it through the focal point of an optical system. Only the wave strictly on the axis will remain parallel to its counterpart in the other arm and hence be able to interfere uniformly and destructively with the latter, while a wave at a small angle of incidence θ, e.g., coming from a nearby companion, will meet its counterpart in the recombination output at an angle 2θ, preventing any effective interference. Moreover, recalling that the effect of passing through a focal point is to introduce a phase shift of π, i.e., a change of sign of the amplitude, the two arms of the interferometer can be of strictly equal length, which simplifies the design, but above all allows destructive interference whatever the wavelength, a significant advantage which justifies calling the device achromatic.[40]

The *visible nulling coronagraph* (Lyon, 2005) is a further variation on the theme of destructive interference. This idea, which also exploits a Michelson interferometer, is analogous to the AIC in many ways. The main difference is that one of the two beams is not reversed but simply translated a distance B (brought to the level of the entrance pupil, i.e., the diameter of the telescope): the pupils overlap significantly in a cat's eye pattern. The phase shift π which ensures destructive interference on one of the two interferometer outputs is introduced by adjusting the optical path difference in one of the arms. The effect of translating one pupil relative to the other is then analogous to the effect obtained in a nulling interferometer comprising two telescopes whose baseline would simply be B. The field of view of the telescope is modulated by a series of fringes, alternately dark (extinguished) and bright (transmissive), placing the bright source that needs to be attenuated on a dark fringe. The further advantage with this setup is that the operation can be repeated by injecting the dark output into a similar interferometer, but applying the separation/translation operation in a perpendicular direction. The extinction profile then goes as θ^4, which attenuates the star more efficiently. Figure 6.57 shows how this works. Note, however, that this kind of optical system is difficult to implement in practice.

To end this discussion of single-pupil interferometric solutions, let us describe two methods which complement the other instruments invented to extinguish the bright source. The main limitation that must be faced by all these solutions when they are eventually set up and required to achieve the extreme levels of performance involved in detecting extrasolar planets, such as those mentioned in the introduction, will be the extreme difficulty in distinguishing a true planet from a speckle due to residual, even very small, defects in the optics. Indeed, even defects of the order of ten picometres on the wavefront of a very large telescope could very easily simulate an Earth-sized planet!

Self-Coherent Camera

This instrument, proposed by Pierre Baudoz, exploits the fact that stellar and planetary photons are not coherent. If the final coronagraphic image is made to interfere with a beam of pure stellar light extracted upstream, the residual speckles will be modulated by a fringe system, while the spot formed by any planet will not be. A careful analysis of these fringes by Fourier transform methods would then provide a way, at least in principle, of removing any ambiguity.

---

[40]This idea could evolve toward a very compact device called the CIAXE. To carry out the required task, this makes judicious use of reflections on the optical interfaces of two thick lenses. However, such an optical system is very difficult to realise and has not yet seen the light of day at the time of writing (2007).

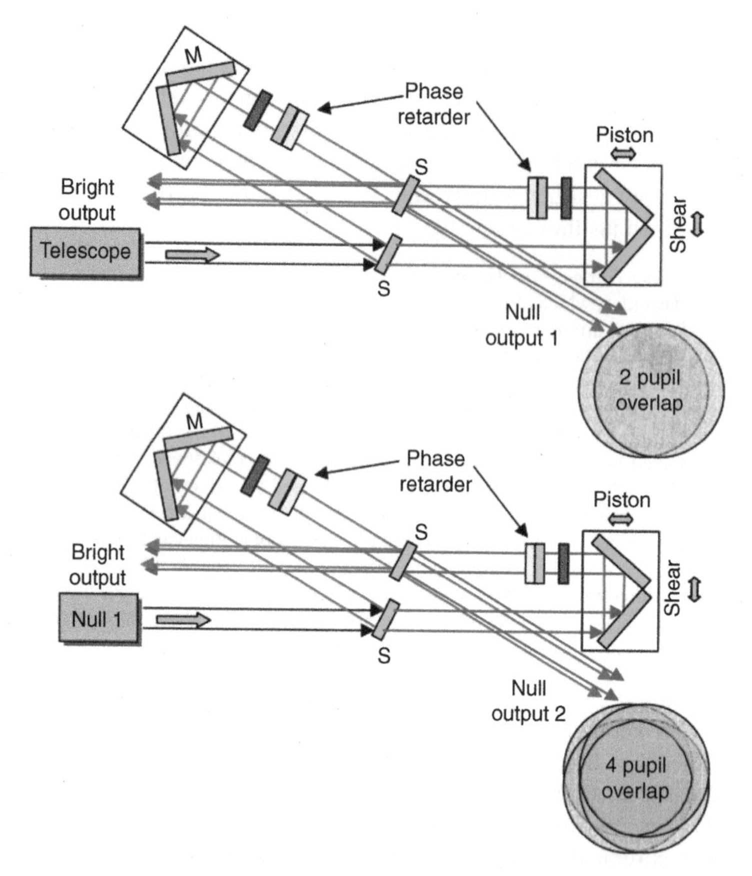

Fig. 6.57 Visible nulling coronagraph due to Lyon and Shao based on a Michelson interferometer with a shear of one beam with respect to the other. S = beam splitter, M = dihedral mirror

Speckle Cancellation

Another way of reducing residual speckles, proposed by Traub and Bordé (2005), would be to measure them, and then cancel them using a deformable mirror. To begin with, a carefully chosen series of very small amplitude wave forms is applied to this mirror, of the same dimensions as residual defects in the wavefront after passing through whatever form of coronagraph has been used. Then, by analysing the resulting series of images and using a little linear algebra, it is possible to reconstruct the right form of wave to apply to the mirror in order to correct these residual defects. After a few iterations, the process should converge.

This solution is based on a remarkable property of the electric field in the pupil after passing through a coronagraph: to a first approximation, it is directly proportional to the perturbed phase Φ, because in the truncated expansion of the input amplitude, viz., $\exp \mathrm{i}\Phi \approx 1 + \mathrm{i}\Phi$, only the first term vanishes as a result of the action of the coronagraph. Put another way, a perfect coronagraph can only extinguish the coherent part of the image of the bright source.

Conclusion

Which of these methods will meet the challenge of forming an image of a planet orbiting around a main sequence star? At the time of writing (2007), no one would be able to reply to this question. The development of such instruments is likely to remain highly active in the coming years. Looking beyond the design of theoretically excellent optical systems, the problem of implementing them in the real world has become a major preoccupation. A clear consensus has already been reached on one key point: only a complete system comprising telescope + extreme adaptive optics + focal instrument, dedicated at the outset to this aim, will ever obtain the image of details of a sister planet to our own.

6.6.2 Nulling Interferometry

Here we discuss a variant of coronagraphy. The idea is still to use interference properties to reduce the extreme intensity contrast between two very close objects, usually a star and a planet, in order to study the fainter of the two. However, in this case, the solution is interferometric recombination of the beams from several telescopes. The term *nulling interferometer* could then be replaced by *coronagraphic interferometer*, as we shall in Sect 9.6.

Basic Idea

The method of nulling interferometry suitably complements conventional coronagraphy. Indeed, the latter is a technique that is essentially applicable to the visible and near infrared regions. This is because it is only at these wavelengths that the angular resolution of a large telescope is of the same order of magnitude as the angle subtended by a typical planetary orbit of 1 AU at a distance of 10 pc. While life does provide spectral signatures in this region of the visible — basically, through oxygen and the chlorophyll function — the star–planet intensity contrast remains extremely high, typically of the order of 10^{10}.

On the other hand, in the mid-infrared region, this contrast is much more favourable, being of the order of $10^{6.5}$. Furthermore, a set of less ambiguous biological signatures than those observable in the visible has been identified. These are provided by ozone at 9.7 μm, carbon dioxide, and water vapour (see Fig. 6.58). It has been shown that the simultaneous presence of these three signatures could constitute a very strong indication that life is present, at least in a form similar to

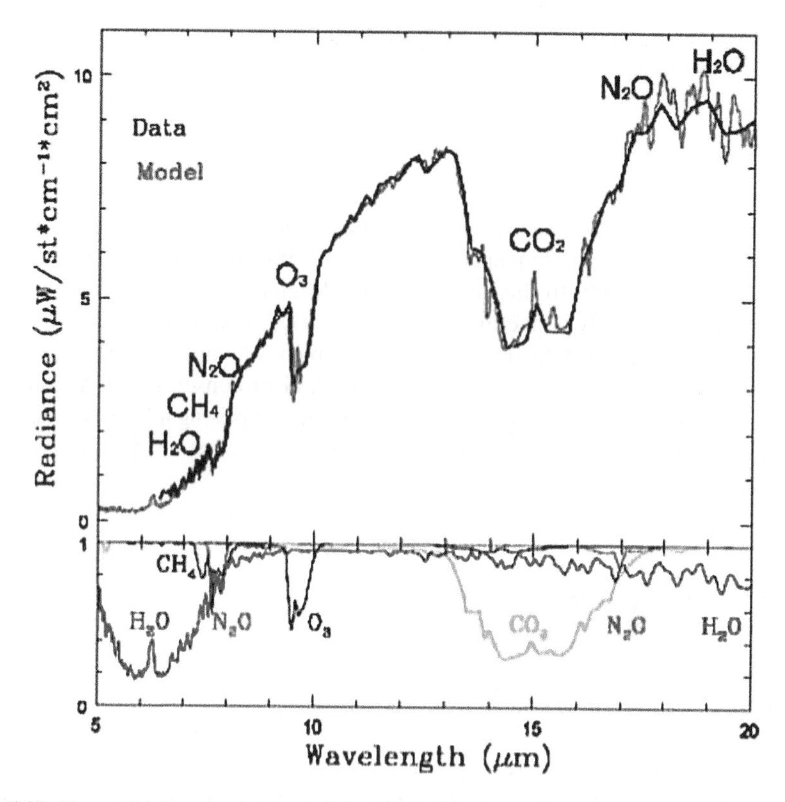

Fig. 6.58 The mid-infrared spectrum of the Earth shows three spectral signals due to ozone, carbon dioxide, and water vapour which, if observed together, would constitute a biosignature that is in principle unambiguous. *Upper*: Observed and modelled infrared spectra of the Earth viewed from a distance. *Lower*: Laboratory spectra of various atmospheric components. See also Fig. 9.20

life on Earth. The quest to identify this trio has been made the cornerstone of ESA's Darwin programme (see above), while NASA's TPF-I programme will pursue a similar aim, but today, both projects have been shelved.

An angular resolution of 0.1 arcsec at 15 μm implies either a monolithic telescope of diameter 30 m, whose feasibility in space is far from obvious, or an interferometer with a baseline longer than 30 m. The second option has been favoured for the Darwin project proposed back in 1996 by Alain Léger and coworkers, based on the assumption that spacecraft could fly in stable and reconfigurable formations without excessive difficulty. However, the problem of dazzle by light from the star remains just as critical as ever with an interferometer, whence it is still essential to apply some blocking procedure analogous to coronagraphy.

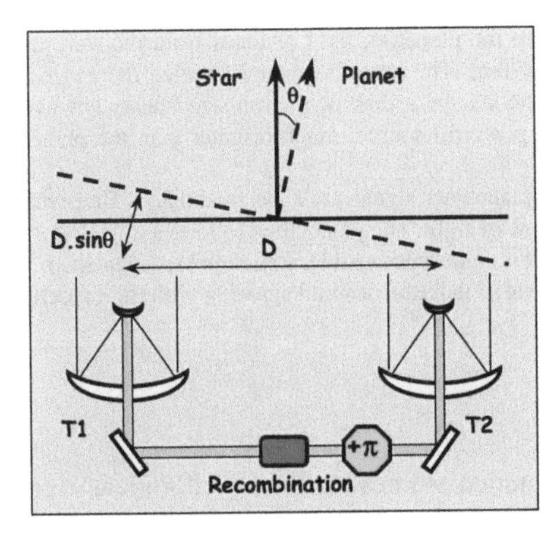

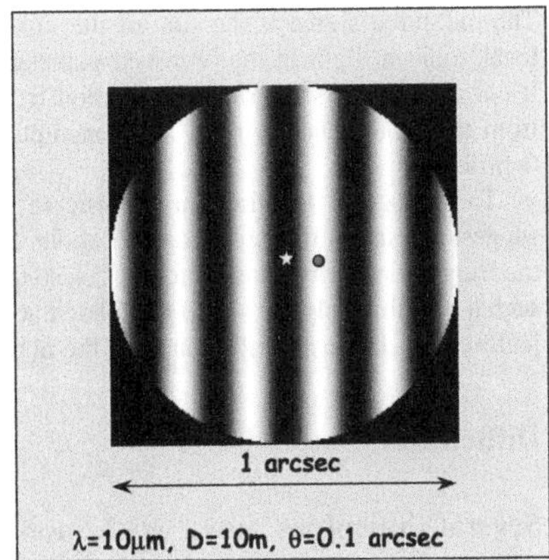

Fig. 6.59 Principle of nulling interferometry, as proposed by Bracewell. Values for the baseline D, the wavelength λ, and the separation θ are given as examples. *Left*: Schematic of the interferometer, with a π phase shift introduced in one arm. *Right*: The on-axis star is extinguished by the dark fringe, while the off-axis planet remains observable. Source: M. Ollivier

The Bracewell Interferometer

The solution to this difficult problem was put forward by Ronald Bracewell[41] as early as 1978, in a seminal paper, where he describes the principle of nulling interferometry. The idea, illustrated in Fig. 6.59, is to produce interference fringes, projected on the sky, by interferometrically recombining the beams from two telescopes a distance B apart. If we can arrange for the star to lie on a dark fringe and for the fringe spacing to correspond to the distance between the star and the supposed planet, then the contrast is optimal.

However, it would be of little use to work at a single wavelength, because the planet is extremely faint and the spectral information required lies in a vast wavelength range, from 5 to 17 μm for the Darwin instrument. In a conventional interferometer, only the bright central fringe, the one corresponding to zero optical path difference, would be common to all wavelengths, while the fringe systems of different wavelengths would overlap and interfere mutually for every other position in the sky. The powerful idea put forward by Ronald Bracewell was to insert a phase shift of π that did not depend on the wavelength, i.e., an achromatic phase shift, in one arm of the interferometer. This produces a dark fringe common to all the fringe systems. If this is placed on the star, efficient extinction becomes possible, without being limited to any spectral band.

[41]The Australian-born radioastronomer Ronald Bracewell (1921–2007), professor at Stanford University in California, made many remarkable contributions to radiofrequency interferometry and signal processing.

The measured signal is the sum of the flux from the planet(s), light residues from the star, and local zodiacal light in the observed planetary system. The latter is generally called the *exozodi*. It corresponds to scattering of radiation from the star by a disk of micron-sized dusts left over from the protoplanetary disk, and constitutes a powerful source, much brighter than the planets themselves.

To distinguish the fraction intrinsic to the planet, its signal must be modulated. Bracewell suggested rotating the interferometer about the line of sight. The planet then follows a trajectory on the fringe system projected onto the sky, so that it is sometimes visible, sometimes extinguished, in such a way that only the signal modulated at the rotation frequency and agreeing with the expected pattern reflects the contribution from the planet.

Difficulties

Several difficulties arose when more detailed studies of Bracewell's ideas were carried out, with a view to detecting exoplanets. It will be instructive to give a brief analysis of these difficulties.

Non-Infinitesimal Angular Size of the Star

Even if it cannot strictly resolve the disk of the photosphere, the nulling interferometer allows photons from its most peripheral regions to escape. Indeed, the extinction zone is not a box shape, but a parabolic function of the angular distance, represented by a truncated expansion of $1 - \cos\theta$. Naturally, very few of these photons get through, but the planet–star contrast is such that they are nevertheless predominant, and their statistical noise becomes a prohibitive limitation. Even more efficient extinction is thus required, going as θ^n with $n > 2$. It can be shown that more than two telescopes are needed to achieve this.

Several configurations have been found, more or less empirically, to achieve this aim. One of the simplest is Angel's cross. This produces a nulling function going as θ^4. It uses four telescopes placed at the ends of a cross, and introduces a π phase shift in one of the two telescopes in each branch of the cross.

Other setups use telescopes of different sizes. Finally, one can also consider configurations where the phase shift is not π but a fraction of 2π, as in the three-telescope configuration studied a few years ago for the Darwin project. Recently, it has been shown that an arbitrary power θ^n can be obtained by combining 2^n telescopes and associating a π phase shift with half of them, chosen using a binary sequence known as the Prouhet–Thué–Morse sequence.[42] This sequence has the property that no finite subsequence is repeated more than twice consecutively in the sequence.

Achromatic π Phase Shift

To obtain a signal of significant magnitude, a broad enough spectral interval must be analysed, and this requires a device that can obtain the π phase shift as

---

[42] The Prouhet–Thué–Morse sequence is defined by the recurrence relations $t_0 = 0$, $t_{2k} = t_k$, and $t_{2k+1} = 1 - t_k$, and begins 0110100110010110100101100110100 1...

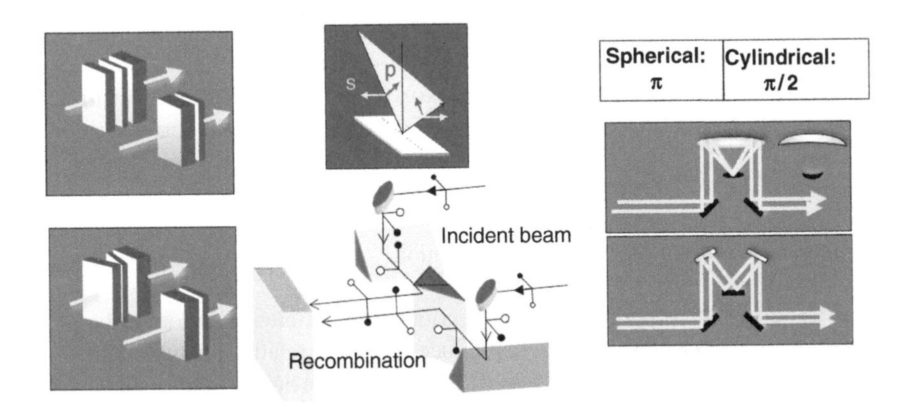

Fig. 6.60 Various methods to produce an achromatic phase shift in a nulling interferometer. Left: using plates of different optical index (see text); center: using a periscopic setup to reverse the electric field direction; right: crossing the focus of a spherical mirror. Source: Y. Rabbia

achromatically as possible throughout the interval. In 2007, several ideas are being explored and have been demonstrated in the laboratory:

- A first solution is to insert a set of transparent plates in each arm of the interferometer, choosing the materials and thicknesses so that the dependence of the path length difference on the wavelength is linear over the whole of the relevant spectral region. The phase difference is then achromatic.
- A second solution exploits a property of Fresnel reflection on a mirror at 45°, viz., only the electric field component lying in the plane of the mirror undergoes a change of sign. The idea is to have the beam reflect on two symmetric pairs of periscopes (set of three successive mirrors at incidence 45°), in such a way that the two components s and p of the electric field end up undergoing a reversal on one channel as compared with the other, as shown in the central part of Fig. 6.60.
- The third idea is to make use of the particular property that an electromagnetic wave undergoes a π phase shift when it passes through a focal point. It is enough to arrange that the beam pass through the focal point of a spherical mirror in just one arm of the interferometer in order to obtain an achromatic π phase shift.

Note that the last two solutions can only achieve phase shifts equal to π.

Separating Planetary and Exozodi Light Signals

The symmetry of revolution of the exozodi light would only correspond to a planetary system viewed face on, which is a highly improbable configuration. Consequently, the elliptical brightness distribution of the exozodi generally produces a modulation by the fringe system, analogous to that of a planet. This is where the choice of interferometer configuration becomes crucial. Depending on the structure of the fringe system, or rather the maxima and minima of the transmission map, the signal modulations produced either by an isolated object like a planet or by a centrally symmetric structure like the exozodi will be more or less distinct. In 2007, investigations aim to identify the configurations that could best discriminate the two.

Temporal Stability of Stellar Flux Cancellation

This last difficulty is perhaps the most serious. Indeed, any deviation of the phase shift from π, even very small, would produce very large relative fluctuations in leaks of stellar light, and these would largely dominate photon noise from the exozodiacal light, the latter fixing the fundamental detection limit. This question is very carefully considered in simulation experiments.

Problems

6.1. Show that, for quasi-monochromatic radiation,

$$|\gamma_{12}(\tau)| = |\gamma_{12}(0)| \quad \text{if} \quad \tau \ll \tau_c \ ,$$

where τ_c is the correlation time. Set $V(t) = f(t)\exp(2\pi i \nu_0 t)$, with $f(t)$ slowly varying.

Answer. With the suggested notation,

$$\Gamma_{12}(\tau) = \langle V_1(t)\, V_2^*(t+\tau)\rangle = \langle f_1(t)\, f_2^*(t+\tau)\exp(-2\pi i \nu_0 \tau)\rangle \ .$$

The quasi-monochromatic approximation, $\tau \ll \tau_c$, then implies $f_2(t+\tau) \approx f_2(t)$, whence

$$\Gamma_{12}(\tau) = \Gamma_{12}(0)\exp(-2\pi i \nu_0 \tau) \ .$$

Taking the modulus and normalising, the result follows.

6.2. Consider an interference experiment with quasi-monochromatic light (Young's slits) and express the visibility $\mathscr{V}$ of the fringes as a function of the fields $V_1(t)$ and $V_2(t)$ of the two sources. Show that

$$|\gamma_{12}(\tau)| = \mathscr{V} \ .$$

Answer. At the point Q observed and at time t, the amplitude of the observed field is obtained from the superposition of the two sources

$$V(Q,t) = V_1(t-\tau_1) + V_2(t-\tau_2) \ ,$$

where τ_1 and τ_2 are the respective times taken by the light between the sources and the point Q.

The observed intensity is $I(Q,t) = \langle V(Q,t)V^*(Q,t)\rangle$, which gives

$$I(Q,t) = I_1(Q,t-\tau_1) + I_2(Q,t-\tau_2) + 2\mathrm{Re}\langle V_1(t-\tau_1)\, V_2^*(t-\tau_2)\rangle \ .$$

The assumption that the light is quasi-monochromatic gives

$$\tau_1 \text{ and } \tau_2 \ll \tau_c = 1/\Delta\nu \ ,$$

which allows simplification of the previous relation. Putting $\tau = \tau_1 - \tau_2$,

$$I(Q,t) = \big[I_1(Q,t) + I_2(Q,t)\big] + 2\sqrt{I_1(Q,t)\, I_2(Q,t)}\,\mathrm{Re}\big(\gamma_{12}(\tau)\big) \ .$$

Finally, as Q moves, the value of τ changes, so that the intensity exhibits fringes of visibility given by

$$\frac{2\sqrt{I_1(Q,t)\, I_2(Q,t)}}{I_1(Q,t) + I_2(Q,t)}\,|\gamma_{12}(\tau)| \ .$$

In the simple case when $I_1(Q,t) = I_2(Q,t)$, this gives visibility $|\gamma_{12}(\tau)|$.

6.3. Check that the coherence degree of order n of a quasi-monochromatic wave obeys

$$|\gamma_n(\tau)| = 1 \ .$$

6.4. A source is in the form of a grid of brightness

$$I(\boldsymbol{\theta}) = I_0 \sin 2\pi\frac{\theta_x}{a} \ .$$

(a) What is the spatial spectrum $\tilde{I}(\boldsymbol{w})$ of this grid?

(b) An astronomical source (a young star forming, or the nucleus of a galaxy) has profile

$$I(\boldsymbol{\theta}) = I_0 \exp\left(-\frac{\theta^2}{2a^2}\right) , \qquad a = 0.1 \text{ arcsec} \ .$$

What is its spatial spectrum?

(c) An optically thin circumstellar envelope around a star, is such that the brightness of the whole ensemble is given by

$$I(\boldsymbol{\theta}) = I_0 \exp\left(-\frac{\theta^2}{2a^2}\right) + I_1\delta(\boldsymbol{\theta} - \boldsymbol{\theta}_0) \ , \qquad a = 0.1 \text{ arcsec} \ .$$

The radius of the star is 10^{12} cm and its distance 400 pc (1 pc $= 3 \times 10^{18}$ cm). The envelope extends out 10^{15} cm all round. What is its spatial spectrum?

(d) A double star is such that one component has twice the intensity of the other. What is its spatial spectrum?

In each of the four cases, analyse the filtering effect on the image produced by pupils of increasing size, for some fixed wavelength.

Answer. The spatial spectra are just the Fourier transforms of the intensities.

(a) For the grid,

$$\tilde{I}(\boldsymbol{w}) = \frac{\mathrm{i}}{2} I_0 \frac{a}{2} \left\{ \delta \left[\frac{a}{2} \left(u - \frac{1}{2} \right), v \right] - \delta \left[\frac{a}{2} \left(u + \frac{1}{2} \right), v \right] \right\} .$$

(b) For the Gaussian source,

$$\tilde{I}(w) = I_0 \sqrt{2\pi} a \exp\left(-2\pi^2 a^2 w^2\right) .$$

(c) For the circumstellar envelope, using the result in (b) and linearity of the Fourier transform,

$$\tilde{I}(w) = I_0 \sqrt{2\pi} a \exp\left(-2\pi^2 a^2 w^2\right) + I_1 \exp\left(-2\mathrm{i}\pi\theta_0 w\right) .$$

(d) For the binary system,

$$I(\boldsymbol{\theta}) = 2I_1 \delta(\boldsymbol{\theta}) + I_1 \delta(\boldsymbol{\theta} - \boldsymbol{\theta}_0) ,$$

which implies

$$\tilde{I}(w) = I_1 \left[2 + \exp(-2\mathrm{i}\pi\theta_0 w)\right] .$$

The modulus thus exhibits fringes.

The effect of the limited size of the pupil is to cut off spatial spectra at the spatial frequency D/λ. In the image plane, the restored image is the convolution of the image source with the PSF, an Airy function which spreads as the pupil diminishes.

6.5. Consider an object $I(\boldsymbol{\theta})$ whose spatial spectrum is circularly symmetric (e.g., a star or a solar granulation). Show that all the spatial information can be obtained by a simplified telescope whose pupil is limited to a slit:

$$G(\boldsymbol{r}) = \Pi\left(\frac{x}{D}\right) \Pi\left(\frac{y}{a}\right) , \quad D \gg a ,$$

replacing a conventional telescope of diameter D.

6.6. A telescope (D=3.6 m) forms an image of a source at wavelength λ=0.5 μm. Its focal ratio is $f/D = 100$. What is the optimal sampling rate x for the image, if it is effectively limited by diffraction? The grain of the detecting photographic plate creates white detector noise up to a frequency corresponding to a grain size of 3 μm.

Show that the choice of a finite sampling rate x has the effect of increasing the noise spectral density in the frequency domain containing the useful information.

6.7. A star appears as a disk, assumed to have uniform surface brightness, at the wavelength λ_0:

$$I(\boldsymbol{\theta}) = I_0 \Pi\left(\frac{\theta}{\theta_0}\right) .$$

Calculate its spatial spectrum. Show that, if the image is formed by a pupil of diameter D such that $\theta_0 = \alpha\lambda_0/D$, ($\alpha < 1$), then the determination of $\tilde{I}(w)$ on $(0, w_c)$ nevertheless gives θ_0, provided that it is assumed *a priori* that the star is a uniformly bright disk (*super-resolution*).

Answer. The spatial spectrum of the star is

$$\tilde{I}(w) = \frac{I_0}{\pi\theta_0 w} J_1(\pi\theta_0 w) .$$

The central peak of the Bessel function J_1 extends to greater spatial frequencies than those sampled by the mirror. Therefore, only a truncated sampling of this function is available, and moreover, it is affected by a certain degree of noise. This sampling can be used to determine the value of θ_0, using least squares or cross-validation fitting methods, provided that the form of $\tilde{I}(\boldsymbol{w})$ is known a priori. In the object plane, this means that the size of the structure is smaller than the diffraction size.

6.8. Michelson Interferometer. Consider a pupil consisting of two circular apertures of diameter d, with separation vector $\boldsymbol{D}$, where $d \gg D$, and an object of spectrum $\tilde{I}(\boldsymbol{w})$, observed at the zenith, at wavelength λ.

(a) Write the transfer function $\tilde{T}(\boldsymbol{w})$ and approximate it in terms of Dirac distributions.
(b) Express the visibility of the fringes as a function of $\tilde{I}(\boldsymbol{w})$.

Answer. The transfer function is the autocorrelation of the pupil. We consider here the limiting case of two pointlike openings,

$$G(\boldsymbol{w}) = \delta(\boldsymbol{0}) + \delta\left(\frac{\boldsymbol{D}}{\lambda}\right) .$$

Then, for any other shift than $\boldsymbol{0}$ and $\pm\boldsymbol{D}/\lambda$, the product $G(\boldsymbol{w} + \boldsymbol{r}/\lambda)\, G^*(\boldsymbol{r}/\lambda)$ is zero, whence the approximation is

$$\tilde{T}(\boldsymbol{w}) = \delta(\boldsymbol{0}) + \frac{1}{2}\delta\left(\boldsymbol{w} - \frac{\boldsymbol{D}}{\lambda}\right) + \frac{1}{2}\delta\left(\boldsymbol{w} + \frac{\boldsymbol{D}}{\lambda}\right) .$$

The spatial spectrum of the object after this filtering is thus

$$\hat{I}(\boldsymbol{0}) + 0.5\hat{I}(\boldsymbol{D}/\lambda) + 0.5\hat{I}(-\boldsymbol{D}/\lambda) ,$$

so that, in the symmetric case, fringes of visibility $\hat{I}(\boldsymbol{D}/\lambda)/\hat{I}(\boldsymbol{0})$ are produced.

6.9. The intensity interferometer (Hanbury-Brown, ARAA **6**, 13, 1968) is based on the following principle: two telescopes, separated by a distance x, receive photons ($\lambda = 0.5\ \mu\text{m}$) from the same source, on two photomultipliers delivering currents $I_1(t)$ and $I_2(t)$.

(a) Show that the cross-correlation function of the two currents, formed in a correlator which introduces a delay τ, and given by

$$C(\tau) = \langle I_1(t)\, I_2(t+\tau)\rangle\ ,$$

contains information about the spatial structure of the observed object. Use the complex degree of coherence of the radiation as an intermediate step.

(b) Before correlation, I_1 and I_2 are filtered by two low-pass electric filters, with transfer functions $H(f) = \Pi\,(f/2\Delta f)$. If $\Delta f \sim 10^6$ Hz, what error is tolerable along the cables bringing each signal to the correlator? Compare with the precision required in a Michelson interferometer, at the same wavelength, and with the spectral interval $\Delta\lambda = 1$ nm.

6.10. Consider a parabolic mirror forming an image on its axis of a point at infinity.

(a) The mirror deviates slightly from a true paraboloid by an error described by $\varepsilon(r) = \varepsilon_0 \cos(2\pi x/a)$, $a = D/n$, $\varepsilon_0 = \lambda/5$, where $n \gg 1$. Show that, as far as the image is concerned, this is as if the pupil function $G(\boldsymbol{r})$ had been replaced by a new complex pupil function $G'(\boldsymbol{r})$ (a phase mask). Deduce the point source response (image of a point) of the imperfect mirror.

(b) Answer the same question if $\varepsilon(\boldsymbol{r})$ is now a random function with constant power spectrum up to some characteristic frequency $s_0 = 1/10\lambda$.

(c) Why is the criterion for mirror surface quality taken as $\sigma_\varepsilon < \lambda/20$, where σ_ε is the standard deviation of the surface error?

(d) Why is the spatial spectrum of the random variable $\varepsilon(\boldsymbol{r})$ important?

6.11. Calculate the coherence length for the electromagnetic signal transmitted by each antenna of a radio interferometer operating at $\lambda = 18$ cm, and receiving a frequency bandwidth $\Delta\nu = 1$ MHz from the incident radiation. Is this length affected by the change in frequency brought about by means of a local oscillator on each antenna?

Answer. The autocorrelation of the field is the Fourier transform of its power spectrum. Hence, if this is truncated over a width $\Delta\nu$, the corresponding coherence time is $\tau_c = \Delta\nu^{-1}$. The coherence length is therefore $l_c = c\tau_c = c/\Delta\nu = 3$ m. This is not affected by any frequency change brought about by local oscillators.

6.12. Consider N antennas, intended to be aligned in an aperture synthesis arrangement. The intervals between antennas are chosen to be multiples of some length a. For each value of N between 2 and 6, inclusive, find the distances which give the non-redundant pupil with the best coverage of spatial frequency, by combination of pairs of antennas. Bearing in mind the rotation of the Earth, should the configuration be arranged east–west or north–south?

6.13. An object is observed in the visible ($\lambda = 0.5\ \mu$m), with spectral resolution $R = \lambda/\Delta\lambda$ increasing from 1 to 10^5. At ground level, the sky background produces a continuous emission of 1 rayleigh Å^{-1} (Sect. 2.3). In a plot of telescope diameter D, from 1 to 20 m, against the magnitude V of the object, trace, for different spectral resolutions R, the loci of the points at which the noise from the sky background is equal to the signal noise. Take the field to be 1 arcsec, and the quantum efficiency multiplied by the total transmission to be 0.5. What conclusions can be drawn for the use of large telescopes in obtaining high spectral resolution?

6.14. An infrared telescope is used at the diffraction limit, at wavelength $\lambda = 100\ \mu$m, with spectral resolution $\lambda/\Delta\lambda = 10$. Calculate the power received at the focus, in the beam étendue corresponding to the diffraction limit, taking $T_{\text{mirror}} = 300$ K, emissivity $\mathscr{E} = 0.1$, and using the Rayleigh–Jeans approximation. Compare this with the sensitivity of a detector with noise equivalent power (NEP) = 10^{-16} W Hz$^{-1/2}$ (the definition of the NEP is given in Sect. 7.1). What can be concluded from this comparison?

Answer. The beam étendue is written $S\Omega$, where the solid angle observed satisfies $\Omega = (\lambda/D)^2$ at the diffraction limit, and this implies a beam étendue of $\pi\lambda^2/4$, regardless of the telescope diameter. Moreover, the flux of a blackbody at temperature T is

$$B_\nu(T) = 2kT/\lambda^2 \qquad [\mathrm{W\,m^{-2}\,sr^{-1}\,Hz^{-1}}]$$

in the Rayleigh–Jeans approximation.

In order to make the comparison with the sensitivity of the detector, calculate the received power per beam étendue, in W Hz$^{-1/2}$, at the wavelength and resolution under consideration ($\Delta\nu = c\Delta\lambda/\lambda^2$):

$$\frac{\varepsilon 2kT}{\lambda^2}\frac{\pi\lambda^2}{4}\sqrt{\frac{c\Delta\lambda}{\lambda^2}} = \frac{\varepsilon\pi kT}{2}\sqrt{\frac{c}{R\lambda}} = 3.4\times 10^{-16}\,\mathrm{W\,Hz^{-1/2}}\,.$$

Therefore, with the given observational characteristics, detector sensitivity is not the limiting factor.

6.15. Invariance of the Complex Amplitude Correlation Function in Fresnel Diffraction. Let $\psi_z(\boldsymbol{x})$ be the complex amplitude of a wavefront at position z, with $\boldsymbol{x}$ the position variable in the wave plane. Show that the complex amplitude at the position 0 is given by the convolution

$$\psi_0(\boldsymbol{x}) = \psi_z(\boldsymbol{x}) \star \frac{1}{\mathrm{i}\lambda z}\exp\left(\mathrm{i}\pi\frac{x^2}{\lambda z}\right)\,.$$

Answer. The method used in this chapter can also be applied here, ensuring the conservation of energy. Show that

$$B_0(x) = \langle \psi_0(\xi)\, \psi_0^*(x+\xi) \rangle = B_z(x) = \left\langle \psi_z(\xi)\, \psi_z^*(x+\xi) \right\rangle .$$

The convolution is converted into a multiplication by Fourier transform.

6.16. *The Fried Parameter.* Assuming the expression $I_{\mathrm{A}}(\boldsymbol{\theta})$, given in this chapter, for the Airy image of a circular pupil, calculate the exact value of the Fried parameter $r_0(\lambda)$, by carrying out the integration

$$\int B_0(\boldsymbol{w})\, \mathrm{d}\boldsymbol{w} = \int \tilde{T}_{\mathrm{A}}(\boldsymbol{w})\, \mathrm{d}\boldsymbol{w} , \qquad \tilde{T}_{\mathrm{A}}(\boldsymbol{w}) = \mathrm{FT}\left[I_{\mathrm{A}}(\boldsymbol{\theta}) \right] ,$$

and with $B_0(\boldsymbol{w})$ equal to the long-exposure MTF. [The resulting $r_0(\lambda)$ is given in the text.]

6.17. The star T Tauri is observed as two sources in the near infrared. One of these sources also emits in the visible, and the other at radiofrequencies. The latter is separated by a distance 0.9 arcsec along the north–south axis, and the intensity ratio is α.

Show that a knowledge of $|\tilde{I}(\boldsymbol{w})|$ from speckle interferometry is not enough in itself to show which source is the more northerly. (Dyck H.M. et al., Ap. J. **255**, L103, 1982.)

Chapter 7
Detectors

This chapter is concerned with devices used to turn information, transported by the incident energy coming from astronomical sources, into signals which the astrophysicist can study. Such devices are called *receivers* or *detectors*.

The terminology depends on which scientific community is using it. Radioastronomers systematically talk about *receivers*, no doubt due to the traditions in telecommunications, i.e., radio and TV. In the spectral range from infrared to X rays, the term *detector* is generally used, in particular to refer to CCDs, which have become the most commonly used detectors, but the word *receiver* is also possible here. Then in the very high energy range of γ radiation, only the term *detector* is used. Here the terminology has naturally been imposed by the traditions of nuclear physics, where particles are considered to be detected.

In every case, the receiver or detector is of course a subsystem achieving the same objective of transforming the incident electromagnetic radiation into a signal, usually electrical, which can then be measured.

Detectors can be based on profoundly different physical principles, depending on which information carrier is involved. Electromagnetic radiation will clearly be the primary concern, occupying most of the chapter. However, detectors associated with the measurement of neutrinos and gravitational waves will also be examined, for these two areas represent a 'new astronomy', with a potential for discovery which, if not for the immediate future, is nonetheless considerable.

Any technical description of the immense variety of detectors used would be rather dull, and would almost immediately become obsolete. Therefore, the present discussion will deal first with general properties, and the main families of physical processes which underlie all detection devices.

The next section will deal with a selection of the main detectors used in astronomy today (2007). These have been classified according to the main wavelength regions, partly because this largely determines the technology, and particle because it determines the range of energies to be detected, bearing in mind that there is a factor of 10^{16} or more between the energies of γ radiation and radiofrequency waves.

An important aim will be to show how closely practical realisations approach the fundamental limitations due to the statistical properties of the radiation, which are outlined at the beginning of the chapter. This will provide an opportunity to measure

P. Léna et al., *Observational Astrophysics*, Astronomy and Astrophysics Library,
DOI 10.1007/978-3-642-21815-6_7, © Springer-Verlag Berlin Heidelberg 2012

the astonishing progress made, particularly in sensitivity, over the last decades of the twentieth century. These developments have often been due to the needs of medicine, defence, and telecommunications, from which astronomy has benefited, without itself being the principal motivation. Lallemand's[1] photomultiplier (1936) shows that this is not a new phenomenon.

Given any observation programme, the experimenter must apply effective criteria in choosing from the great diversity of tools available. For this reason, an attempt will be made to compare the optimal use of the various detectors. In practice, it is often difficult to find the dividing line between the detector itself and the full radiation detection and analysis system, as it was presented in Chap. 1. For example, *spectral selectivity* can be obtained by using a spectrometer before the detector (see Chap 8), but it can also result from an intrinsic selectivity in the detector itself. Finally, the coupling between the detector and the telescope, or radiation collector, can be improved by using devices such as fibre optics, amplifiers, and the like. Several examples will be given, which will show a marked evolution, as mentioned earlier in the book: an astronomical observing instrument has become an *integrated system*, whose various components interact closely and contribute together to the overall performance. This point will be taken up again in Chap 9, where several examples will be discussed for instruments in use today.

7.1 General Properties

By referring to the *detection* subsystem of an observing system, a very general expression can be given to represent the signal coming from one detection element of the detector (which may indeed contain a great number of such elements):

$$x(t) = x_0(t) + f\left[\int_{\Delta\nu} \Phi(\nu)\,\mathrm{d}\nu \int_{\Delta\Omega} I(\boldsymbol{\theta},\nu,t)P(\boldsymbol{\theta})\,\mathrm{d}\boldsymbol{\theta}\right].$$

$I(\boldsymbol{\theta},\nu,t)$ is the specific intensity of the radiation arriving at the detector, $P(\boldsymbol{\theta})$ is the *angular response* of the detector (e.g., field, lobes), and $\Phi(\nu)$ is its *spectral response*. $x_0(t)$ represents any signal delivered by the detector in the absence of an incident signal, and is called the *dark signal*. The function f can be considered to characterise the input–output relation of the detector. The most favourable situation is one in which this function is linear. I can also be considered as the input intensity of the whole observing system, the functions P, Φ, and f then characterising the complete system, and not just the detector.

If the detector is such that $f(I) = 0$ for $I < I_t$, the detector is said to have a *sensitivity threshold* I_t. Similarly, if $f(I)$ is approximately constant for $I > I_s$, the detector is said to *saturate* for $I > I_s$ (non-linearity).

---

[1]André Lallemand (1904–1978) was a French astronomer who developed the idea of the photomultiplier tube for use in astronomy.

It has been assumed, for simplicity, that the variables $\boldsymbol{\theta}$ and ν do not mix together in the responses Φ and P, although this is not always valid (e.g., chromatic effects of diffraction).

7.1.1 Amplitude Detectors. Quadratic Detectors

Depending on which physical process is being exploited, a detector may be linear, with respect to the field amplitude, or it may be quadratic, and thereby linear in the power. In the latter case, the phase information required for coherent detection and image formation in interferometry is generally lost (see Chap. 6).

Amplitude Detectors

This type of detector measures the instantaneous *amplitude* of the electric or magnetic field of a wave of frequency ν,

$$x(t) = \mathrm{Re}\big[E_0 \exp(2\pi \mathrm{i}\nu t + \phi)\big] .$$

There is therefore an instantaneous linearity between the signal and the amplitude, and this justifies calling it a linear, or a coherent detector, although in a more restricted sense than in the general definition given above. In this way, the antenna of a radio receiver converts the field into a current in the antenna, and this current signal is proportional to the field (Fig. 7.1).

In Sect. 7.5, it will be shown how a quadratic detector can be transformed into an amplitude detector by means of a local oscillator.

Quadratic Detectors

This kind of detector delivers a signal proportional to the mean *power* of the wave,

$$x(t) = \frac{1}{\Delta T}\int_t^{t+\Delta T} E(t')E^*(t')\,\mathrm{d}t' = \Pi\left(\frac{t}{\Delta T}\right) \star [E(t)E^*(t)] .$$

where $\star$ denotes convolution and the mean is a running mean, taken over a time ΔT (the *integration time*) much longer than the wave period. An equivalent relation is

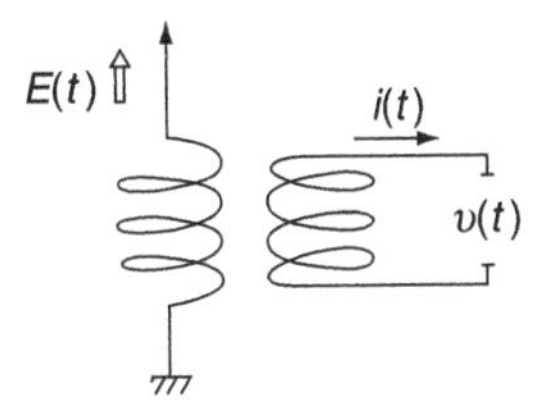

Fig. 7.1 Conversion of an electric field into a current by a receiving aerial

$$x(t) = \frac{1}{\Delta T} \int_t^{t+\Delta T} N(t')\,\mathrm{d}t' \ ,$$

where $N(t)$ describes the arrival of photons, which is usually a Poisson process.

The time ΔT, over which the running mean is taken, defines the temporal *passband* of the detector. Indeed, the relation giving $\tilde{x}(f)$ is

$$\tilde{x}(f) = S_\mathrm{E}(f)\Delta T \,\mathrm{sinc}(f\Delta T) \ ,$$

which makes the low-pass filtering of incident information explicit, defining an equivalent passband by

$$\Delta f = \frac{1}{\mathrm{sinc}^2(0)} \int_{-\infty}^{+\infty} \mathrm{sinc}^2(f\Delta T)\,\mathrm{d}f = \frac{1}{\Delta T} \ .$$

This type of quadratic detector, also known as a power, or an intensity, or again an *incoherent* detector, is therefore linear in the incident power (unless it exhibits threshold or saturation effects). These are the most commonly used detectors, operating by detection and/or counting of photons (the eye, photomultipliers, television cameras, channeltron, photoconductor and photovoltaic detectors, photographic plates, scintillators, spark chambers, etc.), and receiving radio diodes (video mode). In the following, an expression of the type

$$x(t) = \langle E(t')E^*(t') \rangle_{(t-\Delta T/2,\, t+\Delta T/2)}$$

will be assumed to represent the signal. This is a simplification, because in practice several parameters should be taken into account: the solid angle $\Delta\boldsymbol{\theta}$, frequency $\Delta\nu$, and so on, as already mentioned. The mean $\langle\ \rangle$ is taken over the interval ΔT around the instant t.

It is interesting to compare performances within the category of *coherent detectors*, which necessarily isolate a narrow spectral band $\Delta\nu$, with those within the category of *incoherent detectors*, which are not limited in this way. Figure 7.2 illustrates such a comparison, made for the spectral region between $\lambda = 3$ mm and $\lambda = 30$ μm, a region in which both types of detector can be made, using the appropriate technology. It can be seen that the domain of incoherent detection is the one with lower spectral resolutions, and this all the more so as the frequency is low.

7.1.2 Spatial Structure of Detectors

Single-Channel Detectors

These are detectors which, at any given instant, can only measure a single point (x, y) of the image, spectrum, etc. After an interval of time, called the integration time, the telescope or array, or whatever, is moved, so as to bring a new point to

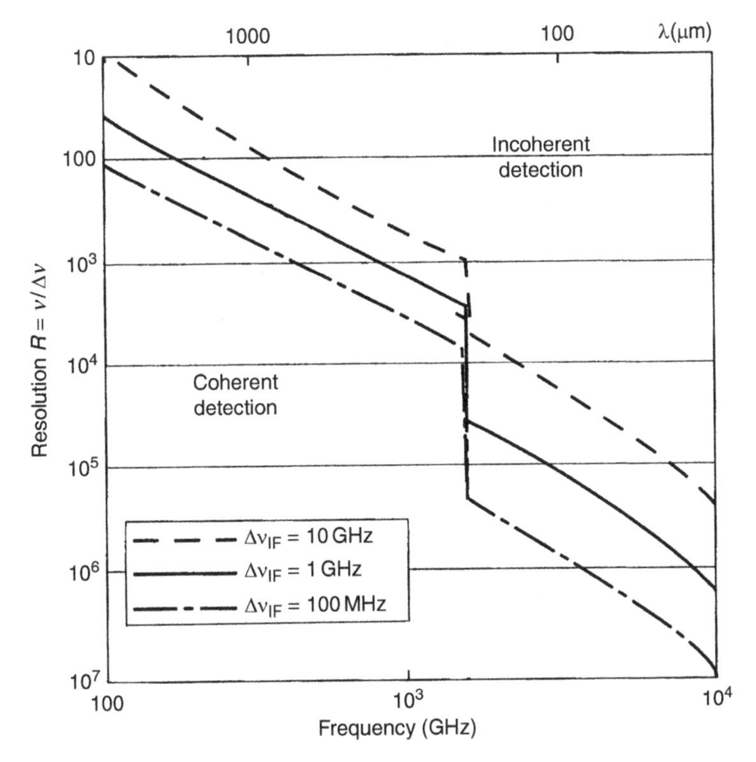

Fig. 7.2 Comparison of coherent and incoherent detection. *Abscissa*: frequency (*lower scale*) and wavelength in μm (*upper scale*). *Ordinate*: Spectral resolution $R = \nu/\Delta\nu$. Curves, established for three different spectral widths from 100 MHz to 10 GHz, give the locus of points at which the two methods give the same signal-to-noise ratio, equal to 3 on the curves. Above, it is incoherent detection which has the advantage with regard to the signal-to-noise ratio, and below, it is coherent detection. The discontinuity at $\lambda \sim 200$ μm comes from the fact that, for incoherent detection, bolometers (Sect. 7.4.10) are used in the submillimeter region and photoconductors (Sect. 7.3.1) at shorter wavelengths, the latter having better noise characteristics. Note that coherent detection has the advantage with regard to the signal-to-noise ratio when the spectral band is narrow. From Philips T., cited by McWatson D., Physica Scripta, 1986

the centre of the detector (Fig. 7.3). This *scanning* can be continuous or made up of discrete steps. If it is continuous, the spatial frequencies of the intensity distribution are transformed into temporal frequencies. The temporal filtering effected by the detector, and the scanning rate, must then be chosen according to whatever spatial frequencies are being sought in the image. If the scanning is step-by-step, the sampling rate is chosen by reference to the sampling theorem.

This type of image analysis, known as *raster scanning*, is similar to the movement of the electron beam in a television cathode ray tube. Its advantage is that the whole image is analysed with a single detector, which occupies the same position relative to the instrumental setup, if it is the telescope which is displaced, and this guarantees a strictly uniform instrumental response. Its drawback is clearly

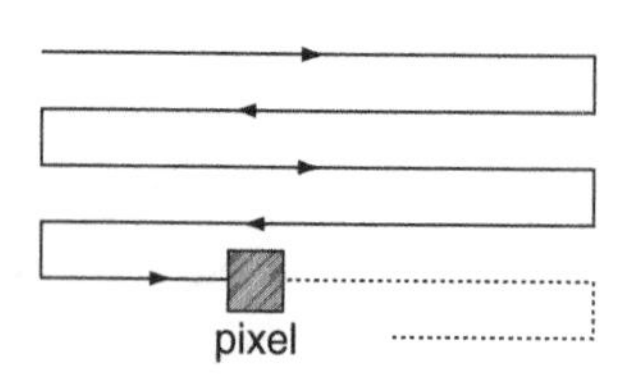

Fig. 7.3 Television (*raster*) scanning of a field

the concomitant inefficiency, for the greater part of the energy of the image, or the spectrum, at any given instant, is simply lost.

Radiofrequency receivers are still usually single channel, although it becomes possible in the millimetre region to build arrays containing a few tens of elements.

Multichannel Detectors

These detectors comprise N independent detectors (*pixels*) which receive photons simultaneously. If each detector is assigned a different direction $\boldsymbol{\theta}$, the setup is an imaging detector (e.g., a photoconducting array in the focal plane of a telescope). If each pixel receives a different wavelength, the setup is a spectral multichannel detector, as exemplified by a CCD array placed at the focus of a grating spectrometer.

Because of their high efficiency, multichannel detectors have been developed for use right across the electromagnetic spectrum. The photographic plate covered only the regions between X rays and the visible, but modern array detectors (the CCD and its spin-offs, and bolometers) already extend from X rays to the far infrared, and are gradually moving into the millimetre region.

Response of a Multichannel Detector

Consider first the simple case of a detector which is continuous in an ideal sense, such as a photographic plate with uniform quantum efficiency η, and hence no grain. Within its domain of linearity, the signal $S(x,y)$ is found from the incident intensity by a convolution

$$S(x,y) = G(x,y) \star I(x,y) ,$$

where the function $G(x,y)$ characterises the *spatial response* of the plate. The Fourier transform $\tilde{G}(u,v) = \tilde{G}(\boldsymbol{w})$ of G is referred to as the *modulation transfer function* (MTF) (Fig. 7.4). The MTF relates the spatial spectrum of the signal to that of the incident information by

$$\tilde{S}(\boldsymbol{w}) = \tilde{G}(\boldsymbol{w})\,\tilde{I}(\boldsymbol{w}) .$$

In a real photographic plate, for which grain is taken into account by assuming that the quantum efficiency can be modelled by a random variable $\eta(x,y)$, the MTF itself becomes random, at least for high spatial frequencies $\sim 1/g$, where g is the

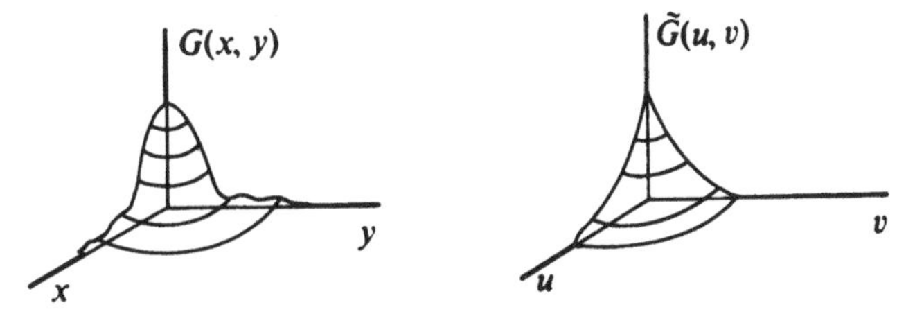

Fig. 7.4 Spatial point source response $G(x, y)$ and its Fourier transform $\widetilde{G}(u, v)$. In the case shown, $\widetilde{G}$ is real. $\widetilde{G}(u, v)$ is the modulation transfer function of the detector

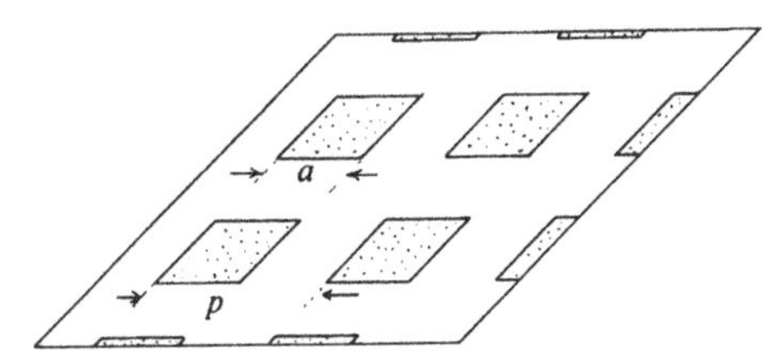

Fig. 7.5 Coverage of the image plane by a multichannel detector. Each detector pixel is shaded. Note, in this example, the relatively low *geometrical filling factor*. Progress in making CCD substrates means that this factor can now be made close to unity. In other cases, a microlens array in the focal plane can improve the exploitation of incident photons

mean grain size. If the detector has isotropic properties, the MTF will clearly exhibit circular symmetry.

A multichannel detector may also be composed of discrete elements (e.g., CCDs) which cover the image plane (Fig. 7.5). Let p be the distance between the centres of the elements, and a the length of their sides, with $a < p$. The signal is then obtained from:

$I(x, y)$ incident intensity,
$I'(x, y) = I(x, y) \star \Pi(x/a,\ y/a)$ convolution with individual elements,
$S(x, y) = ⧢(x/p,\ y/p)\, I'(x,\ y)$ sampling with rate p .

Hence, after normalisation at the origin, the MTF is given by

$$\mathrm{MTF} \propto \mathrm{FT}\left[\Pi\left(\frac{x}{a}, \frac{y}{a}\right)\right] = \mathrm{sinc}(ax)\mathrm{sinc}(ay)\ ,$$

and it can be observed that the spacing p does not necessarily represent a sampling adequate to the task of completely reconstructing $I(x,\ y)$, since $\widetilde{G}(\boldsymbol{w})$ does not have bounded support. The appropriate value of p thus depends on the spectrum $\widetilde{I}(\boldsymbol{w})$: if $\widetilde{I}(\boldsymbol{w}) = 0$ for $|\boldsymbol{w}| > w_\mathrm{c}$, w_c being a cutoff frequency for the system, then a suitable choice is simply $p = 1/2w_\mathrm{c}$.

7.1.3 Temporal Response

In the study of rapidly varying phenomena, such as a pulsar, by rapid photometry, the temporal capabilities of a detector must be determined. To this end, the temporal

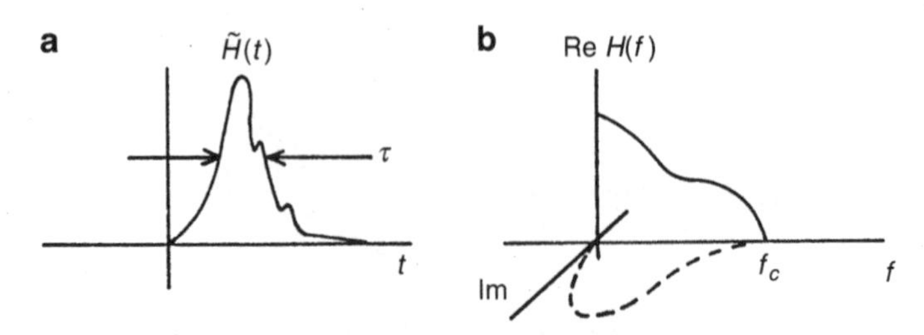

Fig. 7.6 (**a**) Point source response, and (**b**) complex transfer function of a low-pass filter. τ is the temporal width and f_c the frequency width, so that $\tau f_c \approx 1$

response of a detector is defined by a time filter $H(f)$, or by the impulse response $\tilde{H}(t)$ of the detector. The *time constant* of the detector is found from $\tilde{H}(t)$ (Fig. 7.6).

Integrating Detectors

The duration T of any measurement is always finite (so as to involve only a finite quantity of energy). If the incident signal is $x(t)$, the detector delivers

$$y(t) = \int \Pi\left(\frac{t'-t}{2T}\right) x(t')\,\mathrm{d}t' = \Pi\left(\frac{t}{2T}\right) \star x(t) .$$

The integration is thus a case of filtering, by a sinc function (the FT of Π) in the frequency space.

The photographic plate is an example of an integrator, since the instant t is fixed and T is the exposure time. The signal is then the *density* (blackening) of the photographic emulsion which, under certain conditions, is indeed proportional to the intensity (in the domain of linearity of the emulsion). The signal, or blackening of the emulsion, is then defined as the quantity

$$X(x, y) = i(x, y)/i_0 ,$$

or as

$$-\log_{10} X(x, y) ,$$

where $i(x, y)$ denotes the intensity transmitted by the plate at the point (x, y) when it is illuminated by an intensity i_0 (*densitometry* of a photographic image).

7.1.4 Noise

Below we shall examine in some detail the issue of physical fluctuations, which affect the detection of any form of radiation. A detector can augment these by adding further fluctuations, due to the way it is constituted or the way it is used. For example, this may be the *dark current*, *amplifier noise*, *readout noise*, or the *grain* of a photographic plate, etc. We shall return to these different forms of noise intrinsic to detectors after studying fundamental fluctuations. Ways of taking them into account and correcting for them, insofar as this is possible, will be examined in Sect. 9.6.

7.1.5 Characterisation of Detectors

Detectors are too complex to be characterised by a few simple parameters. Table 7.1 summarises all the properties that must be specified when two detectors, operating in the same spectral range, are to be compared, or when a detector must be chosen for some specific observation, or use with a specific instrument (spectrometer, camera, etc.).

The nomenclature here is very general. In any particular case, it can be simplified. The single-channel detector has only one pixel; the modulation transfer function or power spectrum of spatial noise are specific to two-dimensional imaging detectors and are irrelevant at radiofrequencies; readout noise is a specific property of solid-state imagers (CCDs); and so on.

Table 7.1 Detector characteristics

Dynamic range	Sensitivity threshold Domain of linearity Saturation level
Spectral properties	Quantum efficiency η Domain of sensitivity $\Delta\nu$ (or $\Delta\lambda$, or ΔE) Spectral response $\eta(\nu)$ Spectral resolution $\overline{\nu}/\Delta\nu$
Temporal response	Impulse response $H(t)$ Response time Temporal filter function $\tilde{H}(s)$
Geometrical properties	Geometrical dimensions Modulation transfer function $\tilde{G}(\boldsymbol{w})$ Homogeneity of spatial properties Multichannel: pixel number or format $N \times M$
Noise (see below)	Amplifier noise: gain g, fluctuations σ_g Readout noise: NEC or σ_{R} Intrinsic noise: NEP or noise temperature Spatial noise characteristics: PSD Temporal noise characteristics: PSD Detector quantum efficiency DQE
Polarisation	Polarisation selectivity

7.2 Fundamental Fluctuations

In principle, in order to give a complete treatment of light and its interactions with matter, the most correct physical model available should be used, namely the *quantisation* of the electromagnetic field, and the associated concept of the photon (quantum electrodynamics QED). However, the great complexity of this model is hardly ever necessary for the cases studied in this text. Indeed, a designer of optical instrumentation (see Chap. 5) can sometimes even make do with the ultra-simple model which uses light rays. Consequently, the *semi-classical model,* will serve the purposes of this book. This describes field propagation (e.g., diffraction, see Chap. 6) by means of the classical electromagnetic wave, and the interaction with matter (e.g., detection, see below) by means of random Poisson processes. It is not the quantised field, but rather its observables, namely photoelectrons, which are described. Nevertheless, the terminology refers to 'detected photons', and is not quite appropriate when using the semi-classical model, as J.C. Dainty has pointed out. It is so common that the same term will be used here.

The semi-classical model is presented in Mandel L. and Wolf E., Rev. Mod. Phys. **37**, 231, 1965. The discussion here is based on Dainty J.C., in *Modern Technology and Its Influence on Astronomy*.

In this model, light is subject to irreducible *fluctuations* when interacting with matter (*detection*), and this whatever the nature of the detector. The general features of these fluctuations will be presented here, followed by two limiting cases, the quantum and the thermal limits, which are sufficient in practice to give a simple description of the fluctuations of an electromagnetic signal.

Consider a vacuum enclosure at a uniform temperature T, within which the radiation is in thermal equilibrium. The time-averaged electromagnetic energy density per unit frequency interval is given by

$$\langle u(\nu)\rangle = \frac{8\pi h\nu^3}{c^3}\left[\exp\left(\frac{h\nu}{kT}\right)-1\right]^{-1} \qquad \left[\mathrm{J\,Hz^{-1}\,m^{-3}}\right] .$$

At any point inside the enclosure, the electromagnetic field $\boldsymbol{a}(t)$ results from a superposition of the fields radiated by a large number of incoherent elementary oscillators. This field is an ergodic stationary random (or stochastic) process, with zero mean. (For the terminology and results concerning stochastic processes, see Appendix B.) The central limit theorem (see Sect. 9.1) shows that this process is Gaussian (or normal), since $\boldsymbol{a}(t)$ is the sum of a large number of independent elementary fields. (It can also be shown that the normal distribution is the one which maximises the entropy for a given mean power.

In order to homogenise the physical dimensions, $\boldsymbol{a}(t)$ is not the electric field of the wave with wave vector $\boldsymbol{k}$ and frequency ν, but is proportional to it:

$$\boldsymbol{a}(t) = (\varepsilon_0 c\lambda^2)^{1/2}\boldsymbol{E}(t) .$$

The square is then a power [W], and this is also the dimension of the autocorrelation $R(\tau)$, of which an estimator is

$$R(\tau) = \frac{1}{T}\int_0^T a(t)\,a(t+\tau)\,\mathrm{d}t\;.$$

The quantity $P(\nu) = R(\tau)\times$ time is measured in joules, or equivalently, in W Hz^{-1}, as a power per unit frequency interval.

The autocorrelation of this stochastic process, identical to its covariance up to a scale factor, is determined from the power spectrum. The spectral density of $\boldsymbol{a}(t)$ is the electromagnetic power associated with the field of frequency ν, in a *coherence étendue* equal to λ^2. In order to define a finite monochromatic power, a solid angle Ω and an associated surface element S must be defined (Fig. 7.7). We refer to Sect. 3.2 for the demonstration that the radiation can be considered as coherent when attention is limited to the étendue $S\Omega = \lambda^2$. For either of the two polarisations, the associated monochromatic power is

$$P(\nu) = \frac{1}{2}\overline{u}(\nu)\frac{\lambda^2}{4\pi}c = h\nu\left[\exp\left(\frac{h\nu}{kT}\right) - 1\right]^{-1} \qquad [\mathrm{W\,Hz^{-1}}]\;.$$

The 4π factor just normalises the solid angle, and the presence here of the speed of light is due to an integration of the energy contained in a cylinder of length c and unit cross-section.

Thermodynamics shows that a system of mean energy $\langle W\rangle$ fluctuates about this mean, over time, with a variance given by

$$\langle \Delta W^2\rangle = \langle W^2\rangle - \langle W\rangle^2 = kT^2\frac{\mathrm{d}\langle W\rangle}{\mathrm{d}T}\;.$$

Applying this result to the mean power calculated above, it follows that

$$\left\langle [\Delta P(\nu)]^2\right\rangle = kT^2\frac{\mathrm{d}P(\nu)}{\mathrm{d}T} = P(\nu)\,h\nu\left\{1 + \left[\exp\left(\frac{h\nu}{kT}\right) - 1\right]^{-1}\right\}\;.$$

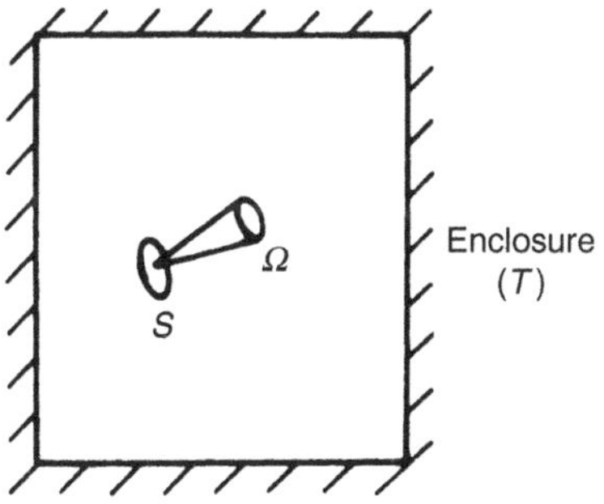

Fig. 7.7 A vacuum enclosure, with perfectly reflecting walls, contains a small quantity of matter at temperature T, in equilibrium with the radiation, at the same temperature. An antenna, symbolised by the circular area S, receives radiation in a cone of solid angle Ω. The coherence étendue is then $S\Omega = \lambda^2$

Inside a blackbody cavity in thermal equilibrium, the fluctuations of the radiation field[2] are manifested by the presence of two terms:

- A term $h\nu\, P(\nu)$ which can be interpreted as a *photon number fluctuation*. Letting $\langle n \rangle$ denote the mean number of photons received per unit time, Poisson statistics implies

$$\langle \Delta n^2 \rangle = \langle n \rangle .$$

- A term

$$h\nu \left[\exp\left(\frac{h\nu}{kT}\right) - 1 \right]^{-1} P(\nu) = \left\langle [P(\nu)]^2 \right\rangle ,$$

  in which the radiation temperature appears explicitly. This fluctuation term can be interpreted as being due to the random phase differences of the wave fields, and the beats between field components of different frequency. These fluctuations are low frequency, because of the exponential term. This phenomenon is analogous to the production of *speckles* (see Sect. 6.2) by interference between waves of random relative phase.

The structure of these two terms means that, to a first approximation, and depending on the value of $h\nu/kT$, one or other of them will dominate. For this reason, they can be studied separately.

If the first term dominates, the fluctuations are referred to as *photon noise*, *quantum noise*, or *shot noise*, and correspond to the quantum limit.

The discussion here will be limited to quantum noise associated with incoherent radiation, which is the type most frequently encountered in astrophysics (for example, in blackbody radiation). In Sect. 3.2, the quantum fluctuations of partially coherent radiation were examined, in the context of temporal and spatial coherence.

If the second term dominates, the fluctuations are referred to as *thermal noise*. In the general case when both fluctuations coexist,

$$\langle \Delta n^2 \rangle = \langle n \rangle (1 + \delta) , \quad \text{with} \quad \delta = (\mathrm{e}^{h\nu/kT} - 1)^{-1} ,$$

where δ is called the *degeneracy factor* and results directly from the Bose–Einstein statistics satisfied by photons. δ is the number of photons which are coherent with each other (refer to the discussion on temporal coherence in Chap. 3).

Remark 1

The expression for the fluctuations applies only to the radiation *inside* a blackbody. If this condition is not fulfilled, the dominant term in the fluctuations may be different.

---

[2]This result was demonstrated in 1909 by Albert Einstein, who used it as an argument to conclude the wave–particle duality of light. Einstein, A., Phys. Zeit. **10**, 817 (1909).

Consider an observation of a star, whose spectrum is assumed similar to that of a blackbody of temperature T. The radiation is observed at some frequency ν such that $h\nu \ll kT$. According to the general formula, if it were applicable, the thermal noise should dominate. Imagine, however, that the star is observed from far enough away to ensure that the photons ($h\nu$) arrive at the detector at well-separated times. Clearly, quantum noise should then dominate. Indeed, the radiation is unidirectional here, the beam étendue sampled by the observation much less than λ^2, and the number of photons sampled in $P(\nu)$ remains very small.

Remark 2

This second term triggered the idea of *intensity interferometry* discussed earlier (see Sect. 6.4). Indeed, when an extended light source is observed simultaneously from two distant points, the intensity fluctuations remain correlated if the separation between these two points is rather small. More precisely, the distance at which the correlation disappears depends on the coherence étendue of the source, whose angular size can thus be determined. This possibility, put forward by the physicist Robert Hanbury-Brown in the 1940s, raised much skepticism until he demonstrated it, first in the laboratory, and then on the stars.

Remark 3

A simple argument explains why quantum noise dominates at high frequencies. The exponential decrease in $P(\nu)$, as ν increases, implies that the power required to transmit a given quantity of information could be made to tend to zero, by using a high enough frequency ν. In the limit, less than one quantum $h\nu$ of energy would suffice. Quantisation must therefore create some other source of fluctuation, to prevent such a physically unacceptable state of affairs, and indeed it is exactly what was referred to as quantum or photon noise which forbids the programme described here.

Figure 7.8 shows the characteristic domains of these two fundamental fluctuations, which are independent of the type of detector used. It is worth noting that the detector does not detect radiation only from the source, but also from its environment (the laboratory, the telescope, the atmosphere, the ground, and so on), whose temperature is generally around 300 K, unless it is located in space and cooled.

- In the visible and ultraviolet regions, quantum noise dominates.
- At radiofrequencies, thermal noise dominates, even if the system can be cooled, so as to reduce its *noise temperature*.

By contrast, in the infrared and submillimetre regions, it may be necessary to take both types of fluctuation into account.

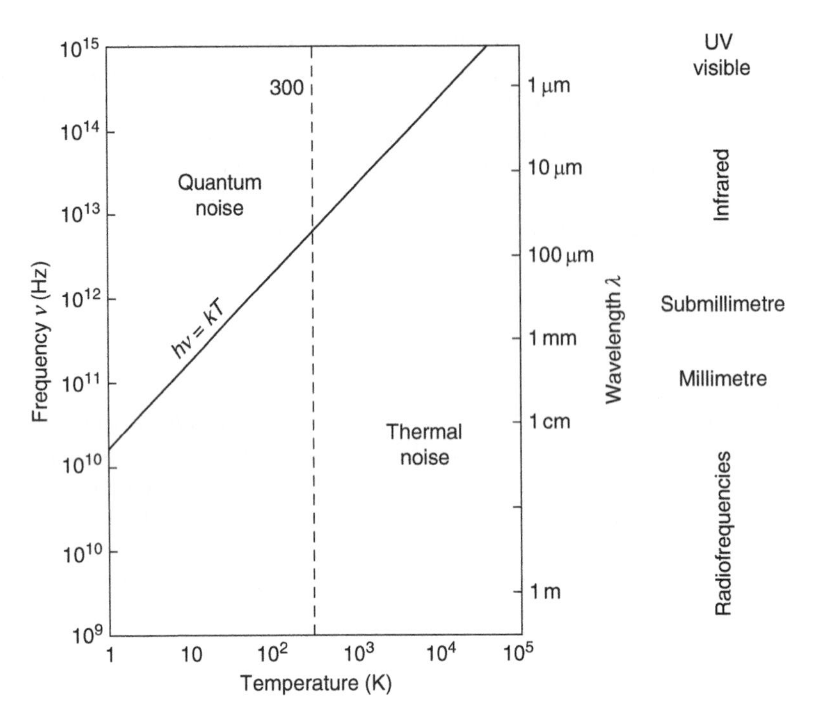

Fig. 7.8 Regions of predominant quantum or thermal noise, separated by the straight line $h\nu = kT$

7.2.1 *Quantum Noise*

Consider a flux of photons defining a stochastic process $\mathbf{x}(t)$ made up of Poisson impulses. The parameter is $\overline{n}$, the mean number of photons received per unit time, within the étendue $S\Omega$ defined by the detection system, e.g., the telescope optics. Then the probability of obtaining p events during a time T is given by

$$P\left[\int_T \mathbf{x}(t)\,\mathrm{d}t = p\right] = \frac{(\overline{n}T)^p\,\mathrm{e}^{-\overline{n}T}}{p!}\,,$$

which leads to a mean number of events n_T during the time T given by

$$n_T = \overline{n}T\,,$$

with variance given by

$$\sigma^2 = \langle n_T^2\rangle - \langle n_T\rangle^2 = \overline{n}T\,.$$

When $n_T \gg 1$, the random variable

$$\delta n_T = n_T - \langle n_T \rangle$$

is normal (Gaussian), with standard deviation $(\overline{n}T)^{1/2}$.

If the measurement involves counting the number of photons received in time T, the signal-to-noise ratio is

$$R = \frac{\overline{n}T}{(\overline{n}T)^{1/2}} = (\overline{n}T)^{1/2} ,$$

and it varies as the square root of the measurement time.

Example

Photon counting in γ-ray astronomy. The COS-B satellite, put into orbit in 1975, detected the brightest source in the sky, the Vela pulsar, at 50 MeV. This pulsar has period 89.2 ms. Using an accurate clock, the signal-to-noise ratio can be improved, by accumulation of successive cycles. With 89 bins, a total of 2 963 detected events were accumulated, over 38 days of observation (Fig. 7.9). (A *bin* is the interval Δx_i of the variable x, of constant width, over which the average value of some function is measured.) Although there is a modulation in the intensity over the cycle, the average number of events per bin, and the standard deviation giving the confidence interval, are

$$\overline{N}_T = \frac{2\,963}{89} = 33.29 , \qquad \sigma_{N_T} = 5.8 .$$

Conversion of Photons into Photocurrent

A photoelectric detector (see below) converts photons into current impulses. The *quantum efficiency* η of the detector is defined as the ratio of the mean number of photoelectrons produced to the mean number of incident photons over some period T.

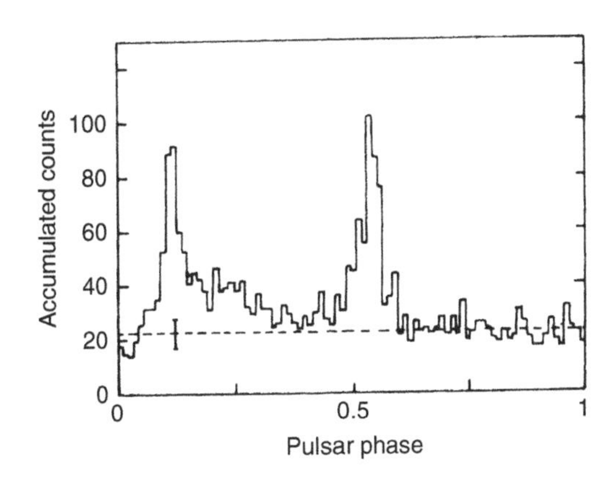

Fig. 7.9 Accumulated γ-ray photon counts, during an observation of the pulsar PSR 0833-45 by the satellite COS-B, at high energy ($h\nu > 50$ MeV). (Benett K. et al., Astr. Ap. **61**, 279, 1977.) The period was divided into 100 bins, and successive cycles were added bin by bin, to improve the signal-to-noise ratio

If $\mathbf{x}(t)$ describes the arrival process of the photons, the process $\mathbf{i}(t)$ describing the photocurrent is

$$\mathbf{i}(t) = \begin{cases} e\,\mathbf{x}(t) & \text{probability } \eta \,, \\ 0 & \text{probability } 1-\eta \,. \end{cases}$$

$\mathbf{i}(t)$ is a new process, constructed from the process $\mathbf{x}(t)$, and

$$\mathbf{i}(t) = G[\mathbf{x}(t)]$$

would be a correct representation, if the passband of the detector were infinite. Real detectors (photomultiplier, photoconductor, etc.) always have a finite frequency passband. The *temporal filtering* property of the detector can be represented by the idealised low-pass transfer function

$$\tilde{H}(f) = \Pi\left(\frac{f}{2f_c}\right) ,$$

which cuts off the frequency at f_c (Fig. 7.10). The photocurrent is now

$$\mathbf{i}(t) = G[\mathbf{x}(t)] \star H(t) ,$$

where $H(t) \rightleftarrows \tilde{H}(f)$ (Fourier transform). Its spectral density is

$$S_i(f) = \left[\bar{n}^2\,\delta(f) + \bar{n}\right]\eta\tilde{H}(f) ,$$

where the general expression for the spectral density of a Poisson s.p. has been used (see Appendix B). The variance of the photocurrent can be calculated as

$$\sigma_i^2 = R(0) - \eta\bar{n}^2 = \int_{-\infty}^{+\infty} S_i(f)\,\mathrm{d}f - \eta\bar{n}^2 = 2\eta\bar{n}\,f_c ,$$

and the signal-to-noise ratio S/N for incident power $P(\nu) = \bar{n}h\nu$ is

$$\frac{i^2}{\sigma_i^2} = \eta\frac{P(\nu)}{2h\nu f_c} .$$

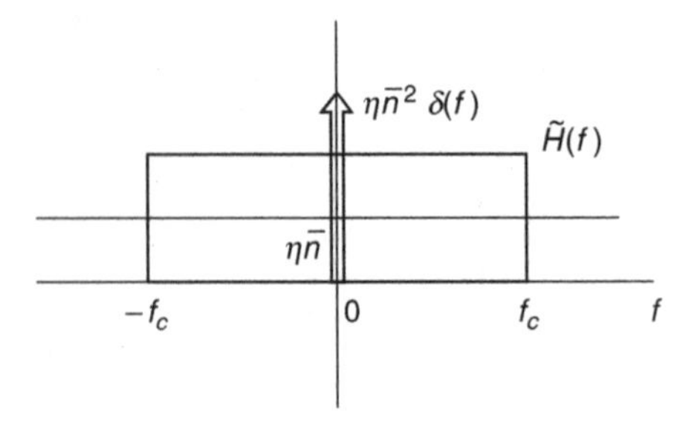

Fig. 7.10 Filtering of the stationary process assumed to describe the photocurrent

The incident power producing a unit signal-to-noise ratio is thus

$$P(\nu) = \frac{2h\nu}{\eta} f_{\mathrm{c}} \, .$$

A simple interpretation of this result can be given. When photon noise dominates the signal, the signal-to-noise ratio is unity for two photoelectrons per hertz of passband, or one photoelectron per polarisation per hertz of passband.

The S/N ratio can evidently be improved by imposing a narrow passband, which means a long integration time (smoothing by *running mean*). Some recent ideas (1990) suggest that the above noise limit could be reduced by *quantum noise compression*.

Background Noise

It often happens that the detector receives simultaneously:

- source photons, leading to a photocurrent $i_{\mathrm{s}}(t)$,
- background photons, leading to a photocurrent $i_{\mathrm{b}}(t)$.

This background emission may be due to the thermal environment of the detector (at infrared or millimetre frequencies), the dark current of the detector (e.g., for photocathodes), or to the secondary particle environment (for high energy astronomy).

The independence of the two signals implies

$$\sigma_{\mathrm{i}}^2 = \sigma_{\mathrm{s}}^2 + \sigma_{\mathrm{b}}^2 \, ,$$

and when the source is faint,

$$i_{\mathrm{b}} \gg i_{\mathrm{s}} \, , \qquad \sigma_{\mathrm{i}}^2 \sim \sigma_{\mathrm{b}}^2 \, .$$

The minimum detectable signal (S/N $= 1$) is then

$$i_{\mathrm{s}} = \sigma_{\mathrm{b}} \, ,$$

which implies an incident power at the detector given by

$$P_{\mathrm{s}} = h\nu \left(\frac{2}{\eta} \frac{P_{\mathrm{b}}}{h\nu} f_{\mathrm{c}} \right)^{1/2} \, .$$

This power is called the *noise equivalent power* (NEP), when referred to a unit passband $f_{\mathrm{c}} = 1$ Hz. It is measured in W Hz$^{-1/2}$.

Given the characteristics of the reception system, namely the collecting area A, the transmission t, the spectral passband $\Delta\nu \ll \nu$ (a prerequisite for the definition), we can further define the *noise equivalent flux* (NEF) to be

$$\mathrm{NEF} = \frac{P_{\mathrm{s}}}{A \Delta\nu \, f_{\mathrm{c}}^{1/2} \, t} \qquad [\mathrm{W\,m^{-2}\,Hz^{-1}\,Hz^{-1/2}}] \, .$$

The NEF is often given in Jy $\mathrm{Hz}^{-1/2}$ (1 Jy $= 10^{-26}$ W m^{-2} Hz^{-1}). It is a convenient unit for determining the detectability of source, by comparing the flux it emits with the efficiency of the system.

At the high energies of γ radiation, the background signal will be expressed by the number of counts in a given spectral interval of the detector, i.e., in counts per keV, and its fluctuations will once again be of Poisson type for low fluxes and approximately Gaussian for high fluxes.

7.2.2 *Thermal Noise*

We return now to the approximate expression for the radiated power density, at low frequency, of a blackbody at temperature T, under the assumption that $h\nu \ll kT$:

$$P(\nu) \approx \frac{h\nu}{\mathrm{e}^{h\nu/kT} - 1} \quad [\mathrm{W\,Hz^{-1}}] \,.$$

The process $\mathbf{a}(t)$, proportional to the electric field, thus has an almost-white power spectral density (PSD) (Fig. 7.11)

$$P(\nu) \approx kT \quad [\mathrm{W\,Hz^{-1}}] \,,$$

up to a frequency $\nu \sim kT/h$. In practice, the receiving system (antenna, mixer, diode, etc.) always has a cutoff frequency ν_c less than kT/h. If $T = 300$ K, the condition $h\nu = kT$ corresponds to a wavelength of 50 μm ($\nu = 6 \times 10^{12}$ Hz), shorter than that of the sensitivity threshold. The effect of the receiving system can thus be represented by a filter, modelled by the transfer function

$$\tilde{H}(\nu) = \Pi\left(\frac{\nu}{2\nu_\mathrm{c}}\right) \,.$$

Let $\mathbf{a}_1(t)$ be the stochastic process resulting from this filtering. It has zero mean, so its autocorrelation and autocovariance are equal, and its variance can be written

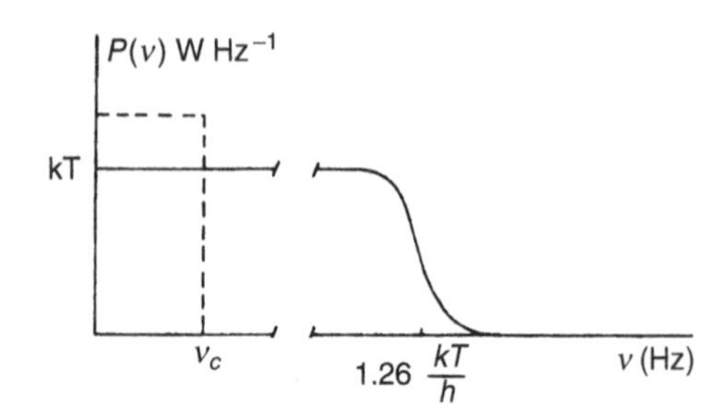

Fig. 7.11 Spectral density of thermal noise. Note that the curve is asymptotic to zero, at frequencies where quantum noise must be added. The *dashed lines* show the filtering produced by a low-pass filter associated with a detector

$$\sigma_{\mathbf{a}_1}^2 = R_{\mathbf{a}_1}(0) = \int_{-\infty}^{+\infty} P(\nu)\,\mathrm{d}\nu = 2kT\nu_{\mathrm{c}}\;.$$

The numerical coefficient relating $\sigma_{\mathbf{a}_1}^2$ to the characteristic quantity $kT\nu_{\mathrm{c}}$ depends on the exact form of $\tilde{H}(\nu)$.

An Important Special Case

The detector can be treated as a simple low-pass resistance–capacitance filter, whose power transfer function is known to be

$$\tilde{H}(\nu) = \frac{1}{1+(2\pi RC\nu)^2}\,, \qquad f_{\mathrm{c}} = (2\pi RC)^{-1}\;.$$

Consequently,

$$P_{\mathbf{a}_1}(\nu) = \frac{kT}{1+(2\pi RC\nu)^2}\,,$$

and also

$$\sigma_{\mathbf{a}_1}^2 = \frac{kT}{2RC} = \pi kTf_{\mathrm{c}}\;.$$

By Fourier transformation (the appropriate Fourier pairs are given in tables of functions, see also Appendix A), the autocorrelation (or, equivalently, the autocovariance) of $\mathbf{a}_1(t)$ is given by

$$R_{\mathbf{a}_1}(\tau) = C_{\mathbf{a}_1}(\tau) = \frac{kT}{2RC}\exp\left(-\frac{|\tau|}{RC}\right) = \pi kTf_{\mathrm{c}}\exp(-2\pi f_{\mathrm{c}}|\tau|)\;.$$

The normal process is therefore completely specified. A different filtering $\tilde{H}(\nu)$ leads to a normal process with different autocorrelation and spectrum.

Thermal Noise of a Resistor

This simple case is of interest because it provides an invaluable relation between the power associated with noise and the temperature of a physical system (a simple resistor, for this example), in a state of statistical equilibrium. In a resistor at temperature T, the free electrons are subject to thermal excitation. A random voltage $\mathbf{v}(t)$ therefore appears across the resistor, and this voltage depends on the temperature T. $\mathbf{v}(t)$ is a zero mean, stationary, normal random process. The non-zero capacitance C associated with the resistance R determines, as above, the form of the spectral density of $\mathbf{v}(t)$:

$$S_{\mathbf{v}}(\nu) = A\frac{1}{1+4\pi^2R^2C^2\nu^2}\,,$$

where A is determined by the equipartition of energy in thermodynamic equilibrium. The latter implies, at low frequencies $\nu \ll (RC)^{-1}$,

$$\frac{S_{\mathbf{v}}(\nu)}{R} = 2kT ,$$

and hence

$$A = 2kTR ,$$

where $2kT$ is the power per unit frequency in unpolarised radiation from a blackbody at temperature T, provided the frequencies are low (Rayleigh–Jeans law). The *cutoff frequency* $(RC)^{-1}$ of this thermal noise is associated with the mean free path of the electrons between atoms, and is of the order of 10^{13} Hz at 300 K. After filtering $\mathbf{v}(t)$ with a filter of frequency bandwidth $\Delta\nu \ll (RC)^{-1}$, the noise power of the filtered voltage is then given by

$$\langle v^2 \rangle = \int_{-\infty}^{+\infty} S_{\mathbf{v}}(\nu)\, \mathrm{d}\nu = 4kTR\Delta\nu ,$$

which is the *Nyquist formula*.

This formula is particularly useful in determining the thermal noise associated with a detector which behaves like a resistance at temperature T.

Quadratic Detectors and Thermal Noise

A quadratic detector has been defined as one which delivers, for example, the quantity proportional to the square $\mathbf{b}(t)$ of the process $\mathbf{a}_1(t)$ resulting from the filtering of the electric field $\mathbf{a}(t)$.

The autocorrelation of the stochastic process $\mathbf{b}(t)$ is given by the expression (see Appendix B)

$$R_{\mathbf{b}}(\tau) = R_{\mathbf{a}_1}^2(0) + 2R_{\mathbf{a}_1}^2(\tau) .$$

Taking the Fourier transform of this relation,

$$S_{\mathbf{b}}(\nu) = R_{\mathbf{a}_1}^2(0)\delta(\nu) + 2S_{\mathbf{a}_1}(\nu) \star S_{\mathbf{a}_1}(\nu) .$$

Letting $\mathbf{a}_1(t)$ be a stochastic process filtered by a low-pass box-function filter, with cutoff at ν_c, it follows that

$$S_{\mathbf{a}_1}(\nu) = kT\, \Pi\left(\frac{\nu}{2\nu_c}\right) , \qquad R_{\mathbf{a}_1}(\tau) = 2kT\nu_c\, \mathrm{sinc}(2\nu_c\tau) ,$$

$$S_{\mathbf{b}}(\nu) = (2kT)^2\, \nu_c^2\delta(\nu) + 2(kT)^2\, 2\nu_c\left(1 - \frac{\nu}{2\nu_c}\right) ,$$

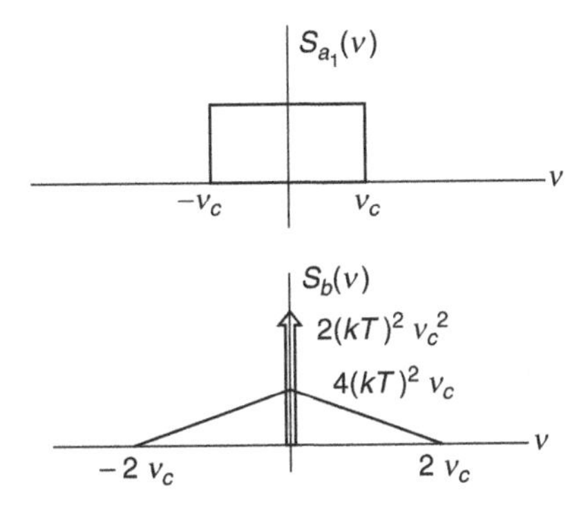

Fig. 7.12 Spectral density of a filtered stochastic process $\mathbf{a}(t)$, and of its square $\mathbf{b}(t)$

as illustrated in Fig. 7.12. The PSD of the stochastic process $\mathbf{b}(t)$ is no longer white, and its mean no longer zero, as is shown by the non-zero value of $S_{\mathbf{b}}(0)$.

Note. A similar method could be used to study the probability distributions of the amplitude and phase of $\mathbf{a}_1(t)$, and of its instantaneous mean power, etc., in those experimental arrangements which involve these quantities.

We now have at our disposal all the tools needed to find variances and autocorrelations associated with thermal noise. These will be used in the next chapter to establish the ultimate limits of receiving systems with respect to this type of noise.

7.3 Physical Principles of the Detection of Electromagnetic Radiation

Detection involves a *matter–radiation interaction* and a transfer of energy towards the matter, which causes it to change state in a measurable way. Clearly, this change of state, which constitutes the signal, must be chosen to correspond to the mean energy $h\nu$ of the detected photons. If it involves too low an energy, the signal will be non-existent or undetectable, but if its energy is too high, the detector may be damaged or destroyed. Figure 7.13 gives an overview of the main interactions used in the detection of photons.

The photoelectric detection of radiation quanta is possible provided that their energies exceed several meV, which means wavelengths less than about 200 μm. At longer wavelengths, thermal fluctuations dominate, so that radiofrequency quanta are hard to detect. This is not, however, a fundamental limitation, and in practice, radiofrequency detectors (in the millimetre and centimetre ranges, and beyond) detect the *electric field* of the wave directly. Finally, *calorimetric detection* of radiant energy constitutes a third approach, which is useful in the submillimetre and γ ray regions of the spectrum.

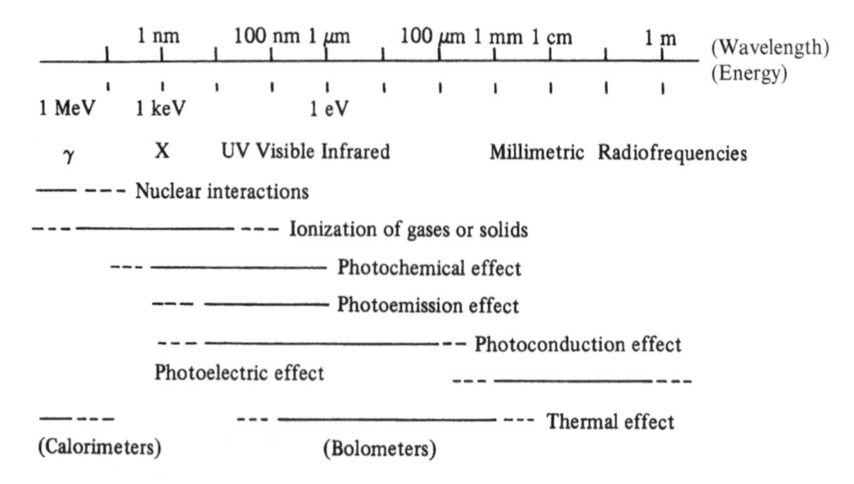

Fig. 7.13 Physical interactions and detection of radiation

7.3.1 *Detection of Quanta*

Here we discuss the physical principles applied in the detection of individual photons and the production of a measurable signal. Recent developments in micro-electronics have profoundly changed, improved, and diversified the processes used by astronomers in their dectectors.

Photoelectric Emission in Vacuum

The surface of a conducting solid presents a potential barrier which opposes the extraction of electrons from that solid. The height E_s of this barrier is called the *work function* of the surface. Only photons with energy greater than this threshold can produce a photoelectron. The first photoelectric cell was devised in 1890 by Elster and Geitel. The work function E_s depends on the crystal structure and the state of the surface of the material, but is generally around 1 eV, which limits the use of the photoelectric effect in vacuum to detection of wavelengths shorter than the micron (visible, UV, and X ray).

When the photon energy is greater than this threshold, the probability of photoelectron production is the *quantum efficiency* η, defined by

$$\eta = \frac{\text{mean number of photoelectrons produced}}{\text{mean number of incident photons}} .$$

This probability (Fig. 7.14) can be determined theoretically from the atomic structure of the material and the thickness of the emitting layer. It is generally

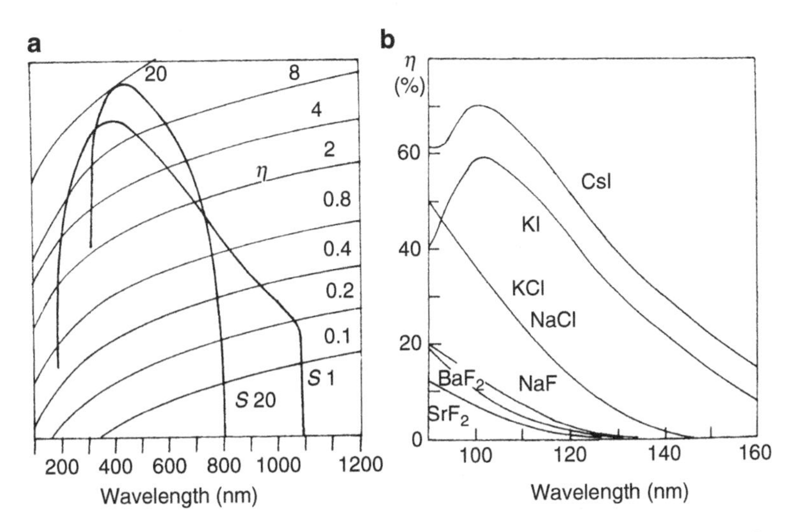

Fig. 7.14 Quantum efficiency η of photocathodes (in %): (**a**) in the visible and near ultraviolet and (**b**) in the far ultraviolet. The S1 layer is a Ag–O–Cs photocathode (1930), and the S20 layer is tri-alkali Na_2KSb (Sommer, 1955). Note the extremely poor response in the red, and excellent response in the far UV

less than unity, unless the photon energies become large compared with the energy differences between levels in the material (i.e., at UV or X wavelengths).

Detectors using the photoelectric effect in vacuum are the *photomultiplier*, the *electronic camera*, and *image intensifiers*.

The Photoelectric Effect in Solids

Even when the energy of a photon is insufficient to eject an electron from the solid surface, it may nevertheless cause the ionisation of an atom in the solid, thus releasing charge carriers within the crystal lattice, and modifying the electrical conductivity of the material. This *photoconducting effect* can be used to detect the incident photon (Smith W., 1873).

Photoconducting materials are semiconducting crystals, and electrons within the crystal lattice have a characteristic energy level structure (Fig. 7.15). The conduction band is separated from the valence band by a gap, and the energy difference E_s associated with this gap is exactly the *photoelectric threshold* of the material.

The width of this gap depends on the wave vector $\boldsymbol{k}$ of the electrons. Let $E(\boldsymbol{k})$ be this width. Its minimum value is E_s, which is usually between a half and a quarter of its mean value. A photon of energy E_γ can transfer an electron of wave vector $\boldsymbol{k}$ into the conduction band, and thus leave a hole in the valence band, if $E_\gamma > E(\boldsymbol{k})$. This pair of charge carriers will move under the effect of an externally applied electric field, producing a current, or photocurrent. A pure crystal which manifests this effect is known as an *intrinsic photoconductor* (Table 7.2).

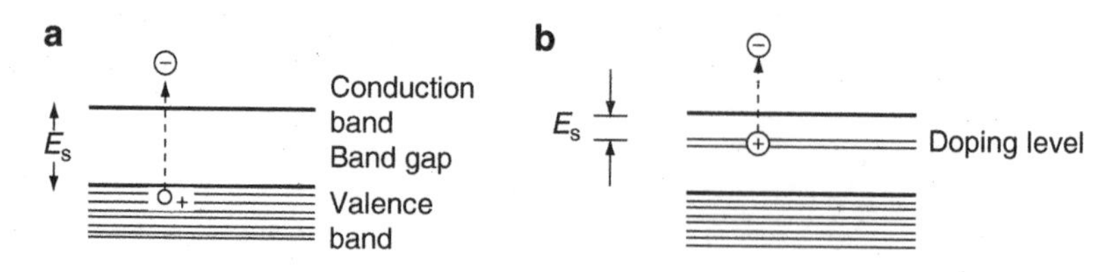

Fig. 7.15 Energy levels in a semiconducting solid. (**a**) Intrinsic photoconductivity. (**b**) Extrinsic photoconductivity of a doped semiconductor

Table 7.2 Intrinsic photoconductivity

Material	Threshold wavelength [μm]	Working temperature [K]	Applications
Germanium Ge	1.88	77	Solar energy
Silicon Si	1.14	300	Astronomical detectors
Cadmium sulphide CdS	0.52	77 to 150	Photography (exposure meters)
Indium antinomide InSb	5.50 (77 K)	4 to 77	Astronomical detectors
$Hg_{1-x}Cd_xTe$	Varies with x but $\lesssim 20$ μm	20 to 300	Thermal surveillance and astronomical detectors

In an *extrinsic photoconductor* (Fig. 7.15b), further energy levels appear in the band gap. These are caused by adjoining atoms of impurities within the crystal lattice, at very low concentrations ($\lesssim 10^{-8}$). As these atoms can trap charge carriers, the energy threshold is lowered, and this explains why all infrared detectors, beyond about $\lambda = 5$ μm, make use of extrinsic photoconduction. For example, replacing one silicon atom (group IV of the periodic table) by one arsenic atom (group V of the same) in the silicon lattice, means adding an atom of similar electronic structure to the other atoms in the lattice, but with one extra, easily removed electron, and it is this extra electron that is responsible for the extrinsic photoconductivity (n-type electron conductivity). Group III atoms inserted into a lattice of group IV atoms will lead to p-type hole conductivity. The threshold can be adjusted, by making a suitable choice of impurity, or *dopant*, and this selects the wavelength region to which the detector is sensitive (Table 7.3). This idealised picture of photoconductivity should also take into account the effects of temperature and impurities.

Photoconductors are used in two ways: firstly, to measure electric current (photoconduction), and secondly, to measure pulses. In the first case, a flux is measured (photometry), and in the second, the energy of quanta is measured (spectrometry). In both cases, the performance of the photoconductor depends on its conductivity.

The conductivity, in turn, depends on the concentration of charge carriers in the conduction band. At low temperatures, this concentration depends on the temperature, and also the level of impurities, whereas at high temperature, it no longer

Table 7.3 Extrinsic photoconductivity

Matrix material: dopant	Threshold wavelength [μm]	Working temperature [K]
Si:As	24	< 20
Si:In	7.4	77
Si:Bi	18.7	29
Si:Ga	17.8	27
Ge:Ga	120	4
Ge:Ga[a]	195	4

[a] Threshold decreased by mechanical pressure on the crystal (stressed photoconductor).

depends on the level of impurities. For an intrinsic semiconductor, the concentration of charge carriers follows from the law of mass action, and can be written

$$n_{\mathrm{i}} = p_{\mathrm{i}} \approx T^{3/2} \exp\left(-\frac{E_{\mathrm{s}}}{2kT}\right) .$$

For $kT > 2 \times 10^{-2} E_{\mathrm{s}}$, the current is dominated by thermal excitation of the electrons near the conduction band, and not by photoionisation. This explains why a photoconductor must be cooled to a temperature $T \ll E_{\mathrm{s}}/k$, and the temperature must decrease as the energy of the quanta decrease, or if their their flux is low (photometry), or again when higher accuracy is required in the measurement of their energy (spectrometry).

Moreover, the crystals can never be perfectly pure, and it is difficult to go below an impurity level of 10^{11} cm^{-3}. Some atoms, occurring as an impurity, create energy levels which are very close to the conduction band, and which can be excited by very low temperatures. Even for $T \ll T_{\mathrm{s}}$, and in the absence of photons, the current will be non-zero (dark current). Hence, the quality of a photoconducting detector can be represented by the value of its impedance at its working temperature, and in the absence of any illumination (typically, $R \gg 1\,\mathrm{G}\Omega$).

Only a fraction of incident photons create electron–hole pairs. Indeed, some photons lose their energy by exciting vibrations in the crystal lattice (phonons). This heats the crystal (*optical absorption*). When the optical depth of the material is sufficient, taking into account the photoabsorption cross-section ($\tau_\lambda \sim 1$), the quantum efficiency η of the material is then determined.

The photoelectric effect in solids has become the basis for a large family of detectors used in astronomy: *avalanche diodes*, *photoconducting* or *photovoltaic infrared detectors*, *solid-state imagers* (also called *Reticons*, CCDs, or CIDs), *blocked impurity band* (BIB) *detectors*, *television tubes*, and *photodiodes*.

The Photovoltaic Effect. Junctions

Consider two doped semiconductors, one with impurities giving n-type (electron) conductivity, the other with impurities giving p-type (hole) conductivity. This would

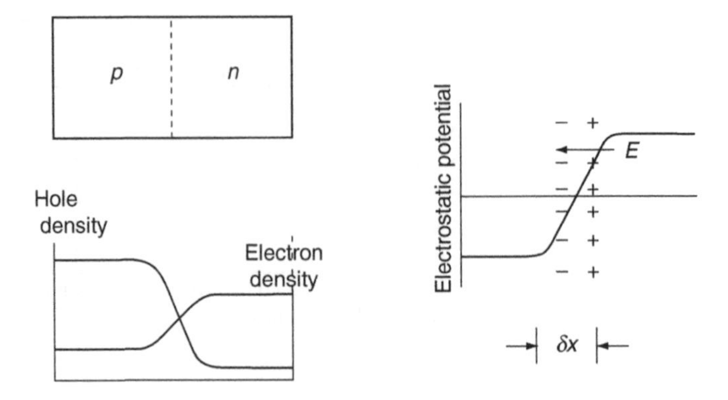

Fig. 7.16 The principle of a junction between two semiconductors

be the case for a silicon (group IV) matrix doped respectively with phosporous (group V) and gallium (group III). When put into contact with each other, these form a *junction* (Fig. 7.16). If two charge carriers, an electron and a hole, are released by an incident photon on the unbiased junction, then the voltage across it changes by an amount

$$\Delta V = \frac{e\delta x}{\varepsilon S} ,$$

where δx is the thickness of the junction, S its area, and ε the electrical permittivity of the material. Absorption of a photon thereby results in a voltage across the junction, and this voltage can be measured by a high impedance circuit. This is the *photovoltaic effect*, discovered by A. Becquerel in 1839.

Quantum Wells

Junctions can be made either from the same semiconducting material, but with different dopants in each part, or from two different materials (*heterojunctions*), in which case the band gap will differ from one side to the other. Starting from gallium arsenide AsGa, aluminium atoms can be inserted in the lattice to make $Al_xGa_{1-x}As$. This is compatible with the crystal structure, and the result can be used to make junctions. A thin layer of AsGa sandwiched between two layers of $Al_xGa_{1-x}As$ has the effect of pushing the conduction band down to the level of AsGa. An electron crossing the junction remains trapped in the AsGa, which is thus a potential well. In such a *quantum well*, the electron can only have certain discrete and well-defined energies.

Infrared detectors can be made using this principle to construct a series of quantum wells, whose levels correspond in such a way that a photoelectron can pass from well to well by tunnelling, as far as the external electrodes.

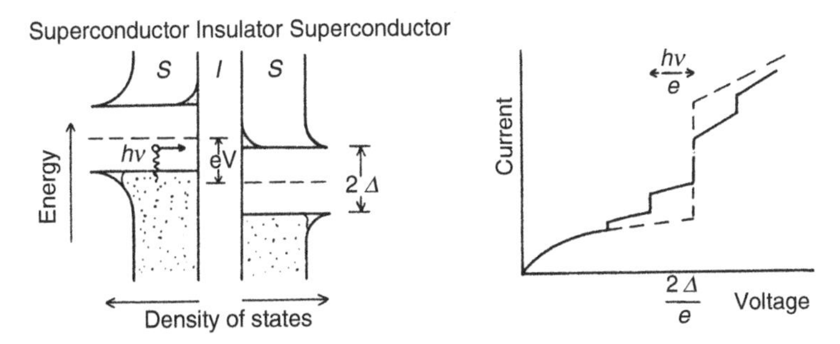

Fig. 7.17 Superconductor–insulator–superconductor tunnel effect junction. (**a**) Energy level structure on either side of the insulator. Occupied states are *shaded*. The *ordinate* gives the energy and the *abscissa* the density of states. The superconductor energy gap is 2Δ. When V is large enough to place occupied states on the left opposite vacant states on the right, a current can easily flow by the tunnel effect. If $eV < 2\Delta$, a photon may nevertheless excite a charge carrier to the energy at which the tunnel effect occurs (assisted tunnel effect). (**b**) Current–voltage characteristic of the junction. *Continuous line*: no flux. *Dashed line*: photons arriving. (Philips T.G., Woody D.P., ARAA **20**, 285, 1982. Reproduced with the kind permission of the Annual Review of Astronomy and Astrophysics, 1982, Annual Reviews Inc.)

Photon Counting by Tunnel Effect

Consider a device made by sandwiching a thin insulating film ($\approx$10 nm) between two superconducting electrodes. At low temperatures, this forms a junction called a superconductor–insulator–superconductor (SIS) junction. These junctions are commonly used as *mixers* in radiofrequency receivers (see Sect. 7.5). However, they also have promising applications for direct photon counting from X rays to the near infrared, where they are referred to as superconducting tunnel junctions (STJ).

Figure 7.17 shows how photons can be detected by such a junction.[3] When there is no light flux and at temperature zero, a quasiparticle current can only pass through the insulating layer when the applied potential difference exceeds the sum of the gaps of the two superconductors, e.g., niobium, hafnium. If $T > 0$, a dark current circulates whenever $\delta V \neq 0$. Absorption of a photon in the superconductor breaks a number of Cooper pairs that is directly proportional to the energy of the photon. This generates quasiparticles, which are then collected as current passing through the junction. The charges collected in this way can be amplified. Furthermore, an avalanche effect occurs because a quasiparticle induced by the photon can couple with a Cooper pair after crossing the insulating barrier, cross it again and be counted once more. Photon counting is a quantum mechanism, contains intrinsic amplification without noise (avalanche), and exhibits a certain spectral selectivity ($\leq$ 100).

[3]See, for example, Delaet, B., Feautrier, P., Villégier, J.-C., Benoît, A. X-ray and optical photon counting detector using superconducting tunnel junctions, IEEE Trans. **11**, 2001.

The Photochemical Effect

This effect is the basis of photography (developed by Nicephore Niepce, 1822, and Charles Daguerre, 1829). An incident photon excites an electron of the emulsion into the conduction band, and leaves a positively charged hole. The released electron is recaptured by a defect in the crystal lattice, and attracts a silver ion Ag^+ to the site, thereby forming a neutral silver atom. The process repeats until a small cloud of silver atoms is produced, constituting one pixel, or element, of the *latent image*. The residual positively charged holes are able to move through thermal effects and then recombine with the silver atoms, which gives back silver ions. The processes of sensitising and hypersensitising photographic plates aim to destroy these residual holes. The production process for the emulsion creates the required defects in the crystal structure of silver chloride (AgCl) grains, by adding a suitable reducing agent (*Gurney–Mott theory*).

During development of the plate, a considerable amplification is produced, and a single silver atom released by a photon can give up to 10^9 atoms in the final image. The noise, or *grain*, of the plate is due to the formation of silver grains at the lattice defects without the action of an incident photon.

The quantum efficiency of a photographic emulsion is low, being of the order of 8×10^{-3}. This is indeed the main drawback of the *photochemical effect*. The threshold of the photochemical effect is governed by the value of the electron excitation energy in AgCl, or some other halide of this kind. This limits the sensitivity of plates at the infrared boundary of the visible ($\lambda \leq 800$ nm).

The detector using the photochemical effect is, of course, the *photographic plate*. As the bombardment of an emulsion by electrons can produce a latent image by the same process as the one described above, *nuclear emulsions* have been developed for the purpose of making latent images from electron impacts, and this is the principle of the *electronic camera*.

Photoionisation of Gases

If the energy of incident photons is high enough ($\gtrsim$ 10 eV, ultraviolet or X-ray regions), an atom may be photoionised. The released electrons are then accelerated by an external electric field, and cause further ionisations by collision. The charges collected on the accelerating electrode are measured by a charge amplifier, or *electrometer*. The operating limit is reached when the multiplier effect leads to a discharge in the gas. Below this limit, the response is linear. It is not strictly correct to speak here of a quantum efficiency in the sense described earlier, for a photon may produce more than one electron, by double ionisation. However, extending the notion, η will be used to denote the mean number of electrons obtained per incident photon. This quantity, which may exceed unity, is referred to as the *quantum yield*. The photoionisation cross-section in a gas is a function of the incident energy E, varying as $E^{-7/2}$. At around 10 keV, it becomes smaller than the Thomson scattering cross-section, and photoionisation is no longer usable.

The noble gases, neon, argon, and xenon, are used because of their high gain, the other common excitation mechanisms, such as rotation or vibration, being absent. An impurity, opaque to the UV radiation emitted by radiative de-excitation of the excited gas atoms, is added to the noble gas, in order to be able to use the energy for further ionisations. These impurities also de-excite the metastable states of the noble gas, by collisions.

The *ionisation of gases* is the basis of *proportional counters*. *Wire chambers*, invented by Georges Charpak,[4] are used to detect events in the vicinity of particle accelerators. They are also based upon the photoionisation effect. The difference is that, in this case, the ionisation is not exclusively due to photons.

Photon–Matter Interaction at High Energy (>10 keV)

The interaction of gamma rays with matter produces charged particles (e^+ and e^-), which carry off the photon energy in kinetic form. The information concerning the energy and direction of the gamma photon is transferred to these particles, which interact, in turn, with other matter by ionisation, bremsstrahlung, annihilation, and so on. At low energies, loss of energy by ionisation is the dominant effect, the charged particles creating a series of electron–ion and electron–hole pairs. Most detectors use these charged pairs to detect the passage of a gamma photon, or to measure its energy by the number of charges produced.

The three main interaction processes are *photoelectric absorption*, *Compton scattering*, and *pair creation*.

Photoelectric Absorption

In this process, all the photon energy E_γ goes to an electron in some atom. This interaction tends to occur with the inner shells (which have higher energy). The electron is ejected with kinetic energy equal to the difference between the incident photon energy and the energy required to extract the electron from its shell. As it propagates, the electron ionises atoms along its path. All the ions so formed then de-excite in a series of recombinations. In particular, the ion formed by the initial interaction of the photon de-excites, by emitting an X-ray photon which is characteristic of the shell of the ejected electron (K, L). If a thin layer of material is used for detection, some of these X-ray photons escape without further interaction, but in the majority of cases, they are not able to escape, and the totality of the energy of the gamma photon is then transmitted to the medium. The cross-section for the photoelectric interaction thus depends on the incident photon energy, and on the energy of atomic electrons in the medium. Discontinuities are observed at energies corresponding to the various shells of atomic electrons (see Fig. 7.18). However, to

[4]The French physicist Georges Charpak (1924–2010), who won the Nobel Prize for Physics in 1992, invented this experimental device for visualising elementary particles at CERN.

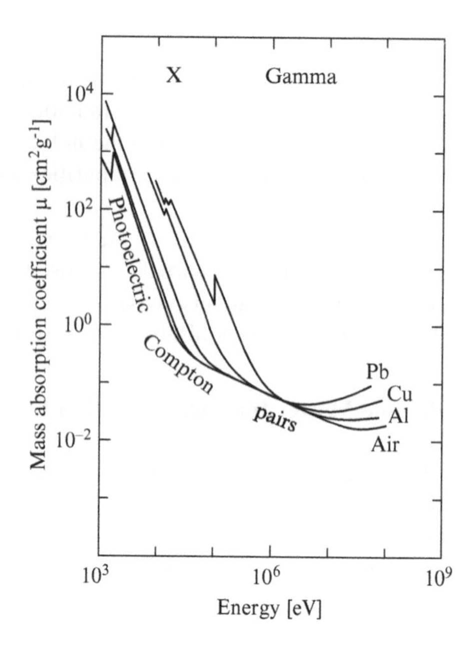

Fig. 7.18 Mass absorption coefficient as a function of incident photon energy. Note the K shell ionisation discontinuities at low energy, and the energy dependence of the cross-section. (After Gursky H., Schwartz D., *X-ray Astronomy*, Reidel, Dordrecht, 1974. With the kind permission of D. Reidel Publishing Company)

a first approximation, this cross-section is given by

$$\sigma_\mathrm{p} = Z^{4.5} E_\gamma^{-3.5} \, .$$

Compton Scattering

In Compton scattering, only a part of the incident photon energy is transferred to the electron. The rest of the energy is carried away by the scattered photon, while the scattered electron, as in the case of the photoelectric effect, loses its energy mainly by ionisation of atoms along its trajectory. The scattered photon may also interact with the medium by a further Compton scattering, or by a photoelectric absorption. The directions of the incident and the scattered photon are determined by conservation of energy and momentum. The angle θ between them is related to the energy E_γ of the incident photon, and the energy E_s of the scattered photon by

$$\cos\theta = 1 - \frac{m_0 c^2}{E_\mathrm{s}} + \frac{m_0 c^2}{E_\gamma} \, .$$

Figure 7.19 shows that the various scattering directions are not equiprobable. The dependence of the differential cross-section for the Compton interaction on the scattering angle θ is given by the *Klein–Nishina formula*

$$\frac{\mathrm{d}\sigma_\mathrm{c}}{\mathrm{d}\Omega} = \frac{1}{2} r_\mathrm{e}^2 \frac{E_\mathrm{s}}{E_\gamma} \left(\frac{E_\gamma}{E_\mathrm{s}} + \frac{E_\mathrm{s}}{E_\gamma} - \sin^2\theta \right) \, ,$$

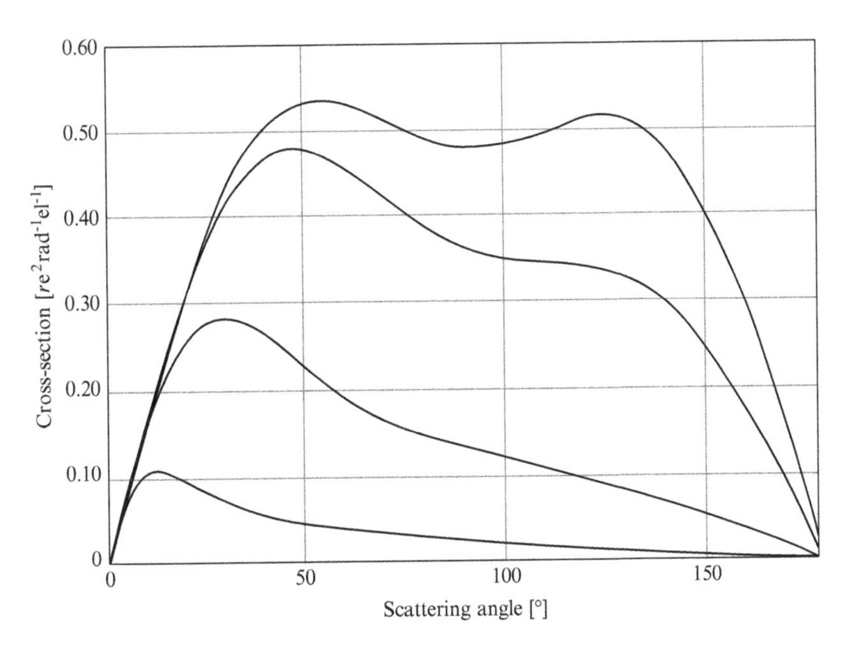

Fig. 7.19 Microscopic cross-section for Compton scattering, as a function of the scattering angle, at various energies. r_e is the classical electron radius. The cross-section is given in rad^{-1}, and is therefore an angular differential quantity

where r_e is the classical electron radius ($r_e = 2.82 \times 10^{-15}$ m), and $d\Omega$ the element of solid angle around direction θ. The higher the energy, the more likely it is that the photon will scatter forwards. The total cross-section for the Compton interaction goes approximately as

$$\sigma_c \propto ZE^{-1} .$$

Pair Creation

In this process, a high energy photon ($E \gg 1$ MeV) interacts with the electric field of a heavy nucleus to create an electron–positron pair. The photon energy is divided between the mass energy of the pair, equal to 1.22 MeV, and their subsequent kinetic energy. Although the momentum carried by the nucleus ensures conservation of momentum for the reaction, its recoil energy can be neglected. In the centre of mass frame, the electron and positron are emitted in opposite directions. During propagation, they gradually lose their energy, mainly by ionisation, bremsstrahlung, and the Cherenkov effect. When the positron energy is low enough (≈ 1 keV), it will annihilate with an electron in the medium, by the inverse reaction, emitting two photons at 511 keV in opposite directions. These photons may escape, or they may in turn interact with the medium by Compton scattering or photoelectric absorption.

The cross-section for pair creation varies roughly as Z^2, and is characterised by its threshold of 1.22 MeV.

Comparison of Processes

The energy dependence of the three interaction processes mentioned above means that each one will be predominant in some energy domain determined by the interaction medium. At low energies, photoelectric absorption dominates, and at high energies, pair creation is the main channel. The energy domain for Compton scattering is situated between the two. Its width is highly dependent on the atomic number Z of the material (Fig. 7.20), increasing for lighter materials. Generally speaking, for equal mass surface densities, light materials tend to scatter, and heavy materials tend to absorb.

Compton scattering and pair creation are the physical processes exploited in *scintillation counters* or *scintillators*, as well as in *semiconductor detectors*. The applications to γ ray astronomy will be discussed in Sect. 7.6.

Thermal Effects. Calorimetry

Once the photon energy has been transferred to the solid, which constitutes the detector, and transformed into thermal excitations, the incident radiation is revealed

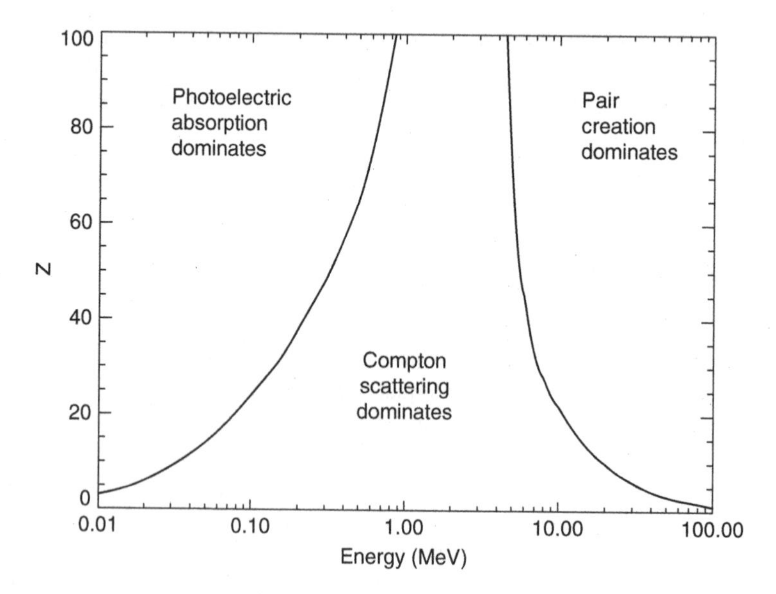

Fig. 7.20 Energy domains associated with each interaction process between γ rays and a material, as a function of its atomic number Z. Curves correspond to loci of equal cross-sections for the two processes they separate

by the slight temperature change it has produced, and which can be measured by the consequent variation in resistivity. In order for this to work correctly, the only condition is that the solid should be as absorbent to photons as possible, not only absorbing incident photons, but also any products of secondary events.

The thermal effect is used in detectors called *bolometers*, mainly in the infrared and submillimetre regions. It is also used in *calorimeters*, which measure the energy of γ rays.

7.3.2 Detection of the Electromagnetic Field

At wavelengths above about 0.2 mm, no quantum detector in the sense described above, involving release of one photocharge by one incident photon, has yet been devised. Across the whole of the radiofrequency region, other detection processes must be employed to measure the electric field of the incident wave. Its electromagnetic field, or the current it induces in an antenna, are applied to a non-linear element called a *diode* or a *mixer*, usually superposing this field on the field of a *local oscillator*. The choice of non-linear element and mixer determines the final detection performance. We shall thus postpone discussion of these issues to a full presentation of the various elements.

7.4 Astronomical Detectors from X Ray to Submillimetre

Since the 1970s and 1980s, a genuine revolution has transformed the kind of detectors used in astronomy. The driving force here has been the emergence of microtechnologies, contributing to the design of light detectors across the board from X ray to submillimetre wavelengths with levels of sensitivity (quantum efficiencies) that have quickly superseded traditional photography, along with the various systems that had been developed to supplement it. In addition to this new level of performance, direct digitisation of the signal has greatly facilitated transmission, storage, and processing, and it has also brought mechanical advantages by making it easier to equip space observatories or remote missions.

In this context of rapid progress, yesterday's detectors are often mentioned only for completeness, to describe the historical role they have played and explain references to them in the literature. However, one must not forget the extraordinary wealth of data contained in photographic astronomical archives, collected throughout the twentieth century and available today for the main part in digitised form (see Chap. 10). In the first section here, we consider more traditional detectors, mainly exploiting the generation and transfer of charge in semiconductors and appealing to microelectronics for their fabrication.

Outside this broad spectral range from X ray to submillimetre wavelengths, detectors in the γ ray region on the one hand and radiofrequency receivers on the other have not witnessed similar revolutions. Figure 7.21 gives an overview.

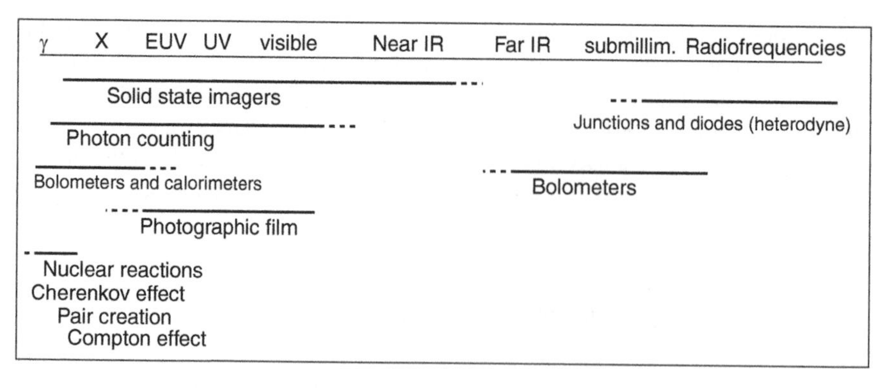

Fig. 7.21 Astronomical detectors

7.4.1 Noise Performance

At the beginning of this chapter, we discussed fundamental physical noise due to the Poissonian nature of the light–matter interaction or the thermodynamics of a thermal equilibrium. Of course, a given detector rarely achieves this ultimate level of performance. The true performance of a detector thus depends on what fluctuations it may add to those of the signal itself (signal noise) and to those of the detector environment (background noise). It is therefore important to qualify these real performance levels since they will influence the final quality of an observation.

Detector Quantum Efficiency

The *detector quantum efficiency* (DQE) δ quantifies this notion. It extends the idea of quantum efficiency η already introduced, but should not be confused with it. The definition is

$$\delta = \text{DQE} = \frac{\left(\langle \Delta N^2 \rangle / \overline{N}^2\right)_{\text{real detector}}}{\left(\langle \Delta N^2 \rangle / \overline{N}^2\right)_{\text{ideal detector}}} .$$

It is straightforward to show that, if the detector, of quantum efficiency η, introduces no further fluctuations, only the Poisson statistics of the incident photons remains, and

$$\langle \Delta N^2 \rangle_{\text{ideal}} = \overline{N} ,$$

$$\langle \Delta N^2 \rangle_{\text{real}} = \eta \overline{N} , \qquad \overline{N}_{\text{real}} = \eta \overline{N}_{\text{ideal}} .$$

In this case, $\delta = \eta$, which justifies the name given to δ.

As well as the fundamental fluctuations linked to the radiation itself, a detector may be subject to *amplifier noise* and *readout noise*.

Amplifier Noise

Some photoelectric detectors (photomultipliers, the avalanche effect in avalanche diodes, or CCDs in the X-ray region) exploit the following sequence of events upon impact by a photon: 1 photon $\rightarrow \eta$ photoelectrons $\rightarrow$ amplification $\rightarrow \eta g$ electrons $\rightarrow$ measurement. The amplification gain g is a random variable of mean $\overline{g}$ and standard deviation σ_g, from one photon impact to another. The noise this creates in the signal is not additive but multiplicative, degrading the DQE by the factor

$$\left(\frac{1+\sigma_g^2}{\overline{g}^2}\right)^{-1} .$$

Readout Noise

Some detectors (e.g., CCDs, see Sect. 7.4.6) measure at regular intervals the photocharges accumulated since the last measurement. Each charge transfer is then accompanied by a fluctuation, whose standard deviation is measured in numbers of electrons per reading. This number is the *readout noise*, or *noise equivalent charge* (NEC). Being a standard deviation, it will be denoted σ_{R}, given in rms electron number.

The value of σ_{R} for a given device may depend on the rate at which readings are made, and therefore on the integration time T. In general, σ_{R} increases as T decreases. However, σ_{R} no longer exhibits the standard $T^{1/2}$ dependence of integrated noise (see below).

The DQE is degraded by a factor

$$\left(\frac{\sigma_{\mathrm{R}}}{\overline{g}\eta\sqrt{\overline{N}T}}\right)^2 ,$$

so that, in the most general case,

$$\delta = \eta\left[\left(1+\frac{\sigma_g^2}{\overline{g}^2}\right) + \frac{\sigma_{\mathrm{R}}^2}{\overline{g}^2\eta^2\overline{N}T}\right]^{-1} ,$$

where $\overline{N}$ is the mean photon flux per second, T the readout time, the fluctuations being quadratically additive.

Problem 7.4 concerns the expression for the DQE when the dark current is included, and also an extension of the definition of δ to the case of an imaging detector, including spatial frequencies.

7.4.2 Photographic Plates

Considerable progress has been made in the sensitivity of photographic plates, over the period since their invention, going from a flux of 10^{-4} J cm^{-2} as required to

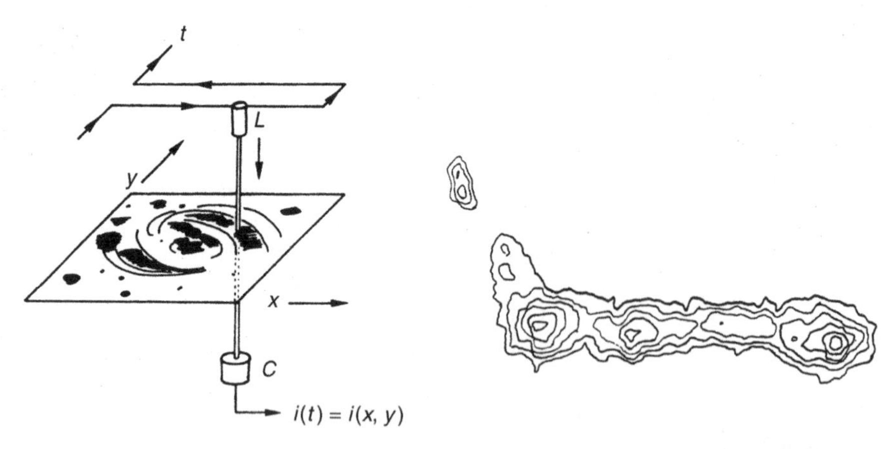

Fig. 7.22 *Left*: Microphotometry of a photographic or electronographic plate. The light source L emits a ray, received by the photosensitive cell C. The intensity i measured, together with the x and y coordinates, read by a position sensor, are digitised. L performs a two-dimensional raster scan. *Right*: Isophotes of the jet emerging from the quasar 3C 273 obtained by digitising the plate. Note the oscillation of the jet around its mean direction. Picture courtesy of Wlérick G., Lelièvre G., Canada–France–Hawaii telescope, wide field electronic camera, U filter (near ultraviolet), exposure 165 min, 1984

print a plate in 1850, to a flux 10^6 times lower a century later. We can nevertheless consider that it has reached its limit today. The main advantages are simplicity of use, ability to integrate low fluxes over long periods, and, above all, the high number of pixels that can be simultaneously exposed (up to 10^{10}). It is its own mass memory.

The drawbacks are its non-linearity, its rather restricted dynamic range, the grain effect, and also the need to use a *microdensitometer* (Fig. 7.22) to transform analogue information, stored in the form of blackening on the plate, into digital information, which can then be processed quantitatively. This point by point scanning process is still used by the modern *scanner* to digitise documents.

In order to set up digital archives of photographic records like those described in Chap. 10, machines have been built to measure and digitise astronomical photographic plates. These machines, which were developed in the 1990s, are capable of processing enormous numbers of pixels. Examples are MAMA, or *Machine à Mesurer pour l'Astronomie*, in France, and COSMOS, in the United Kingdom.

The registered signal is measured by the *density* of the emulsion. If the pixel being examined by microdensitometer transmits a fraction I/I_0 of the incident radiation, the density d is defined by

$$d = -\log_{10} \frac{I}{I_0} .$$

The photographic plate remains a useful detector but, despite its high number of pixels (up to $\sim 10^{10}$), it has now been superseded by CCD arrays which, at the time of writing (2007), can count up to 10^8 pixels.

7.4.3 Photomultipliers and Classical Cameras: X Ray, UV, and Visible

Once again, these detectors played an important role over the period 1950–1990, but were then completely superseded by CCDs and similar detectors. They worked by photon counting.

Photomultipliers (PM)

For many years regarded as the ideal detector, the photomultiplier (abbreviated to PM) consists of an evacuated tube containing a series of electrodes, the first of which is exposed to the radiation. When a photoelectron is released, it is extracted from the photocathode by means of a first dynode, or positive electrode, creating secondary electrons by impact. These are in turn accelerated, causing further electrons to be released, and the process continues. The final current is measured with a charge amplifier or electrometer. The photomultiplier nevertheless remains widely used in high energy (γ-ray) detectors, where it detects the photons induced by interactions and counts them (see below).

Based on the photoelectric effect, there is no sensitivity threshold, and indeed the photomultiplier can detect a single quantum. It is perfectly linear up to the point where the flux is high enough to cause an electron cascade, which destroys the electrodes. Its sensitivity is limited to the region 20 to 1 200 nm, by cascade effects in the far UV, and the photocathode response in the near IR. Figure 7.14 gives the response and the quantum efficiency. The PM has almost no spectral selectivity, and its response time is extremely short (the transit time of the electrons between dynodes, which is several nanoseconds). It clearly has no multichannel capabilities (except for certain configurations, which channel the photoelectrons and can discriminate between points of the photocathode), and the response of the photocathode may vary somewhat across its surface. These characteristics are summarized in Table 7.4.

At non-zero temperatures, thermoelectronic emission from the cathode produces accelerated electrons, even when there are no incident photons. *Richardson's law* gives the value of the current density j extracted at temperature T, for an extraction potential of E_s/e :

$$j = \frac{4\pi e k^2}{h^3}(1-\rho_e)\,T^2 \exp\left(-\frac{E_s}{kT}\right)$$

$$= 1.2\times 10^6\,(1-\rho_e)\,T^2 \exp\left(-\frac{E_s}{kT}\right)\ \mathrm{A\,m^{-2}}\,,$$

where ρ_e is the coefficient of reflection, measuring the probability of the electrodes recapturing an emitted electron. At $T = 300$ K, and for $S = 1\ \mathrm{cm}^2$, $E_s = 1$ eV, it follows that $I = jS = 0.2$ nA. This is a large *dark current*. For comparison, with $\eta = 0.5$, a flux of one photon per second would produce a current of 0.8×10^{-19} A (Table 7.4).

The dark current decreases rapidly with T (if the photomultiplier is cooled with dry ice or a Peltier refrigerator). It also decreases when E_s increases (this concerns UV photocathodes). Although the mean value $\langle jS\rangle = \langle i_B\rangle$ of the dark current can be subtracted from the measurements, its fluctuations $\langle \delta i_B^2\rangle = 2e\langle i_B\rangle\Delta f$, where Δf is the passband of the electric filter which follows the PM, set a fundamental limit to the sensitivity (Problem 7.1).

Table 7.4 Photomultipliers

Threshold	One quantum
Format	Up to 8 × 8 pixels
Quantum efficiency	0.43 (Ag–O–Cs at 800 nm) to 0.1
Spectral response	Wide
Temporal response	Nanosecond
Size	About 10 cm^2 (cathode)
Noise	Dark current $\sigma_\text{B} \approx 10$ counts $\text{Hz}^{-1/2}$
Polarisation	Insensitive

With the above current, and taking $\Delta f = 1$ Hz,

$$\langle \delta i_\text{B}^2 \rangle^{1/2} = 9 \times 10^{-15}\,\text{A}\,\text{Hz}^{-1/2} .$$

The signal can be measured in various ways.

- By integration of charge: a filter circuit (resistor and capacitor), or an integrating circuit (running mean), provides a continuously variable analogue voltage.
- By counting photon arrivals: a counter registers each burst of electrons, whatever their number and whatever the duration of the burst. The signal is digitised, and the intensity and duration of the pulses can be used to protect against unwanted signals.

Other Photon Counting Detectors

Much effort has been made to combine the advantages of the photoelectric effect in vacuum and the detection of 2D images. Here we give a brief overview of the sometimes imaginative systems devised through 1950–1990, but which have for the main part been superseded by CCDs (see below).

Electronic Camera (or Electronographic Tube)

Invented and first exploited by A. Lallemand, the electronic camera consists of a photocathode on which an image $I(x, y)$ is formed. The photoelectrons produced are accelerated by electron optics with electrostatic or magnetic focusing, and then an electron image is formed on a high contrast photographic plate (*nuclear emulsion*). The camera can operate at extremely low levels of intensity. Indeed, it does not exhibit the threshold effect of the normal photographic plate, and can almost detect single photoelectrons. The time integration capacity in long exposures is excellent (there is no fogging), and the advantages for astronomical photography are clear, particularly for faint objects, such as quasars, which require an accumulation of long exposures (Fig. 7.22).

Image Intensifiers Using Microchannel Plates

The input for an *image intensifier* is an image, formed by a low flux of photons, at a certain wavelength. The output from such device consists of an image with the

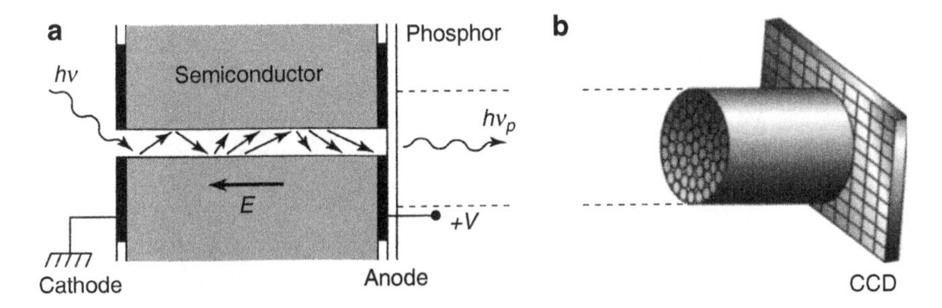

Fig. 7.23 (**a**) Microchannel in a semiconductor, constituting one element of a microchannel plate. The incident photon releases a photoelectron, which is accelerated by the field E, and multiplied by successive electronic collisions. The phosphor is excited at the exit and emits a large number of photons. (**b**) The phosphor emits across a wide solid angle and this requires fibre optic coupling. Each fibre is in contact with the phosphor and transmits the photons corresponding to one pixel to the CCD detector. The figure shows straight channels to simplify, but in general they are curved in order to prevent ion movement

same number of pixels. Each output pixel behaves as a source of photons, whose number is proportional to the number of incident photons, but whose wavelength is different. The system is therefore characterised by a *gain* g.

The *microchannel plate* or *channeltron* is an intensifier, exploiting the photoelectric effect in solids. The photoelectrons are guided and multiplied in a channel with accelerating voltage gradient (Fig. 7.23). The final impact is on a phosphor, which emits in the visible. The incident photons may be visible, ultraviolet, or soft X rays. Several microchannel plates can be arranged in series, in order to increase the amplification factor.

A microchannel plate has no *information storage capacity*, and so could not be used alone. Hence it is coupled to a CCD, and the two together are referred to as an *intensified* CCD.

Photon Counting Television

Inspired by the tubes developed for commercial television in the 1950s, many improvements were brought to bear to increase their sensitivity, to arrive at a genuine photon counting device (Fig. 7.24). A whole series of astronomical detectors was thereby developed.

Vidicon Tubes and Derivatives. Television tubes use the photoelectric effect in silicon. Released photoelectrons are stored in a layer of microcapacitors, which are periodically recharged (25 times a second) by a scanning electron beam, the Vidicon tube. The recharge current of the microcondenser is modulated by the light intensity. Passing through the charge resistance of the target, it produces a potential difference, which is the *video signal*, at a frequency of a few tens of MHz. The readout noise is high ($\sigma \sim 2 \times 10^3$ electrons pixel^{-1} Hz$^{-1/2}$), and is easily observed on the TV

Fig. 7.24 Photon counts, as they appear on the TV screen of a Nocticon tube (exposure 25 ms). Note the variable gain, manifested by the varying brightness of the events. Note also the noise events, fainter and easy to discriminate. The image is read on a 512 × 512 CCD array. Each event is digitally centered. (Observatoire de Marseille, image due to Boulesteix J. in the 1980s)

screen: the background, in the absence of any signal, appears as a random pattern of white on black, like snow.

In the *secondary electron conduction tube* or SEC tube, the photoelectrons are accelerated onto a target where they produce a secondary emission of electrons. The latter are read by standard TV scanning. The gain of around 1 000 is obtained at the cost of high non-linearity.

In the *silicon intensified target* (SIT) tube, or the *electron bombarded silicon* (EBS) tube, the photoelectrons are accelerated onto a silicon target at sufficiently high energy to create several thousand electron–hole pairs upon each impact. These are then collected by an array of junctions or diodes. The electric field of the diode separates them and they discharge a microcapacitor, which is then recharged by the scanning electron beam. An analogous principle is used in the *electron bombarded CCD* (EB-CCD).

The Photon Counting Camera. The *image photon counting system* (IPCS) combines the linearity and sensitivity of the electronographic camera with the convenience of the TV camera, and a capacity for digital storage of information, which can be processed at a later date. It was developed by Alec Boksenberg[5] in Cambridge (UK), and was soon being widely used in astronomical observatories, until it was replaced by CCDs. A photon counting camera was designed and made in Europe to equip the Hubble Space Telescope in the 1990s. This camera, the Faint Object Camera (FOC), was used to make the first deep sky observations with this telescope, before being replaced in orbit by CCD detectors, installed by astronauts during servicing missions.

This camera includes an *image intensifier*: photons incident on the photocathode generate photoelectrons, which are accelerated and focused magnetically onto a phosphor. The latter

[5]Boksenberg A. and Coleman C.I., Advances in Electronics and Electron Physics **52**, 355, 1979.

re-emits photons upon each impact. The amplification process is repeated in three successive stages, to give a final gain of 1.3×10^5.

The intensifier is followed by optics. The image, formed on the output phosphor of the intensifier, is re-formed by an objective on the entry side of a cylindrical bundle of optical fibres, which define each pixel with great precision. The photons are then channelled onto the photocathode of an EBS television tube. Each photon impact is converted into a pulse in the output video signal, and the timing of each pulse is used to determine exactly which event should be assigned to which pixel in the image. The digitised video signal increases by one the number of photons so far counted in the pixel.

This sequence of devices had many advantages: the system involved almost no noise, reading was non-destructive, the 512×512 pixel format was good at the time, and photon impacts could be accurately localised. Figure 7.24 gives an example of the instantaneous image obtained using a photon counting camera.

Other versions of the photon counting camera were developed during the 1980s, such as the connection of two image intensifiers in series, followed by a CCD. Another idea couples an electronic camera (photocathode with acceleration of the photoelectrons) to a CCD operating in an unusual but effective way: bombarded by the accelerated electrons, each pixel releases a large number of charges which are then read by the readout circuit (EB-CCD, or electron bombarded CCD).

Avalanche Photodiodes

Consider a junction which is strongly biased in the opposite direction, and hence non-conducting. The diode material being silicon, an incident photon generates a photoelectron. The very high bias (several V μm^{-1}) accelerates this electron, which photoionises in series the Si atoms it encounters. The process is amplified by avalanche multiplication until a current pulse is created across the poles of the diode. This current pulse may reach 10^9 charges, which is easy to detect above the intrinsic noise of the diode. Count rates can reach 10^6 photocharges per second.

This device has a very fast response, reaching several nanoseconds. The photoimpulses can thus be counted individually, a pause of several hundred nanoseconds being required to restore the charges (*quenching*). A promising application may be the study of radiation coherence by correlation.

The great advantage of these 'solid-state photomultipliers' is that they can work in counting mode (and therefore with the limit imposed by the Poisson noise of the signal), whilst benefiting from the excellent quantum efficiency of silicon (> 0.6). An extension of this idea to the near infrared, using germanium, can be envisaged. Their limitation is two-fold. Firstly, the diodes are cumbersome and it is difficult to assemble them into arrays, although arrays containing several units or several tens of units became available by the 2000s, and it may be that $1\,000 \times 1\,000$ arrays will appear one day. Secondly, the recombination of charges produces a flash of light, and although this presents no problem when limited to a single diode, it soon becomes problematic in the case of arrays, because it leads to *cross talk* between neighbouring pixels.

Compared with photomultipliers, which are their close relatives, they are still limited by the smaller dimensions of their photosensitive zone. Nevertheless, they are fast developing, and sizes up to 200 mm^2 have been achieved. In 2007, these detectors face serious competition from L3-CCDs (see below).

7.4.4 X-Ray Detection (0.1–10 keV)

Here we discuss a family of photon counting detectors developed hand in hand with the emergence of X-ray astronomy, and which in the 2000s have gradually been superseded by CCD detectors, presented later.

Proportional Counters

Figure 7.25 gives a schematic view of a proportional counter. The window enclosing the gas must be thin, and it is the specific combination of the absorption properties of this window and those of the gas which determines the spectral response of the counter. Table 7.5, which is not exhaustive, thus specifies the spectral ranges that can be covered. Note the extreme thinness of the windows, and the properties of the ionisation thresholds in the absence of the window.

The energy resolution is, like the quantum efficiency, determined by the combination of the gas properties (photoelectric cross-section of the gas) and those of the window (Fig. 7.26). This resolution is determined by the Poisson statistics of the photoelectrons, and is thus proportional to $\sqrt{E}$. In general,

$$\Delta E_{\mathrm{keV}} \approx 0.4\sqrt{E_{\mathrm{keV}}} \, , \tag{7.1}$$

the proportionality factor dropping to 0.2 in *gas scintillation proportional counters* (GSPC).

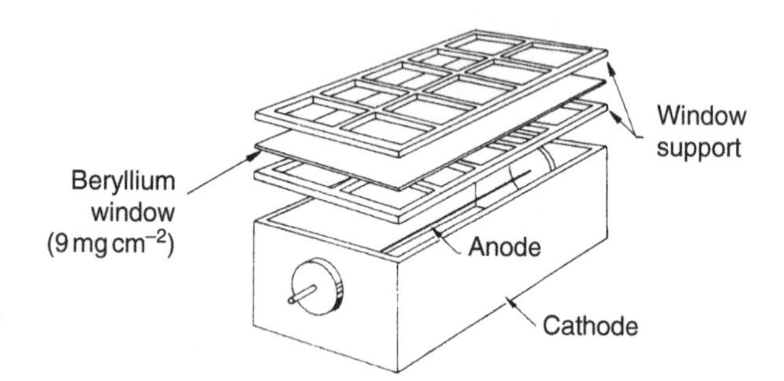

Fig. 7.25 Exploded view of a proportional counter

Table 7.5 Proportional Counters

Window	Thickness [μm]	Gas filling	Spectral response [nm]
Beryllium	125	Neon or argon	0.02–0.8
Aluminium	6	Neon	0.02–0.6 and 0.8–1.6
Mylar	6	Nitrogen or helium	0.02–1.5 and 4.4–6
Nitrocellulose	0.1	Argon	< 30
No window	–	Helium	< 50.43
No window	–	Xenon	< 102.21
Lithium fluoride	1 000	Ethyl bromide	104–120
Quartz (silicon)	1 000	Tri-*n*-propyl amine	160–171.5

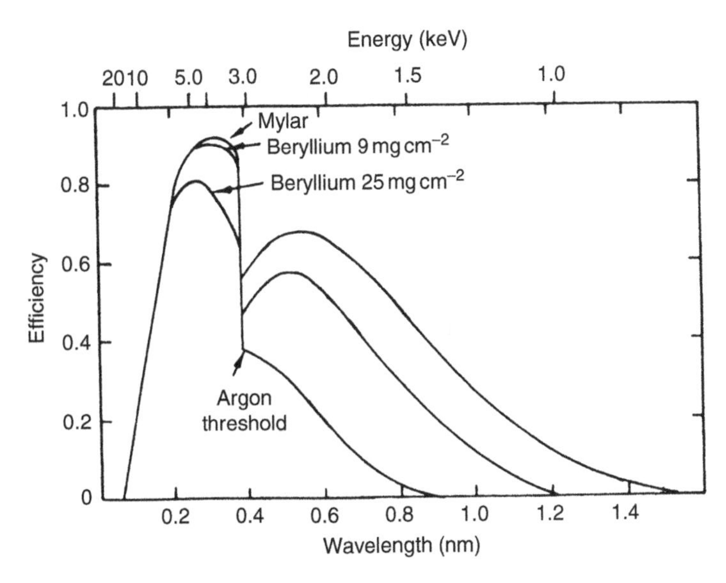

Fig. 7.26 Quantum efficiency of a proportional counter filled with argon (column density 5.4 mg cm^{-2}). The effect of different windows is shown. From Giacconi R. and Gusky H., *X-Ray Astronomy*, Reidel, Dordrecht, 1974. With the kind permission of D. Reidel Publishing Company

The proportional counter has no intrinsic noise, but suffers from amplification noise (gains of the order of 10^3–10^5). Only the statistics of the signal, whether it be the source or a background signal, can therefore create fluctuations, and this alone will determine the signal-to-noise ratio. This idea was extended in the 1970s to the *position sensitive proportional counter*, which equipped the camera of the Einstein satellite, launched in 1978.

Microchannel Plates

Already described in the context of image intensifiers used in the detection of visible and ultraviolet radiation (see Sect. 7.4.3), this device can be used to make multichannel detectors, without energy resolution, for energies up to several keV. Figure 7.27 shows an image obtained with such a detector on the Exosat satellite, launched in 1982, the first X-ray satellite of the European Space Agency.

7.4.5 Solid-State Imagers

These detectors exploit the photoelectric effect in solids. Their multichannel character results from integrating elementary photoconducting pixels into an integrated *array*, which includes a *readout circuit*. State-of-the-art microelectronics is used

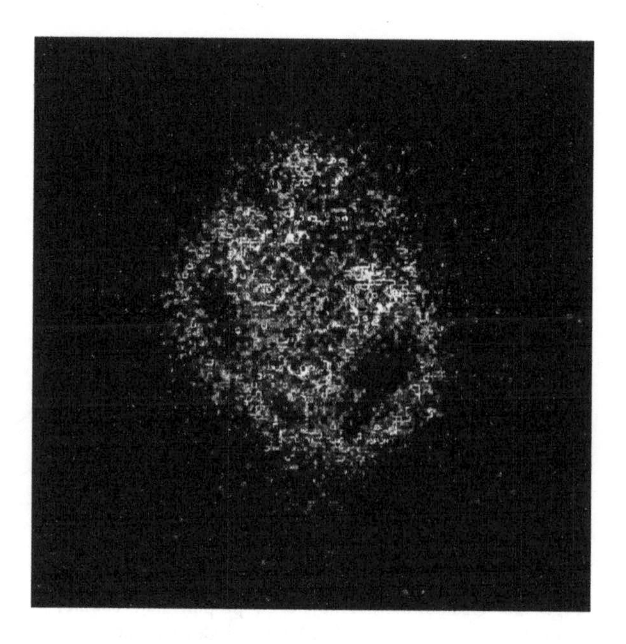

Fig. 7.27 Historical X-ray image of the supernova remnant Cassiopea A, obtained with the microchannel camera (*Channel Multiplier Array*, or CMA) on the European satellite Exosat. Spectral band 0.1–2 keV. Spatial resolution 18 arcsec, on the optical axis. (Picture provided by the Exosat Observatory, European Space Agency, 1985)

in their manufacture. They have developed considerably since their introduction in 1970 and the trend continues, with improved performance in sensitivity, dynamic range, format, miniaturisation, reliability, and integrated intelligence. They produce a digital output signal which is easy to process and have a temporal integration capacity.

These detectors are known generically as *solid-state imagers*. The term covers many variations, such as Reticons, charge coupled devices (CCDs), and charge injection devices (CIDs). At visible wavelengths, it is the CCD exploiting the photoconductivity of silicon which has taken the lead since the end of the 1980s. For infrared detection, microelectronic integration techniques have made it possible to superpose a suitable photosensitive material for the given wavelength and readout circuits, thus constructing the equivalent of a CCD at these wavelengths. Finally, recent developments, in the 2000s, have created a competitor for the CCD, namely the CMOS (complementary metal oxide semiconductor) detector.

The strategy in solid-state imagery is to accept a gain close to unity, but to reduce the readout noise σ_R to near zero, so that the detector quantum efficiency δ attains the value of the photoelectric quantum efficiency η. The trend is also to make the latter tend toward unity.

7.4.6 Charge Coupled Device (CCD)

The basic structure of a charge coupled device is a metal–oxide–semiconductor (MOS) capacitance. It is a capacitance consisting of an insulator sandwiched

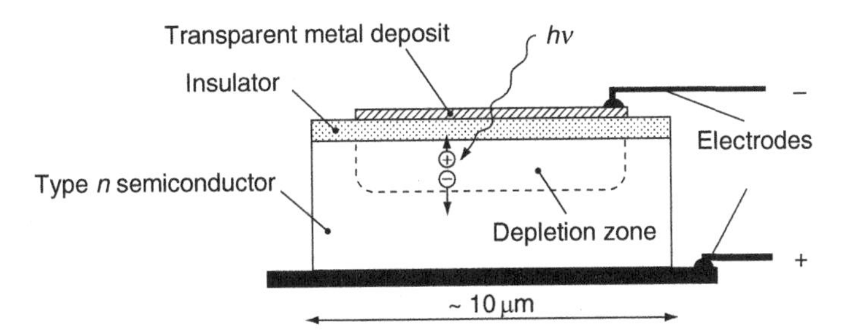

Fig. 7.28 Metal–oxide–semiconductor (MOS) capacitance. The voltage applied at the upper electrode acts on mobile charges in the n-type semiconductor. For $V > 0$, the majority carriers (e^-) accumulate at the surface under the insulator. For $V < 0$, the majority carriers (e^-) are repelled far from the surface, creating a depopulated or *depletion zone*, without mobile charges, which is thus insulating. For $V \ll 0$, a positively charged inversion layer is formed at the surface, and this is a potential well for the minority carriers (+ holes). The photocharges caused by incident photons ($h\nu$) modify the surface potential. The lower electrode can be made transparent to radiation, freeing the upper part for the electrical connections of the readout circuit

between a metal electrode and a semiconductor exhibiting the photoelectric effect (Fig. 7.28). The insulator may be formed by direct oxidation of the semiconductor (SiO_2 on silicon), or it may be deposited by vaporisation, sputtering, or condensation in the vapour phase. The semiconductor is usually silicon.

The applied potential difference creates an electric field which modifies the charge distribution in the semiconductor. A potential well appears, capable of storing in the capacitance any photocharges released in the semiconductor by an incident photon, as well as any thermally excited mobile charges. This last phenomenon makes it necessary to cool solid-state imagers, in order to reduce the thermally generated signal.

Each pixel of the CCD comprises a capacitance of this kind, and the full set of pixels is organised in a 2D array forming the detector array of $N \times M$ pixels which specifies the $N \times M$ format of the CCD. Once the charges have accumulated, the CCD contains a readout circuit, using *field effect transistors* (FET) that function as switches and also a clock, which transfers the charges of a given column one after the other into an output series register and then, having done this, moves on to the next column, until all the stored charges have been read. The charges are transformed into successive voltage pulses by the series register, and these are amplified then digitised in a synchronised way by the clock controlling readout. The CCD controller oversees the electronic functions. Photosites have been gradually reduced in size with progress in silicon etching techniques, to reach a few micrometres.

Figure 7.29 illustrates the idea of charge transfer driven by a clock-controlled variation of potential at successive electrodes in rows and columns. The output signal is thus a video-type signal, in which each pulse is proportional to the stored photocharge. The image can then be generated at the standard TV rate of 25 frames per second (in Europe), but it can also be read at an arbitrarily slow rate (long exposures).

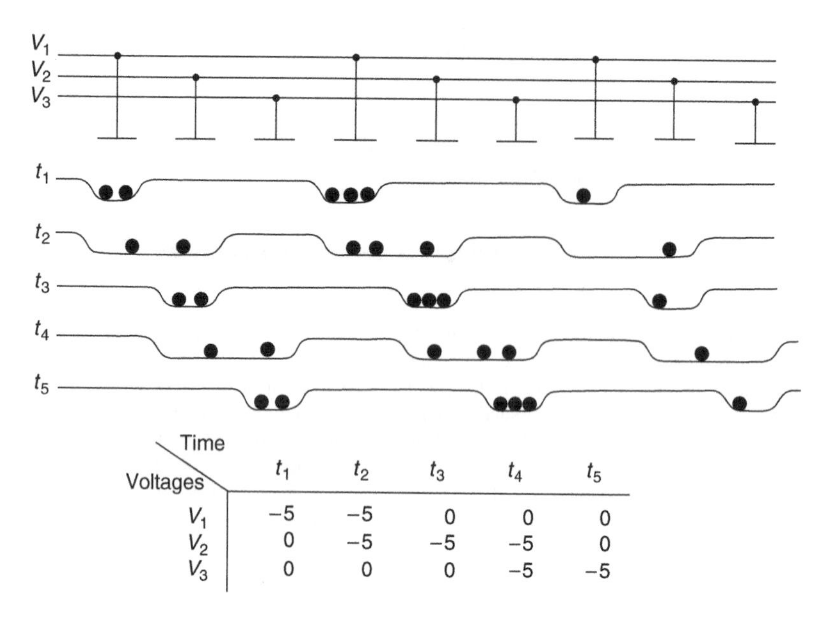

Voltages \ Time	t_1	t_2	t_3	t_4	t_5
V_1	−5	−5	0	0	0
V_2	0	−5	−5	−5	0
V_3	0	0	0	−5	−5

Fig. 7.29 Principle of charge transfer in the CCD detector. Each pixel is defined by three electrodes. Electrons are represented by balls, which move within the potential wells created by the electrodes, and evolve in time. (After Fauconnier T., doctoral thesis, University of Paris VII, 1983)

When all functions, i.e., release of photocharges, storage, and transfer to the output register, are handled within the same pixel, the CCD is said to be *full frame*. However, it also possible to separate these functions in the so-called *frame transfer CCD*. A $2N \times N$ array is divided into two parts, with one $N \times N$ piece used to generate the charges and the other $N \times N$ piece storing them. If the transfer from one to the other is very fast (< 10 ms for a 1k × 1k format, the time depending on the number of rows), the use of time is optimal, since the charges can be read out while the exposure is still underway. This doubles the size of the final array, which may complicate image formation when a large focal plane is to be covered by CCDs placed side by side.

Quantum Efficiency. Thinned CCDs

Impressive progress has been made with the quantum efficiency of CCDs in the visible wavelengths (400–900 nm). Figure 7.30b shows efficiencies almost equal to unity in the extreme red, and greater than 50% in the blue. Most of the loss of efficiency is due to reflection losses. There is a great reduction in sensitivity when CCDs are used in the blue (< 400 nm) and ultraviolet, because silicon can absorb photons at these wavelengths.

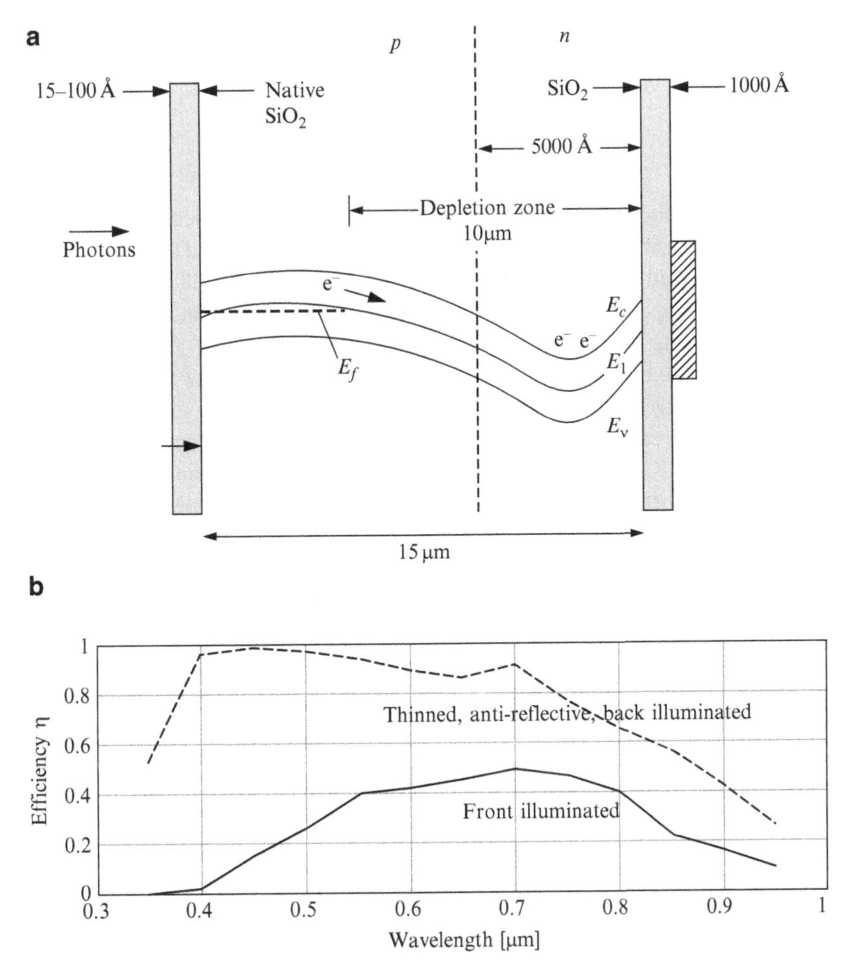

Fig. 7.30 (**a**) Schematic cross-section of a thinned CCD. The limited extent of the potential wells is shown, as is the depletion zone that the photocharges must reach. (**b**) DQEs for front illumination and for anti-reflectance treatment. (After Lesser M., 1994)

The electrodes placed on the front face absorb the radiation. This led to the idea of illuminating the CCD from behind, so that the radiation would only encounter silicon immediately effective in photoelectric conversion. However, as a CCD was around 300 μm thick, the photocharges had to cover such a distance before encountering the potential wells, whose extension under the electrodes barely exceeds 10 μm, that the charges would never be able to recombine. Illuminating the CCD from behind therefore required the CCD to be thinned down, to a thickness of about 15 μm. A compromise must be found between the capture of photocharges and their generation, which requires a photon path in the silicon that increases with the wavelength. Thinning is thus necessary in the blue, and of course for UV and X-ray detection, but unfavourable for the high quantum efficiency in the red. It is impossible to make silicon wafers of this thickness directly. A standard CCD is thus turned round, with its front face stuck onto a suitable substrate, and then its back face is thinned by mechanical or chemical processes, passivated (antioxidant treatment), and treated to reduce reflection losses (Fig. 7.30a).

Readout Noise

The transfer of photocharges and ensuing amplification are accompanied by two sources of noise, whose (quadratic) sum is generally observed. On the one hand, there is a fluctuation in this number of charges. This has thermal origins and depends on the capacitance storing the charges. On the other hand, there is a further fluctuation σ_R due to transfer and amplification of the signal, called the *readout noise* of the CCD. Each of these terms is reduced by cooling the CCD, generally to the temperature 77 K of liquid nitrogen.

The ultimate readout noise depends on the read rate. Indeed, the noise level decreases as the CCD is read more slowly, going as $1/f$ in the amplification stages. But increasing formats implies reading a larger number of pixels, and this has consequences for the readout noise, which has decreased at the time of writing (2007) to a few photocharges (1–2 electrons rms) at rates of 50 kilopixels per second (or 20 s to read a $1\,024 \times 1\,024$ CCD). Very high readout rates ($\leq$ 1 Mpixels per second) are sought for certain applications, such as image or wavefront reading in times less than the coherence time of atmospheric turbulence (e.g., in adaptive optics). The objective is a noise $\sigma_R \approx 2$ electrons rms for a readout rate of the order of 1 Mpixel per second. Another solution to the problem of increasing the readout rate, without increasing the readout noise, consists in having parallel exits, by dividing the CCD into 2, 4, or 8, with a separate exit for each part.

For a fixed number of generated photocharges, let us evaluate the number of charges read. Just as it was possible to characterise the thermal noise output by a resistance R at temperature T (see Sect. 7.2.2), in terms of a power $4kTR\Delta f$, it can be shown that, across a capacitance C, the charge q has variance

$$\sigma_q^2 = kTC\Delta f \ .$$

This is therefore the fundamental limit of the noise across C. Assuming a capacitance of 50 femtoF, below which it is difficult to go, this gives $\sigma_q = 46\,\mathrm{e}^-$ at 77 K, and $10\,\mathrm{e}^-$ at 4 K. It is thus useful to reduce this capacitance, but that in turn would reduce the maximum amount of charge that could be stored, whence a lower flux would saturate the CCD.

Concerning the maximum charge that can be stored by this capacitor under a voltage V, namely CV, a value of $V = 1$ gives $6.25 \times 10^6\mathrm{e}^-$ per pF. There is thus a simple relation for the dynamic range of this capacitor, in the absence of any other source of noise which would reduce it:

$$\text{dynamic range} = V\sqrt{\frac{C}{kT}} \ .$$

An important property of CCDs is their linearity up to a high charge storage capacity, around 10^6 charges per pixel in the best cases. This makes possible very long exposures of faint objects, with no increase in readout noise (Problem 7.11).

Table 7.6 CCD detectors. The table gives mean values, open to improvement in the future

Threshold	One quantum
Dynamic range	$\lesssim 10^6$ charges pixel^{-1}
Quantum efficiency	$\eta \geq 0.8$ between 500 and 900 nm
Spectral response	Normal CCD: 400–1 100 nm
	Thinned CCD: soft X ray (0.1–50 keV)
Temporal response	Around 50 ms
Format	Up to 2 048 × 4 096 pixels
Noise	Readout $\sigma_R \sim$ 2–5 e$^-$ rms depending on rate
Information storage	Digital, destructive readout

Very long exposures nevertheless increase the chances of superposing a further type of noise in addition to readout noise, for example, due to the impact of cosmic rays, which is random and may saturate a pixel. Such events can occur both in space and at ground level, particularly during solar flares, or during a crossing of the South Atlantic anomaly (see Sect. 2.9.2). They then contribute to *background noise*. A specific treatment of the pixels can reduce this noise for spaceborne CCDs (see below).

CCD Formats

At the time of writing (2007), the 2 048 × 4 096 format is a technological limit that would appear to be rather hard to improve upon, given the great difficulty involved in maintaining uniformity over surfaces including several tens of millions of pixels. In 2007, it is commonplace to make CCDs with format 1 024 × 1 024 (usually denoted by 1k × 1k), but 4k × 4k is possible. Beyond this, it is easy to place CCDs side by side in order to construct large arrays. The present trend is therefore towards the tiling of focal planes by juxtaposition, which can give formats up to 8k × 8k, or 8 192 × 8 192. There is, of course, a dead space between the CCDs, which can be reduced to the area of a few dozen pixels. If one of the four sides of the CCD is reserved for electrical contacts, then juxtaposition can make a 4 × 2 array.

For example, the Megacam camera of the Canada–France–Hawaii telescope comprises 40 thinned CCDs, each with format 2k × 4.5k, laid side by side to cover a total field of 1.4 degrees with 3.6×10^8 pixels. The size of the CCDs is 13.5 μm. Each exposure lasts at most 20 s (to avoid risk of saturation) and provides 0.77 Gbits of data. A specific processing centre called Terapix (see Chap. 10) has been set up to process this considerable volume of information (Table 7.6).

Dark Current

Thermal generation of charges in the CCD creates a signal, even when there is no illumination. This signal, mainly produced by traps at the Si–SiO_2 interface, must therefore be calibrated on average and subtracted from the signal, an exercise referred to as doing a dark frame subtraction. The dark current decreases with temperature. It can be significantly reduced by means of a *charge sloshing* or *dithering* mode, which modulates readout voltages faster than the lifetime of surface states.

The Low Light Level CCD (L3-CCD)

As the name suggests, the *low light level CCD* or L3-CCD is one that operates at low light levels. Here the readout noise becomes negligible, up to high readout rates ($\sim$ 1 kHz). This is achieved by virtue of a multiplicative avalanche effect in the transfer register which transforms the incident photoelectron into a charge packet. A gain of around 500 is possible, making this L3-CCD a promising detector, operating in photon counting mode. It looks like becoming a serious competitor for the *avalanche photodiode*.

CCDs for X-Ray Detection

By thinning, the sensitivity of the CCD can be extended as far as X rays. The European mission XMM–Newton (1999–2010) is equipped with two cameras, each using 7 CCDs with format 600×600. A third camera uses 12 CCDs, laid side by side, each containing 64×189 pixels. The CCDs are used either in full-frame mode or in frame transfer mode, so as to minimise the readout noise and maximise the useful exposure time.[6] Analogous setups are in preparation for the future German mission ROSITA (around 2010), with quantum efficiencies exceeding 0.8 for X rays in the range 0.2–20 keV.

The pn-CCD differs from conventional CCDs by the structure of its transfer registers, here formed by p–n junctions in the silicon. This ensures low sensitivity to undesirable hard radiation like γ rays or cosmic rays, but also fast charge transfer and high charge storage capacity.[7]

Finally, the CCDs used in the X-ray region have a certain spectral selectivity, i.e., an intrinsic energy resolution, of $\Delta E \sim 100$ eV for $E < 1$ keV, and $\Delta E \sim 140$ eV for $E \sim 6$ keV. Indeed, a photon with energy in the keV range releases a large number of photocharges, a number which then becomes a measure of the incident photon energy. To measure the energy of each photon, it is thus essential to read the CCD before another photon causes a release of photocharges at the same place. A summary of the CCD performances is given in Table 7.6.

The CCD Controller

This electronic device is an essential element in the use of CCDs. It generates the clock signals and controls the charges and their transfer:

---

[6] The website www.src.le.ac.uk/projects/xmm/instrument/index.html provides a detailed description of cameras on board the XMM mission.

[7] See, for example, www.pnsensor.de/Welcome/Detector/pn-CCD/index.html.

- It modifies the *readout frequency* (also called the *pixel frequency*), depending on whether the readout noise must be minimised (slow frequency) or not.
- It modifies the gain, that is the number of electrons per level of digitisation, depending on the strength of the signal and the required dynamic range.
- It can read the CCDs in *binning mode*, in which adjacent pixels are summed according to a predefined geometry (2×2, 4×4, etc.). In this case, spatial resolution is lost, but the signal-to-noise ratio is improved, since the signal is increased for the same level of readout noise. This mode, which reduces the volume of data collected and the readout time, is useful for making adjustments in order to take into account variable image quality in photometry, if a high spatial resolution is not required, and so on.
- It also provides a *windowing mode*, in which a subset of pixels is selected within a window of the array.

As an example of the way CCD pixels are manipulated, let us consider a device proposed for the detection of an exoplanet which measures the tiny differential polarisation over time of the radiation recieved from the star–planet system when the planet orbits around the star. Indeed, the stellar radiation scattered by the atmosphere of the exoplanet is polarised, and depends on the phase angle. The polarisation of the incident signal is analysed very rapidly (1 kHz), then transformed into an intensity variation. The image is formed on a CCD in which every other column is masked. The masked columns are used to store the photocharges generated in the first column and transferred to the second at the fast rate of the modulation. After many cycles, the difference between the two columns indicates the presence of a possible polarisation. This is the so-called ZIMPOL technique, proposed by Schmid, H.M. et al., and described at the website saturn.ethz.ch/instrument/zimpol.

7.4.7 The Hybrid CMOS Detector

CMOS stands for complementary metal oxide semiconductor and refers to a whole family of 2D detectors, also fabricated by conventional techniques of microelectronics, but based on a different idea to the CCD. In this case, each pixel comes with its own readout and amplification transistors (3 or 4 in all). These transistors take up some space on the sensitive surface and this reduces the filling factor of the array, whence it may be necessary to use microlens arrays to improve photon gathering in the image. Furthermore, the CMOS technology is not amenable to rear illumination, which would avoid absorption of a fraction of the radiation by the surface layers. Instead of transferring the charges to registers as in a CCD, a clock device addresses the pixels of the array one by one to read the signal.

Many ideas were tested during the period 1980–2000 for making infrared detectors. For the record, since they have now been abandoned, let us just mention *charge injection devices* (CID), CCDs made entirely from InSb rather than Si, and hybridisation of photoconductors such as HgCdTe on silicon CCDs.

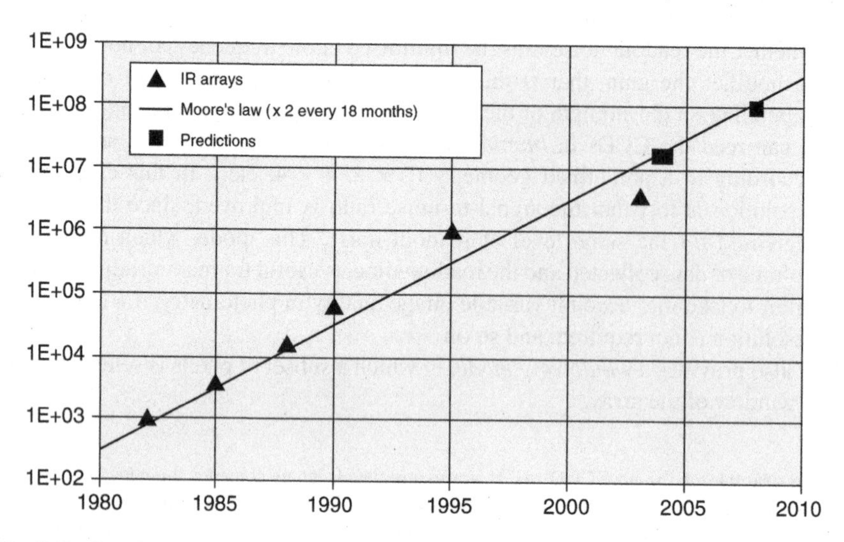

Fig. 7.31 Growing formats of imaging detectors in the near and mid-infrared using CMOS technology. *Ordinate*: Number of pixels on the detector. The *straight line* shows a doubling every 18 months (known as Moore's law). Figure established in 2004, courtesy of Hodapp K.W., Hall D.N., University of Hawaii

This technique was first developed for the near infrared, in order to make imaging detectors beyond the photoconductivity of silicon. It was subsequently extended to the mid- and far infrared in the middle of the 1990s. Based on the photovoltaic effect, these detectors couple a photodiode, made from a material that is photosensitive at the desired wavelength, with a readout circuit made from silicon, the whole thing being in a large format array. They are sometimes called *hybrid arrays* since they employ two different materials hybridised together. Since the 1980s, these arrays led to the rapid development of astronomical imaging in the infrared region (Fig. 7.31).

It is interesting to situate these advances with respect to the history of astronomical exploration in the infrared region, which developed slowly at first, then leapt forward, largely due to the quality of the available detectors, as can be seen from Fig. 7.32, for the period up to 2000.

The very low energies involved (a fraction of an electron-volt) make it necessary to cool the devices, in order to avoid the thermal generation of charges (dark current). The temperature ranges from 77 K (liquid nitrogen), at the lower wavelengths ($\leq 2.5\ \mu$m), and from 4 to 20 K, at the longer wavelengths, where the band gap is narrower (Fig. 7.33). The use of consumable cryogenic liquids, such as N_2 and He, is gradually giving way, on Earth and in space alike, to closed circuit cryogenic systems (*Joule–Thomson refrigerators*).

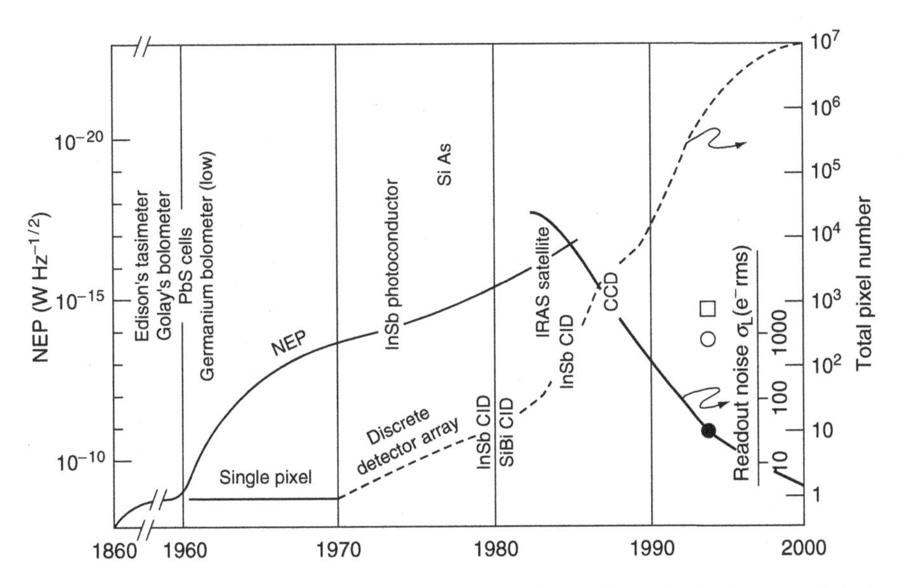

Fig. 7.32 Progress in infrared detection. The Edison tasimeter (Eddy A., J. Hist. Astr. **3**, 187, 1972) was remarkably sensitive for its day (NEP $\approx 10^{-8}$ W Hz$^{-1/2}$). The *thin continuous curve* represents a mean sensitivity obtained with single pixel detectors. The *dashed curve* shows the arrival of array detectors (CID and CCD), and the formats are given. The readout noise σ_R is given by the *thick continuous curve*, with a brief indication of trends in performance levels beyond 2000 for different wavelength regions: • = 1–2.5 μm, ○ = 2.5–5 μm, □ = 5–20 μm

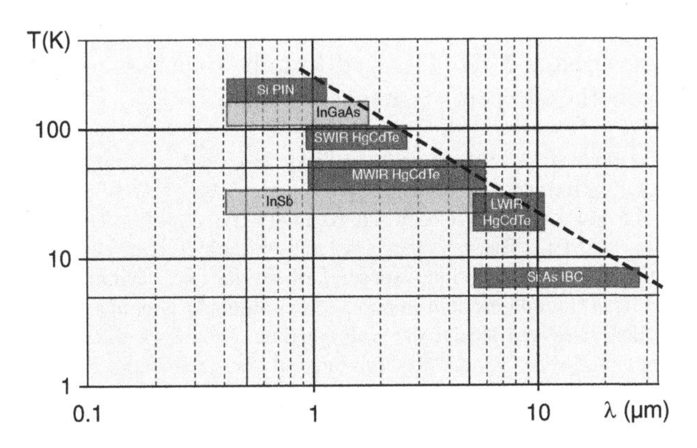

Fig. 7.33 Operating temperature of CMOS arrays at different wavelengths. Different regions are covered by silicon Si, then InGaAs, HgCdTe, InSb, and S:As. Courtesy of Hodapp K.W., Hall D.N., University of Hawaii

Visible Region

CMOS technology competes with CCDs in this region, coupling a silicon diode with the pixel transistors, also fabricated in the same silicon substrate (see Table 7.7).

Table 7.7 Comparing CCDs and CMOS devices. From Hodapp K.W., Hall D.N.B, Institute of Astronomy, Hawaii, 2005

Property	CCD	Hybrid CMOS
Resolution	> 4k × 4k	4k × 4k (under development)
Pixel (μm)	10–20	18–40 (< 10 under development)
Domaine spectral	400–1000 nm	400–1000 (Si PIN)
Noise	Several e^- rms	Same with multiple sampling
Shutter	Manual	Electronic
Consumption	High	10 times lower
Sensitivity to radiation	Sensitive	Barely sensitive
Control electronics	High voltage clock	Low voltage
Special modes	Binning	Windowing
	Adaptive optics	Random access
		High dynamic range

The popular invention known as the *webcam* is equipped with detectors using this technology.

Near and Mid-Infrared

The hybrid HgCdTe arrays originally commercialised under the trade name NICMOS, which stands for *Near Infrared Camera and Multi-Object Spectrometer*, were designed for the Hubble Space Telescope with a 1 024 × 1 024 format in the 1990s (Fig. 7.33). They are now called Hawaii detectors and have become more or less universal for the spectral bands J, H, and K, to be found on all the major ground-based and spaceborne telescopes (see Figs. 7.34 and 7.35).

The format and readout noise of these detectors have seen constant improvement and they now have excellent noise performance in the region 1–2.5 μm, viz., $\sigma_R \approx 10\,e^-$. At the time of writing (2006), formats of 6 144 × 6 144 have been achieved with 10 μm pixels. They contain 248 million field effect transistors (FET). Thermal charge generation (dark current) is low, which means that, in the absence of background radiation, exposures of several tens of minutes are feasible. It can also be read quickly, in several tens of milliseconds, an invaluable asset in adaptive optics or optical interferometry, where rates are linked to the evolution times of atmospheric turbulence. The storage capacity of the potential wells is not very high (around 10^5 photocharges), and this limits use to long exposures with high incident fluxes or in the presence of a significant background signal (spectral bands J and H).

Hybrid InSb (indium antimonide) arrays, commercialised under the trade name *Aladdin*, pick up from there, up to the photoconductivity limit of thise material, in formats reaching 512 × 4 096 on one of the CRIRES focal instruments of the European Very Large Telescope.

Hybrid Si:As arrays, in a 256 × 256 format, then bridge the gap to the ground-based observation window of the N band around $\lambda = 10$ μm (the VISIR instrument of the VLT). They work in BIB mode described below. The IRAC camera of the Spitzer space mission launched in 2003 is also equipped with this detector, operating

Fig. 7.34 The Orion nebula observed in 2001 with the European Very Large Telescope and the ISAAC instrument. This image is put together using images taken in three different spectral bands (J, H, and K) and is in fact a 3×3 array of individual images, with the $1\,024 \times 1\,024$ format of the HgCdTe detector. Field $7' \times 7'$, pixel $0.15''$. North is at the top and east on the left. Source ESO and M. McCaughrean

at 5.8–8μm. The imaging photometer (MIP) of this mission comprises a 128×128 Si:As array, observing around $\lambda = 24$ μm. It also works on the blocked impurity band (BIB) principle discussed below.

The same hybridisation scheme can be used with Si:Sb, for which photoconductivity extends up to 40 μm, in 256×256 formats, for use in space. The IRS spectrometer of this same mission is equipped with a 128×128 format camera.

Blocked Impurity Band (BIB) Detectors

Two opposing requirements must be balanced when an extrinsic photoconductor is used in a detector: the level of dopant should be low in order to ensure a high resistivity, but it should be sufficient to maintain an adequate quantum efficiency when absorbing photons. The principle of the *blocked impurity band* (BIB) is to separate these two requirements spatially, in the material of the photoconductor,

Fig. 7.35 An image obtained with a NICMOS detector, on the 1 m telescope at the European Southern Observatory (La Silla, Chile). Field $12' \times 12'$ in the central region of Orion, showing molecular cloud and star formation region. The same region has been shown at radiofrequencies in Fig. 5.5. Wavelength and spectral band J (1.2 μm). Image format 780×780, although the solid-state device includes only 256×256 pixels, the telescope being displaced by *offsets* of one third of a pixel and the image reconstructed by interpolation. Total exposure time 9×1 s. (Photo by Copet E., DENIS programme, 1995)

so that each can be optimised. An added benefit is an amplification effect, which improves noise performance.

Consider the material Si:As, silicon doped with arsenic, with a band gap which situates the photoelectric effect in the region $\lambda \leq 23$ μm. The photosensitive layer (Fig. 7.36) is highly n-doped, implying a good quantum efficiency, and also a small volume, which makes the detector less sensitive in an environment of high energy particles (e.g., in space, see Sect. 2.9.2). This layer is preceded by a thin blocking layer, made of a pure material, which ensures a high resistivity between the two electrodes. A photoelectron released in the sensitive layer drifts in the electric field, towards the blocking layer. A strong field can then transmit enough energy for the photoelectron to produce secondary electrons, by ionising collisions. This does not happen in the blocking layer, for the material there is too pure, and the ionisation energy is therefore high (around 1 eV). Amplification can reach a factor of ten. This is presented in detail by Rieke G.H., *Detection of Light: from the Ultraviolet to the Submillimetric*, 2nd edn., Cambridge 2003, and the discussion here follows closely.

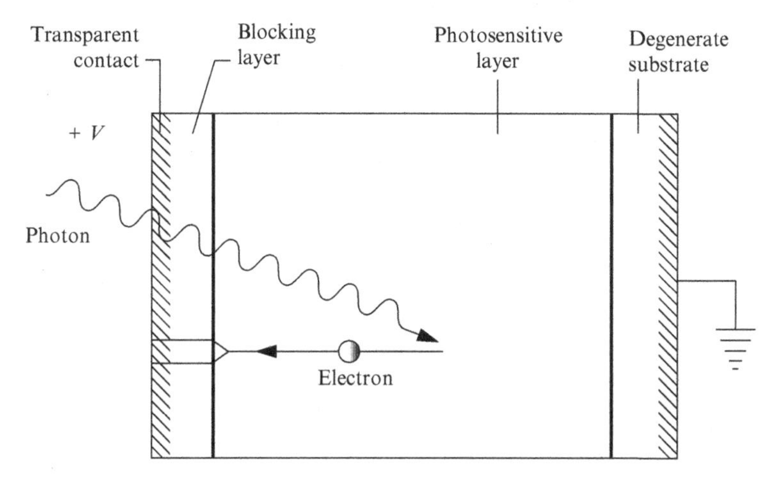

Fig. 7.36 Schematic cross-section of a BIB detector. The electron is released in the photosensitive layer and drifts towards the blocking layer. The electrodes are on either side. Here, the BIB is illuminated from the front. (After Rieke G.H., 1994)

The amplification technique in the BIB detector is pushed a step further in the *solid-state photomultiplier* or SSPM. One difficulty with the BIB is that the amplification zone, situated at the interface of the blocking and photosensitive layers, is not very clearly defined. If this role is assigned specifically to a third layer, a significant cascade effect is obtained, in which a single electron may generate a current peak of 40 000 charges. Further, this current pulse is very brief, lasting only a few nanoseconds, if followed by high speed electronics. The detector then operates as a photon counter, and is able to discriminate events which are highly fluctuating in time. This constitutes, in the infrared domain, a property already mentioned for silicon *avalanche diodes*.

Detection in the Far Infrared

The only photoconductor that can be used beyond 40 μm is the material Ge:Ga, with a sensitivity region extendable to 210 μm by applying pressure (*stressed detector*). However, the stressing device requires space and microelectronic techniques are no longer applicable, whence hybridisation is carried out pixel by pixel. This in turn reduces the available format sizes. These detectors equip the imaging photometer of the Spitzer mission (format 2×20), the PACS instrument of the European space mission Herschel (format 5×5), and the FIFI instrument of the airborne telescope SOFIA (format 25×16).

Quasi-Optical Coupling

At wavelengths lying between the far infrared and millimetre, diffraction phenomena become significant, for the size of the detectors becomes comparable with the

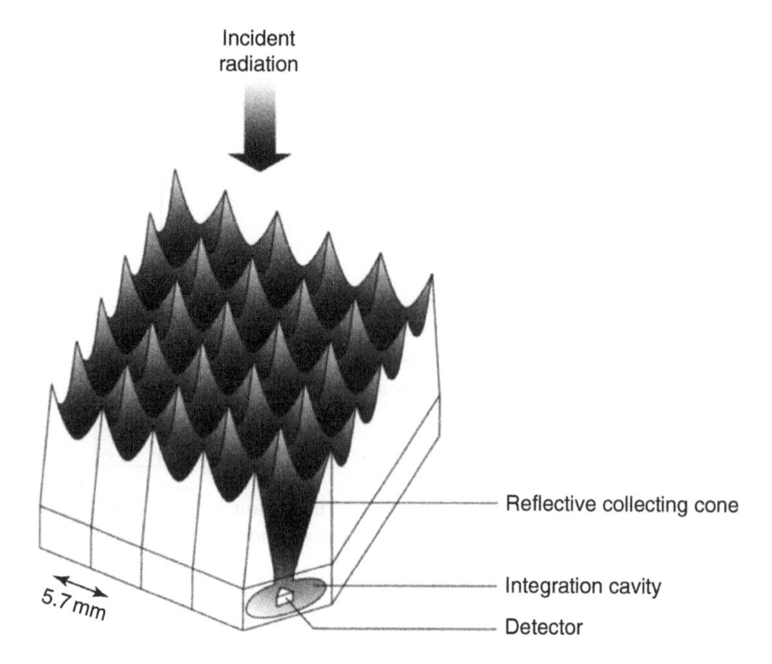

Fig. 7.37 Quasi-optical arrangement of a detector array (5 × 5), in gallium-doped germanium, operating at $\lambda < 240\,\mu$m. (After Geis N., doctoral thesis, Max Planck Institut für Extraterrestrische Physik, 1991)

order of magnitude of the wavelength. In order to better concentrate the energy on the detector, guides are used in *quasi-optical mode*, which means that they operate somewhere between the classical image formation by a lens (see Chap. 5) and the standard waveguide (see Sect. 6.4). This is illustrated in Fig. 7.37, for detection in the range 100 to 240 μm.

7.4.8 Observing Conditions in the Infrared

The radiation received at the detector includes the signal from the source and also the signal from the thermal background. The latter may be due to the optics of the telescope itself, if it is not cooled, or to the thermal background of the atmosphere. These two components can be reduced or even totally eliminated, when the detector is placed in space and equipped with cooled optics. Except in the latter case, the readout noise of the detector does not need to be reduced below the thermal noise generated by the background signal.

Two spectral regions thus stand out clearly, particularly for ground-based observations: wavelengths less than about 1.8 μm, where the background signal is negligible and the reduction of readout noise can be pushed to its limit. Beyond this point, the fluorescence of atmospheric OH radical enters the picture, as discussed in

Sect. 2.3, followed by a significant rise in the thermal background radiation. This region does not therefore require extreme performance with regard to the readout noise σ_R.

In space, the absence of thermal background, and the very limited background due to a fully cooled telescope and optics, require the readout noise to be as low as possible.

An Instrument from the 1990s: The ISO Satellite. It is interesting to consider the progress achieved by describing the performance of the long wavelength observations (5–17.8 μm) of the ISO satellite, which was designed in the mid-1980s, launched in 1995, and exhausted its cryogenic fluid in 1998 (a joint project backed by the USA, GB, and ND). This camera (ISOCAM) was equipped with a 32 × 32 array made with Si:Ga hybridised on Si, and had a readout device that prefigured the CMOS system. (This was called *direct voltage readout* or DVR, which no longer transferred the charges, but read the amplified voltages on each pixel in sequence.) The readout noise σ_R was approximately 180e$^-$ rms, lower than the background noise, the integration time was at most 100 ms before saturation by the residual thermal background, and the pixels had a storage capacity of 10^6 charges. This camera, which provided the first quality images in the near infrared with the 32 × 32 format (with two arrays, HgCdTe and Si:Ga), produced an extraordinary crop of new results.

The important development of the ISO satellite camera in France between 1985 and 1993 had many spinoffs. It is interesting to see how a single idea can be adapted to different contexts. At ground level, the thermal background is very much higher than that which prevails in space, for a cooled telescope. A solid-state device, based on the same principles, but with a storage capacity of 10^7 charges per pixel, was devised to equip the camera of a ground-based telescope (the TIMMI camera of the 3.5 m NTT at the European Southern Observatory). It originally had a format of 128 × 128 but, after extension to a format of 256 × 256 and a change of dopant from Ga to As, it equipped the *Very Large Telescope* (VLT), operating in the atmospheric transmission window between 8 and 12 μm. It is interesting to compare the performance of the same detector at ground level (high background signal) and in space, with totally cooled optics (low background signal).

Even earlier, before the advent of detector arrays, NASA's *Infrared Astronomical Satellite* (IRAS), which was the first to map the infrared sky, throughout the year 1983, carried individual detectors arranged side by side (Fig. 7.38).

7.4.9 Development of Solid-State Imaging Arrays

As in the case of many other astronomical detectors, many parameters are required to completely characterise these detectors. Three examples are listed here:

- *The Maximum Storage Capacity of Photocharges per Pixel.* This is an important quantity when faint sources are sought in the vicinity of some bright object. *Saturation* by the bright object determines the dynamic range of the detector. Furthermore, as already mentioned, the wells can be very quickly saturated, in a few milliseconds, by any thermal background radiation, so that it is pointless trying to extract the signal, after subtraction of this background. The solution is therefore to increase the capacity, which may reach a few times 10^6 or 10^7 photocharges per pixel.
- *The Dependence of σ_R on the Readout Rate.* The faster the readout, the larger σ_R becomes, and this may be harmful in rapid readout applications, such as optical

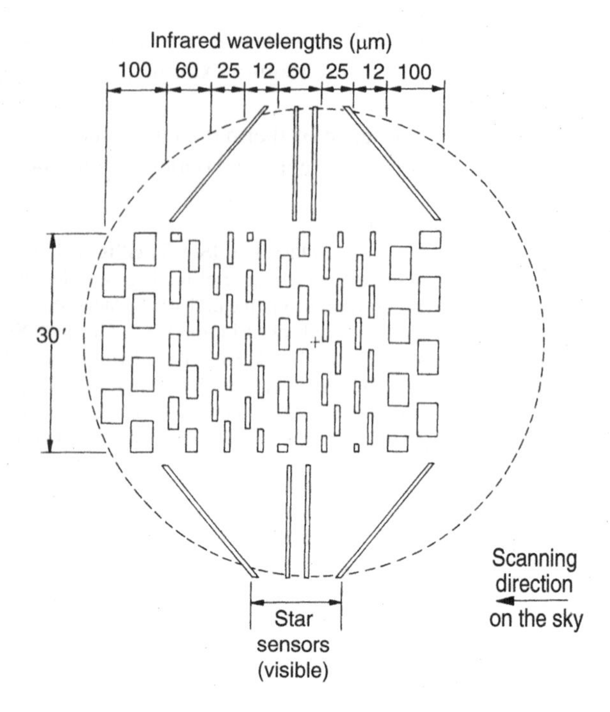

Fig. 7.38 A historic mission: the focal plane of the IRAS satellite 60 cm telescope (1983). The individual detectors are arranged to obtain the maximum accuracy in locating a source as it crosses the focal plane. The visible sensors locate the field by detecting stars known in the visible, as they pass across. From Neugebauer G. et al., Ap. J. **278**, L1, 1984. With the kind permission of the Astrophysical Journal

interferometry, adaptive optics, and speckle interferometry, in which wavefront and image vary on the time-scale of atmospheric turbulence. The same is true for rapidly varying sources, such as pulsars and bursts.

- *The Format*. The trend is towards larger formats. Although technology is unlikely to provide formats greater than $4\,096 \times 4\,096$ per unit of a detector array at affordable prices,[8] the technique of juxtaposing solid-state imagers to form arrays can increase the useful format, and thus give a better use of the field at the telescope focus.

Noise Equivalent Power

A photoconductor, operating at a temperature T, has a non-zero resistance, even in the absence of illumination, because of the thermal release of charge carriers. The thermodynamic fluctuations in this resistance at temperature T, called *Johnson noise*, are generally negligible compared with other noise sources.

Each incident photon produces η electron–hole pairs, rather than η electrons, as happens in the vacuum photoelectric effect, and so the quantum noise, due to the signal or the thermal background, has twice the variance.

[8] The high cost of these detectors is largely due to the complex quality control required by the many steps in the microelectronic fabrication process, and sorting out those detectors with no defective pixels and transistors from the resulting series.

The dominant source of noise from the detector itself is usually due to the transistor in the first stage of amplification. Denoting the detector response by R [V W^{-1}], and the noise of this transistor by σ_R [V Hz$^{-1/2}$], the noise equivalent power of the transistor, often used to characterise the detector (see Sect. 7.2.1), is given by

$$\mathrm{NEP} = \sigma_{\mathrm{R}}/R \qquad [\mathrm{W\,Hz^{-1/2}}] .$$

Typical values are 10^{-16}–10^{-17} W Hz$^{-1/2}$. Table 7.8 summarises some common properties of solid-state devices used in the near and mid-infrared.

7.4.10 Bolometers

In this type of detector, the energy of the incident photon is no longer used to excite the quantum jump of an electron into the conduction band. Instead, it is directly converted into thermal agitation, either of the crystal lattice, or of the free electron gas, if this is indeed decoupled from the lattice (the *hot electron bolometer*). These detectors therefore measure the temperature variation in either the crystal lattice or the electron gas, by a change in electrical resistance. They played a very important role in the development of infrared astronomy over the period 1960–1990, and then, considerably improved, continued to do so in the 2000s in the submillimetre region, where the fact that they have no intrinsic spectral selectivity gives them a significant advantage over heterodyne detection. With the use of microelectronic and multiplexing techniques, large format arrays can be made for both ground-based and spaceborne telescopes.

Standard Germanium Bolometer

The *germanium bolometer* doped with gallium (Ga) is shown schematically in Fig. 7.39. The idea is to measure a slight change in resistance due to a rise in

Table 7.8 Infrared solid-state devices

Spectral range	Depends on material and dopant
Threshold	One quantum
Dynamic range	10^4–10^7 photocharges
Quantum efficiency	Generally > 0.5
Temporal response	≤ 1 ms
Integration time	About 10 min if limited by thermal generation
Readout noise	About 10 ms if limited by thermal background
Storage	10–500 e$^-$ depending on device
Format	$1\,024 \times 1\,024 \rightarrow 4\,096 \times 4\,096$ (4k × 4k)
Operating temperature	77 K ($\leq 2.5\,\mu$m) → 4 K (long λ)
Special features	Addressable pixels

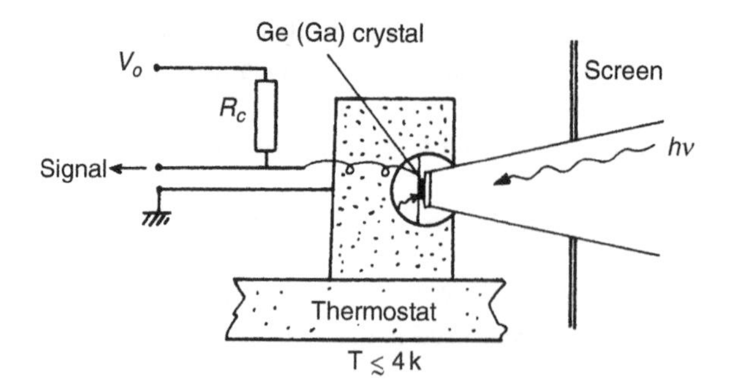

Fig. 7.39 The germanium bolometer (Ge:Ga). The resistivity of the semiconductor crystal varies rapidly with temperature ($\rho \propto T^{-2}$). A blackened shield efficiently (~100%) absorbs the incident photons. The cross-section is improved by surrounding the crystal by a reflecting sphere. The crystal (of heat capacity C) is linked to the thermostat by wires (of heat conductance G). The time constant is $\tau \sim C/G$. A bias V_0 is applied across the resistor R_C. The variation of the resistance ΔR_B of the bolometer, caused by the heating, is transformed into a signal $\Delta V = \Delta R_B/R_C + R_B$. The solid angle within which radiation is received is controlled by a low temperature screen

temperature of the material. Developed in the 1960s by F. Low,[9] it covered the near and mid-infrared (infrared photometry or spectrophotometry) and led to many discoveries. A semiconductor crystal is used here, cooled to 2–4 K by liquid helium.

The bolometer can be considered to comprise two thermally coupled parts which are to be optimised separately. One is a radiation absorber, and the other a resistance serving as thermometer. The noise in this type of bolometer is thermal, and corresponds to its operating temperature T (Johnson noise of the resistance and phonon noise leading to thermal energy). Noise equivalent powers from NEP $\sim 10^{-15}$ W Hz$^{-1/2}$ at $T = 2$ K to 10^{-16} W Hz$^{-1/2}$ at $T = 0.3$ K are achieved with this type of device.

This conventional bolometer has witnessed considerable improvements.[10] An example is the array equipping the European space mission Herschel (launch 2008) to cover the spectral region 60–210 μm. This array contains 2 560 silicon bolometers, arranged into 16×16 subsystems and operating at a temperature of 0.3 K, obtained by a ^{3}He dilution refrigerator. The NEP reaches 10^{-16} W Hz$^{-1/2}$, comparable with the noise induced by the thermal background in the observed beam étendue, whence there is no point in attempting further improvements. The signals, produced by individual amplifiers, are multiplexed in order to reduce the number of connecting wires.

---

[9]Frank James Low was an American solid-state physicist and astronomer, born in 1933, who pioneered infrared astronomy. He discovered the emission of the Kleinmann–Low nebula in Orion. See fr.wikipedia.org/wiki/Frank-Low.

[10]See Billot, N. et al. Recent achievements on the development of the Herschel/PACS bolometer arrays, arXiv:astro-ph/0603086 (2006).

Superconducting Bolometer

Here a superconducting material such as an Nb–Si alloy is held very close to the critical temperature at which the resistivity depends sensitively on the temperature. If energy in input by absorption of incident radiation, the resistance is altered. The resulting change in the current induces a magnetic field which is measured by a *superconducting quantum interference device* (SQUID). The operating temperature is close to 0.1 K, obtained using a ^{3}He refrigerator. These cooling systems can operate in closed circuit, and provide an elegant solution when long lifetimes are required for space missions, without depending on a cryogenic fluid which would eventually be used up. The NEP can be of the order of 10^{-17} W Hz$^{-1/2}$, and improves as the temperature is lowered.

One difficulty lies in coupling the bolometer surface with the incident radiation. Indeed, it must be similar in size to the wavelength, while having a heat capacity, which determines the time constant of the bolometer, as low as possible. Finally, it is desirable to construct $N \times N$ bolometer arrays to obtain images efficiently.

The 10.4 m telescope of the *Caltech Submillimeter Observatory* (CSO)[11] in Hawaii is equipped with a linear array of 24 thermal (Si) bolometers operating at a temperature of 0.3 K obtained by a ^{3}He dilution refrigerator. Atmospheric windows at 0.35, 0.45, and 0.87 mm are accessible in broad band photometry ($\Delta\lambda/\lambda \sim 0.1$). The CSO is also equipped with a hexagonal bolometer array (11×11) operating at 1.2 and 2.1 mm (BOLOCAM).

The European space mission Planck, launched in 2008, is designed to study the cosmological blackbody radiation and its spatial structure. It seeks maximal sensitivity and imaging capacity at submillimetre wavelengths, using a 1.5 m telescope cooled to 60 K. The *High Frequency Instrument*[12] (HIFI) uses 4×4 and 8×8 bolometer arrays cooled to 0.1 K to cover the range 100–857 GHz in six bands at low spectral resolution (~ 0.3). Cooling is carried out in several stages: hydrogen desorption, then Joule–Thomson expansion to reach 4 K, then ^{3}He dilution. The NEP can reach around 10^{-17} W Hz$^{-1/2}$.

In less than three decades, the spectacular progress made in bolometer design has made them invaluable submillimetre detectors, both on the ground and in space (see Fig. 7.40).

Hot Electron Bolometer

In a very pure material like InSb ($\sim 3 \times 10^{13}$ carriers/cm^3) at low temperatures (~ 4 K), a fraction of the conduction electrons remain free and interact weakly with the crystal lattice. Photon absorption by electrons raises their temperature and quickly modifies the resistivity of the material. This is photoconductivity by free carriers, exploited in the *hot electron bolometer* (HEB).[13] However, the resistivity is

---

[11]This telescope started operations in 1988 at the top of Mauna Kea. Details can be found at www.submm.caltech.edu/cso/.

[12]See www.rssd.esa.int/SA/PLANCK/include/payl/node7.html for a description.

[13]The idea of the hot electron bolometer was originally put forward by Kinch M.A. and Rollin B.V., Br. J. Appl. Phys. **14**, 672, 1963.

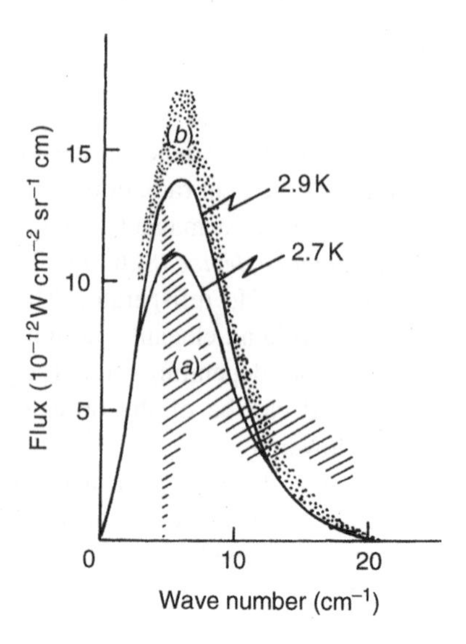

Fig. 7.40 A particularly difficult measurement using a bolometer: the continuum emission of the 2.7 K cosmological background near its maximum. The 2.7 K and 2.9 K Planck functions have been plotted with the following measurements: (**a**) using a 0.3 K bolometer (NEP $\sim 10^{-15}\,\mathrm{W\,Hz^{-1/2}}$), Woody and Richards measured the spectrum from a balloon at altitude 40 km; (**b**) the result of a similar measurement, made by Gush from a rocket between 150 and 370 km altitudes. The *shaded areas* correspond to the error bars given by the authors. (After Weiss R., ARAA **18**, 489, 1980. Reproduced with the kind permission of the *Annual Review of Astronomy and Astrophysics*, © 1980, Annual Review Inc.) These historical results have been considerably improved by the COBE satellite, since 1990, then by the BOOMERANG and WMAP missions at the beginning of the 2000s

proportional to the incident energy, hence also to the square of the electric field, whence the system functions as a *mixer*, and is frequently used as such in the radiofrequency range (see Sect. 7.5).

X-Ray Bolometer

At the high energies characteristic of X rays (> 0.1 keV), the energy delivered is enough to produce a temperature pulse in the bolometer, indicating the arrival of individual photons. The amplitude of the pulse can be used to determine the energy. This is therefore a detector with an intrinsic spectral selectivity, invaluable for spectroscopic applications. Since the energy threshold and spectral resolution are proportional to the volume of the detector, small detectors are favoured. This is also advantageous in imaging, which requires small pixels. On the other hand, detection efficiency depends on the thickness of the detector.

A diamond crystal of several mm^3, cooled to below 0.1 K, is thermally coupled to a germanium component, whose conductivity varies sensitively with temperature (Fig. 7.41). If X-ray photon arrivals are separated by time intervals greater than the time constant of the bolometer, which is of millisecond order, individual events

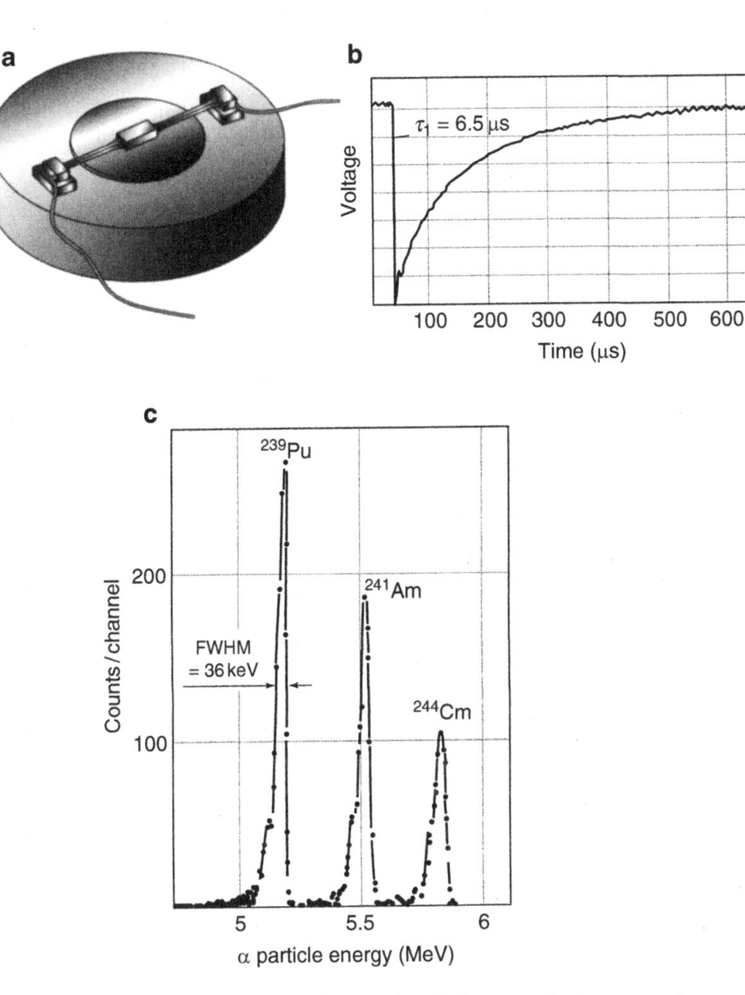

Fig. 7.41 (**a**) Schematic view of a doped germanium bolometer. At the centre, the germanium resistance, stuck on the absorbing diamond disk. (**b**) Display of the time constant of the bolometer, after absorbing the energy of a photon (or any other absorbed particle). (**c**) Spectrum for emission of α particles by a radioactive material, with $T_{\mathrm{bol}} = 1.3$ K. Data kindly provided by Coron N., 1985

can be discriminated. An energy resolution of a few eV can be achieved with these detectors in the laboratory at the time of writing (2007).

7.5 Astronomical Detectors: Radiofrequencies

The principle of radiofrequency detection using *non-linear elements* as *mixers*, with a frequency change, then detection, was first developed at metre wavelengths, and then extended gradually down to centimetre, millimetre, and finally submillimetre wavelengths, at which point it joined up with quantum and thermal detection techniques (photoconductors and bolometers, respectively, which were discussed earlier). Frequency-changing techniques are sometimes successfully applied in infrared spectroscopy, using photoconductors as mixers (see Sect. 8.4).

In this section, we begin by discussing concepts which apply across the whole radiofrequency region, before examining some practical realisations and the performance of detection systems they use.

7.5.1 General Features

Here we discuss how to convert the incident electric field into current, either directly or by frequency change, then describe the noise characteristics affecting the signal produced by the measurement, taking into account the thermal nature of the received radiation, both from the source and from the environment of the telescope.

The Conversion of Field into Current

The wave, focused at the focal point of a telescope, enters a *horn*, which matches the vacuum impedance to that of a *waveguide*. The latter selects a polarisation. The wave is guided into a *resonant cavity*, which defines a passband $\Delta\nu$, through its selectivity $Q = \nu/\Delta\nu$, close to the frequency ν of the radiation (Fig. 7.42).

Received Power. Antenna Temperature

Consider first the simple case of a thermal source of radiation, of emissivity ε_ν and temperature T. The power reaching the non-linear element is given by

$$\frac{\mathrm{d}P_\nu}{\mathrm{d}\nu} = \frac{1}{2}\varepsilon_\nu B_\nu(T)\eta_\nu \mathscr{E} ,$$

where $\mathscr{E}$ is the beam étendue sampled, and the factor of $1/2$ comes from the selection of one polarisation. η_ν denotes the total transmission, and $B_\nu(T)$ is the Planck function. It is assumed that the arrangement is as in Fig. 7.42a above.

It was shown, in Chap. 6, that the beam étendue $\mathscr{E}$ is limited by the coherence of the radiation, and must be taken as equal to λ^2. In other words, for a given telescope area S, the field on the sky, which is imposed by coherence, is given by the solid angle $\omega = \lambda^2/S$. Any radiofrequency receiver is intrinsically limited to analysis of a single pixel, with field equal to ω. (In aperture synthesis, the receiver is limited to one 'Fourier space pixel', rather than one pixel in the image plane.)

The power per frequency interval is then

$$\frac{\mathrm{d}P_\nu}{\mathrm{d}\nu} = \frac{1}{2}\varepsilon_\nu \eta_\nu \lambda^2 B_\nu(T) ,$$

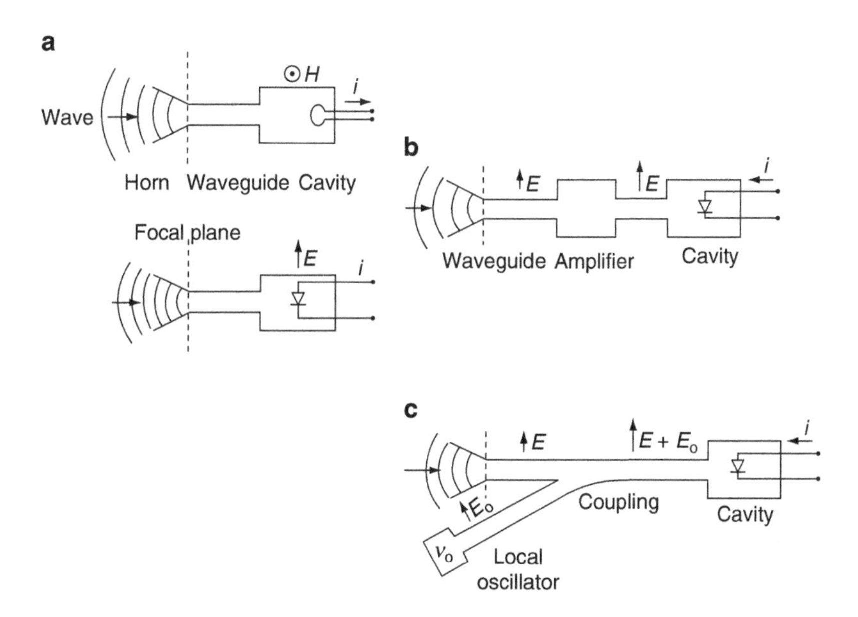

Fig. 7.42 Three receiver configurations for radio waves. In the first arrangement (**a**), a coil, normal to the magnetic field vector for example, is excited by induction and transforms the field into a current, which is transported by coaxial cable. A variation on this arrangement consists in directly exciting the non-linear element by the electric field. In the second arrangement (**b**), an *amplifier* is placed on the waveguide axis and amplifies the wave before it excites the coil, or is detected. In the third arrangement (**c**), a *local oscillator*, of frequency ν_0, superposes its field on that of the incident signal, by means of a *coupler*. The summed field is then applied to the non-linear element. An amplifier can be incorporated into this arrangement

and, in the case of blackbody radiation ($\varepsilon_\nu = 1$), in the Rayleigh–Jeans approximation,

$$\frac{\mathrm{d}P_\nu}{\mathrm{d}\nu} = \eta_\nu kT \ .$$

The received power is then simply given by a temperature, known as the *antenna temperature*, which is independent of the transmission of the system.

If the source is:

1. optically thick at the wavelength considered,
2. at a temperature justifying use of the Rayleigh–Jeans approximation for the frequency considered ($h\nu \ll kT$),
3. sufficiently extended to fill the field of *lobe* ω of the instrument,
4. in local thermodynamic equilibrium,

then the antenna temperature T_A is also the physical temperature of the radiation from the source. In the case of a non-thermal source, or one with an arbitrary intensity function, the power received at the telescope is clearly given by

$$dP_\nu = kT_A(\nu)\,d\nu ,$$

where T_A is the antenna temperature measured at the frequency ν.

Noise Temperature

In practice, even when an instrument is pointed towards a region containing no sources of radiation, and at zero temperature, a non-zero current is measured, whose power is also non-zero. There are several causes of this residual signal:

- Residual thermal emission from the atmosphere (see Sect. 2.3.2), the telescope, and the waveguide.
- Contribution from the thermal emission of the ground, detected in the sidelobes of the diffraction pattern.
- Thermal noise generated in the detector itself (for example, fluctuations in the tunnelling current, in the AsGa diode).

The fields of these different sources are incoherent, and therefore add quadratically, so that the equation for the total noise power as the sum of the powers gives

$$T_{\text{noise}} = T_{\text{atmosphere}} + T_{\text{lobes}} + T_{\text{detector}} ,$$

even if the 'temperatures' do not correspond to actual physical temperatures. They merely represent the power from which the signal must be distinguished. The first two terms are highly dependent on wavelength, for the same is true of atmospheric emission and diffraction. They are usually around 100 K for Earth-based telescopes, and around 10 K for space telescopes. In the best of cases, the limit will be imposed by T_{detector}, also known as the *system temperature*, which cannot be less than the physical temperature of the detector, and indeed, it is generally much higher than this value.

Consider now each of the configurations in Fig. 7.42a, b, and c, and their total noise temperatures, excluding contributions from the antenna and atmospheric emission:

(a) $T_{\text{noise}} = T_{\text{quadratic detector}}$.
(b) $T_{\text{noise}} = T_{\text{amplifier}} + T_{\text{mixer}}/\text{gain}$.
(c) $T_{\text{noise}} = T_{\text{RF amplifier}} + T_{\text{mixer}}/\text{RF amplifier gain}$
$+T_{\text{IF amplifier}} + T_{\text{IF mixer}}/\text{IF amplifier gain}$.

Case (c) has been written in the most general case of heterodyne detection.

The critical factor in each case is the mixer, whose noise temperature varies from a few tens to a few thousands of degrees kelvin (Fig. 7.43). The key features will be discussed below.

Note that the best receivers approach the quantum limit (Problem 7.9). At frequencies below 100 GHz (centimetre or metre), the RF signal must be amplified, and this use of an amplifier reduces T_{noise} to between 50 K (6 to 21 cm) and

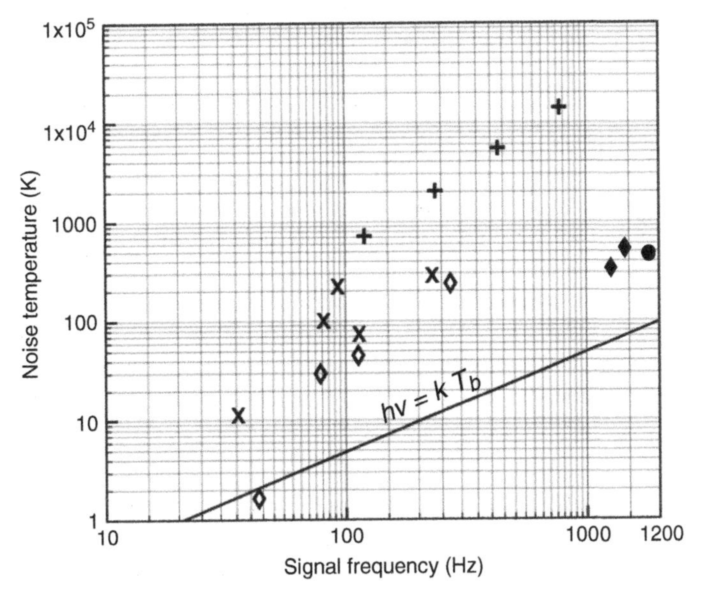

Fig. 7.43 Noise temperature of mixers used in radioastronomy. (+) Schottky diodes at 300 K, (×) cooled Schottky diodes, (◇) SIS junctions, (o) hot electron bolometer (HEB). All these detectors have an intrinsically wide passband (> 100 MHz), except the bolometer. The *straight line* indicates the limit $h\nu = kT_n$. (After Philips T.G., Woody D.P., ARAA **20**, 285, 1982. Reproduced with the kind permission of the *Annual Review of Astronomy and Astrophysics*, Vol. 20, ©1980, Annual Reviews Inc.) The initial values given in the figure ($T_n \geq 100$ K) are those of the mid-1980s. By the mid-1990s, little progress had been made in the region below 500 GHz, the greatest being a factor of 2 for SIS junctions. By the mid-2000s (*black symbols*), SIS junctions are still being improved, e.g., T_n in the range 50–150 K for the detectors of the 30 m IRAM telescope or those of the HIFI instrument on board the European mission Herschel (launch 2008) beyond 1 THz, whereas the hot electron bolometer takes this mission beyond 1 THz, viz., $\lambda = 330\ \mu$m. (Values for 2007 with the courtesy of G. Beaudin)

300 K (1.3 cm). These values are typical of receivers for the aperture synthesis radiotelescope known as the Very Large Array (VLA) in New Mexico.

By pointing the receiver alternately at the source and at a neighbouring point of the sky, assumed empty of sources, two values of the antenna temperature can be measured:

$$T_{\text{system}} + T_{\text{source}} \quad \text{and} \quad T_{\text{system}} \, .$$

On averaging, the difference of these gives the desired quantity, subject to fluctuations in each of the terms.

Minimum Detectable Power and Signal-to-Noise Ratio

Figure 7.44 shows the simplest possible detection scheme. It can be used at metre wavelengths for low signal frequencies (< 100 MHz). Following the antenna, the

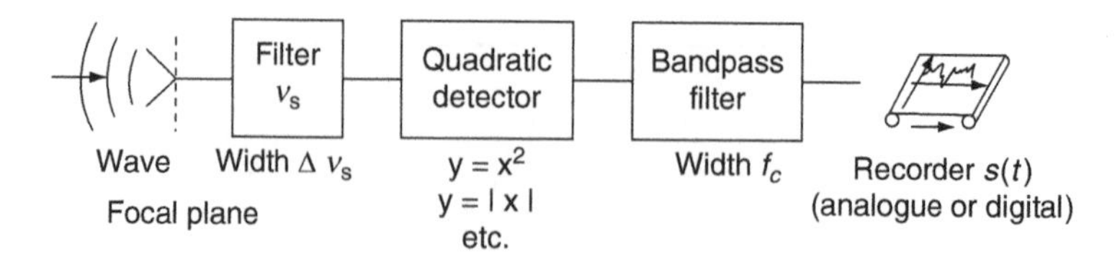

Fig. 7.44 Simple detection scheme (referred to as video detection) for the field of an electromagnetic wave

cavity, behaving as a bandpass filter, isolates a frequency band $\Delta\nu_s$ in the signal, centered on a frequency ν_s.

It is assumed, to simplify the calculation, that the non-linear element has the effect of transforming the signal $x(t)$ into the quantity

$$y(t) = x(t)^2 , \qquad \text{current } i(t) \propto E(t)^2 .$$

In practice, the transformation may be slightly different, for example,

$$y(t) = |x(t)| , \qquad \text{current } i(t) \propto |E(t)| ,$$

but the conclusions would be similar. It will suffice to consider the appropriate transformation of the random variable $\mathbf{x}(t)$, which is assumed to be a Gaussian process, with white spectral density, characterised by a noise temperature $T(\text{noise}) = T_n$. The current $i(t)$ is then filtered by a filter (Fig. 7.45) of passband (ε, f_c), which excludes zero. Finally, the signal is recorded in digital or analogue form.

The result giving the power spectrum of the square of a random process, which is described in Appendix B, can be applied, noting that $\Delta\nu_s \ll \nu_s$ (for example, in the case of a centimetre receiver, $\nu_s \sim 10^{10}$ Hz, and $\Delta\nu_s \sim 10^7$ Hz). The PSD of the signal is then given by the convolution

$$S_s(\nu) \simeq (2kT_n)^2 \Delta\nu_s \left[\delta\left(\nu - \frac{f_c}{2}\right) + \delta\left(\nu + \frac{f_c}{2}\right)\right] \star \Pi\left(\frac{\nu}{f_c}\right) ,$$

for $f_c \ll \Delta\nu_s$. Fourier transforming gives the autocovariance $C_s(\tau)$, and finally

$$\sigma_s^2 = C_s(0) = (2kT_n)^2\, 2\Delta\nu_s f_c .$$

The signal due to the source is obtained simply by putting $T_n = 0$,

$$\text{signal} = \langle s(t) \rangle = \langle E^2_{\text{source}}(t) \rangle = \sigma_E^2 = 2kT_{\text{source}} \Delta\nu_s ,$$

and the signal-to-noise ratio is then

$$\frac{2kT_{\text{source}}\Delta\nu_s}{2kT_n\,(2\Delta\nu_s f_c)^{1/2}} = \frac{T_{\text{source}}}{T_n}\left(\frac{\Delta\nu_s}{2f_c}\right)^{1/2} .$$

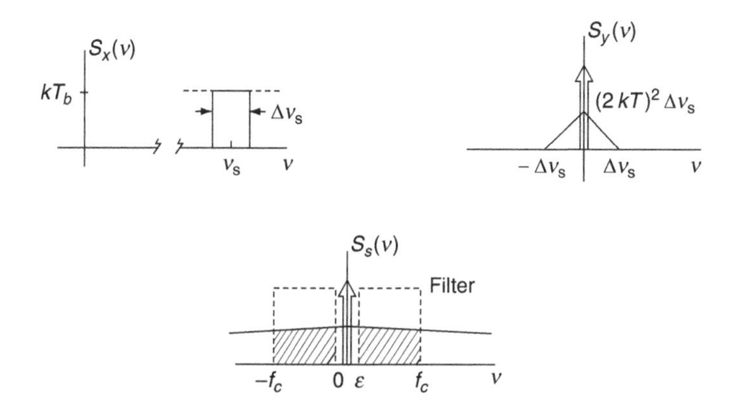

Fig. 7.45 Spectral densities: (**a**) of the filtered incident signal at ν_s, (**b**) after quadratic detection, only the low frequency part being shown, (**c**) after low frequency filtering, on an expanded scale. The bandpass filter cuts out zero frequency, and hence the impulse at the origin. Since $f_c \ll \Delta\nu_s$, the PSD is almost constant across the whole filter width

The bandpass f_c of the filter is related to the *integration time* τ by an equation of the form

$$f_c\tau = \alpha\ , \qquad \alpha \sim 1\ .$$

The exact value of the coefficient α depends on the *transfer function* of the filter used. The general rule stating that the signal-to-noise ratio only increases as the square root of the measurement time is once again satisfied. Moreover, this ratio only increases as the square root of the passband $\Delta\nu_s$. Indeed, when $\Delta\nu_s$ increases, the signal increases, but then so does the noise power.

The unwanted contributions from atmospheric emission and instrumental noise can be eliminated by pointing the antenna at two neighbouring points of the sky, and taking the difference, so that the minimum detectable power is, for a signal-to-noise ratio equal to one,

$$P_{\text{min}} = kT_{\text{n}}\left(\frac{2f_{\text{c}}}{\Delta\nu_{\text{s}}}\right)^{1/2}\Delta\nu_{\text{s}}\quad [\text{W}]\ .$$

In the model considered here, it was assumed that the same filter $\Delta\nu_s$ was applied to all components of $i(t)$, namely, the signal due to the source, the signal from instrumental emission, and the signal due to the fluctuations of the detector. If this were not the case, it would be a simple matter to modify the argument.

7.5.2 Heterodyne Detection

This is the method now most commonly used on radiotelescopes, both on the ground and in space, from the submillimetre to the metre wave region. It was presented briefly in Fig. 7.42c. Here we shall discuss the different components required, together with the underlying physical principles and the way they are implemented.

Figure 7.46 gives a more detailed overview of the successive operations to which the incident signal is subjected. Its frequency is first changed by superposition, without loss of information, at a quadratic detector, of the signal field at frequency ν_s, and the field of a *local oscillator* at frequency ν_0. The current output by the *mixer* is then, up to proportionality,

$$i(t) \propto E_0(t)E_s(t) .$$

It therefore depends linearly on the field $E_s(t)$, and has a component at the frequency $(\nu_0 - \nu_s)$. This current, which is called the *intermediate frequency* (IF) *signal*, is then amplified and filtered, before being detected again and averaged by the exit filter of width f_c. The first advantage of this setup (Fig. 7.46) is the possibility of precisely controlling the filtering after the IF stage, which allows a spectral analysis of the signal. The second advantage is the amplification of the signal due to the mixing of the two frequencies. If the local oscillator produces a strictly coherent field, the current $i(t)$ can be made large compared with all noise sources within the receiver, leaving only those fluctuations originating in the signal itself. This argument fails for wavelengths at which *quantum noise* in the local oscillator can no longer be neglected. In the end, the noise properties of the system as a whole are determined by those of the amplifier(s) and the mixer. The spectral properties are determined by the feasibility and stability of the local oscillator, the operating frequency and frequency bandwidth of the mixer, and then the properties of the spectral analysis stages to be described later (see Sect. 8.4).

The mixer does not in itself differentiate between two signal frequencies which are symmetrical about the frequency ν_0 of the local oscillator. This confusion can perturb the measurement (analysis of a spectral line), in which case a filter can be placed before the mixer to block frequencies $\nu < \nu_0$. This is referred to as *single sideband reception* or SSB reception, as opposed to *double sideband reception* or DSB reception. In the latter case, the signal-to-noise ratio is clearly reduced by a factor of $\sqrt{2}$ (Problem 7.8), but the spectral coverage is doubled.

The intermediate frequency stage, after the mixer, and the ensuing stages are independent of ν_s and ν_0. They depend only on the difference $\nu_0 - \nu_s$. The heterodyne setup is thus applicable at all wavelengths, provided that one has a suitable local oscillator, amplifier, and mixer for the chosen frequency, sensitivity, and spectral characteristics of the source under investigation.

Local Oscillator

The role of this subsystem is to provide a coherent signal, hence with stable frequency, without noise, at a frequency ν_0 very close to the frequency ν_s of the signal. In addition, the power supplied must be large enough to ensure if possible that the signal after mixing exceeds the noise of the mixer. The three requirements

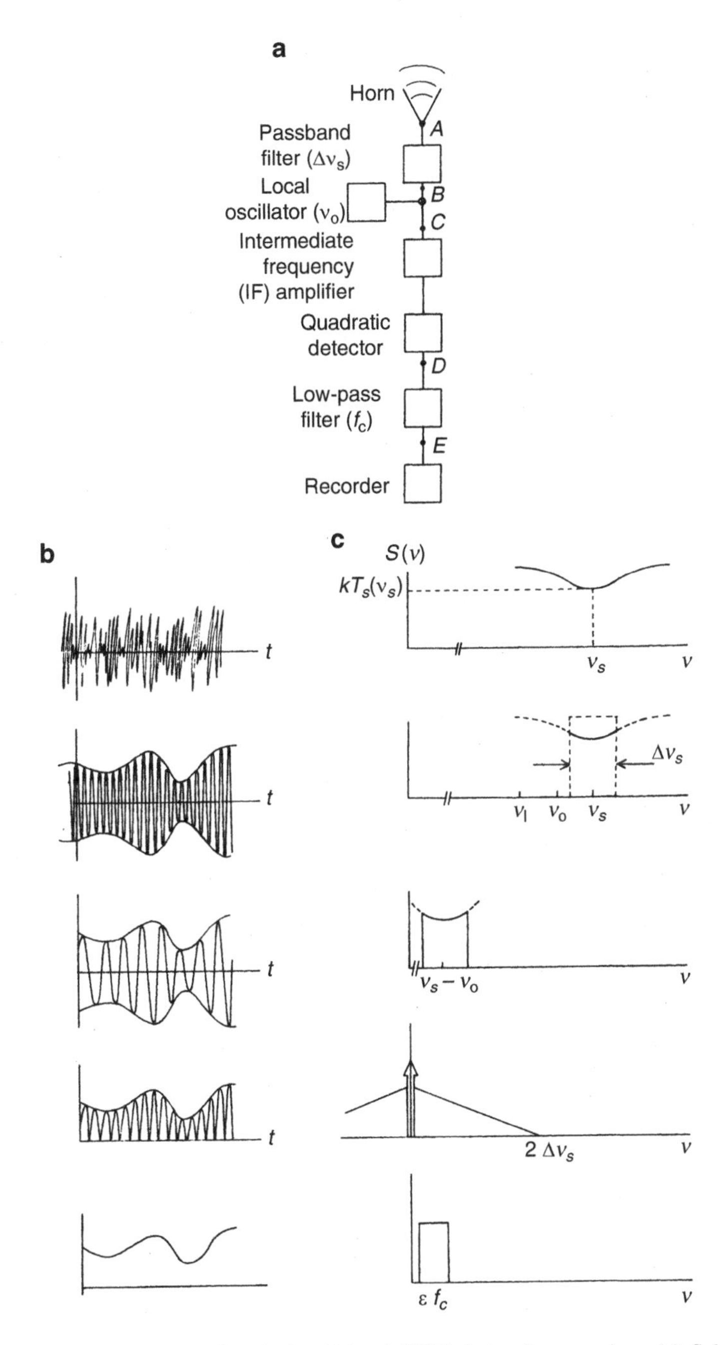

Fig. 7.46 Successive stages of a single sideband (SSB) heterodyne receiver. (**a**) Subsystems. (**b**) The time dependent signal. (**c**) Spectral density of the signals. Note the image frequency ν_I eliminated by the filter $\Delta\nu$

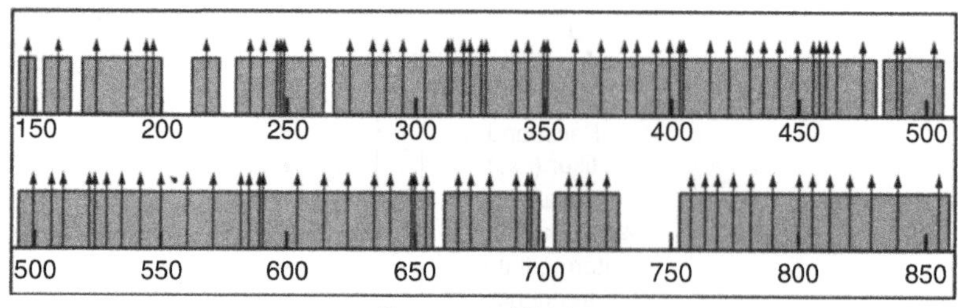

Fig. 7.47 Frequency regions covered by molecular lasers between 150 and 850 GHz (0.35–2 mm) (frequencies indicated by *vertical arrows*) and mixers whose bandwidth would be ±5 GHz. The *shaded area* shows frequencies that can be reached by using these lasers as local oscillators. (After Roeser H.P. et al., Astron. Astrophys. **165**, 287, 1986)

made of a local oscillator are thus: matched frequency, sufficient power, and absence of noise.

Let us mention briefly two types of local oscillator which played an important role in the development of radioastronomy throughout the second half of the twentieth century.

The *klystron* is a wave generator that was used for $\nu_0 \lesssim 180$ GHz. Capable of supplying high power (25–100 mW), it was developed mainly for telecommunications and radar.

The *carcinotron* is a wave generator, working on the following principle. A beam of electrons, of the same kinetic energy, move parallel to a grill. Modulation of the grill potential creates a variable electric field, whose frequency is related to the grill spacing. This can emit a few tens of mW up to 1 000 GHz, and is a high quality source, but cumbersome and power-consuming, and impossible to send into space, for example.

Harmonic generators are *frequency doublers* or *multipliers*, using a non-linear element, such as a Schottky diode, excited by a klystron. The non-linear element then radiates the harmonics 2, 3, ..., of the excitation frequency. The power obtained can reach several mW, up to around 500 GHz.

Molecular lasers offer great spectral purity and stability, together with a very large number of lines, covering not only the submillimetre, but also the far and middle infrared (20 μm and beyond). The available power is from 0.1 to 1 mW. The main drawback is that they provide only spectral lines at discrete frequencies, which may not be close to the frequency ν under study (see Fig. 7.47).

Modern oscillators (2007) are *Gunn effect generators*. A solid-state diode is designed so that its current–voltage characteristic in a certain region has negative resistance ($\mathrm{d}V/\mathrm{d}I < 0$). With direct current, the diode begins to oscillate and the desired frequency ν_0 is obtained by placing the diode in a resonant cavity at this frequency, from which it is extracted by a waveguide. These diodes are made with semiconducting materials like GaAs (from a few GHz to 100 GHz), then gallium nitride (up to THz). Once they had become the standard local sources, they greatly facilitated the development of submillimetre astronomy, e.g., on the *California Submillimeter Observatory*[14] (CSO) in Hawaii, or the HIFI instrument on board the Herschel space mission.

[14] See www.submm.caltech.edu/cso/receivers.

Another possible configuration has a powerful amplifier after the diode, in fact a *high electron mobility transistor* (HEMT) described further below, then a frequency multiplier producing a harmonic at the desired frequency. Noise temperatures lower than 100 K can be achieved at 300 GHz, and lower than 1 000 K at 700 GHz.

Other methods for making oscillators are under development at the time of writing (2007), using Josephson junctions,[15] laser beats, up to very high frequencies (THz), or beyond and up to the mid-infrared, the *quantum cascade laser* (QCL).

In conclusion, we have presented a rather general radiofrequency detection setup, and we have shown that, in practice, a wide range of configurations is possible on this basis (see Table 7.9).

Amplifiers

In most radiofrequency detection systems, the non-linear element is a significant source of noise. It is useful, therefore, to amplify the signal of frequency ν_s, or the IF signal $\nu_{IF} = \nu_0 - \nu_s$, before detection.

The principle of the various types of amplifiers is as follows. Energy is extracted from a local radiation source and transferred to the incident wave, in proportion with the incident energy, and in phase with the incident field. A coherent amplification effect is obtained, and the signal-to-noise ratio is improved as long as the amplifier gain remains less than

$$g = \frac{T_n}{T_a} ,$$

where T_a is the noise temperature at the amplifier input, and T_n the noise temperature of the detection element.

Table 7.9 Radiofrequency reception. RF radiofrequency, IF intermediate frequency

Component parts	RF amplifier
	Local oscillator
	Non-linear RF element
	IF amplifier, filter(s), and correlator
	IF detector (transistor)
Geometrical properties	Single pixel, fixed étendue λ^2
	Small arrays ($N \leq 10$)
Polarisation	One polarisation detected
Spectral properties	Intrinsically wideband components
	Post-IF spectral selectivity
Noise	Close to the thermodynamic limit imposed by the system temperature
Temporal response	No particular limit

[15]In the Swedish *Submillimetron* project (2007), a long-lived autonomous module carrying a submillimetre telescope, serviced by the International Space Station.

The modern method of amplification, which can be applied either directly to the incident RF signal or to the intermediate frequency (IF) signal, uses so-called *high electron mobility transistors* (HEMT). These are field-effect transistors based on a heterojunction, i.e., two semiconducing materials with different bandgaps. For example, the ODIN space mission (Sweden, Canada, Finland, and France), launched in 2001 and still running in 2007, includes a 1.1 m radiotelescope and focal systems operating between 180 and 520 GHz, with IF amplifiers based on this idea.

Here we briefly mention some of the systems developed over the period 1950–2000.

The *parametric amplifier* exploits the following idea. Consider an oscillator represented, for example, by a circuit with a resistor R, an inductance L, and a capacitance C, in which one of the elements determining the resonant frequency, e.g., C, can be varied. Let ω_0 be the original frequency, and $2\omega_0$ the frequency of modulation of C. Assuming low damping, that is

$$Q = \frac{1}{\omega_0 RC} \gg 1 ,$$

the oscillator equation is

$$\ddot{x} + \frac{1}{\tau}\dot{x} + \omega_0^2(1 - \varepsilon \sin 2\omega_0 t)x = 0 ,$$

with

$$C = C_0(1 - \varepsilon \sin 2\omega_0 t) , \quad \varepsilon \ll 1 , \quad LC_0\omega_0^2 = 1 , \quad \tau = \frac{L}{R} .$$

Seeking a solution of the form

$$x = f(t) \sin \omega_0 t ,$$

it follows that

$$f(t) = K \exp\left(\frac{\varepsilon\omega_0}{2} - \frac{1}{2\tau}t - \frac{\varepsilon}{4}\sin 2\omega_0 t\right) \sim K \exp\left(\frac{\varepsilon\omega_0}{2} - \frac{1}{2\tau}\right) .$$

Amplification occurs if the two conditions

$$\varepsilon > \frac{1}{\omega_0\tau} \quad \text{and} \quad \varepsilon \ll 2 + \frac{1}{\omega_0\tau}$$

are simultaneously satisfied. This is indeed possible, when the selectivity Q is much greater than unity, as we assumed.

The system is only linear in a restricted amplitude range. An appreciable gain can be obtained. Power is transferred to the signal by modulation of the capacitance C. It therefore comes from the oscillator which modulates this capacitance at the frequency $2\omega_0$. In practice, the variable capacitance, sometimes called a *Varactor*, is a Schottky junction barrier (see below), whose capacitance depends on the applied voltage by the relation

$$C = C_0\left(1 - \frac{V}{V_0}\right)^{-n} ,$$

where $0.3 \lesssim n \lesssim 0.5$. The typical noise temperature of a cooled parametric amplifier is 10 to 20 K.

The idea of the *maser amplifier* is to use an external power source to populate the excited levels of a solid or gas beyond the population which would correspond to thermodynamic equilibrium. The stimulated emission produced by the incident radiation is in phase with that radiation, and in the same direction, so that *amplification* occurs.

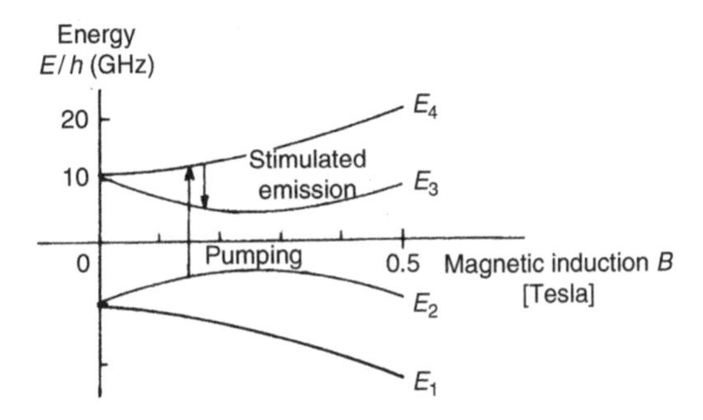

Fig. 7.48 Energy states in a ruby maser, as a function of the applied magnetic field

A *ruby maser* is a crystal of alumina Al_2O_3, in which 0.05% of the Al^{3+} ions are replaced by chromium ions Cr^{3+}. The combined effect of the Al_2O_3 lattice electric field and an external magnetic field creates four energy levels (Fig. 7.48) for the electron spin $S = 3/2$ of the Cr^{3+} ion.

Magnetic dipole transitions are possible between these levels. The maser can be pumped by a local oscillator (Fig. 7.48) at the frequency $(E_4 - E_2)/h$, and it will then emit the energy $(E_4 - E_3)$ (stimulated emission). The crystal must be cooled so that the differences of population are significant:

$$\frac{E_3 - E_1}{h} \quad (\text{at } B = 0) \quad = 11.46 \text{ GHz}$$

corresponds to a temperature

$$T = \frac{E_3 - E_1}{k} = 0.55 \text{ K} .$$

The typical noise temperature of a maser amplifier lies between 15 and 50 K.

The Rydberg maser uses states close to the ionisation continuum of the hydrogen atom. The energy difference between two of these states is given by

$$\Delta E = R\left[\frac{1}{n^2} - \frac{1}{(n+1)^2}\right] \sim Rn^{-3} ,$$

for $n \gg 1$. For $n \sim 30$, the energy difference corresponds to millimetre wavelengths. Electrical dipole transitions between these levels are easy to excite, and the amplification principle is as before.

Mixers

We said above that there were two key factors in the design of a mixer, i.e., the spectral bandwidth B it accepts and the noise temperature that characterises it. All things being equal, the desired value of B increases with the frequency ν_s of the incident signal. This observation means that simpler mixing techniques are possible at frequencies below 80 GHz ($\lambda \gtrsim 4$ mm). Today, HEMT transistors can be used, operating at ever higher frequencies and able to amplify then detect IF signals, while maintaining a sufficiently broad spectral bandwidth B.

At higher frequencies, viz., millimetre and submillimetre, the predominant mixer is the SIS superconducting junction and the hot electron mixer (HEB). The Schottky diode is still used, but performs less well with regard to noise.

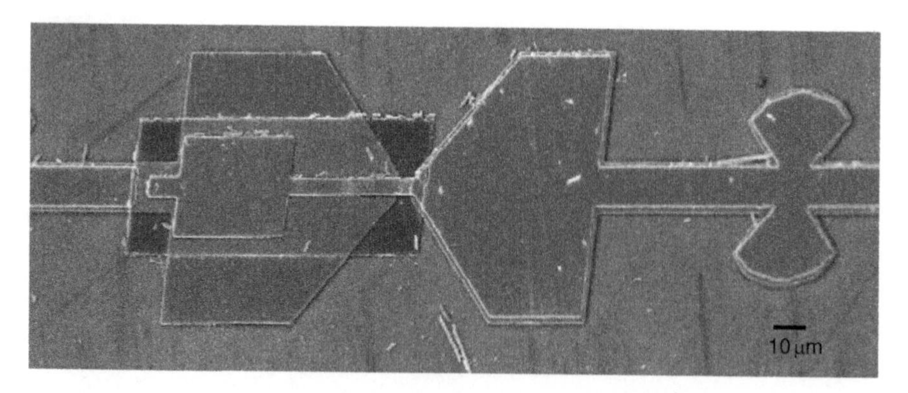

Fig. 7.49 Electron microscope image of a SIS junction made at the Paris Observatory (LERMA) for the frequency band 385–500 GHz of the European telescope ALMA in Chile. Document courtesy of LERMA

SIS Junction

The underlying principle of the superconductor–insulator–superconductor (SIS) junction has already been mentioned in Sect. 7.3. The circulation of a quasiparticle current in the insulator is a non-linear effect, as can be seen from the current–voltage characteristic of the junction shown in Fig. 7.17. It can thus be used as a frequency mixer.[16] Since the interaction between the radiation and the material occurs by quanta, the theoretical noise temperature limit is very low, viz., $T_0 = h\nu/k$. At 100 GHz, we have $T_0 = 5$ K, whereas the best diodes reach $T_n \approx 20$ K, and at 1 THz, $T_n \approx 1\,000$ K. This favourable factor means that a low power local oscillator can be used, and it can even be integrated into the diode. However, the intrinsic capacitance of the geometry (see Fig. 7.49) does not authorise a use exceeding the THz range.

The high frequency behaviour depends on the product of the resistance and the capacitance of the junction. At the time of writing (2007), Nb(niobium)–AlO_2(alumina)–PbBi junctions are used up to 1.2 THz ($\lambda = 0.35$ mm).

SIS junctions are small and can thus be arranged in small format 2D arrays to make multipixel image detectors at radiofrequencies. For example, the 30 m IRAM telescope is equipped with a 3×3 array, fed by juxtaposed horns (which reduces the filling factor), and the James Clerk Maxwell Telescope (JCMT) in Hawaii has a 4×4 array, operating at $\lambda = 800$ μm.

Hot Electron Mixer

We have already seen how the free electron gas in a very pure InSb semiconductor is heated by incident radiation, modifying its resistivity when this energy is transferred

[16]The mixing effect of a quasiparticle current is described by Richards P.L. et al., Appl. Phys. Lett. **34**, 345, 1979.

to the crystal lattice, and thereby constituting a *hot electron bolometer* (HEB). However, this change in resistivity, proportional to the incident energy, is thus proportional to the square of the electric field incident on the material, whence it can function as a *mixer*.

In this case, a nanoscale superconducting bridge made from niobium nitride (NbN) and connecting two gold electrodes[17] The signal (source and local oscillator) is focused on this bridge by a lens or quasi-optical waveguide. Heterodyne mixing results from the purely resistive transition between the superconducting and normal states, induced when the RF signal heats the material. A broad IF bandwidth (up to $B = 2$ GHz) is obtained by reducing the volume of the NbN film to give the lowest possible thermal time constant. A noise temperature of $T_\mathrm{n} = 750$ K is obtained at 1.5 THz, rising to 3 000 K at 3 THz in double sideband (DSB) mode.

Schottky Diode

This non-linear component, also called a *Schottky barrier*,[18] was long considered the ideal radiofrequency mixer. It has gradually been superseded by detectors coming closer to the quantum noise limit, such as those we have just described. However, it remains widely used when the signal is strong enough not to be significantly degraded by the mixer noise. This is the case, for example, when a millimetre wave instrument is used to observe the Earth atmosphere from space, or the surface of Mars from a probe in orbit around the planet. The signal, similar to a blackbody at around 250 K, is then strong enough.

The aim here is to outline the effects of contact between a metal and a semiconductor. Figure 7.50 illustrates the phenomena resulting from such a contact. In equilibrium, the majority charge carriers in the semiconductor move away from the contact area, in such a way that the Fermi levels in the metal coincide with those in the semiconductor. A *barrier*, or *depletion zone*, emptied of majority charge carriers, appears in the semiconductor. A simple calculation shows that the width of this barrier is given by the expression

$$x_\mathrm{b} = \left(\frac{2\varepsilon |V_0|}{Ne} \right)^{1/2} ,$$

where ε is the permittivity of the material, V_0 is the difference between Fermi levels, and N is the carrier number density. For $\varepsilon = 16\varepsilon_0$, $V_0 = 0.5\ V$, $N = 10^{16}\ \mathrm{cm}^{-3}$, this gives $x_\mathrm{b} = 0.3\ \mu\mathrm{m}$.

The *current–voltage characteristic* of the junction or diode is shown in Fig. 7.51. In the absence of any applied voltage, a current may nevertheless cross the junction, either because the thermal energy of the electrons is large enough to allow them to cross the potential barrier, or because, having insufficient energy, they still manage to cross it by *tunnelling*. At low temperatures ($T \ll 300$ K), this latter effect dominates, and the characteristic can be written

[17] See www.sron.nl/index.php?option=com-content&task=view&id=44&Itemid=111.

[18] The German physicist Walter Schottky (1886–1976) was one of the first to study the electronics of semiconductors and build devices with them.

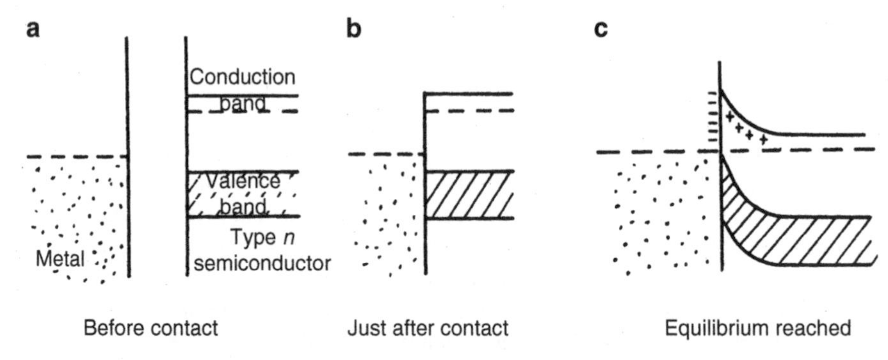

Fig. 7.50 Contact between a metal and an n-type semiconductor. The *dashed line* represents the Fermi level in each material. The *ordinate* is the electrostatic potential

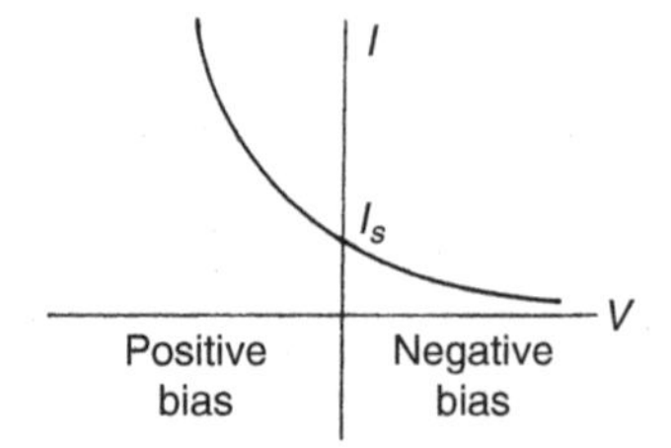

Fig. 7.51 Current–voltage behaviour of a metal/n-type semiconductor. A positive voltage increases the potential barrier, and the current is decreased

$$I \sim I_s \exp\left(\frac{eV}{E_\infty}\right) ,$$

where I_s is the saturation current, and $E_\infty \propto x_b^{-1} \propto N^{1/2}$.

If Δf is the frequency passband of the filter which follows the diode, the current i will produce a fluctuation (see Sect. 7.2)

$$\langle \Delta i^2 \rangle = 2ei\Delta f ,$$

and the corresponding noise power is (Problem 7.2)

$$P = \frac{R\langle \Delta i^2 \rangle}{4} = \frac{\mathrm{d}V}{\mathrm{d}I}\frac{2ei\Delta f}{4} .$$

This power can be characterised by a *noise temperature*, using the power spectrum of thermal noise (see Sect. 7.2). It follows that

$$P = kT_\mathrm{d}\Delta f = \frac{1}{2}\frac{\mathrm{d}V}{\mathrm{d}I}ei\Delta f ,$$

implying

$$T_\mathrm{d} = \frac{E_\infty}{2k} .$$

The noise temperature T_d characterises the intrinsic quality of the diode.

In practice, the semiconductor used is highly doped gallium arsenide (AsGa). A thin layer of lower dopant concentration is formed on the surface by *molecular epitaxy*, and this surface is then put into contact with the metal.

Epitaxy is the name for the process which consists in growing a thin layer of different composition on the surface of a monocrystalline structure, in such a way that the layer has the same crystal structure as the substrate. This process is commonly used in microelectronics, for example, to deposit doped silicon onto pure silicon. The layer is deposited by condensation of a vapour, or crystallisation of a liquid.

The drive to reduce noise requires low operating temperatures ($\lesssim$ 20 K), and low dopant levels in the epitaxial layer, in order to reduce tunnelling. Moreover, the thickness of this layer must not exceed that of the depletion zone, for this would introduce extra resistance.

There is a high frequency limit to the response of these diodes. Beyond 300 GHz ($\lambda < 1$ mm), the intrinsic capacitance of the diode creates a resistance–capacitance filter which reduces the response. Certain processes have been devised to reduce this capacitance to 1.5×10^{-15} F, thereby extending the domain of operation up to 700 GHz.

Wide Band Receiver

Heterodyne detection has intrinsic spectral selectivity by spectral analysis of the IF signal (see Chap. 8). However, the total received power, limited only by the passband B of the mixer, can be measured directly at the mixer output, hence with very low spectral selectivity. An example is shown in Fig. 7.52.

7.5.3 *The Diversity of Radioastronomy*

Radiofrequency detection covers a vast spectral region which today links up with the infrared via the submillimetre region. Needless to say, this involves a wide range of different technologies for the components used in observing systems. Table 7.9

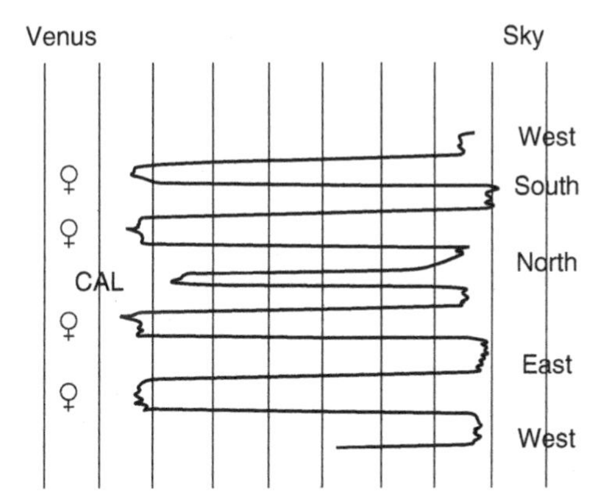

Fig. 7.52 Wide band receiver. A measurement of the power radiated by the planet Venus at $\lambda = 6$ cm, with the 42.6 m radiotelescope at Kitt Peak (Arizona). The planet is alternated with the sky background in its neighbourhood (E, W, N, S). CAL denotes a calibration (Sect. 3.5.1). The selected polarisation is in the N–S direction. The size θ of the planet is small compared with the field of coherence, and the deviation is proportional to $\int_{\Delta\nu} B_\nu(T_{\text{venus}})\, \pi\theta^2\, \mathrm{d}\nu/4$

sums up the basic properties of detectors, on the basis of which many different observational combinations are possible.

As in other wavelength regions, the observing system couples an Earth-based or spaceborne telescope, sometimes part of an array (see Sect. 6.5.1), with these receivers, associated with a spectroscopic capacity to be examined in detail in Chap. 8. These different components are arranged and combined as dictated by the desired observation and the given astrophysical problem. By visiting the websites of radioastronomy observatories, the reader will be able to confirm the extreme diversity and wealth of possible combinations, put together in each case to optimise the observation. Here are some examples:

- Detection of a very faint extragalactic radio source, using a large collecting area, a spectral band $\Delta\nu_s$ as wide as possible, and an integration time as long as possible.
- Analysis of the very rapid variations of a pulsar, in order to determine the period, the spectral band again being wide, but the integration time short (less than one millisecond).
- Analysis of solar flares, in which the rapid variations of the emission depend on the frequency, and high spectral selectivity and temporal response must be achieved, whilst maintaining the signal-to-noise ratio.
- Analysis of the spectral line profile of an interstellar molecule, which requires very high spectral resolution ($\nu_s/\Delta\nu_s \sim 10^5$), using the kind of spectrometer described in more detail in Sect. 8.4.
- Measurement of the angular size and imaging of a radiosource. This requires the use of an interferometer array, with comparison of the phase of a wave at two spatially separated receivers, in order to obtain information about the size of the source. The spectral resolution is not necessarily high, unless the emission is spectrally narrow, e.g., an emission line.

Figure 7.53 shows schematically the various configurations that can be associated with such problems. A fourth dimension could be added, to show the frequency of the received signal.

7.6 Observing Systems for Gamma-Ray Astronomy

In this high energy region, which we shall define as having a lower limit of the order of 10 keV ($\lambda \sim 0.1$ nm), continuing on from the X-ray region, the notions of telescope, detector, and spectrometer considered up to now to be independent entities tend to merge into a single system which fulfills all these functions. This is mainly because, in contrast to other wavelengths, there are no mirrors available that could form images from incident γ radiation. The lower limit, clearly fixed in the years 1970–2000 at around 10 keV by the limits to the possible use of mirrors, is tending to move toward ever higher energies, and will no doubt exceed a few hundred keV in the near future, while the X-ray region moves up to higher energies

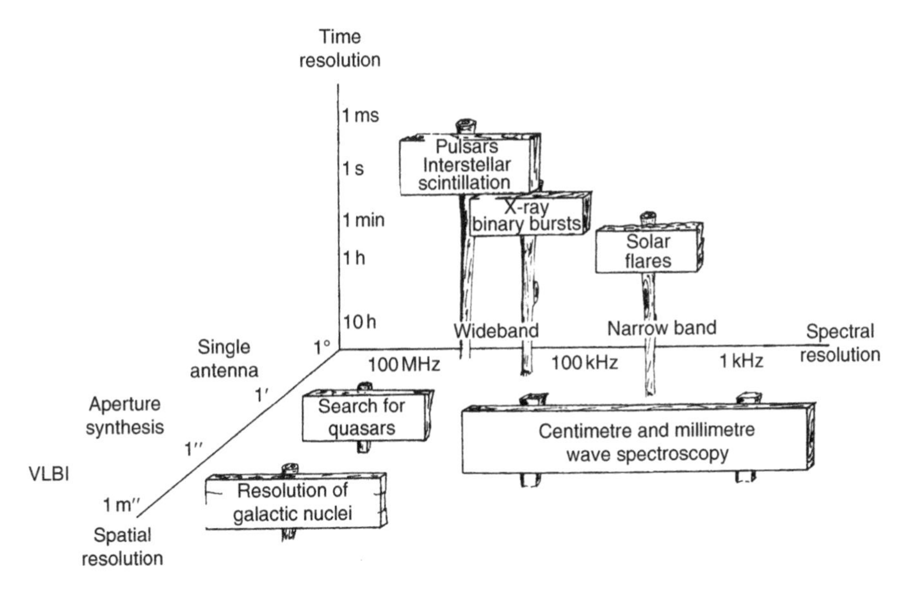

Fig. 7.53 The diversity of observing combinations in radioastronomy. Domains of application are only given up to an order of magnitude. A fourth dimension is needed to represent frequency, but each subject identified is treated successively over the whole frequency range from 1 THz to at least 100 MHz

by means of its own techniques. This illustrates the somewhat arbitrary but merely practical nature of the way the electromagnetic spectrum has been divided up for the purposes of this book. At the very highest energies, the γ ray region continues to move upwards, and now extends beyond 10 TeV.

In Chap. 5, we presented certain features of image formation at γ ray wavelengths which tended rather to resemble those of conventional telescopes. Here we bring together the notions which for other wavelength regions have been distributed throughout the first part of this chapter on detectors and Chap. 8 on spectrometers. In addition, we shall retain the term *telescope*, which is commonly used to refer to the whole system receiving, detecting, and analysing the γ radiation energy from astronomical sources. Such systems began to appear naturally enough at the beginning of the 1960s, in particle or nuclear physics laboratories which have their own traditions of instrumentation. In particular, since the 1990s, a field known as *astroparticle physics* has come into being, bringing together the problems of high energy astrophysics, where the relevant mechanisms involve the nuclei of atoms rather than their electronic shells. Instrumentation for observing astronomical sources emitting at such energies has also been developing at a considerable rate.

It should be said, however, that, despite immense progress accomplished since the 1970s in the γ ray region, angular resolutions remain less good than in other spectral regions at the time of writing (2007), and most sources must still be considered as unresolved by today's instruments. But considerable progress has been made in levels of sensitivity for the detection of faint sources, spectral

Table 7.10 Detection in γ ray astronomy

Interaction	Type of detector	Energy range	Imaging
Photoelectric absorption	CdTe crystal	10 keV–1 MeV	Coded mask and segmented imager (10^4 pixels)
	Scintillators	100 keV–10 MeV	Coded mask and segmented imager (10^3 pixels)
	Ge crystal	10 keV–10 MeV	Coded mask and segmented imager (10 pixels)
Compton scattering	Scintillators or CdTe + CsI Si + Ge Si + CdTe	100 keV–10 MeV	Intrinsic and/or coded mask
Pair effect	Spark chamber then Si trackers	20 MeV–300 GeV	Intrinsic
Pair effect (air shower formation and Cherenkov radiation)	Photomultiplier tubes	100 GeV–100 TeV	Intrinsic

Table 7.11 Main instruments for γ ray astronomy

Mission	Date	Energy range	Instrument	Separating power	Spectral resolution $E/\Delta E$
HEAO-1	1977–1979	80 keV–2 MeV	A4-MED	17°	4
HEAO-3	1979–1981	50 keV–10 MeV	HRGRS	30°	500
COS-B	1975–1982	50 MeV–10 GeV		0°	2
GRANAT	1989–1998	30 keV–1 MeV	SIGMA	20′	5–10
CGRO	1991–2000	20 keV–10 MeV	BATSE	5°	10
CGRO	1991–2000	20 keV–10 MeV	OSSE	3.8° × 11.4°	10
CGRO	1991–2000	1–30 MeV	Comptel	2–4°	10–20
CGRO	1991–2000	20 MeV–30 GeV	EGRET	1–5°	5
INTEGRAL	2002–	15 keV–1 MeV	IBIS/ISGRI	12′	3–30
INTEGRAL	2002–	15 keV–8 MeV	SPI	2.6°	50–500
AGILE	2007–	10 MeV–10 GeV	LAT	0.1–3°	10
GLAST	2008–	10 MeV–100 GeV	LAT	0.1–3°	10
Telescope					
Whipple	1989–2000	500 GeV–5 TeV	–	0.5°	5
HEGRA	1993–2002	500 GeV–10 TeV	–	20′	5
CAT	1996–2002	200 GeV–10 TeV	–	20′	5
HESS	2003–	100 GeV–100 TeV	–	10′	7
MAGIC	2004–	50 GeV–50 TeV	–	18′	7

resolution, and the very high energy frontier. We shall thus discuss the question of *spatial resolution*, then *spectral resolution*, both being closely related to the type of detector used. Tables 7.10 and 7.11 give an overview. Each of these devices is then used in a specific instrument, generally carried in space, although occasionally exploiting the Earth atmosphere as detector.

7.6.1 Spatial Resolution of Gamma-Ray Sources

Here we discuss several types of γ ray telescopes, in the sense of generalised system using different physical processes to determine the direction of the incident radiation.

Compton Telescope

In the energy region extending from several hundred keV to several MeV, the interactions of γ rays with matter are dominated by *Compton scattering*[19] (see Sect. 7.3.1). An acceptable measurement of the incident photon energy can only generally be made in this region by instigating a series of interactions which end in a photoelectric absorption, and such a succession of interactions can only be guaranteed by a detector of a certain thickness. This kind of detector, bombarded by radiation belt protons (see Sect. 2.9.2) and cosmic rays, suffers from a great deal of noise. If it is split into two separate parts and only those events simultaneously triggering the two parts are retained, almost all background events are eliminated, because these events generally only trigger one of the two detectors. On the other hand, a large proportion of events undergoing Compton scattering is conserved. Consequently, a significant background reduction is obtained, usually by more than an order of magnitude, and this can procure a considerable gain in terms of signal-to-noise ratio if the system is designed to favour the Compton scattering of photons along the telescope axis. The nature and size of the first part of the detector must be chosen so as to maximise the probability of single Compton scattering (low atomic number Z of the material). In contrast, the second detector must maximise absorption (high Z, high mass), in order to ensure total absorption of energy.

Equipped with a collimator, such a Compton telescope can give an image of the whole celestial sphere by scanning. This is analogous to the technique used in the HEAO1-A4 experiment (1977–1979), but operating at higher energies. If the detectors are position sensitive, such scanning is not necessary, but one is faced with a classic dilemma: a large aperture results in high sensitivity but poor angular resolution, while a small aperture guarantees good angular resolution at the expense of the sensitivity. As we saw in Sect. 5.2.5, the coded mask provides a solution to this problem and can be profitably applied here. The idea is then analogous to the Sigma experiment: imaging would be of comparable quality, being linked to coded mask techniques, but at higher energies (greater than 1 MeV).

Still considering the case of position sensitive detectors, it is also possible to use the *ballistics* of the Compton interaction to restrict the direction at which the photons arrive (scattering angle) and deduce the position of the source statistically. In this case, imaging performance depends on the spectral resolution of the two detectors and the distance between them. However, efficiency decreases rapidly with this distance and a compromise must be found between sensitivity and angular resolution. This latter method is used in the COMPTEL experiment aboard the *Compton*

---

[19] Arthur Compton was an American physicist (1892–1962) who won the Nobel Prize for Physics in 1927 for discovering the inelastic scattering of light by a particle of matter, an effect which now carries his name.

Gamma-Ray Observatory (GRO) which flew from 1991 to 2000. The distance of about 1.2 m between the two detectors leads to a separating power of around 5 degrees. It is also possible in the case of very closely positioned detectors to use both a coded mask for imaging and also Compton ballistics for a further background reduction. One then selects only those events for which all calculated incident directions lie within the solid angle subtended by the mask at the point of first interaction. This method is used by the IBIS instrument carried on board the INTEGRAL mission (launched in 2002).

The spectral performance of this type of telescope with regard to sensitivity and imaging are governed by the energy resolution of the weakest of the two detectors.

Pair Effect Telescope

Beyond energies of a few MeV, the pair effect dominates the interactions between photons and matter (see Sect. 7.3.1). The photons are detected in two stages: first conversion, i.e., pair creation, then measurement of the electron and positron trajectories, from which the direction of the incident γ ray can be deduced.

Indeed, in the center of mass system, the electron and positron are emitted in opposite directions, while in the laboratory system, they are emitted in the direction of the photon, modified by the recoil motion of the nucleus responsible for the electric field in which electron–positron production occurred. Up to this recoil, the directions of the electron and positron are related to the direction of the incident photon.

Pair creation becomes more efficient in stronger electric fields. In matter, it occurs in the electric fields of atomic nuclei, and the strength of these fields depends on the number of protons in the nucleus. This is why materials with high atomic number (high Z) provide the most efficient conversion (see Fig. 7.20), and lead is usually used for this purpose. On the other hand, good conversion efficiency also requires a thick converter, and the electrons will tend to scatter more (Molière scattering) as the atomic number of the converter material and thickness of the converter are increased. This means that electrons leaving a thick converter are likely to have lost all memory of the direction of the incident γ ray.

This difficulty has given rise to the basic design of the pair effect telescope. In order to limit electron scattering, thin plates of lead are used, inserted between devices able to measure the electron trajectories. Good conversion efficiency is guaranteed by using many plates, while the fact that they are thin ensures an accurate measurement of electron and positron trajectories (Fig. 7.54). The latter are measured using techniques developed for particle accelerators, such as spark chambers up until the mid-2000s in missions like SAS-2, COS-B, Compton/EGRET (*Energetic Gamma Ray Telescope*), and silicon strip trackers in the future, e.g., in the *Gamma ray Large Area Space Telescope* (GLAST) mission (launch 2008).

The angular resolution of such a telescope thus depends on the photon energies, since the error resulting from electron scattering increases as the energy decreases, being a few degrees at 50 MeV, and only a few tens of arcmin in the GeV range and beyond. The photon energies, generally measured using a calorimeter absorbing

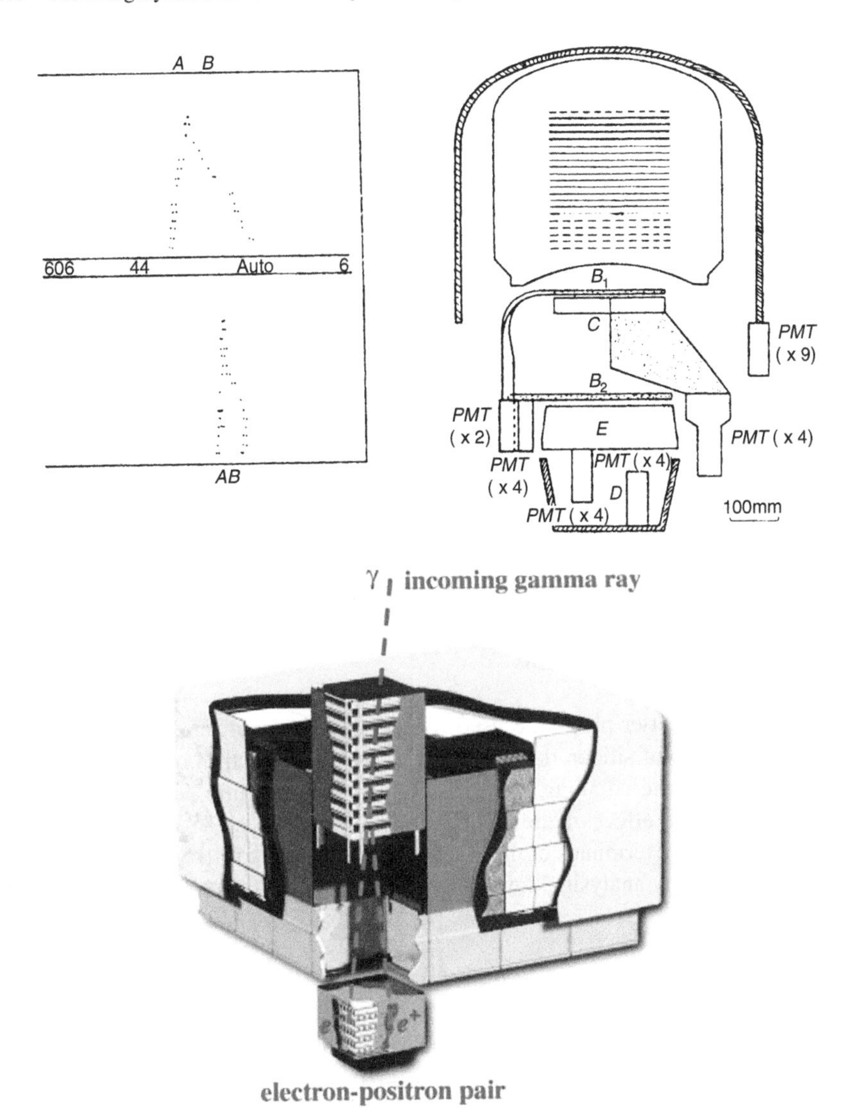

Fig. 7.54 Telescopes for γ-ray observation. *Top right*: The telescope carried by COS-B, one of the first γ-ray observation satellites (1975–1982). The plates of the spark chamber can be seen, and also the scintillators (*shaded*) placed around the main detector as *anticoincidence* devices, which are designed to detect and eliminate the background signal due to charged particles. The scintillators B_1, C and B_2 trigger the chamber, and the calorimeter E measures the energy of the particles. PMT stands for photomultiplier tube. *Top left*: View of an event (pair formation) in the spark chamber. From Paul J., 1979, private communication. See also Bignami et al., Sp. Sci. Instrum. **1**, 245, 1975. With the kind permission of D. Reidel Publishing Company. *Bottom*: Schematic diagram of the Fermi *Large Area Telescope* (LAT). The telescope has dimensions 1.8 m × 1.8 m × 0.72 m. The required power and the mass are 650 W and 2 789 kg, respectively. From Atwood W.B. et al., Ap. J. **697**, 1071, 2009

the pair, can also be estimated from the observed scattering of the electron and positron. The event rate is high, but events due to celestial γ rays are rather rare, and observation times and data processing are thus very long (of the order of a month). The production of e^+e^- pairs is an unambiguous signature of γ rays with energies greater than a few MeV. This signature provides an efficient way of rejecting the background, whence pair effect telescopes can claim relatively good sensitivity: $\approx 10^{-7}\,cm^{-2}\,s^{-1}$ in one month of observation by EGRET.

Spark Chambers

In a *spark chamber*, γ rays interact by pair effect in plates made from a high Z material. Alternatively, charged sheets of wires are inserted between the plates. A created electron and positron ionise a gas (neon) and a spark is produced in the gas between the electrodes formed by the sheets of wires. The trajectories of the particles are thus materialised (Fig. 7.54) and recorded either photographically or using a vidicon tube (see Sect. 7.4).

Silicon Strip Tracking Systems

Once again, converter plates are used for pair creation, but the gas and wires are replaced by grids of silicon detectors (Fig. 7.54). The electron or positron ionises the silicon encountered along their trajectories. The resulting electrons and holes migrate under the effect of the electric field toward the electrodes and induce a current pulse at the terminals of these electrodes. The track followed by the pair can be reconstructed by analysing the coincidence signals in the tracker ensemble.

Atmospheric Cherenkov Telescope

A 1 TeV photon entering the Earth atmosphere interacts with it at an altitude of around 10 kilometres, producing an electron–positron pair. The electron quickly generates high energy γ rays by inverse Compton scattering, while the positron annihilates to emit a further two high energy γ rays. These γ rays will in turn create e^+e^- pairs and the process repeats, generating an *air shower*. The electrons in the shower are relativistic, moving a speed v close to the speed of light in vacuum, and in fact faster than the speed of light in the atmosphere (c/n, where n is the refractive index of the atmosphere). The result is a very brief flash of blue Cherenkov emission,[20] lasting only a few nanoseconds, emitted along a cone with axis close to

[20]The Russian physicist Pavel Cherenkov (1904–1990) won the Nobel Prize for Physics for explaining in 1937 the origin of the light radiation emitted by a liquid irradiated by a radioactive source, the effect which now carries his name.

Fig. 7.55 Two of the four light collectors forming the HESS instrument in the Namibian desert. Photo courtesy of the HESS collaboration and CEA/DAPNIA

the direction of the γ ray which originated the shower and with greater apex angle θ for higher values of the speed v of the electrons. In fact, $\cos\theta = c/vn$. At an altitude of around 1 000 m, an electromagnetic shower contains only Cherenkov photons and extends over about a kilometre.

There are two techniques for measuring Cherenkov air showers. In the more popular imaging technique — used, for example, in the *High Energy Stereoscopic System* (HESS), set up in Namibia in 2004[21] (see Fig. 7.55) — each Cherenkov cone, reflected by a parabolic mirror, forms an ellipse in the focal plane of the mirror. The dispersion in the directions and speeds of the electrons produces ellipses with different centers and different sizes, forming a relatively uniform quasi-elliptical spot. This feature is used to distinguish electromagnetic showers from hadronic showers, which produce much less regular spots. The orientation of the ellipse is related to the direction of the initial γ ray. By measuring the angular distribution of the Cherenkov light, 99% of the cosmic ray protons can be rejected, and at the same time angular resolutions of a tenth of a degree can be achieved. The detection threshold with this technique is of the order of a few hundred GeV.

The second, so-called sampling technique consists in measuring as accurately as possible the shape of the blue light wavefront by means of fine spatial sampling and accurate time measurements. This wavefront is almost spherical for an electromagnetic shower and more spread out for a hadronic shower. This technique, which requires an array of optical reflectors with large collecting area, has been developed by reusing former solar energy production units. This is exemplified by the CELESTE experiment, set up since 2000 at the Thémis power generating site in the French Pyrénées. At the present time, sampling systems are less sensitive than imagers, but have an energy detection threshold ten times lower.

Whatever technique is used, the brevity and colour of the Cherenkov emission necessitates photomultiplier tubes as detecting element (see Sect. 7.4.3).

[21] See www-dapnia.cea.fr/Sap/ for a description.

Between 1996 and 2007, decisive progress was made with this method. On the one hand, the imaging technique, developed in particular in the United States for the Whipple instrument in the 1980s, is now in a dominant position. On the other hand, new instruments are improving this technique. The French CAT experiment, set up at the Thémis site in France in 1996, uses a 600 pixel high resolution camera able to carry out detailed analysis of Cherenkov images, with a consequently improved separation between γ photons and hadrons. At the same time, the German *High Energy Gamma Ray Astronomy* (HEGRA) experiment[22] in the Canaries has implemented an innovative stereoscopic technique, able to observe a single air shower simultaneously with four telescopes. This robust method can achieve angular resolutions of the order of one tenth of a degree and further reduces the hadron background noise.

In 2004, a new generation of detectors came on the scene with the French–German *High Energy Stereoscopic System* (HESS).[23] This experiment associates the methods used successfully in CAT and HEGRA, viz., high resolution Cherenkov imaging and stereoscopy.

HESS comprises 4 large mirrors of 100 m^2 each, placed at the corners of a square of side 120 m. The focal length of the mirrors is 15 m. A camera consisting of 960 small photomultiplier tubes is placed at the focal point of each mirror. These PMTs are particularly well suited to detecting the flash of Cherenkov light thanks to their fast response, sensitivity in the blue region, and low noise levels. The sensitivity achieved above 1 TeV is 10^{-13} cm^{-2} s^{-1}. The gain in sensitivity over previous experiments (an order of magnitude) is such that, between 2004 and 2006, about thirty new sources of very high energy γ radiation were discovered, mainly in the galactic plane. The experiment is set up in Namibia, giving access to the southern sky and hence to the central region of our Galaxy, a region known to contain many γ ray sources. For the first time it has become possible to resolve extended sources, and in particular to study the morphology of supernova remnants, by virtue of the field of view of 5 degrees and resolution of around 10 arcmin (FWHM).

7.6.2 Spectral Analysis of Gamma-Ray Sources

Spatial resolution and detection efficiency are clearly the criteria for quality in an image-forming (or imaging) detector. But those for a spectrometer are more subtle. Indeed, although spectral resolution is an obvious criterion, the total efficiency, on the other hand, is not in itself a good criterion in the γ ray region. This is because Compton scattering, in which the scattered photon escapes, does not lead to a determination of the incident photon. In fact, it is the efficiency in the peak of the total absorption that must be considered, and this increases with the charge number Z, and the mass (volume $\times$ density) of the detector. The spatial resolution of a γ ray detector depends on the energies involved, being limited by the Compton scattering events. It will improve if the scattering is minimal, which implies a high Z and dense

---

[22] See www.mpi-hd.mpg.de/hfm/CT/CT.html.

[23] The acronymn also remembers the Austrian physicist Viktor Hess (1883–1964) who won the Nobel Prize for Physics in 1936, for his discovery of cosmic radiation in 1912.

Table 7.12 Materials for γ-ray detection. S scintillator, SC semiconductor

Material	Type	Z	Density [g cm^{-3}]	Volume [cm^3]		Spectral Resolution at	
				Min	Max	122 keV	511 keV
NaI(Tl)	S	23–53	3.7	1	10^4	0.12	0.09
CsI(Tl)	S	50–53	4.5	1	10^4	0.16	0.12
Ge	SC	32	5.3	1	10^2	0.015	0.004
CdTe	SC	48–52	6.1	10^{-3}	1	0.016	0.010
HgI_2	SC	53–80	6.4	10^{-3}	10^{-1}	0.1	0.04

material, and these are therefore the working criteria for selecting materials, whether for an imager or for a spectrometer. In contrast, high volume is advantageous in spectrometry, but not in segmented imaging, in which each pixel is an independent detector. These simple remarks can be used to classify some materials (Table 7.12).

It would have made perfect sense to place the following in previous sections, in particular, the one dealing with the physical principles of detection (see Sect. 7.3). This necessarily somewhat arbitrary choice reflects once again the extent to which the boundaries between wavelength regions, physical principles, and methods for making detectors are fuzzy and mutable, depending as they do on advances made in physics and technology.

Scintillators (100 keV–10 MeV)

Photoelectric absorption or Compton scattering of X or γ rays by the electrons in a solid transfers a large fraction of the incident photon energy to those electrons. This electron energy is progressively passed on to the atoms of the solid, either by excitation of electrons and molecules in bound states, or by excitation of acoustic vibrations (*phonons*) in the crystal lattice. These excited states decay radiatively, by emitting, in particular, visible and near UV photons. If the material is transparent at these wavelengths, it suffices to measure the intensity of the observed *scintillation*, in order to deduce the energy loss of the incident photon, and hence, within certain limits, its energy. A photomultiplier is used, with very low, or zero, intrinsic noise, high spectral sensitivity in the de-excitation wavelengths (blue), and very short time constant, which allows excellent time discrimination (around 10^{-3} s, Problem 7.7).

Two classes of scintillators are *inorganic* and *organic scintillators*. In the first class, the material is an alkali halide, such as NaI, or CsI, doped with impurities, such as thallium. These produce interstitial centres in the crystal lattice, which are easily ionised and luminescent when recombination occurs. The *conversion yield* depends on the temperature: around 20% of the incident X or γ-ray energy is converted into visible photons, a 100 keV photon producing something like 4 000 photons at around 420 nm.

Many scintillators were developed following the use of NaI:T1 in 1948, mainly to meet the needs of nuclear medicine. Table 7.13 gives the characteristics of the most

Table 7.13 Properties of scintillators. BGO $Bi_4Ge_3O_12$, GSO Gd_2SiO_5, LSO Lu_2SiO_5, YAP $YAlO_3$

Crystals	NaI (Tl)	CsI (Tl)	BGO	GSO	LSO	YAP
Density	3.67	4.5	7.1	6.7	7.4	5.5
Effective Z	51	54	75	59	66	33.5
Attenuation length at 500 keV [mm]	29.1	22	10.4	14.1	11.4	21.3
Number of scintillation photons for 100 keV	4 100	1 800	900	800	3 000	1 700
Response time (μs)	0.23	1	0.3	0.06	0.04	0.03
Hygroscopic	Yes	≈ no	No	No	No	No
Refractive index	1.85	1.79	2.15	1.85	1.82	1.95

commonly used scintillators. Scintillators have no intrinsic directivity and must be combined with optics (for X rays), and collimators or masks (for γ rays), if the aim is to form and detect an image (see Chap. 5).

Anticoincidence Devices

The effects of charged particles can be limited, at least as regards their direct effects and spontaneous de-excitation emission, by equipping γ ray telescopes with anticoincidence systems. The idea is to surround the main detector with another detector. If this is triggered, a 'veto' signal is generated, which prevents the main detector from being triggered in its turn, but for a very short lapse of time, although longer than the response time of the main detector and the anticoincidence detector. Used to limit the field of view of an experiment, this system has the advantage over the collimator of eliminating its own spontaneous de-excitation emission. In addition, it can improve the spectral response of spectrometers by eliminating a large proportion of Compton scattering events taking place in the main detector. However, this system has some disadvantages, too. For example, it increases the production of neutrons and the contribution from delayed de-excitation. Moreover, it limits the useful experiment time.

The anticoincidence device is probably the most delicate part in the design of a γ ray experiment in astronomy, because it is essential, but at the same time highly complex. The design of such a system involves *Monte Carlo simulations*, as well as measurements in proton accelerators. Due to their fast response and simple implementation, scintillators are usually chosen to make anticoincidence detectors. Those designed to reject charged particles use a plastic scintillator. Those that must also stop γ rays off the telescope axis use heavy scintillators like CsI or BGO (see Table 7.13), which have photon detection efficiencies as high as 100% up to energies of several hundred keV.

Phoswich Detectors

A phoswich detector is a sandwich of two scintillators, which are chosen so that their characteristic times are very different (see Table 7.13). Light emitted by the two detectors is collected by a photomultiplier (or a photodiode), and a selection made on the basis of the *pulse shape discrimination* (PSD) signal. This selection distinguishes those events having interacted in both scintillators (charged particles or Compton interactions) from those which have interacted only in the upper scintillator. The phoswich detector is equivalent to a detector–anticoincidence ensemble. This requires more complicated electronics, but the readout system for the upper scintillator is no longer needed. The significant reduction in mass within the anticoincidence device implies a large decrease in detector background, which, in the γ ray region, is predominantly due to de-excitation of the surrounding matter. NaI–CsI sandwiches are often used, as they give good spectral results and effective selection of events.

Semiconductor Detectors (10 keV–10 MeV)

In this energy domain, semiconductor detectors are no longer used for photoconduction (measurement of a direct current as in Sect. 7.4), but rather as spectrometers, that is, measuring the energy deposits induced by the interaction of photons with the matter (see Sect. 7.3.1). The electrons produced by these interactions generally have a much higher energy than the energy levels of atoms in the detector. They lose this energy very quickly while propagating, by ionising atoms on their path, thereby creating electron–hole pairs or charge carriers which themselves propagate, following the electric field lines. This drift under the effect of the electric field, until they reach the collecting electrodes (Fig. 7.56), constitutes the current charging up the capacitor. The collected signal Q (the charge) is then the product of the

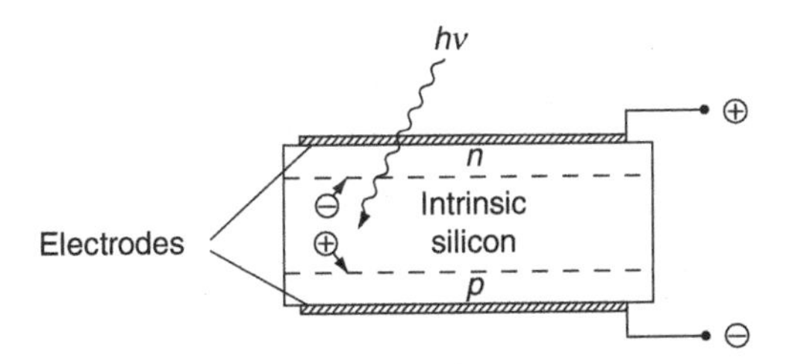

Fig. 7.56 PIN diode. The base material is p-doped silicon (or germanium). Supplementary doping with lithium compensates for the p-doping in the volume, making the silicon intrinsic by compensation. The surface layer has an excess of lithium (n). The pair created by the X-ray photon moves in the applied electric field

elementary charge and the sum over the electrons and holes of the path elements, relative to the interelectrode distance x_i/d (*Ramo's theorem*):

$$Q = \frac{e}{d}\sum x_i \ .$$

Charge Losses

In general, this migration involves charge losses, since each electron and hole has a certain probability of capture per unit time, characterised by its mean lifetime (τ_e and τ_h, respectively). If the transit time of a type of charge carrier is not negligible compared with the lifetime of these carriers, part of the signal will be lost, and this is referred to as *charge loss*. The transit time of carriers depends on their distance from the collecting electrode, the applied electric field E, and also on their *mobility*, which is a characteristic of the semiconductor (μ_e and μ_h, respectively, for electrons and holes). Charge losses therefore depend on the mean free paths of the carriers, $\lambda_e = \mu_e E \tau_e$ and $\lambda_h = \mu_h E \tau_h$. For n electron–hole pairs, produced at distance x from the negative electrode, the charge collected is given by *Hecht's equation* (see Problem 7.13)

$$Q = \frac{ne}{d}\left\{\lambda_e\left[1-\exp\left(-\frac{d-x}{\lambda_e}\right)\right]+\lambda_h\left[1-\exp\left(-\frac{x}{\lambda_h}\right)\right]\right\} \ .$$

The number of electron–hole pairs created in semiconductors is well above the number of photoelectrons created in scintillators, which explains their superior spectral resolution (Fig. 7.57).

It is therefore possible to minimise the charge loss in a given detector by increasing the electric field, that is, by increasing the voltage across the detector. In practice, too high a voltage generates a lot of noise, which can then decrease the resolution of the detector. If all interactions of the photons were to take place at the same distance from the collecting electrode, the charge loss would always be the same, and, as the loss of gain could be corrected for, its only effect would be a reduction in spectral resolution.

In reality, the interactions occur at a range of distances from the electrode, and a continuous range of losses is observed. These losses are manifested by the appearance of a continuum in the amplitude spectrum of pulses produced by a radioactive source (Fig. 7.58). They therefore amount to an energy loss, analogous to the one observed when Compton scattering occurs. However, when the charge loss is high, this means that the carriers must have travelled over a longer period, and if it is possible to accurately observe the pulse shape, it will be found that the growth time is much longer. A measurement of the pulse growth time can therefore lead to a correction for charge loss, and this correction allows for a symmetrisation of the observed lines (Fig. 7.58). However, charge loss always results in line broadening, which means a loss of resolution.

Table 7.14 gives some characteristics of semiconductors susceptible to charge loss effects.

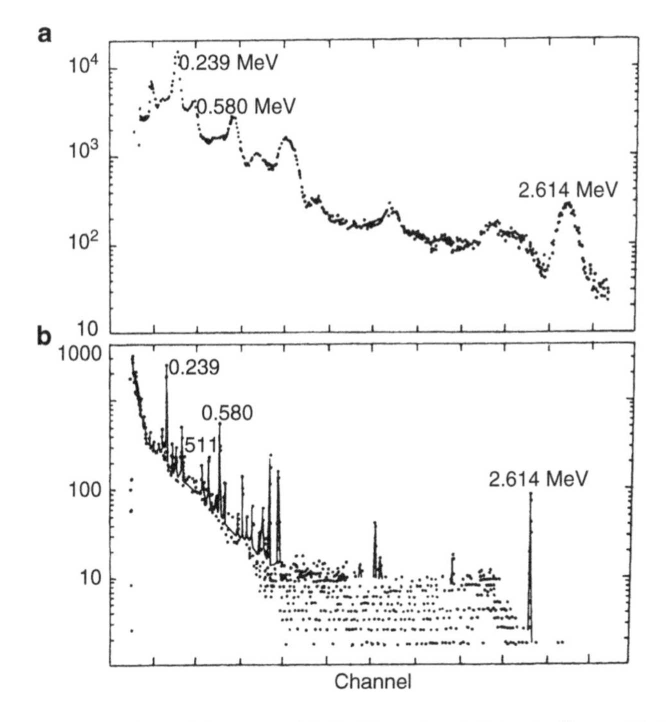

Fig. 7.57 Spectral resolution of detectors. (**a**) Thallium-doped NaI scintillator. (**b**) Ge–Li semiconductor detector. *Ordinates* are in units of 5×10^{-4} counts per second

Fano Factor

If all the energy deposited in a semiconductor were converted into electron–hole pairs, the number of these produced would depend only on the total energy deposited. On the other hand, if the energy was distributed in a completely uncorrelated way between electron–hole pairs and phonons, the number of electron–hole pairs would obey Poisson statistics, and its variance would be equal to its average value. In fact, it is found that the variance of this number is always less than the average value. The ratio of the variance and the average is called the *Fano factor*. This factor is of order 0.1 for most semiconductors.

Germanium

Cooled germanium offers by far the best spectral resolution. Charge losses are low, but can nevertheless be corrected for, so improving the resolution even further. The feasibility of obtaining large volume crystals undoubtedly makes it the best possible choice for a laboratory spectrometer. In contrast, the need to cool it down to 77 K considerably limits its feasibility for a space spectrometer. In addition, it is relatively

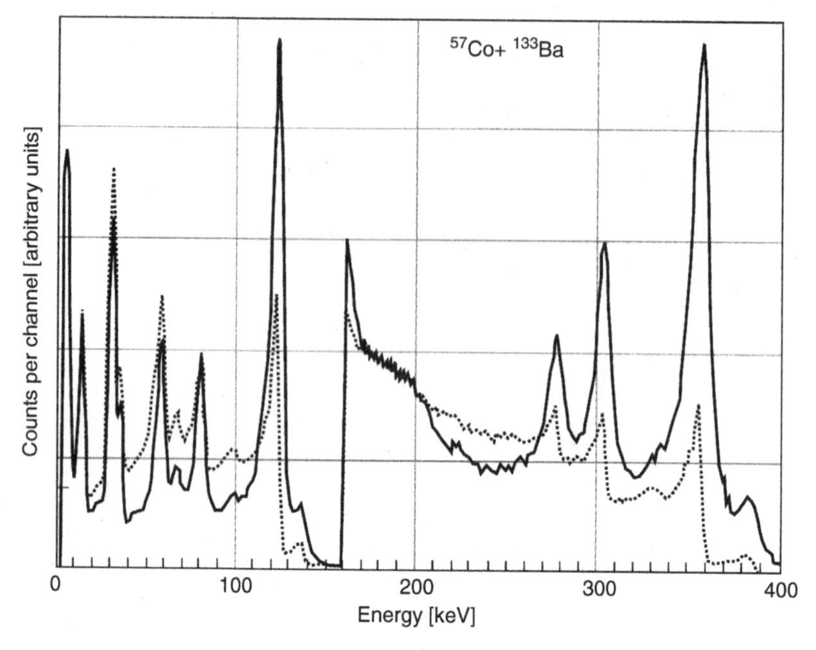

Fig. 7.58 Correction for the phenomenon of charge loss in the material CdTe. The count per energy channel is given as a function of energy, for laboratory radioactive sources (^{57}Co and ^{133}Ba). This spectrum, obtained in 1995, is amplified beyond 160 keV. *Dotted curve*: amplitude spectrum of pulses. *Continuous curve*: spectrum corrected for charge loss

Table 7.14 Semiconductor characteristics

	Si (300 K)	Ge (77 K)	CdTe (300 K)	HgI_2 (300 K)
Band gap width [eV]	1.12	0.74	1.45	2.1
Pair creation energy [eV]	3.61	2.96	4.43	4.22
μ_e [cm^2 V^{-1} s^{-1}]	1 300	36 000	1 100	100
μ_h [cm^2 V^{-1} s^{-1}]	500	42 000	80	4
τ_e [μs]	3 000	1 000	1	1
τ_h [μs]	3 000	1 000	1	1
E_{max} [V cm^{-1}]	1 000	100	1 000	10 000

sensitive to irradiation by charged particles, its spectral resolution degrading at a rate of 50% per year. However, the example of the INTEGRAL spectrometer[24] shows that the semiconductor can be periodically annealed in orbit to repair radiation damage and recover spectral performance levels. Annealing involves heating the detector to 300 K for several days, then cooling it back down to its operating temperature of 77 K. This procedure removes most of the traps created by the

[24]The site sigma-2.cesr.fr/spi/index.php3 gives a description of this spectrometer.

passage of protons. As far as imaging is concerned, the crystal can be divided into several pixels, of about one centimetre in size, by depositing conducting bands on opposite faces of it. This gives a spectral resolution of centimetre order, although limited at high energies by Compton scattering in this material, which has a very high atomic number ($Z = 32$).

Cadmium Telluride

This material exhibits significant charge loss, due essentially to low hole mobility. These losses can be corrected for by measuring the rise time of the pulses or the ratio of the signals on each electrode, but these measurements are difficult to make and somewhat inaccurate. A more effective method is to segment the anode into small pixels in such a way that the signal induced by the holes becomes negligible while that induced by the electrons depends only weakly on the interaction depth. With this configuration and thanks to progress in microelectronics and hybridisation, its spectral performance at the time of writing (2007) is close to what can be achieved with germanium (see Fig. 7.59), but with the considerable advantage that it can work at room temperature, i.e., 20°C. However, it does perform significantly better at lower temperatures, in the range −40 to 0°C. On the other hand, the resulting detectors remain relatively small and limit the application of this semiconductor to energies below MeV.

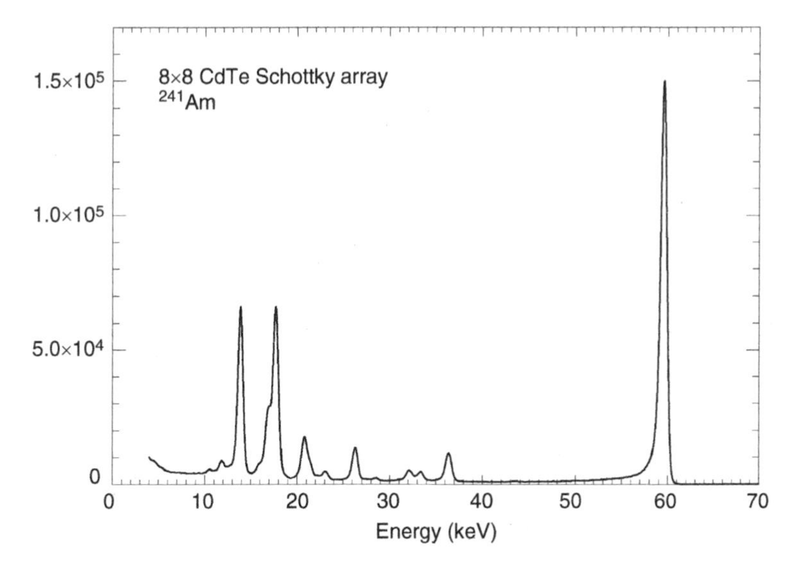

Fig. 7.59 Amplitude spectrum obtained in 2007 by summing the response of pixels in an 8×8 format CdTe Schottky array, illuminated by a radioactive source (^{241}Am). Pixel size 1 mm^2. Crystal thickness 2 mm. The array is hybridised on a substrate connected to ASIC readout components (IDEF-X). Credit: O. Limousin, CEA/DAPNIA/SAP

In orbit, irradiation by protons in cosmic rays and solar flares slightly degrades the electron transport properties. For the INTEGRAL *Soft Gamma Ray Imager* (ISGRI), carried on board the European INTEGRAL mission, a 3% loss of gain per year has been observed. In contrast to germanium, it is not therefore necessary to anneal CdTe detectors.

CdTe is now produced on an industrial scale, usually being obtained by zone refining. Crystals produced by this method exhibit limited resistivity of the order of $10^9\ \Omega\,\text{cm}$, which precludes the possibility of applying the high voltages required to minimise charge loss, since the leakage current, which generates noise, becomes too large. But a new method has been available since the mid-1990s, where growth occurs under high pressure (100 atm), including 20% zinc. This gives larger and more resistant crystals, which can be used in higher fields (up to $10\,000\ \text{V cm}^{-1}$). CdTe crystals produced by zone refining can be equipped with an indium electrode to create a Schottky barrier (see Sect. 7.5.2). These diodes exhibit extremely low leakage currents, whence it is possible to apply very high fields. Charge loss is then minimised (see Fig. 7.59).

As regards imaging, the small size of these detectors is not a disadvantage, and CdTe seems to be a good choice for the construction of a segmented imager, operating up to a few hundred keV.

Mercury Iodide

Roughly speaking, the same advantages and disadvantages as CdTe are present in mercury iodide (HgI_2), but it is more difficult to produce. It is highly resistant to irradiation. The thickness of present detectors does not exceed one millimetre, and this limits use to energies below 100 keV, despite its having a better intrinsic stopping power due to the properties of the mercury atoms. Moreover, the deposition process of the electrodes is rather delicate, and rules out industrial production for the present (2007). Despite these drawbacks, mercury iodide has been used in X-ray astronomy. Trials carried by stratospheric balloons have shown the advantage of using a high Z detector. Indeed, for the same stopping power, there is around five times less noise than with a germanium or scintillation detector, working been 40 and 80 keV.

7.7 Neutrino Observing Systems

In Chap. 1 we described how neutrinos have recently joined the panoply of information carriers exploited by astrophysics, alongside electromagnetic radiation.[25] Even more so than for γ radiation, it is no longer possible to separate the notions of telescope, spectrometer, and detector. In this section, we shall thus discuss all these functions together. We shall consider neutrinos with energies from 100 keV (solar neutrinos) up to neutrinos from supernovas ($\sim$ 10 MeV) or active galactic nuclei (TeV and beyond).

---

[25] In this section, the authors are grateful for the invaluable assistance of Michel Cribier.

7.7.1 *Radiochemical Detection of Solar Neutrinos*

Historically, the first encounter between neutrinos and astrophysics occurred when John Bahcall[26] contacted Raymond Davis to enquire about the means available for detecting solar neutrinos. The basic idea of these radiochemical techniques is the so-called inverse β reaction, which leads to the transmutation of a target nucleus (A, Z) to a generally unstable nucleus $(A, Z+1)$:

$$\nu_e + {}^{A}X_Z \longrightarrow e^- + {}^{A}X_{Z+1} \,.$$

The great advantage with this detection method is that it is virtually insensitive to undesirable effects. Several targets have been successfully used here, such as chlorine and gallium. The following conditions must be satisfied for a target nucleus to be viable:

1. A threshold for the transmutation reaction that is compatible with the neutrinos one hopes to detect.
2. A sufficiently high probability of inducing this reaction.
3. A sufficient isotopic abundance of the target nucleus.
4. The resulting nucleus must produce an exploitable signature.

Detection Using Chlorine

In the mid-1960s, Raymond Davis constructed the first experiment for detecting solar neutrinos. He used the inverse β reaction proposed by Pontecorvo[27] in 1946:

$$\nu_e + {}^{37}\mathrm{Cl} \longrightarrow e^- + {}^{37}\mathrm{Ar} \,.$$

Five or six times a year, the isotope ^{37}Ar is extracted by flushing the system with helium. This radioactive isotope of argon (Ar) decays with a half-life of 35 days, emitting characteristic radiation. The cross-section of the transmutation reaction is very small, and the reaction only occurs when the neutrino energy is greater than 814 keV. Davis put 615 tonnes of carbon tetrachloride C_2Cl_4 at the bottom of a gold mine in South Dakota (USA). Applied to the Sun, the theories of nuclear physics predict that a certain number of neutrinos should be emitted per unit time. Then, given the mass of the detector and the reaction cross-section, one ^{37}Ar atom was expected to be produced every day. However, the experiment produced only one such atom every three days. This significant difference was the basis of the

[26]The Princeton astrophysicist John Bahcall (1934–2005) was a leading expert on the solar neutrino problem, along with the American chemist and physicist Raymond Davis (1914–2006), winner of the Nobel Prize for Physics in 2002.

[27]Bruno Pontecorvo (1913–1993) was an atomic physicist. Born in Italy, he was an assistant of Enrico Fermi in Chicago, then emigrated to the Soviet Union in 1950.

solar neutrino enigma, only solved in 2002, and today attributed to the property of neutrino oscillation between different neutrino species. Raymond Davis received the Nobel Prize for Physics for this result in 2002.

Detection Using Gallium

The idea of using a gallium target to detect solar neutrinos is due to Kuz'min, who suggested the transmutation of the gallium isotope ^{71}Ga into the radioactive isotope ^{71}Ge of germanium by the reaction

$$\nu_e + {}^{71}\mathrm{Ga} \longrightarrow \mathrm{e}^- + {}^{71}\mathrm{Ge}\ .$$

The advantage compared with the experiment using ^{37}Cl is that, thanks to the lower threshold of 233 keV, ^{71}Ga can be used to detect so-called primordial neutrinos, not emitted by the Sun. The GALLEX detector (1991–1997), later called the *Gallium Neutrino Observatory* (GNO), was set up in the underground Gran Sasso laboratory, in the Abruzzo region of central Italy, with 30 tonnes of gallium. The *Soviet–American Gallium Experiment* (SAGE) set up in the Baksan laboratory in the Russian Caucasus in the 1990s operates on the same principle.

The isotope ^{71}Ge decays with a half-life of 11.4 days. After exposing the gallium to solar neutrinos for several weeks, the germanium atoms that have been produced must be recovered, and their decay observed. Given the mass of gallium used and the theoretical prediction for solar production, the expected rate of production by solar neutrinos is a little over one germanium atom per day.

The gallium is in the form of a liquid compound, gallium trichloride in solution in hydrochloric acid, in a large cylindrical tank 8 m high and 4 m in diameter. The germanium produced is then in the form of germanium tetrachloride, which is extremely volatile in the presence of hydrochloric acid. Every three or four weeks, several thousand cubic metres of nitrogen is passed through the tank to flush out the germanium tetrachloride. A trap is placed at the tank outlet, comprising a glass column containing tiny glass threads, designed to capture the germanium tetrachloride. Pure water is passed through from top to bottom while the gas flow goes from bottom to top. The water dissolves the germanium tetrachloride and allows the nitrogen to escape.

The next step is to form germanium hydride GeH_4. This gas, very similar to methane, is placed in a small proportional counter. In order not to swamp the tenuous signal of the ^{71}Ge under a signal due to tritium decay, this chemical reaction is carried out with water taken from 3 000 metres underground in the Negev desert, which contains no tritium atoms. The decay of the germanium by electron capture is observed in the proportional counter. This yields a characteristic electrical signal. Each counter is left for several months in its lead shielding (dating from ancient Rome), itself placed in a Faraday cage where undesirable radiation is reduced to a minimum. The counters are made from quartz, sculpted and worked by hand and selected to produce the least possible unwanted noise.

Detection Using Heavy Water

The *Sudbury Neutrino Observatory* (SNO) is a detector operating in real time, comprising 1 000 tonnes of heavy water D_2O, placed in a nickel mine in Canada, some 2 000 m underground. The detection of electron neutrinos exploits the following reactions:

$$\nu_e + {}^2\mathrm{H} \longrightarrow e^- + p + p ,$$

$$\nu_e + e^- \longrightarrow \nu_e + e^- .$$

In the first reaction, the electron energy provides a direct determination of the neutrino energy, but with rather limited information about the direction of the incident neutrino. In the second, which is elastic scattering, information about the neutrino energy is limited, while more can be found out about the direction of the incident neutrino. But for the first time, this experiment can explicitly measure the flux of other neutrino *flavours*, via the decay of the deuteron induced by the neutrino, to produce a neutron and a proton:

$$\nu_x + {}^2\mathrm{H} \longrightarrow \nu_x + p + n , \qquad x = e, \mu, \tau .$$

Thanks to the complementarity of these reactions, the SNO experiment has made a major contribution to solving the solar neutrino problem, showing that all the neutrinos from the Sun which seemed to be missing were transformed during their trip into neutrinos of another flavour by the oscillation phenomenon.

The detector comprises a sphere of acrylic material of diameter 12 m, containing 1 000 tonnes of ultrapure heavy water, itself immersed in a tank of diameter 22 m and height 34 m containing very pure ordinary water, the whole thing subjected to the permanent scrutiny of 9 600 photomultipliers. The radioactive purity requirements are very strict indeed.

To detect the neutrons that are the signature of the third reaction, two tonnes of highly purified sodium chloride have to be dissolved in the heavy water, while maintaining the same level of purity. But it has to be possible to extract all the dissolved salt in a few days to restore the initial purity of the heavy water, in order to proceed with another neutron detection method using ^{3}He proportional counters immersed in the water.

The Prospects for Neutrino Detection Using Indium

Indium was proposed as neutrino target by the physicist Raju Raghavan in 1968. The inverse β reaction used here results in an excited state of the final nucleus, with a mean lifetime of 4.7 μs, decaying with the emission of two photons:

$$\nu_e + {}^{115}\mathrm{In} \longrightarrow e^- + {}^{115}\mathrm{Sn}^{**} ,$$

followed by

$$^{115}\mathrm{Sn}^{**} \longrightarrow {}^{115}\mathrm{Sn} + \gamma_1\,(116\,\mathrm{keV}) + \gamma_2\,(496\,\mathrm{keV})\,.$$

What is interesting here is to obtain a direct detection, with a threshold of 117 keV which makes it possible to detect primordial neutrinos, determining their energies individually by measuring the energy of the resulting electrons, and hopefully improving discrimination thanks to the different spatial coincidences (e^-, γ_1), temporal coincidences (γ_1, γ_2), and the characteristic energies of the photons.

Despite ingenious experimental ideas, the problem shared by all these detectors remains the lack of selectivity of the final radiation, which makes it difficult to deal efficiently with intrinsic background noise (the natural radioactivity of indium) or external noise sources. In 2011, this approach has been more or less abandoned, particularly since the initial motivation of explaining the mystery of solar neutrinos has since been removed.

7.7.2 Neutrino Detection by Cherenkov Radiation

To detect the particles resulting from the interactions of incident neutrinos, the Super-Kamiokande detector and similar detection systems use Cherenkov radiation, already encountered earlier in the context of detection via the Earth atmosphere. Charged particles, and only charged particles, moving through water at speeds 75% greater than the speed of light in water will radiate blue light in a conical configuration around the direction of their trajectory. This so-called Cherenkov light will cross the pure water of the reservoir without absorption, and can then be detected on its inner wall, lined with photomultiplier tubes operating in counting mode.

Each photomultiplier measures the total amount of light reaching it, and the time of arrival of each photon. These measurements are used to reconstruct the trajectory and energy of the particles crossing the water. The Cherenkov light cone produces a more or less regular ring of light on the walls. If this ring appears distinctly, the particle was a muon, but if the edges of the ring are fuzzy, the particle was an electron whose multiple scattering has modified the direction. This latter feature is used to distinguish between muon neutrino and electron neutrino interactions.

This technique has been successfully used for neutrinos with energies of a few MeV (Super-Kamiokande), but also for high energy neutrinos (AMANDA, ANTARES, etc.).

Since 1987, the Kamioka Nucleon Decay Experiment (Kamiokande), and since May 1996, Super-Kamiokande, developed by Masatoshi Koshiba,[28] have been measuring the upper region of the solar neutrino spectrum with a Cherenkov detector using water. The Super-Kamiokande

---

[28]The Japanese physicist Masatoshi Koshiba, born in 1926, won the Nobel Prize for Physics in 2002 for his contribution to demonstrating the existence of neutrino oscillations.

experiment comprises 50 000 m^3 of pure water, monitored by 11 000 photomultiplier tubes (PMT). The neutrinos are scattered by electrons in the target ($\nu_e + e^- \rightarrow \nu_e + e^-$) and transmit a large part of their energy to them. The Cherenkov radiation produced by the scattered electron provides information about the energy and direction of the incident neutrino. The rate of production of photons by solar neutrinos is around 10 per day. The energy threshold of Super-Kamiokande is around 5.5 MeV. Below this threshold, the Cherenkov light is too faint to exceed the background signal, produced by undesirable phenomena. Super-Kamiokande can measure the energies of neutrinos and also the direction from which they arrive. These detectors have confirmed the solar origin of the observed neutrinos, and also detected neutrinos produced by the supernova SN 1987A.

Neutrinos from the Supernova SN 1987A

This type of water Cherenkov detector, e.g., Kamiokande and IMB in the USA, recorded neutrinos produced in the supernova SN 1987A, using the same idea of Cherenkov emission, on 23 February 1987. Experimentally, this is easier than detecting solar neutrinos, owing to the generally higher energy ($\approx$ 20 MeV) and the narrow time window ($\approx$ 12 s) in which the 20 detected events were concentrated.

7.7.3 High Energy Neutrino Astronomy

Since the 1990s, enormous detectors have been built to make neutrino telescopes for neutrinos reaching energies of several TeV. Indeed, very high energy γ rays ($\geq 10^{12}$ eV) have been detected from active galactic nuclei and the Crab nebula. The underlying mechanisms in these compact sources simultaneously produce neutrinos which can escape and cross intergalactic space without deviating from their initial trajectory.

Detection of these high energy but rare neutrinos uses the Earth both to filter unwanted radiation and as a target to convert the neutrinos into muons travelling in the same direction as their progenitors. These upwardly moving muons travel hundreds of metres, and in water, emit electromagnetic radiation by the Cherenkov effect. This radiation can in turn be measured by optical detectors (called optical modules) distributed along detection strings several hundred metres long. The first attempt to build a high-energy neutrino telescope goes back to the end of the 1980s with the *Deep Underwater Muon and Neutrino Detector* (DUMAND), just off the coast of Hawaii. A prototype string attached by cable to a vessel at the surface was deployed in 1987. Following this success, a project was put forward to build a network of 9 strings, anchored to the seabed at a depth of 4 800 m. Although only the first detection string was set up, it was able to detect the signatures of atmospheric muons. However, it had only been operating for 10 hours when a leak occurred in one of the electrical units. The experiment was eventually stopped in 1996 due to lack of funds. In the meantime, other projects had been set up, some persevering with the idea of using a liquid medium (the sea or a lake), others preferring ice. Today, the telescope known as IceCube, set up under the Antarctic ice and hence sensitive to sources in the northern hemisphere, and the ANTARES telescope, operating from 2 500 m under the Mediterranean sea, just off the French coast at Toulon, form a complementary pair that can observe the whole sky.

Southern Hemisphere

AMANDA

The *Antarctic Muon and Neutrino Detector Array* (AMANDA) was started in the 1990s, under the Antarctic ice at the geographical South Pole. It comprises a network of 676 optical modules, distributed over twenty detection strings. These were deployed by drilling a vertical holes in the ice by means of high pressure hot water jets, setting up the detection strings, then letting the water freeze them rapidly into place. Apart from the relative ease of deployment, ice has the advantage of inducing a low optical background noise, of the order of 1 kHz per optical module, making it possible to detect very low energy neutrinos (tens of MeV) as might be emitted by supernova explosions. However, the ice layer traps air bubbles and these induce light scattering that makes it difficult to reconstruct the particle trajectories. The angular resolution is thus only a few degrees, so it is a delicate matter to study point sources (see Fig. 7.60). By increasing the size of the detector, and hence the data available for reconstructing trajectories, it should be possible to improve the angular resolution. This explains how AMANDA's successor IceCube can achieve angular accuracies better than one degree.

After 7 years of data acquisition, the AMANDA telescope accumulated a total of 6 595 candidate neutrinos (compatible with the expected background), and thereby measured the energy spectrum of atmospheric neutrinos down to two orders of

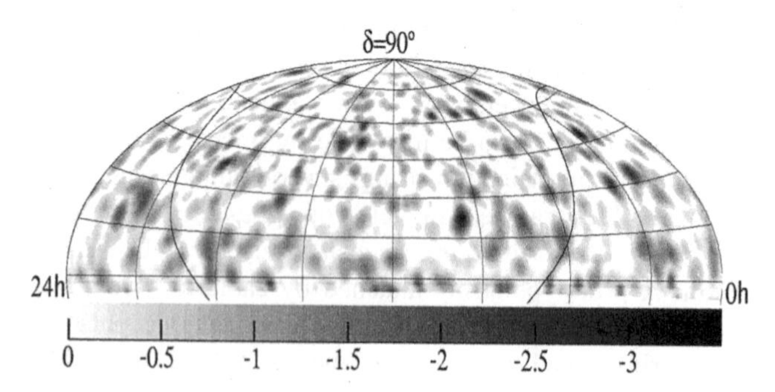

Fig. 7.60 Sky map of the northern hemisphere showing the results of observations made by AMANDA since 2000. The *grey scale* at the bottom, in units of 10^{-8} cm^{-2} s^{-1}, gives the upper limits of the high energy neutrino flux, above 10 GeV, with a spectrum assumed to go as E^{-2}, measured from muon events over a total of 607 days. No correlation has yet been observed with potential high energy neutrino sources such as quasars or supernovas. With the kind permission of the IceCube collaboration

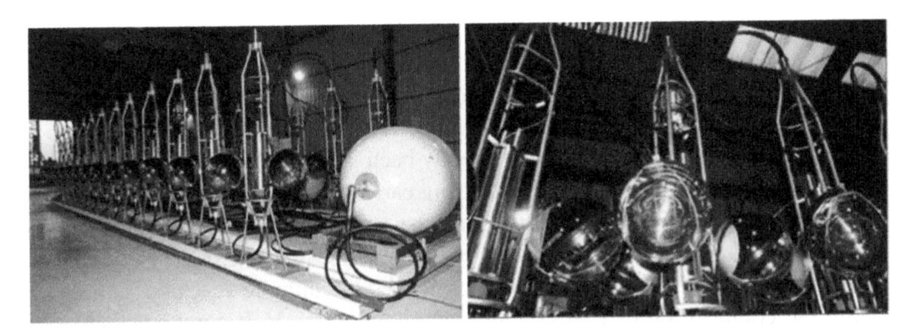

Fig. 7.61 *Left*: Photograph of a detection string in the assembly room, folded up on a pallet, just prior to immersion. *Right*: Photograph of one storey of a detector string, showing the photodetectors inside their protective spheres. The associated electronics is housed in a central titanium container

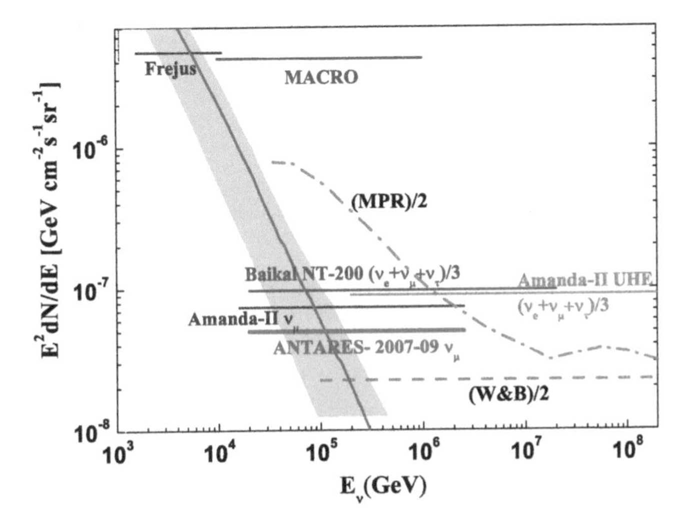

Fig. 7.62 Upper bounds on a diffuse flux of high energy cosmic neutrinos, compared with theoretical predictions (W&B and MPR). The most restrictive bound yet published comes from ANTARES, but the sensitivity expected from analysis of recent data gathered by IceCube will exceed it by almost an order of magnitude

magnitude below previous investigations carried out in underground detectors, i.e., down to about 100 TeV. Regarding the high energy diffuse cosmic neutrino flux, an upper limit was established by studying the selected batch of neutrinos, and on this basis it was possible to reject a certain number of production models, e.g., by active galactic nuclei (see Figs. 7.60 and 7.62).

IceCube Neutrino Observatory

The IceCube detector has taken over from its predecessor AMANDA. It is the largest neutrino telescope in the world. Most of the results obtained after the 7 years of data acquisition by AMANDA have already been superseded by data from the first few detector strings of IceCube. Each string carries 60 optical modules frozen into the ice at a depth of 1500–2500 m. The first string was set up in January 2005, and the detector has had its full complement of 86 strings since January 2011. At the surface is the IceTop detector, which serves to calibrate IceCube and also to study the composition of cosmic radiation above 10^{17} eV. It comprises tanks for detecting Cherenkov light emitted by charged particles reaching ground level. When data was taken with the 40 string detector, a total of 14 139 candidate upwardly moving neutrinos were selected over 375.5 days. A search for point sources was then carried out, but no significant accumulation was observed. The limits obtained with this batch of data are now tighter by a factor of 7 than those laid down by AMANDA, and they should be improved in the near future through analysis of data from the final configuration.

Northern Hemisphere

Baikal

Located in Lake Baikal in Siberia, this neutrino telescope was the first to go into operation. Deployment and maintenance of the sensor strings are carried out in the winter in order to take advantage of the thick ice on the lake. After several research and development stages, beginning in 1980, followed by construction, beginning in 1990, the Baikal telescope now runs with 192 photomultipliers distributed over 8 strings. The modest size of the detector and the poor transparency of the water in the lake limit sensitivity to cosmic neutrinos. An extension of the detector has nevertheless been planned to increase sensitivity to the highest energies. This will involve deploying 3 further strings, each 140 m long and carrying 36 optical modules, with operations beginning in April 2005. This new configuration should provide a fourfold increase in sensitivity over results already published, and it may itself constitute the first step toward the construction of a detector on the kilometre scale.

NESTOR

The *Neutrino Extended Submarine Telescope with Oceanographic Research* (NESTOR) project was the first to have begun R & D studies in the Mediterannean sea, in 1989. The original aim was to deploy a set of towers, 410 m high, to detect Cherenkov light, eventually forming a kilometre-sized detector made up of over 1 000 photodetectors, at a depth of almost 4 000 m, just off the bay of Navarino, near Pylos in Greece. A first, reduced-size prototype storey, 5 m in radius and equipped with 12 optical modules, was set up temporarily in March 2003. Several events were then recorded which triggered 4 of the 12 optical modules in coincidence. A reconstruction algorithm was used to extract the zenithal angular dependence of atmospheric muons, and also to measure the muon flux at a depth of about 4 000 m. The Nestor team are now concentrating their activities within the Km3NeT consortium, which aims to deploy a kilometre-sized neutrino telescope in the Mediterranean.

NEMO

The NEMO collaboration, created in 1999, originally planned to build a neutrino telescope with kilometre dimensions in the Mediterranean sea at a site about 80 km off the coast of Capo Passero in Sicily, at a depth of about 3 500 m. This site seems to ensure a low concentration of bioluminescent bacteria, which are a source of background noise for the detector. The design is based on semi-rigid towers. The underlying idea is that these towers will require a smaller number of in situ connections than the detector strings used elsewhere. A mini prototype tower comprising 4 storeys was deployed at a depth of 2 000 m, only 25 km away from Catania, and operated between 18 December 2006 and 18 May 2007, leading to about 200 hours of data acquisition. Tracks of atmospheric muons were thereby reconstructed and their angular distribution compared with simulations, revealing a good agreement with predictions. The project has now entered a new phase, integrated into the common programme of the Km3NeT consortium. In July 2007, an electro-optical cable 100 km long was laid between the Capo Passero site and a completely new coastal station in the port of Portopalo.

ANTARES

The ANTARES project, begun in 1996, is the most advanced Mediterranean project. The site chosen for the telescope is in the Porquerolles trench, about 40 km off the coast of Toulon (France). The water is pure with limited scattering of Cherenkov light, so it is possible to achieve good accuracy in measuring the direction of the muon, and hence also the neutrino ($< 0.5°$ at 10 TeV). The 2 475 m of water above the detector ensure partial shielding against atmospheric muons. The telescope comprises 12 independent detection strings, some 60 to 75 m apart, distances comparable with the absorption length of Cherenkov light. This distance also satisfies constraints imposed by safety and maintenance. In particular, a submarine must be able to move between the detector strings to ensure connection to the main junction box. The latter controls the electricity supply to the detectors and centralises the optical fibres coming from the various strings before data is sent to the land station by means of an electro-optical cable. The junction box contains an optical distributor to connect the clock to each string through a signal sent from the coastal station.

Each string is made up of 5 identical sectors, each comprising 5 storeys with a vertical spacing of 14.5 m (see Figs. 7.61 and 7.63). Each storey in turn holds three optical modules linked to the control module for that storey. As well as ensuring voltage distribution and signal transmission, this module is equipped with a compass and a tiltmeter to determine the orientation of the storey. Each string is anchored to the seabed, while its upper end is attached to a buoy which moves around according to the sea currents, together with the all the optical modules. The position and orientation of each optical module must be continuously monitored. The tiltmeters measure the inclination of a storey with respect to the horizontal plane, while the compass measures the components of the Earth's magnetic field in three directions. Each string also carries five acoustic detectors (hydrophones), spread out along its length. Moreover, the anchor holds an acoustic transponder. The hydrophones receive signals emitted by one of the three beacons arranged on the sea floor around the detector or by one of the emitters carried by the strings. By measuring the different travel times between acoustic emitters and receivers, the position of each hydrophone can be obtained by triangulation. The combination of acoustic data and information from the tiltmeters is used to fix the positions of the optical modules to an accuracy of 10 cm. This level of accuracy is needed to guarantee the desired angular resolution.

The detector was gradually deployed over the period from March 2006, when the first detector string was immersed, to May 2008. Since then, ANTARES has been the largest neutrino telescope in the northern hemisphere, with unequalled sensitivity over a large region of the galactic plane and in particular toward its centre. The data already recorded have been analysed to search for cosmic neutrinos, among other things, in the form of an isotropic diffuse signal. This study has established an upper bound comparable with theoretical predictions (see Fig. 7.62). Further data is currently under analysis.

Fig. 7.63 Artist's representation of the strings of light sensors making up the ANTARES detector in the Mediterranean sea. A total of 1 000 photomultipliers are distributed over the 12 strings, covering a horizontal area of 0.1 km^2 over a height of 350 m. Image courtesy of the ANTARES collaboration

Km3Net

The ANTARES, NEMO, and NESTOR groups have joined together to form a European consortium called Km3NeT, with a view to designing and determining the site for the next generation of cosmic neutrino telescope in the Mediterranean. The European Union will fund this operation in two stages. The first, called *Design Study*, began in February 2006 and was completed in 2009 with the publication of a *Technical Design Report* (TDR) detailing the proposed technical solutions. The fundamental requirements for such a detector are first of all an instrumented volume of more than one cubic kilometre and in addition an angular precision of the order of $0.1°$ around 10 TeV. Furthermore, the detector must be sensitive to the three flavours of neutrino. The second stage, or *Preparatory Phase* (PP), began in March 2008. It focuses on strategic and financial issues but nevertheless includes prototyping activities. Construction of the detector should begin in 2013. The Km3NeT project forms part of a broad cross-disciplinary field bringing together a wide range of scientific activities from oceanography, to seismology, geochemistry, and others. The site will be an entirely novel submarine observatory, accessible to a varied scientific community.

7.8 Gravitational Wave Detection

Gravitational waves were briefly presented in Chap. 1, in the context of astronomical information carriers.[29] We defined the dimensionless quantity $h = \delta L/L$, which characterises the amplitude of a gravitational wave and allows an assessment of detector capabilities. Detection involves measuring the relative displacement of two *free masses* (which means that non-gravitational forces can be neglected in the relevant temporal frequency band). Maximal sensitivity is achieved when the masses are separated by a distance equal to half a wavelength, and this property fixes the optimal scale for detection, although it is not necessarily attained, or even possible. Baselines of the order of a million kilometres would be required for the lower frequency waves (less than 1 Hz), and a hundred kilometres for the higher frequency waves (10 kHz).

It is still not clear what to call such instruments. The terms *gravitational wave telescope*, *gravitational antenna*, or *gravitational wave detector* are all sometimes used.

Mechanical Resonance Detectors

The aim here is to measure the deformation of a solid as the wave passes through it. This was the approach followed by Weber in the first attempt at detection

[29] In this section, the authors are grateful for the invaluable assistance of Philippe Laurent.

(1969). The sensitivity was already quite remarkable ($h \sim 3 \times 10^{-16}$), but much too low for the predicted signals. A metal cylinder is insulated from terrestrial seismic perturbations by a high-pass mechanical filter. Excitation of the mechanical resonance of the cylinder produces a relative variation of its length (this being defined by the measurement itself as a mean value, which smooths out local atomic effects), measured by piezoelectric sensors. Various improvements, such as cooling the cylinder down to a temperature of a few kelvin, so as to reduce the effects of Brownian motion, have increased sensitivity by about two orders of magnitude in the kHz frequency domain, by using a 4 800 kg aluminium cylinder.[30] This type of detector, of intrinsically narrow spectral range, will ultimately be limited by quantum uncertainty (square root of spectral density $\approx 1.5 \times 10^{-22}\,\mathrm{Hz}^{-1/2}$), assuming that the antenna can be cooled sufficiently to remove thermal noise.

Interferometric Detectors

The aim is to build a detector, operating over a broad frequency band, and capable of reaching a spectral sensitivity

$$\lesssim 3 \times 10^{-23}\,\mathrm{Hz}^{-1/2}\,.$$

It can be shown[31] that a Michelson interferometer, with arms of length $L > 1$ km, illuminated by a 10 W continuous-wave laser, would achieve this sensitivity in the frequency range between 50 Hz and several kHz (Fig. 7.64).

The great length of the arms pushes back the quantum uncertainty limit. The photon noise of a laser of wavelength λ and power P leads to a spectral uncertainty in the path length difference x, viz.,

$$\Delta x = \frac{\lambda}{\pi}\sqrt{\frac{h\nu}{2P}} \approx 10^{-16}\,\mathrm{m\,Hz}^{-1/2} \quad \text{for } P = 1\,\mathrm{W},\ \lambda = 0.5\,\mu\mathrm{m}\,.$$

The required sensitivity would impose $P = 500$ W, which is quite unrealistic. The light must therefore be 'recycled'. This means reinjecting the light which leaves the interferometer, in order to increase the stored energy without increasing the equivalent length of the arms. The desired spectral sensitivity is then within reach (Fig. 7.65), provided that laser fluctuations in frequency ($\Delta\nu/\nu < 10^{-17}\,\mathrm{Hz}^{-1/2}$), in power ($\Delta P/P < 10^{-6}\,\mathrm{Hz}^{-1/2}$), and in angular spread ($\Delta\phi < 10^{-7}$ rad), can be reduced.

In the mid-1990s, construction began on two interferometers of this type: LIGO in the USA, and VIRGO in Pisa, Italy. They have baselines around 3 km long.

---

[30]Boughin S.P. et al., Ap. J. **261,** L19, 1982.

[31]See the very complete discussion by Vinet, J.-Y. Optical detection of gravitational waves, Compt. R. Acad. Sci. **8**, 69–84, 2007.

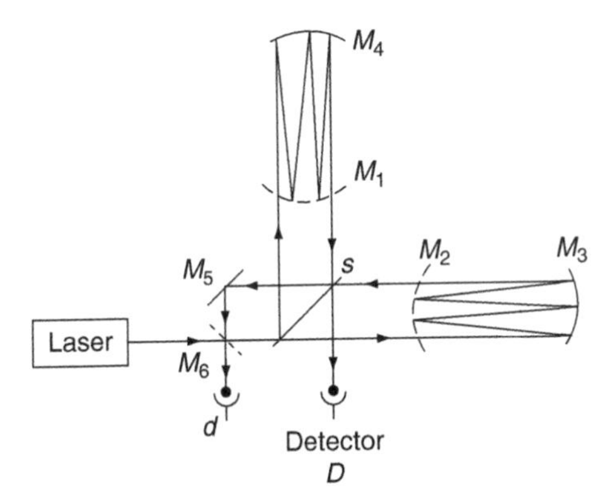

Fig. 7.64 Interferometric detector of gravitational waves. The stabilised single frequency laser illuminates, via a separator S, the two arms of the interferometer, and these are arranged so that the light passes several times, in order to increase sensitivity (the sensitivity is multiplied by n for n passes). Recycling light by mirrors M_5 and M_6 involves adjusting the interferometer to give destructive interference at the detector D, and then adjusting the laser frequency and the mirrors M_1 and M_2 to minimise the signal at the auxiliary detector d. The pairs of mirrors M_1, M_4 and M_2, M_3 are suspended and insulated from seismic vibrations, so that they behave like free masses. The passage of any gravitational wave, whose wave plane coincided with the arms of the interferometer, would cause a variation in the flux of light received at D. (After Brillet A., Virgo project, private communication, 1985)

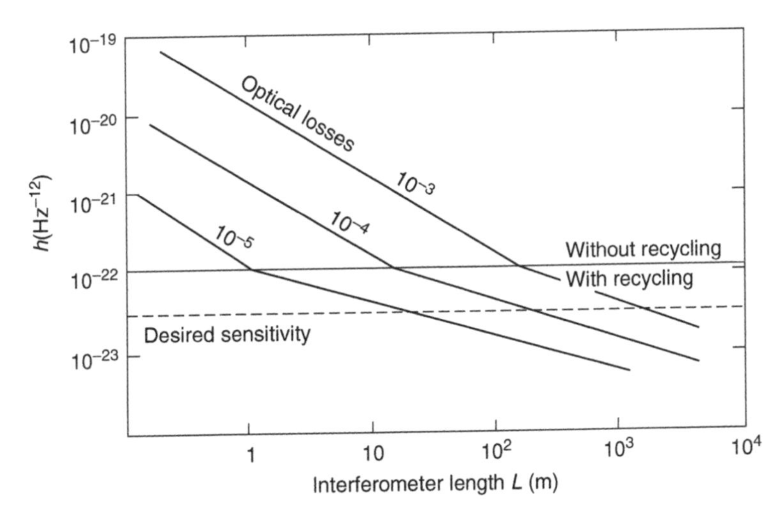

Fig. 7.65 Sensitivity limit imposed by photon noise, for an interferometer of length L, illuminated by a laser of power $P = 10$ W. Optical loss factors are shown, in a reasonable range of values, given the present state of optical surface technology. The desired sensitivity requires recycling and low losses. (After Brillet A., Virgo project, private communication, 1985)

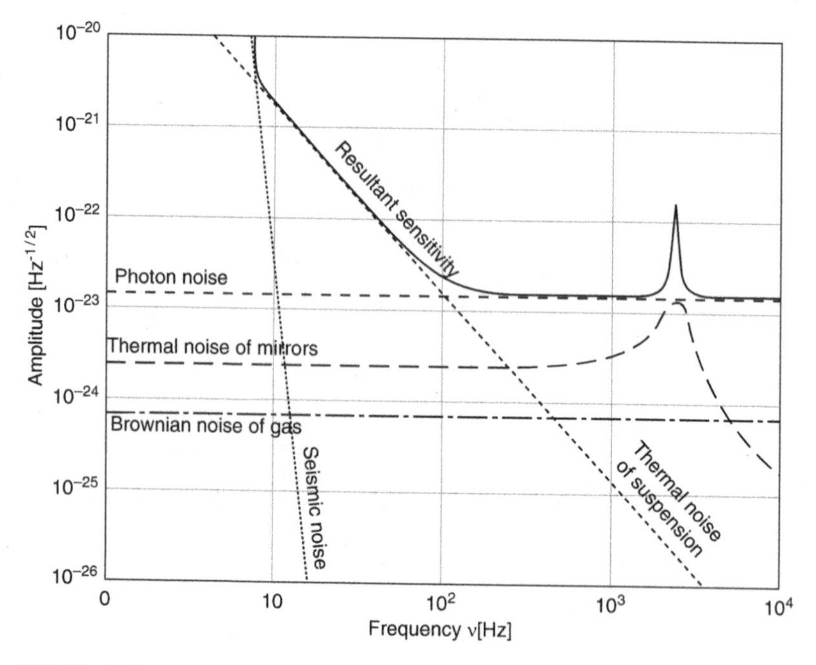

Fig. 7.66 Sensitivity, expressed by the amplitude of the power spectrum of h [$\mathrm{Hz}^{-1/2}$], as a function of the frequency ν [Hz] of the gravitational wave to be detected. The different noise sources and their amplitude in a VIRGO-type interferometer are indicated between 1 and 10^4 Hz. The *continuous curve* gives the resultant, and hence, the best possible performance

Figure 7.66 shows the various noise sources affecting the ultimate sensitivity of an interferometer:

- *Seismic noise*, reduced by the anti-vibration suspension of the mirrors.
- *Thermal noise of this suspension*, which manifests macroscopically the Brownian fluctuations in the position of their centre of gravity.
- *Thermal noise of the mirrors*, whose surface also contributes a position uncertainty, due to the thermal motions of its atoms, despite a certain degree of cooling.
- *Photon noise*, due to the finite power of the laser.
- *Collision noise*, due to residual atoms in the vacuum surrounding the mirrors and beams.

The VIRGO and LIGO instruments[32] are extremely difficult to realise in practice. Results are expected in the decade 2010–2020 and beyond.

The size of the Earth, and seismic effects, set limits which can only be surmounted by placing instruments in space. The baseline could then be increased, using two or three satellites, separated by distances of 10^6 km or more. Low

[32]See virgo.web.lal.in2p3.fr and www.ligo.caltech.edu to follow progress with these instruments.

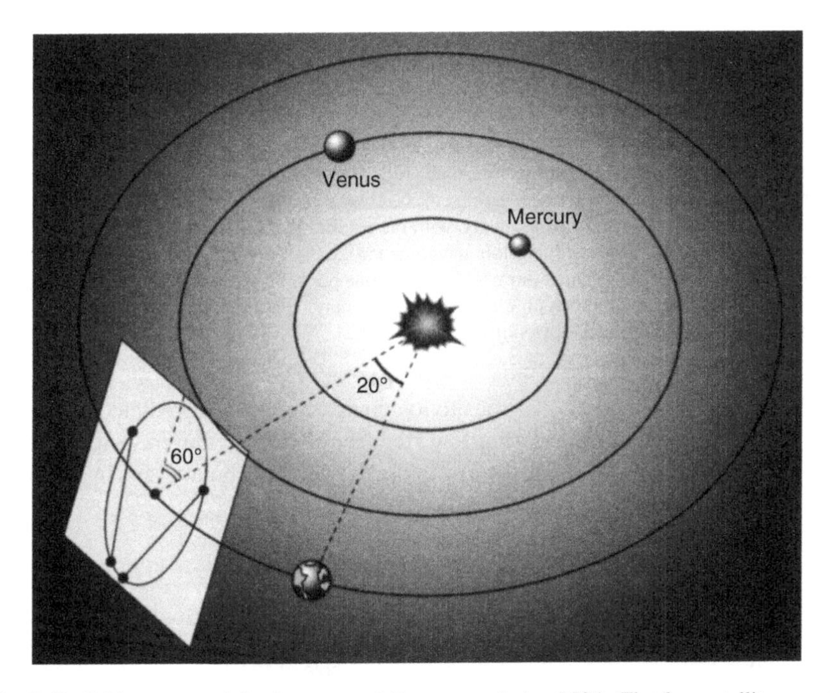

Fig. 7.67 Orbits suggested for the proposed European mission LISA. The four satellites constituting the interferometer orbit in a plane at 60° to the ecliptic, and 20° behind the Earth

frequency gravitational waves would then become detectable ($\nu = 10^{-1}$ to 10^{-4} Hz), with spectral sensitivity up to $10^{-23}\,\text{Hz}^{-1/2}$. The distance between satellites can be measured using laser beams, and the detection limit will very likely be set by perturbing accelerations, due to radiation pressure from solar photons, and impacts of solar wind protons and electrons.

The *Laser Interferometer Space Antenna* (LISA) is an extraordinarily ambitious joint project by NASA and ESA for observing gravitational waves. The project was first considered in the 1990s, with a possible launch in the decade 2010–2020. It will comprise three spacecraft, separated from one another by distances of around 5 million kilometres, but keeping their relative positions to an extremely high level of accuracy, of micrometric order. When a gravitational wave goes by, the spacecraft will be displaced relative to one another, just like a fleet of ships when an ocean wave goes by.

The orbits of the three satellites will be situated on a circle, which will itself lie in a plane tilted at an angle of 60° to the ecliptic, with center trailing the Earth by 20° in its orbit around the Sun (Fig. 7.67). Each satellite will carry two test masses held in free fall, by virtue of an electrostatic servosystem which compensates for the drag effects due to the interaction of the satellite with the interplanetary medium. The distance between the test masses of each satellite will be continually monitored by laser interferometry (Fig. 7.67).

The space interferometer LISA will complement Earth-based detectors for a different frequency range, viz., 10^{-4} to 0.1 Hz. LISA is planned for launch in 2012 or 2013. The very advanced technology used for LISA will have to be tested during a preparatory mission called LISA Pathfinder, which should be launched in 2008.

LISA should be able to detect gravitational waves emitted by very close compact binary systems. One or other of these objects may be a black hole, a neutron star, or a white dwarf. Since these binary systems have small masses, observations of such systems are limited to our own Galaxy and nearby galaxies. However, most galaxies probably contain a black hole at their centre, with a mass several million times the solar mass. It may be that two galaxies sufficiently close to one another could attract gravitationally, with the result that their black holes eventually merge into one. Our best hope of studying this phenomenon lies in gravitational radiation. The rate of occurrence of this type of event is estimated at about one per galaxy per million years. Considering the millions of galaxies in the visible Universe, LISA may be able to detect several such events, associated with these supermassive objects, every year.

Figure 7.68 gives the expected sensitivity, comparing it with predicted amplitudes for several particularly interesting low frequency sources: the gravitational background of cosmological origin, the low frequency signal of binary stars in the galaxy, which are many, and a black hole collision.

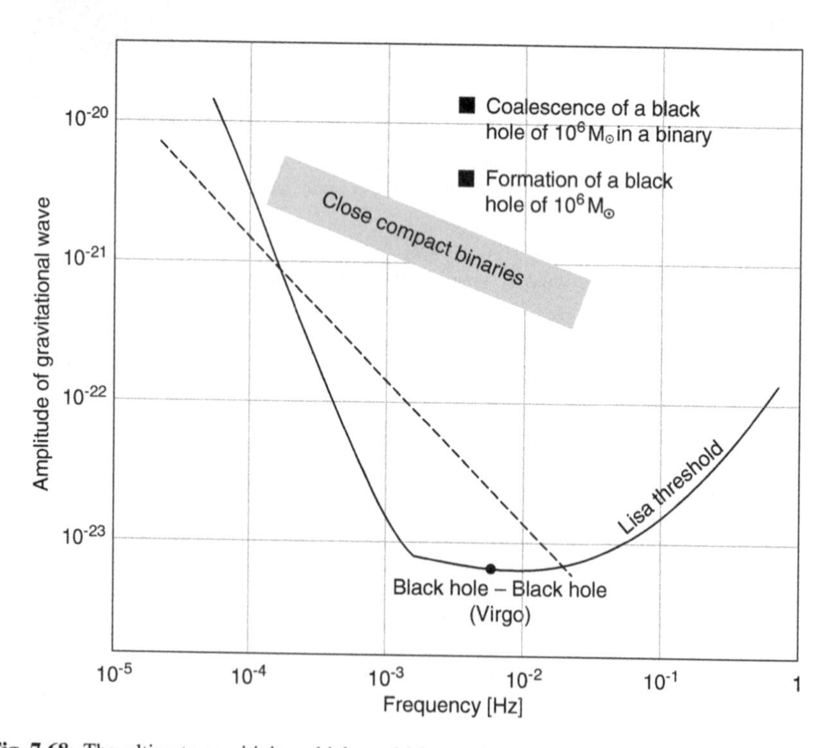

Fig. 7.68 The ultimate sensitivity which could be envisaged for a low frequency gravitational wave detector in space (European mission LISA). The calculated amplitudes of various sources are given. The *dashed line* shows the theoretical contribution of the *cosmological background of gravitational waves*. Integration period 1 year, and signal-to-noise equal to 5. (LISA Working Group project and European Space Agency, 1994)

Problems

7.1. A photomultiplier (PM) receives a flux of N photons per second. Its quantum efficiency is η. Describe the Poisson process for the photocurrent $i(t)$ (mean density, power spectral density). The PM is assumed to filter $i(t)$ by the transfer function $\Pi(f/f_c)$. Calculate the variance of the photocurrent.

Answer. $\sigma^2 = 2\eta N e^2 f_c = 2e\langle i \rangle f_c$.

7.2. Starting from the noise power at the ends of a resistor R, at temperature T (see Sect. 7.2), calculate the variance $\langle \Delta i^2 \rangle$ of the current in the resistor.

Answer. $\langle \Delta i^2 \rangle = 4kT\Delta f/R$.

7.3. Consider a sighting, formed by strips of equal width, alternately opaque and transparent, with n strips per mm. Describe the image observed in a system for which the MTF is zero beyond the spatial frequency: (a) $w_c < n$ mm^{-1}, (b) $n < w_c < 2n$ mm^{-1}, (c) $w_c \gg n$ mm^{-1}. Answer the same question for an MTF such that $\tilde{G}(w_c) = 0.5G(0)$ (value at 3 dB). The magnification of the system is 1.

7.4. DQE in the Presence of a Dark Current. If the photocathode of the receiver (or its equivalent in the case of a semiconductor detector) emits γ electrons cm^{-2} s^{-1}, the pixel area being a, show that the DQE is

$$\delta = \left[\left(1 + \frac{\sigma_g^2}{\overline{g}^2}\right)\left(1 + \frac{a\gamma}{\eta \overline{N}}\right) + \frac{\sigma_L^2}{\overline{g}^2 \eta \overline{N} T}\right]^{-1} .$$

7.5. Signal-to-Noise Ratio and Spatial Frequencies. Consider a multichannel detector, with MTF $\tilde{G}(\boldsymbol{w})$, assumed isotropic to simplify. The detector gives an image $I(\boldsymbol{r})$. In the absence of a signal, the 'dark image' (i.e., the grain of the unexposed photographic plate) is the random function $I_0(\boldsymbol{r})$, with spectral density $\tilde{I}_0(\boldsymbol{w})$. This noise is assumed additive. The signal noise is the photon noise on each pixel, expressed by the spectral density $S_0(\boldsymbol{w})$. Show that the DQE is

$$\delta(\boldsymbol{w}) = \eta \frac{S_0(\boldsymbol{w})}{|\tilde{I}_0(\boldsymbol{w})|^2/|\tilde{G}(\boldsymbol{w})|^2 + S_0(\boldsymbol{w})} .$$

Deduce that, in the presence of grain, the signal-to-noise ratio S/N is generally smaller when the high frequencies of an image are studied than when the low frequencies are studied.

7.6. High-Speed Television. Calculate the passband needed to transmit a high resolution TV image (1 000 lines), at a rate of 50 images per second. The intensity of each pixel is defined on 12 bits.

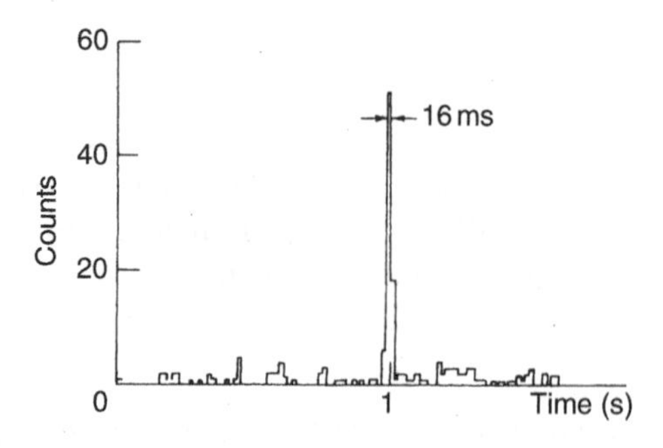

Fig. 7.69 Structure of an intense γ-ray emission

7.7. Gamma-Ray Bursts. Figure 7.69 shows the structure of an intense γ-ray burst, detected on 13 June 1979 by the Venera II probe. The time resolution is less than 1 ms (scintillator). Deduce the maximum possible size of the unknown source which produced this emission. (Vedrenne G., *Les sursauts* γ, La Recherche **12**, 536, 1981.)

7.8. Noise in Heterodyne Detection. Show that the general expression

$$\frac{S}{N} = \frac{T_{\text{source}}}{T_{\text{n}}} \left(\frac{\Delta\nu_{\text{s}}}{2 f_{\text{c}}} \right)^{1/2},$$

which gives the signal-to-noise ratio in a direct detection, applies equally to heterodyne detection, where $\Delta\nu_{\text{s}}$ is then the bandwidth of the IF filter, and T_{n} the total noise temperature. When is the result independent of the power of the local oscillator (assumed noise-free)? At fixed ν_0, how should the signal ν_{s} be filtered before the mixer to avoid superposition of frequencies $\nu_{\text{s}} + \alpha$ and $\nu_{\text{s}} - \alpha$? Why is a single-band receiver used in spectroscopy, when double-band receivers are used to measure continuum spectra? Why is the signal-to-noise ratio S/N increased by a factor of $\sqrt{2}$ in the latter case?

7.9. Quantum Noise Limit. Starting from the analysis of quantum noise in Sect. 7.2, show that the quantum fluctuation of blackbody radiation is reached when the noise temperature T_{n} of a detector, sensitive to only one polarisation of radiation at frequency ν, is such that $kT_{\text{n}} = h\nu$. Reconsider Fig. 7.43 in the light of this result.

7.10. The Structure of the Universe at Large Distances. Figure 7.70 (after Kellerman and Toth, ARAA **19**, 373, 1981) shows the number $N(S)$ of radiogalaxies as a function of their flux S. The abscissa is the flux density S at $\lambda = 6$ cm, and the ordinate is the difference $\Delta N = \log[N(S)/60]$ between the observed distribution $N(S)$ and a uniform distribution in volume in a static Euclidean universe, as it is observed over short distances. All the galaxies are assumed to have the same luminosity. Determine the antenna temperature T_{A} corresponding to

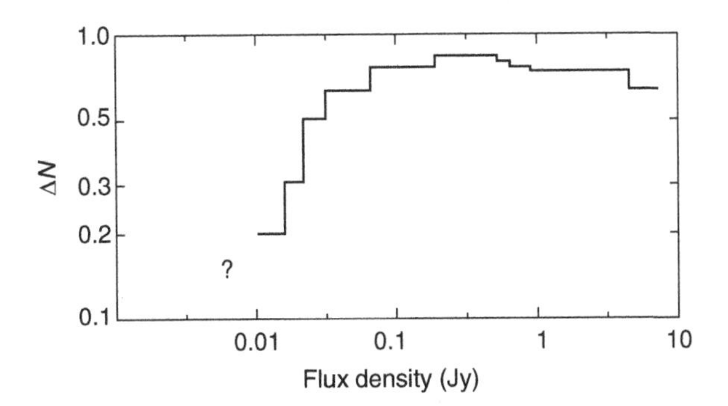

Fig. 7.70 Density of radiogalaxies as a function of their flux

the faintest sources (10 mJy). Determine the telescope time τ required to observe a source with S/N = 10, $\Delta\nu_s = 100$ kHz, $T_n = 300$ K.

Answer. $\tau \sim 10^4$ s.

7.11. A field contains two galaxies, of magnitudes $m_V = 18$ and 28. An image is formed in such a way that each one only occupies a single pixel of the detector. The telescope, of diameter 3.6 m, is assumed to have unit transmission, and a standard V filter is used (of effective bandwidth $\Delta\lambda = 0.089$ μm). Assuming reasonable quantum efficiencies for (a) a CCD detector, (b) a photon counting camera, study how the signal-to-noise ratio varies for each object, as a function of exposure time. The CCD can be assumed to have readout noise 30 electrons rms.

Answer. The flux from the source is, in W m^{-2} μm^{-1},

$$e = e_0 10^{-m/2.5} ,$$

and, in terms of counts at the detector,

$$n = \frac{eS\Delta\lambda}{(hc/\lambda)\eta t} ,$$

where $S \approx 10$ m^2 is the collecting area, $\Delta\lambda = 0.089$ μm is the effective bandwidth of the filter, $\lambda = 0.55$ μm is the wavelength of the filter, η is the quantum efficiency, and t is the exposure time.

(a) CCD detector. The readout noise is a fixed, additive noise, independent of the read time, hence the expression

$$\frac{\mathrm{S}}{\mathrm{N}} = \frac{n}{\sqrt{n} + 30}$$

Fig. 7.71 Hecht's formula

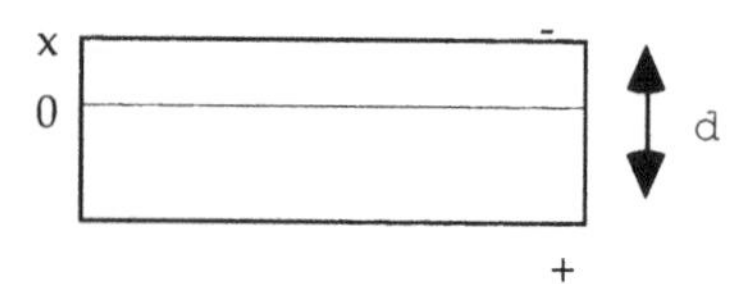

for the signal-to-noise ratio. The readout noise becomes significant when the number of counts accumulated during an exposure is low (in this case, less than 1 000). For an efficiency of 70%, and $t = 1$ ms, S/N $= 2 \times 10^3$ for $m_{\mathrm{v}} = 18$, and S/N $= 8.5$ for $m_{\mathrm{v}} = 28$.

(b) Photon counting camera. S/N $= \sqrt{n}$ (proportional to $\sqrt{t}$). For an efficiency of 10%, and $t = 1$ ms, S/N $= 7.9 \times 10^2$ for $m_{\mathrm{v}} = 18$, and S/N $= 7.9$ for $m_{\mathrm{v}} = 28$. These values must be corrected for a real transmission of less than unity.

7.12. The photon counting camera is rapidly saturated by bright objects in the field. Calculate the magnitude of a quasar, which saturates such a camera, used with a 3.6 m telescope of unit transmission, in the B photometric band, using the fact that saturation occurs at 100 counts per pixel per second.

7.13. Prove Hecht's equation from Ramo's theorem, which gives the charge collected in a semiconductor detector (Fig. 7.71) at energies above a few tens of keV.

Answer. The number of charges trapped in dx, after travelling a path length x (Fig. 7.71), is

$$\mathrm{d}n = \mathrm{d}x \frac{n_0}{\lambda} \mathrm{e}^{-x/\lambda} .$$

Calculating then the total charge Q deposited, for a given type of charge carrier,

$$Q = e \left(\int_0^l \frac{x}{d} \,\mathrm{d}n + \frac{l}{d} n_0 \mathrm{e}^{-l/\lambda} \right) ,$$

Hecht's formula is obtained by integration.

Chapter 8
Spectral Analysis

Spectroscopy is at the heart of astrophysics, and indeed, it is usually spectroscopic observations and their interpretation which provide the strongest constraints on the models proposed by astrophysicists. There are a great many examples: energy balance, abundance of molecules, atoms, ions, or other particles, macroscopic or microscopic velocity fields, local physical conditions such as the temperature, the density, the magnetic or electric fields, states of equilibrium or of deviation from equipartition of energy, and so on.

Whether the subject of study is planetary or stellar atmospheres, hot and dilute interstellar media or cool and dense molecular clouds, the intergalactic medium, accretion or thermonuclear processes occurring near the surface of very compact objects, energy generation processes in galactic nuclei, the expansion of the Universe, or the cosmic background radiation, spectroscopy is almost always the main diagnostic tool.

For this reason, the development of *spectrometers*, using the available photons to maximum advantage, is a constant priority. The combined efforts of astrophysicists and physicists, who are equally interested in this area, have brought spectrographic instrumentation close to perfection, in the sense that it attains or exceeds the limits of resolution imposed by the physical conditions in which the radiation is produced. The main advances today stem from better coupling between spectrometers and detectors, so that spectrometry can now benefit from any progress made in detector development.

The aim of this chapter is therefore to discuss the fairly classical techniques of spectroscopy proper. It begins with a brief survey of the use of spectra in astrophysics, extending the study of radiation in Chap. 3, and introducing the vocabulary and general ideas basic to all spectroscopy. This is followed by some general comments about spectrometers. Then we describe the spectrometry based on interferometric techniques, which plays a fundamental role, right across from the X-ray to the far-infrared region. Its application to the design of astronomical spectrometers is discussed. Radiofrequency spectrometry falls into a category

P. Léna et al., *Observational Astrophysics*, Astronomy and Astrophysics Library,
DOI 10.1007/978-3-642-21815-6_8, © Springer-Verlag Berlin Heidelberg 2012

of its own and is treated separately. Gamma-Ray spectrometry (energies above 10 keV), together with certain X-ray spectrometers (bolometers), use intrinsic energy discrimination by detectors, and this subject is therefore dealt with in Chap. 7 which examines detectors.

8.1 Astrophysical Spectra

The radiation received from a source is characterised by its specific intensity $I(\nu, \boldsymbol{\theta})$, at frequency ν, and in the direction of observation $\boldsymbol{\theta}$. The polarisation of the radiation can also be specified (linear, circular, elliptical, etc., see Sect. 3.1). The spectrum of the source is given by the ν dependence of $I(\nu, \boldsymbol{\theta})$. Strictly speaking, the term *spectrum*, which is used here and is common usage, refers to the spectral density of the electromagnetic field, regarded as time dependent variable, and more often than not, a random time variable.

8.1.1 Formation of Spectra

The elementary interaction between radiation and matter (atomic nuclei, atoms, molecules, solid particles, etc.) results in an energy exchange. When the system makes a transition from one state to another, with energies E_1 and E_2, and the probability of photon emission or absorption varies greatly with frequency ν in the neighbourhood of the transition frequency,

$$\nu_0 = \frac{E_2 - E_1}{h},$$

where h is Planck's constant, the microscopic conditions for the formation of a spectral line are satisfied. This line may be an emission line or an absorption line, depending on the direction of the energy exchange, and the fundamental probabilities are given in detail by nuclear, atomic, molecular, or solid state physics, depending on the nature of the system.

The radiation $I(\nu)$ received by an observer is the result of a macroscopic summation over a large number of elementary interactions situated either inside the source itself, or, especially in the case of a more distant object, along the whole length of the line of sight. These radiation–matter interactions cause emission, absorption, and elastic or inelastic scattering, depending on the various media encountered, their physical state (solid, gaseous, plasma, etc.), their physical parameters (temperature, pressure, velocity field, etc.), and the local fields (magnetic, gravitational, etc.). The spectrum finally observed thus results from a complex process of radiative transfer. The general *transfer equation* can be written

$$I(\nu) = \int_0^{\infty} S(\nu, x) e^{-\tau(\nu,x)} \, dx,$$

where $S(\nu, x)$ is the *source function*, which characterises the elementary emission at the point x, measured along the line of sight, ν is the frequency, and τ is the *optical depth* between 0 and x,

$$\tau(\nu, x) = \int_0^{x} \kappa(\nu, \xi) \, d\xi,$$

with $\kappa(\nu, \xi)$ the local absorption coefficient for the radiation. *Inversion* of the transfer equation,[1] establishing the local source function $S(\nu, x)$ at the object, is a fundamental problem in astrophysics, leading as it does from the measured spectrum $I(\nu)$ to the local conditions which created S.

The great diversity of observed spectra is a consequence of the complexity of radiative transfer processes, in media such as stars, interstellar gases, and molecular clouds, which are themselves diverse.

The spectrum is said to be *continuous* if the observed function $I(\nu)$ varies slowly with ν. The blackbody sky background radiation, the continuous radiation of the Sun, or synchrotron radiation emitted by active galactic nuclei, are all good examples of continuous spectra, in which radiative transfer has attenuated or smoothed away the emissions of discrete transitions on the microscopic scale.

The spectrum is referred to as a *line spectrum* if it exhibits emission or absorption lines, which means that $I(\nu)$ varies rapidly over some narrow frequency interval $\Delta\nu \ll \nu$ (Fig. 8.1). The absorption spectrum of a star of advanced spectral type, and hence cold, or the submillimetre emission spectrum of a molecular cloud, are typical examples of line spectra.

The Main Transitions and Corresponding Frequencies

The quantification of matter systems assigns energy levels E_i, and hence associates spectral transitions with those systems. Depending on its complexity, a semi-classical or quantum mechanical treatment is required to evaluate the probability of a transition between energy levels, or the cross-section of an interaction with radiation. The main transitions which concern astronomical spectroscopy are reviewed in this section, and summarised in Table 8.1. When the radiative transfer is sufficiently simple, the spectrum $I(\nu)$ received from a source is dominated, at least in a restricted spectral region, by a local source function expressing specific spectral transitions.

---

[1] Methods for inverting the transfer equation are discussed in Rybicki G. B., Lightman P., *Radiative Process in Astrophysics*, Wiley, 1979 and also Kourganoff V., *Introduction to Advanced Astrophysics*, Reidel, 1980.

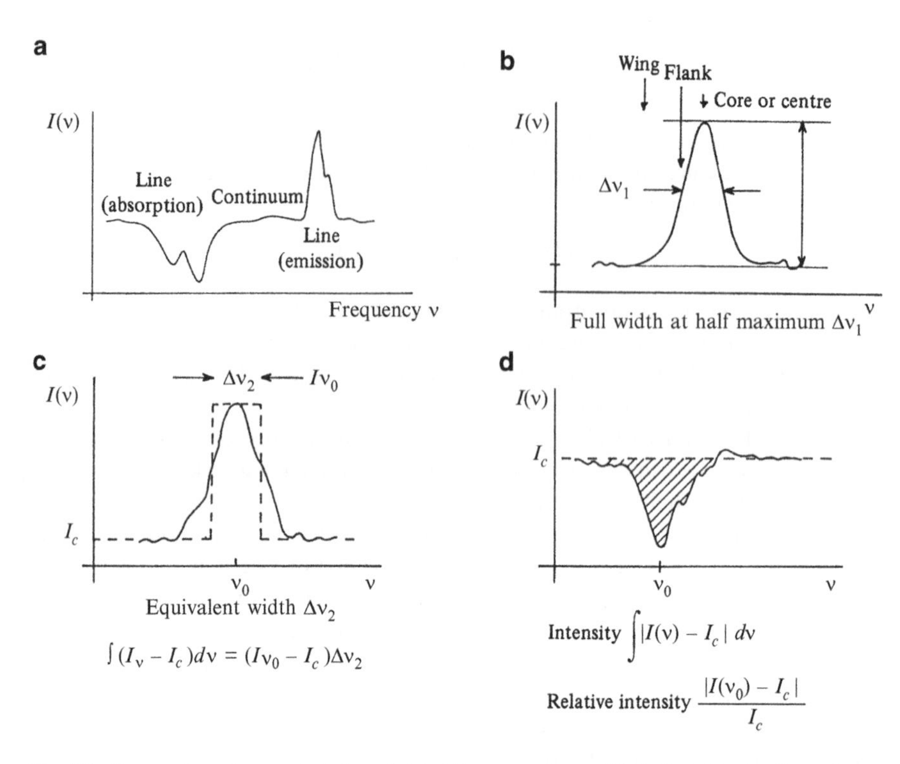

Fig. 8.1 Terminology for spectral lines. (**a**) Arbitrary spectrum. (**b**) Full width at half maximum of an emission line. (**c**) Equivalent width of a line. (**d**) Relative intensity of an absorption line

Electronic Transitions

In the case of the hydrogen atom, or any hydrogen-like atom, the frequencies ν_0 radiated when the atom changes energy state are given by the *Rydberg formula*

$$\nu_0 = 2cRZ^2 \frac{m-n}{n^3},$$

where n and m are the principal quantum numbers characterising the two energy states, $R = 109\,737.31\ \mathrm{cm}^{-1}$ is the Rydberg constant, and Z is the effective charge as seen by the electron. This simple expression will now be used to specify some systems undergoing transition, and the energy ranges they involve.

- **The Hydrogen Atom.** The transitions with $n, m < 5$, whose energies $h\nu$ lie between 10 and 0.1 eV for $Z = 1$, are radiated in the near UV, the visible and the near infrared. These energies correspond to temperatures $T = h\nu/k$ between 10^5 and 10^3 K, and are therefore characteristic of the outer layers of stars (stellar atmospheres), or ionised hydrogen regions in interstellar space. The spectra of these objects are thus dominated by such transitions. For example, the Lyman

Table 8.1 Examples of discrete transitions

Transition	Energy [eV]	Spectral Region	Example
Hyperfine structure	10^{-5}	Radiofrequencies	21 cm hydrogen line
Spin–orbit coupling	10^{-5}	Radiofrequencies	1667 MHz transitions of OH molecule
Molecular rotation	10^{-2}–10^{-4}	Millimetre and infrared	1–0 transition of CO molecule at 2.6 mm
Molecular rotation–vibration	1–10^{-1}	Infrared	H_2 lines near 2 μm
Atomic fine structure	1–10^{-3}	Infrared	Ne II line at 12.8 μm
Electronic transitions of atoms, molecules and ions	10^{-2}–10	Ultraviolet, visible, infrared	Lyman, Balmer series, etc., of H, resonance lines of C I, He I, and K, L shell electron lines (Fe XV, O VI)
Nuclear transitions	$> 10^4$	X and γ rays	^{12}C line at 15.11 keV
Annihilations	$\gtrsim 10^4$	γ rays	Positronium line at 511 keV

α line ($\lambda = 121.6$ nm) of hydrogen appears in the spectra of hot stars, and the Balmer α line ($\lambda = 656.3$ nm) appears in the spectra of ionised nebulas (H II regions), and so on.

- **Hydrogen-like Atoms.** The Rydberg formula shows that spectra are shifted towards the near or extreme ultraviolet as Z increases, which raises the related energies and temperatures accordingly. It is worth recalling here the special notation used by astrophysicists: a neutral atom carries the label I, e.g., C I, H I, while a once-ionised atom carries the label II, e.g., H II (H^+), Fe II, and so on. This group includes the important lines of helium (the $1s^2$–$1s2p$ transition of He at $\lambda = 58.4$ nm), of carbon (the $2p3s$–$2p3p$ transition of the neutral atom, or C I, at $\lambda = 1069.1$ nm), and many others. Ions such as He II, C II, Fe^{n+}, etc., also undergo this type of transition, the effective charge Z increasing with the degree of ionisation. Their spectral lines are characteristic of hotter regions, including accretion disks in binary systems, the solar corona, and eruptions at the surfaces of stars.
- **Molecules.** The frequencies of electronic transitions in molecules consisting of several atoms like H_2, CO, and NH_3, radicals like OH and CN, or molecular ions like H_2O^+, are given by a more complicated expression than the Rydberg formula, but nevertheless fall into the same energy range, from the near ultraviolet (10 eV) to the near infrared (0.1 eV). The principal sites generating such spectra would be the atmospheres ($T \approx 3\,000$ K) and circumstellar envelopes ($T = 1\,000$ to $3\,000$ K) of cold stars, rich in molecules and radicals that are stable at low temperatures, and also dense interstellar regions, excited by shock waves.
- **Deep Atomic Shells.** Transitions involving low electron shells, and hence high effective charges Z, close to the total charge of the nucleus ($Z = 10$ to 100), lead to energies as high as 10 keV, given the Z^2 dependence of ν_0. Such radiation therefore occurs in the EUV and X-ray regions, and the associated temperature

can reach 10^8 K. One of the advantages in observing this radiation is that it is indifferent to any relationship the atom may have with other atoms, for example, whether or not it belongs to a molecule or a solid, and it is also irrelevant whether the atom is neutral or ionised. X-ray radiation is observed in the solar corona, accretion phenomena (close binaries, quasars), and thermonuclear outbursts on stellar surfaces (cataclysmic variable stars).
- **Rydberg States.** This term refers to high quantum number (n) states of the hydrogen atom or hydrogen-like atoms. Transitions between such states have energies of the order of 10^{-5} eV, and are observed at radiofrequencies, for sufficiently dilute media in which the temperature pushes the atoms close to ionisation.

Electronic Fine Structure Transitions

The electron levels have a fine structure springing from the coupling between orbital angular momentum and electron spin. The extra levels produced have characteristic energy differences between 10^{-2} and 10^{-5} eV. The principal lines $n \rightarrow m$ are split into several components, hence the term 'fine structure', and there are also transitions between the new sublevels, which are observable in the far infrared or at radiofrequencies. The relative populations of energetically close sublevels are virtually independent of temperature, but highly sensitive to density, and this is the great importance of these lines. Examples are the carbon C I lines at 609 and 370 μm (submillimetre region), C II lines at 158 μm, and the oxygen O I lines at 63 and 145 μm, all of which are important in the study of the galactic interstellar medium.

Electronic Hyperfine Structure Transitions

The coupling of electron spin and nuclear spin also leads to the appearance of sub- levels, the transitions here having energies between 10^{-6} and 10^{-5} eV, which radiate at centimetre radiofrequencies. The best known example is the splitting of the ground state ($n = 1$) of hydrogen into two sublevels, separated by 5.9×10^{-6} eV. These sublevels are populated even in the cold phase of the interstellar medium, and lead to the important emission line at 21 cm, corresponding to the radiofrequency 1420.406 MHz. In the same way, the hyperfine structure of the molecular radical OH, present in the interstellar medium and circumstellar media, leads to several lines in the neighbourhood of 1 667 MHz ($\lambda \approx 18$ cm).

Molecular Transitions

The movements of molecular structures (rotation, vibration) require less energy than those associated with electronic transitions. Their spectral signatures are thus located in the infrared, submillimetre, and millimetre regions.

The simplest case is a rotating dissymmetric diatomic molecule, possessing therefore a non-zero dipole moment, and moment of inertia I. When quantised, this motion is characterised by a quantum number J and a corresponding energy

$$E(J) = \hbar^2 \frac{J(J+1)}{2I}.$$

Radiative transitions $J \rightarrow J-1$, referred to as pure rotational transitions, produce photons of energy $\hbar^2 J/I$.

As an example, the molecule CO exhibits a fundamental transition 1–0 at wavelength $\lambda = 2.6$ mm (230.538 GHz), as well as all the transitions for higher values of J, which occur in the submillimetre and far infrared. The same is true for a great number of molecules and radicals to be found in molecular clouds.

Quantising the vibratory motions of the atoms which make up a molecule gives more widely separated energy levels, usually in the range 1 to 10^{-2} eV. Each of these levels is split into sublevels by the molecular rotation. A whole range of transitions becomes possible, and this gives an appearance of complexity to the vibrational–rotational spectra of molecules, situated in the near and middle infrared (1–100 μm). Only moderate excitation temperatures are required, in the range 100 to 3 000 K, and these exist in circumstellar envelopes, and the atmospheres of cold stars, when molecules or radicals like SiO or OH are present. Emission from the OH radical is also present in the Earth's atmosphere. The hydrogen molecule H_2 deserves a special mention: it occurs abundantly, but is symmetrical, and thus displays no pure rotational spectrum. On the other hand, its vibrational–rotational spectrum can be observed in the near infrared ($\lambda \sim 2$ μm), in particular during shock wave excitation.

Nuclear Lines

The quantification of the energy states of atomic nuclei leads to levels separated by very large energies, which means that the associated transitions are observed in the gamma-ray region. For example, the excited states of ^{14}N, ^{12}C, and ^{16}O produce spectral de-excitation lines around 2.31, 4.43, and 6.14 MeV, respectively.

Other mechanisms may also produce spectral lines. A case in point is the nucleosynthesis of elements, exemplified by cobalt ^{60}C lines at 1.33 and 1.17 MeV. These can be observed during supernova explosions, but are also produced in the interstellar medium by cosmic rays. Another case is the so-called *positronium line*, produced at 511 keV by electron–positron annihilation.

Transitions in Solids

In a crystalline solid comprising a very great number of atoms (at least a thousand), the regularity from a large distance partially removes the individual properties of its atoms or molecules. The matter–radiation interaction manifests the collective energy transfers of the crystalline array, whose quantised vibrations are called *phonons*. These energies are necessarily less than the bonding energies within the crystal, which are of the order of 1 eV. For this reason, the characteristic spectra of solids lie in the near or far infrared (between 1 and 100 μm). Any given solid has characteristic frequencies depending on the mass of its atoms, the type of bonds between them (ionic or covalent), and so on. In the case of amorphous solids, which do not display the same ordered appearance as crystals, when viewed from a great distance, spectral lines can nevertheless be identified, and indeed, this is because there is some order over short distances. Hence, H_2O ice, in crystalline form, has a characteristic emission wavelength near 45 μm and another near 3.05 μm, both of which are slightly displaced in amorphous ice.

Some of the transitions associated with solids are particularly important in determining the physical composition of small solid particles or grains, produced in comets, the atmospheres of cold stars, and also in novas. They are abundant in the interstellar medium. Silicates, like SiO_4Mg, have been identified by their transition at 9.6 μm, and so, too, has water ice, as mentioned earlier.

Between isolated molecules and macroscopic solids are aggregates, which may comprise from ten to several hundred molecules. They may exhibit the beginnings of a crystal structure, but without completely losing the spectral features of their constituent atoms or molecules. Polycyclic aromatic hydrocarbons (PAH), formed by juxtaposition of benzene rings, represent an intermediate stage in the progression towards the crystalline structure of graphite, and are present in the interstellar medium (Fig. 8.2). Their characteristic spectrum consists of lines in the near infrared (3.31 μm, and a few others).

8.1.2 Information in Spectrometry

In this section we discuss the relation between the spectrum as measured and the information it contains about the physical conditions governing its emission. The aim is not to study how spectra are formed, which involves the transfer equation, but a few elementary notions are useful in understanding the constraints that spectroscopic observation and spectrometers must satisfy.

Qualitative Features of a Spectrum

The great complexity of the process linking microscopic interactions of well-defined frequency ν_0, and observed spectra $I(\nu)$, has already been underlined. Figure 8.1 shows some local views of a spectrum $I(\nu)$, and highlights the main empirical characteristics of a spectral line, which must be physically interpreted later.

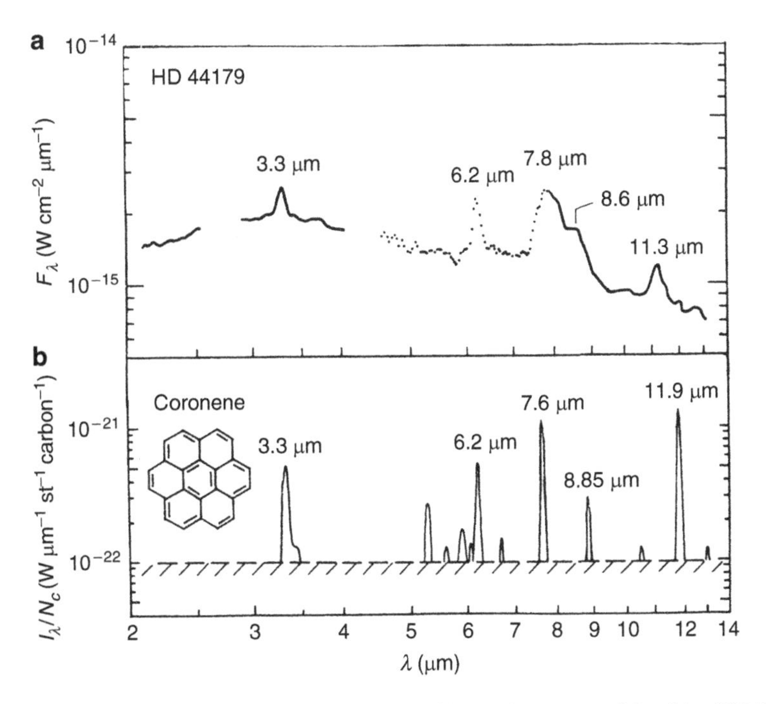

Fig. 8.2 The spectra of molecular aggregates. *Upper*: Observed spectrum of the object HD 44179, a galactic reflection nebula, known as the Red Rectangle. The spectrum was obtained in the near infrared, using a variable passband interference filter. Those lines not originating in atoms or elementary molecules are indicated by their wavelength. (After Russell, 1978). *Lower*: Emission spectrum of coronene $C_{24}H_{12}$. The probable identification of the lines as originating from coronene, or a PAH of this type, is made by comparison. From Léger A., Puget J.L., Astron. Ap. **137** L5, 1984. With the kind permission of Astronomy and Astrophysics

A spectrum consists of a continuous part $I_c(\nu)$, varying slowly with ν (for example, blackbody radiation), and superposed lines. The following definitions will be useful:

- **Line Profile.** Several parameters are used to represent complex profiles in a simplified way (Fig. 8.1): the *full width at half maximum*, or at *half depth*, $\Delta\nu_1$, the *half power beam width*, or HPBW, and the *equivalent width*, $\Delta\nu_2$. These convenient parameters are only of use if the line profile is relatively simple. Also used are the *core* or *centre*, the *flanks* or *edges*, and the *wings*.
- **Line Intensity.** This denotes the total power contained within the line, excluding any contribution from the continuum. The *relative intensity* measures the contrast between the line and the continuum (Fig. 8.1d).
- **Line Position**. This quantity is just the position of the maximum of $I(\nu)$, for emission, and the position of the minimum of $I(\nu)$, for absorption. It can be given as a frequency ν (Hz), a wavelength λ, a wave number $\sigma = 1/\lambda$ (cm^{-1} or kayser), or an energy $h\nu$ (eV). If the line is broad or highly asymmetric, it may be

preferable to define its position as the frequency corresponding to the full width at half maximum or the HPBW.

- **Polarisation of $I(\nu)$.** The spectrum may not be the same when observed in different polarisations. For example, the spectra $I_+(\nu)$ and $I_-(\nu)$, observed in two opposite circular polarisations, will be different if the radiation is produced by the Zeeman effect.

Each of these empirical parameters can be used to describe an arbitrary spectrum, in any spectral region, from the gamma-ray region to radiofrequencies.

Physical Parameters Associated with Lines

The first and most fundamental application of spectroscopy in astrophysics is to identify the presence of some element, molecule, or ion in the emitting region. This science began in 1868, with the spectroscopic discovery of helium in the solar spectrum (Lockyer). Measurement of macroscopic velocity fields is often made concurrently. Smaller spectral shifts, due to gravity or magnetic fields, are supplementary sources of information.

Position of the Line

The position ν_0 of the line is related to the transition which produced it, and thus, taking into account the accuracy with which spectroscopic frequencies can be determined in the laboratory, to the well-identified presence of some given constituent. However, a frequency change may have occurred somewhere between emission and reception, and this complicates the identification. We now consider some of the possible causes of such a change.

Doppler Effect. The emitter is in motion relative to the observer, with relative velocity v. The resulting change in frequency $\Delta\nu$, referred to as the Doppler effect, is given by:

$$\Delta\nu = \begin{cases} \nu_0 \dfrac{v_\parallel}{c}, & \text{non-relativistic case } (v \ll c), \\[2ex] \nu_0 \left[1 - \dfrac{\left(1 - v_\parallel^2/c^2\right)^{1/2}}{1 - v_\parallel/c} \right], & \text{relativistic case,} \end{cases}$$

where $v_\parallel$ is the projection of the velocity along the line of sight. Some types of motion are relatively simple, well understood, and easy to correct for, such as the relative motion of the Earth (diurnal or orbital rotation), the motion of a space observatory relative to the Sun, or the motion of the Sun relative to the centre of mass of the Galaxy. Other types of motion are less well understood, such as the relative speed at which a gas jet is ejected from a galaxy, or the local speed of rotation of some element within a molecular cloud. The *Local Standard of Rest*, or LSR, is therefore defined, choosing if possible the centre of mass of the object under study. After correcting for any trivial effects, the speed v_{LSR} of the LSR relative to the Earth can be defined, and also the local velocity field relative to the LSR:

$$v_{\mathrm{local}\|} + v_{\mathrm{LSR}\|} = c\frac{\Delta\nu}{\nu_0}.$$

For small speeds ($v \ll c$), corrections are correspondingly small, and transitions are easy to identify. The speeds can then be deduced. By contrast, for the observation of distant objects, with large cosmological redshift $z = \Delta\nu/\nu$, $z \sim 1$, such as quasars or galaxies, it may be difficult to identify the spectrum, knowing neither the value of z nor the transition frequency in the laboratory frame.

An original application of the Doppler effect is *Doppler imaging*. Consider the surface of a rotating star, as seen by a distant observer, who is assumed for simplicity to be in the equatorial plane of the star. The photosphere of the star may contain spots, characterised by their lower temperature and the presence of a magnetic field. Both of these factors influence the profiles of spectral lines emitted by such spots, which thus differ from the profiles of lines emitted in calmer photospheric regions. The speed of rotation, projected along the line of sight, depends on the latitude and longitude of the emitting point, and varies right through the rotation, unless the spot is located at one of the poles. In general, the observer does not have available an instrument capable of resolving such details on the stellar surface, and a single spectrum is obtained which mixes together all the information coming from the various points of the surface. However, this spectrum exhibits an evolution in time, and the profile of any particular line will also evolve. It is then possible to invert, in some sense, the set of profiles obtained over a period of time, and reconstruct the configuration of spots over the stellar surface (their extent, temperature, and even their magnetic field), with the help of several reasonably simple assumptions. A kind of image is thus obtained, without possessing any other information than the spectrum (Fig. 8.3).

Zeeman Effect. When the line ν_0 is produced in the presence of a magnetic field, the existence of Zeeman sublevels causes several components to appear, with different frequencies and polarisations. In the simplest case (the normal Zeeman effect), three components appear, with frequencies ν_0 (the π component) and $\nu_0 \pm \Delta\nu$ (the σ components). The formula for $\Delta\nu$ is

$$\Delta\nu = \frac{eB}{4\pi m} = 1.4 \times 10^{10}\, B,$$

where B [T] is the magnetic field. The π component is linearly polarised in the plane containing the line of sight and the direction of the magnetic field, whereas the σ components are elliptically polarised. If B is directed along the line of sight, the π component disappears, and the σ components are circularly polarised. The Zeeman shift may be much less (by a factor of 5×10^{-3}) than the width $\Delta\nu$ of the line, as described above, but nevertheless detectable through its polarisation property. The

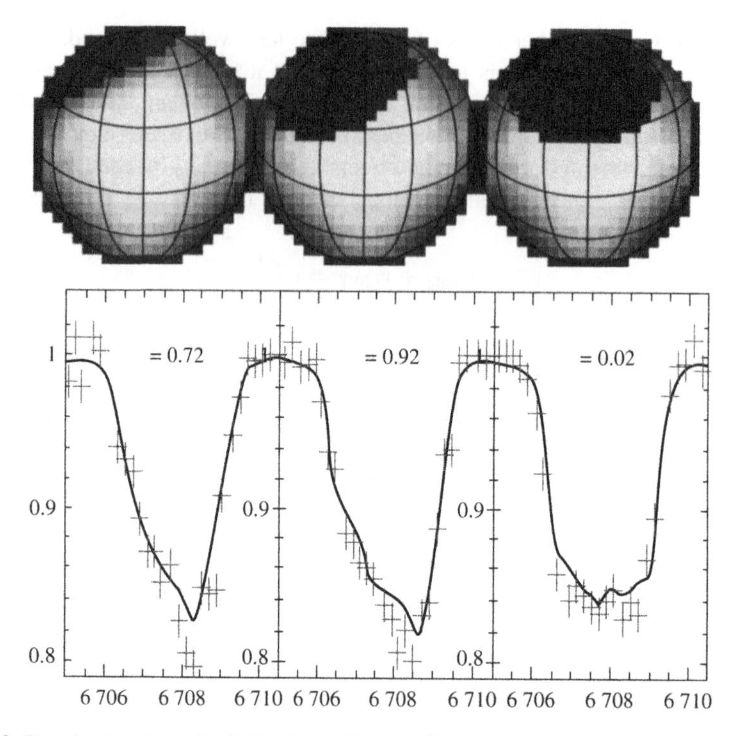

Fig. 8.3 Doppler imaging of a T Tauri star. The star V410 Tau is observed in the lithium line ($\lambda = 670.7$ nm) at different instants. The line profile (*bottom*) varies in time and is measured by the phase ϕ of the rotation of the star, which has period 1.52 days. Using the profiles for various values of ϕ, a unique surface brightness distribution can be deduced (*top*), which clearly shows the spots, at temperature 3 000 K, on a photosphere at temperature 4 800 K. The Aurélie spectrometer with the 1.2 m telescope at the Observatoire de Haute-Provence (France). Document courtesy of Joncour I., 1994

magnetic field of sunspots has been measured using a magnetograph to detect the Zeeman shift of lines at visible frequencies. The interstellar galactic magnetic field has been measured by its Zeeman splitting of the hydrogen 21 cm line, using the polarisation selectivity of radiofrequency detectors.

Einstein Effect. When a photon of frequency ν_0 leaves the surface of a massive object, its frequency changes, and this *gravitational redshift* is given by

$$\frac{\nu_0 - \nu}{\nu} = \left(1 - \frac{2GM}{Rc^2}\right)^{-1/2} - 1,$$

or, in a weak gravitational field,

$$\frac{\Delta\nu}{\nu} \sim \frac{GM}{Rc^2}.$$

This effect is very slight, but nevertheless measurable, in the case of the Sun. It becomes more significant for white dwarfs, and even more so again for neutron stars.

Line Intensity

The integrated intensity (Fig. 8.1c), which measures the total absorption or emission in a line, is a function of the number N of nuclei, atoms, ions, molecules, or solid grains taking part in the transition at ν_0:

$$\int_{\text{line}} [I(\nu) - I_c] \, d\nu = f(N).$$

This is a linear relation when N is small enough, but ceases to be so when N becomes large and the line is saturated. The theory of the *growth curve* of lines determines $f(N)$, and in particular its linear part, as a function of the characteristic atomic parameters of the ν_0 transition. Indeed, we have now arrived at the second fundamental application of spectroscopy, namely, determination of the *abundance* of a given constituent in the emitting region.

Line Profile

The profile $\phi(\nu)$ is defined here as a function describing as closely as possible the shape of the line, and normalised to unity:

$$\int \phi(\nu) \, d\nu = 1.$$

The profile $\phi(\nu)$ of a line reflects exactly the local physical conditions which produced it.

Doppler Lines

When the atoms, ions, or molecules undergoing transitions are in Maxwellian thermal agitation, at some temperature T, the line profile is primarily determined by the Doppler effect due to that agitation. It is given by a Gaussian function

$$\phi(\nu) = \frac{1}{\sigma(2\pi)^{1/2}} \exp\left[-\frac{(\nu - \nu_0)^2}{2\sigma^2}\right],$$

where the frequency dispersion σ is given by

$$\sigma = \frac{\nu_0}{c\sqrt{2}}\left(\frac{2kT}{m} + V^2\right)^{1/2},$$

and m is the mass of an individual atom, ion, or whatever type of particle is involved. This general expression for σ includes both the kinetic effect of the temperature T, and also those effects due to any isotropic microturbulence speeds, whose most likely value is denoted by V.

The full width at half maximum of this profile is called the *Doppler width* of the profile and is given by

$$\Delta\nu_{\mathrm{D}} = \frac{2\nu_0}{c}\left[\ln 2\left(\frac{2kT}{m} + V^2\right)\right]^{1/2} = 2.3556\,\tau.$$

A *maser line* is formed when a self-amplified stimulated emission process is predominant, a situation which arises, for example, with H_2O lines and radical OH lines in the interstellar medium. In this case, the width of the line is reduced compared with the expression given above, the amplification being more intense for frequencies at which the line is already intense. If τ_0 is the optical depth at the centre of the line, assumed unsaturated, the width is given by

$$\Delta\nu = \frac{\Delta\nu_{\mathrm{D}}}{\sqrt{\tau_0}}.$$

Table 8.2 gives some typical relative widths, for later comparison with the resolution of spectrometers.

When the medium emitting the line manifests additional macroscopic motions, such as large scale rotation, macroturbulence, spherically symmetric expansion or collapse, etc., the line profile carries the signature of these motions and can be interpreted.

Lines Dominated by Collisions

At sufficiently high pressures, collisions between those particles undergoing transitions and other particles of the medium lead to a further damping effect, which

Table 8.2 Doppler spectral widths (hydrogen atom)

V [km s^{-1}]	T [K]	$\Delta\nu/\nu_0$
0	10	1.9×10^{-6}
0	10^4	6×10^{-5}
0.3	10	2.7×10^{-6}
30	10	2×10^{-4}

Table 8.3 Physical and spectroscopic parameters

Type of element	Line position
Abundance	Intensity or equivalent width
Macroscopic velocity field	Position and profile
Temperature, pressure, gravity	Intensity
Microscopic velocity field	Profile
Magnetic field	Zeeman components, polarisation

broadens the line. The line profile, in this case, is referred to as a *Lorentz profile* and is given by

$$\phi(\nu) = \frac{1}{2\pi} \frac{\Delta\nu_L}{(\nu - \nu_0)^2 + (\Delta\nu_L/2)^2},$$

where $\Delta\nu_L = 1/\pi\tau$, and τ is the mean time between collisions, directly related to the physical conditions described by agitation speed and collision cross-section.

General Case

A more complex profile, called the *Voigt profile*, results when the Lorentz and Doppler widths are of comparable order. This profile is the convolution of the two previous profiles.

This brief review shows that the profiles of spectral lines are a rich source of information about the physical conditions that prevail in the emitting region (Table 8.3).

8.2 Spectrometers and Their Properties

Ideally, a *spectrometer* is an instrument that can measure and deliver the quantity

$$I_0(\nu, \boldsymbol{\theta})$$

which describes the intensity received from the source, as a function of the frequency ν, for each direction $\boldsymbol{\theta}$ of the object. If the observation is made simultaneously for several values of $\boldsymbol{\theta}$, and therefore in several pixels of the image, the instrument is referred to as an *imaging spectrometer.*

In practice, the spectrometer delivers a quantity $I(\nu, \boldsymbol{\theta})$ differing from $I_0(\nu, \boldsymbol{\theta})$. Indeed, its ability to discriminate between two neighbouring frequencies is never perfect. The detector associated with it, which measures the received intensity, suffers from fluctuations or noise, and the image (or $\boldsymbol{\theta}$ dependence), if such is provided by the spectrometer, is subject to distortion, aberration, and so on. And, of course, the spectrometer can only provide the values of $I(\nu, \boldsymbol{\theta})$ in the limited frequency range for which it was designed.

A spectrometer and the detector associated with it must therefore be adapted to the type of source (point-like or extended), and the type of spectrum (spectral region, width and intensity of lines), in order to obtain the maximum amount of information in the shortest possible measurement time.

8.2.1 Quantities Characterising a Spectrometer

Spectral Resolution

This fundamental quantity refers to the capacity of the spectrometer to give the frequency dependence of $I(\nu, \boldsymbol{\theta})$, and is illustrated in Fig. 8.4. It is a direct consequence of the way the spectrometer is designed. We consider here an instrument with no imaging capacity, and therefore limited to one direction $\boldsymbol{\theta}$, or to a single image pixel.

The *instrumental profile* $P(\nu)$ is the frequency point response of the spectrometer, that is, the quantity $I(\nu)$ delivered when the incident radiation $I_0(\nu)$ consists of one infinitely narrow line $I_0(\nu) = \delta(\nu - \nu_0)$.

The observed spectrum of a source with general spectrum I_0 is given, for a linear detector, by the convolution

$$I(\nu) = P(\nu) \star I_0(\nu).$$

The equivalent width $\Delta\nu_\text{P}$ of the instrumental profile is defined by

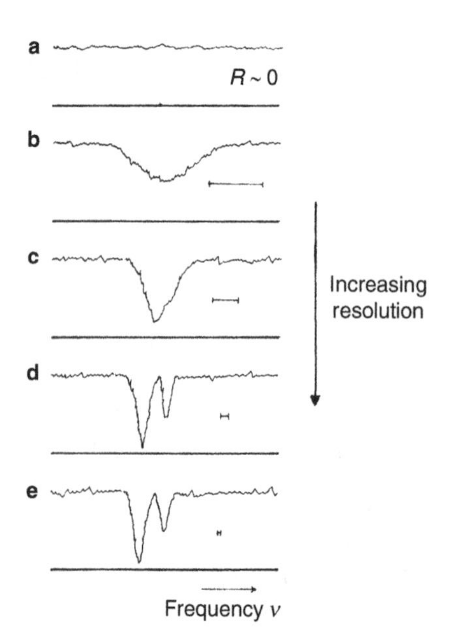

Fig. 8.4 Observation of a spectral line at increasing resolution. (**a**) $R \sim 0$. (**b**) The line appears. (**c**) The line appears double, but unresolved (*blended*). (**d**) The two lines are resolved. (**e**) The true width of the lines is attained, and line purity can be no further improved by increasing R. The *horizontal line* shows the instrumental spectral width $\Delta\nu = \nu_0/R$ used in each observation

$$\Delta\nu_{\mathrm{P}} = \frac{\int P(\nu)\,\mathrm{d}\nu}{P(\nu_0)}.$$

It is clear that two instrumental profiles of the same equivalent width may have very different shapes, and hence specific advantages and disadvantages. The presence of extended wings would not obstruct the search for faint lines around some principal line, but could impede the exact determination of a continuum.

The resolution criterion is defined as the ability of the spectrometer to distinguish two lines $\delta(\nu-\nu_1)$ and $\delta(\nu-\nu_2)$, when the signal-to-noise ratio is assumed infinite. It depends on the exact shape of the profile $P(\nu)$. It is assumed, as a convenient approximation, that the minimum separation which can be resolved is

$$\nu_2 - \nu_1 = \Delta\nu_{\mathrm{P}},$$

and then the *spectral resolving power*, or simply the *spectral resolution*, is defined to be the quantity

$$R = \frac{\nu_0}{\Delta\nu_{\mathrm{P}}}.$$

A measurement of $I(\nu)$, sampled at an interval $\Delta\nu$, possibly much greater than $\Delta\nu_{\mathrm{P}}$, is called a *spectral element*, or *bin*. The precise sampling interval to adopt depends on the shape of the instrumental profile $P(\nu)$, and a correct application of the sampling theorem (see Sect. 9.1.3). Recall that, applying the latter, there is no point in sampling the measurement points of $I(\nu)$ at closer frequency intervals, provided that any noise has been adequately filtered.

The distinction between photometry and spectrometry is rather arbitrary, since in both cases a specific part of the spectrum is singled out. There is a general consensus that spectrometry (or spectroscopy) refers to situations in which the spectral resolution R exceeds a few units.

Table 8.4 gives some typical resolutions R of those spectrometers considered later in the chapter. It is instructive to compare these values with the intrinsic widths of spectral lines given in Table 8.2. Observe that, apart from exceptional cases (for example, measurement of the gravitational redshift by the Sun), it serves no purpose to push the resolution beyond 10^6. Another special case of major astrophysical importance is the measurement of the radial velocities of stars in order to detect exoplanets. Here, the accuracy required for the velocity is of the order of $1\ \mathrm{m\,s^{-1}}$, i.e., a resolution R of the order of 3×10^8 (see Sect. 8.3).

Beam Etendue (Throughput)

A spectrometer can accept a maximum beam étendue or throughput $E = A\Omega$, whilst preserving its maximum resolution, where A is the area of its dispersive element, and Ω is the solid angle of the beam it accepts. The larger the value of the accepted beam étendue, the greater the *light gathering power* of the

Table 8.4 Resolution of spectrometers. The values given in the table are typical orders of magnitude, and should not be considered as absolute

Spectrometer	Region	Typical resolution
Interference filter	Visible, IR	10^2–10^3
Grating	IR, visible, UV	10^3–10^6
Bragg crystal	X ray	10^3
Atomic resonance	Visible, UV	10^7
Fabry–Pérot	Visible, IR	10^4–10^6
Fourier transform	Visible, IR	10^4–10^6
Heterodyne	Radiofrequencies	$> 10^6$
	IR, submillimetre	$> 10^5$
Bolometer (detector)	X ray	10^2
Scintillator (detector)	γ ray	10^3

spectrometer. Since the beam étendue is conserved during propagation through an optical instrument, the condition for the spectrometer to operate with a telescope of collecting area A_T, gathering the radiation from a source in the sky subtending a solid angle ω, is simply

$$A_T\omega = A\Omega .$$

This simple relation shows that a spectrometer with a wide field of view on the sky (ω) must have either a large dispersive element (large A), or highly inclined beams (large Ω). All things being equal, the size of the spectrometer must increase with that of the telescope (A_T), and this imposes a considerable constraint for optical telescopes exceeding 8 or 10 m in diameter.

When radiation is detected coherently, at radiofrequencies, the étendue of the spectrometer is necessarily equal to λ^2 (see Sect. 6.1.1).

Transmission

This is defined for each frequency by

$$t(\nu) = \frac{I_{\mathrm{exit}}(\nu)}{I_{\mathrm{entrance}}(\nu)} ,$$

and is an important factor in determining the throughput of the spectrometer. It can generally be optimised by adequate preparation of all the optical surfaces, in order to minimise reflection or transmission losses. Despite all these precautions, the total transmission of a spectrometer placed at the Coudé focus of a telescope, rarely exceeds 2–5%.

Signal-to-Noise Ratio

When the spectrometer has been fitted with a detector, the final spectrum $I(\nu, \boldsymbol{\theta})$ is characterised by a signal-to-noise ratio for each spectral element, and in the final result this will determine a confidence interval for any measurements of spectral line parameters.

If the instrumental profile $P(\nu)$ is accurately known by calibrating the spectrometer with quasi-monochromatic radiation, $I(\nu)$ can in principle be extracted by *deconvolution* (see Sect. 9.6), or, at least, the resolution can be somewhat improved. In practice, this cannot usefully be done unless the measurement has a high signal-to-noise ratio ($>10^2$).

8.2.2 Spectral Discrimination

Spectrometers can be based on a range of different physical principles depending on the spectral region and the desired resolution R. The aim is always to isolate a spectral element of fixed width with minimal contamination by neighbouring spectral elements. This can be achieved in various ways, outlined here and further discussed later:

- **Interference Method.** The spectral element is isolated by imposing a phase change on the incident wave, then arranging for constructive interference of this wave at the desired frequency and destructive interference at other frequencies. This method is applied over a very broad spectral range from X rays to the far infrared. It uses optical components like interference filters, gratings, or more elaborate systems like Michelson or Fabry–Pérot interferometers.
- **Resonant Electrical Filters.** After reducing the frequency, the electrical signal generated by the incident wave is distributed over a set of filters, each of which selects a spectral interval or element of resolution. This method is used at radiofrequencies on the intermediate frequency (IF) signal, or directly on the signal at very long wavelengths.
- **Digital Autocorrelation.** This method exploits the integral transformation property which, by Fourier transform, relates spectral density and autocorrelation (see Appendix A). At radiofrequencies, high performance correlators can be designed because it is easy to digitise the intermediate frequency (IF) signal, thereby allowing very flexible spectral selectivity.
- **Atomic Resonance.** In this original approach, highly efficient spectral selectivity is obtained by the response of an atomic vapour to excitation by the incident radiation, since this response occurs in a very narrow spectral region corresponding to the resonance line.
- **Detector Selectivity.** As we have seen in Chap. 7, a detector has a certain spectral selectivity, whereby it will be sensitive only in a limited spectral region. Some detectors based on semiconductors have spectral selectivity due to interband

transitions and are used in X-ray spectrometry. Likewise, in γ ray astronomy, it is the properties of the detector itself that incorporate the spectroscopic function. For this reason, γ ray spectroscopy was discussed in Chap. 7.

8.2.3 *The Modes of a Spectrometer*

The way the spectrometer is coupled to the associated detector(s) leads to the following different arrangements for determining the spectrum $I(\nu, \boldsymbol{\theta})$. Each has its own specific advantages and is designed to meet specific objectives: different resolutions, different spectral range, different fields of view to be covered, different numbers of objects to be studied simultaneously, and so on. There may also be other constraints, such as mass and volume for a spaceborne instrument, stability of frequency calibration, and so on. There are therefore many different configurations which we shall now review.

Sequential Spectrometer. This analyses, sequentially in time, the successive spectral elements of a single image pixel (i.e., a single point in the field). This is the case in a grating spectrometer, whose rotation scans the spectrum on a single detector. It thus provides a continuous spectrum. In the case of a Fourier spectrometer, it is the Fourier transform of the spectrum or interferogram which is explored step by step, and the spectrum itself is obtained at the end by a Fourier transform of the interferogram.

Multichannel Spectrometer. This combines a dispersive element and a multi-element detector. Many spectral elements are recorded simultaneously, spectral scanning is unnecessary, and the gain in time is considerable. However, this only provides the spectrum of a single element of the source. An example is a grating spectrometer in conjunction with a CCD array (e.g., 2048 pixels), each pixel receiving a distinct spectral element. To cover a broad spectral range, *echelle gratings* are used. This is also the case for a heterodyne spectrometer equipped with a series of filters, receiving the IF signal in parallel, at radiofrequencies, or with a digital autocorrelator.

Imaging Spectrometer. This simultaneously records the same spectral element, of frequency ν, or even a set of spectral elements covering the relevant spectral range, for all directions $\boldsymbol{\theta}$ of the observed field. This method is invaluable for efficient observation of extended objects. Various instrumental arrangements lead to a whole range of different ways of achieving this. The method is mainly used in the ultraviolet, visible, and infrared spectral regions, where multipixel detectors are available. The aim is always to maximise efficiency, that is, obtain the largest possible number M of spectral elements across the largest possible number of image pixels, and this in the given measurement time T.

For example, with a Fabry–Pérot spectrometer, a single spectral element is obtained at a given instant of time in two dimensions of the image (on a CCD detector), then the successive spectral elements are obtained by scanning as time goes by. Alternatively, if the source has well separated emission lines, the two dimensions of the image can be obtained simultaneously with the spectral dispersion of the image. With the Fourier spectrometer, the imaging mode obtains an interferogram, hence a spectrum, for each point of the field of view.

In long-slit spectroscopy, one dimension of the image is explored along the slit, showing the spectral dispersion perpendicular to the slit. This idea is used to give spectral images of the Sun with a *spectroheliograph*. In a different version, the light from different points of the field is carried optically to a slit by an *image slicer*.

Finally, this method finds its most developed form in *integral field spectroscopy* (IFS). The image is sliced up using a microlens array, with one microlens per pixel of the field, and then each micro-image of the pupil is used as the source for a grism spectrograph, or else injected into an optical fibre which transfers the light to the entrance slit of a conventional spectrograph (the so-called *argus* method, named after the mythological shepherd with a hundred eyes). Another solution is to juxtapose image slicers which redistribute the light over the spectrograph slit, deflecting the light rays from each element of the image in a specific way. These optical components are generally extremely delicate to make. The dispersion angle is judiciously chosen so that the spectra do not overlap on the final CCD detector.

Multi-Object Spectrometer. In an extended field, when there are distinct objects such as stars and galaxies, it is interesting to build a spectrometer that can focus on the radiation emitted by each object in order to carry out spectral analysis, while ignoring the rest of the field. What makes this difficult is that no field is exactly like any other field, so the sampling of the field has to be reconfigured for each new observation. The development of cosmological observations and sky surveys (see Chap. 10) has stimulated this concept of *multiple object* or *multi-object spectroscopy* (MOS). It has been implemented on the very large telescopes with a high level of automation and remarkable efficiency. The field can be sampled by optical fibres or slits cut into a mask.

Spectrometers used in astronomy are usually coupled to a telescope which collects light and forms an image. Of particular importance are the focal ratio f/D of the telescope, which can vary from 1 to 100, depending on the type of telescope and the focal point used, and the focal scale a (in μm or mm per arcsec or per arcmin), which follows directly from the equivalent focal length f of the telescope.

The spectrometer is thus designed to be part of a system, as has already been pointed out. Choosing which specific spectrometric combination to use is a complicated process. It depends on the spectral region, the resolution, the number M of spectral elements required, the field to be covered, and, in certain cases, the simultaneous imaging capacity desired, the type of detector available for the relevant spectral region and the kind of noise affecting it, and finally, the size of the telescope to which the spectrometer must be coupled. We shall now give some examples.

8.3 Interferometric Spectrometers

This vast category of spectrometers uses interference devices. Spectral discrimination is obtained by introducing phase changes in the incident wave, and arranging for constructive interference to occur at the desired frequency, and destructive interference at all other frequencies.

8.3.1 General Criteria

The following parameters are used to compare the various configurations.

- The resolution $R = \nu/\Delta\nu = \lambda/\Delta\lambda$, where $\Delta\nu$ (or $\Delta\lambda$) is the width of the spectral element.
- The total number M of spectral elements, which gives the number of independent values of $I(\nu)$ obtained in the measured frequency interval (ν_1, ν_2).
- The total measurement time T, which, for a given detector, fixes the signal-to-noise ratio for the measurement of the spectrum.
- The power received from the source. Two cases arise: a point source, giving a monochromatic flux $e(\nu)$ [W m^{-2} Hz^{-1}], and an extended source, characterised by its specific intensity $B(\nu)$ [W m^{-2} sr^{-1} Hz^{-1}].
- The detector, characterised by its noise, the beam étendue it will accept, its number of pixels, and the physical dimension of those pixels.
- The beam étendue. It can be shown that, for any interferometric instrument, the maximum beam étendue that such a spectrometer can accept, while maintaining a resolution R, is

$$U = Q\frac{A}{R},$$

  where Q is a dimensionless geometric factor, characteristic of the type of instrument, and A is the area of the dispersive element. The value of U should be compared with the coherence étendue λ^2, when a point source is the subject of observation. But when the source is extended, the spectrometer has greater light gathering power as U increases, for any given resolution.

By combining these factors, a *figure of merit* can be defined for each configuration.[2] For example, for a given observation time and resolution, the question as to which instrument gives the maximum number of spectral elements with the best signal-to-noise ratio can be answered, or which instrument gives the maximal signal-to-noise ratio and resolution for a single spectral element, and so on.

[2]The introduction of the figure of merit and comparison between grating, Fourier, and Fabry–Pérot spectrographs is very clearly explained in Jacquinot P., J.O.S.A. **44**, 761, 1954.

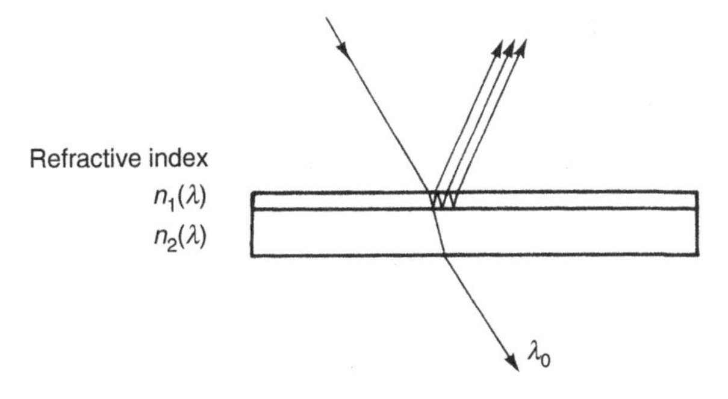

Fig. 8.5 Principle of the interference filter. A single coating is shown

8.3.2 *Interference Filters*

This device (Fig. 8.5), used from the near ultraviolet to the far infrared, well illustrates the principle of spectral discrimination by multiple wave interference. The filter transmission is maximal for a given wavelength when the interference is destructive for the reflected waves. This implies that, for normal incidence,

$$\frac{2n_1(\lambda_0)\,e}{\lambda_0} + \frac{\pi}{2} = 2k\pi .$$

The condition is satisfied for a narrow band around λ_0 and for certain other values λ_i, which are eliminated by an absorbing filter (transmission or absorption). Interference filters, which may contain more than ten layers, can give resolutions R up to 1 000. This is illustrated by the spectra in Fig. 8.2. In the far infrared, the coatings would become too thick, and are replaced by metal grids.

Thin metal films, combining the interference and the absorption by the material, are used as transmission filters in the extreme ultraviolet (λ between 10 and 100 nm). Layers of aluminium, indium, or silicon, several tens of nanometres thick, constitute a narrow filter ($\lambda/\Delta\lambda \approx 1$), whose peak transmission can exceed 60%.

8.3.3 *Grating Spectrometers*

The grating spectrometer also uses multiple interference, the difference being that here all the waves, diffracted from a periodic structure, have the same amplitude.

General Characteristics

Resolution

Let a be the period of the diffracting element, N the number of periods, i and i' the angles of incidence and diffraction, respectively, m the order, and λ the wavelength. The condition for constructive interference is (Fig. 8.6a)

$$\sin i \pm \sin i' = m\frac{\lambda}{a} \qquad \text{(grating relation)},$$

where the sign on the left-hand side is + for a reflection grating and − for a transmission grating.

The angular dispersion is obtained by differentiating the grating relation, for fixed i, which gives

$$\frac{\mathrm{d}i'}{\mathrm{d}\lambda} = \frac{m}{a\cos i'}.$$

The resolution can be deduced from the angular width of the diffraction pattern of the grating, which forms the entrance pupil of the spectrometer,

$$(\mathrm{d}i')_{\text{pupil}} = \frac{\lambda}{Na\cos i'}.$$

Equating $(\mathrm{d}i')_{\text{pupil}}$ with $(\mathrm{d}i')_{\text{dispersion}}$ gives

$$\frac{m\mathrm{d}\lambda}{a\cos i'} = \frac{\lambda}{Na\cos i'},$$

which implies, finally,

$$R = mN.$$

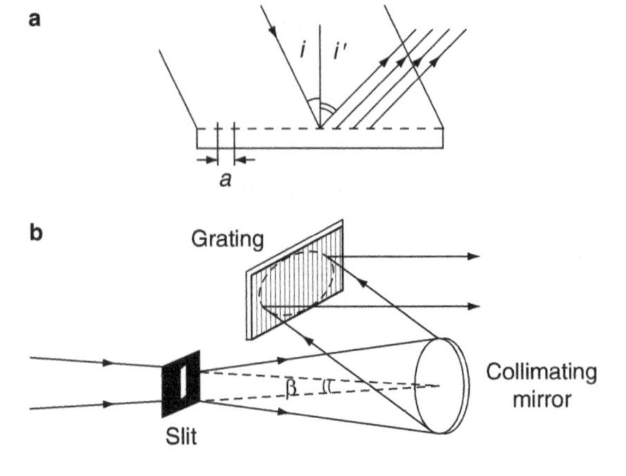

Fig. 8.6 Principle of the diffraction grating

This value, $R = mN$, is a maximum resolution. Often the actual resolution R' is less, being determined by the width L of the entrance slit of the spectrometer. This is placed at the focal point of a collimating mirror of focal length f_{coll} (Fig. 8.6b). In the end, for ground-based observation, the choice of L is determined by the seeing, i.e., the width of the image spot at the telescope focus. This is why the collimator of a high-resolution spectrograph must have a very long focal length. The actual resolution is then

$$R' = \frac{\sin i \pm \sin i'}{\cos i'} \frac{f_{\text{coll}}}{L}.$$

Another parameter often used is the *inverse linear dispersion*, expressed in Å mm^{-1},

$$D = \frac{\mathrm{d}\lambda}{f_{\text{cam}} \mathrm{d}i'} = \frac{m}{a \cos i'},$$

where f_{cam} is the focal length of the camera mirror (Fig. 8.11).

Calculation of the Factor Q

Because a slit must be used to illuminate the grating (Fig. 8.6b), the beam étendue is

$$U = \beta A \cos i \,\mathrm{d}i,$$

where A is the area of the grating and β is the angular height of the slit. At fixed wavelength, the two relations

$$\cos i \,\mathrm{d}i + \cos i' \,\mathrm{d}i' = 0$$

and

$$\cos i' \,\mathrm{d}i' = m \frac{\lambda}{Ra}$$

imply

$$U = \beta \frac{m\lambda}{a} \frac{A}{R}.$$

Since

$$U = Q \frac{A}{R},$$

we deduce that $Q \sim \beta$.

Aberration by the collimating mirror means that β has to be limited to small angles (10^{-2}–10^{-1} rad), thus limiting the usable étendue, and hence the energy accepted by the spectrometer.

Blazed Grating

For an arbitrary shape of periodic diffracting structure, energy is diffracted in an essentially uniform manner into the different orders m, and restricting to just one order is inefficient. The directions of constructive interference and specular reflection from the *blazed* faces of the grating can be made to coincide, for given wavelength and order, by choosing a suitable shape for the periodic structure (Fig. 8.7). The condition to the blaze is (reflection grating)

$$i - \theta = i' + \theta, \quad \theta = \frac{i - i'}{2}, \quad m\lambda_b = 2a \sin\theta \cos(i - \theta).$$

The blaze angle θ is fixed by construction, so that just one blaze wavelength λ_b is associated with each order.

Echelle Grating

The idea here is to use a very high order of interference m, with a long period a ($\gg\lambda$) and a large angle of incidence, such that the *Littrow condition* $i = i' = \theta$ is satisfied. At the blaze wavelength, the grating relation is (Fig. 8.8)

$$m\lambda_b = 2a \sin i = 2t,$$

and the resolution R is, for a slit of width L and a collimator of focal length f_{coll},

$$R = 2 \tan\theta \frac{f_{coll}}{L}.$$

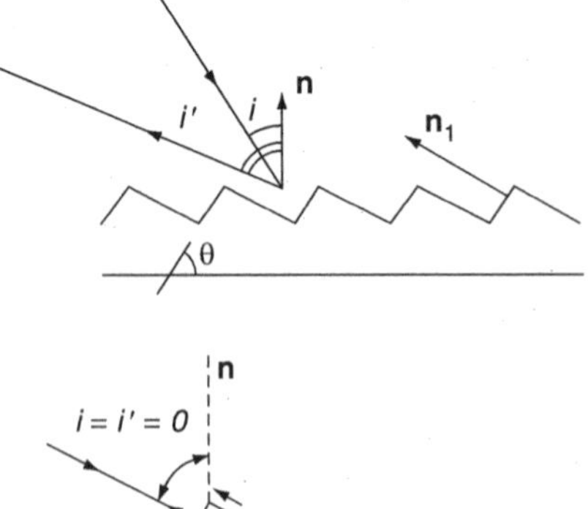

Fig. 8.7 Blazed grating. $\boldsymbol{n}$ is normal to the grating, $\boldsymbol{n}_1$ is normal to the blazed faces, and θ is the blaze angle

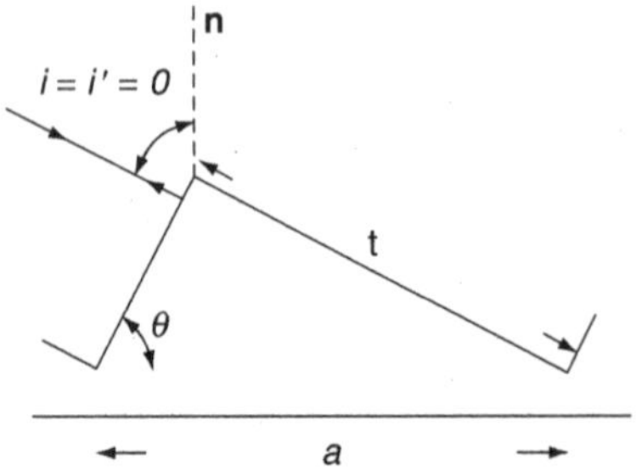

Fig. 8.8 Echelle Grating

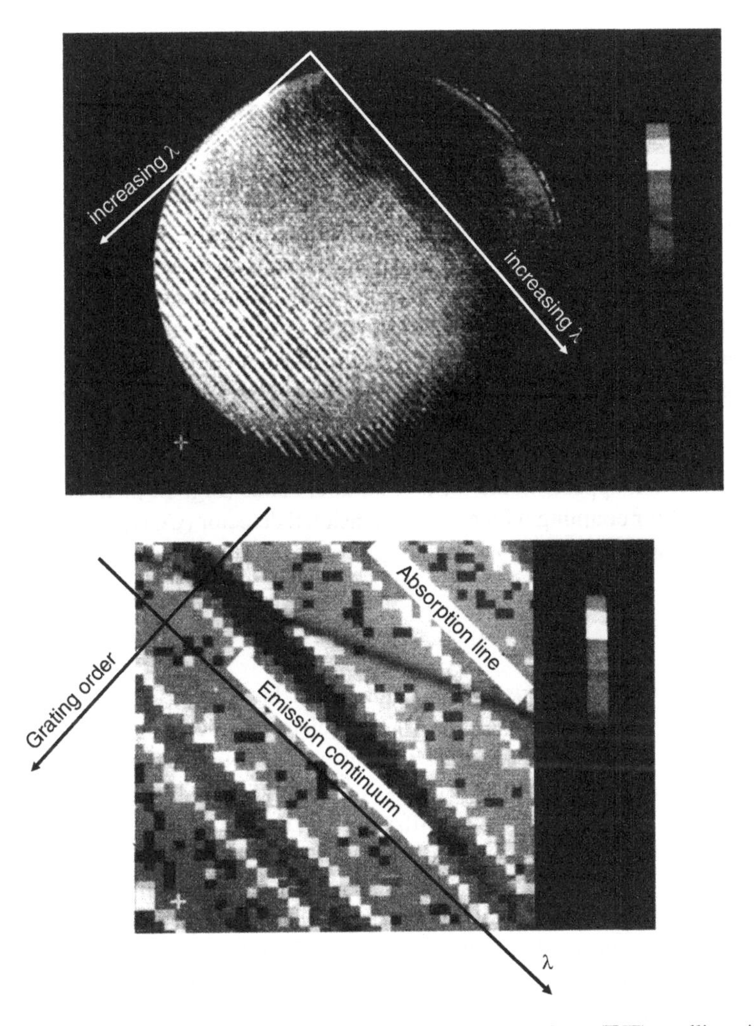

Fig. 8.9 Echellogram obtained by the International Ultraviolet Explorer (IUE) satellite mission. Spectrum of the A-type supergiant α Cyg, in the range 200–330 nm. Spectral resolution 0.02 nm. The detector is a Vidicon tube, with 768 × 768 pixels. The dynamic range is 0–256 counts, each level corresponding to an increment of 16 counts. (**a**) Photo of the whole field. (**b**) Enlargement showing an emission continuum and an absorption line. (European Space Agency and Praderie F., 1984, private communication)

As a is relatively large, the periodicity can be accurately maintained over a large number of grooves, which leads to high resolution ($R = mN$) and high dispersion ($\propto m/a$).

Example. A square grating of side 5 cm has 40 grooves per mm ($a = 25\ \mu$m), blazed at $\lambda_b = 430$ nm. For an angle of incidence $i = 60°$,

$$N = 2\,000, \quad t = 21.6\,\mu\text{m}, \quad m = 100, \quad R = 2 \times 10^5.$$

The orders tend to overlap. The quantity $\Delta\lambda$ defined by

$$(m+1)\lambda = m(\lambda + \Delta\lambda),$$

which implies

$$\Delta\lambda = \frac{\lambda}{m},$$

is called the *free spectral interval.* This spectral interval becomes smaller as m increases, so that an extended spectral region cannot be studied without confusion arising between orders. For this reason, a *predisperser* is introduced. This is a grating or prism whose direction of dispersion is orthogonal to that of the principal grating. Each order is then separated from adjacent orders, and the entire spectrum, or *echellogram*, appears in the form of parallel bands (Fig. 8.9). An echelle grating is thus ideal for coupling with a two-dimensional detector (CCD), and we shall give several examples below.

Making Gratings

The requirements here are high resolution and broad spectral coverage. Echelle gratings with $a \gg \lambda$ can cover the far ultraviolet ($\lambda \gg 50$ nm) and are only limited by the reflection power of the metal deposits used to make them. Ordinary gratings with $a \gtrsim \lambda$ can be made on large scales $Na \sim 0.2$–0.5 m. The grating can be made directly from a photograph of the diffraction pattern by the process of *holographic etching*. Extremely regular spacings can be obtained. The price to pay is a sinusoidal profile focusing the energy less efficiently than a blazed grating. If necessary, the grating can be deposited on a spherical surface, or even on an aspherical surface, which optimises the quality of the image formed on the detector.

In the infrared (large λ), there is no difficulty in making gratings, but the requirement of high resolution leads to prohibitive sizes, e.g., $R = 10^6$ at 100 μm would lead to $Na = 100$ m!

From the 1990s, it has been known how to make *grating arrays* in order to increase the effective value of Na. For example, the *Ultraviolet Visible Echelle Spectrograph* (UVES) of the European VLT uses two plane gratings with a blaze angle of 76°. Holographic gratings can now be made with a non-sinusoidal profile by means of reactive ion etching, whereby the initial sinusoidal profile is modified by ion bombardment.

At X-ray wavelengths (0.5–10 nm), it is still possible to use reflection gratings etched on silicon carbide (SiC) and coated with gold. The European space mission XMM–Newton (1999–2010) carries a reflection grating spectrometer (RGS), whose grating comprises 182 independent and suitably aligned etched plates.

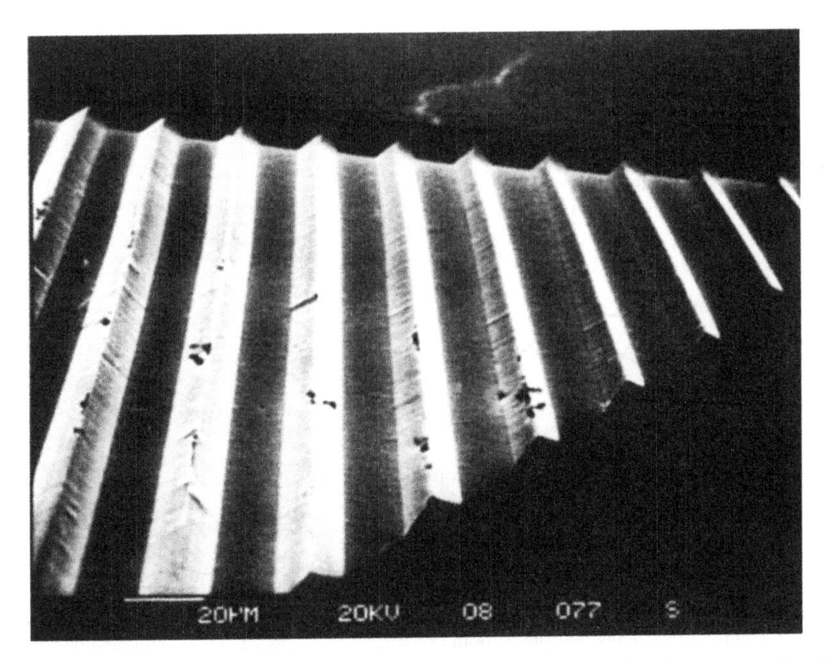

Fig. 8.10 Transmission grating with spacing 35 μm for a grism, made from germanium by anisotropic etching. Mid-infrared camera project for the European VLT. Electron microscope image courtesy of H.U. Kaüfl and the Fraunhofer Institut, Münich, 1994

Grisms

The *objective prism* is a small angle prism (a few arcmin) placed on the entrance pupil. Each star or galaxy image is dispersed in the focal plane on the surface of a CCD detector, whence it becomes possible to carry out low resolution spectral analysis of each object over a wide field. The *grism* exploits a similar idea, but combines a transmission grating and a prism (Fig. 8.10). Note that only the grating really contributes to dispersion by a grism, the prismatic effect being entirely due to the structure of the grating. A grism is in fact nothing other than a blazed transmission grating.

Volume Holographic Gratings

However carefully attempts are made to concentrate light in a single order, the efficiency of etched gratings is always reduced by the existence of multiple orders, which it is difficult to control fully. In the *volume holographic grating*, a thin film of transparent material is placed between two glass plates, and the refractive index of this material is modulated in intensity over a thickness that can also be adjusted. As in a surface grating, the spatial frequency of the modulation determines the

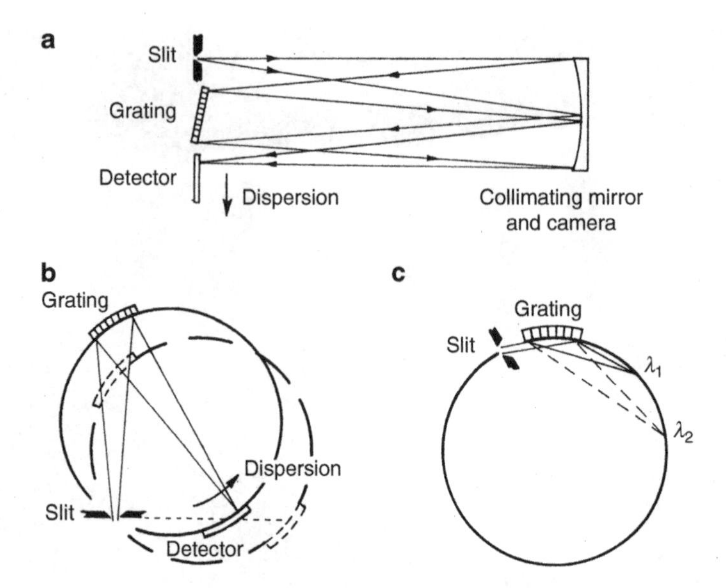

Fig. 8.11 Grating spectrometer arrangements. (**a**) *Ebert–Fastie mounting* with a planar grating. Here, a single mirror plays the role of both *collimator* and *camera*. This could also be achieved using two mirrors. (**b**) *Rowland* mounting with a concave grating. Rotating the assembly around the entrance slit varies the wavelength of radiation received by the detector. (**c**) *Rowland* mounting for the far ultraviolet. The grating operates at grazing incidence ($i = 82$ to $89°$)

diffraction angle (Bragg condition), while the diffracted energy fraction, which depends critically on the properties of the modulation (intensity and thickness), can be made closer to 100% by making the relevant spectral interval narrower. Such a grating can also be used as a spectral filter, adjusted by varying the angle of incidence.

Astronomical Grating Spectrometers

Figure 8.11 shows some simple arrangements, in which the classic components of a grating spectrometer can be identified: the *entrance slit*, the *collimating mirror*, or any equivalent optical device which gives a beam of parallel light (an *afocal* or *collimated* beam), the dispersive element, and the mirror, or any other optical device referred to as the *camera*, which recombines the dispersed image of the field on a one- or two-dimensional detector. It is sometimes possible to combine the collimator and camera functions into a single optical device. A certain number of additional elements can be joined to this basic configuration, such as a predisperser, cross-dispersion, and so on. Having obtained the required resolution and field, the main concern is the quality of the optical image, and especially the total transmission, since spectrometers may contain a large number of reflecting surfaces.

A grating spectrometer must be coupled to a detector. Several coupling arrangements exist between dispersive element(s) and detector, depending on the desired resolution R, and whether the detector is one- or two-dimensional (CCD). Some of these arrangements combine spectroscopy and imaging (*spectro-imaging*), and greatly improve the efficient use of the photons collected.

The various arrangements are as follows:

- **One-Dimensional Detector.** Only one source and one order. If the image of the source is bigger than the spectrometer slit (high R or large seeing disk), an image slicer is required (fibre optics) and the spectrum obtained will mix together the radiation coming from all points of the source.
- **Two-Dimensional Detector.** At high resolution, echelle mode (cross dispersion): only one source but several orders of dispersion. At low resolution ($R < 10^3$), single-order dispersion: one of the detector dimensions can then be used for λ dependence, and the other for spatial dependence along one of the dimensions of the source, i.e., (x, λ)-mode. If spatial information is not required, e.g., as happens for a source which is known a priori to be unresolved, the possibility of binning the pixels of a CCD, i.e., associating several pixels together, provides a way of summing several pixels in the direction of the slit and thereby increasing the signal-to-noise ratio (Sect. 7.4.6).
- **Two-Dimensional Detector and Dispersion.** In this mode, referred to as *integral field spectrometry*, spectral information and two-dimensional spatial information (x, y, λ) are obtained simultaneously. This arrangement, originally due to the French astronomer G. Courtès, is clearly the one which makes the best use of the photons. We shall describe it in much greater detail later.
- **Two-Dimensional Detector.** One order, low resolution ($R < 10^3$), or even high resolution, *multi-object*. By suitably dividing up the image plane (slit mask, fibre optics), the spectrometer samples the radiation at a limited number of image points, juxtaposing the spectra on the detector.

This subsystem must be adapted to the telescope, both from a mechanical point of view (choosing a focal point to suit its mass and stresses) and from an optical point of view (focal ratio, adaptive optics where necessary).

Associated Detectors

Let f be the equivalent focal length of the exit mirror of the spectrometer. The pixel size of the detector must be no larger than

$$f \frac{\lambda}{Na \cos i'}$$

in the direction of dispersion. Photographic plates have now been replaced either by linear arrays of photoelectric detectors, which contain as many sensitive elements as there are spectral elements to be measured simultaneously, or by 2D detectors (CCDs or infrared arrays) in the imaging modes described below.

The sampling condition (Shannon) requires at least two pixels of size p per resolved spectral element, and hence states that

$$2p \le f_{\text{cam}} \mathrm{d}i' = \frac{f_{\text{cam}}}{R} \frac{\sin i \pm \sin i'}{\cos i'} = L \frac{f_{\text{cam}}}{f_{\text{coll}}} \frac{\cos i}{\cos i'}.$$

The slit width L thus appears, modified by the ratio of focal lengths and multiplied by an *anamorphic factor* which depends on the grating.

This can be expressed in another way: if α denotes the angular size of the slit on the sky (for example, the dimension of the seeing), D the diameter of the primary mirror of the telescope, and D_s the diameter of the spectrometer pupil (equal to the useful size of the grating), then

$$2p \le \alpha D \frac{f_{\text{cam}}}{D_s} \frac{\cos i}{\cos i'}.$$

This leads to the expression

$$R = 2 \tan \theta \frac{1}{\alpha} \frac{D_s}{D}$$

for the resolution, assuming the Littrow condition for simplicity.

Application. Returning to the grating used above, with a $D = 4$ m telescope, CCD pixels of $p = 15\ \mu$m, a Littrow mounting ($i = i'$), and a camera mirror of $f_{\text{cam}} \approx 4$ cm, the resolution is $R = 4\,500$ for a slit of dimension $\alpha = 1''$, which is the dimension of an average seeing. This is a long way from the intrinsic resolution of the grating. With $D_s = 25$ cm (which implies a large grating, of length 50 cm), a resolution of $R = 45\,000$ can be attained.

In order to cover a free spectral interval, the detector must contain n_p pixels, where

$$n_p = \frac{\lambda_b/m}{\lambda_b/R} = \frac{R}{m}, \qquad n_p = 2\,000.$$

As the free spectral interval is $\lambda/m = 4.3$ nm, a $2\,000 \times 500$ pixel detector would allow coverage of the whole visible spectrum. Note that the information content of a single spectrum is considerable (10^6 pixels, 12 bits per pixel, and hence more than 10 Mbits per spectrum).

Beam Etendue

The elongated shape of a grating spectrometer slit creates difficulties in matching it to a large telescope. Indeed, consider an image formed on the slit. At ground level, atmospheric turbulence defines a pixel size of around $\theta = 0.5$ arcsec, in good conditions. If A_T is the aperture of the telescope, the corresponding étendue is

$$U = A_T \theta^2.$$

This étendue is accepted by the spectrometer if the slit width corresponds to θ, implying

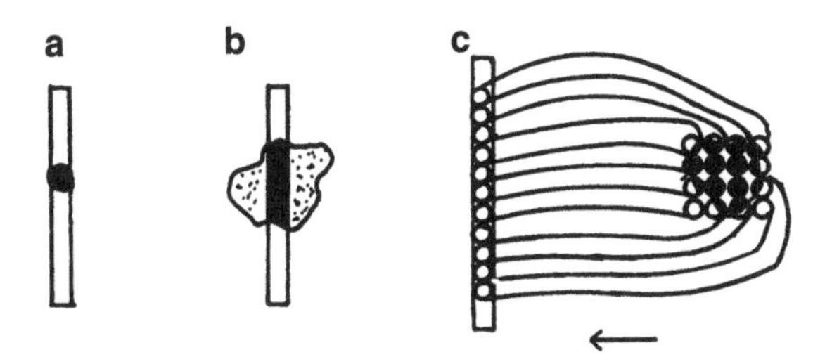

Fig. 8.12 Image slicer. (**a**) Seeing disk and spectrometer slit of comparable size. (**b**) Seeing disk more smeared, so that part of the energy misses the slit. (**c**) A bundle of optical fibres brings all the energy back to the slit

$$A_{\mathrm{T}}\theta^2 \sim \frac{A}{R}\,\mathrm{d}i \sim \frac{A}{R^2}.$$

Putting $A_{\mathrm{T}} = 10\ \mathrm{m}^2$, $A = 1$ m, the result is

$$R \lesssim \frac{1}{\theta}\sqrt{\frac{A}{A_{\mathrm{T}}}} = 10^6.$$

At ground level, high resolutions thus require very good images and a telescope whose size does not exceed a few metres. For larger telescopes, an *image slicer* must be used (Figs. 8.12 and 8.17), in order to bring all the energy to the spectrometer slit. In space, or when using adaptive optics on the ground, the image size is limited by diffraction, and not by turbulence, so the problem disappears.

At first sight, grating spectrometers would not appear well suited to obtaining the spectra of extended objects (i.e., several arcsec or arcmin across, depending on the area A_{T} of the telescope), if a high resolution, between 10^4 and 10^6, is sought. Below we shall see how imaging spectroscopy has solved this problem in several different ways.

Echelle Spectrometers

We have already described the underlying idea of the echelle spectrometer. It is important enough in astronomy to justify giving several examples.

High Resolution Spectrograph (HRS)

Figure 8.13 shows the arrangement of the high-resolution spectrograph (HRS) of the Hubble Space Telescope (since 1989), for which the spatial resolution in the image reaches the diffraction limit with $D = 2.4$ m, while the spectral resolution can reach 10^5.

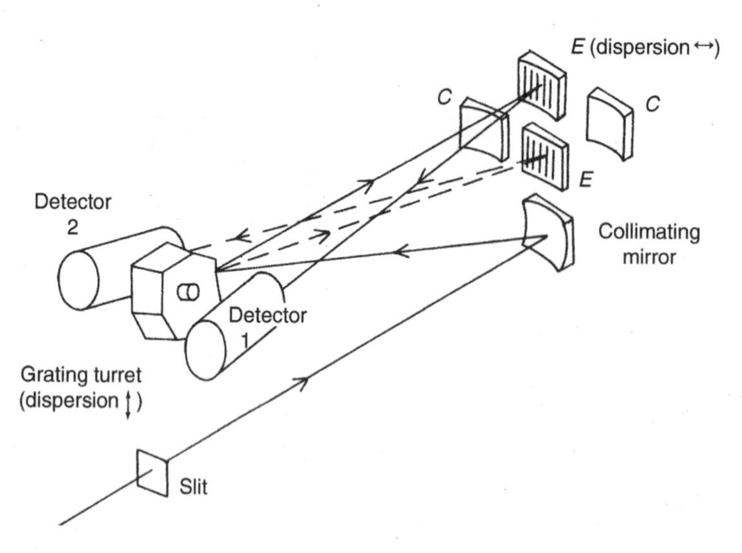

Fig. 8.13 Optical arrangement of the high resolution spectrograph (HRS) of the Hubble Space Telescope. The turret holds a set of six planar holographic gratings ($a = 0.16$ to $3.16\ \mu$m). In normal operation, the mirrors C form the spectrum on detector 1 or 2 (linear Digicon arrays of 512 pixels), the mirrors E are gratings giving the cross-dispersion for the echelle mode. Wavelengths are from 105 to 320 nm, and resolution from $R = 2\,000$ to 10^5. The useful spectral interval is 2.5–29 nm

HARPS and the Search for Exoplanets

The Doppler shift of the spectral lines of a source such as a star or galaxy determines its radial velocity v relative to the Earth. Every spectral line of the source is subject to the same relative shift:

$$\frac{\Delta\lambda}{\lambda} = \frac{v}{c},$$

and this means that the signal-to-noise ratio of a measurement of the velocity v can be considerably improved by simultaneously measuring all the lines over a broad spectral range. This nevertheless requires the object to be bright enough to ensure that all the lines can be detected. Developed by the Geneva Observatory (Switzerland)and the Observatoire de Haute-Provence (France) to make systematic measurements of stellar motions in the Galaxy (CORAVEL and CORALIE) in the period 1980–2000), this technique has been widely used in the search for exoplanets associated with other stars, measuring the time dependence of their radial velocities with respect to the Earth. Indeed, the star orbits the common centre of mass of the system. In order to be able to detect planetary companions with the smallest possible masses (a few dozen times the Earth mass), the velocity sensitivity must clearly be as high as possible.

The *High Accuracy Radial velocity Planet Searcher* (HARPS)[3] has been operating on the 3.6 m ESO telescope at La Silla (Chile) since 2001. A significant part of the operating time of this telescope is devoted to a systematic search for exoplanets. This is an echelle spectrometer with spectral resolution reaching 90 000 in the broad spectral range from 380 to 690 nm. In order to maximise stability over time, the spectrometer is held in a fixed position, in a pressure-controlled container, with thermostatic control (Coudé room), supplied by two optical fibres from the Cassegrain focus of the telescope, one carrying light from the star sampled in the focal plane, and the other light from a standard (thorium–argon) spectral lamp ensuring permanent wavelength calibration. These two sources are dispersed on two CCD arrays (2 000 × 4 000 pixels), in echelle mode, with 68 orders. The broad spectral interval here requires correction for atmospheric dispersion whenever the star is not at the zenith, and this is done by means of a correcting optical component placed upstream of the transfer fibre.

The HARPS instrument performs exceptionally well for the measurement of radial velocities (Fig. 8.14), and no doubt represents the ultimate limit of what is possible from an Earth-based observatory. The limiting velocity modulation that can be detected is $1\ \mathrm{m\,s^{-1}}$, while the presence of the planet Jupiter modulates the radial velocity of the Sun by about $13\ \mathrm{m\,s^{-1}}$ for an observer located outside the Solar System. Over five years, this instrument has already discovered or confirmed the existence of 12 exoplanets, and results are still coming in.

Multi-Object Spectroscopy (MOS)

This is a multiplex arrangement, where the spectra of several objects in the field are obtained simultaneously, with a considerable gain in observing time and the possibility of common calibration. For a long time, *objective prisms* were used for stellar studies. A large prism placed on the entrance pupil, or an image of it, disperses each star image on the detector. There are two drawbacks: one is that the spectra may overlap, and the other is that the sky background adds to each spectrum. The selection of objects prior to dispersion is carried out by a suitable process for slicing up the image plane, viz., a slit mask or optical fibres. This delivers a spectrum $I_j(\lambda)$ for each object j, although the latter must be small, i.e., no more than a few arcsec. Multi-object spectrometers (MOS) built in this way have extraordinarily high performance levels.

Slit Masks

For image slicing using a *mask*, a non-dispersed image of the field is first made on the CCD detector. Then a mask is produced, containing a series of slits

[3] See www.unige.ch/sciences/astro/ for HARPS.

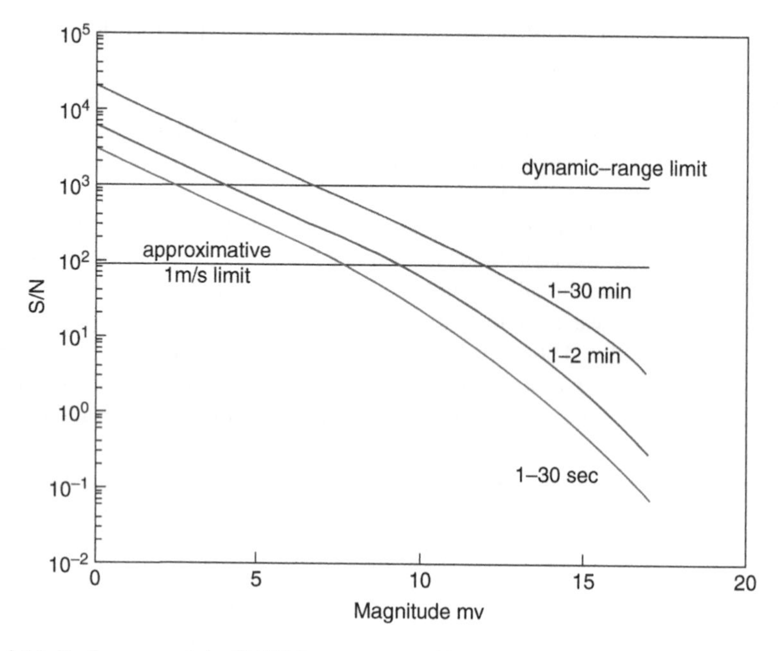

Fig. 8.14 Performance of the HARPS spectrometer. The signal-to-noise ratio, depending on an exposure of 0.5 to 30 min, is shown on the *ordinate* as a function of the magnitude m_V of a star, here of spectral type G8. The *horizontal line* is the limiting detectable velocity of 1 m s^{-1}, below which measurements are no longer significant owing to various fluctuations, including fluctuations in the velocity field of the star's photosphere. The saturation limit of the CCD due to excess photons is also shown. With the kind permission of the Geneva Observatory

adequate for the desired resolution, on the basis of this image. This can be done by photoengraving, for example. The mask is inserted in the image focal plane, and a grism in the following pupil plane. Spectroscopy can then be carried out on a hundred or so sources simultaneously, e.g., with the CFH telescope (Fig. 8.15). The method is well suited to faint sources (galaxy surveys) in fields containing many sources, for example, several hundred sources in a field of 10 arcmin. Sky subtraction is good, so a magnitude limit of around 1% of the sky magnitude per spatial resolution element is possible (since the sky brightness is measured in square arcsec). There are two drawbacks with this solution: one is that it is not possible to obtain the spectrum of two objects close to one another in the direction of the dispersion, and the other is that the spectra of objects at the edge of the field are truncated at the two ends of the spectral region. Masking was chosen for the *Visible Multi-Object Spectrograph* (VIMOS) of the VLT.[4]

[4]See www.eso.org/instruments/vimos/.

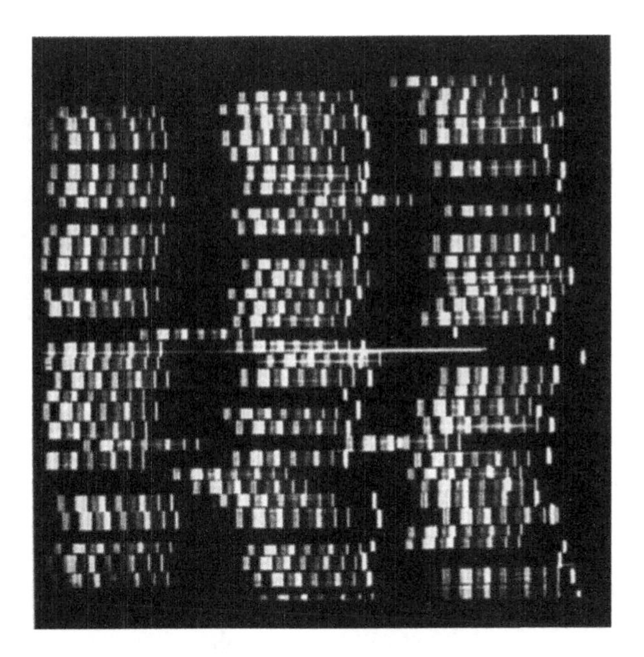

Fig. 8.15 Multi-object spectroscopy using slits. One hour exposure with the Canada–France–Hawaii Telescope (CFHT) in a field of 10′. 80 masks were used. Each vertical rectangle corresponds to the sky (slit = 12″) plus source spectrum, covering 450 to 850 nm. Sky emission lines and bands (*horizontal lines*) predominate. The fainter objects ($m_I = 22$, $m_B = 24$) are only visible after processing and recombining 8 exposures of 1 hour each. (Document due to LeFèvre O., Canada–France Redshift Survey, 1995)

Optical Fibres

The idea here is straightforward. One end of each multimode fibre is positioned on the selected object in the focal plane, while the other is placed on the entrance slit of the spectrograph. The latter does not therefore need to be fixed to the telescope and move around with it, whence a gain in stability. Some fibres are reserved for measuring the sky background, others for field stars which may in some cases be used for finely tuned pointing. The detector pixels are thus exploited to their maximal potential by juxtaposing individual spectra (see Figs. 8.16 and 8.17).

The fibres can be placed in various ways, either by making holes at suitable positions in a plate and linking the fibres to them, or by displacing each fibre head using a robotic arm, commanded from positions selected on a previously made image of the field (Fig. 8.18). The density of sources accessible in the field is low, several hundred in a field of the order of one degree, because of the space required for the positioning arms. Sky subtraction is less good than for the slit technique, because in that approach the background is measured locally at the ends of the slit, whereas background measurement fibres must be positioned further from the sources. Small scale heterogeneity in atmospheric emission (Sect. 2.6) thus

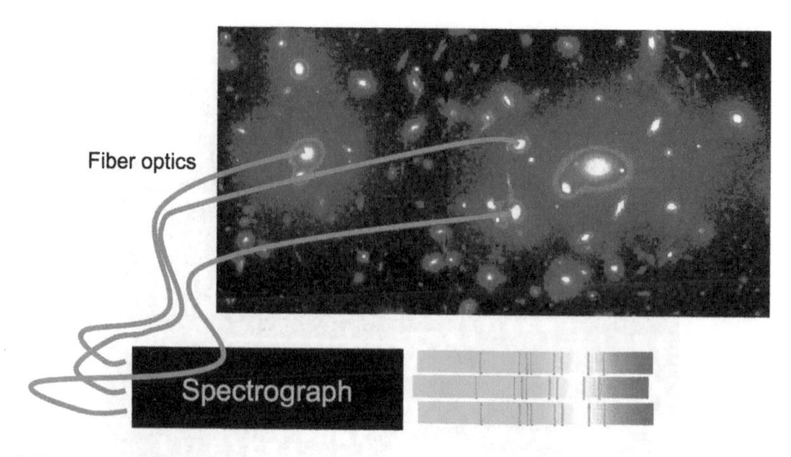

Fig. 8.16 Multi-object spectroscopy

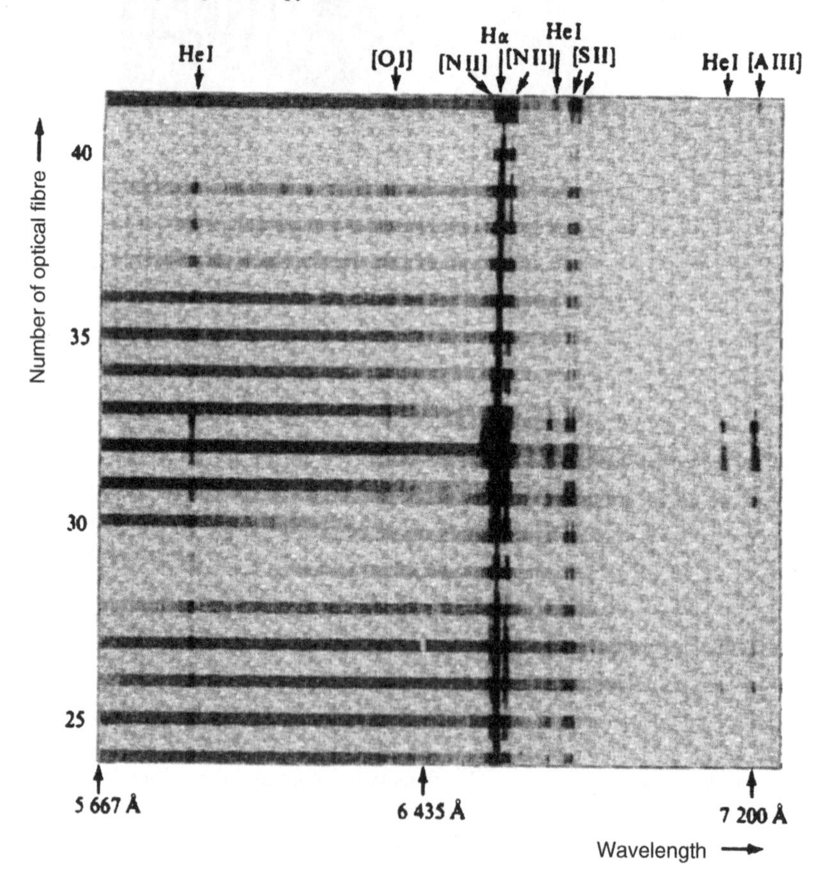

Fig. 8.17 Multi-object spectroscopy. Spectrograms of various regions in the Orion Nebula M42, showing lines characteristic of H II regions. The detector is a CCD. (After Enard D., 3.6 m telescope at the European Southern Observatory)

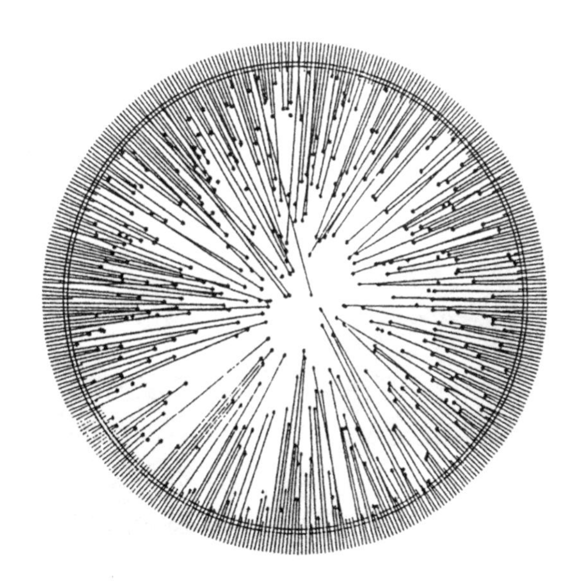

Fig. 8.18 Positioning of optical fibres in a field, using straight robot arms. 400 fibres are fitted to 400 sources, according to two different geometries. (After Lewis et al., in *Fiber Optics in Astronomy II*, ASP Conf. Ser. **37**, 1993)

introduces fluctuations in the background and its spectra. The limiting magnitudes that can be reached lie between 17 and 21 at visible wavelengths.

An interesting idea has been implemented in the GIRAFFE spectrometer set up on the Very Large Telescope in 2006. It combines the multi-object feature, selecting up to 132 galaxies in the field, with the wide-field feature, since each light-sampling head comprises 20 fibres which sample different points of the field. It is thus possible to obtain the spectrum of different regions of a galaxy, useful for making elemental abundance or rotation measurements, for example, and at the same time a large number of galaxies in a total field of 25 arcmin. It can also be combined with low order adaptive optics, guaranteeing a stable image quality, with a PSF of a few hundred milli arcsec.

In the mid-1990s, the slit spectrometer of the 4.2 m Anglo-Australian Telescope (AAT) (Mount Stromlo, Australia) can position 400 fibres, connected to two spectrographs, in just 2 minutes. The Sloan Digital Survey (see Sect. 10.2) is preparing the spectrometry of all 10^6 galaxies and 10^5 quasars in a π steradian fraction of the sky. The instrument, which will use 3 000 photographic plates with holes pierced in them for manual installation of the fibres, and has a field of 3° and 600 fibres. It can reach magnitude $m_\mathrm{v} = 17$.

Integral Field Spectroscopy

The aim here is to measure the spectral information $I(\nu, \boldsymbol{\theta})$ at every point of the field, rather than at a limited number of previously selected points, as happens in multi-object spectroscopy. This approach is not only relevant to grating spectrometers. Later we shall encounter further illustrations with the Fabry–Pérot spectrometer and the Fourier transform imaging spectrometer. A typical setup couples a dispersing element with a microlens array or a long slit (for studying the Sun).

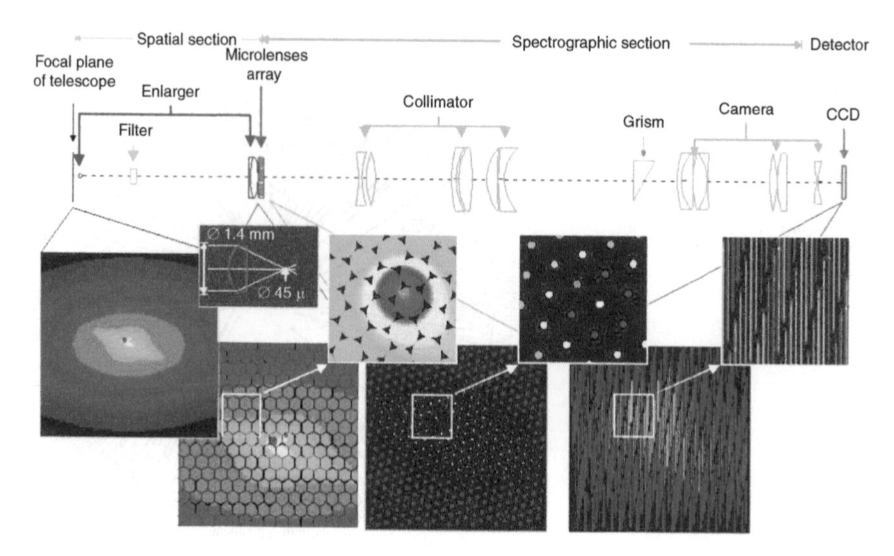

Fig. 8.19 Integral field spectrometer with microlens array. Each microlens, placed in a focal plane, slices the field up into elements whose spectrum is then given by a grism. The dispersion angle must be chosen judiciously so that the spectra do not overlap

Microlens Array Spectrometer

An array of microlenses, typically a few hundred micrometres across, is placed in an image plane conjugate to the focal plane with suitable magnification. This forms a set of images of each pixel of the field thereby sampled. These images are dispersed by a grism (transmission grating) and the spectra are detected on a CCD (Fig. 8.19).

There are alternatives to using microlenses to slice up and reconfigure the focal plane. A remarkable example is the SINFONI instrument, set up on the Very Large Telescope in 2004. This operates in the near infrared (1.1–2.45 μm) and can be coupled with adaptive optics.[5] It is designed to work at the diffraction limit of an 8 m telescope so each pixel is necessarily small, and the total field can be anything from $8'' \times 8''$ to $0.8'' \times 0.8''$, depending on the available configuration. This field is divided into 32 slices (*image slicing*), each of which is then dispersed onto 64 pixels of a HgCdTe detector (see Chap. 7). We thereby obtain $32 \times 64 = 2\,048$ spectra of the given sky region. The associated adaptive optics module can use either a natural guide star (NGS), or an artificial guide star generated by laser (LGS) in the upper atmosphere (see Chap. 6), whence it is possible to observe in a sky region without bright reference star.

This idea of slicing up and reconfiguring the field can also be used at submillimetre wavelengths. In the *Photodetector Array Camera and Spectrometer* (PACS), designed for the Herschel space observatory,[6] a field of 5×5 pixels, each one $9.4''$, is sliced and dispersed to

---

[5] See www.eso.org/instruments/sinfoni/ for more details.

[6] Herschel is a European space mission for 2008–2012, designed to study submillimetre radiation and equipped with three instruments, including the PACS. The latter is described in more detail at pacs.mpe.mpg.de.

form 16 spectral elements on a 16×25 array of individual detectors (Ge:Ga, see Sect. 7.4) in the submillimetre bands 57–105 μm and 105–210 μm.

Solar Spectroscopy

The aim is to obtain a monochromatic image of an extended object, in this case the surface of the Sun, using the high spectral resolution of a grating spectrograph. Replacing the exit slit by a one-dimensional detector (for example, a Reticon or CCD array, which usually has 512 pixels, see Sect. 7.4.5), a slice of the object is obtained at wavelength λ_0. Displacing the image of the Sun on the entrance slit by an amount equal to the slit width, a second such slice is obtained, and so on. The image of the object at wavelength λ_0, known as a *spectroheliogram*, is thus reconstructed (Figs. 8.20 and 8.21).

The *magnetograph* combines the imaging capacity of the spectroheliograph with analysis of the polarisation. In a chosen spectral line, the Zeeman effect is measured at each point of the image. These measurements can be used to deduce the component of the magnetic field along the line of sight, for the layer of the atmosphere in which the line is formed.

8.3.4 Fourier Transform Spectrometer

Principles and Properties

The Fourier Transform Spectrometer (FTS), also called the *Michelson interferometer*, is a two-wave interferometer, as opposed to the grating, which causes N waves to interfere (one from each groove of the grating). We shall not employ the term Michelson interferometer here, although it is common, to avoid confusion with the *spatial* Michelson interferometer used in imaging and described in Sect. 6.4.

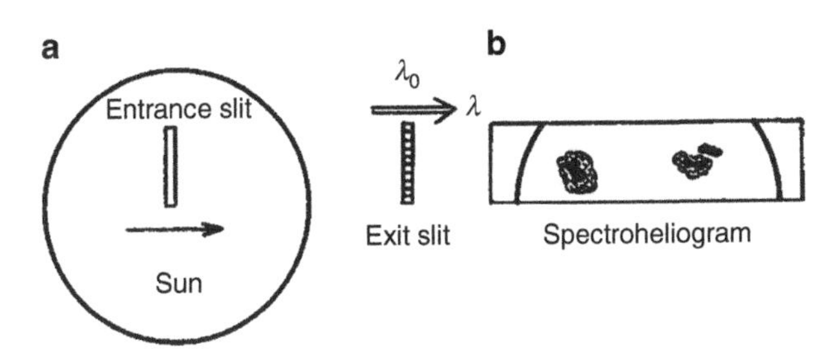

Fig. 8.20 Principle of the spectroheliograph. (**a**) The image of the Sun is scanned by the entrance slit, and the exit slit selects the wavelength λ_0 on a detector array. (**b**) Monochromatic image at wavelength λ_0

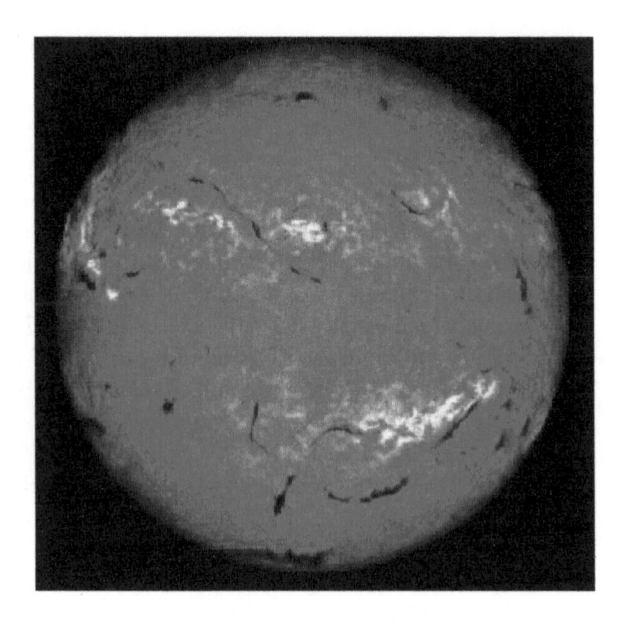

Fig. 8.21 Solar spectroheliogram obtained in the Hα line. Image courtesy of Henoux J.-C., Observatoire de Paris Meudon, 1985

If x is the difference in the path length (Fig. 8.22) of the two waves, the exit intensity is given by

$$\tilde{I}(x) = \frac{I_0}{2}(1 + \cos 2\pi\sigma x),$$

for monochromatic incident intensity I_0 of wave number σ. An arbitrary source whose intensity has a spectral distribution $I_0(\sigma)$ limited to the range $[\sigma_1, \sigma_2]$, produces the signal

$$\tilde{I}(x) = \frac{1}{2}\int_{\sigma_1}^{\sigma_2} I_0(\sigma)(1 + \cos 2\pi\sigma x)\,\mathrm{d}\sigma.$$

This signal, referred to as an *interferogram*, is measured by the detector placed at the exit. Up to a constant, it is the Fourier cosine transform of the spectrum of the object. This is *spectral multiplexing* since, for each value of x, all the spectral elements of the incident spectrum contribute to the signal, each one being *coded* by a cosine function orthogonal to the others.

Recovering the Source Spectrum

$I_0(\sigma)$ can be recovered by inverse Fourier transformation. After subtracting the mean value

$$\langle\tilde{I}(x)\rangle = \frac{1}{2}\int_{\sigma_1}^{\sigma_2} I_0(\sigma)\,\mathrm{d}\sigma,$$

we obtain

$$I_0'(\sigma) = \mathrm{FT}\left[\tilde{I}(x) - \langle\tilde{I}(x)\rangle\right].$$

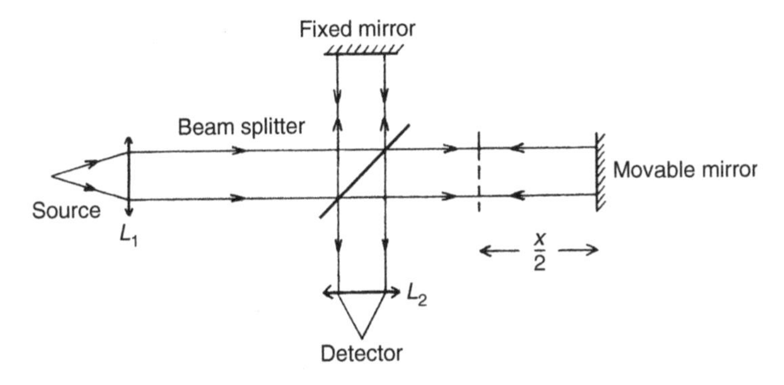

Fig. 8.22 Optics of a Fourier transform spectrometer

If $\tilde{I}(x)$ were known over the whole interval $x \in (-\infty, +\infty)$, I_0' would equal I_0. When $\tilde{I}(x)$ is only known on an interval $[-x_m/2, +x_m/2]$, it can be shown (Problem 8.1) that

$$I_0'(\sigma) = I_0(\sigma) \star \mathrm{sinc}(x_m \sigma),$$

and the recovered spectrum differs from I_0 by a degradation in resolution due to this convolution.

Resolution

The interferogram of a monochromatic source $I_0(\sigma) = \delta(\sigma - \sigma_0)$ is the *instrumental profile*

$$\mathrm{sinc}(x_m \sigma) \star \delta(\sigma - \sigma_0),$$

where x_m is the maximum optical path length difference. The first zero of the instrumental profile occurs at

$$\sigma - \sigma_0 = \pm \frac{1}{x_m},$$

and it is assumed that this value gives the separation at which it is just possible to discriminate between two lines. The resolution is therefore

$$R = \frac{\sigma_0}{\Delta\sigma} = x_m \sigma_0.$$

When high resolutions are sought, the size of the interferometer is determined by the displacement amplitude (equal to $x_m/2$) of a moveable mirror. This size is therefore $\lambda R/2$, and comparable to the size of a grating spectrometer with

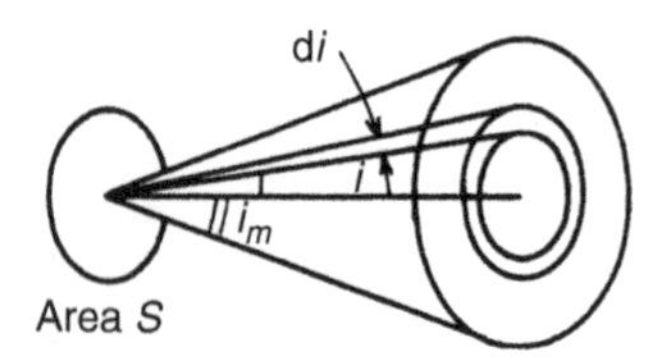

Fig. 8.23 Field of the interferometer. The area S and the angle i_m define the beam étendue. S is the area of the beam at the lens L_1 shown in the previous figure

the same resolution. In a typical application, $\lambda = 0.5\ \mu\mathrm{m}$, $x_m = 2$ m, and $R = x_m/\lambda = 4 \times 10^6$.

It can be shown (Problem 8.2) that the optimum sampling rate for the interferogram is an increment $\Delta x = 1/2\sigma_m$, where σ_m is the maximum wave number contained in the source spectrum. A finite number of samples of $\tilde{I}_0(x)$ is thus obtained, and a numerical Fourier transform can easily be carried out on them. Note, furthermore, that it is possible to control the sampling by using the interference pattern $\tilde{I}_1(x)$ given by a reference laser whose wave number is precisely known. The wavelength scale of the spectrum $I_0(\sigma_m)$ is then determined with the same precision, a considerable advantage over the grating spectrometer, which must be calibrated using reference spectra.[7]

Throughput. The aim here is to establish the maximal beam étendue, beyond which the resolution would be degraded (determination of the Q factor). Referring to Fig. 8.23, let

$$\Omega_{\max} = \pi\, i_m^2 .$$

Those rays inclined at an angle i to the axis contribute

$$\mathrm{d}\Omega = 2\pi\, i\, \mathrm{d}i$$

to the solid angle, and the associated variation in path difference is

$$|\mathrm{d}x| = x(0)\, i\, \mathrm{d}i, \quad \text{using} \quad x(i) = x(0)\left(1 - \frac{i^2}{2}\right),$$

so that

$$\mathrm{d}\Omega = \frac{2\pi}{x(0)}\, \mathrm{d}x .$$

The power delivered by a source of uniform brightness is thus

$$\mathrm{d}\Phi = S \frac{2\pi}{x(0)}\, \mathrm{d}x(i)\, \mathscr{C}[x(i)],$$

[7]Some authors make a distinction between the *spectrograph*, which requires a reference for calibration, and the *spectrometer*, which, as in this case, is intrinsically accurate in its wavelength calibration.

where the factor $\mathscr{C}$ expresses the modulation due to the interference of the two waves. For a monochromatic source of unit brightness, the modulated part of the flux $\tilde{I}(x) - \langle \tilde{I}(x) \rangle$ can be written

$$\Phi[x(0)] = S\frac{2\pi}{x(0)} \int_0^\infty D[x(i)] \cos[2\pi\sigma_0 x(i)]\, \mathrm{d}x(i),$$

where D is a 'diaphragm' function, giving a first approximation to $\mathscr{C}$:

$$D = \begin{cases} 1, & x(0) \le x(i) \le \left(1 - \frac{i_\mathrm{m}^2}{2}\right) x(0), \\ 0, & x(i) \quad \text{outside this interval.} \end{cases}$$

It follows that

$$\begin{aligned} \Phi[x(0)] &= \frac{2\pi S}{x(0)} \frac{1}{2} \mathrm{FT}\big[D[x(i)]\big] \\ &= \frac{\pi S}{x(0)} \operatorname{sinc}\left[\pi\sigma_0 x(0) \frac{i_\mathrm{m}^2}{2}\right] \cos 2\pi\sigma_0 \left(1 - \frac{i_\mathrm{m}^2}{4}\right) x(0), \end{aligned}$$

or simply

$$\Phi(x) = \frac{\pi S}{x} \operatorname{sinc}\left(\pi\sigma_0 x \frac{i_\mathrm{m}^2}{2}\right) \cos 2\pi\sigma_0 x.$$

The modulation term $\cos 2\pi\sigma_0 x$ is therefore subject to a damping effect which tends to remove the modulation, either for large path differences at given i_m, or for large fields i_m at given x_m.

The modulation is zero for

$$\pi\sigma_0 x \frac{i_\mathrm{m}^2}{2} = k\pi, \qquad \text{i.e.,} \quad x = k\frac{2\pi\lambda_0}{\Omega}.$$

The resolution is $R = x_\mathrm{m}/\lambda$, so the beam étendue is

$$U = S\Omega = 2\pi\frac{S}{R}, \qquad \text{i.e.,} \quad Q = 2\pi.$$

This result reveals the *throughput advantage* of the Fourier transform spectrometer: at a resolution equal to that of a grating spectrometer, the throughput is increased by the factor

$$\frac{2\pi}{\beta} \sim 10^2 \text{ to } 10^3.$$

This greatly facilitates the observation of extended sources (that is, sources whose angular dimensions were incompatible with the slit width of the grating spectrometer). Note that this advantage is clearly related to the axisymmetry of this interference arrangement, possessed also by the Fabry–Pérot spectrometer.

The Holographic Fourier Spectrometer. The moveable mirror is one of the less convenient aspects of the above setup. It has been suggested that one (or both) of the mirrors could be replaced by a grating, tilted in such a way that it acts as a plane mirror for some wavelength λ_0 (Littrow reflection of the grating). Neighbouring wavelengths λ are reflected at different angles, and each one of them produces straight-line interference fringes, like the ones produced by an air wedge, on the two-dimensional detector. The amplitude is proportional to the intensity $I(\lambda)$ and Fourier analysis of the signal restores $I(\lambda)$. The name comes from the fact that the frequency ν_0 is reduced to zero (uniform flux at the detector) and so produces a holographic effect. (Douglas N. et al., Astroph. Sp. Sci. **171**, 307, 1980.)

The Multiplex Advantage

A simple argument readily establishes the principle here, due to Felgett. Suppose we wish to determine, over a total measurement time T, the spectrum $I_0(\sigma)$ of a source, in the interval (σ_1, σ_2), and with resolution R. The number of spectral elements is M. Suppose that the intensity is comparable in each spectral element (absorption or continuum spectrum, the argument requiring modification in the case of a spectrum comprising a small number of emission lines). Suppose, finally, that the measurement is made by a detector whose intrinsic noise σ_D dominates the other sources. For sequential scanning of the spectrum, the signal-to-noise ratio for each spectral element is

$$r_1 = \frac{I_0(\sigma)\,(T/M)}{\sigma_D\sqrt{T/M}}.$$

For FT spectroscopy, the whole time T is used for each spectral element, and the corresponding ratio is

$$r_2 = \frac{1}{2}\frac{I_0(\sigma)\,T}{\sigma_D\sqrt{T}},$$

the factor $1/2$ coming from the light lost at the beam splitter. The *multiplex gain* is thus $\sqrt{M}/2$, and can be very large for extended spectra at very high resolution (Fig. 8.24). Historically, this gain made the FT spectrometer the optimal instrument for infrared observation, in which detector noise dominates, and for which, until the 1990s, no multichannel detectors were available to avoid sequential scanning.

Even though this advantage is now disappearing, with the progress made in detector technology, FT spectroscopy maintains its lead in throughput, in very high resolution (of the order of 10^5), and in rigorous wavelength calibration.

It is worth pointing out that the spectroscopic capacities of the FT spectrometer can be combined with imaging, by placing an array of detectors in the focal plane. The difficulties of data handling, whether related to the storage of measurement data or the calculation of spectra (Problem 8.3), are daunting.

Astronomical Applications

The FT spectrometer was widely used in the near infrared over the period 1960–2000, but progress in the sensitivity of 2D detectors has gradually eliminated it from ground-based observation. However, it is still used in space because of its high level of accuracy, compactness, and multiplex facilities.

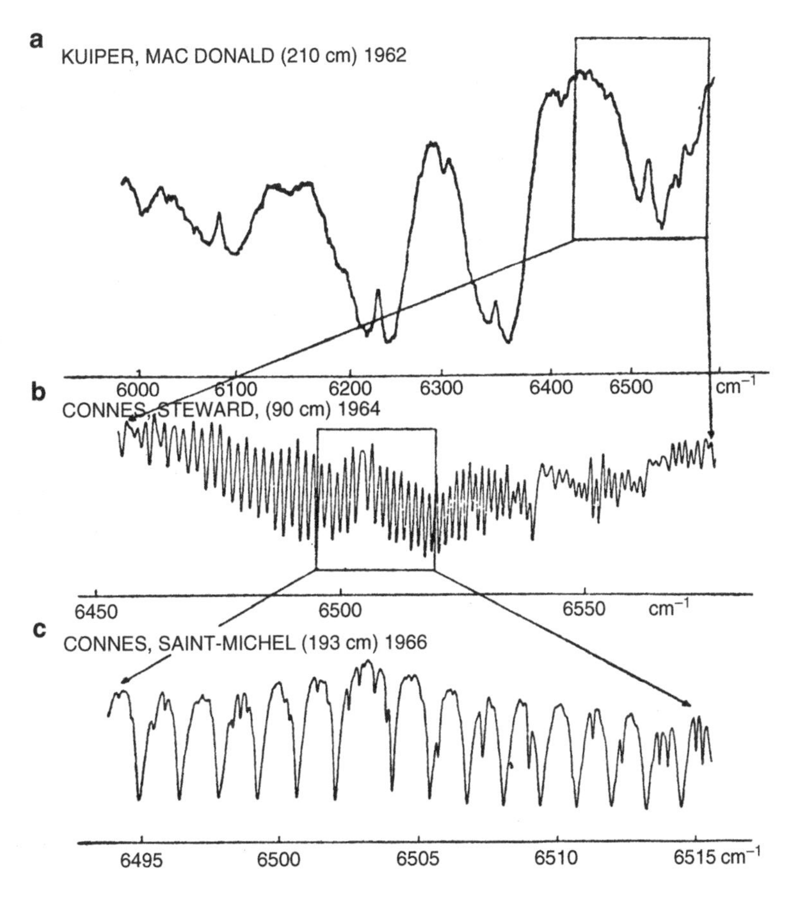

Fig. 8.24 Progress in resolving the infrared spectrum of Venus, during the historical emergence of Fourier spectroscopy. (**a**) Grating spectrometer in the atmospheric window around 1.6 μm. Four intense rotational–vibrational bands of the CO_2 molecule are visible. (**b**) Fourier transform spectroscopy with $\Delta\sigma = 0.7\,\text{cm}^{-1}$. (**c**) The same, with $\Delta\sigma = 0.08\ \text{cm}^{-1}$. The CO_2 bands are resolved. The apertures of the telescopes used are given. From Connes P., ARAA **8**, 209, 1970. Reproduced with the kind permission of the *Annual Review of Astronomy and Astrophysics*, Vol. 8, ©1970, Annual Reviews Inc

Ground-Based Applications

In the infrared region observable from the ground, a background of thermal or OH emission is the predominant noise for $\lambda \gtrsim 1.5\ \mu\text{m}$, thanks to progress in detector technology. The multiplex advantage then disappears (Problem 8.4), but the throughput advantage remains, when the observed infrared sources are extended ($\gg$1 arcsec). Moreover, as the wavelength is long, the need for mechanical precision is less restrictive, and the FTS is relatively easy to build. For this reason, they now equip several large telescopes (Fig. 8.25). When coupled to an array, it then operates

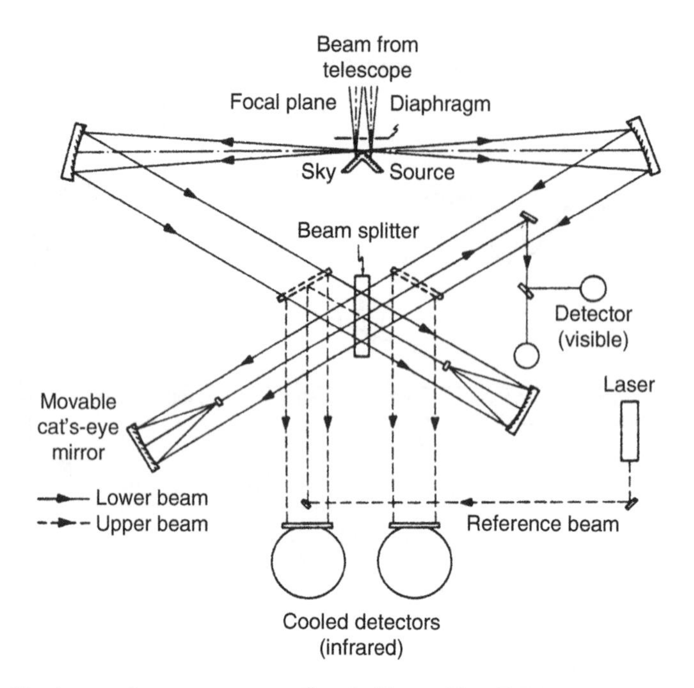

Fig. 8.25 Fourier transform spectrometer, Canada–France–Hawaii Telescope. The instrument is symmetrical, with signals, $I_1(x)$ and $I_2(x)$, received at the two infrared detectors. The quantity $(I_1 - I_2)/(I_1 + I_2)$ is independent of atmospheric fluctuations. The beam from the visible He–Ne laser provides reference fringes which control the sampling of the infrared signal. Upper and lower beams are spatially separated. The whole setup is mounted at the Cassegrain focus of the telescope. (After Maillard J.P., Michel G., I.A.U. Colloquium, **67**, Reidel, Dordrecht, 1982. With the kind permission of D. Reidel Publishing Company)

as an integral field spectrometer (the BEAR instrument of the CFH telescope in Hawaii, Fig. 8.26).

Spaceborne Applications

A nice example here is the ESA's Herschel mission,[8] observing the sky in the far infrared and submillimetre regions from Lagrange point L2 using a cooled 3.5 m telescope from 2008. The *Spectral and Photometric Imaging Receiver* (SPIRE) is a Fourier transform integral field spectrometer (2.6′), operating simultaneously in the two bands 200–300 μm and 300–670 μm, with spectral resolution as high as 1 000.

[8] See sci.esa.int/science-e/www/area/index.cfm?fareaid=16.

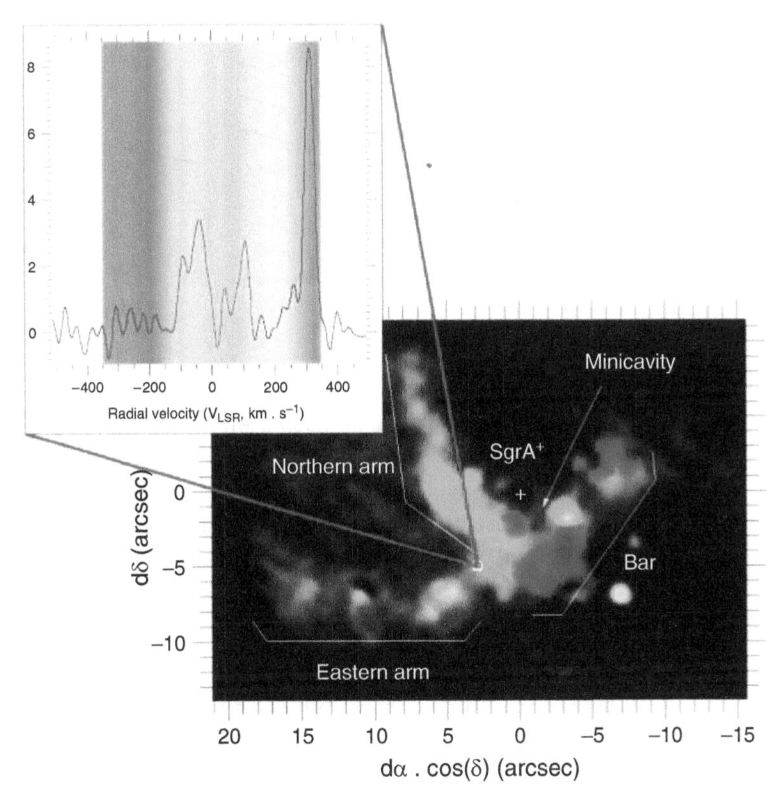

Fig. 8.26 Integral field spectroscopy. High spatial and spectral resolution image of helium streamers, obtained with the Fourier spectrometer BEAR on the CFHT (Hawaii) in the HI Brackett γ line at $\lambda = 2.058$ μm, from the minispiral located in the immediate vicinity of the galactic centre. Stellar contributions have been subtracted. The position of SgrA* is marked by +. The color coding of velocity is explicit in the *insert*, where it clearly appears that the spectral line in this particular pixel shows several radial velocities, proving the presence of several clouds with different velocities along the line of sight. Paumard, T. et al., 2001, courtesy of Astronomy & Astrophysics

8.3.5 The Fabry–Perot Spectrometer

This spectrometer (or interferometer) combines the throughput advantage of the FT spectrometer ($Q = 2\pi$) with a very high spectral resolution, whilst at the same time being far less bulky than the two previous instruments for the same resolution.[9]

[9]Charles Fabry (1867–1945), French physicist and optician, and Alfred Perot (1863–1925), French physicist, worked together at the Marseille Observatory to build a multiple-wave interferometer, and started a strong tradition of optical development there. They both taught at the Ecole polytechnique in France.

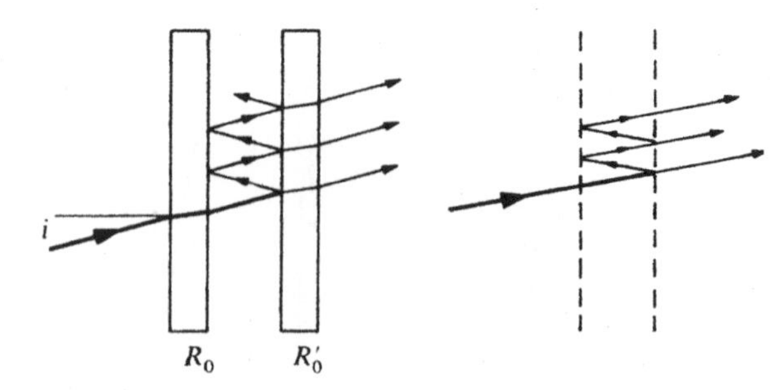

Fig. 8.27 Arrangement of the Fabry–Perot interferometer. (**a**) In the UV, visible, and near infrared, the plates are made of transparent material. (**b**) In the far infrared, the plates are self-supporting metallic grids

Principle

Two parallel plates, supposed non-absorbent, have a high reflection coefficient r ($r \lesssim 1$), and transmission $t = 1 - r$ (Fig. 8.27). This is called the *Fabry–Perot etalon* (or FP etalon), and is so named because it allows a wavelength calibration by physical measurement of the distance between the plates, or conversely. Neglecting reflection by the faces R_0 and R_0', the transmission of the FP etalon, illuminated by monochromatic radiation of wave number σ at incidence angle i, results from multiple wave interference and can be expressed by Airy's formula

$$I = I_0 \left(\frac{r}{1-r} \right)^2 \left[1 + \frac{4r}{(1-r)^2} \sin^2(2\pi e\sigma \cos i) \right]^{-1} .$$

At zero incidence, the FP etalon has transmission peaks when

$$2\pi e\sigma = m\pi, \quad m \text{ an integer,}$$

which means

$$\sigma = \frac{m}{2e}, \quad m \text{ an integer.}$$

These peaks are therefore more widely separated as the thickness e is increased. The whole number m is called the *order* of the interferometer, and $\Delta\sigma = 1/2e$ is the *free spectral interval*.

At zero incidence, the full width at half maximum $\delta\sigma$ of a transmission peak is given by $I = I_0/2$, which implies

$$\sin \left[2\pi e \left(\sigma + \frac{\delta\sigma}{2} \right) \right] \sim \pi e \delta\sigma = \frac{1-r}{2\sqrt{r}} .$$

This width decreases as the reflection coefficient r increases. Introducing the *finesse* of the FP by

$$\mathscr{F} = \frac{\pi\sqrt{r}}{1-r},$$

it follows that

$$\delta\sigma = \frac{1}{2e\mathscr{F}}.$$

The resolution is then given by

$$R = \frac{\sigma}{\delta\sigma} = 2e\mathscr{F}\sigma = m\mathscr{F}.$$

This increases with both the finesse and the order.

As the FP has rotational symmetry about its axis, the interference pattern at infinity, when it is illuminated by monochromatic radiation of wave number σ_0, is a set of concentric rings. Let δi be their angular size at half maximum. Then the ring maximum is given by

$$\cos i = \frac{m\pi}{2\pi e\sigma_0},$$

and the width δi is found by solving

$$\frac{4r}{(1-r)^2}\sin^2\left[2\pi e\sigma_0\cos\left(i+\frac{\delta i}{2}\right)\right] = 1.$$

A lens or mirror placed behind the FP forms an image of these rings in its focal plane. The maximum throughput for an FP without loss of resolution follows from an identical calculation to the one carried out above in the case of the FT spectrometer. Once again, if S is the area of the etalon, the result is

$$U = 2\pi\frac{S}{R}.$$

Use as a Spectrometer

The FP etalon can be used in different ways (Fig. 8.28). In any arrangement, it must be preceded by a *predisperser* (another etalon, interference filter, grating, or other), in order to isolate a single free spectral interval $\Delta\sigma$ on the FP.

In a scanning mode, and illuminated by arbitrary monochromatic radiation of spectrum $I_0(\sigma)$, the FP is followed by a diaphragm isolating the central ring, which is received on the detector. The thickness e is then varied, and the detector signal is proportional to the incident intensity $I_0(\sigma)$. It is also possible to vary the pressure, and hence the refractive index, of a gas introduced between the plates.

8.3.6 The Bragg Crystal Spectrometer (X-Ray Region)

The first extragalactic spectral line observed in the X-ray region was the $2p$–$1s$ line of iron XXIV, in the Virgo, Coma, and Perseus clusters of galaxies by NASA's OSO-8 satellite. This observation (Fig. 8.29) used the low resolving power of

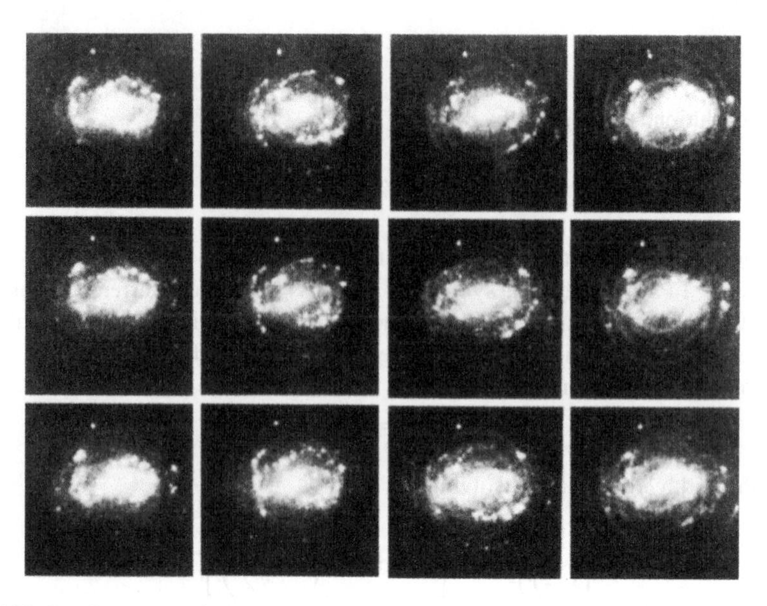

Fig. 8.28 Interferograms of the galaxy NGC 2903, obtained using a scanning Fabry–Perot interferometer in the Hα line (656.3 nm). The variation in order is obtained using piezoelectric wedges, and the free spectral interval is scanned in 24 steps (of which 12 are shown on the figure). Each photon is recorded on a 512 $\times$ 512 matrix by a photon counting camera. The number of photons measured is proportional to the intensity emitted at a given Doppler speed. The radial velocity of hydrogen in the galaxy is thus mapped. The integration time per step was 5 min. (Observatoire de Marseille, 3.6 m CFH telescope, Hawaii, Marcelin M. et al., Astron. Ap. **128**, 140, 1983. With the kind permission of Astronomy and Astrophysics)

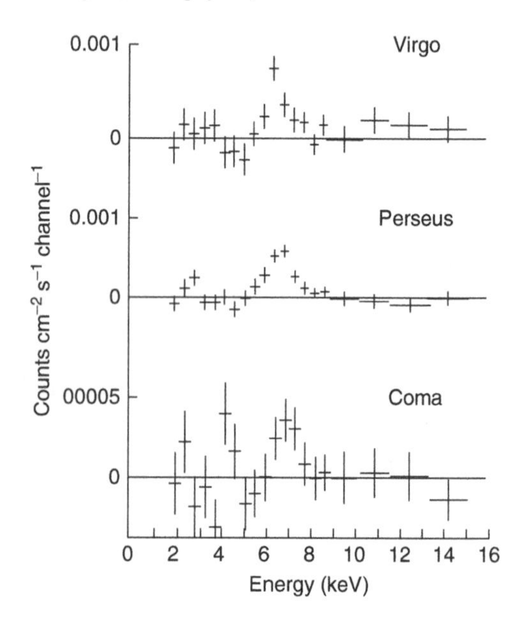

Fig. 8.29 X-ray emission from the Virgo, Perseus and Coma clusters of galaxies. The spectral resolution, obtained by a *proportional counter*, is less than 10, but nevertheless sufficient to show an iron line at 7 keV. From Serlemitsos P.J. et al., Ap. J. **211**, L63, 1977. With the kind permission of the Astrophysical Journal

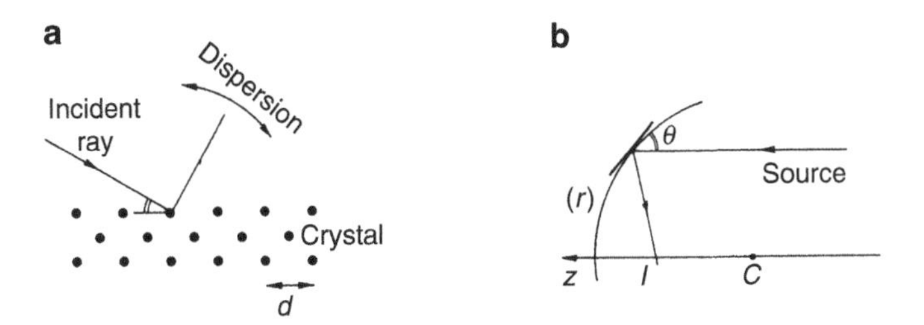

Fig. 8.30 Bragg diffraction. (**a**) Lattice plane of the crystal, with period d. (**b**) Concave crystal forming the pseudo-images of a source at infinity

the *proportional counter*, but demonstrated the potential of X-ray spectroscopy. Developments continued using crystals to produce diffraction, but in the 1980s, these were gradually replaced by conventional diffraction gratings with very close rulings (600 grooves/mm), first by transmission (the European EXOSAT mission), then by reflection (the European mission XMM–Newton, 1999), thereby covering the energy range 0.1–2 keV. In NASA's Chandra X-ray observatory launched in 1999, the slitless high resolution spectrometer or *High Energy Transmission Grating spectrometer* (HETG)[10] uses a transmission grating formed by depositing gold microrods on a plastic substrate. Whether the disperser is a crystal or a grating, the X-ray spectrometer also includes a camera on which the dispersed image is formed (Sect. 7.4.6).

A high resolving power ($R \sim 1\,000$) can be obtained by means of a crystal, such as lithium fluoride LiF, which functions as a diffraction grating. *Bragg diffraction* occurs on a crystal of spacing d (Fig. 8.30a), at an angle θ given by

$$\sin\theta = \frac{\lambda}{2a}.$$

Just as a Rowland mounting (Fig. 8.11) forms an image with a concave grating, a curved crystal produces a series of quasi-anastigmatic monochromatic images from a point source at infinity (Fig. 8.30b). If the crystal surface is part of a spherical surface of centre C, the lattice planes being tangent to the sphere, dispersion is along the z axis and the distance CI to the image I of radiation with wavelength λ is given by

$$CI = \frac{r}{2\sin\theta}.$$

The resolution R at which an energy interval $\mathrm{d}E$ can be separated is given by

$$R = \frac{E}{\mathrm{d}E} = \frac{\lambda}{\mathrm{d}\lambda} = \frac{r}{2\sin\theta\,\mathrm{d}z}.$$

The spacing of the LiF crystal is 2.01 Å, giving $\theta = 25°$ for an energy $E = 7.3$ keV. The resolution $R = 1\,000$ corresponds to $\mathrm{d}z = 1$ mm for $r = 1$ m. A proportional counter is thus placed on the z axis, to measure both the z coordinate, and hence the energy, of the photon, as well

[10]See space.mit.edu/CSR/hetg.

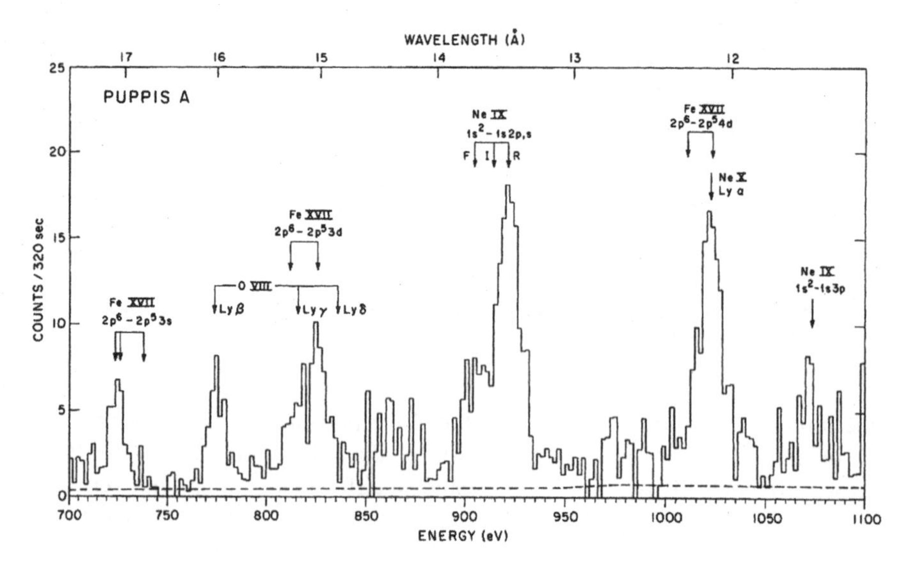

Fig. 8.31 Spectrum of the *supernova remnant* (SNR) Puppis A, observed with the curved grating spectrometer on the Einstein satellite. Lines due to various ions are identified. Note the profile of the rays, due to the limited resolution of the spectrometer. From Winkler P.F. et al., Ap. J. **246**, L27, 1981. With the kind permission of the Astrophysical Journal

as the corresponding intensity. The spatial resolution of the proportional counter may indeed limit the spectral resolution of the instrument. As the system is not perfectly anastigmatic for a point source at infinity, extended objects give a series of pseudo-images at different z.

Note that such a spectrometer is sensitive to polarisation, since only the component of the electric field parallel to the crystal surface is reflected at the Bragg angle. Between 1978 and 1981, the Einstein satellite (High Energy Astronomical Observatory 2) was equipped with such a spectrometer, and provided a resolution of 50 to 100 up to 0.4 keV, and 100 to 1 000 beyond (Fig. 8.31). Figure 8.32 shows a spectrum obtained 20 years later with the RGS instrument aboard the European XMM–Newton mission.

Finally, note that, as in other wavelength regions, a certain spectral selectivity can be introduced by the detector itself (Sect. 7.4.6), whence no further dispersive system is required.

8.4 Radiofrequency Spectrometry

Interference spectrometers use the spectral selectivity which results from interference of a wave with itself. *Heterodyne spectrometers* achieve this selectivity by interfering the incoming wave with a locally produced wave, and analysing the result in a set of electric filters, or by other methods outlined below. These methods cover the conventional radiofrequency region, from long wavelengths (centimetre to

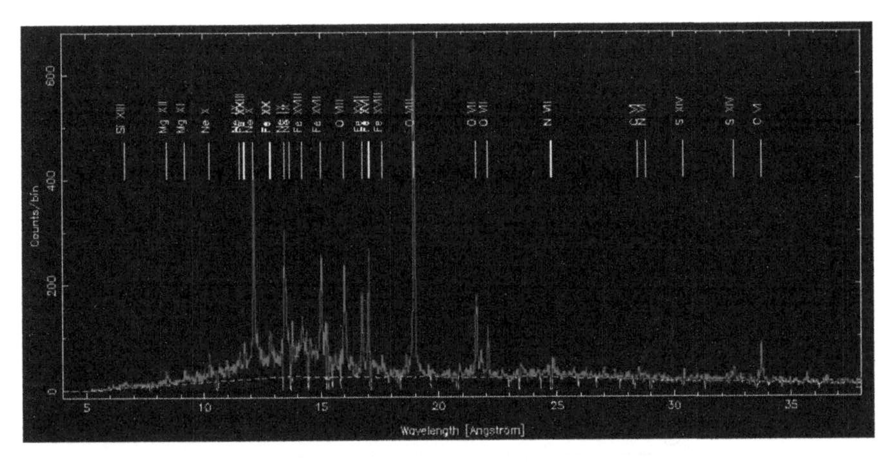

Fig. 8.32 Spectrum obtained with the reflection grating spectrometer (RGS) of the Newton mission. Spectrum of the young, active star AB Doradus between 2 and 0.3 keV. There are many ionised iron, magnesium, and neon lines. Image courtesy of J. Sanz-Forcada ESA XMM–Newton mission

hectometre) down to the shortest, in the millimetre and submillimetre region which we first began to seriously explore at the beginning of the 2000s.

The idea of heterodyne detection, from radiofrequencies to mid-infrared, was discussed in detail in Sect. 7.5. It involves a local oscillator (LO) with frequency very close to that of the incident radiation. The frequency change by *heterodyning* transfers the information carried by the incident wave to a lower frequency region called the intermediate frequency (IF), and this is where spectral analysis and final detection can be carried out. The detector remains limited to a coherence étendue λ^2 and one polarisation direction, but detectors can be juxtaposed to form small arrays (3×3 or 4×4), thereby speeding up image reception efficiency. Single wavelength maps are produced either in the spatial frequency plane $\boldsymbol{w}$ or in the image plane $\boldsymbol{\theta}$ by combining spectral discrimination, the image formation properties discussed in Chaps. 5 and 6, and detector properties discussed in Chap. 7.

8.4.1 Spectral Discrimination Methods

Spectral discrimination on the intermediate frequency (IF) signal can be achieved in several ways, which we shall discuss in detail here:

- Using electrical filters connected in parallel and identifying each spectral band.
- Using acousto-optical techniques forming a diffraction grating.
- By purely digital processing of the signal, either by digital correlator or by fast Fourier transform.

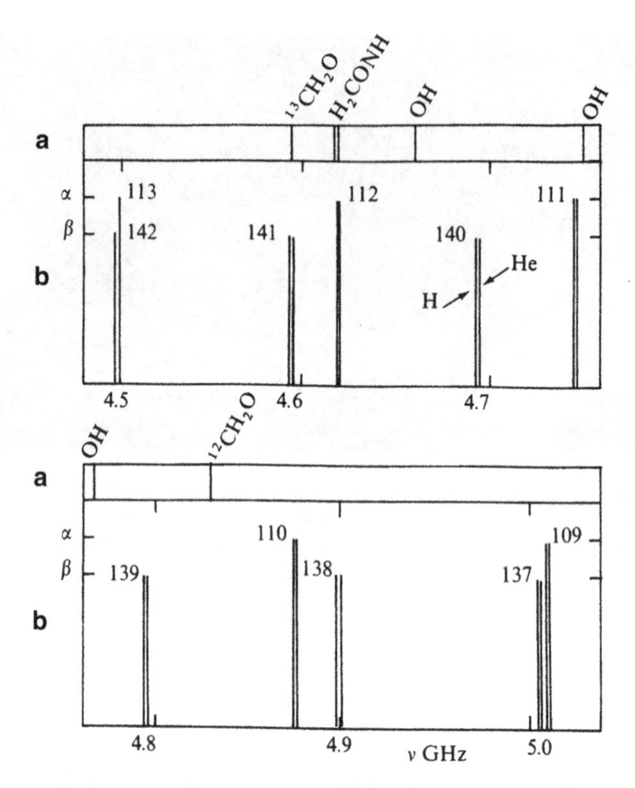

Fig. 8.33 Typical lines in the radio region 4.5 to 5 GHz. (**a**) Molecular lines of formaldehyde H_2CO, and formamide H_2CONH, and fine structure lines of the OH radical. (**b**) Recombination lines of hydrogen and helium, α and β transition series. Values of the quantum number n are given. A list of transitions and their frequencies is given in AQ, Sects. 2.12, 2.14

Any method will do, whatever the spectral region from submillimetre to metre wavelengths, since they all apply to the IF signal, and not directly to the incident signal. Only the mixing stage can differ here.

Considering the regions from $\lambda = 100\ \mu m$ ($\nu = 3$ THz) to $\lambda = 10$ cm ($\nu = 3$ GHz), one seeks passbands for the IF stage able to exceed 1 GHz. What is the desired resolution? When a gas has thermal agitation speeds of the order of several km s^{-1}, an emission or absorption line (Fig. 8.33) will have relative width $\Delta\nu/\nu \sim 10^{-5}$, implying $\Delta\nu = 14$ kHz at $\lambda = 21$ cm, or 1.5 MHz at $\lambda = 200\ \mu m$. There is little point in pushing the spectral resolution to values much above 10^5–10^6, since for one thing the spectral lines are broader, and for another, a very high level of stability would have to be guaranteed for the local oscillator.

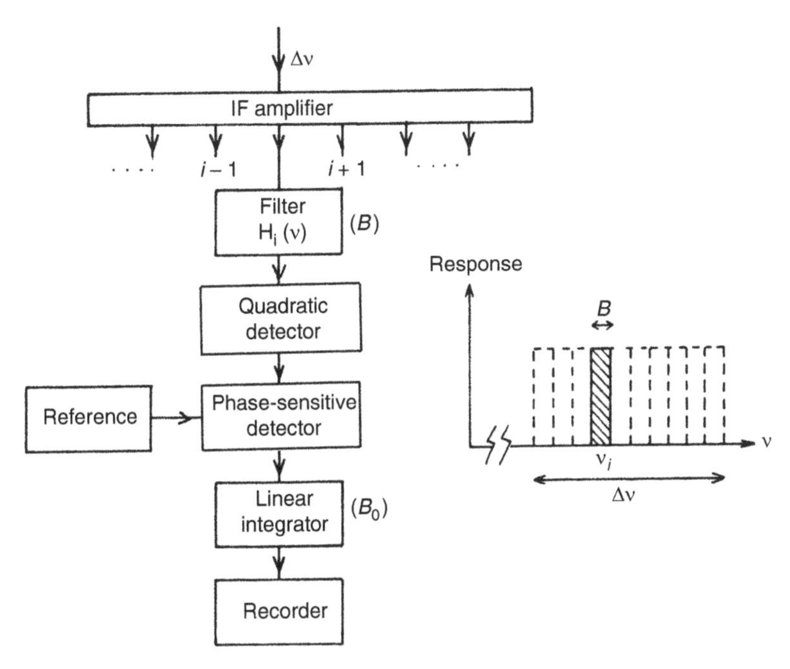

Fig. 8.34 Set of filters in parallel, and the resulting transfer function. The intermediate frequency signal from the mixer is amplified, then distributed through a series of electrical filters arranged in parallel, each one discriminating a spectral band B and thereby fixing the spectral resolution

Multichannel Filter Spectrometers

In the second half of the twentieth century, radioastronomy developed on the basis of this method of spectral discrimination, which is now tending to disappear (2007).

Consider the signal at intermediate frequency (IF) leaving the mixer (Fig. 8.34). A set of filters, with transfer functions $H_i(\nu)$ and passband B, slice the frequency band $\Delta\nu$ into $N = \Delta\nu/B$ channels, over which the IF signal is distributed in parallel. Each filter is followed by a quadratic detector, then a commutator, which compares the signal with a reference signal (local calibration source, see Sect. 3.4). A linear integrator (e.g., low-pass filter, digital running mean) of equivalent passband B_0 reduces the variance of the fluctuations, and the resulting noise temperature is given by

$$T = T_{\text{system}} \left(\frac{B_0}{B} \right)^{1/2} ,$$

where T_{system} is the system noise temperature defined for $B = B_0$. Note that half of the total measurement time is used for calibration, which therefore doubles the noise power. Each of the above operations is carried out digitally.

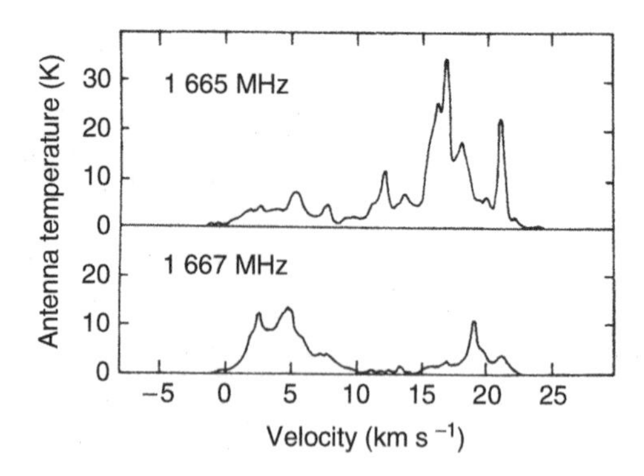

Fig. 8.35 Spectrum of the galactic object W49 OH (a very intense H II region). Spectral element $B = 2$ kHz. Number of channels $N = 100$. Central frequencies $\nu_0 = 1\,665$ and 1 667 MHz, respectively (radical OH lines). Note the complex radial velocity structure of the various components of the object

In brief, the source spectrum is found in the frequency interval $\Delta\nu$ around the frequency ν_0, with resolution $R = \nu_0/B$, averaged over a time interval of about $1/B_0$.

Multichannel spectrometers (Fig. 8.35) can include up to 512 channels operating in parallel, and the width B of each channel can be made almost arbitrarily narrow, but rarely less than 1 kHz in practice.

The Acoustic Spectrometer

This second method appeals to a different idea. The intermediate frequency signal excites vibrations in a piezoelectric crystal, which in turn generates an ultrasonic sound wave in a lithium niobate crystal ($LiNbO_3$). Other materials (tellurium dioxide or gallium phosphide) can be used to vary the useful band between 0.2 and 2 GHz. Promising tests (2007) aim to replace acoustic excitation of the crystal by laser excitation, keeping the crystal at low temperature (~ 4 K) in order to extend the lifetime of metastable states created by the laser radiation. The band could then reach 20 GHz.

The fluctuations of the refractive index at successive wavefronts are used to diffract the beam of an optical laser (for example, He–Ne, or a solid-state laser diode). The diffracted energy is distributed across a multichannel linear detector (a 1024 or 2048 pixel CCD array), in which each pixel has a size corresponding to the frequency spread B, or less, if over-sampling is required. The frequency resolution is limited by diffraction and attenuation in the acoustic medium. If the IF signal is monochromatic, all the diffracted energy is concentrated on a single pixel (*Bragg reflection*), and the spectrum produced by the sequential readout of the diodes is indeed non-zero only on a single element. In the case of a more complicated spectrum, the energy distribution on the diodes is simply proportional to the IF spectrum (Fig. 8.36).

This instrument is particularly compact and low in power consumption (a few watts), factors which make it ideal for use on board a satellite. For example, the European Herschel mission

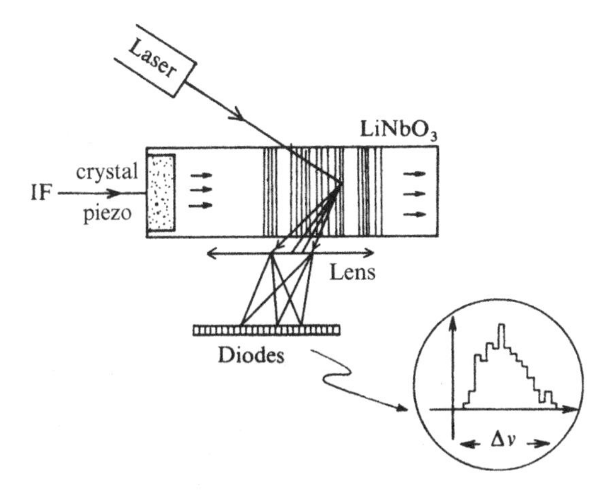

Fig. 8.36 Principle of the acoustic spectrometer. The *inset* shows the intensity received by each of the diodes, namely, the spectral density (spectrum) of the IF signal applied to the piezoelectric crystal. For the purposes of clarity, the angle between the incident direction (laser) and the diffracted direction has been grossly exaggerated. In reality, it would not be more than a few degrees. Note that the lens forms the Fourier transform of the diffracted field on the detector (see Sect. 6.1)

(launch 2008) carries an acousto-optical spectrometer, the *Wide Band Spectrometer* (HIFI) with four juxtaposed units of $B = 1$ GHz each, to observe in the range 0.5–1.9 THz.

Other spectrometers with similar inspiration may come to light in the future, e.g., using *surface waves* produced on a solid by interference between two lasers, and used to diffract the radiofrequency signal.

The Autocorrelation Spectrometer

This has become the main method used in radioastronomy observatories, thanks to progress in speed and capacity of digital processing, and the correlator has become a standard tool for radiofrequency spectrometry.

The *autocorrelation spectrometer* makes use of the integral transformation property relating the Fourier transform of the spectral density to the autocorrelation function (see Appendix A). If $I(\nu)$ is the spectrum (in fact, a spectral density, as already pointed out) of the time-dependent IF signal $x(t)$, then $I(\nu)$ is also the Fourier transform of the autocorrelation $R(\tau)$ of $x(t)$:

$$R(\tau) = \langle x(t)x(t+\tau) \rangle .$$

At the expense of a slight loss of efficiency, when $x(t)$ is digitised on a small number of bits, the digitisation of $x(t)$ followed by an autocorrelation calculation can give

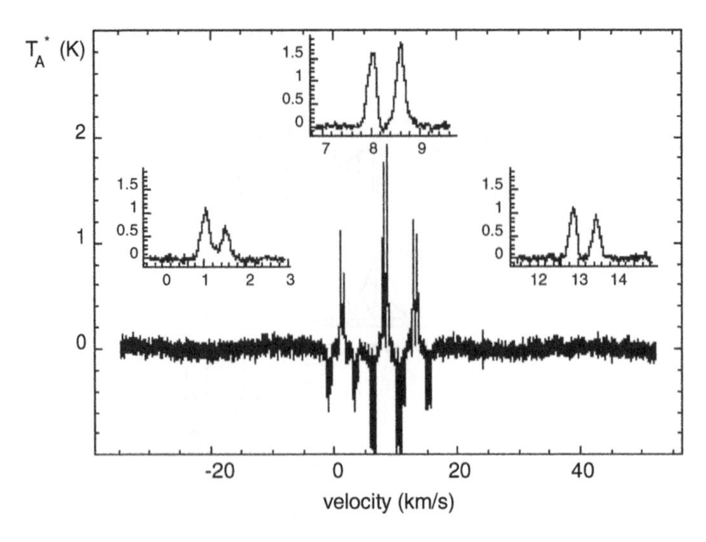

Fig. 8.37 Spectrum of a cold molecular cloud, obtained using the VESPA instrument on the IRAM 30 m radiotelescope (Spain) at 88 GHz ($\lambda = 3.4$ mm). *Ordinate*: Antenna temperature. *Abcissa*: Frequency converted to radial velocity. Resolution $B = 30$ MHz, and 3.3 kHz (corresponding to a gas velocity width of 11 m s^{-1}) in the *inserts*, showing the hyperfine structure of the line. Document courtesy of G. Paubert and C. Thum

$R(\tau)$ in real time, and hence also $I(\nu)$. Most correlators work on two bits (1 and 0) with efficiency reduced to 0.6, but sometimes on three (1,0,-1), and up to eight bits, whereupon the efficiency reaches 0.85. A correlator can cover up to 250 MHz, and several correlators can work in parallel to cover a broader range. The spectral resolution $B = 1/\tau_{\max}$ is determined by the width of the support of R, hence by the amplitude of the maximum chosen shift τ.

Note that the technique can be generalised to the cross-correlation of two signals $x_1(t)$ and $x_2(t)$, viz.,

$$R_{12}(\tau) = \langle x_1(t)\, x_2(t+\tau) \rangle ,$$

and this makes it possible to study two perpendicular polarisations, hence giving direct access to the *Stokes parameters*.

An example is the *Versatile Spectrometer Assembly* (VESPA) correlator,[11] set up on the IRAM 30 m millimetre wave radiotelescope in Spain in 2002. It covers an IF band of 480 MHz, which can be sliced into spectral elements of width $B = 40$ kHz, while specific modes can increase the spectral resolution up to a maximum of 10 kHz, i.e., a velocity resolution of 10 m s^{-1} at $\nu = 100$ MHz, as shown in Fig. 8.37. Spectra are often represented as a function of the LSR radial velocity (Sect. 8.1).

[11]See www.iram.fr/IRAMFR/ARN/dec02/node6.html.

Fast Fourier Transform (FFT)

The spectrum of the intermediate frequency (IF) signal can now be calculated directly almost in real time, thanks to the increasing speed of computer calculation. This spectrum is just the source spectrum but translated in frequency by the heterodyne setup. The 12 m millimetre wave radiotelescope *Atacama Pathfinder EXperiment* (APEX), set up at the future site of the ALMA millimetre array in Chile (Sect. 6.4), has a processing mode[12] with 16 384 spectral channels in a total band of 1 GHz: the resolved bandwidth is thus $B = 61$ kHz. The 100×110 m metre and centimetre wave radiotelescope at Greenbank (USA) is also equipped with this kind of processing system. The efficiency is excellent in principle since the digitised signal can be treated directly.

8.4.2 Submillimetre Spectroscopy

The submillimetre region (50 μm to 1 mm), still largely unexplored at the time of writing (2007), lies between the more conventional optical regions (mid-infrared) and the radiofrequencies. Both for radiation detection and spectroscopy, it usually borrows its observational methods from one or other of these regions. Very rich in spectral lines, particularly of molecular origin, in the interstellar medium, it also contains some quite fundamental lines, such as the neutral carbon C I fine structure line at 492 GHz and also the fine structure line of O I at 63 μm.

Observation in this region thus uses optical filters and grating spectrometers with moderate resolution ($< 1\,000$), as in the PACS instrument of the European Herschel mission (2008–2012) or in the IRS instrument (IRS) of the Spitzer mission (2003–2008), operating up to wavelengths close to 200 μm. However, the Herschel mission also carries a very high resolution heterodyne spectrometer, the HIFI instrument, covering the range 280–1 910 GHz (157–1070 μm). It works on the principle, typical to radiofrequency observation, of a local oscillator and mixers with a superconductor–insulator–superconductor (SIS) junction or hot electron bolometers (see Sect. 7.3). Likewise, the airborne telescope SOFIA[13] is equipped with a heterodyne spectrometer covering the range 250–600 μm to study the interstellar medium.[14]

Consider again the arrangement of a heterodyne instrument (Fig. 8.38), with a laser as local source and using a quantum quadratic detector of efficiency η to mix. The IF exit signal has frequency $\nu_s - \nu_0$ and the exit current at the detector is

[12]See www.apex-telescope.org/instruments for details of the fast Fourier transform (FFT) mode.

[13]This telescope, carried by a B 747 aircraft, is operated by NASA and the German space agency *Deutsches Luft- und Raumfahrt* (DLR). It made its first flight in 2007 (see Sect. 5.2).

[14]CASIMIR is described at www.sofia.usra.edu/Science/instruments/.

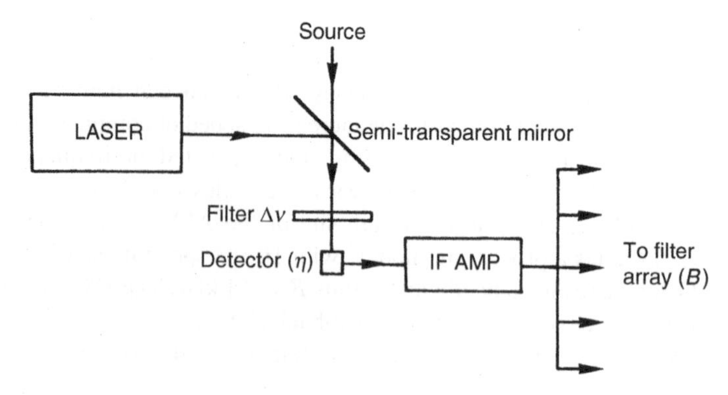

Fig. 8.38 Heterodyne spectrometer, with a laser as local oscillator

$$i(t) \propto \eta \left[\frac{E_0^2}{2} + E_\mathrm{s}\, E_0 \cos 2\pi(\nu_\mathrm{s} - \nu_0)t \right],$$

neglecting the term in E_s^2 ($E_\mathrm{s} \ll E_0$), and also the sum frequency term, which is too high for the frequency response of the detector. After filtering (bandwidth B) and linear integration (equivalent bandwidth B_0), the signal-to-noise ratio is, in power,

$$\eta \frac{P_\mathrm{s}}{h\nu\,(BB_0)^{1/2}}.$$

Note that, if $B = B_0$, the expression $\eta P_\mathrm{s}/h\nu B$ is recovered, which means that the minimal detectable power P (signal-to-noise ratio $= 1$) corresponds to one photoelectron per hertz of the passband (see Sect. 7.2.2).

In the special case when the source is a blackbody of temperature T, and thus of intensity

$$\frac{2h\nu^3}{c^2}\left(\mathrm{e}^{h\nu/kT} - 1\right)^{-1} \qquad [\mathrm{W\,m^{-2}\,sr^{-1}\,Hz^{-1}}],$$

the beam étendue is λ^2, because coherence of the incident field is required in this method of measurement. The ideal signal-to-noise ratio is then deduced to be

$$2\eta\left(\mathrm{e}^{h\nu/kT} - 1\right)^{-1}\left(\frac{B}{B_0}\right)^{1/2}.$$

The resolution of such a spectrometer is only limited by the frequency stability of the local oscillator and by the passband B of the filter.

Extension to Mid-Infrared. The boundary between the submillimetre and infrared regions is fuzzy as far as observational techniques are concerned, as already emphasised. As an example, consider an infrared heterodyne spectrometer made in 1983 (Fig. 8.39). The local oscillator is a

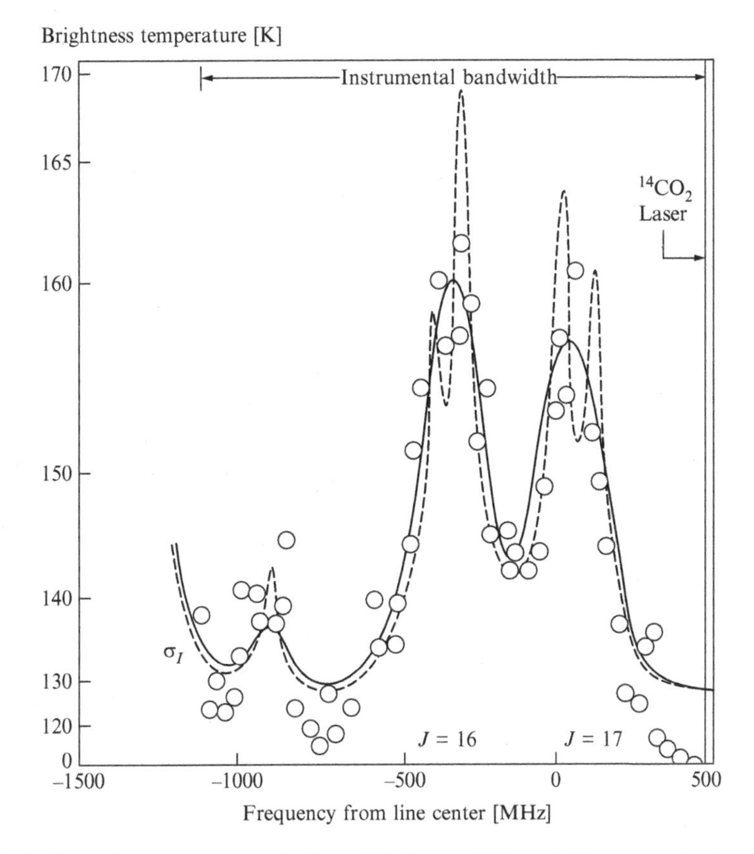

Fig. 8.39 Heterodyne spectrometry in the mid-infrared (10.6 μm). Measurement of an ethane (C_2H_6) line emitted by the atmosphere of Jupiter. Beam étendue λ^2, i.e., a field of 2″ for a 1.5 m telescope. Rotational lines corresponding to levels $J = 16$ and 17 are fitted to a model (*dashed line*), which is corrected for line broadening due to rotation of the planet (*continuous line*). The estimated σ is shown *on the left*. From Kostiuk T. et al., Ap. J. **265**, 564, 1983. With the kind permission of the Astrophysical Journal

$^{14}C^{16}O_2$ laser, emitting at 829.93 cm^{-1}, and the detector is a HgCdTe photoconductor mixer, with passband $\Delta\nu = 1.6$ GHz. The intermediate bandwidth is $B_{IF} = 25$ MHz, and the resolution is 10^6.

8.5 Resonance Spectrometers

A particularly elegant method of spectrometry is based upon the following principle. An atomic vapour is illuminated by the radiation under study. If this radiation contains photons whose frequency is equal to the characteristic frequency of some atomic transition in the vapour, then those photons will be absorbed. If a transition is chosen such that the lower level is a ground state, therefore highly populated,

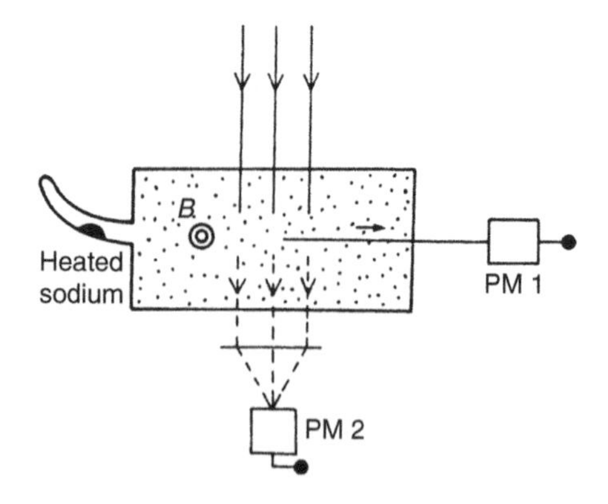

Fig. 8.40 Resonance spectrometer. The photons re-emitted by the vapour are measured at photomultiplier 1, while the signal proportional to the incident radiation is measured at photomultiplier 2. The ratio of the two is independent of atmospheric and instrumental fluctuations

and the excited level a resonance, so that the probability of excitation is high, the number of absorbable photons can be maximised. This number is then measured by detecting the photons re-emitted in spontaneous de-excitation of the upper level, and the number of these is proportional to the number of incident photons (Fig. 8.40).

A quite general tuning relation is

$$\frac{V_{\text{source/cell}}}{c} + \frac{V_{\text{source}}}{c} = \frac{V_{\text{vapour}}}{c} + \frac{eB}{4\pi m},$$

which includes the thermal speeds of atoms in the source and in the vapour (both small relative to c), the motion of the source (e.g., the Sun) relative to the detecting cell (fixed to the Earth or a satellite in space), and also a Zeeman shift which may be imposed on the atoms of the vapour by a magnetic field B. By varying B, the profile of the excitation line can be scanned. A vapour at low temperature T has $V_{\text{vapour}} \ll V_{\text{source}}$, and the maximum resolving power of such a spectrometer is given by

$$R \approx \frac{c}{V_{\text{vapour}}} \approx c\left(\frac{m}{3kT}\right)^{1/2}$$

where m is the atomic mass of the vapour atoms. At $T = 500$ K, $R \approx 2 \times 10^5$ for sodium vapour.

Although limited to certain favourable circumstances, this type of *resonance spectrometer* has been successfully used to measure the photospheric oscillations of the Sun and other stars (in the 1980s) when the velocities to be detected are of order cm s^{-1}, and also to study the velocity field of interstellar hydrogen.

Problems

8.1. Resolution of a Fourier Transform Spectrometer. Consider an interferogram $I(x)$, taken over the optical path differences 0 to x_m.

(a) Show that the observed interferogram is simply the product of the full interferogram (x varying from $-\infty$ to $+\infty$) and a suitable box function. Deduce the instrumental profile of the spectrometer.

(b) Show that, by multiplying the observed interferogram by a suitably chosen function, it is possible to modify the instrumental profile. For example, giving up some resolution in order to decrease the wings of the profile. Examine the use of the multiplying functions

$$\cos\left(\frac{\pi x}{2x_{\mathrm{m}}}\right)$$

and the sawtooth function

$$1-\frac{x}{x_{\mathrm{m}}} \quad \text{if } x \le x_{\mathrm{m}}, \qquad \text{and} \quad 0 \quad \text{if } x > x_{\mathrm{m}}.$$

This weighting of the interferogram is called *apodisation*, because it suppresses the 'feet' of the instrumental response. It is also possible to apodise a pupil by the use of an appropriate amplitude or phase mask (see Sect. 6.6).

Answer. (a) The observed interferogram is truncated

$$I(x) = I_{\mathrm{c}}(x)\Pi(x/x_{\mathrm{m}}),$$

and in the frequency space,

$$\tilde{I}(\sigma) = \tilde{I}_{\mathrm{c}}(\sigma) \star \mathrm{sinc}(x_{\mathrm{m}}\sigma).$$

The instrumental profile is therefore $\mathrm{sinc}(x_{\mathrm{m}}\sigma)$.

(b) Multiplying the interferogram by a cosine amounts to modifying the instrumental profile of the spectrometer by

$$\mathrm{FT}\left[P(x/x_{\mathrm{m}})\cos\left(\frac{\pi x}{2x_{\mathrm{m}}}\right)\right].$$

The resulting profile is wider because the high frequencies are filtered, but the feet of the profile are cleaner (Fig. 8.41).

8.2. Sampling in Fourier Transform Spectroscopy. Let σ_{M} be the highest wave number contained in the source spectrum. What step Δ should be chosen for the moveable mirror so that the interferogram for this radiation is optimally sampled (in the sense of Shannon's theorem)? (It is easily shown that the question amounts to finding the minimum sampling for a sinusoidal function.)

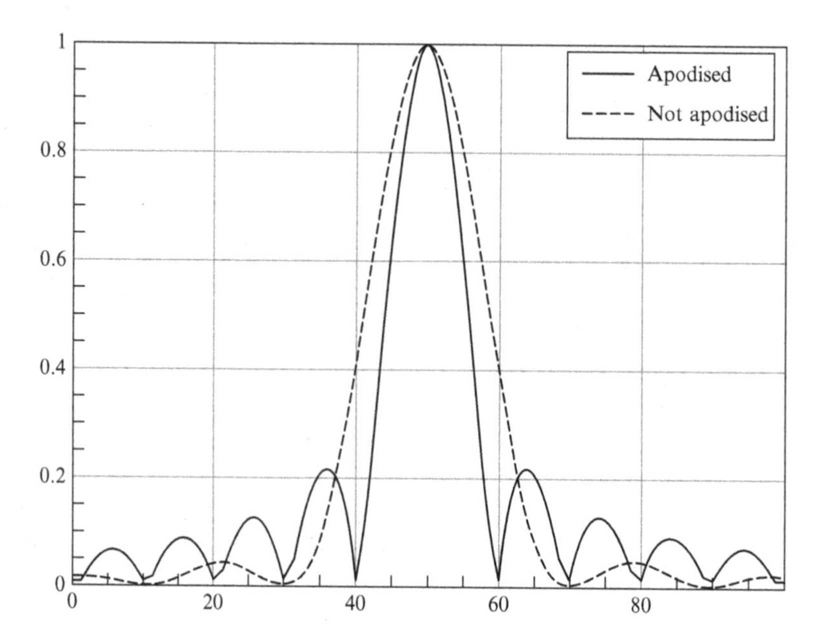

Fig. 8.41 Example of apodisation. *Continuous line*: Point source response of a circular pupil (Airy function). *Dashed line*: Point source response of a circular pupil after apodisation by a sawtooth function. Note the broadening of the central spike and the significant reduction of the wings

8.3. Imaging and Fourier Transform Spectroscopy. The multiplex capability of an FT spectrometer, combined with a digital detector with many pixels, can in principle be used to obtain a large number of monochromatic images of a source simultaneously, at adjacent frequencies. If the aim is to obtain such images over a 1×1 arcmin field, with an image quality limited by the atmosphere to 0.5 arcsec, and a resolution of 5×10^4, calculate the total number of bits of information accumulated in the interferograms, each data point being digitised on 16 bits.

Answer. The resolution requirement imposes on x_m the condition: $R = \sigma x_\mathrm{m}$. The sampling of the interferogram requires $\Delta x = 1/2\sigma$, whence the number of images $N = x_\mathrm{m}/\Delta x$. Moreover, the information in an image is coded on

$$(2 \times 60)^2 \times 16 = 2.3 \times 10^5 \,\text{bits}.$$

The digital dynamic range thus obtained with 16 bits is then $2^{16} \sim 65\,000$, which corresponds to the usual dynamic range of CCD detectors. So this cube of information is coded on 23 Gbits.

8.4. An FT spectrograph is used in the photometric band M (see Sect. 3.4), with a high-performance detector (NEP $= 10^{-16}\,\mathrm{W\,Hz^{-1/2}}$). Calculate the thermal photon noise due to atmospheric emission. Take its emissivity to be 0.05, and the beam size fixed by a 4 m telescope with a 10 arcsec field of view at the detector. Show that the broadening of the spectral band allowed by the spectrograph (multiplex advantage) would not improve the signal-to-noise ratio.

Part III
Data Analysis

Chapter 9
The Signal in Astronomy

Astronomy has often developed by pushing instruments to the limits of their capacities. Many discoveries spring from a combination of noise reduction, by careful design of instrumentation, and efficient signal processing.

Light, or any other information carrier, sent out from a source, crosses various media which influence it on its path, including in some cases the Earth's atmosphere. It then reaches the observing system and its components: an image is formed by a telescope, the spectrum is analysed by a spectrometer, and a measurable and storable signal is finally produced by a detector. For the astrophysicist, the signal represents a translation of the information emitted by the source, and it is on this signal that the work of interpreting observations will be based.

Since the beginning of the 1990s, there has been a genuine revolution in signal processing thanks to developments in computing. It is now possible to digitise all the information provided by an instrument, which can itself be controlled precisely and automatically by computer. That information can then be stored in practically unlimited quantities, while nevertheless keeping it easily accessible, and it can be processed by extremely sophisticated algorithms, without requiring exorbitant computation times. Observing instruments and systems, comparison of data with models, use of a priori known information to reduce uncertainties due to noise or the limitations of a given instrument, here are some of the novelties made available by extremely elaborate processing of astronomical signals. Clearly this exploits tools developed in a wide range of different applications, but it can also inspire the scientific community to original ideas, born from particular features specific to astronomical observations, such as stringent sensitivity or image quality requirements.

After a brief review of the basic elements of signal processing, this chapter shows how an observing system can be completely modelled by computer, whence its operation can be simulated in every conceivable configuration. This is typical of a common practice today when dealing with complex systems. For example, an aircraft is fully modelled in this way before it makes its first real flight. The previous chapters have dealt individually with the various subsystems, such as imaging devices, spectrographs, and detectors, but the aim here will be to provide

P. Léna et al., *Observational Astrophysics*, Astronomy and Astrophysics Library,
DOI 10.1007/978-3-642-21815-6_9, © Springer-Verlag Berlin Heidelberg 2012

a more global point of view of several complete systems through several specific examples, describing how they actually work when carrying out real observations: a radiotelescope, infrared adaptive optics, a photometric satellite, and a γ ray observatory. We shall then be able to determine whether the desired observation of the given astronomical source will in fact be possible.

The maximal correction for all effects due to known properties of the instrument, referred to as the *instrumental signature*, is discussed in the following section. Then comes a more mathematical section, describing the ideas most frequently used in the problem of *estimation*, where the aim is to deduce the properties of the observed object from the acquired data, using where necessary any information that looks a priori as though it is likely to be true on the basis of the available models of the object.

The last section deals with the tools used to solve the *inverse problem*, presenting a certain number of applications to the problem of *image processing*, central to the interpretation and use of observations in astrophysics.

Questions of archiving the immense volume of data produced by all the astronomical instruments that have ever operated or are still operating today, both on Earth and in space, and in particular the methods used to maintain accessibility to this data, are postponed until Chap. 10.

Some of the mathematical tools required in this chapter and throughout the book, such as Fourier transforms, probability, and statistics, are reviewed in Appendices A and B, to ensure consistency of notation insofar as possible.

9.1 The Signal and Its Fluctuations

In this section, the subjects will be the role of the detector, the relation between the signal and information from the source, and, finally, the perturbation or noise to which the signal is subjected during an observation.

9.1.1 Observing System and Signal

A *detector* (the general term *transducer* is also used) transforms the photon flux into a current, a voltage, a charge, or some other measurable or storable quantity, which is usually digitised and processed immediately by computer.

If the detector has imaging capabilities, it provides N distinct pieces of information, each corresponding to an image element of field $\mathrm{d}\omega$ (solid angle), called a *pixel*.

One particularly important class of detection systems in astronomy (interferometry, see Sect. 6.4) has the remarkable property of sampling information from the source, not in the $\boldsymbol{\theta}$ plane of the sky, but rather in the conjugate Fourier transform space. Hence, the information element is not defined by its direction of observation,

Table 9.1 Examples of information sampling

• Direction • Temporal frequency e.g., mirror + field diaphragm + photographic plate + spectral filter	• Direction • Time e.g., imaging detector + FT spectrometer
• Spatial frequency • Temporal frequency e.g., pair of radiotelescopes + electric filter	• Spatial frequency • Time e.g., pair of IR telescopes + FT spectrometer
• Direction • Energy e.g., γ coded mask + scintillator	• Direction • Polarisation e.g., submillimetre telescope + heterodyne detector

but rather by its position in the conjugate space of *spatial frequencies*. Instead of measuring the quantity $I(\boldsymbol{\theta})\mathrm{d}\omega$, which is the specific intensity received in the direction $\boldsymbol{\theta}$, in the solid angle $\mathrm{d}\omega$, we measure

$$\tilde{I}(\boldsymbol{w}) = \iint I(\boldsymbol{\theta})\mathrm{e}^{-2\mathrm{i}\pi\boldsymbol{\theta}\cdot\boldsymbol{w}}\,\mathrm{d}\boldsymbol{\theta}$$

in a spatial frequency interval $\mathrm{d}\boldsymbol{w}$.

Since Fourier transforms are one-to-one, knowing $\tilde{I}(\boldsymbol{w})$ on the whole $\boldsymbol{w}$ space is completely equivalent to knowing $I(\boldsymbol{\theta})$ in all directions $\boldsymbol{\theta}$. Aperture synthesis methods (see Chap. 8) make use of this approach.

In the same way, *spectral filtering* involves picking out from the spectral density $S(\nu)$ of the incident radiation a spectral element of frequency width $\Delta\nu$, and this is equivalent to sampling the information in a time interval Δt and measuring the Fourier transform of the spectral distribution of the radiation (see Chap. 8).

It is also possible to combine these two approaches, as shown schematically in Table 9.1. In each example, the symbol (•) marks the two variables sampled by the system. Clearly, the signal may take a wide variety of forms, extracting information from the source via some specific filter.

9.1.2 Signal and Fluctuations. Noise

In most cases, the raw signal delivered by a detector has the form

$$x(t) = x_{\mathrm{s}}(t) + n(t),$$

where $x_{\mathrm{s}}(t)$ is the signal from the source and $n(t)$ may be a background signal, background noise, detector noise, or a combination of all three. The problem is, firstly, to understand the causes and characteristics of the noise $n(t)$, and, secondly, to optimise the acquisition and recording of $x(t)$.

For a source of constant intensity, the quantity $x(t)$ measured by the detection system would have a perfectly well-defined value, but in the reality of physical observation, it has the character of a *random variable* or *process*. In the following, therefore, it will be denoted by a bold-type letter $\mathbf{x}(t)$.

The physical measurement made of $\mathbf{x}(t)$ consists in observing one realisation ζ of this random or stochastic process (abbreviated to s.p. in the following). A measurement carried out over the time interval ΔT will only be indicative of a process if certain restrictive assumptions can be made about that process, for example, stationarity, or ergodicity.

The causes of fluctuation, which give the signal its random character, are various. Some are due to interference from the measurement itself, and can be compensated for, or eliminated. Some are of a fundamental nature and connected to the physics of the detection process itself, or to the quantum and discrete nature of the electromagnetic interaction (see Sect. 7.2).

The term *noise* will mean those fluctuations whose effect is to introduce a random deviation between the *estimators* of the signal, that is, measurements made over a time ΔT, and the true mean values of the signal. The quantity called the *signal-to-noise ratio* is a relative measure of this deviation between the desired quantity and its estimator.

This idea of *signal-to-noise ratio*, which is indeed a ratio of two quantities, denoted by S/N, always refers to a measurement of a specific quantity. There is no such thing as intrinsic signal-to-noise. If $\mathbf{x}$ is a positive random variable, aiming to estimate the physical quantity x, then the S/N of this measurement is defined as the average of $\mathbf{x}$ divided by its standard deviation.

Background Noise

Even when there is no signal from the source, the detector may nevertheless deliver a signal $\mathbf{x}_0(t)$, with the character of an s.p. A few examples are given here. Other noise phenomena typical of detectors were introduced in Chap. 7.

Dark Current (Detector Noise)

Even when no photons are incident on its photocathode, a photomultiplier or the pixels of a CCD will produce a non-zero photocurrent, resulting from random electron impacts due to charges whose thermal agitation energies exceed the work function of the metal (the potential holding them in the metal) or the photoconductor gap. The statistics of these fluctuations was discussed in Sect. 7.2. $\mathbf{x}_0(t)$ is called the *dark current*, and adds to the source signal. Its fluctuations lead to fluctuations in the measured signal.

Grain of a Photographic Plate (Detector Noise)

When an unexposed plate is developed, it exhibits a random blackening $X_0(x, y)$, at points (x, y) of the plate. This quantity has a non-zero mean value, the continuous background or veiling, and a variance which is the background noise of the plate, caused by the *grain of the emulsion*.

Gamma-Ray Background

When used as a detector in a γ-ray telescope, a spark chamber detects events even in the absence of radiation from the source. These events are due to γ photons produced by the interaction of cosmic rays with the Earth's atmosphere (see Sect. 2.9), or with the material of the detector itself, and also natural radiation, or other sources. They are random events and give a signal, whose mean value, evaluated as the number of parasitic counts per unit time, is the background signal, and whose variance is the background noise (Fig. 9.1).

Cosmic Rays on CCDs

The impact of a high energy particle on the photoconducting material of a CCD creates a large number of charges by photo-ionisation. These are detected as if they came from incident photons. This extraneous signal will affect one or more pixels and must be eliminated from the final image. The level of events depends on the location of the detector (on the Earth, at a high altitude, in space, in a radiation belt, etc.). At ground level, it is usually around one event per frame (i.e., over all the pixels), for a 256×256 pixel format, and per minute of observation time.

Thermal Background Noise in the Infrared

A detector placed at the focus of an infrared or submillimetre telescope will receive photons even in the absence of any source in the field of view. There are thermal photons emitted by the Earth's atmosphere, by the telescope mirrors, and by the optical components encountered (windows, filters, and so on). This flux of photons gives an average background signal, whose fluctuations are called the *thermal background noise* (see Sect. 7.2).

In each of these cases, the aim will be to understand the nature and statistics of the process $\mathbf{x}_0(t)$, in order to deduce its effect on measurement of the signal $\mathbf{x}(t)$.

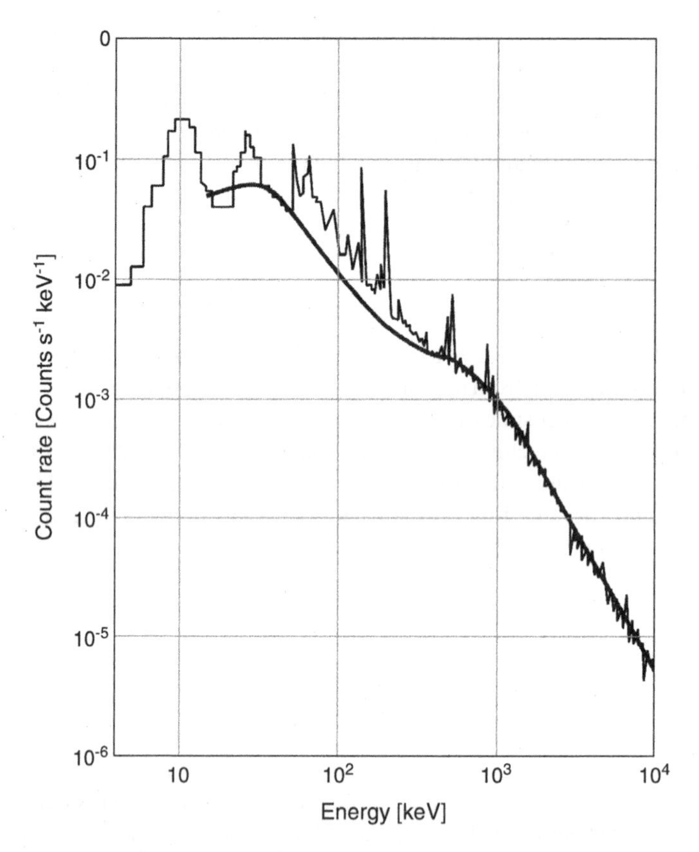

Fig. 9.1 Background noise measured during observation of gamma radiation in the Earth's stratosphere. The *smooth curve* gives the background modelled as a superposition of local atmospheric background, elastic scattering of neutrons, and the induced β radioactivity. The noisy curve, in good agreement with the model, results from measurements made in the spectral range 5 keV to 10 MeV. Any astronomical signal would be superposed on this background. (Hexagone instrument, on a stratospheric balloon, altitude 3.6 mbar, Alice Springs, Australia, 1989. Document courtesy of Védrenne G.)

Signal Noise

The signal from the source itself may also exhibit random fluctuations.

Transmission Noise (Multiplicative Noise)

If the absorption properties of the medium vary randomly in time, the transmitted signal will have the form

$$\mathbf{x}(t) = A(t)\,\mathbf{x}_s(t),$$

where $\mathbf{x}_s(t)$ is the source signal and $A(t)$ a fluctuating transmission. This is the case when there is *scintillation* at visible or radio frequencies, this phenomenon being related to fluctuations in the refractive index of the air (see Sect. 6.2), or of the electron density in the ionosphere, respectively. It is also the case in the infrared, when there is variation in the water vapour content of the atmosphere.

Signal Noise (Thermodynamic or Photon Noise)

An electromagnetic wave emitted by a thermal process (e.g., blackbody) is incoherent. The field $\boldsymbol{E}(t)$ is a random process, whose fluctuations are explained by a lack of phase correlation over the atoms radiating the macroscopic field. A photon description can also be given. The rate of arrival of photons from the source is a random function, whose mean value is the mean received flux.

In brief, these phenomena can be classified into three main categories, according to the various causes of fluctuation:

- *Fundamental limitations*, of thermodynamic origin, related to the physical nature of the radiation (photon noise of the signal), to its incoherence, or to its interaction with matter in the detection process (irreducible detector noise).
- *Practical limitations*, resulting from the quality of the detectors, whose design is constrained by the present state of understanding and technology. Such limitations are constantly being pushed back: thus, the intrinsic noise of near infrared detectors (1–10 μm) is characterised by a variance whose value has been reduced by a factor of 10^8, a gain of eight orders of magnitude, between 1960 and 1990, while the format progressed from 1 pixel to almost 10^7 pixels in 2007.
- *Observational limitations*, where the signal is affected by the local environment in which observations are made. These observation conditions can almost always be improved, by changing the instrumental arrangement, or the site, going outside the Earth's atmosphere, and so on.

Recall that the first category here, of fundamental limitations, was discussed in Sect. 7.2, along with the practical performance of detectors. It is quite clear that these fundamental limitations will determine the performance that can ultimately be attained by any measurement system.

Signal-to-Noise Ratio. Detection Limits

The signal $x(t)$ is one realisation of a random process $\mathbf{x}(t)$, and extraction of the information it contains constitutes an *estimate* (see Appendix B and Sect. 9.5). The degree of confidence which can be accorded to this estimate will depend on the type of information contained in the signal, as well as the statistics of the fluctuations it manifests.

The signal is written in the form

$$x(t) = x_0(t) + x_s(t).$$

In the case of statistically independent additive noise, the variance of the signal is written

$$\sigma_x^2 = \sigma_{x0}^2 + \sigma_{x_s}^2.$$

We examine the simple case in which the desired quantity is the mean $\langle x_s \rangle$. An example would be the measurement of a spectrum by making a best estimate of the intensity in each spectral element. Assume that $\langle x_0(t) \rangle$ can be estimated with relatively low uncertainty by means of an independent experiment (determining the statistical properties of the noise). The signal estimate is then[1]

$$\langle x(t) \rangle_T - \langle x_0(t) \rangle.$$

If the statistics of $x_s(t)$ and $x_0(t)$ are known, then so are σ_{x0} and σ_{x_s}, and the estimate of the signal-to-noise ratio is

$$\frac{\mathrm{S}}{\mathrm{N}} = \frac{\langle x(t) \rangle_T - \langle x_0(t) \rangle}{\left(\sigma_{x0}^2 + \sigma_{x_s}^2\right)^{1/2}}.$$

If the statistics are not known, then the quantities σ_{x0} and σ_{x_s} must be estimated, by repeating the measurements, so that the next order moment (the variance of the variance) is small (see Sect. 9.5).

If $\langle x_0(t) \rangle$ could not be determined without error, the signal estimate would be

$$x(t) - \text{estimate of } \langle x_0(t) \rangle,$$

and the signal-to-noise ratio would be

$$\frac{\text{signal estimate}}{\left(\sigma_{x0}^2 + 2\sigma_{x_s}^2\right)^{1/2}}.$$

It must be decided what value of this ratio allows us to assert that the source is detectable. The question can only be answered by making some hypothesis concerning the statistical properties of the random process which constitutes the noise.

---

[1]In this book, the average of a random quantity is denoted by the operator $\langle\ \rangle$, a standard notation for physicists. If necessary, the angle brackets will carry a subscript to indicate the quantity on which the random variable depends, e.g., $\langle\ \rangle_t$ for the time t. We shall also sometimes use the mathematician's notation $E\{\ \}$ for expectation or expected value.

Fig. 9.2 Spectral characteristics of a particular example of Gaussian noise. The spectral density $S_N(f)$ is almost constant (white) at frequencies $f \lesssim f_c$

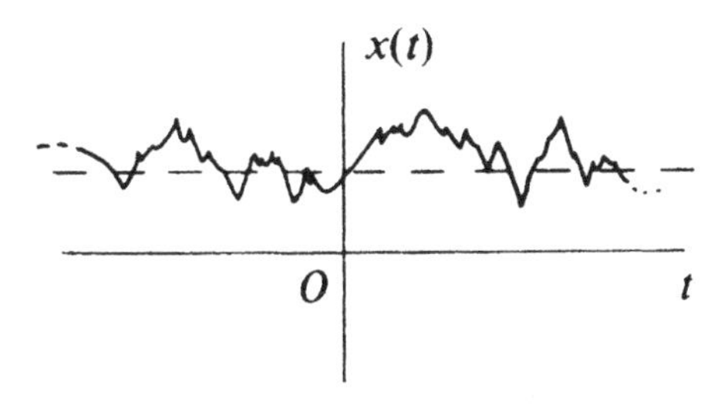

The determination of a signal from measurements is therefore a process of estimation. Care must be taken to use unbiased estimators, or else to correct for any bias (see Appendix B and Sect. 9.5.4). For example, in order to estimate the variance of the noise, with a view to evaluating the signal-to-noise ratio, a simple estimate of the variance on the measurement sample gives a biased value, which must be corrected for.

The term *bias* is used in a more general context, when the aim is to measure a physical quantity X using some measuring instrument. The latter gives, in reality, $X + B$, where B is a measurement bias due to some extraneous signal, a background, or the like. B must then be estimated independently in order to subtract the bias.

Gaussian Noise

Consider a signal represented by

$$x(t) = x_s + N(t),$$

where $\mathbf{N}(t)$ is a Gaussian s.p. with zero mean, autocorrelation $R(\tau)$, and spectral density $S(f)$, supposed approximately constant (almost-white noise)[2] for $f \lesssim f_c$ (Fig. 9.2). Consider a measurement made over a time T.

We wish to determine the degree of confidence that can be attributed to the estimate $\hat{x}$ of $\langle x(t) \rangle$ given by

$$\hat{x} = \frac{1}{T} \int_0^T x(t)\, \mathrm{d}t.$$

It can be shown (see Appendix B) that

$$\sigma_{\hat{x}}^2 = \frac{R(0)}{2 f_c T}.$$

[2]White noise has constant spectral density whatever the frequency. This can only ever be a limiting case, never actually attained in reality, because such noise would involve infinite power.

The Gaussian probability distribution gives a degree of confidence for intervals centred on x_s, with extent $\pm\sigma$, $\pm 3\sigma$, etc. (see Table B.1 in Appendix B).

The central limit theorem[3] shows that many of the fluctuation phenomena encountered in measurement can be considered as Gaussian processes. In the presence of Gaussian noise, a confidence interval of $\pm 1.5\sigma$ is often taken as satisfactory.

In order to be intelligible, every curve and every measurement result published in the literature must be accompanied by an error bar $\pm n\sigma$ based on the variance σ.

Signal-to-Noise Ratio and Integration Times

The relation

$$\sigma_{\hat{x}}^2 = \frac{R(0)}{2 f_c T}$$

shows that the variance of the noise decreases as the $-1/2$ power of the measurement time, although exceptions are sometimes encountered. This very general result fixes a practical limit on how much the signal-to-noise ratio can be improved by increasing the measurement time, taking into account the average lifespan of an astronomer. Indeed, it is altogether feasible to gain a factor of 10^3 by extending the time from 1 μs to 1 s, and a further factor of 300 in increasing the time to one day. However, the next factor of 200 would require extension from one day to one century.

Averaging is the most straightforward procedure for improving the signal-to-noise ratio, as shown in Fig. 9.3. It is implicit in this analysis that the noise can be assumed stationary, and that the required quantity x_s does not itself fluctuate, during the period considered.

It is sometimes observed that this rule no longer holds for faint sources superposed on a strong sky emission background. Subtraction of the sky background (see Sect. 2.3) then creates a noise increasing as $T^{1/2}$, when this background is intense. The signal-to-noise ratio, instead of continuing to increase with T, tends to some constant value as T increases. In some approaches using CCDs, it is possible to subtract the background at frequent and regularly spaced intervals of time, so avoiding the saturation effect (*sloshing mode*).

Figure 9.3 illustrates the gain in the signal-to-noise ratio by averaging, both in a 1D scan and in an image. The latter shows the superposition of 3 000 individual images, obtained with a TV camera (exposure $1/30$ s), using a broad band filter (410–760 nm), of a magnitude 16 quasar. First, the thirty images taken each second are cumulated, and *realigned* on the brightest pixel, and then all these one second images are again cumulated. Image realignment techniques are important and we shall return to this matter in Sect. 9.6. This simultaneously improves the S/N ratio, by the averaging effect, and corrects for atmospheric activity (see Sect. 6.2), hence improving the *resolution* of the image. The brightest image was extended (i.e., had two barely resolved components), and was

[3]This is a translation of the original term used in German by the Hungarian mathematician G. Polya, viz., *Zentralen Grenzversatz*, which means 'theorem establishing a limit that plays a fundamental role in probability theory'. Translations are often misleading.

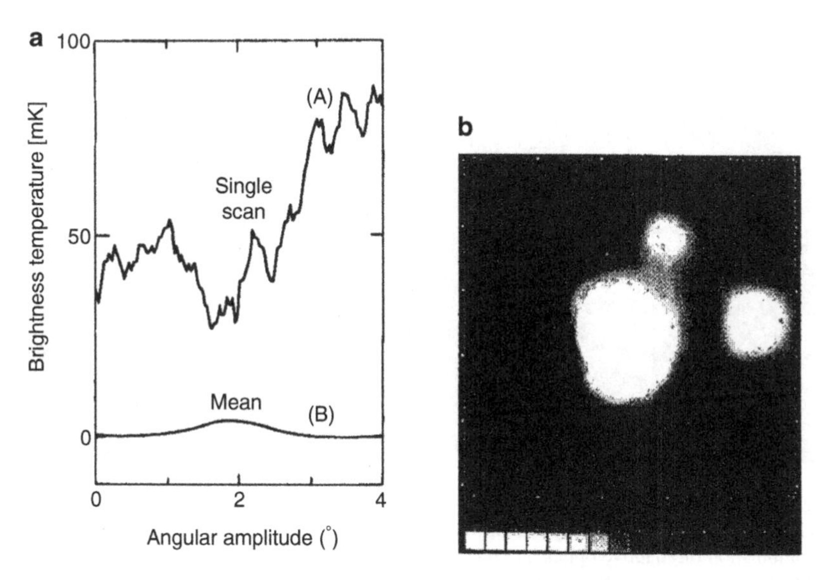

Fig. 9.3 Extraction of a signal from noise by averaging. (**a**) Thermal emission from cold dust in the plane of the Galaxy, at a wavelength $\lambda = 0.9$ mm. The brightness temperature is in millikelvin [mK], measured as a function of galactic latitude, at a given longitude. A = single scan, $T = 0.7$ s, and B = mean over 1 400 successive scans. From Pajot F., doctoral thesis, Paris VII, 1983. See also Problem 2.8. (**b**) The triple quasar QSO PG 1115 + 08. The second quasar was discovered in the form of a gravitational mirage. This observation was made using a 1.8 m telescope (one element of the Multi-Mirror Telescope). Scale 512 × 512 pixels, 1 pixel = 10^{-2} arcsec. Image courtesy of Hege E.K. For the speckle observing method, see Hege E.K. et al., Ap. J. **248**, L1, 1981. With the kind permission of the Astrophysical Journal

later fully resolved by speckle interferometry (see Sect. 5.2), then by adaptive optics. The four components constitute a *gravitational mirage* of the same quasar, but the galaxy causing it is too faint to be extracted from the noise. It was observed in the 1990s from the ground and also by the Hubble Telescope. More recently, this same quasar was observed in the infrared by the Japanese Subaru telescope in Hawaii (see Fig. 9.4). The image shows not only the four components, but also the whole gravitational mirage in the form of a beautiful *Einstein ring*.

9.1.3 Elementary Signal Processing

Digitisation

Although certain operations, such as filtering, can be carried out by analogue techniques, the trend in modern methods of signal processing is towards digital processing, in which the signal $x(t)$ is simultaneously sampled and transformed into a binary-coded numerical value by means of an *analogue-to-digital converter* (ADC). The number of bits of information used to represent the maximum value of the analogue variable $x(t)$ is called the *dynamic range* of the converter. The value

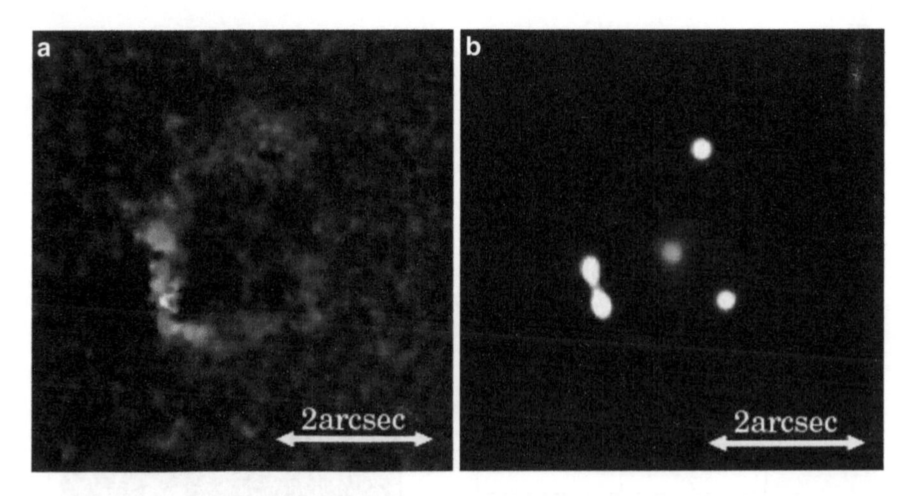

Fig. 9.4 The quasar QSO PG 1115+08, observed in the near infrared (J and K′ spectral bands) with the Japanese Subaru lescope in Hawaii. The resolution of the image is 0.12 arcsec, thanks to a combination of adaptive optics and deconvolution. (**a**) An Einstein ring (gravitational mirage). (**b**) Composite image (J+K′) showing the four components. Compare with Fig. 9.3, obtained two decades earlier. Reproduced with the kind permission of Iwamuro, F. et al., Publ. Astr. Soc. Jap. **52**, 25–32, 2000

of the analogue variable corresponding to one bit is called the *digitisation step*. The operation thus amounts to replacing $x(t)$ by $\hat{x}$, if $x(t)$ lies between $\hat{x} - q/2$ and $\hat{x} + q/2$, where q is the digitisation step.

The moments of the new random variable $\hat{x}$ are clearly different from those of $x(t)$. It can be shown, for example, that the error in taking the autocorrelation function $\hat{C}(\tau)$, rather than the autocorrelation function $C(\tau)$ of $x(t)$, is negligible, provided that q is chosen less than σ, where σ^2 is the variance of $x(t)$. For certain operations, it is enough to digitise using just one bit. For example, if $\mathbf{x}(t)$ is a r.v. with zero mean, digitising can be limited to specifying the sign of $x(t)$. There is a significant gain in calculation time (Problem 9.11).

Sampling

Sampling means reading off the value of the signal at times t (or, more generally, at discrete values of the variable labelling the signal, such as x and y coordinates on a photographic plate). The interval between two successive readings is called the *sampling rate*.

This rate is usually constant, but it need not be so. Sometimes it is random. Suppose the aim is to study the magnitude variations of a star, possibly periodic, in a site affected by uncertain weather conditions. The nights when measurement is possible are distributed randomly in time, and the brightness $b(t)$ of the star is thus randomly sampled. An analogous problem occurs in the sampling of a signal due to the photospheric oscillations of the Sun (*helioseismology*), or a star (*astroseismology*).

Using the Dirac comb distribution, defined in Appendix A, to describe sampling at rate Δt,

$$x(t) \text{ sampled } = ⧢\left(\frac{t}{\Delta t}\right) x(t).$$

Choice of Sampling Rate. Shannon's Theorem

Consider a function $f(x)$ whose spectrum $\tilde{f}(s)$ has bounded support $[-s_\mathrm{m}, +s_\mathrm{m}]$. The sampling of $f(x)$ defined by the distribution

$$F(x) = f(x) ⧢\left(\frac{x}{\Delta x}\right),$$

with the sampling rate

$$\Delta x = \frac{1}{2s_\mathrm{m}},$$

is enough to be able to reconstruct $f(x)$ for all x, by the convolution

$$f(x) = F(x) \star \operatorname{sinc}\left(\frac{x}{\Delta x}\right).$$

A sufficient condition to ensure the converse is $\tilde{f}(s_\mathrm{m}) = 0$. For sampling in which $\Delta x = 1/2s_\mathrm{m}$ holds exactly, information can be lost (Fig. 9.5).

Therefore, with the exception of this component, $f(x)$ is completely specified by the discrete sampling $F(x)$. This is confirmed by

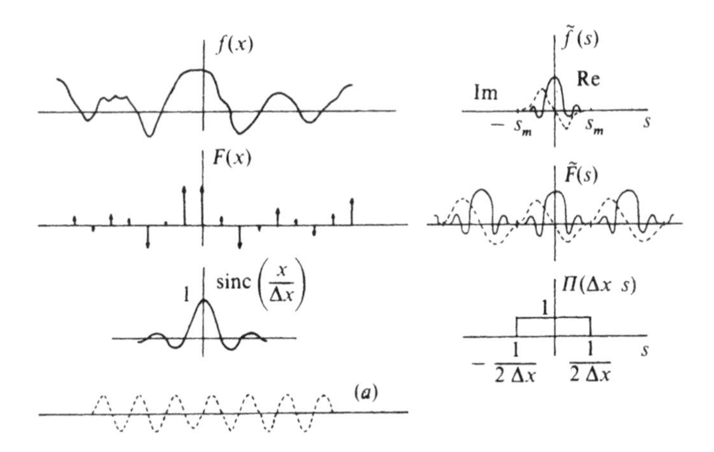

Fig. 9.5 Shannon's theorem, or sampling theorem. The representation of $F(x)$ uses a sequence of Dirac distributions, symbolised by an *arrow* whose height is proportional to the weight. In (**a**) is shown a component of $f(x)$ which would be lost in the sampling

$$F(x) * \left(\frac{x}{\Delta x}\right) \rightleftarrows \tilde{F}(s)\Pi(s\Delta x),$$

which is equivalent to

$$f(x) \rightleftarrows \tilde{f}(s), \quad \text{if} \quad \Delta x = \frac{1}{2s_\mathrm{m}},$$

where the arrows denote Fourier transform.

The great importance of this theorem, known as *Shannon's theorem*,[4] springs from the physical nature of the functions sampled. A receiver does not respond to infinitely large frequencies, and always behaves as a low-pass filter, with a cutoff at some temporal or spatial frequency s_m.

Sampling at any rate above or below the *Shannon rate* is called *oversampling* or *undersampling*, respectively. The only drawback of oversampling is the amount of redundant measurement data, whilst undersampling leads to the phenomenon of *aliasing*. The latter can have a considerable adverse effect on the results of measurement, by re-injecting into the spectrum, and at low frequencies, Fourier components of frequency above the cutoff frequency (Problem 9.10).

Filtering

In some cases it is useful to be able to examine a signal in *real time*, that is, without waiting until the end of the measurement. The optimal signal processing may then differ from the one that is most appropriate when the signal is available in its totality, over the whole measurement period.

Filtering a signal implies keeping only a part of the information it carries, by reducing or modifying its initial frequency spectrum. Only frequencies containing relevant information are conserved, if they are known beforehand, or else those at which the noise has a tolerable level, or those satisfying a combination of these criteria.

Optimising the way the signal is processed, whether in real time or otherwise, leads to the idea of *optimal filtering*. An analogue signal (that is, in the form of a continuous physical quantity, such as a voltage, or the blackening of a photographic plate) is *digitised*. This means that it is transformed into a finite set of numerical values, and hence coded by a finite number of bits of information. If this is a very small electric charge (the pixel of a CCD), or a flux of singly detected photons, digitisation is immediate, since the signal is already a discretised quantity. This digitisation occurs at discrete time intervals: the signal has been *sampled*.

[4]Claude Shannon (1916–2001, USA) was one of the founders of the theory of information.

Let us examine how these operations modify the information content of $x_s(t)$, and how they may improve the signal-to-noise ratio of the measurement.

Optimal Filtering

Suppose first that $x(t)$ is measured over an infinitely long time ($T \to \infty$). Suppose also that the statistics of the stochastic processes $\mathbf{x}_s(t)$ and $\mathbf{n}(t)$ are known, and that they are both stationary processes. This would be the case in the following situation: $\mathbf{x}_s(t)$ could be the arrival of photons from some periodically varying source, with frequency f_0, such as a pulsar, and $\mathbf{n}(t)$ could be the background noise due to temporal fluctuations in the sky background.

The problem is to find the best possible estimate of $\mathbf{x}_s(t)$, which will be denoted $\hat{x}_s(t)$. Set

$$\hat{x}(t) = \int_{-\infty}^{+\infty} x(t-u)\, h(u)\, \mathrm{d}u,$$

where the function $h(u)$ is chosen in such a way as to minimise the expectation value

$$E\left\{|x(t) - \hat{x}(t)|^2\right\}.$$

It can be shown that the *linear filter* $h(t)$ may be found in the following way. Let $S_{\mathbf{x}}(f)$ be the spectral density of $\mathbf{x}(t)$, and $S_{\mathbf{x}\mathbf{x}_s}(f)$ be the cross-spectrum between the observed quantity and the desired signal. (As explained in Appendix A, the cross-spectrum is simply the Fourier transform of the cross-correlation function.) Then, denoting the FT of $h(t)$ by $\tilde{h}(f)$,

$$\tilde{h}(f) = \frac{S_{\mathbf{x}\mathbf{x}_s}(f)}{S_{\mathbf{x}}(f)}.$$

The transfer function $\tilde{h}(f)$ is that of the optimal linear filter, or *Wiener–Kolmogorov filter*.[5] In Sect. 9.6, we shall return in more detail to the question of how this filtering can use a priori information about the source.

- In the particular case when signal and noise are uncorrelated, it follows that

$$S_{\mathbf{x}_s\mathbf{n}} = 0, \quad S_{\mathbf{x}}(f) = S_{\mathbf{x}_s}(f) + S_{\mathbf{n}}(f), \quad S_{\mathbf{x}\mathbf{x}_s}(f) = S_{\mathbf{x}_s}(f),$$

$$\tilde{h}(f) = \frac{S_{\mathbf{x}_s}(f)}{S_{\mathbf{x}_s}(f) + S_{\mathbf{n}}(f)}.$$

[5]The American mathematician Norbert Wiener (1894–1964) was one of the founders of cybernetics and pioneered information theory, while the Soviet mathematician Andreï Kolmogorov (1903–1987) axiomatised probability theory and also did theoretical work on dynamic systems and the complexity algorithm.

If the spectra $S_{\mathbf{x}_s}(f)$ and $S_{\mathbf{n}}(f)$ do not overlap, then $\tilde{h}(f) = 1$ wherever $S_{\mathbf{n}}(f) = 0$, and $\tilde{h}(f) = 0$ wherever $S_{\mathbf{n}}(f) \neq 0$. Note that $\tilde{h}(f)$ is arbitrary outside these intervals. It is also true that the error in the estimate $\hat{x}(t)$ for $\mathbf{x}_s(t)$ is everywhere zero.

- If there is some overlap between the supports of $S_{\mathbf{x}_s}(f)$ and $S_{\mathbf{n}}(f)$, then filtering by $\tilde{h}(f)$ leads to an estimate $\hat{x}(t)$ for $\mathbf{x}_s(t)$ with error, given in least squares by

$$\varepsilon = \frac{1}{2\pi} \int_{-\infty}^{+\infty} \frac{S_{\mathbf{x}_s}(f) S_{\mathbf{n}}(f)}{S_{\mathbf{x}_s}(f) + S_{\mathbf{n}}(f)} \, \mathrm{d}f.$$

The idea of a signal-to-noise ratio is difficult to apply here, for the estimated signal $\hat{x}(t)$ can take a wide range of values on $(-\infty, +\infty)$, whereas ε is a well-defined quantity. On the other hand, since $\mathbf{x}_s(t)$ is periodic in the example chosen, the signal- to-noise ratio is well-defined in the Fourier space, where the signal is concentrated around the frequency f_0.

The results, obtained here for a signal depending on the single parameter t, can easily be extended to the optimal filtering of an image, in which case the signal and the noise are both functions of the two coordinates describing a point in the image plane.

Online Filtering

We have studied the case in which all the information is available before the Wiener filter is applied to it. A different case occurs when the signal must be processed in real time, because of some communications problem. Then, the message must be decoded or extracted from the noise before it has all been transmitted. Indeed, we wish to give the best possible estimate of the message at each instant of time.

An example would be the transmission of an image of Uranus from a probe in flyby. Because of the great distance and the limited power available on board, the image is transmitted pixel by pixel, and the transmission is affected by various noise sources. After receiving the whole image, the ground base can process it by means of an appropriate filter, eliminating noise as far as possible and producing a filtered image. The astronomer may also wish to have an incomplete image but in real time, in which case the intensity of each pixel must be estimated online from the raw signal.

Another example can be given, concerning the variability of an X-ray source, observed by an orbiting X-ray telescope. The measured flux is transmitted to Earth, affected by noise of known statistical behaviour. After several months of observation, the raw signal is analysed, for example, in order to determine periodic fluctuations by filtering and Fourier analysis. But, considering the length of the observation, it may also be desirable to make a best estimate, at each instant t, of the true flux of the source.

This type of filtering is particularly suitable for real time telecommunications (e.g., telephone, television), in which there is no question of awaiting the end of a signal before processing it.

Other Applications of Filtering

The *linear filtering* described by the convolution

$$y(t) = h(t) \star x(t),$$

or by its equivalent in the Fourier space

$$\tilde{y}(f) = \tilde{h}(f)\tilde{x}(f),$$

can be used to eliminate the continuous component of a signal [setting $\tilde{h}(0) = 0$], or an extraneous frequency f_0 [setting $\tilde{h}(f_0) = 0$], or high frequencies (using a low-pass filter, cutting off at $f > f_c$). For example, a photometer measuring the flux from a star would be followed up by a low-pass filter, eliminating fluctuations due to scintillation, and providing a photometric measurement of the mean brightness.

A wide range of filters exists for specific applications, and their description can be found in specialised texts.

Here we shall give an example, serving to introduce the methods discussed in Sect. 9.5. It concerns an important method of filtering known as *cross-validation*. It often happens that the best processing of a set of raw data depends on an unknown parameter α. This may be a cutoff frequency, or a whole number, such as the degree n of an interpolating polynomial, or the number n of components (spectral lines) giving the best fit to an observed spectrum. For any given value of this parameter α, the data processing procedure is known (e.g., filtering, regression, fitting, etc.), but the value of α is unknown. The cross-validation method provides a way of estimating a good value for α, precisely by minimising prediction errors.

In order to illustrate this claim, consider the problem of smoothing some experimentally determined points by means of a polynomial of degree n. As the degree is fixed, the smoothing procedure consists in fitting the observational data to a polynomial, by applying the least squares criterion. n is then a parameter controlling the degree of smoothing. If $n = 1$, the experimental points are simply fitted to a straight line by least squares (maximal smoothing, Fig. 9.6a). If $n = N - 1$, where N is the number of points, the polynomial goes through every point, corresponding to a minimal smoothing.

Experience shows that, in the latter case, the polynomial will oscillate a great deal between the experimental points which constrain it (Fig. 9.6b). If a further point were given between two already known points, it would be possible to say whether the prediction given by the polynomial was a good one or not. We could then quantify the error in the prediction, between the interpolated value and the new experimental point. Clearly, such an extra piece of data is not available, for if it had been, it would have been treated in the same way as the others, without obtaining any further benefit from it. However, it is possible to 'simulate' these hidden, and then rediscovered, data points. This is the principle of cross-validation, which tests a certain degree n by 'hiding' the second point, then smoothing and comparing the interpolated value with the hidden value, whence a prediction error denoted $\varepsilon_2^{(n)}$. This procedure is repeated, omitting successively all the points, from the second to the $(N-1)$th. A prediction error $Q(n)$ can then be given, which depends on the degree n,

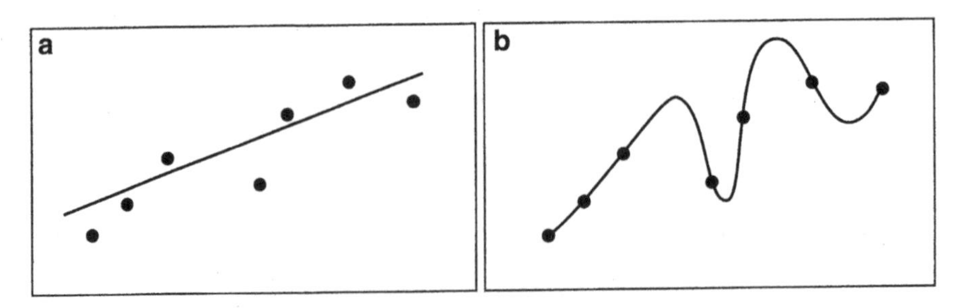

Fig. 9.6 (**a**) Smoothing of N experimental points by a least squares straight line ($n = 1$). (**b**) Smoothing of the same points by a polynomial of degree $N - 1$

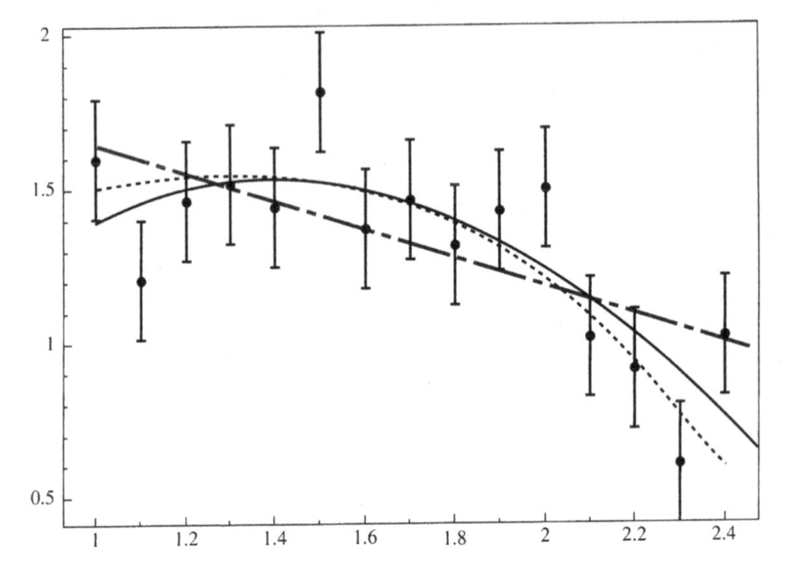

Fig. 9.7 Simulated cross-validation. The 'experimental points' are shown with their variance. The *continuous curve* is the smoothing (or *regression*) obtained for a polynomial whose degree $n = 3$ was itself obtained by cross-validation. The *dashed line* is the least squares straight line ($n = 1$). The *dotted curve* is the theoretical curve used to generate the 'experimental points'. It is well reproduced by the data processing method here

$$Q(n) = \sum_{i=2}^{N-1} \varepsilon_i^{(n)2} .$$

Note that the first and last points have been omitted, as they would not allow interpolation, but rather an extrapolation.

Cross-validation consists in choosing that value of n which minimises the error $Q(n)$. It is mainly a fact of experience that, whatever the nature of the control parameter α, this method gives good results, as can be seen in Fig. 9.7.

Data Compression

Data produced by an instrument can show a high level of internal redundancy. In particular, the value of a pixel in an image is often strongly correlated with the values of the neighbouring pixels. In order to determine the level of *data redundancy* and, hence, how much the data might possibly be compressed during a transmission, it is useful to measure the quantity of information contained in the image. The notion of *entropy* is thus introduced (see also Sect. 9.6).[6]

Consider an image whose pixels are coded on N bits, which implies 2^N possible values. Let P_i be the number of pixels taking the i th value, as i runs from 0 to $2^N - 1$. The histogram giving P_i as a function of i represents the probability that the signal takes a given value on any pixel. Starting from this empirical probability distribution P_i, the *entropy* H of the image can be constructed by

$$H = -\sum_{i=0}^{2^N-1} P_i \log_2 P_i \,.$$

The entropy is expressed in bits per pixel and represents the number of bits necessary to code the image without loss of information. The entropy is calculated for an image whose redundancy has already been reduced, and which therefore requires fewer coding bits. Decorrelation algorithms exist to this effect. For example, a simple method consists in replacing each pixel $p_{i,j}$ by the difference between the value of this pixel and that of its neighbour, namely $p_{i,j} - p_{i,j-1}$.

Applying this to the example of a field of stars, each of which occupies just one or a very small number of pixels, on a uniform background, the operation described amounts to replacing the value of the uniform field by zero almost everywhere. This leads to a considerable saving in coding bits.

If all the available values are equiprobable, then $H = N$ is constant, and no compression can be applied. If on the contrary the image is binary, as is the case for a text of black letters on a white background, then $H < N$, and compression algorithms can be applied with a high level of reduction (≈ 100). Generally speaking, the maximum compression ratio applicable to an image of entropy H coded on N bits, after eliminating redundancy between neighbours, is just N/H.

Application specific integrated circuits (ASIC) carry out compression operations in real time. Apart from their use in satellites for terrestrial observation (e.g., SPOT satellites), compression methods are used mainly for transmissions within computer networks. They were also used during an observation mission by the satellite Clementine (1993), in lunar orbit, which mapped the whole surface of the moon in more than ten colours and at high resolution.

---

[6]The remarks here are based on a text by G. Moury, in *L'Instrumentation scientifique dans le domaine spatial*, Centre National d'Etudes Spatiales, 1992.

9.1.4 A Specific Example of Data Processing

Data such as spectra, images, and so on, can be treated in many ways. Analysis into *principal components* (or *factor analysis*) aims to expose any linear structure underlying large amounts of data. Karhunen–Loève series are used when this data is a realisation of a stochastic process.

In order to illustrate the method in a specific case, consider a random vector $\boldsymbol{X} = (X_1, X_2, \ldots, X_n)$, which will be analysed into its principal components.

Any particular realisation of $\boldsymbol{X}$ can be represented by a point in an n-dimensional space. Note that this analysis into principal components only really becomes useful when n is large. A set of N realisations of the random vector, $N > n$, will be spread out in the space like a sort of 'cloud' of points. It may happen, in so-called degenerate cases, that these points do not fill out all the dimensions available, but rather remain within some affine submanifold of $\mathbb{R}^n$ (a hyperspace, for example), of dimension $k < n$. If this is the case, the data can be compressed by describing it using only the k parameters of the submanifold, rather than all the n components of the canonical basis. Along the normal to the submanifold, the variance of the data is zero, and this immediately eliminates an axis which supplies no information about the way the cloud of points is distributed in the space. The new basis of dimension k gives a better description of the data than did the original basis.

This ideal case is rarely encountered in practice. More common is the case in which the spread of data along a certain axis is not exactly zero, but is nevertheless small. The points are *almost* distributed on a k-dimensional submanifold. If the cloud of points is described using the k-dimensional basis, then the mean squared error incurred will correspond to the spread of the data along the neglected axis. Finding this axis amounts to identifying the direction along which the variance of the data, as a measure of its spread, is a minimum.

The problem is solved by placing the origin of the space at the mean of the data, and then choosing as basis the eigenvectors of the variance/covariance matrix of the vector $\boldsymbol{X}$. The notation $\boldsymbol{X}$ will hereafter refer to the components in this new, centred basis. The corresponding eigenvalues are exactly the data variances along these axes. If a loss of information can be accepted up to a certain level, then the proper axes can be neglected until the sum of their associated eigenvalues, classed in increasing order, reaches that level.

In order to formulate this, the data can be written

$$\boldsymbol{X} = \sum_{i=1}^{n} A_i \boldsymbol{f}_i ,$$

where $\boldsymbol{X}$ is the centered random vector, and A_i the random components of $\boldsymbol{X}$ in the fixed basis $\boldsymbol{f}_i$, $i = 1, \ldots, n$. The eigenvalue equation is

$$E\left\{XX^{\mathrm{T}}\right\} f_i = \lambda_i^2 f_i ,$$

where $E\left\{XX^{\mathrm{T}}\right\}$ is the expected value of the variance–covariance matrix and λ_i^2 is the variance of $\boldsymbol{X}$ in the direction of $\boldsymbol{f}_i$. This basis is orthonormal, being the set of eigenvectors of a symmetric matrix, and the A_i are uncorrelated random variables, in the sense that

$$E\left\{A_i A_j\right\} = \lambda_i^2 \delta_{ij} .$$

In the special case when $\boldsymbol{X}$ is a normal random variable, the A_i are, in addition, independent random variables.

In this way, the stochastic process is replaced by a set of random variables, ideally independent. Apart from the linear analysis already mentioned, this leads to a simplification of the data processing.

9.2 Complete Model of an Observing System

Since astronomical instruments are ever more sophisticated and expensive, it is easy to understand the importance of carrying out studies before they are actually put together for real. For one thing, there must be no risk of error in predicted performance when millions of euros are involved, not to mention the work of dozens of people over several years. For another, the increasing complexity of instruments also multiplies the risk of malfunction, or unexpected interactions between subsystems or between a priori unconnected functions.

A good instrumental model is now an essential tool for several reasons:

- To obtain the best possible design for the given scientific requirements and to make good choices, or at least the best possible compromise, e.g., when working out the error budget for the various subsystems.
- To predict performance in order to convince the scientific community and funding organisations that a new instrument is worthwhile and that the chosen solutions are the best.
- To provide users with a *virtual instrument*, whereby they may estimate the feasibility of the astrophysical programme they have in mind and begin to plan their observations.

In the latter case, forward planning is absolutely necessary, both on Earth and in space, because observations are highly automated for reasons of efficiency and safety.

This instrumental model is generally put together during the project, on the basis of submodels which are each linked to some specific technical area and/or a specific type of knowhow. Ever more sophisticated computational tools now allow increasingly accurate simulations of the behaviour intrinsic to each of these areas. When an instrument is designed today, several different submodels are generally made, which we shall now discuss.

Thermal Model

The aim is to calculate the temperature distribution, energy exchange, e.g., by conduction, radiation, or convection, transient departures from equilibrium and their evolution, and so on. This model is particularly important for spaceborne instrumentation, because of the extreme environmental conditions, and also in the infrared to submillimetre region where background radiation is a key problem and optical systems and especially detectors must be cooled to low temperatures.

Mechanical and Thermoelastic Model

This is the model most familiar to us, since it uses computer aided design (CAD) software to produce usually 3D depictions, sometimes animated, which provide a concrete view of the instrument. The model is used to build a consistent mechanical architecture, choose materials, predict relative motions, calculate bending, distortion, and thermal expansion, and to estimate mechanical errors and tolerances in the manufacture of parts. Impressive software is available to visualise mechanisms and even to produce the digital codes required to command machine tools making the basic parts.

Optical Model

The aim here is to calculate the paths of light rays from one optical component to the next, e.g., mirrors, lenses, prisms, etc., using the laws of geometrical optics, and generate the final point spread functions (PSF). Optical aberrations can be minimised and the best compromises achieved by optimising different characteristics of the optical components, such as refractive index, surface curvature, and so on. The model also helps to take into account the phenomena of physical optics, such as diffraction, scattering, dispersion, and reflections on optical surfaces, and to assess their various impacts. Many programs are available commercially. Among the most widely used are *Zemax* and *Opticad*. More subtle effects, or at least effects that are difficult to estimate roughly, can now be taken into account in models. An example is light pollution, often a critical issue in astronomy, which can now be modelled in a more and more realistic way using software, admittedly still somewhat cumbersome, to simulate secondary sources and reflections from the surfaces and edges of the mechanical structure (as described by textures and their microstructure), and to optimise the geometry of protective screens or baffles.

Model of Detectors and Detection Chain

The aim here is to take into account, as realistically as possible, the pixel geometry, electrical response to the photon flux, efficiency as a function of wavelength, nonlinearity, pixel saturation and overflow effects, detector noise, dark current, pixel cross-talk, electrical biases, and so on. Such models are often tailor-made, injecting a detailed knowledge of the detector built up from experience.

Electrical Model

The development of both digital and analogue electronic circuits is more and more often accompanied by detailed simulations of their electrical behaviour. This is done using software exploiting a universal data bank that contains detailed characteristics of most components now on the market. In this way, without connecting up any real components, it then becomes possible to create a circuit, test it, refine the choice of components, generate realistic dynamic responses, and simulate its behaviour under extreme conditions, such as extreme temperature conditions, for example. An example of this kind of software is *Spice*, which is in the public domain and widely used. The special case of servosystems is often handled more functionally using dedicated software.

Computing Model

This is rather a model for the computer architecture controlling and piloting the instrument. Software systems like *Labview*, are able to design genuine virtual instruments carrying out all the functions of a measuring instrument or combinations of several sub-instruments, along with their control by virtual control panels, i.e., screens with indicators, graphics, buttons, and cursors. Such software systems can assemble well defined functionalities and join them together by data exchange connections. The parameters of the instrument can be optimised by clicking on the *buttons*, and carry out data acquisition operations, while piloting the various functions of the instrument.

Astrophysical Model

The characteristics of sources, such as their morphology, size, emission spectrum, environment, and so on, are generally supplied by an ad hoc model. This serves to check in a genuinely convincing way that, with the estimated level of performance, it really will be possible to achieve the initial astrophysical objectives. It also provides a way for future observers to prepare for their observations by working with a model of the source that is close to what is expected.

Environmental Model

In some cases, this kind of model is also necessary in order to achieve the desired level of realism. This is the case in particular for instruments using an adaptive optics system, or optical interferometric arrays. In order to simulate their operation

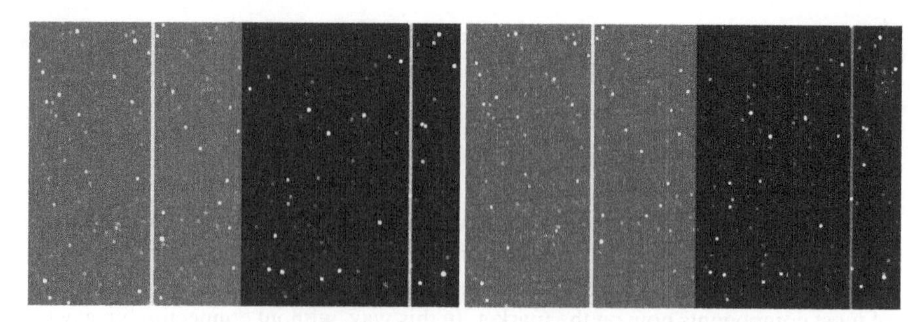

Fig. 9.8 Raw CCD image of part of the first field of view observed by the COROT satellite in January 2007. *Left*: CCD image actually obtained. *Right*: Simulation on the basis of a detailed model of the detector, optics, and star catalogues. The striking likeness between the two images illustrates the quality that can now be achieved in instrument simulation. Images kindly provided by M. Auvergne, Lesia, Paris Observatory

and estimate their performance, it is essential to use a model reproducing the spatial and temporal behaviour of atmospheric turbulence, along with vibrations of the telescope, the effects of wind on the outer structure, and so on.

Finally, all these submodels are put together into a global or *end-to-end model*. This general involves a command metalanguage that can interface the software building blocks described above, or rather their inputs and outputs, and put together a complete measurement chain from the celestial object emitting its radiation right down to correction for instrumental signatures and data compression. This model is often the basis for an *exposure time calculator*, provided for future observers to assess the feasibility of their scientific programme and the time required to carry out a specific observation in order to obtain a given level of performance, given the various constraints and random factors.

The level of realism that can be achieved today in global simulations of the measurement chain is illustrated in Fig. 9.8. This compares the simulated image of the first field of view observed by the photometric satellite COROT in January 2007 with the image that was actually obtained, without any correction. It is hard to find any difference at all between the simulation and the reality!

9.3 Overall Performance of an Observing System

This book has aimed at a systematic presentation of the different aspects of an observing system:

- Telescopes which collect light and form images, the phenomena of diffraction and coherence loss which affect the latter, and the way in which telescope arrays can improve them.

- Spectrometers which decompose and analyse radiation.
- Detectors which transform the electromagnetic radiation received from celestial bodies into a signal that can be processed digitally.

As discussed in Chap. 1 and stressed again above, observation combines all these features into a single *observing system*, which may be exceedingly complex, whether Earth-based or spaceborne. While the performance and fundamental limitations of each subsystem, e.g., atmosphere, telescope, spectrometer, detector, etc., will have been carefully investigated in its turn, the final performance, which is after all the main concern of the astrophysicist, nevertheless results from the combination and compatibility of the subsystems. This is the role of the *integration process* leading to *qualification* of the instrument.

Performance estimates generally lead to a quantitative response to the following question: what is the faintest object that a particular system can observe, in a given time, with given spectral and angular resolutions, for a given polarisation, and with a given signal-to-noise ratio? However, the question may take other forms, such as: what is the fastest temporal fluctuation of a given object that can be observed with this system?

In this section, we shall give several examples to illustrate the way the overall performance of an instrument appears to the user. This is important because the user rarely has an intimate knowledge of the many complex subsystems making up the instrument, or even of their individual performance levels. The important thing is the final result.

9.3.1 Observing with the IRAM Millimetre Interferometer

Here we consider the millimetre interferometer of the *Institut de radioastronomie millimétrique* on the Plateau-de-Bure in France, already presented in Sect. 6.5.1. What sensitivity can be obtained in spectral and angular resolution using this system, which consists of a high-altitude station (2 560 m) equipped with 6 radiotelescopes of 15 m, detectors, and digital correlators, e.g., with a view to observing quasars? The first step is to visit the internet site[7] which gives the characteristics in some detail. Table 9.2 provides a summary.

[7]Almost all instruments in the world's observatories are accurately described in this way. In this case, see the website of the *Institut de Radio Astronomie Millimétrique* (IRAM) at iram.fr/IRAMFR/GILDAS/doc/html/pdbi-intro-html/node2.html (Guilloteau, S., Lucas, R., Dutrey, A.), from which the quoted values were extracted (in 2007).

Table 9.2 Characteristics of the Plateau-de-Bure interferometer in 2007

Number of telescopes	6
Telescope diameters	$D = 15$ m
Altitude	2 560 m
Maximal baselines	N–S 230 m, E–W 408 m
Receivers (per antenna)	81–115 GHz and 205–250 GHz
Correlators	8 units, resolution 0.39–2.5 MHz

Detection Mode

For observation of a point source, i.e., unresolved with the chosen baseline configuration, and in the detection mode, i.e., aiming to extract a point source from the background, the sensitivity, defined as the signal leading to unit signal-to-noise ratio, is a value expressed in jansky (Jy) and given by the following expression, supplied by the IRAM website and easily derived from the considerations of Sect. 7.5:

$$\delta S \ \ [\mathrm{Jy}] = \frac{\rho_{\mathrm{e}} T_{\mathrm{sys}}}{\eta_{\mathrm{c}}\eta_{\mathrm{p}}\eta_{\mathrm{j}}\sqrt{N(N-1)}\sqrt{\delta\nu\delta t}}, \tag{9.1}$$

where the notation is explained as follows:

- N is the number of antennas used in the chosen interferometric configuration.
- $\delta\nu$ is the passband used, which determines the noise level.
- δt is the observation time.
- T_{sys} is the system temperature (see Sect. 7.5), taking into account atmospheric transmission [$T_{\mathrm{sys}} \approx 150$ K below a frequency of 110 GHz, both in summer and winter].
- $\rho_{\mathrm{e}} = 2k/\eta A$ is the efficiency of an antenna (or mirror), measured in Jy K^{-1}, with A (m^2) its area, η its reflection efficiency, k is Boltzmann's constant [$\rho_{\mathrm{e}} \approx$ 22 Jy K^{-1}].
- η_{c} is the correlator efficiency [$\eta_{\mathrm{c}} \approx 0.88$].
- $\eta_{\mathrm{j}} = \mathrm{e}^{-\sigma_{\mathrm{j}}^2/2}$ is the coherence loss, or *instrumental decorrelation coefficient*, due to phase instability of the local oscillator [$\sigma_{\mathrm{j}} \approx 4°$].
- $\eta_{\mathrm{p}} = \mathrm{e}^{-\sigma_{\mathrm{p}}^2/2}$ is the coherence loss due to phase instability (or phase noise) of the atmosphere above the instrument [σ_{p} depends on the baseline B and atmospheric humidity, and is about 15° rms for $B \leq 100$ m].

The characteristic numerical values for the system around the frequency 90 Hz (corresponding to $\lambda = 3.3$ mm), as provided by IRAM, are specified in square brackets. We then obtain the following result. Using the full interferometer configuration ($N = 6$ baselines) for one hour of integration and with a spectral width $\delta\nu = 580$ MHz, the detection threshold $\delta S_{\mathrm{threshold}}$ (unit signal-to-noise ratio) at $\lambda = 3$ mm is between 0.4 and 0.8 mJy depending on the atmospheric conditions.

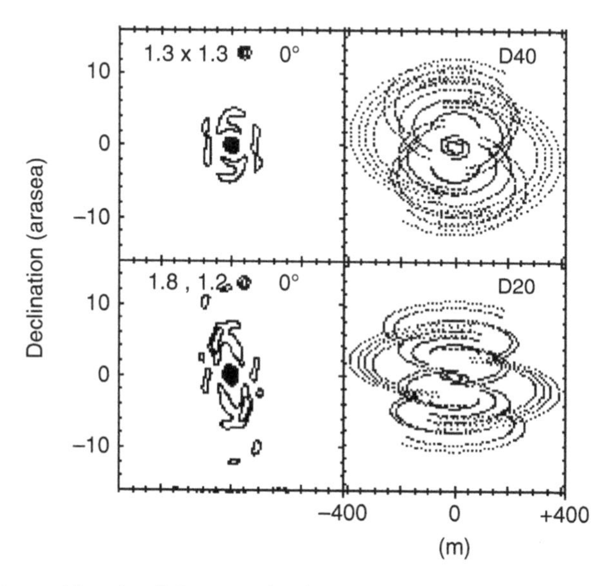

Fig. 9.9 Point spread function (*left*, on a scale of arcsec) and spatial frequency coverage (*right*, on a scale of metres, dividing by λ to obtain spatial frequencies u and v) for the Bure interferometer in its highest resolution configuration, for two values (+40 and +20°) of the declination D of the source. Figure kindly provided by IRAM

Imaging Mode

Now consider the interferometer operating in imaging mode. The choice of baseline configuration is dictated by the desired angular resolution. Assuming we select the highest value possible, i.e., around 0.6 arcsec at $\nu = 230$ GHz, Fig. 9.9 shows the interferometer point spread function (PSF) and the coverage of the u–v plane obtained with the corresponding arrangement of the six telescopes, depending on the declination of the observed source.

What sensitivity can be obtained in mapping mode for the same observation time? From the sensitivity δS to a point source calculated above, the noise (rms value) in the brightness temperature (see Sect. 7.5) is given by

$$\delta T_{\mathrm{m}} = \frac{\lambda^2}{2k(f\theta_1\theta_2)}\delta S, \tag{9.2}$$

where k is Boltzmann's constant, θ_1 and θ_2 are the values in arcsec of the PSF axes (assuming the PSF to be close to an ellipse), and f is a numerical factor close to unity and depending on the choice of apodisation (or *tapering*) mode used on the antenna lobes. With δS as calculated above, λ given in mm, and angles in arcsec (at

Table 9.3 Infrared imaging with the VLT (2007)

UT4 telescope (Yepun)	$D = 8.2$ m
Location	24° 40′ S, 70° 25′ W, $H = 2\,535$ m
Adaptive optics (NAOS)	Cassegrain focus
Wavefront sensors	Visible (0.4–1.0 μm) and infrarouge (0.8–2.5 μm)
Reference star	$m_V \leq 16.7$ and $m_K \leq 12$
Maximal correction field	55 arcsec
Camera (CONICA)	1 024 × 1 024 InSb array
Spectral region	1–5 μm
Modes	Imaging, polarimetry, spectroscopy, coronography

90 GHz, $\theta_1 = \theta_2 = 1.5$ arcsec), the result is

$$\delta T_\mathrm{m} = 15 \frac{\lambda^2}{f \theta_1 \theta_2} \delta S_\mathrm{threshold} = 0.06\ \mathrm{mK}. \tag{9.3}$$

Once the observed sky region has been mapped, this value $\delta T_\mathrm{m} = 0.06$ mK indicates the noise level, hence the confidence interval relevant to the details of the map.

Spectroscopic Mode

Let us now consider the interferometer operating in spectroscopic mode. If it observes an unresolved source, the bandwidth $\delta\nu = 580$ MHz used for the above calculation of $\delta S_\mathrm{threshold}$ must be reduced to the spectral resolution used by the correlator (up to the maximal resolution of 0.39 MHz). The same was true in imaging mode.

9.3.2 *Observing with NAOS Adaptive Optics*

Consider the imaging system operating at the diffraction limit in the near infrared, including one of the telescopes of the Very Large Telescope (C. Paranal, Chile), with its adaptive optics system NAOS, and the CONICA camera that follows it. The main characteristics of this system,[8] which is just one of the many possible configurations of the VLT, are summarised in Table 9.3. However, this table only gives a very limited idea of the wide range of observation modes possible with the VLT–NAOS–CONICA system, since the observer manual contains more than a hundred pages, and it would not be possible to summarise all that here. The layout of the manual

[8]For more detailed information, see www.eso.org/instruments/naco/inst/. The numerical values used here are taken from *NAOS–CONICA User Manual VLT-MAN-ESO-14200-2761*, Ageorges, N., Lidman, C. (2007).

is typical of such a complex system, containing a description of the performance of each subsystem, and a presentation of all the instructions the observer must prepare prior to using the telescope, in the form of computer information sheets or *templates*, in order to minimise the decisions that need to be taken in real time, which can lead to handling errors, and in order to automate the observation procedure as far as possible. The aim is to reduce dead time and obtain the best possible use of nights when the weather is favourable, bearing in mind the high cost of telescope time.

The observation procedure must be prepared in the greatest detail, with the help of automated *observing blocks*, which break down into: acquisition of the source, i.e., locating the sky field that contains it, the observation itself, and calibration of the properties of the instrument and the atmosphere at the measurement time.

Since each subsystem involves many variables or possible configurations, e.g., atmospheric seeing, pixel size, spectral filters, dichroic elements separating the light between wavefront sensor and imaging camera, polarisers, and so on, many examples would be required to illustrate all these possibilities and their corresponding performance. We shall expand on just one of these here.

Consider the following project: to observe the galactic center (SgrA*) and obtain an image of its immediate neighbourhood, in order to monitor the motions of the stars there in the gravitational field of the massive black hole, in the spectral band K_s (2.2 μm). What is the faintest magnitude m_K of such a star that can be observed?

- **Adaptive Optics.** While extinction eliminates any reference star in the visible, by pure chance there is a bright supergiant (IRS7) of magnitude $m_K = 6.4$ at 9″ from SgrA* (see Fig. 6.24 in Sect. 6.3 for an image of this field). This can be taken as reference by using the infrared wavefront sensor (WFS), since the guide tells us that this has a radius of 55″ and limiting magnitude $m_K = 12$. The Strehl ratio $\mathscr{S} \approx 40$ is given as accessible at a distance 9″, and this is a good correction value. Since the reference star is bright, it suffices to use 20% of its flux for the WFS and to send 80% to the camera via a suitable dichroic component.
- **Camera.** There are two important choices to make: the field and the spectral band. Given the good value of $\mathscr{S} \approx 40$, we may choose a pixel equal to the diffraction spot in the K band, with $\lambda/D = 54$ milliarcsec (mas). One of the available cameras (S54) has a magnification leading to 54 mas per pixel and covering a field 56″×56″. Since the aim is to detect the faintest stars in the field of view, we use the filter with the greatest possible width $\Delta\lambda$. Indeed, the sensitivity is determined here by the existence of atmospheric background radiation and thus varies as $\sqrt{\Delta\lambda}$. In order to obtain the spectral type of these stars once detected, it could be interesting to use a spectroscopic mode with narrower filters, or a *grism*, while increasing the exposure time.
- **Background Radiation.** In the K band, the thermal background is negligible and it is OH emission that dominates. Under the conditions prevailing at the Paranal site, with the chosen dichroic, this background has brightness $m_{K_s} = 12.5$ arcsec^{-2}. It will thus have to be subtracted for observation of faint stars,

by alternating observation of the field containing them with observation of a neighbouring field that is as empty as possible, in order to be able to carry out a sky subtraction during data compression. With the chosen pixel size, the background signal per pixel is reduced by a factor of $(1\,000/54)^2 = 343$, corresponding to $2.5 \log_{10}(343) = 6.34$ magnitudes. It is thus equivalent to a star of magnitude $m_{K_s} = 18.8$ with flux concentrated on this single pixel.

- **Detector.** This InSb array (1 024×1 026 pixels), first implemented in 2004, has an excellent quantum efficiency $\eta \approx 0.8$–0.9. The dark current (superposition of a slight thermal background produced by the instrument itself and the dark current of the photoconductor proper) is 0.05–0.015 ADU $\mathrm{s^{-1}}$ $\mathrm{pixel^{-1}}$. Dark frames must therefore be carried out, i.e., exposures with no signal, in order to subtract the pseudo-signal due to this current.[9]
- **Signal.** The photometric values of the standard star α Lyrae (Vega) are a monochromatic flux of $e_\lambda = 4.3 \times 10^{-10}\ \mathrm{W\,m^{-2}\,\mu m^{-1}}$ for a magnitude $m_{K_s} = 1.86$ at $\lambda = 2.149\ \mu\mathrm{m}$ (K_s band). With a telescope of area $S = \pi(8.2)^2/4\ \mathrm{m}^2$ in the band $\Delta\lambda = 0.35\ \mu\mathrm{m}$, a star with the same magnitude as the background, i.e., 18.8, will thus give a signal of $1.36 \times 10^4 t$ photons $\mathrm{s^{-1}}$, where t is the overall transmission (atmosphere + telescope + instrument). With $t \approx 0.3$ and $\eta = 0.85$, the current will be 3 500$\mathrm{e^-}$ $\mathrm{s^{-1}}$. With a gain of 11$\mathrm{e^-}$ per ADU, the background signal will be 318 ADU $\mathrm{s^{-1}}$ and its quadratic fluctuation 18 $\mathrm{Hz^{-1/2}}$.

With the Strehl ratio obtained here, about 50% of the energy of a star is focused on the pixel of the chosen size. For an exposure time of 1 min, a star of magnitude 18.8 will give a signal of 159×60 ADU, whereas the background noise corresponds to $18 \times \sqrt{60}$, or S/N = 68. With this exposure time by integration on the CCD, and for S/N = 5, the detection limit is then an object 13.6 times fainter, i.e., a magnitude $m_{K_s} = 21.6$. This could be further improved by a longer exposure, up to a limit determined by pixel saturation due to the background.

9.3.3 *Observing with the Photometric Satellite COROT*

The satellite COROT, launched by the French space agency CNES in December 2007, is designed for very high accuracy photometry with the specific aim, for one of its two channels, of identifying extrasolar planets by the transit technique. The idea is to measure the reduced brightness of a star when one of the planets in its system passes between the star and the observer during a transit. Naturally, this assumes the rare situation in which the observer (COROT) is almost exactly in the orbital plane of the planet. What performance may we expect from COROT for the detection of extrasolar planets by this method? For example, what is the smallest

---

[9] The unit here is the widely used *analog digital unit* (ADU), which is a conversion factor between a current, i.e., a number of electrons per second, and the bits of the digitised signal.

Table 9.4 Main characteristics of the COROT satellite

Telescope diameter	27 cm
Focal length	1.2 m
Detector	2 CCD $2\,048^2$
Pixel	13.5 μm
Readout noise (RON)	14 electrons
Elementary exposure time	32 s
Global efficiency	0.7
Relative dispersion of efficiency	.01
Pointing fluctuations	0.4 pixels

planet that we could hope to detect in this way? The main features of the COROT system are summarised in Table 9.4.

The sensor is a frame transfer CCD, so no time is lost during readout and the CCD detects photons all the time (see Sect. 7.4.6). There is no wavelength filtering, so in order to evaluate the signal, it is assumed that this is equivalent to simultaneously observing with the three juxtaposed filters B, V, and R of the Johnson photometric system (see Sect. 3.1), with central wavelengths λ_{B}, λ_{V}, and λ_{R}, respectively, and widths $\Delta\lambda_{\mathrm{B}}$, $\Delta\lambda_{\mathrm{V}}$, and $\Delta\lambda_{\mathrm{R}}$, respectively. To simplify, it is also assumed that the overall efficiency η (optical + quantum) of photon–electron conversion does not depend on the wavelength.

Consider two stars of solar spectral type G2V and apparent magnitude $m_V = 10$ and 15, i.e., in the magnitude range covered by COROT. The Johnson colour indices for this spectral type are $\mathrm{B}-\mathrm{V} = 0.63$ and $\mathrm{V}-\mathrm{R} = 0.53$. We begin by calculating the fluxes f_{B}, f_{V}, and f_{R} of the star in each filter:

$$
\begin{aligned}
f_{\mathrm{B}} &= F_{\mathrm{B}} \times 10^{-0.4(m_{\mathrm{V}}+\mathrm{B}-\mathrm{V})}, \\
f_{\mathrm{V}} &= F_{\mathrm{V}} \times 10^{-0.4 m_{\mathrm{V}}}, \\
f_{\mathrm{R}} &= F_{\mathrm{R}} \times 10^{-0.4(m_{\mathrm{V}}-\mathrm{V}+\mathrm{R})},
\end{aligned}
$$

where F_{B}, F_{V}, and F_{R} are the reference fluxes in $\mathrm{W\,m^{-2}\,\mu m^{-1}}$ of a star of magnitude $m_{\mathrm{V}} = 0$. The next task is to evaluate the total number N_{tot} of photoelectrons produced during an exposure time τ by making the reasonable approximation that the filters are narrow enough to ensure that, in a filter, the energy of each photon corresponds to the average wavelength of the filter. Then

$$
N_\lambda = \pi \frac{D^2}{4} \tau \gamma F_\lambda \Delta\lambda \frac{c\lambda}{h},
$$

and

$$
N_{\mathrm{tot}} = \pi \frac{D^2}{4} \tau \gamma \frac{F_{\mathrm{B}}\lambda_{\mathrm{B}}\Delta\lambda_{\mathrm{B}} + F_{\mathrm{V}}\lambda_{\mathrm{V}}\Delta\lambda_{\mathrm{V}} + F_{\mathrm{R}}\lambda_{\mathrm{R}}\Delta\lambda_{\mathrm{R}}}{h/c}. \tag{9.4}
$$

We thus obtain $N_{\mathrm{tot}} = 3.16 \times 10^6$ for $m_{\mathrm{V}} = 10$ and $N_{\mathrm{tot}} = 3.16 \times 10^4$ for $m_{\mathrm{V}} = 15$.

Table 9.5 Change in the number of photoelectrons during a transit for two different planet sizes and two cumulative exposure times

m_V	Jupiter, 32 s	Earth, 32 s	Noise	Jupiter, 4 h	Earth, 4 h	Noise
10	2.32×10^4	232	1.78×10^3	1.04×10^7	1.04×10^5	3.78×10^4
15	232	2.3	177	1.04×10^5	1.04×10^3	3.78×10^3

We now estimate the change ΔN in the number of photoelectrons received during an elementary exposure when the planet passes in front of the stellar disk. If the radius of the planet is R_{P} and the radius of the star is R_{S}, we have

$$\Delta N = N_{\mathrm{tot}} \left(\frac{R_{\mathrm{P}}}{R_{\mathrm{S}}} \right)^2 . \tag{9.5}$$

Table 9.5 sums up the results for $R_{\mathrm{S}} = 7.0 \times 10^8$ km, $R_{\mathrm{P}} = 60\,000$ km (an exoplanet like Jupiter) and $R_{\mathrm{P}} = 6\,000$ km (an exoplanet like Earth), comparing then with the photon noise of the star alone, which is simply $b_{\mathrm{S}} = N_{\mathrm{tot}}^{1/2}$.

Let us now examine the various sources of noise or error affecting the measurement. Note first that the image spot of a star is spread over about 80 pixels, due to a slight defocusing, designed to limit problems of saturation and reduce the noise due to small pointing fluctuations.

Readout Noise

If for reasons of convenience we choose to sum the pixels over a square of 10×10 pixels, centered on the star spot, the readout noise n_{R} of the measurement will be $n_{\mathrm{R}} = (10 \times 10)^{1/2} \times RON = 140\,\mathrm{e}^-$.

Background Photon Noise

For reasons associated with the choice of orbit, COROT's pointing direction must lie in a plane close to the ecliptic, so the zodiacal light (light from the Sun reflected by tiny particles in the plane of the Solar System, see Sect. 2.9) is an important source of spurious radiation. Its specific intensity I_λ is $260 \times 10^{-8}\,\mathrm{W\,m^{-2}\mu m^{-1}\,sr^{-1}}$. The number N_{Z} of photoelectrons due to zodiacal light detected in the 10×10 square during the elementary exposure time is given by $N_{\mathrm{Z}} = 36\,800\mathrm{e}^-$ and the corresponding photon noise is $n_{\mathrm{Z}} = N_{\mathrm{Z}}^{1/2} = 191\mathrm{e}^-$.

Pointing Noise

A further source of noise can be ascribed to small changes in the satellite pointing and this must also be taken into account. The image spot moves over the CCD which does not have uniform response, and this induces changes in the measurements. The effect can be reduced by the defocusing of the image on the CCD already mentioned.

Table 9.6 Noise assessment for the COROT satellite. The table sums up the various contributions (in e^- rms) to the total noise for two different stars (m_V) and an integration time of 4 hr

Readout noise	Pointing noise	Zodiacal noise	Star photon noise	Jupiter signal	Earth signal
$m_V = 10$					
2.97×10^3	1.70×10^4	4.05×10^3	3.77×10^4	1.31×10^7	1.05×10^5
Source photon noise dominates and all planets are detectable					
$m_V = 15$					
2.97×10^3	1.70×10^2	4.05×10^3	3.77×10^3	1.31×10^5	1.05×10^3
Zodiacal photon noise dominates and only planets of Jovian dimensions are detectable					

Consider a small displacement δ (a fraction of a pixel) of the spot. The surface area not shared by the two spots is then $4R\delta$, where R is the radius of the spot in pixels. On average, the relative change $\Delta N/N$ in the number of photoelectrons for a pixel at the edge of the spot is then

$$\frac{\Delta N}{N} = \frac{4\delta}{(\pi R)^2} .$$

The normalised quantum efficiency or *flat field* of the pixels can be treated as a normal random variable centered on 1 and with standard deviation σ. The corresponding noise n_D, i.e., the standard deviation of the change in the number of photoelectrons due to pointing fluctuations is then

$$n_D = \delta N \sigma (2\pi R)^{1/2} ,$$

so that, with $R = (80/\pi)^{1/2}$, $\sigma = 0.01$, and $\delta = 0.4$ pixel, we have $n_D = 801e^-$.

Table 9.6 sums up the different noise sources n_S, n_Z, n_R, and n_D, for the cases $m_V = 10$ and $m_V = 15$, for a cumulative measurement time of 4 hr. It is thus affirmed that COROT has the capacities required to detect planets much smaller than Jupiter, and indeed it did as proven by the detection of Corot-7b a planet with $R = 1.7 R_{Earth}$.

9.3.4 Observing with a Coded Mask Gamma-Ray Instrument

Consider the IBIS/ISGRI instrument on board the INTEGRAL satellite.[10] This uses a coded mask to form images, as described in Sect. 5.2.5. The image, which is recorded on board, comprises a background and the superposition of shadows of

[10]The INTEGRAL mission, launched by the ESA in 2002 and still operating in 2009, is a gamma-ray observatory. The IBIS/ISGRI instrument is designed to form images (*Imager on Board the Integral Satellite*) using the ISGRI camera (*Integral Soft Gamma-Ray Imager*).

the mask projected by all the sources in the field of view. A pseudo-image of the sky and the γ ray sources it contains is thereby formed by correlation between the recorded image and a decoding or inverse matrix deduced from the pattern of the coded mask.

Let $\mathscr{M}$ be the matrix representing the mask, made up solely of ones (transparent elements) and zeros (opaque elements). The image formed on the detector is the convolution of the sky image $\mathscr{S}$ and $\mathscr{M}$, to which the background noise $\mathscr{B}$ must be added:

$$\mathscr{D} = S \star \mathscr{M} + \mathscr{B}.$$

If $\mathscr{M}$ has an inverse $\mathscr{G}$ such that $\mathscr{M} \star \mathscr{G} = \delta$, we find

$$\mathscr{S}' = \mathscr{D} \star \mathscr{G} = \mathscr{S} \star \mathscr{M} \star \mathscr{G} + \mathscr{B} \star \mathscr{G} = \mathscr{S} \star \delta + \mathscr{B} \star \mathscr{G} = \mathscr{S} + \mathscr{B} \star \mathscr{G},$$

and $\mathscr{S}'$ only differs from $\mathscr{S}$ by addition of the term $\mathscr{B} \star \mathscr{G}$. When the mask carries a cyclic repetition of the same basic pattern and the background $\mathscr{B}$ is uniform, the term $\mathscr{B} \star \mathscr{G}$ is a constant that can be ignored. In practice, if the background is not uniform but its structure is known, the image $\mathscr{D}$ can be corrected to make the background uniform. Replication of the basic pattern, which allows one to obtain a fully coded field, implies the existence of a strong correlation between the partial patterns projected by sources separated by an angle equal to the one subtended by the basic pattern of the mask at the detector.

Deconvolution will therefore produce a *ghost* of the main source at each of these positions. For a basic pattern that is square and a fully coded source, there will be 8 main ghosts[11] located in the middle of the sides and at the corners of a square surrounding the actual position of the source. If the source is partially coded, only a part of the ghosts will be visible. Since each source produces up to 9 peaks (8 ghosts plus the source) in a deconvoluted image, a field containing more than 3 sources will produce an extremely confused image in which it will be hard to detect faint sources. It may also be that one source is hidden by the ghost of another. It is thus important to exorcise the ghosts.

This is done using an iterative process, similar to the CLEAN method used in radioastronomy (see Sect. 6.4.5). This process, shown in Fig. 9.10, consists in identifying the brightest source, modelling the point spread function at the position of the source, and subtracting it from the image. This process is repeated until there is no longer any significant source. The main peak of the point spread function is then added at the position of each of the sources found in this way.

Since INTEGRAL is an observatory, there is a committee in charge of allocating observing time to those who respond to a yearly call for tender. The latter, published by the European Space Agency, comes with documentation and computing tools to help the astronomer work out an appropriate duration and strategy for observation

[11]This number would be different if the mask had different symmetry, e.g., hexagonal, as discussed in Sect. 5.2.5.

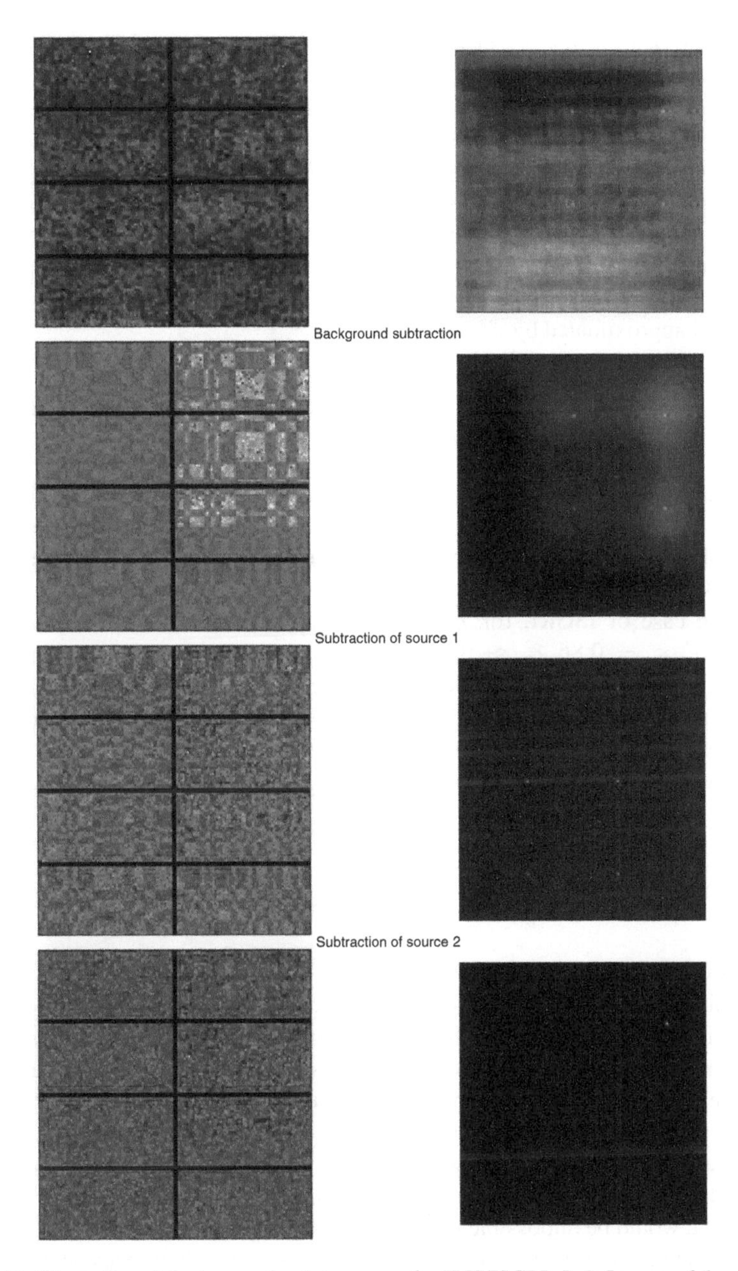

Fig. 9.10 Illustration of the image cleaning process for IBIS/ISGRI. *Left*: Images of the detector at different stages in the process, going from top to bottom. *Right*: Deconvoluted images corresponding to the images of the detector. *Top line*: Images from the detector and sky background at the beginning of the process. *Second line*: Images obtained after background substraction. *Third line*: Images resulting from subtraction of the brightest source (Cygnus X-1). *Bottom left*: Detector image after subtracting the second brightest source (Cygnus X-3). *Bottom right*: Final image obtained by deconvoluting the detector image on the left and adding the central peak of the point spread function at the position of each of the sources

in order to achieve his or her objectives. With the exception of γ ray bursts and extremely bright sources, the count rate of coded mask instruments is dominated by the instrumental background. The sensitivity, or more precisely, the minimum detectable flux, is thus limited by background noise. Let $B(E)$ be the background count rate in an energy band E of width δE, δt the observing time, $A(E)$ the detection area, ϵ_{i} the imaging efficiency, $\epsilon_{\mathrm{p}}(E)$ the detection efficiency in the total absorption peak, $\eta_0(E)$ the transparency of the mask and $\eta_1(E)$ the transparency of the holes in the mask. The detection sensitivity for a point source, expressed in $\mathrm{cm}^{-2}\,\mathrm{s}^{-1}\,\mathrm{keV}^{-1}$, for a signal level n times stronger than noise of standard deviation σ, is well approximated by

$$S = \frac{2n\sqrt{B(E)\delta t}}{\epsilon_{\mathrm{i}}\epsilon_{\mathrm{p}}(E)(1-\eta_0)\eta_1 A\delta t\delta E}. \tag{9.6}$$

The imaging efficiency ϵ_{i} results from imperfect sampling of the mask shadow. In the best possible case, it is given by $(1 - 1/3R)$, where R is the ratio between the size of the basic element of the mask and the size of a detector pixel ($R = 2.43$ for IBIS/ISGRI).

In the case of ISGRI, for E in the range 20–40 keV, $B \approx 150\ \mathrm{s}^{-1}$, $A \approx 2\,000\ \mathrm{cm}^2$, $\epsilon_{\mathrm{i}} = 0.86$, $\epsilon_{\mathrm{p}} \approx 0.9$, and $\eta_0 = 0$, $\eta_1 \approx 0.8$. For an exposure time of 10^6 s (about 10 days), at the 3σ detection level, we may thus expect a sensitivity of the order of $3 \times 10^{-6}\ \mathrm{cm}^{-2}\,\mathrm{s}^{-1}\,\mathrm{keV}^{-1}$, or around 3 000 times less than the flux from the γ ray source in the Crab nebula, a supernova remnant. This is sometimes expressed by referring to a sensitivity equal to one third of a milliCrab.

9.4 Removing Instrumental Signatures

An astrophysical measurement is always affected by several forms of bias and uncertainty. Some are intrinsic to the measurement, such as photon noise, and are almost impossible to control other than by increasing the integration time, while others are related to imperfections in the instrumental chain and can usually be taken into account to improve the accuracy of the measurement and thereby make it as close as possible to the fundamental limits. In this section, we discuss the set of operations designed to deal with the latter problem during data reduction, and referred to as *correction for the instrumental signature*.

Since it would be impossible to give an exhaustive account for the whole panoply of instrumentation used in observational astrophysics, we shall focus on instruments used in the visible and near infrared, a spectral region in which the strategy employed to remove instrumental signatures has been well established and appeals to relatively standardised tools. Similarly inspired processes are implemented in the observational protocols applied at radiofrequencies and high energies (X and γ rays).

The term *bias* refers to any spurious signal adding systematically to the photosignal from the source. This may be sky background emission, detector dark current, detector voltage interference, light pollution, ghost images, diffraction spikes, and so on.

For their part, uncertainties correspond to some random feature affecting each measurement. This may be statistical noise, scintillation, temporal variations in atmospheric transmission and emission, variations in quantum efficiency between detector pixels or transmission over the field, or impacts by cosmic rays. In addition, these two families of phenomena are not mutually independent. For example, there is random noise associated with a dark current.

9.4.1 Intrinsic Emission from the Instrument

While the night sky is not actually truly dark at infrared and millimetre wavelengths (see Sect. 2.3), the situation gets worse due to the instrument itself, and in particular its thermal emission, which makes a major contribution to the background radiation picked up by the detector. This background can even become much brighter than the target sources. Such emission is spatially structured owing to non-uniformities in the beam and the different emissivities of the various parts of the instrument, and it is also time variable. It must therefore be included in the instrumental signature.

To limit this signature, at least to first order, the solution is to carry out another measurement on an *empty* sky region and then subtract it. In doing so, it should not be forgotten that the statistical noise of the photoelectrons due to the sky or the instrument cannot be subtracted in this operation. Quite the opposite, in fact: the variance of this noise will be doubled. In order to measure the reference sky region faster than the typical time of its variation, which is generally a fraction of a second, the beam is switched between the two positions, either by offsetting the telescope, or by using an oscillating secondary mirror, which is faster. This is known as *chopping*.

In the latter case, since the telescope optics is also emissive, displacing the beam on the primary mirror will induce a further undesirable signal due to dust or other inhomogeneities. A structured background is thus superposed on the sky image. One then evaluates this contribution, which is due only to the rapid chopping, by regularly offsetting the telescope itself to some other region of the sky. This is the basic principle of *nodding*. Correction for this particular signature is thus achieved by carrying out a double subtraction, as shown in Fig. 9.11.

9.4.2 Dark Current

This signal corresponds to the charges produced in the detector by thermal excitation, even in the absence of any photons. For example, in a photosensitive semiconducting material, the probability of producing an electron–hole pair is

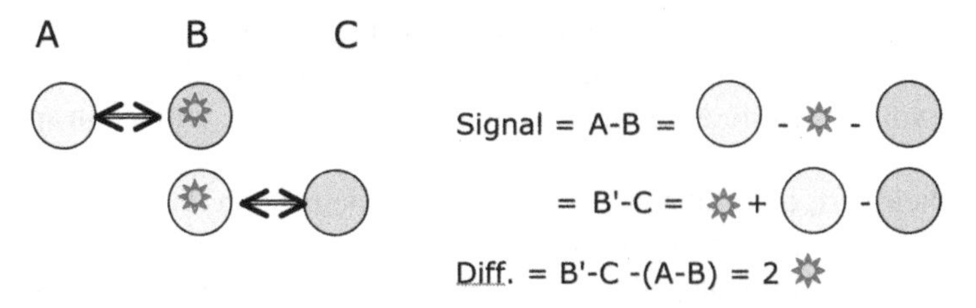

Fig. 9.11 Chopping. The shaded background in the detector field, represented by a *disk*, represents variable emission from the sky background, from which the star signal must be extracted. A, B, and C are different positions of the field on the sky. (A − B) and (B′ − C) represent chopping, while their difference (B′ − C) − (A − B) leaves only twice the source signal, having eliminated the sky background (nodding)

proportional to the quantity $T^{3/2}\exp(-E_{\text{gap}}/2kT)$, where E_{gap} is the energy difference or gap between the valence band and the conduction band, i.e., $E_{\text{gap}} = h\nu_{\text{min}} = hc/\lambda_{\text{max}}$, where λ_{max} is the wavelength beyond which the detector is no longer sensitive. For silicon, the photosensitive material used to make CCDs, $\lambda_{\text{max}} = 1.15\,\mu\text{m}$ (see Sect. 7.3). More charges are thus produced as T increases and the threshold decreases, whence the cutoff wavelength of the detector is longer. This is why infrared detectors always have larger dark current than visible wavelength detectors. The longer the operating wavelength, the more they must be cooled (see Sect. 7.4.7). Since the dark current produced after a given integration time differs slightly from one pixel to another, a structure *signature* appears, superimposed on the photon signal. It is generally reproducible if the detector temperature is well stabilised. This signature, which may be quite significant in the infrared, is simply removed by carrying out measurements with the shutter closed for the same time as the sky observation, then subtracting the resulting signal.

There are two things to note here. First of all, it is important to remember that subtracting the *dark charge* does not remove the statistical uncertainty associated with it. It is thus always useful to reduce this effect as much as possible by cooling the detector. The second point is that an accurate value for the dark charge is obtained by accumulating many measurements, and thereby reducing statistical noise.

9.4.3 Non-Linearity Defects

The non-linearity of the detector is another example of instrumental signature which must often be taken into account. The response of a detector always saturates above a certain level of photon signal called the *saturation threshold*. It is thus important never to exceed this threshold. However, the response generally starts to become non-linear at a lower level than this threshold. The aim is usually to remain well below this limit to ensure that the measurement can be given a simple interpretation in terms of light flux. If the detector is in fact non-linear over its whole operating range, the measurement must then be corrected for this non-linearity. This is therefore established by prior laboratory measurements, e.g., by observing a source, usually calibrated, but most importantly, highly stable, and by increasing the exposure time in precise steps in order to explore the whole signal range.

9.4.4 Bias

In general, even for zero exposure time and closed shutter, a non-zero signal is measured at the detector output, varying from one pixel to another. This bias has several origins. It depends on the controlling electronics and/or effects within the detector, such as impurity levels, voltages arising at the junctions, and characteristics of the readout circuit integrated within the detector.

The bias is measured by carrying almost zero exposures which are averaged and can be subtracted from the image made on the object. In fact, this operation is rarely necessary, because this undesirable signal is necessarily included in the measurement of the dark charge and is therefore subtracted along with that.

9.4.5 Light Interference

Bright sources in the field or close to the sighting axis, such as the Moon or stars, can produce ghost images, caused by unwanted reflections on the optical surfaces, in particular in dioptric systems. These images, which correspond to longer optical paths, are generally out of focus and appear in the form of a halo reproducing the shape of the pupil. Figure 9.12 gives an example of this type of artifact in an image obtained using the MEGACAM camera at the prime focus of the Canada–France–Hawaii telescope.

Ghost images arising from off-axis sources are difficult to anticipate and remove because they vary over time (relative motions of celestial objects or field rotation) and they depend on the finer details of the pointing direction. For example, microscanning can introduce ghost images which move significantly on the field.

Diffraction by the arms of the spider holding the secondary mirror of the telescope in place produces spikes on the brighter star images which can also perturb measurements (see Fig. 9.12). In an equatorially mounted telescope, the resulting pattern is fixed and can be subtracted from one exposure to another using a measurement on a bright star. However, for a telescope with altazimuth mounting, rotation of the field and pupil renders this operation impossible, or at least much less efficient.

Fig. 9.12 Part of an image of a very deep field obtained with the MEGACAM camera at the focus of the Canada–France–Hawaii telescope. Note two types of artifacts around the bright stars. First, there are two extended and overlapping halos due to unwanted reflections on the lenses of the focal reducer. These are circular and reproduce an image of the telescope pupil with its central obstruction. Then there are spikes in the shape of a cross. These result from the diffraction of light on the arms of the spider which holds the mechanical structure in which the camera is installed at the primary focus

9.4.6 Flat Field Corrections

This instrumental signature reflects a combination of non-uniformities in the detector quantum efficiency and the optical transmission over the field. It is also sometimes called *fixed spatial noise*. Indeed, if we observe a scene with perfectly uniform brightness, the measurement will nevertheless produce a spatial fluctuation in the signal from one pixel to another, and generally in a reproducible way. This is in fact a multiplicative effect, as indeed are non-linearity defects, since this noise is proportional to the flux of the uniformly bright source. To correct for this, several techniques can be combined. The first is to observer a scene of highly uniform brightness and then to divide by its median in order to produce the normalised *flat field* by which to divide all our images.

Certain precautions must be taken. Firstly, a good flat field must be affected by much less noise than the noise it is supposed to correct for. It must therefore be built up from a very large number of photocharges per pixel. This will involve adding up a large number of exposures, each with an average value close to the maximum number of charges that can be stored per pixel. Secondly, the flat field must be obtained using the whole optical train, including the telescope, so that non-uniformities in the optical transmission can be eliminated. A region of the sky observed at sunrise or sunset is generally considered to provide a good flat field source.

In order to limit the effects of differential response, a second technique is to carry out a *microscan*, i.e., displacing the telescope by a few pixels, usually in a random manner, from one exposure to the next. The image is reconstructed by recentering numerically to the nearest fraction of a pixel before adding up the exposures. A given region of the sky is thereby viewed by several different pixels and the average of their response reduces the variance of the spatial noise.

Another technique is *time delay integration*, specific to CCD detectors. In this mode, the scene is allowed to drift across the detector in the direction of the CCD rows, e.g., stopping the telescope motion, while ensuring charge transfer from pixel to pixel at exactly the same speed as the image drift. In this way, a source contributes to the construction of a photocharge throughout the scan on the CCD, but having successively illuminated a very large number of different pixels whose responses are thus averaged. This technique is used in the SLOANE sky survey and it is the one planned for the astrometric satellite GAIA.

9.4.7 Defective Pixels

A panoramic detector is never perfect. Owing to microscopic pollution during the manufacturing process or following deterioration by high energy particles, several pixels may become inoperative (dead pixels) or noisier, or they may produce a large dark current (hot pixels). These defects can be corrected for by exploiting the microscanning technique mentioned above. During an exposure, when the flux from one sky region has not been measured because it corresponds to a dead pixel, a good approximation can nevertheless be obtained by averaging the other spatially shifted measurements, i.e., having displaced the position of the dead pixel. If the microscanning technique cannot be implemented, an acceptable estimate of this flux can still be given by averaging the eight pixels surrounding the defective pixel. This does of course assume that image sampling has been correctly chosen with regard to its spatial frequency content (see Sect. 9.1.3).

9.4.8 Effects of High Energy Particle Impacts

High energy cosmic rays continually bombard our environment with a frequency that increases with altitude and hence becomes significant for many modern observatories. Indeed, it is always significant for satellites (see Sect. 2.9). When these particles deposit their energy in the bulk of the detector substrate, they create thousands of unwanted charges that are quite indistinguishable from photocharges. This effect increases with the thickness of the photosensitive material. In general, a particle only affects a single pixel on a CCD, but with the thicker infrared detectors, cosmic particle impacts can leave a trail across several pixels. The image produced by a very long exposure is thus scattered with bright pixels, namely those that have been impacted. The most effective solution here is to divide the exposure into several short exposures in order to be able to identify the measurements that must be rejected for a given pixel before addition. It is nevertheless important not to fall into the trap of making the exposures too short, for they would then be limited by readout noise.

9.5 The Problem of Estimation

Chance and randomness are everywhere. Humans would certainly have noticed this very early on in their evolution, and it must have been commonplace to note that, during a series of apparently identical experiments, different results were always obtained. Indeed, this same observation would have inspired games of chance, so widespread in all civilisations. On the other hand, it seems that the idea that chance could itself obey laws was one that long escaped the wisdom of our ancestors. There does appear to be a component to natural phenomena that fluctuates about an unchanging part, but it was not until the advent of probability theory and its laws that it became possible to estimate this invariable part.

When probability theory came upon the experimental scene, it imposed its way of viewing things: an observation is a realisation of one or more random variables which obey an underlying distribution, e.g., the normal distribution. This distribution is governed by unknown parameters θ_k, e.g., its average and standard deviation. And it is these parameters that often form the 'unchanging part' of the phenomenon mentioned above, and which constitute the information to be extracted from the noise. This is the approach of statistical data processing, and more precisely the theory of estimation to be discussed in this chapter.

9.5.1 Samples and Statistics

Theoretically, statistical data processing is based on probability theory, but it uses a different vocabulary to refer to its basic objects. From a statistical standpoint, a set $(X_1, \ldots, X_n)$ of n random variables is a *sample* and n is its *size*. We then say that we have a sample of size n, or an n-sample. A set $(x_1, \ldots, x_n)$ of n numbers from this n-sample is called an *realisation* or an *observation* of this sample. The probability distribution of the random variables X_i is called the *parent population*. Finally, in order to support their decisions, statisticians use functions $g(X_1, \ldots, X_n)$ of the random variables X_i which they call *statistics*.

Example: Typical Problem Expressed in the Language of the Statistician

An experiment has led to the observation of n numbers $(x_1, \ldots, x_n)$, identified as one realisation of an n-sample $(X_1, \ldots, X_n)$. We assume that the n-sample comprises independent random variables resulting from the same parent population with mean μ. In order to estimate μ, we propose to use the statistic

$$g(X_1, \ldots, X_n) = \frac{1}{n} \sum_{i=1}^{n} X_i . \tag{9.7}$$

This statistic, often denoted by $\overline{X}$, leads to an estimate for the mean μ by the arithmetic mean of the x_i, viz.,

$$\overline{x} = \frac{1}{n} \sum_{i=1}^{n} x_i . \tag{9.8}$$

This is a reasonable choice since a result from probability theory, namely the law of large numbers, asserts that $\overline{X}$ converges almost surely to μ when the sample size tends to infinity.

9.5.2 *Point Estimation*

In order to estimate the unknown parameters of a parent population, we use functions $\delta(X_1, \ldots, X_n)$ of the sample called *estimators* No? These are statistics. An estimator for θ constructed using a sample of size n is traditionally denoted by $\widehat{\theta}_n$, while an estimator of $\tau(\theta)$ is denoted by T_n, but note that in general, $T_n \neq \tau(\widehat{\theta}_n)$. A realisation of these estimators will be called an *estimate*. An estimator is thus a random variable, whereas an estimate is a number. However, they will be denoted by the same symbol. Finally, to simplify, the sample $(X_1, \ldots, X_n)$ will often be denoted simply by X and its realisation $(x_1, \ldots, x_n)$ simply by x.

The problem of point estimation is to specify what is to be understood by a *good estimator* of the unknown parameter θ or a function τ of this parameter. This amounts to choosing, among all the different ways $\delta(X)$ of combining the data, the one for which there are good reasons for preferring it above all others. Since the estimation of $\tau(\theta)$ covers the case of θ itself, we shall consider only the first case, and let $\delta(X)$ denote the estimator of $\tau(\theta)$.

9.5.3 *Elements of Decision Theory*

In order to motivate the above considerations, it is useful to introduce some objective criteria for comparing estimators. These criteria are taken from decision theory, which we shall outline below.

Estimation Error

The first thing to do is to define an estimation error. The error involved in estimating $\tau(\theta)$ by $\delta(X)$ is $\delta(X) - \tau(\theta)$. If the error is positive, then $\tau(\theta)$ is *overestimated*, and otherwise it is *underestimated*.

Loss Function

One must then face the consequences of estimation errors. This is the role of the *loss function*, which specifies the cost associated with these errors. This function $L(\theta, t)$ is positive and its value is the 'price to pay' for the estimation errors. It is zero if and only if $t = \theta$. For largely practical reasons, the quadratic loss function is often chosen:

$$L\big(\tau(\theta), \delta(X)\big) = \big[\delta(X) - \tau(\theta)\big]^2. \tag{9.9}$$

Risk

Estimators are random variables and as such are likely to yield different outcomes from one experiment to another. In order to be able to compare them, one must have a global measurement that takes into account all the possible outcomes of these variables, and *risk* is the notion that fulfills this demand. The risk R is the expectation value of the loss function:

$$R(\theta, \delta) = \mathrm{E}\big\{L\big(\theta, \delta(X)\big)\big\}. \tag{9.10}$$

The expectation refers to all possible values of the sample X, so the risk is a function of θ. If the loss function is quadratic, the risk attached to the estimator $\delta(X)$ for $\tau(\theta)$ is the mean squared error of $\delta(X)$ about $\tau(\theta)$, or

$$R\big(\tau(\theta), \delta(X)\big) = \mathrm{Var}\big(\delta(X)\big) + \Big[\mathrm{E}\{\delta(X) - \tau(\theta)\}\Big]^2. \tag{9.11}$$

The quantity $\mathrm{Var}(\delta(X))$ is the variance of $\delta(X)$ and $\mathrm{E}\{\delta(X) - \tau(\theta)\}$ is called the *bias*.

Partial Ordering of Estimators

When comparing two estimators for $\tau(\theta)$, we shall say that δ_1 is better than δ_2 *in the sense of risk* if

$$\forall\theta, \quad R\big(\tau(\theta), \delta_1\big) < R\big(\tau(\theta), \delta_2\big).$$

This ordering relation 'is better than' is only a partial ordering. The condition cannot be satisfied for all θ and the estimators are not then comparable. The obvious question here is this: for a given loss function, is there a best estimator, better than all others in the sense of risk?

In fact, no estimator can fulfill this requirement. Indeed, consider the trivial estimator δ_0 which, regardless of the value of X, declares that $\delta_0(X) = \tau(\theta_0)$, where θ_0 is an arbitrary value. The risk attached to δ_0 will generally be very high, unless we should happen to have $\theta = \theta_0$, in which case it will be zero. No other estimator can then be better than this one for θ_0, so there is no estimator that is preferable to all the others in the sense of the above partial ordering.

The problem is that we were trying to compare a candidate optimal estimator with a class containing all the others and for all θ. This class is much too big and the problem is too general to hope to find an estimator that stands out above all the rest. If we are to have any chance of finding optimal estimators, we must be less ambitious and restrict the class of estimators. To do this, there are three conventional methods which, for a given problem, supply objective reasons for preferring one estimator in a class which nevertheless remains extremely large.

Principle of Impartiality

The problem with the trivial estimator is that it favours the value θ_0 to the detriment of all the others, whence the idea of seeking estimators in a class that does not possess this shortcoming. The most conventional way is to restrict the search for the best estimator to the class of unbiased estimators and impose $\mathrm{E}\{\delta(X)\} = \tau(\theta)$. Comparing estimators in the sense of risk then amounts to comparing their variances. Effectively, it can be shown that, in this class, optimal estimators can be found. These are called *uniform minimum variance unbiased (UMVU) estimators*, with the understanding that the loss function is the quadratic function.

Bayes' Principle

Another, less ambitious approach is to abandon any hope of minimising the risk uniformly, i.e., for all θ, and to seek the estimator which minimises a risk calculated on average over all θ. For this average to have any meaning, it must be calculated using a weighting function $\pi(\theta)$. We then calculate the average risk or *Bayes risk* R_π from

$$R_\pi(\delta) = \int_\theta R\big(\tau(\theta), \delta\big)\pi(\theta)\mathrm{d}\theta. \tag{9.12}$$

In the Bayesian interpretation of this average, θ is a random variable with a priori probability density $\pi(\theta)$. A Bayes estimator is one that minimises the Bayes risk.

When taken in this way, the above notion is not controversial, but some practitioners criticise it when it is not conceivable for θ to be a random variable. They then reject the idea accepted by the Bayesian school and interpret $\pi(\theta)$ as a measure of the subjective confidence level attributed to the possible values of θ. In any case, the Bayesian approach has the advantage that it gives a framework within which it becomes possible on the basis of acquired experience to refine the a priori density $\pi(\theta)$ by replacing it by the a posteriori density $\psi(\theta|X)$. Indeed, according to the Bayes relation,

$$\psi(\theta|X) = \frac{l(X|\theta)\pi(\theta)}{\int_\theta l(X|\theta)\pi(\theta)}. \tag{9.13}$$

In this expression, $l(X|\theta)$ is the *likelihood function*, i.e., the probability density for observing X, knowing θ. The function ψ is then taken as an improvement on π in view of the experimental results X.

Minimax Principle

According to this principle, the overall performance of an estimator is accounted for by noting the maximum of its attached risk, and this maximum risk is then minimised. Hence, an estimator δ^* will be a minimax estimator if

$$\max_{\theta} R\big(\tau(\theta), \delta^*\big) = \min_{\delta} \max_{\theta} R\big(\tau(\theta, \delta)\big). \tag{9.14}$$

The minimax risk R_{m} is defined as

$$R_{\mathrm{m}}(\delta) = \max_{\theta} R\big(\tau(\theta, \delta)\big).$$

It can be shown that the minimax risk is always greater than or equal to the risk of an arbitrary Bayes estimator. If θ has finite domain of definition, it can be shown that the minimax estimator corresponds to the Bayes estimator of the most unfavourable a priori density π. The minimax principle turns out to be a precautionary principle, which may even be qualified as pessimistic.

There are other principles leading to interesting estimators. An example is provided by non-deterministic estimates in which the same sample does not necessarily lead to the same estimate.

9.5.4 Properties of Estimators

The ideas discussed above suggest a more careful investigation of the properties that should be satisfied by *good* estimators. It is convenient to begin with convergence, which is a property that all estimators should possess, whether they are UMVU, Bayesian, or something else. The next issue is the bias and variance of estimators, and the methods available for reducing them. Finally, we discuss a fundamental result in the class of unbiased estimators, namely the non-zero lower bound of their variance.

Convergence

Definition. A sequence of estimators $\{\widehat{\theta}\}$ or, more concisely, an estimator $\widehat{\theta}_n$, is convergent when it converges in probability to the value it claims to estimate. According to this definition, an estimator is convergent if

$$\forall \epsilon > 0, \qquad \lim_{n \to \infty} \Pr\left\{\left|\hat{\theta}_n - \theta\right| \geq \epsilon\right\} = 0. \tag{9.15}$$

This convergence is expressed by writing $\widehat{\theta}_n \xrightarrow{\Pr} \theta$.

The notion of convergence in probability is used to state the *law of large numbers*, which stipulates that the arithmetic mean M of n independent realisations $(X_1, \ldots, X_n)$ of the same random variable X is a convergent estimator of its average μ (provided it exists):

$$M = \frac{1}{n} \sum_{i=1}^{n} X_i \xrightarrow{\Pr} \mu. \tag{9.16}$$

This result, first discovered at the beginning of the seventeenth century, was the point of departure for probability theory in the field of experiment.

Unbiasedness

Definition. An estimator $\widehat{\theta}_n$ for θ or T_n for $\tau(\theta)$ is said to be *unbiased* if

$$\mathrm{E}\{\widehat{\theta}_n\} = \theta \quad \text{or} \quad \mathrm{E}\{T_n\} = \tau(\theta), \tag{9.17}$$

for all possible values of θ. If the bias is defined to be the quantity $\mathrm{E}\{\widehat{\theta}_n\} - \theta$, an unbiased estimator is one with zero bias.

Example. The arithmetic mean

$$M = \frac{1}{n} \sum_{i=1}^{n} X_i$$

is an unbiased estimator of the average μ and of the variance σ^2 of the parent population (provided it exists):

$$\mathrm{E}\{M\} = \mu. \tag{9.18}$$

This is a simple consequence of the linearity of mathematical expectation.

The notion of bias, introduced by Gauss in 1821, corresponds to the idea of *systematic error* as used by physicists. Apart from the theoretical considerations discussed above, it seems reasonable to regard it as a good property if the average of the estimator is equal to the parameter that it is supposed to estimate. However, it is not necessarily a good idea to insist on dealing only with unbiased estimators, if only because there are cases where no such estimator actually exists. The following counterexample is a classic.

Example of a Situation with No Unbiased Estimator. Let K be a random variable distributed according to the binomial distribution $\mathscr{B}(n, p)$. The number of trials n is given, but the probability p of success in a given trial is unknown. However, it is not p that we are trying to estimate, but rather $\tau(p) = 1/p$. Let $t(k)$ by the values taken by the estimator T for $1/p$ when we observe k successes in n trials. By definition, for T to be unbiased, we must have

$$\sum_{k=0}^{n} t(k) C_n^k p^k (1-p)^{n-k} = \frac{1}{p} \quad \text{for} \quad 0 < p \leq 1.$$

Suppose now that for a certain p, the estimator T is unbiased. Then the sum on the left is finite, which implies in particular that $t(0)$ is finite. Now when $p \to 0$, the term on the left tends to $t(0)$, whereas the term $1/p$ on the right can be as large as we like. Hence, T cannot be unbiased for all p.

This is not such an artificial example. It corresponds to the so-called *capture–recapture method*, which aims to estimate an animal population. Let N be the population of animal species, e.g., fish in a lake. To estimate N, we capture m animals which are then tagged and released. After tagging, the probability of recapturing an animal is $p = m/N$, and estimating the population $N = m/p$ amounts to estimating $1/p$.

In fact, to estimate $1/p$ in an unbiased way, the animals must be recaptured up to the point where the number of tagged individuals is equal to a prechosen number r. If $r + Y$ is the total number of animals that must be recaptured in order to obtain r tagged animals, it can be shown that $(r+Y)/r$ is the best unbiased estimator for $1/p$. In this procedure, it is assumed either that N is very large or that animals are recaptured and released immediately (drawing with replacement).

Asymptotically Unbiased Estimators

When defining asymptotic properties, it is essential to specify the order of operations when taking the limit. In the case that concerns us here, will a sequence of estimators $\widehat{\theta}_n$ be declared asymptotically unbiased if the limit of the expectation tends to zero, or if the expectation of the limit tends to zero? There are thus two possible definitions.

Definition: Limit of Expectation. A sequence of estimators $\{\widehat{\theta}\}$ for θ will be called asymptotically unbiased if the sequence of expectations of $\widehat{\theta}_n$ tends to θ, i.e., if

$$\lim_{n\to\infty} \mathrm{E}\{\widehat{\theta}_n\} = \theta.$$

No particular comment is needed here, because this is just a sequence of scalars.

Definition: Expectation of the Limit. A sequence of estimators $\{\widehat{\theta}\}$ for θ will be called *asymptotically unbiased* if the sequence of differences $\widehat{\theta}_n - \theta$, suitably normalised, tends in probability to a random variable of zero expectation, i.e., if there is a sequence k_n such that

$$k_n(\widehat{\theta}_n - \theta) \xrightarrow{\text{Dist}} Y \quad \text{and} \quad \mathrm{E}\{Y\} = 0.$$

The normalisation constant k_n is often proportional to $\sqrt{n}$, as for the central limit theorem. Convergence in distribution means that, at the limit, the two random variables obey the same distribution function, except perhaps on a set of zero measure.

An estimator can be asymptotically unbiased according to one definition and not for the other. In practice, the first definition, i.e., the limit of expectations, is assumed.

Convergence and Absence of Bias

Convergence refers to the distribution function while absence of bias refers to the first order moment of this function. There is no particular reason for the properties of convergence and absence of bias to be related. They are independent notions. In addition, a convergent estimator is not necessarily asymptotically unbiased by either definition of the absence of asymptotic bias. The following example shows this for the limit of expectations definition. (For the expectation of the limit definition, it is easy to find a counterexample in which the limiting distribution has no average, e.g., the Cauchy distribution.)

Example. Let $\{\widehat{\theta}\}$ be a sequence of estimators for zero, where the estimator $\widehat{\theta}_n$ is a discrete random variable equal to 0 with probability $1-1/n$ and n^2 with probability $1/n$. The sequence of estimators is obviously convergent since the probability associated with the point which is not at zero tends to 0 as $1/n$ when $n \to \infty$. However, it is biased, and its bias does not tend to zero, but rather to infinity:

$$\mathrm{E}\{\widehat{\theta}_n\} = 0(1-1/n) + n^2/n = n.$$

Reduction of Bias

Unbiased estimators are of great theoretical importance since they define a class where optimal estimators can exist, but they have another advantage outside this class. If we attempt to apply the law of large numbers by taking the arithmetic mean of a set of biased estimators, we will converge toward the average of the underlying distribution, but this will not be the parameter we hoped to estimate. It will entail an error equal to the bias. This is the whole point about trying to correct any bias, or at least trying to reduce it.

Note, however, that, if we try to reduce the bias using the methods presented below when there is no unbiased estimator, we may reduce the bias of the estimator, but we will increase its variance in such a way that its mean squared error will end up being large.

When the Bias is Easy to Calculate

In certain situations, a simple calculation can correct a bias. This happens when estimating the variance of a population we shall use here as an example. Suppose we have an independent sample $(X_1, \ldots, X_n)$ for a distribution with average μ and variance σ^2. The parameter to be estimated is σ^2, e.g., a noise power. There are two possible cases: either the average μ is known or it is not.

If the population average is known, it can be shown that the estimator

$$S'^2 = \frac{1}{n}\sum_{i=1}^{n}(X_i - \mu)^2,$$

called the sample variance, is unbiased, i.e., $\mathrm{E}\{S'^2\} = \sigma^2$. However, if we estimate the population average by the sample mean

$$M = \frac{1}{n}\sum_{i=1}^{n}X_i,$$

the estimator S'^2 is only asymptotical unbiased. Indeed, it can be shown that for

$$S'^2 = \frac{1}{n}\sum_{i=1}^{n}(X_i - M)^2,$$

we have

$$\mathrm{E}\{S'^2\} = \sigma^2 - \frac{\sigma^2}{n}.$$

The sample variance is thus systematically smaller than the population variance.

In this case, it is easy to correct for the bias by considering the estimator

$$S^2 = \frac{n}{n-1}S'^2.$$

It follows that

$$\mathrm{E}\{S^2\} = \mathrm{E}\left\{\frac{n}{n-1}S'^2\right\} = \frac{n}{n-1}(\sigma^2 - \sigma^2/n) = \sigma^2.$$

In order to estimate the variance of a population without bias, it is thus worth considering the estimator

$$S^2 = \frac{1}{n-1}\sum_{i=1}^{n}(x_i - M)^2. \tag{9.19}$$

Note, however, that it has higher variance than the biased estimator S'^2. Indeed, we have

$$\mathrm{Var}(S^2) = \frac{2\sigma^4}{n-1} > \mathrm{Var}(S'^2) = 2\sigma^4\frac{n-1}{n^2}. \tag{9.20}$$

If an estimator $\widehat{\theta}$ is unbiased for θ, there is no reason why $g(\widehat{\theta})$ should be an unbiased estimator for $g(\theta)$. For example, if g is a convex function, the Jensen inequality holds: $\mathrm{E}\{g(X)\} \geq g(\mathrm{E}\{X\})$, which shows that we should expect to observe a bias.

This is the case when estimating the standard deviation σ. We have $\mathrm{E}\{S^2\} > (\mathrm{E}\{S\})^2$, so we should expect the estimator $S = \sqrt{S^2}$ to be a biased estimator for σ. For a normal variable $\mathscr{N}\mu, \sigma^2$, for example, the unbiased estimator for the standard deviation σ is

$$S^* = k_n\left[\frac{1}{n-1}\sum_i (X_i - M)^2\right]^{1/2}, \quad k_n = \sqrt{\frac{n-1}{2}}\frac{\Gamma\left(\frac{n-1}{2}\right)}{\Gamma\left(\frac{n}{2}\right)}, \quad n \geq 2, \tag{9.21}$$

where Γ is the Euler function of the second kind.

When the Bias is Hard to Calculate

We then resort to resampling methods or *bootstrap methods*. The archetypal example is the following.

Quenouille's Method

If there is a bias of order $1/n$, this method can reduce it to order $1/n^2$. We assume that the mean value of the estimator can be expanded in powers of $1/n$. Half the sample is extracted, having removed one point if the sample size was odd to begin with. We then have

$$\mathrm{E}\{\widehat{\theta}_n\} = \theta + \frac{1}{n}\beta + O(1/n^2),$$

$$\mathrm{E}\{\widehat{\theta}_{2n}\} = \theta + \frac{1}{2n}\beta + O(1/n^2),$$

whence

$$\mathrm{E}\{2\widehat{\theta}_{2n} - \widehat{\theta}_n\} = \theta + O(1/n^2).$$

The bias going as $1/n$ has disappeared, but in general the variance of this new estimator will increase by a factor of order $1/n$. A better method would be to divide the $2n$-sample at random into two equal parts, evaluate the corresponding estimators $\widehat{\theta}_n$ and $\widehat{\theta}'_n$, and calculate the new estimator

$$2\widehat{\theta}_{2n} - \frac{1}{2}\left(\widehat{\theta}_n + \widehat{\theta}'_n\right).$$

Jackknife Method

There is another method which only increases the variance by a term in $1/n^2$. This is the *jackknife method*. (For more details, see *The Jackknife. A Review*, Miller, 1974.) This involves more sophisticated calculation, but with today's ever faster and ever cheaper electronic means of computation, the jackknife is superior to the Quenouille method described above.

Let $\widehat{\theta}_n$ be an estimator for θ, calculated from an n-sample. Once again, expand its mean value in powers of $1/n$:

$$\mathrm{E}\{\widehat{\theta}_n\} = \theta + \sum_{k=1}^{\infty} \frac{a_k}{n^k}. \tag{9.22}$$

Now recalculate the n estimators $\widehat{\theta}_{-i}$, each time removing a point i from the n-sample, and calculate the arithmetic mean of these estimators, viz.,

$$\overline{\overline{\theta}}_{n-1} = \frac{1}{n} \sum_{i=1}^{n} \widehat{\theta}_{-i}. \tag{9.23}$$

Finally, define the jackknife $\widehat{\theta}'_n$ as

$$\widehat{\theta}'_n = n\widehat{\theta}_n - (n-1)\overline{\overline{\theta}}_{n-1} = \widehat{\theta}_n + (n-1)\left(\widehat{\theta}_n - \overline{\overline{\theta}}_{n-1}\right). \tag{9.24}$$

It can be shown that, for this estimator, the order $1/n$ bias has also vanished.

Efficiency

We have seen that, for unbiased estimators and for a quadratic loss function, these can be ordered according to the values of their variances. According to this idea, the estimator $\widehat{\theta}_1$ will be preferable to $\widehat{\theta}_2$ if $\mathrm{Var}\big(\widehat{\theta}_1\big) < \mathrm{Var}\big(\widehat{\theta}_2\big)$. We then say that $\widehat{\theta}_1$ is more efficient than $\widehat{\theta}_2$.

This raises two questions:

1. For a given unbiased estimator, can its variance be reduced?
2. Is it possible, at least in principle, to make its variance arbitrarily small?

The answer to the first question is affirmative in some cases, but for this to be possible, there must be an *exhaustive statistic*. We shall see below what this involves. The answer to the second question is negative, if there is a lower bound below which the variance cannot be reduced. In general, this lower bound is not known, but an approximation can be given, viz., the Rao–Cramér bound. These considerations are developed in the following.

Reducing the Variance

Reduction of the variance involves the idea of an exhaustive summary, i.e., statistics containing everything we can hope to know about the parameter we are trying to estimate. Let us make this more precise.

Definition. A statistic T is said to be *exhaustive* if the conditional distribution of the sample X given that $T(X) = t$ does not depend on the parameter θ. If $f\big(x|T(x) = t\big)$ denotes this conditional probability density of the sample X, we have

$$\frac{\partial}{\partial\theta} f\big(x|T(x) = t\big) = 0. \tag{9.25}$$

Although abstract, this definition corresponds to what is usually done when compressing data. By keeping only the value $T(X)$ from the original sample $X = (X_1, \ldots, X_n)$, we keep only what depended on θ. Indeed, the equation $T(x) = t$ specifies a hypersurface in the sample space $\mathbb{R}^n$, and saying that T provides an exhaustive summary amounts to saying that only the value t of this statistic is important. The distribution of 'points' X_i actually on this hypersurface is of no importance, since it does not depend on what we seek here. If we had to reconstruct a sample with the same characteristics as the original sample with regard to estimation of θ, the points could be distributed at random over the hypersurface $T = t$ according to the corresponding conditional distribution.

In order to show that a statistic is exhaustive, it is often simpler to appeal to the following criterion.

Theorem. Factorisation or Fisher–Neyman Criterion. A statistic T is exhaustive for estimating θ if and only if the likelihood function of the sample can be factorised in the form

$$L(x, \theta) = l(t, \theta)h(x), \tag{9.26}$$

where x denotes a value taken by the sample, i.e., $(X_1 = x_1, \ldots, X_n = x_n)$, and L is the probability distribution at this point (or just the probability if the variables are discrete). The functions l and h must be positive.

Example: Poisson Distribution. For an independent and identically distributed sample with Poisson distribution, we have

$$\Pr\{X = x\} = \prod_{i=1}^{n} \frac{\lambda^{x_i}}{x_i!} \mathrm{e}^{-\lambda} = \frac{\lambda^{\sum x_i} \mathrm{e}^{-n\lambda}}{\prod x_i!}. \tag{9.27}$$

This satisfies the factorisation criterion with $h(x) = 1/\prod x_i!$ and the statistic $T = \sum_{i=1}^{n} x_i$ is exhaustive for the parameter λ of the Poisson distribution.

With this idea in hand, we can now state a version of the Rao–Blackwell theorem for the case of quadratic loss functions.

Rao–Blackwell Theorem. If $\widehat{\theta}$ is an unbiased estimator of $\tau(\theta)$ and if the sampling law admits an exhaustive statistic T, then the estimator $\widehat{\theta}_1$ calculated from

$$\widehat{\theta}_1 = \mathrm{E}\big\{\widehat{\theta} | T = t\big\} \tag{9.28}$$

is unbiased and has smaller variance than the original estimator:

$$\forall \theta, \quad \mathrm{Var}\big(\widehat{\theta}_1\big) < \mathrm{Var}\big(\widehat{\theta}\big),$$

unless the estimators turn out to be identical.

The improvement achieved by the Rao–Blackwell theorem does not guarantee that the new estimator has minimal variance, however. For this to be the case, the statistic T must also be *complete*.

Definition. The statistic T is said to be complete if the expectation of all integrable functions of T is non-degenerate, i.e., if

$$\forall \theta, g \in L^1, \quad \mathrm{E}\{g(T)\} = 0 \Longrightarrow g(t) = 0. \tag{9.29}$$

The θ dependence is contained in the sampling distribution used to calculate the expectation.

Example: Binomial Distribution. The sample X of size $k = 1$ is drawn from a binomial distribution with unknown parameter θ, while the parameter n is known. The statistic $T = X$ is trivially exhaustive. Show that it is also complete. For all integrable g,

$$\forall\theta \in (0,1), \quad \mathrm{E}\{g(T)\} = \sum_{x=0}^{n} g(x) C_n^x \theta^x (1-\theta)^{n-x} = 0$$
$$= (1-\theta)^n \sum_{x=0}^{n} g(x) C_n^x \left(\frac{\theta}{1-\theta}\right)^x = 0.$$

The last expression is a polynomial in $\theta/(1-\theta)$ of degree n which vanishes for infinitely many values of its variable. It is therefore identically zero and $g(x) = 0$. A similar result can be proven for a sample of arbitrary size k and for $T = \sum_{i=1}^{k} X_i$.

Lehmann–Scheffé Theorem. If T is an exhaustive and complete statistic, then the improvement of $\widehat{\theta}$ under the conditions of the Rao–Blackwell theorem has smaller variance. The resulting estimator $\widehat{\theta}_1$ is thus the UMVU estimator.

Example. Following Tassi (see the bibliography), consider the case where an object has probability p of possessing some feature considered to be exceptional in some sense. The objects are arranged into r classes and n objects are then drawn independently from each class. The number of objects per class is relatively unimportant, provided it is greater than n. The class itself is said to be exceptional if there is at least one exceptional object among the n objects drawn out, and otherwise it is said to be ordinary. Let X_i be the number of exceptional objects in the sample of size n drawn from class i. The problem is to estimate without bias, given the data $(X_1, \ldots, X_r)$, the probability θ that a class is ordinary.

Consider the variable $\mathbf{1}_{X_i=0}$, which is the indicator function for class i to be ordinary. If we set $\widehat{\theta} = \mathbf{1}_{X_i=0}$, the estimator is a Bernoulli variable with parameter θ. We then have

$$\mathrm{E}\{\widehat{\theta}\} = \mathrm{E}\{\mathbf{1}_{X_i=0}\} = \theta,$$

and this estimator is thus unbiased. In order to improve it using the Rao–Blackwell theorem, we require an exhaustive statistic for estimation of θ. The likelihood function of the data $(X_1, \ldots, X_r)$ is the product of r binomially distributed random variables. We have

$$L(x,\theta) = \prod_{i=1}^{r} C_n^{X_i} p^{X_i} (1-p)^{n-X_i} = \left(\frac{p}{1-p}\right)^{\sum_i X_i} (1-p)^{nr} \prod_{i=1}^{r} C_n^{X_i}.$$

This is indeed in the form (9.26) and the statistic $T = \sum_{i=1}^{r} X_i$ is exhaustive for p. It is also exhaustive for θ, which depends only on p. The estimator $h(\widehat{\theta})$ improved by Rao–Blackwell is then

$$h(\widehat{\theta}) = \mathrm{E}\{\widehat{\theta} | T = t\}.$$

Considering the class $i = 1$ for calculation of $\widehat{\theta}$, it follows that

$$h(\widehat{\theta}) = \mathrm{E}\{\widehat{\theta}|T = t\} = \Pr\{X_1 = 0|T = t\} = \frac{\Pr\{X_1 = 0, T = t\}}{\Pr\{T = t\}}.$$

The sampling model is binomial, whence

$$\frac{\Pr\left\{X_1 = 0, \sum_{i=2}^{r} X_i = t\right\}}{\Pr\left\{\sum_{i=1}^{r} X_i = t\right\}} = \frac{C_{nr-n}^{\mathrm{T}}}{C_{nr}^{\mathrm{T}}}.$$

This estimator depends only on the value of the variable T, denoted by $h(T)$. Then

$$h(T) = \frac{C_{nr-n}^{T}}{C_{nr}^{T}}. \tag{9.30}$$

Since we have seen that this statistic is also complete, $h(T)$ is therefore the UMVU estimator for θ.

9.5.5 *Fréchet or Rao–Cramér Inequality*

Since the variance of an estimator can be reduced under certain conditions, it is then legitimate to ask how far such a reduction can be taken. Clearly, the variance cannot be less than zero, but is there a strictly positive bound below which, for a given problem, it is impossible to find an estimator achieving this level of performance? The problems relating to the search for this bound involve theoretical and practical features of great importance, which we shall discuss in this section.

An unbiased estimator which reaches this limiting variance will be preferable to all others in the sense of quadratic risk. It will be called the optimal estimator and hence will be the UMVU estimator. However, in the general case, there is no way to find the optimal limit. On the other hand, it is possible to give an approximation, namely the Rao–Cramér bound, or *minimum variance bound*. Furthermore, there is a large class of estimators that actually reach this bound, known as *efficient estimators*, which we shall now examine.

Such an estimator T, by assumption unbiased, has mean $\mathrm{E}\{T\} = \tau(\theta)$ and variance $\mathrm{Var}(T) = \mathrm{E}\{[T - \tau(\theta)]^2\}$. To calculate this variance, we use the mathematical expectation operator, which is a linear operator and with which we can associate a scalar product. This scalar product obeys the Cauchy–Schwartz inequality. Let ψ be a function of the n-sample (denoted by X) and the parameter θ to be estimated. By the Cauchy–Schwartz inequality,

$$\mathrm{E}\{[T - \tau(\theta)]^2\}\mathrm{E}\{\psi(X,\theta)^2\} \geq \left[\mathrm{E}\{[T - \tau(\theta)]\psi(X,\theta)\}\right]^2, \tag{9.31}$$

whence, given that $\mathrm{E}\{[T-\tau(\theta)]^2\} = \mathrm{Var}(T)$,

$$\mathrm{Var}(T) \geq \frac{\left[\mathrm{E}\{[T-\tau(\theta)]\psi(X,\theta)\}\right]^2}{\mathrm{E}\{\psi(X,\theta)^2\}}. \tag{9.32}$$

Equality obtains if and only if the function ψ is proportional to $T-\tau(\theta)$, that is, if and only if

$$\psi(X,\theta) = A(\theta)\left[T-\tau(\theta)\right]. \tag{9.33}$$

The inequality (9.32) can in principle be used to find the optimal bound, since

$$\mathrm{Var}(T) \geq \max_{\psi} \frac{\left[\mathrm{E}\{[T-\tau(\theta)]\psi(X,\theta)\}\right]^2}{\mathrm{E}\{\psi(X,\theta)^2\}}. \tag{9.34}$$

However, it is almost impossible for all T and $\tau(\theta)$ and all parent populations to exhibit the optimal bound by calculating the maximum. The best that can be done is to find a function ψ that provides a bound close enough to the optimal bound. If the function ψ is not well chosen, we find a lower bound that is too small, e.g., if ψ does not depend on X, we will find $\mathrm{Var}(T) \geq 0$, which we already knew. In order to find a best lower bound, we must identify a common case where the function ψ has the above form (9.33), so that the bound is attained. For this particular case, the bound obtained will be the optimal one.

Let us consider the case where the n-sample X comes from a normal parent distribution with known mean μ and variance σ^2, and the parameter θ to be estimated is the mean μ, estimated by the empirical average M. Assume also that the sample is made up of independepent and identically distributed random variables.

First we calculate the likelihood function for this n-sample:

$$L(X|\theta) = \prod_{i=1}^{n} \frac{1}{\sigma\sqrt{2\pi}} \exp\left[-\frac{1}{2}\left(\frac{X_i-\theta}{\sigma}\right)^2\right]. \tag{9.35}$$

The product is eliminated by taking the logarithm of the expression to give

$$\ln L = -n\ln(\sigma\sqrt{2\pi}) - \sum_{i=1}^{n} \frac{1}{2}\left(\frac{X_i-\theta}{\sigma}\right)^2. \tag{9.36}$$

We now eliminate the constant and the square by differentiating with respect to θ to obtain

$$\frac{\partial \ln L}{\partial\theta} = \frac{1}{\sigma^2}\sum_{i=1}^{n} x_i - \frac{n\theta}{\sigma^2} = \frac{n}{\sigma^2}(M-\theta). \tag{9.37}$$

We then see that in the particular case of estimation of the mean μ of a normal population with known variance by the arithmetic mean M, the function $\partial \ln L/\partial\theta$ has precisely the form (9.33) which makes the Cauchy–Schwartz inequality into an equality. In this case the resulting bound is the optimal bound.

For an arbitrary population and when estimating a function of an arbitrary parameter θ by a statistic T, we can be sure that the variance of T will always be greater than or equal to the limit found if we replace the function ψ by $\partial \ln L/\partial\theta$ in the Cauchy–Schwartz inequality. It will not be the optimal bound but it is of great practical importance because of the case giving rise to it. It is called the *minimum variance bound* (MVB).

It remains to actually calculate this bound, and to do this, we have to find the various expectation values appearing in (9.32). The calculation is relatively simple and we shall not go into the details here. Note, however, that the final result is only valid when the operations of integration and differentiation with respect to the parameter to be estimated can be legitimately interchanged, e.g., when the integration bounds do not depend on this parameter. With this proviso, if the estimator T is unbiased, we deduce the following inequality for the variance, known as the Rao–Cramér or Fréchet inequality:

$$\boxed{\mathrm{Var}(T) \geq \frac{\left[\tau'(\theta)\right]^2}{\mathrm{E}\{-\partial^2 \ln L/\partial\theta^2\}}.} \tag{9.38}$$

We have seen that the relatively arbitrary choice of function ψ implies that this lower bound is not necessarily attained. For it to be attained, a necessary and sufficient condition is

$$\frac{\partial}{\partial\theta} \ln L = A(\theta)\left[T - \tau(\theta)\right], \tag{9.39}$$

where $A(\theta)$ is an arbitrary function of the parameter θ to be estimated. If the variance of an estimator reaches this bound, it is called the *minimum variance bound* (MVB).

9.5.6 Efficient Estimators

When the MVB is attained, we have

$$\frac{\partial \ln L}{\partial\theta} = A(\theta)\left[T - \tau(\theta)\right]. \tag{9.40}$$

Substituting this directly into (9.38), we obtain

$$\mathrm{Var}(T) = \frac{\tau'(\theta)}{A(\theta)}, \tag{9.41}$$

and if $\tau(\theta) = \theta$,

$$\mathrm{Var}(T) = \frac{1}{A(\theta)}. \tag{9.42}$$

The last two expressions can in some cases be used to find the variance of an MVB estimator rather easily, as illustrated by the following examples.

Example: Variance of the Arithmetic Mean of a Normal Distribution. For an n-sample drawn from a normal distribution with known variance σ^2 and for estimation of the mean θ by the arithmetic mean $T = M$, we saw earlier that we had

$$\frac{\partial}{\partial \theta} \ln L = \frac{n}{\sigma^2}(M - \theta), \quad \text{hence} \quad A(\theta) = \frac{n}{\sigma^2},$$

so that

$$\mathrm{Var}(T) = \frac{\sigma^2}{n}.$$

Example: Variance of the Sample Variance of a Normal Distribution. Consider a normal population with probability distribution

$$f(x) = \frac{1}{\sqrt{2\pi}\theta} \exp\left(-\frac{x^2}{2\theta^2}\right). \tag{9.43}$$

The average is known and can be assumed to be zero, but the variance θ^2 is not known. We then have

$$\frac{\partial \ln L}{\partial \theta} = \frac{n}{\theta^3}\left(\frac{1}{n}\sum X_i^2 - \theta^2\right) = \frac{n}{\theta^3}(S'^2 - \theta^2). \tag{9.44}$$

The form of the function $\partial \ln L/\partial\theta$ naturally suggests choosing the statistic

$$S'^2 = \frac{1}{n}\sum X_i{}^2 \tag{9.45}$$

as estimator for θ^2. This estimator will be MVB provided that it is unbiased. It is easy to demonstrate that this last condition is satisfied since

$$\mathrm{E}\{S'^2\} = \int s^2 L \mathrm{d}\boldsymbol{x} = \frac{1}{n}\sum_i \int X_i{}^2 L \mathrm{d}\boldsymbol{x} = \frac{1}{n}\sum_i \theta^2 = \theta^2. \tag{9.46}$$

Therefore, S'^2 is indeed an MVB estimator for the function $\tau(\theta) = \theta^2$. We can immediately calculate its variance to be

$$\mathrm{Var}(S'^2) = \frac{\tau'(\theta)}{n/\theta^3} = \frac{2\theta}{n/\theta^3} = \frac{2\theta^4}{n}. \tag{9.47}$$

The standard deviation, denoted here by θ, is usually denoted by σ, whence

$$\mathrm{Var}(S'^2) = \frac{2\sigma^4}{n}. \tag{9.48}$$

We thus find the variance of the sample variance for a normal sample with known mean.

9.5.7 *Efficiency of an Estimator*

Let T be a convergent unbiased estimator for estimating $\tau(\theta)$. Its efficiency is measured by the reciprocal of the ratio of its variance to the limiting variance given by the MVB:

$$\mathrm{Eff}(T) = \frac{\mathrm{Var}_{\mathrm{MVB}}(\theta)}{\mathrm{Var}(T)}. \tag{9.49}$$

The estimator T will be said to be *efficient* if it has efficiency equal to 1 (or 100%), or in other words if it attains its MVB. As already noted, the MVB is not necessarily attained and an optimal estimator is not necessarily efficient.

9.5.8 *Biased Estimators*

Let T be a biased estimator for θ. By the definition of bias, we have

$$\mathrm{e}T = \theta + b(\theta). \tag{9.50}$$

If we choose the function $\tau(\theta) = \theta + b(\theta)$, we have

$$\mathrm{e}T = \tau(\theta), \tag{9.51}$$

which shows that T is an unbiased estimator for $\tau(\theta)$. Let us apply the Rao–Cramér formula for this function τ. We find $\tau'(\theta) = 1 + b'(\theta)$, and also

$$\mathrm{Var}(T) \equiv \mathrm{E}\big\{\big[t - \big(\theta + b(\theta)\big)\big]^2\big\} = \mathrm{E}\big\{(T-\theta)^2\big\} - b^2(\theta). \tag{9.52}$$

The term $\mathrm{E}\{(T-\theta)^2\}$ is the *mean squared error* (MSE) in T. We have just seen that it is always greater than or equal to the variance of T. Finally, we may write

$$\mathrm{E}\big\{(T-\theta)^2\big\} = \mathrm{Var}(T) + b^2(\theta) \geq \frac{\big[1 + b'(\theta)\big]^2}{\mathrm{E}\big\{-\partial^2 \ln L/\partial\theta^2\big\}} + b^2(\theta). \tag{9.53}$$

While it is true that the mean squared error in a biased estimator is always greater than its variance, the example below shows that it is possible to find a biased estimator for which the mean squared error is smaller than the smallest variance associated with the class of unbiased estimators.

Example: Estimating the Parameter of an Exponential Distribution. Consider an independent and identically distributed n-sample $(X_1, \ldots, X_n)$ drawn from a population with exponential distribution having mean θ:

$$f(x) = \frac{1}{\theta}\exp\left(-\frac{x}{\theta}\right), \quad \text{where} \quad \mathrm{E}\{X_i\} = \theta, \quad \mathrm{Var}(X_i) = \theta^2. \tag{9.54}$$

Let us choose the arithmetic mean $\overline{X}$ as estimator for θ. We know that it is an unbiased estimator for the population average, and hence $\mathrm{E}\{\overline{X}\} = \theta$. In this case, its mean squared error with respect to θ is equal to its variance:

$$\mathrm{e}(\overline{X} - \theta)^2 = \mathrm{E}\Big\{\big(\overline{X} - \mathrm{E}\{\overline{X}\}\big)^2\Big\} \equiv \mathrm{Var}(\overline{X}). \tag{9.55}$$

We now show that $\overline{X}$ is MVB by calculating $\partial \ln L / \partial\theta$:

$$\ln L = -n \ln\theta - \frac{1}{\theta}\sum_{i=1}^{n} X_i, \tag{9.56}$$

$$\frac{\partial \ln L}{\partial\theta} = -\frac{n}{\theta} + \frac{1}{\theta^2}\sum_{i=1}^{n} X_i = \frac{n}{\theta^2}(\overline{X} - \theta). \tag{9.57}$$

This shows that $\overline{X}$ is indeed MVB for estimating θ, whence it follows immediately that

$$\mathrm{Var}(\overline{X}) = \frac{\theta^2}{n}. \tag{9.58}$$

On the other hand, the estimator $\widehat{X} = n\overline{X}/(n+1)$ is biased but its mean squared error is smaller than the MVB θ^2/n of the unbiased estimators. Indeed, it is biased:

$$\mathrm{E}\{\widehat{X}\} = \frac{n}{n+1}\theta, \qquad b_n(\theta) = -\frac{\theta}{n+1}, \tag{9.59}$$

but we have the inequalities

$$\mathrm{Var}(\widehat{X}) = \frac{n}{(n+1)^2}\theta^2 < \mathrm{E}\{(\widehat{X} - \theta)^2\} = \frac{\theta^2}{n+1} < \mathrm{Var}(\overline{X}) = \frac{\theta^2}{n}. \tag{9.60}$$

This example shows that it can be too restrictive to search for the best estimator only in the class of unbiased estimators.

9.5.9 Minimum Variance Bound and Fisher Information

The non-negative quantity

$$I_n(\theta) \equiv \mathrm{E}\left\{\left(\frac{\partial}{\partial\theta}\ln L\right)^2\right\} = \mathrm{E}\left\{-\frac{\partial^2}{\partial\theta^2}\ln L\right\} \qquad (9.61)$$

is called the *Fisher information* contained in the n-sample. Since the random variable $\partial \ln L/\partial\theta$ has zero mean, the Fisher information is just the reciprocal of the variance of this random variable.

For independent and identically distributed samples, this quantity depends only the size of the n-sample and the probability distribution of the parent population, and we have shown that $I_n(\theta) = nI_1(\theta)$. This means that, in the case where the Rao–Cramér inequality applies, the variance of an estimator T decreases more slowly than $1/n$. The inequality then becomes

$$\mathrm{Var}(T) \geq \frac{\left[\tau'(\theta)\right]^2}{nI_1(\theta)}. \qquad (9.62)$$

When estimating the midpoint of a uniform probability density on $(0, \theta]$, the estimator

$$\frac{1}{2}\left[X_{(n)} + X_{(1)}\right]$$

of the mean $\theta/2$ has asymptotic variance

$$\lim_{n\to\infty} \mathrm{Var}\left(\frac{1}{2}\left[X_{(n)} + X_{(1)}\right]\right) = \frac{1}{2n^2}. \qquad (9.63)$$

This asymptotic variance decreases faster than the Rao–Cramér bound because this is exactly the case where this bound does not apply. Indeed, the mathematical expectation is given by the integral $\int_0^\theta \mathrm{d}\boldsymbol{x}$, and the upper bound is a function of the parameter to be estimated, whence the Rao–Cramér does not apply.

The bounds deduced from the Rao–Cramér inequalities are adapted to the case where measurement errors are *small* around the estimated parameter. By this we mean that these errors have a distribution close to normal. When the errors are large, e.g., close to a Cauchy distribution, there are then better suited bounds, i.e., greater than the Rao–Cramér bounds.

9.5.10 Multidimensional Case

When the experimenter wishes to make joint estimates of a set of k parameters $(\theta_1, \ldots, \theta_k)$, there is a generalisation of the Rao–Cramér inequality. This appeals to the *Fisher information matrix* J, with elements J_{ij} given by

$$J_{ij} = \mathrm{E}\left\{-\frac{\partial^2}{\partial\theta_i\partial\theta_j}\log L\right\}.$$

If V denotes the matrix of variances and covariances for the paramters θ, the inequality is given symbolically by $V \geq J^{-1}$. We use the convention that an inequality sign $>$ between matrices means that their difference is a positive definite matrix, i.e., all its eigenvalues are positive, while $\geq$ means that at least one eigenvalue may be zero.

From a practical standpoint, this means that the error ellipse at the confidence level γ, i.e.,

$$(\theta - \widehat{\theta})^{\mathrm{T}}\mathrm{V}^{-1}(\theta - \widehat{\theta}) = k_\gamma^2,$$

encompasses the confidence region calculated with the matrix J, i.e.,

$$(\theta - \widehat{\theta})^{\mathrm{T}}\mathrm{J}(\theta - \widehat{\theta}) = k_\gamma^2.$$

This last confidence region can be used as an approximation to the actual confidence region. The scalar k_γ^2 is calculated (under the normal distribution hypothesis) using the distribution function F of a χ^2 distribution with k degrees of freedom: $k_\gamma^2 = F^{-1}(1-\gamma)$.

9.5.11 Robust Estimators

As we have already seen, there are generally several estimators for the same parameter θ. Among other things, the arithmetic mean, median, and midpoint of the sample are often considered to estimate the population average. For example, for a normal distribution, the arithmetic mean of the sample is unbiased and MVB. It is thus the best possible estimator and there is no reason to look for any other. But which one should we choose when we do not know the nature of the parent population, or when we do know that but there is contamination by measurement errors of unknown type? To fix ideas, we shall consider three very different parent populations and give the asymptotic variances of the three previous estimators. As these three densities are symmetric, it is in fact the position of the axis of symmetry that we seek to determine. To simplify, we shall take all scale factors equal to unity. We then obtain Table 9.7, in which values in bold type are the smallest possible.

We see in this example that, if we have no idea what kind of underlying parent population we are dealing with, it is safer to use the median, in order to obtain a finite variance in all cases. We may consider Table 9.7 as the loss matrix for a game against nature, where we can in fact only lose in this case. The median is chosen

Table 9.7 Asymptotic variance of three different populations. The uniform distribution is taken here over the interval from 0 to 1

	Uniform distribution	Normal distribution	Cauchy distribution
Median	$\frac{1}{4n}$	$\frac{\pi}{2n}$	$\frac{\pi^2}{4n}$
Mean	$\frac{1}{12n}$	$\frac{\mathbf{1}}{\mathbf{n}}$	∞
Midpoint	$\frac{\mathbf{1}}{2n^2}$	$\frac{\pi^2}{24\ln n}$	∞

according to the minimax strategy, i.e., minimising the maximum loss. The median is said to be a *reliable* or *robust estimator* for this set of populations.

There are several other reliable estimators for the symmetry axis of a population with symmetric distribution. Depending on their type, they can be suitable for a range of different populations. Here are some of them:

- **Truncated or Trimmed Mean.** Consider the symmetric bilateral truncated mean. We remove $2q$ points from the ordered n-sample

$$(X_{(1)} \leq X_{(2)} \leq \ldots \leq X_{(n)}),$$

namely the q smallest and the q largest, to obtain the new sample

$$(X_{(q+1)} \leq \ldots \leq X_{(n-q)}).$$

The truncation parameter is defined here as

$$\alpha = \frac{q}{n}, \tag{9.64}$$

and the arithmetic mean of the remaining $(n-2q)$-sample will be called the mean symmetrically truncated at $100\alpha\%$ and denoted by T_α. We will thus have

$$T_\alpha = \frac{1}{(n-2q)} \sum_{i=q+1}^{n-q} X_{(i)}. \tag{9.65}$$

Note that α is generally fixed between 0 and 1. We then calculate $q = \lfloor \alpha n \rfloor$. In this case, the mean is truncated by at most $100\alpha\%$ and at least $100(\alpha - 1/n)\%$.

- **Winsorized Mean.** The procedure is the same as for the truncated mean, except that we replace the q smallest values by $X_{(q+1)}$ and the q largest by $X_{(n-q)}$, then calculate the mean:

$$W_\alpha = \frac{1}{n}\left[\sum_{i=q+1}^{n-q} X_{(i)} + q\left[X_{(q+1)} + X_{(n-q)}\right]\right]. \tag{9.66}$$

- **Hodges–Lehmann Median.** We construct the new sample $(Y_{11}, Y_{12}, \ldots, Y_{nn})$ from the sample $(X_1, \ldots, X_n)$ by the rule

$$Y_{ij} = \frac{1}{2}(X_i + X_j), \tag{9.67}$$

and we define the Hodges–Lehmann median HL as the median of the Y_{ij}.
- **Midpoint.** This statistic is the point located at equal distance from the two outermost points, and it is thus equal to

$$P = \frac{1}{2}\left[X_{(1)} + X_{(n)}\right]. \tag{9.68}$$

Example: Asymptotic Variances of Several Estimators for the Cauchy Distribution. Consider an independent and identically distributed n-sample drawn from a Cauchy distribution with probability distribution

$$f(x) = \frac{1}{\pi\left[1 + (x-\theta)^2\right]}. \tag{9.69}$$

The asymptotic variances of different estimators for θ are given in Table 9.8. From the table, we see that the mean truncated at 38% is only slightly less efficient, but much easier to calculate than the estimate of the maximum likelihood in this optimal case.

9.5.12 Some Classic Methods

It seems that effective estimators can be obtained by implementing a method that takes into account both the observed data x and what is known about the unknown parameter θ. For this purpose, we might consider looking for the value of θ that maximises the a posteriori probability density $\psi(\theta|x)$, or a functional of this. However, ψ is written as the product of the likelihood $l(x|\theta)$ and the a priori

Table 9.8 Asymptotic variance of various estimators for θ_0

Estimator	Asymptotic variance
Maximum likelihood	$\frac{2}{n}$
Mean truncated at 38%	$\simeq \frac{2.28}{n}$
Median	$\frac{\pi^2}{4n} \simeq \frac{2.47}{n}$

probability density $\pi(\theta)$, and if we do not possess information about θ, $\pi(\theta)$ remains undetermined. We may then attempt to maximise the likelihood and look for $\widehat{\theta}$ as solution of the implicit equation: $l(x|\theta) = \max_\theta l(x|\theta)$. The estimator found in this way is the estimator for the maximum likelihood.

This estimator is often found as a solution to the equations $\partial l/\partial\theta = 0$ with $\partial^2 l/\partial\theta^2 < 0$. Under these conditions, it can be shown that the estimator $\widehat{\theta}$ is convergent. In addition, the quantity

$$\left(-\left.\frac{\partial^2 l}{\partial\theta^2}\right|_{\theta=\widehat{\theta}}\right)^{1/2}(\widehat{\theta}-\theta)$$

behaves roughly like a normal random variable with zero mean and unit variance. This shows that, provided that the associated moments exist, the maximum likelihood estimator is asymptotically unbiased and asymptotically efficient. It is this result that gives the maximum likelihood estimator its special status.

If the problem reduces to estimating the mean $\mu = \mu(\theta)$ of a normal sample, the likelihood function is equal to $\exp(-\chi^2/2)$ up to a multiplicative constant, where χ^2 is a quadratic form equal to $(x-\mu)^{\mathrm{T}}V^{-1}(x-\mu)$, with V the variance–covariance matrix of x and X^{T} the transpose of X. This therefore amounts to looking for the maximum of l or $\log l$ and the maximum likelihood then corresponds to the minimum of the form $(x-\mu)^{\mathrm{T}}V^{-1}(x-\mu)$. This remark introduces the underlying idea of the least squares method, discussed here in the normal case and for measurements of unequal accuracy.

Outside the normal case, the least squares method (or least χ^2 for measurements of unequal accuracy) does not always yield optimal estimators. However, there is another favourable situation, namely the linear case. This refers to a situation in which the mean μ depends linearly on the unknown parameters θ, and no assumptions are made about the type of data apart from the existence of μ and V. In particular, we do not necessarily know the probability density. The decomposition of μ in terms of a polynomial basis or in a Fourier series expansion are two examples of linear situations. The coefficients of these expansions are the unknown parameters θ. If $\mu = X\theta$ is the matrix equation expressing the linear dependence of the model μ on θ and if y denotes the data (it is traditional to use y rather than x in this case), the least squares estimators are given by the expression $\widehat{\theta} = (X^{\mathrm{T}}X)^{-1}X^{\mathrm{T}}y$ in the case where the matrix V is the identity matrix (and we may always reduce to this case by a change of variables). The Gauss–Markov theorem states that these estimators are unbiased and that, for each parameter θ, they have the smallest variance in the class of linear estimators. It is remarkable that the estimator $\widehat{\theta}$ has optimal properties without any reference to the probability density of the measurements y.

The maximum likelihood and least squares methods often provide good estimators, but they can prove to be unacceptable. They may be the best in a class that turns out to be too small, e.g., normal case or class of unbiased estimators. They may also exhibit anomalies, e.g., if the operator $(X^{\mathrm{T}}X)^{-1}$ does not exist. In this case, one must revert to the idea mentioned at the beginning of this section, namely to introduce a certain dose of a priori knowledge about the desired estimator. This aim can be achieved by means of a regularising function g which reflects the faithfulness of the solution with respect to what is expected of θ. The likelihood l reflects the faithfulness to the data. With this in mind, we look for estimators maximising the quantity $f = \lambda_0 l + \lambda_1 g$, where $\lambda_0 + \lambda_1 = 1$. The scalars λ_0 and λ_1 are Lagrange multipliers, accounting respectively for the relative importance of data fidelity and the relative importance of the a priori knowledge of θ. The Bayesian framework provides a natural probabilistic interpretation of this method. Indeed, the maximum of ψ coincides with the maximum of $\log \psi$, which can be written $\log \psi = \log l + \log \pi$, a form which can be identified with f. Many methods for finding estimators are based on this idea whenever the conventional methods fail. They are classified according to the nature of the regularising function, e.g., entropy maximum, Tikhonov regularisation, pseudo-inverse, etc., or the nature of its probabilistic interpretation in the Bayesian framework. Examples of these methods can be found elsewhere in this book. The difficulty is often to find the weighting by the multipliers λ_0 and λ_1 which leads to optimal estimators. Cross-validation is one approach that tackles this very problem.

9.6 From Data to Object: the Inverse Problem

In this section,[12] we discuss the procedures for obtaining the *best* possible estimate of the observed object, using all the information available about the instrument used for the observation and about sources of noise intrinsic to the instrument or to the received signal, but also the information we possess concerning the source itself, which is generally not totally unknown. We use the term *data* to refer to the complete set of measured values resulting from the observation process. By taking into account all this information in a suitable way, not only is the object better reconstructed by data processing, but error bars affecting this reconstruction can be estimated. Figure 9.13 illustrates these procedures schematically.

The developments and tools described in the following can be applied to a very broad range of data types encountered in astrophysical observation: images with any format and at all wavelengths, spectra, polarisation rates as a function of frequency, and so on. However, in order to make the following more concrete and to illustrate it with examples, we have chosen to discuss only wavefront sensing and *image processing*, and to develop examples taken from the areas of adaptive optics, which began in the 1990s (see Sect. 6.3), and interferometry (see Sect. 6.4). Indeed, these two areas have considerably stimulated such processing techniques. Readers will then be able to adapt the ideas discussed here to other cases they may encounter in radioastronomy, high energy observation, etc.

[12]This section was entirely contributed by Laurent Mugnier.

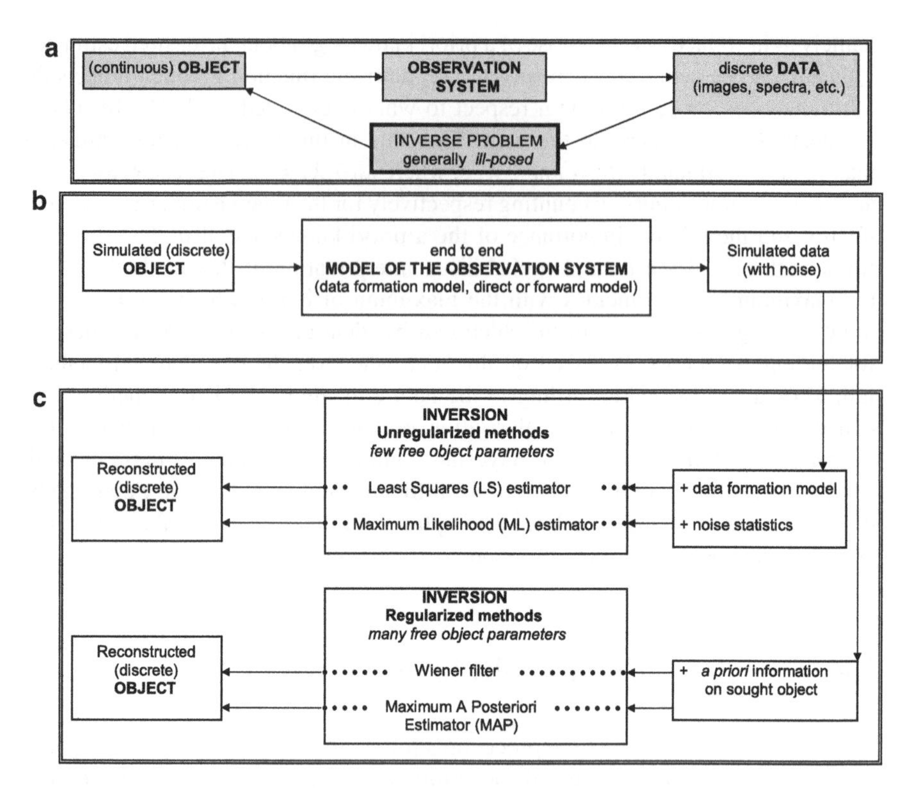

Fig. 9.13 Schematic view of the inverse problem and of the various solutions discussed in this section

9.6.1 Posing the Problem

Forward Problem

Consider the following classic problem in physics. An initial object such as an electromagnetic wave propagates through a system that will affect it in some way, e.g., atmospheric turbulence, an instrument involving aberrations and diffraction, etc. We can deduce the properties of the output wave if we know all the properties of the system, such as would be included in a *forward* or *direct model*, or *data formation model* for this system. This is the basic principle of *end-to-end models*, discussed in Sect. 9.2.

This can be illustrated by wavefront sensing as carried out in adaptive optics, i.e., the measurement of aberrations intrinsic to the telescope or due to the passage through turbulent atmospheric layers, using a (Hartmann–Shack) wavefront sensor. The relevant physical quantity here is the phase $\varphi(\boldsymbol{r}, t)$ on the exit pupil of the telescope, which contains all the information regarding the aberrations suffered by the wave. The measurement data are the average slopes of the phase in two perpendicular directions, viz., $\partial\phi(\boldsymbol{r}, t)/\partial x$ and $\partial\phi(\boldsymbol{r}, t)/\partial y$, on each Hartmann–Shack subaperture, which can be recorded together in a vector $\boldsymbol{i}$. The calculation of the slopes $\boldsymbol{i}$ given the phase φ is a classic *forward problem* in physics. It requires a choice of data formation model.

An end-to-end forward model includes models for data formation right up to detection, and even up to data storage when data are transmitted with compression. It thus takes into account photon noise, detector readout noise, digitiser quantisation noise, and data compression noise, if any (see Sect. 9.2).

Inverse Problem

The problem here is work back up from the data to knowledge of the object that originated the data. This is called the *inverse problem*. It involves inverting the data formation model. Our senses and our brains solve such inverse problems all the time, analysing the information received, e.g., by the retina, and deducing positions in space, 3D shapes, and so on. This is also the most common situation for the observer in astronomy, who makes deductions from observations and attempts to conclude as accurately as possible about some set of properties of the source.

In physics, and especially in astronomy, the processing of experimental data thus consists in solving an inverse problem, in practice, after a data reduction or preprocessing stage which aims to correct for instrumental defects in such a way that the data can be correctly described by the chosen forward model (see Sect. 9.4).

Estimating (see Sect. 9.5) or working back up to the phase φ from the slope vector $\boldsymbol{i}$ is the *inverse problem* of Hartmann–Shack wavefront sensing. It involves inverting the corresponding data formation model.

Let us consider another example, taken from another spectral region. The γ ray astronomy mission INTEGRAL (see Sect. 5.2.5) has three instruments capable of carrying out spectral analysis, namely, JEM-X, IBIS ISGRI+PICSIT, and SPI. Each of these can measure the spectrum of the observed source, here the black hole binary system Cygnus X-1. The data formation model is contained in a software called XSPEC, and an a priori model of the source containing a certain number of free parameters is injected into this. The simulated output of this software is then compared with measurements from each of the instruments, and the free parameters are fitted in the best possible way, in a sense to be made precise shortly, as shown in Fig. 9.14.

We shall see below that inversion can often take advantage of statistical knowledge regarding measurement error, generally modelled as *noise*. Naive inversion methods are often characterised by *instability*, in the sense that the inevitable measurement noise is amplified in an uncontrolled way during inversion, leading to an inacceptable solution. In this case, where the data alone are insufficient to obtain an acceptable solution, more sophisticated inversion methods, called *regularised inversion methods*, are brought to bear. They incorporate further constraints to impose a certain regularity on the solution, compatible with what is known *a priori* about it. This is a key point: in data processing, we introduce supplementary

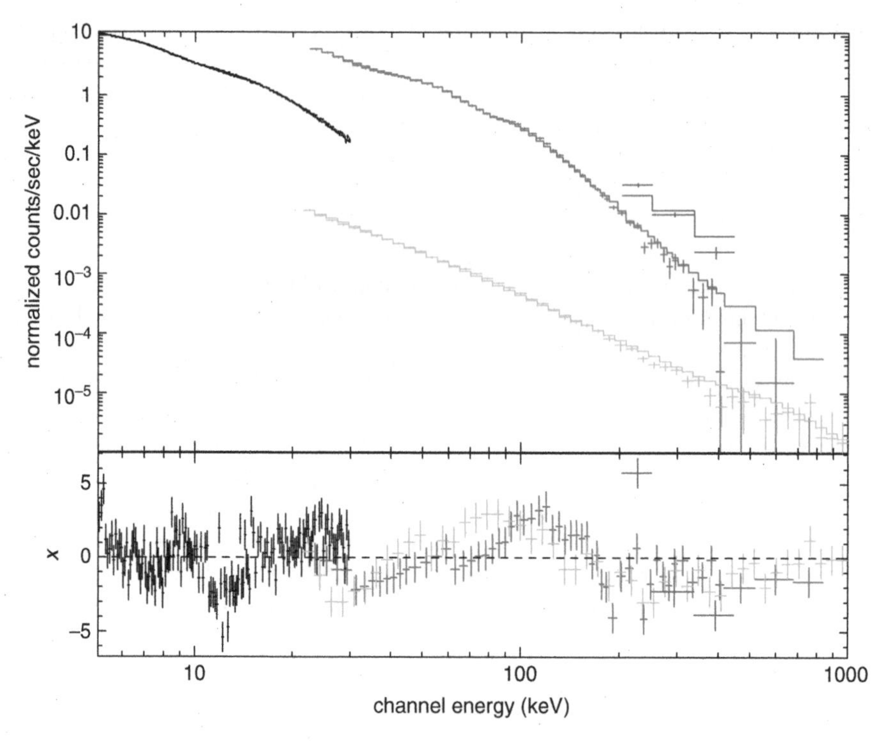

Fig. 9.14 Spectrum of the accreting binary system Cygnus X-1, viewed by the various instruments of the γ-ray observatory INTEGRAL. The model fitted to the data is a hybrid model, in which electrons in the corona have a non-thermal distribution. From Cadolle-Bel M. et al., Soc. Fr. Astron. Astroph., *Scientific Highlights* 2006, p. 191

information as dictated by our prior knowledge of the object. For example, a brightness can only take positive values, an object such as a stellar disk is bounded spatially by a contour, a temporal fluctuation cannot contain frequencies above a certain threshold, and so on.

It is extremely productive to conceive of data processing explicitly as the inversion of a forward problem. To begin with, this forces us to model the whole data formation process, so that each step can be taken into account in the inversion. It also allows us to analyse existing methods, e.g., a typical deconvolution, or filtering, as discussed in Sect. 9.1.3, and to clearly identify the underlying assumptions, or the defects. It is then possible to conceive of methods that take advantage of knowledge concerning the data formation process as well as knowledge of the source or the relevant physical quantity, available *a priori*, i.e., before the measurements are made.

In the following, we discuss the basic notions and tools required to tackle the resolution of an inverse problem, a subject that has been deeply transformed by the advent of powerful computational tools. For the purposes of this textbook,[13] the tools discussed are illustrated on several relatively simple, i.e., linear, inverse problems encountered in astronomy: wavefront reconstruction from measurements by a Hartmann–Shack sensor, adaptive optics corrected image restoration and multispectral image reconstruction in nulling interferometry (see Sect. 6.6). We shall not discuss extensions to non-linear problems, which do exist, however.

9.6.2 Well-Posed Problems

Let the relevant physical quantity be o, referred to as the *object* in what follows. It could be the phase of the wave in the case of a wavefront sensor, a star or galaxy in the case of an image in the focal plane of a telescope, the spectrum of a quasar in the case of a spectrograph, etc. Let i be the *measured data.*[14] This could be the slopes measured by the Hartmann–Shack sensor, an image, a spectrum, etc. We shall consider for the moment that o and i are continuous quantities, i.e., functions of space, time, or both, belonging to Hilbert spaces denoted by X and Y.

The forward model, deduced from physics and from the known properties of the instrument, can be used to calculate a model of the data in the case of observation of a known object. This is what is done in a *data simulation* operation:

$$i = H(o). \tag{9.70}$$

We shall restrict attention here to *linear* forward models, whence

$$i = Ho, \tag{9.71}$$

where H is a continuous linear operator mapping X into Y. It was in this context that Hadamard[15] introduced the concept of *well-posed problems*.

When the forward model is linear and translation invariant, e.g., for imaging by a telescope within the isoplanatic patch (see Sect. 6.2), H is a convolution operator and there is a function h called

---

[13] A good reference for any reader interested in inverse problems is Titterington, D.M. General structure of regularization procedures in image reconstruction, Astron. Astrophys. **144**, 381–387, 1985. See also the extremely complete didactic account by Idier, Bayesian approach to inverse problems, ISTE/John Wiley, London, 2008, which has partly inspired the present introduction.

[14] The notation in this section is as follows: the object o and image i are in italics when they are continuous functions, and in bold type when discretised. (In Chap. 6, the capital letters O and I were used.) Matrices and operators are given as italic capitals like M.

[15] Jacques Hadamard (1865–1963) was a French mathematician who contributed to number theory and the study of well-posed problems.

the point spread function (*PSF*) such that

$$i = Ho = h \star o. \tag{9.72}$$

We seek to invert (9.71), i.e., to find o for a given i. We say that the problem is *well posed* in the sense of Hadamard if the solution o satisfies the usual conditions for existence and uniqueness, but also the less well known condition of *stability*, i.e., the solution depends continuously on the data i. In other words, a small change in the data — in practice, another realisation of the random noise — only brings about a small change in the solution. These three conditions, known as Hadamard's conditions, are expressed mathematically in the following way:

- **Existence.** There exists $o \in X$ such that $i = Ho$, i.e., $i \in \mathrm{Im}(H)$, the image or range of H.
- **Uniqueness.** The kernel of H contains only zero, i.e., $\mathrm{Ker}(H) = \{0\}$.
- **Stability.** The inverse of H on $\mathrm{Im}(H)$ is continuous.

Note that for the stability condition the inverse of H on $\mathrm{Im}(H)$, denoted by H^{-1}, is well defined, because we assume that $\mathrm{Ker}(H) = \{0\}$. Note also that the stability condition is equivalent to requiring the set $\mathrm{Im}(H)$ to be closed, i.e., $\mathrm{Im}(H) = \overline{\mathrm{Im}(H)}$.

For many inverse problems, even the first two of these conditions are not satisfied, let alone the third. Indeed, on the one hand $\mathrm{Im}(H)$ is the set of possible images when there is no noise, a smaller set than the space of noisy data to which i belongs. For example, in imaging, $\mathrm{Im}(H)$ is a vector space that contains no frequency greater than the optical cutoff frequency of the telescope (D/λ), whereas the noise will contain such frequencies. In general, the existence of a solution is therefore not guaranteed. On the other hand, the kernel of H contains all objects *unseen* by the instrument. So for a Hartmann–Shack sensor, these are the unseen spatial modes, e.g., described in terms of Zernike polynomials, such as the piston mode (Z_1) or the so-called waffle mode. For an imager, these are the spatial frequencies of the object above the optical cutoff frequency of the telescope. The kernel is therefore generally not just $\{0\}$ and uniqueness is not granted.

The mathematician Nashed introduced the idea of a well-posed problem in the sense of least squares, which provides a way to ensure existence (in practice) and uniqueness of the solution and then to show that inversion in the least squares sense does not lead to a good solution of the inverse problem owing to its instability, also called *non-robustness* to noise.[16] It thus remains ill-posed.

We say that $\hat{o}_{\mathrm{LS}}$ is a least-squares solution to the problem (9.71) if

$$\|i - H\hat{o}_{\mathrm{LS}}\|^2 = \inf_o \|i - Ho\|^2, \tag{9.73}$$

where $\| \ \|$ is the Euclidean or L_2 norm. Nashed then showed the following:

[16] The term *robustness to noise*, often used when discussing inversion methods, means that the given method does not amplify the noise power in an exaggerated way.

- **Existence.** A least-squares solution exists if and only if $i \in \mathrm{Im}(H) + \mathrm{Im}(H)^{\perp}$. This condition is always satisfied if $\mathrm{Im}(H)$ is closed.
- **Uniqueness.** If several solutions exist, we choose the unique solution with minimal norm, i.e., we project the solution on to the orthogonal complement $\mathrm{Ker}(H)^{\perp}$ of the kernel. We denote this by $H^{\dagger} i$, and call it the *generalised inverse*. The operator $H^{\dagger}$ associates the least-squares solution of $i = Ho$ of minimal norm, i.e., the only solution in $\mathrm{Ker}(H)^{\perp}$, to all $i \in \mathrm{Im}(H) + \mathrm{Im}(H)^{\perp}$.

Moreover, it can be shown that $H^{\dagger}$ is continuous if and only if $\mathrm{Im}(H)$ is closed.

We say that the problem (9.71) is well-posed in the least-squares sense if there exists a unique least-squares solution (with minimal norm) and it is stable, i.e., it depends continuously on the data i. We then see that the problem (9.71) is indeed well posed in the sense of least squares if and only if $H^{\dagger}$ is continuous, i.e., if and only if $\mathrm{Im}(H)$ is closed. And for many operators, e.g., when H is the convolution with a square-integrable response h, this condition is not satisfied.

We can understand the meaning of such a solution intuitively by characterising the least-squares solutions. Let H be a continuous linear operator from a Hilbert space X to the Hilbert space Y. Then the following three propositions are equivalent:

1. $\|i - H\hat{o}_{\mathrm{LS}}\|^2 = \inf_{o \in X} \|i - Ho\|^2$.
2. $H^* H\hat{o}_{\mathrm{LS}} = H^* i$, where H^* is the adjoint of H (normal equation).
3. $H\hat{o}_{\mathrm{LS}} = Pi$, where P is the (orthogonal) projection operator of Y on $\overline{\mathrm{Im}(H)}$.

In particular, characterisation (3) tells us that the least-squares solution *exactly* solves the original equation (9.71) when the data i are projected onto the (closure of the) set of all possible data in the absence of noise, i.e., $\overline{\mathrm{Im}(H)}$.

In finite dimensions, i.e., for all practical (discretised) problems, a vector subspace is always closed. As a consequence, we can be sure of both the existence and the continuity of the generalised inverse. However, the ill-posed nature of the continuous problem does not disappear by discretisation. It simply looks different: the mathematical instability of the continuous problem, reflected by the non-continuity of the generalised inverse in infinite dimensions, resurfaces as a numerical instability of the discretised problem. The discretised inverse problem in finite dimensions is *ill-conditioned*, as we shall explain shortly. The conditioning of a discretised inverse problem characterises the robustness to noise during inversion. It is related to the dynamic range of the eigenvalues of $H^* H$ (a matrix in finite dimensions), and worsens as the dynamic range increases.

9.6.3 Conventional Inversion Methods

In the following, we shall assume that the data, which have been digitised (see Sect. 9.1.3), are discrete, finite in number, and gathered together into a vector $\boldsymbol{i}$. In imaging, for an image of size $N \times N$, $\boldsymbol{i}$ is a vector of dimension N^2 which concatenates the rows or the columns of the image (in a conceptual rather than computational sense).

The first step in solving the inverse problem is to discretise also the sought object $\boldsymbol{o}$ by expanding it on to a finite basis, that is, a basis of pixels or cardinal sine functions for an image, or a basis of Zernike polynomials for a phase representing aberrations. The model relating $\boldsymbol{o}$ and $\boldsymbol{i}$ is thus an approximation to the continuous forward model of (9.70) or (9.71). By explicitly incorporating the measurement errors in the form of additive noise $\boldsymbol{n}$ (a vector whose components are random variables), this can be written

$$\boldsymbol{i} = H(\boldsymbol{o}) + \boldsymbol{n}, \tag{9.74}$$

in the general case, and

$$\boldsymbol{i} = H\boldsymbol{o} + \boldsymbol{n}, \tag{9.75}$$

in the linear case, where H is a matrix. In the special case where H represents a discrete convolution, the forward model can be written

$$\boldsymbol{i} = \boldsymbol{h} \star \boldsymbol{o} + \boldsymbol{n}, \tag{9.76}$$

where $\boldsymbol{h}$ is the PSF of the system and $\star$ denotes the discrete convolution.

Note that, in the case of photon noise, the noise is not additive in the sense that it depends on non-noisy data $H\boldsymbol{o}$. Equation (9.74) then abuses notation somewhat.[17]

Least-Squares Method

The most widely used method for estimating the parameters $\boldsymbol{o}$ from the data $\boldsymbol{i}$ is the least-squares method. The idea is to look for $\hat{\boldsymbol{o}}_{\mathrm{LS}}$ which minimises the mean squared deviation between the data $\boldsymbol{i}$ and the data model $H(\boldsymbol{o})$:

$$\hat{\boldsymbol{o}}_{\mathrm{LS}} = \arg\min_{\boldsymbol{o}} \|\boldsymbol{i} - H(\boldsymbol{o})\|^2, \tag{9.77}$$

where $\arg\min$ is the argument of the minimum and $\|\ \|$ is the Euclidean norm. This method was first published by Legendre[18] in 1805, and was very likely discovered by Gauss[19] a few years earlier but without publishing. Legendre used the

[17] Some authors write $\boldsymbol{i} = H\boldsymbol{o} \diamond \boldsymbol{n}$ for the noisy forward model to indicate a noise contamination operation which may depend, at each data value, on the value of the non-noisy data.

[18] Adrien-Marie Legendre (1752–1833) was a French mathematician who made important contributions to statistics, algebra, and analysis.

[19] Carl Gauss (1777–1855), sometimes called *Princeps mathematicorum* or the prince of mathematicians, was a German astronomer and physicist who made deep contributions to a very broad range of areas.

least-squares method to estimate the ellipticity of the Earth from arc measurements, with a view to defining the metre.

When the measurement model is linear and given by (9.75), the solution is analytic and can be obtained by setting the gradient of the criterion (9.77) equal to zero:

$$H^{\mathrm{T}} H \hat{\boldsymbol{o}}_{\mathrm{LS}} = H^{\mathrm{T}} \boldsymbol{i} . \tag{9.78}$$

If $H^{\mathrm{T}} H$ can be inverted, i.e., if the rank of the matrix H is equal to the dimension of the unknown vector $\boldsymbol{o}$, then the solution is unique and given by

$$\hat{\boldsymbol{o}}_{\mathrm{LS}} = \left(H^{\mathrm{T}} H\right)^{-1} H^{\mathrm{T}} \boldsymbol{i} . \tag{9.79}$$

Otherwise, as in infinite dimension (see Sect. 9.6.2), there are infinitely many solutions, but only one of them has minimal norm (or 'energy'). This is the *generalised inverse*, written $H^{\dagger} \boldsymbol{i}$.

Relation Between Least Squares and Inverse Filter

When the image formation process can be modelled by a convolution, the translation invariance of the imaging leads to a particular structure of the matrix H. This structure is approximately the structure of a circulant matrix (for a 1D convolution), or block circulant with circulant blocks (for a 2D convolution). In this approximation, which amounts to making the PSF $\boldsymbol{h}$ periodic, the matrix H is diagonalised by discrete Fourier transform (DFT), which can be calculated by an FFT algorithm, and its eigenvalues are the values of the transfer function $\tilde{\boldsymbol{h}}$ (defined as the DFT of $\boldsymbol{h}$). The minimal norm least-squares solution of the last section can then be written in the discrete Fourier domain:[20]

$$\tilde{\hat{o}}_{\mathrm{LS}}(\boldsymbol{u}) = \frac{\tilde{\boldsymbol{h}}^*(\boldsymbol{u}) \tilde{\boldsymbol{i}}(\boldsymbol{u})}{|\tilde{\boldsymbol{h}}(\boldsymbol{u})|^2} = \frac{\tilde{\boldsymbol{i}}}{\tilde{\boldsymbol{h}}}(\boldsymbol{u}) \qquad \forall\, \boldsymbol{u} \text{ such that } \tilde{\boldsymbol{h}}(\boldsymbol{u}) \neq 0, \quad \text{and } 0 \text{ if } \tilde{\boldsymbol{h}}(\boldsymbol{u}) = 0, \tag{9.80}$$

where the tilde represents the discrete Fourier transform. In the case of a convolutive data model, the least-squares solution is thus identical to the *inverse filter*, up to the above-mentioned approximation.

Maximum Likelihood Method

In the least-squares method, the choice of a quadratic measure of the deviation between the data $\boldsymbol{i}$ and the data model $H(\boldsymbol{o})$ is only justified by the fact that the

---

[20]The transfer function $\tilde{\boldsymbol{h}}^*$ in the discrete Fourier domain corresponds to the matrix H^{T} and the matrix inverses give simple inverses.

solution can then be obtained by analytical calculation. Furthermore, this method makes no use whatever of any knowledge one may possess about the statistical properties of the noise. But this information about the noise can be used to interpret the least-squares method, and more importantly to extend it.

We model the measurement error $\boldsymbol{n}$ as noise with probability distribution $p_{\rm n}(\boldsymbol{n})$. According to (9.74), the distribution of the data $\boldsymbol{i}$ conditional to the object, i.e., for a given object $\boldsymbol{o}$ (hence supposed known), is then[21]

$$p(\boldsymbol{i}\,|\boldsymbol{o}) = p_{\rm n}\big(\boldsymbol{i} - H(\boldsymbol{o})\big). \tag{9.81}$$

Equation (9.81) can be used to draw realisations of noisy data knowing the object, i.e., to *simulate* data. On the other hand, in an inverse problem, one has only one realisation of the data, namely those actually measured, and the aim is to estimate the object. The maximum likelihood (ML) method consists precisely in reversing the point of view on $p(\boldsymbol{i}\,|\boldsymbol{o})$ by treating $\boldsymbol{o}$ as variable, with $\boldsymbol{i}$ fixed equal to the data, and seeking the object $\boldsymbol{o}$ that maximises $p(\boldsymbol{i}\,|\boldsymbol{o})$. The quantity $p(\boldsymbol{i}\,|\boldsymbol{o})$ viewed as a function of $\boldsymbol{o}$ is then called the *likelihood* of the data, and the object $\hat{\boldsymbol{o}}_{\rm ML}$ which maximises it is the one which makes the actually observed data the most likely:[22]

$$\hat{\boldsymbol{o}}_{\rm ML} = \arg\max_{\boldsymbol{o}} p(\boldsymbol{i}\,|\boldsymbol{o}). \tag{9.82}$$

The most widely used model for noise is without doubt the centered (i.e. zero mean) Gaussian model, characterised by its covariance matrix $C_{\rm n}$:

$$p(i\,|o) \propto \exp\left\{-\frac{1}{2}\big[i - H(o)\big]^{\rm T} C_{\rm n}^{-1}\big[i - H(o)\big]\right\}. \tag{9.83}$$

The noise is called *white noise* (see Appendix B) if its covariance matrix is diagonal. If this matrix is also proportional to the identity matrix, then the noise is called *stationary* or *homogeneous* (*white noise*). The readout noise of a CCD detector (see Sect. 7.4.6) is often modelled by such a stationary centered Gaussian noise. Photon noise is white, but it has a Poisson distribution (see Appendix B), which can be shown, for high fluxes, to tend towards a non-stationary Gaussian distribution with variance equal to the signal detected on each pixel.

Maximising the likelihood is obviously equivalent to minimising a criterion defined to be the negative of the logarithm of the likelihood and called the *neg-log-likelihood*:

$$J_i(o) = -\ln p(i\,|o). \tag{9.84}$$

In the case of Gaussian noise, the neg-log-likelihood is

$$J_i(o) = \frac{1}{2}\big[i - H(o)\big]^{\rm T} C_{\rm n}^{-1}\big[i - H(o)\big]. \tag{9.85}$$

---

[21]The equation reads: the probability of $\boldsymbol{i}$ given $\boldsymbol{o}$ is the noise probability distribution at $\boldsymbol{i} - H(\boldsymbol{o})$.

[22]The notation arg max refers to the *argument of the maximum*, i.e., the value of the variable which maximises the ensuing quantity, in this case $p(\boldsymbol{i}\,|\boldsymbol{o})$.

If the noise is also white, we have

$$J_i(\boldsymbol{o}) = \frac{1}{2}\sum_k \frac{|\boldsymbol{i}(k) - H(\boldsymbol{o})(k)|^2}{\sigma_n^2(k)}, \tag{9.86}$$

where $\sigma_n^2(k)$ are the elements on the diagonal of the matrix C_n. $J_i(\boldsymbol{o})$ is a *weighted least squares criterion*. If the noise is also stationary with variance σ_n^2, then $J_i(\boldsymbol{o}) = \|\boldsymbol{i} - H(\boldsymbol{o})\|^2/2\sigma_n^2$ is precisely the *ordinary*, as opposed to weighted, least squares criterion.

The least squares method can thus be interpreted as a maximum likelihood method in the case of stationary white Gaussian noise. Conversely, if the noise distribution is known but different, the maximum likelihood method can take this knowledge of the noise into account and then generalises the least squares method.

Example: Wavefront Reconstruction by the Maximum Likelihood Method

Consider a Hartmann–Shack wavefront sensor that measures aberrations due to atmospheric turbulence on a ground-based telescope. The phase at the exit pupil is expanded on a basis of Zernike polynomials Z_k, the degree of which is necessarily limited in practice to some maximal value $k_{\max}$:

$$\varphi(x, y) = \sum_{k=1}^{k_{\max}} o_k Z_k(x, y). \tag{9.87}$$

We thus seek $\boldsymbol{o}$, the set of coefficients o_k of this expansion, by measuring the average slopes and inserting them into a measurement vector $\boldsymbol{i}$. The data model is precisely (9.75), viz., $\boldsymbol{i} = H\boldsymbol{o} + \boldsymbol{n}$, where H is basically the differentiation operator and is known as the interaction matrix.

In the simulation presented below, the sensor contains 20×20 subapertures of which only 276 receive light, owing to a central obstruction of 33% by the secondary mirror. This gives 552 slope measurements. The true turbulent phase φ is a linear combination of $k_{\max} = 861$ Zernike polynomials, drawn from a Kolmogorov distribution. The matrix H thus has dimensions 552×861. The noise on the slopes is stationary white Gaussian noise with variance equal to the variance of the non-noisy slopes, implying a signal-to-noise ratio of 1 on each measurement. Under such conditions, the maximum likelihood estimate of the phase is identical to the least squares solution (see last subsection). The matrix $H^{\mathrm{T}}H$ in (9.79) has dimensions $k_{\max} \times k_{\max}$ and cannot be inverted because we only have 552 measurements. The generalised inverse solution cannot be used because it is completely dominated by noise. A remedy often adopted here is to reduce the dimension $k_{\max}^{\mathrm{rec}}$ of the space of unknowns $\boldsymbol{o}$ during reconstruction of the wavefront. This remedy is one of the known *regularisation* methods, referred to as *regularisation by dimension control*. An example of reconstruction for different values of $k_{\max}^{\mathrm{rec}}$ is shown in Fig. 9.15.

For $k_{\max}^{\mathrm{rec}} = 210$, a value well below the number of measurements, the reconstructed phase is already unacceptable. The particular shape is due to the fact that the telescope has a central obstruction, whence there are no data at the centre of the pupil.

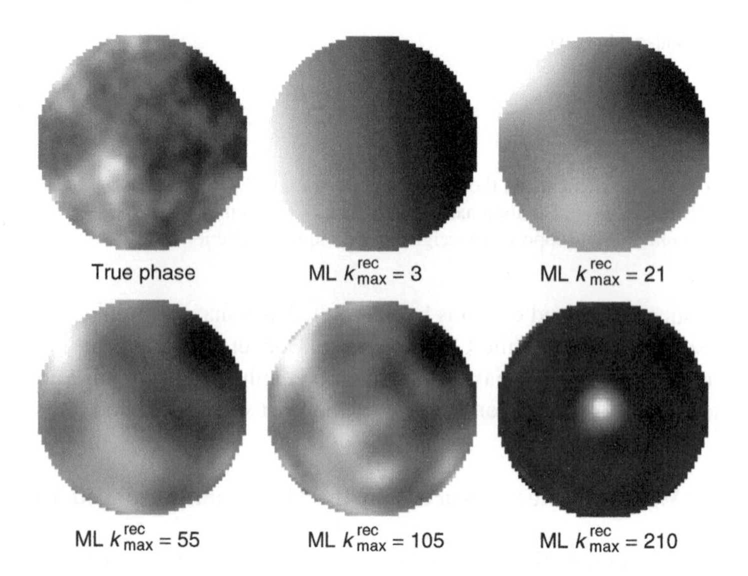

Fig. 9.15 Wavefront sensing on a pupil. *Top left*: Simulated phase (called *true phase*). *Other figures*: Phases reconstructed by maximum likelihood (ML) by varying the number of reconstructed modes

This example clearly illustrates the case where, although the matrix $H^{\mathrm{T}}H$ can be inverted, it is ill-conditioned, i.e., it has eigenvalues very close to 0, leading to uncontrolled noise amplification.

Truncating the solution space to lower values of $k^{\mathrm{rec}}_{\mathrm{max}}$ produces more reasonable solutions, but on the one hand the optimal choice of $k^{\mathrm{rec}}_{\mathrm{max}}$ depends on the levels of turbulence and noise and this adjustment is difficult to achieve in practice, while on the other hand this truncation introduces a modelling error, since it amounts to neglecting all the components of the turbulent phase beyond $k^{\mathrm{rec}}_{\mathrm{max}}$. We shall see below how the introduction of a priori knowledge regarding the spatial regularity of the phase (Kolmogorov spectrum) can lead to a better solution without this *ad hoc* reduction of the space of unknowns.

Interpreting the Failure of Non-Regularised Methods

The failure of the maximum likelihood method illustrated in the above example may seem surprising given the good statistical properties of this method, which is an estimator that converges towards the true value of the parameters when the number of data tends to infinity, which is asymptotically efficient, etc. However, these good properties are only asymptotic, i.e., they only concern situations where there is good *statistical contrast*, defined simply as the ratio of the number of measurements to the number of unknowns.

In practice, and in particular in the previous example, the problem is often to estimate a number of parameters of the same order of magnitude as the number of measurements, and sometimes greater, in which case these asymptotic properties are of little help. In this commonly encountered situation of unfavourable statistical contrast, the inversion is unstable, i.e., highly sensitive to noise, which can often be interpreted as arising due to the ill-posed nature of the underlying infinite dimensional problem.

Image restoration is another typical case of this situation. Obviously, for an image containing $N \times N$ pixels, we try to reconstruct an object of the same size, and if we increase the image size, we also increase the number of unknown parameters relating to the object, without improving the statistical contrast, which remains of order unity. We have seen that the least-squares solution was in this case given by the inverse filter, and we know that this is highly unstable with regard to noise. This instability is easily understood by reinserting the measurement equation (9.76), in the Fourier domain, into the solution (9.80):

$$\tilde{\hat{o}}_{\mathrm{LS}} = \frac{\tilde{\boldsymbol{i}}}{\tilde{\boldsymbol{h}}} = \tilde{o} + \frac{\tilde{\boldsymbol{m}}}{\tilde{\boldsymbol{h}}}. \tag{9.88}$$

According to this, it is clear that the noise is highly amplified for all frequencies for which the value of the transfer function $\tilde{\boldsymbol{h}}$ is close to zero. One way to reduce noise amplification would be to modify (9.80) and divide $\tilde{\boldsymbol{i}}$ by $\tilde{\boldsymbol{h}}$ only for frequencies $\boldsymbol{u}$ where the transfer function $\tilde{\boldsymbol{h}}$ is not too small. This is also a form of regularisation by controlling the dimension of the solution, very similar to the choice of $k_{\max}^{\mathrm{rec}}$ when reconstructing a wavefront, and it suffers from the same ad hoc character and the same difficulty in its tuning.

To sum up, simple inversion methods like least squares or maximum likelihood only give satisfactory results if there is good statistical contrast. For example, maximum likelihood can be successfully applied to problems like image registration, where we seek a 2D vector in order to register two images comprising a large number of pixels. More generally, it applies to the search for the few variables of a parsimonious parametric model from a large enough data set, e.g., estimating a star's diameter from visibilities in optical interferometry, estimating the Earth's ellipticity from arc measurements, etc.

In many problems where the statistical contrast is not favourable, the problem is ill-conditioned and regularisation, i.e., the addition during inversion of a priori knowledge and constraints on the solution, becomes highly profitable, as we shall now show.

9.6.4 *Inversion Methods with Regularisation*

Regularisation of an inverse problem corresponds to the idea that the data alone cannot lead to an acceptable solution, so that a priori information about the regularity of the considered observed object must necessarily be introduced. Here the term *regularity* implies that, for physical reasons intrinsic to the object, it must have certain properties, or obey certain rules regarding sign, size, or frequency, for example. The solution then results from a compromise between the requirements of the object's postulated regularity and of the data fidelity.

Indeed, several very different solutions, some very poor and some rather good, may be compatible with the data. For instance, in the previous example of wavefront reconstruction, the true and reconstructed phases for $k_{\max}^{\text{rec}} = 210$ (Fig. 9.15) give very similar likelihood values, whence they are both faithful to the data. In addition, the *smoother*, more regular solution obtained for $k_{\max}^{\text{rec}} = 55$ is less well fitted, i.e., less faithful to the data than the one obtained for $k_{\max}^{\text{rec}} = 210$, since it was obtained by optimising in fewer degrees of freedom, and yet it is much closer to the true phase.

Bayesian Estimation and the Maximum a Posteriori (MAP) Estimator

Bayesian estimation, presented briefly here, provides a natural way of combining the information brought by measurements and information available a priori. Let us assume that we have expressed our a priori knowledge of the observed object in the form of a probability distribution $p(\boldsymbol{o})$, called the a priori distribution. The idea is not to assume that the observed object *really* is the realisation of a random phenomenon with distribution $p(\boldsymbol{o})$. This distribution is simply taken to be representative of our a priori knowledge, i.e., it takes small values for reconstructed objects that would be barely compatible with the latter and larger values for highly compatible objects.

The Bayes rule[23] provides a way of expressing the probability $p(\boldsymbol{o} \mid \boldsymbol{i})$ for the object $\boldsymbol{o}$ that is conditional upon the measurement data $\boldsymbol{i}$ as a function of the *prior probability distribution* $p(\boldsymbol{o})$ of the object and the probability $p(\boldsymbol{i} \mid \boldsymbol{o})$ of the measurements conditional upon the object. The latter contains our knowledge of the data formation model, including the noise model. This probability $p(\boldsymbol{o} \mid \boldsymbol{i})$ is called the *posterior distribution*, since it is the probability of the object given the results of the measurements. The Bayes rule is (see Sect. 9.5)

$$p(\boldsymbol{o} \mid \boldsymbol{i}) = \frac{p(\boldsymbol{i} \mid \boldsymbol{o}) \times p(\boldsymbol{o})}{p(\boldsymbol{i})} \propto p(\boldsymbol{i} \mid \boldsymbol{o}) \times p(\boldsymbol{o}). \tag{9.89}$$

Using the last expression, how should we choose the reconstructed object $\hat{\boldsymbol{o}}$ that would best estimate the true object? One commonly made choice is the maximum a posteriori (MAP) estimator. The idea is to define as solution the object that maximises the posterior distribution $p(\boldsymbol{o}|\boldsymbol{i})$, so that

[23]Thomas Bayes (1702–1761) was a British mathematician whose main work concerning inverse probabilities was only published after his death.

$$\boldsymbol{o}_{\mathrm{MAP}} = \arg\max_{\boldsymbol{o}} p(\boldsymbol{o} \mid \boldsymbol{i}) = \arg\max_{\boldsymbol{o}} p(\boldsymbol{i} \mid \boldsymbol{o}) \times p(\boldsymbol{o}). \tag{9.90}$$

This is the most likely object given the data and our prior knowledge. The posterior distribution, viewed as a function the object, is called the *posterior likelihood*. Through (9.89), it takes into account the measurement data $\boldsymbol{i}$, the data formation model, and any prior knowledge of the object. In particular, it should be noted that, when the prior distribution $p(\boldsymbol{o})$ is constant, i.e., when we have no information about the object, the MAP estimator reduces to the maximum likelihood method.

It can be shown that choosing the MAP estimator amounts to minimising the mean risk (see Sect. 9.5) for a particular cost function called the *all-or-nothing cost function*. However, this goes beyond the scope of this introduction to inverse problems. Other choices of cost function, such as a choice of $\hat{\boldsymbol{o}}$ minimising the mean squared error with respect to the true object $\boldsymbol{o}$ under the posterior probability distribution, can also be envisaged, but they generally lead to longer computation times for obtaining the solution.[24]

Equivalence with Minimisation of a Regularised Criterion

Maximising the posterior likelihood is equivalent to minimising a criterion $J_{\mathrm{MAP}}(\boldsymbol{o})$ defined as minus its logarithm. According to (9.90), this criterion can be written as the sum of two terms, viz.,

$$J_{\mathrm{MAP}}(\boldsymbol{o}) = -\ln p(\boldsymbol{i} \mid \boldsymbol{o}) - \ln p(\boldsymbol{o}) = J_{\mathrm{i}}(\boldsymbol{o}) + J_{\mathrm{o}}(\boldsymbol{o}), \tag{9.91}$$

where J_{i} is the data fidelity criterion deduced from the likelihood [see (9.84)], which is often a least-squares criterion, while $J_{\mathrm{o}}(\boldsymbol{o}) \triangleq -\ln p(\boldsymbol{o})$ is a regularisation or penalisation criterion (for the likelihood) which reflects faithfulness to prior knowledge.

The expression for the MAP solution as the object that minimises the criterion (9.91) shows clearly that it achieves the compromise between faithfulness to the data and faithfulness to prior knowledge asserted at the beginning of this section.

When $\boldsymbol{o}$ is not a realisation of a random phenomenon with some probability distribution $p(\boldsymbol{o})$, e.g., in the case of image restoration, $J_{\mathrm{o}}(\boldsymbol{o})$ generally includes a multiplicative factor called the regularisation coefficient or hyperparameter.[25] Its value controls the exact position of the compromise. Unsupervised, i.e., automatic,

---

[24] A good reference for the theory of estimation is Lehmann, E. *Theory of Point Estimation*, John Wiley, 1983.

[25] The term *hyperparameter* refers to all the parameters of the data model (e.g., the noise variance) or of the a priori model (e.g., atmospheric turbulence and hence the parameter r_0) which are kept fixed during the inversion.

fitting methods exist for this coefficient, but they go beyond the scope of this introductory account.[26]

The Linear and Gaussian Case. Relation with the Wiener Filter

Here we consider the case where the data model is linear [see (9.75)], the noise is assumed to be Gaussian, and the prior probability distribution adopted for the object is also Gaussian,[27] with mean $\overline{\boldsymbol{o}}$ and covariance matrix C_{o} :

$$p(\boldsymbol{o}) \propto \exp\left[-\frac{1}{2}(\boldsymbol{o}-\overline{\boldsymbol{o}})^{\mathrm{T}} C_{\mathrm{o}}^{-1}(\boldsymbol{o}-\overline{\boldsymbol{o}})\right]. \tag{9.92}$$

Using (9.85), we see that the criterion to minimise in order to obtain the MAP solution is

$$J_{\mathrm{MAP}}(\boldsymbol{o}) = \frac{1}{2}(\boldsymbol{i}-H\boldsymbol{o})^{\mathrm{T}} C_{\mathrm{n}}^{-1}(\boldsymbol{i}-H\boldsymbol{o}) + \frac{1}{2}(\boldsymbol{o}-\overline{\boldsymbol{o}})^{\mathrm{T}} C_{\mathrm{o}}^{-1}(\boldsymbol{o}-\overline{\boldsymbol{o}}). \tag{9.93}$$

This criterion is quadratic in $\boldsymbol{o}$ and thus has analytic minimum, obtained by setting the gradient equal to zero:

$$\hat{\boldsymbol{o}}_{\mathrm{MAP}} = \left(H^{\mathrm{T}} C_{\mathrm{n}}^{-1} H + C_{\mathrm{o}}^{-1}\right)^{-1}\left(H^{\mathrm{T}} C_{\mathrm{n}}^{-1}\boldsymbol{i} + C_{\mathrm{o}}^{-1}\overline{\boldsymbol{o}}\right). \tag{9.94}$$

Several remarks can throw light upon this otherwise slightly dull looking result. To begin with, the maximum likelihood solution is obtained by taking $C_{\mathrm{o}}^{-1} = 0$ in this equation. Incidentally, this shows that this solution corresponds in the Bayesian framework to assuming an infinite energy for $\boldsymbol{o}$. Then if we also take C_{n} proportional to the identity matrix, we see that we recover precisely the least-squares solution of (9.79).

Finally, the case of deconvolution is particularly enlightening. The noise is assumed stationary with power spectral density (PSD) S_{n}, while the prior probability distribution for the object is also assumed stationary with PSD S_{o}. For all the relevant quantities, we make the same periodicity approximation as on p. 583 when examining the relation between least squares and inverse filter. All the matrices in (9.94) are then diagonalised in the same discrete Fourier basis, and the solution can be written in this basis, with ordinary rather than matrix multiplications and inverses:

---

[26]See, for example, Idier, J. op. cit, Chaps. 2 and 3.

[27]There is no assumption here that the *shape* of the object is Gaussian, but only that its *probability distribution* is Gaussian. This happens, for example, when $\boldsymbol{o}$ is a turbulent phase obeying Kolmogorov statistics (Sect. 2.6). The central limit theorem (see Appendix B) often leads to Gaussian random objects when the perturbation of the object is due to many independent causes.

$$\tilde{\hat{o}}_{\mathrm{MAP}}(\boldsymbol{u}) = \frac{\tilde{\boldsymbol{h}}^*(\boldsymbol{u})}{|\tilde{\boldsymbol{h}}|^2(\boldsymbol{u}) + S_{\mathrm{n}}/S_{\mathrm{o}}(\boldsymbol{u})}\tilde{\boldsymbol{i}}(\boldsymbol{u}) + \frac{S_{\mathrm{n}}/S_{\mathrm{o}}(\boldsymbol{u})}{|\tilde{\boldsymbol{h}}|^2(\boldsymbol{u}) + S_{\mathrm{n}}/S_{\mathrm{o}}(\boldsymbol{u})}\tilde{\bar{\boldsymbol{o}}}(\boldsymbol{u}). \tag{9.95}$$

In this expression, $S_{\mathrm{n}}/S_{\mathrm{o}}(\boldsymbol{u})$ is the reciprocal of a signal-to-noise ratio at the spatial frequency $\boldsymbol{u}$ and $\tilde{\bar{\boldsymbol{o}}}$ is the Fourier transform of the a priori object, generally taken to be zero or equal to an object of constant value.

This expression is just the *Wiener filter* (see Sect. 9.1.3) for the case where the a priori object is not zero. For frequencies where the signal-to-noise ratio is high, this solution tends toward the inverse filter, and for frequencies where this ratio is low, the solution tends toward the a priori object. It can even be seen that, at each spatial frequency $\boldsymbol{u}$, the solution is on a segment that connects the maximum likelihood solution (inverse filter) to the a priori solution, with the position on the segment given by the signal-to-noise ratio:

$$\tilde{\boldsymbol{o}}_{\mathrm{MAP}}(\boldsymbol{u}) = \alpha\frac{\tilde{\boldsymbol{i}}(\boldsymbol{u})}{\tilde{\boldsymbol{h}}} + (1-\alpha)\tilde{\bar{\boldsymbol{o}}}(\boldsymbol{u}), \tag{9.96}$$

where

$$\alpha = \frac{|\tilde{\boldsymbol{h}}|^2(\boldsymbol{u})}{|\tilde{\boldsymbol{h}}|^2(\boldsymbol{u}) + S_{\mathrm{n}}/S_{\mathrm{o}}(\boldsymbol{u})}. \tag{9.97}$$

This expression clearly illustrates the fact that regularisation achieves a compromise between bias and variance that enables us to reduce the estimation error. The mean squared error on the estimated object is the square root of the sum of the squared bias and the variance of the estimator. The solution of the inverse filter (obtained for $S_{\mathrm{n}}/S_{\mathrm{o}} = 0$) has zero bias but amplifies the noise in an uncontrolled way, i.e., has unbounded variance. Compared with the maximum likelihood solution, the solution (9.95) is biased toward the a priori solution $\bar{\boldsymbol{o}}$. By accepting this bias, we can significantly reduce the variance and globally reduce the mean squared error on the estimated object.

Application to Wavefront Reconstruction

We return to the example of wavefront reconstruction from data provided by a Hartmann–Shack sensor, as discussed in Sect. 9.6.3. The noise is still assumed to be Gaussian and white with covariance matrix $C_{\mathrm{n}} = \sigma^2 I$. We assume that the phase obeys Kolmogorov statistics and is therefore Gaussian, with known covariance matrix C_{o}, and depends only on the Fried parameter r_0 which quantifies the strength of turbulence. The true phase has spatial variance $\sigma_\varphi^2 = 3.0\ \mathrm{rad}^2$. The most likely phase given the measurements and taking into account this a priori information is the MAP solution given by (9.94). This solution is shown in Fig. 9.16. The estimation error corresponds to a spatial variance $\sigma_{\mathrm{err}}^2 = 0.7\ \mathrm{rad}^2$, which is lower than the best solutions obtained previously by truncating the representation of the phase to a small number of Zernike polynomials ($\sigma_{\mathrm{err}}^2 = 0.8\ \mathrm{rad}^2$, obtained for $k_{\mathrm{max}}^{\mathrm{rec}} = 55$).

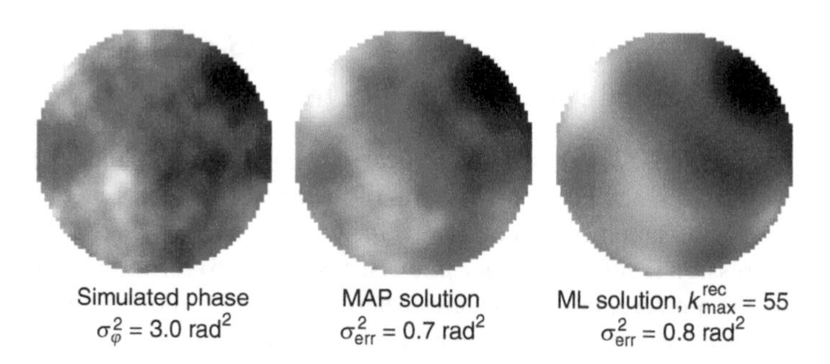

Fig. 9.16 Comparing reconstructed phases on a pupil. *Left*: Simulated turbulent phase, to be estimated. *Centre*: Phase estimated by MAP. *Right*: Best phase estimated by maximum likelihood on a truncated basis

The MAP solution takes advantage of a priori knowledge of the spatial statistics of the turbulence. To use this in adaptive optics, where the sample rate is generally well above $1/\tau_0$, it is judicious to opt for a natural extension of this estimator that also exploits a priori knowledge of the temporal statistics of the turbulence. This extension is the optimal estimator of Kalman filtering.

9.6.5 *Application to Adaptive Optics Imaging*

Here we apply these tools to solve a specific inverse problem, namely, adaptive optics imaging. We shall illustrate with several examples, either by simulation or by the processing of genuine astronomical observations.

Ingredients of the Deconvolution

Long exposure images corrected by adaptive optics (AO) must be deconvolved, because the correction is only partial. Considering that the PSF $\boldsymbol{h}$ is known, the object $\hat{\boldsymbol{o}}_{\mathrm{MAP}}$ estimated by MAP is given by (9.90), i.e., it minimises the criterion (9.91). Let us see how to define the two components of this criterion.

To be able to measure objects with high dynamic range, which are common in astronomy, the data fidelity term J_{i} must incorporate a fine model of the noise, taking into account both photon noise and electronic noise. This can be done by treating the photon noise as a non-stationary Gaussian noise, and it leads to a weighted least-squares criterion J_{i} [see (9.86)] rather than an ordinary least-squares term.

For objects with sharp edges, such as artificial satellites, asteroids, or planets, a Gaussian prior (like the one leading to the Wiener filter on p. 590), or equivalently, a quadratic regularisation criterion, tends to smooth the edges and introduce spurious oscillations or *ringing* in their vicinity. One interpretation of this effect is that, when minimising the criterion $J_{\mathrm{MAP}}(\boldsymbol{o})$, a quadratic regularisation attributes to a step a cost proportional to the square of its value, e.g., at the edge of an object, where there is a large difference in value between adjacent pixels. One solution is then to use an *edge-preserving criterion*, such as the so-called quadratic–linear or L_2L_1 criteria. These are quadratic for small discontinuities and linear for large ones. The quadratic part ensures good noise smoothing and the linear part cancels edge penalisation.

In addition, for many different reasons, we are often led to treat the PSF $\boldsymbol{h}$ as imperfectly known. Carrying out a *classic* deconvolution, i.e., assuming that the point spread function is known but using an incorrect point spread function, can lead to disastrous results. Conversely, a so-called *blind* deconvolution, where the same criterion (9.91) is minimised but simultaneously seeking $\boldsymbol{o}$ and $\boldsymbol{h}$, is highly unstable, rather like unregularised methods. A *myopic deconvolution* consists in jointly estimating both $\boldsymbol{o}$ and $\boldsymbol{h}$ in a Bayesian framework with a natural regularisation for the point spread function and without having to fit an additional hyperparameter. The joint MAP estimator is given by

$$(\hat{\boldsymbol{o}}, \hat{\boldsymbol{h}}) = \arg\max_{\boldsymbol{o},\boldsymbol{h}} p(\boldsymbol{o}, \boldsymbol{h}|\boldsymbol{i}) = \arg\max_{\boldsymbol{o},\boldsymbol{h}} p(\boldsymbol{i}\,|\boldsymbol{o}, \boldsymbol{h}) \times p(\boldsymbol{o}) \times p(\boldsymbol{h})$$

$$= \arg\min_{\boldsymbol{o},\boldsymbol{h}} \Big[J_i(\boldsymbol{o}, \boldsymbol{h}) + J_o(\boldsymbol{o}) + J_h(\boldsymbol{h}) \Big],$$

where J_h is a regularisation criterion for $\boldsymbol{h}$, which introduces constraints on the possible variability of the PSF.

The next section presents experimental results obtained by the MISTRAL restoration method,[28] which combines the three ingredients discussed above: fine noise modelling, non-quadratic regularisation, and myopic deconvolution.

Image Restoration from Experimental Astronomical Data

Figure 9.17a shows a long exposure image of Ganymede, a natural satellite of Jupiter, corrected by adaptive optics. This image was made on 28/09/1997 on ONERA's adaptive optics testbed installed on the 1.52 m telescope at the Observatoire de Haute-Provence in France. The imaging wavelength is $\lambda = 0.85\ \mu$m and the exposure time is 100 s. The estimated total flux is 8×10^7 photons and the estimated ratio D/r_0 is 23. The total field of view is 7.9 arcsec, of which only half is shown here. The mean and variability of the point spread function were estimated from recordings of 50 images of a bright star located nearby. Figures 9.17b and c show restorations obtained using the Richardson–Lucy algorithm (ML for Poisson

[28] This stands for *Myopic Iterative STep-preserving Restoration ALgorithm*. This method is described in Mugnier, L.M. et al., MISTRAL: A myopic edge-preserving image restoration method, with application to astronomical adaptive-optics-corrected long-exposure images, J. Opt. Soc. Am. A **21**, 1841–1854, 2004.

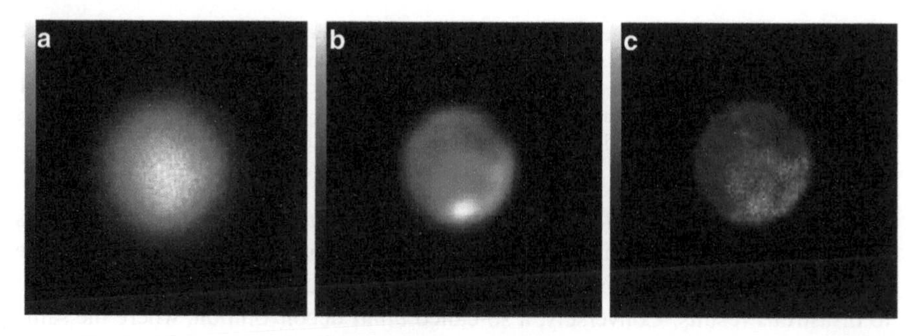

Fig. 9.17 (**a**) Corrected image of Ganymede obtained using the ONERA adaptive optics testbed on 28 September 1997. (**b**) Restoration using the Richardson–Lucy algorithm, stopped after 200 iterations. (**c**) Likewise, but stopped after 3 000 iterations. From Mugnier, L. et al., Chap. 10 of Idier, J. op. cit.

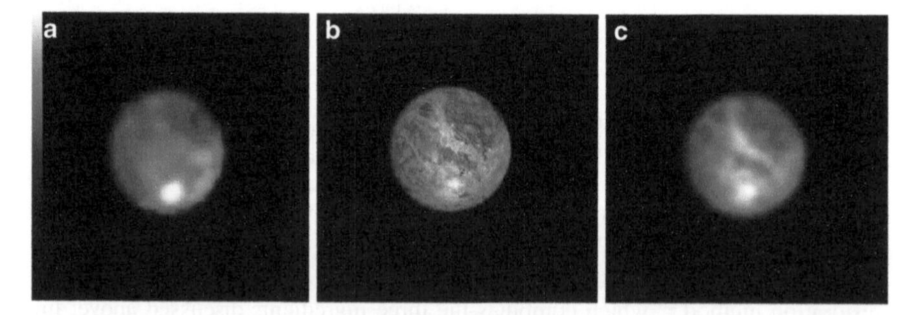

Fig. 9.18 (**a**) Deconvolution of the Ganymede image in Fig. 9.17 by MISTRAL. (**b**) Comparison with a wide band synthetic image obtained using the NASA/JPL database. (**c**) The same synthetic image, but convolved by the perfect point spread function of a 1.52 m telescope. From Mugnier, L.M. et al., MISTRAL: A myopic edge-preserving image restoration method, with application to astronomical adaptive-optics-corrected long-exposure images, J. Opt. Soc. Am. A **21**, 1841–1854, 2004

noise), stopped after 200 and 3 000 iterations, respectively.[29] In the first case, similar to restoration with quadratic regularisation, the restored image is somewhat blurred and displays ringing, while in the second, very similar to the result of inverse filtering, noise dominates the restoration.

Figure 9.18a shows a myopic deconvolution implementing an edge-preserving prior, while Fig. 9.18b is a wide band synthetic image obtained by a NASA/JPL space probe[30] during a Ganymede flyby. Comparing the two, we find that many

[29]The idea of stopping a non-regularised algorithm before convergence is still a widespread method of regularisation, but extremely ad hoc.

[30]See space.jpl.nasa.gov/ for more details.

features of the moon are correctly restored. A fairer comparison consists in jointly examining the myopic deconvolution carried out by MISTRAL with the image of Fig. 9.18b convolved by the perfect PSF of a 1.52 m telescope, as shown in Fig. 9.18c.

9.6.6 Application to Nulling Interferometry

We now discuss a second example of the inversion problem, this time relating to the detection of extrasolar planets by means of a nulling interferometer (see Sect. 6.6). With the Darwin instrument, or NASA's *Terrestrial Planet Finder Interferometer*, both under study during the 2000s, data will be very different from an image in the conventional sense of the term, and their exploitation will require implementation of a specific reconstruction process. They will consist, at each measurement time t, of an intensity in each spectral channel λ. This intensity can be modelled as the integral over a certain angular region of the instantaneous transmission map of the interferometer, denoted by $R_{t,\lambda}(\boldsymbol{\theta})$, which depends on the time t owing to rotation of the interferometer relative to the plane of the sky, multiplied by the intensity distribution $o_\lambda(\boldsymbol{\theta})$ of the observed object. The data model is thus linear, but notably non-convolutive, thus very different from the one used in imaging. The transmission map is a simple sinusoidal function in the case of a Bracewell interferometer, but becomes more complex when more than two telescopes interfere simultaneously.

By judiciously combining the data, and with asymmetrical transmission maps, the contribution to the measured signal of the components of the observed object with even spatial distribution can be eliminated. These components are stellar leakage, exozodiacal light, and a fortiori zodiacal light and thermal emission from the instrument itself (which have constant level in the field of view). It is then possible to seek out only planets during image reconstruction, which corresponds to the following object model:

$$o_\lambda(\boldsymbol{\theta}) = \sum_{k=1}^{N_{\mathrm{src}}} F_{k,\lambda}\,\delta(\boldsymbol{\theta} - \boldsymbol{\theta}_k), \tag{9.98}$$

where N_{src} is the number of planets, assumed known here, and $F_{k,\lambda}$ is the spectrum of the k th planet in a spectral interval $[\lambda_{\mathrm{min}}, \lambda_{\mathrm{max}}]$ fixed by the instrument. This parametric model can substantially constrain the inversion in so as to counterbalance the fact that the data are distinctly poorer than an image.

With this model of the object, the data formation model is

$$i_{t,\lambda} = \sum_{k=1}^{N_{\mathrm{src}}} R_{t,\lambda}(\boldsymbol{\theta}_k) F_{k,\lambda} + n_{t,\lambda}, \tag{9.99}$$

where $n_{t,\lambda}$ is the noise, assumed to be white Gaussian, whose variance $\sigma^2_{t,\lambda}$ can be estimated from the data and is assumed known here. The inverse problem to be solved is to estimate the positions $\boldsymbol{\theta}_k$ and spectra $F_{k,\lambda}$ of the planets, these being grouped together into two vectors $(\boldsymbol{\theta}, \boldsymbol{F})$. The ML solution is the one that minimises the following weighted least-squares criterion, given the assumptions made about the noise:

$$J_{\mathrm{i}}(\boldsymbol{\theta}, \boldsymbol{F}) = \sum_{t,\lambda} \frac{1}{\sigma_{t,\lambda}^2} \left[i_{t,\lambda} - \sum_{k=1}^{N_{\mathrm{src}}} R_{t,\lambda}(\boldsymbol{\theta}_k) F_{k,\lambda} \right]^2 . \tag{9.100}$$

As we shall see from the reconstruction results, the inversion remains difficult under the high noise conditions considered here. The object model (9.98), separable into spatial and spectral variables, already contains all spatial prior information concerning the object. It is nevertheless possible to constrain the inversion even more by including the further knowledge that the spectra we seek are positive quantities (at all wavelengths), and furthermore that they are relatively *smooth functions* of the wavelength. The latter fact is taken into account by incorporating a spectral regularisation into the criterion to be minimised, which measures the roughness of the spectrum:

$$J_{\mathrm{o}}(\boldsymbol{F}) = \sum_{k=1}^{N_{\mathrm{src}}} \mu_k \sum_{\lambda=\lambda_{\mathrm{min}}}^{\lambda_{\mathrm{max}}} \left(\frac{\partial^m F_{k,\lambda}}{\partial \lambda^m} \right)^2 , \tag{9.101}$$

where the m th derivative of the spectrum ($m = 1$ or 2 in practice) is calculated by finite differences and where the μ_k are hyperparameters used to adjust the weight allocated to the regularisation. The MAP solution is the one minimising the composite criterion $J_{\mathrm{MAP}}(\boldsymbol{\theta}, \boldsymbol{F}) = J_{\mathrm{i}}(\boldsymbol{\theta}, \boldsymbol{F}) + J_{\mathrm{o}}(\boldsymbol{F})$. It is a rather delicate matter to implement this minimisation because there are many local minima. We use the fact that, for each assumed position $\boldsymbol{\theta}$ of the planets, the MAP estimate of the spectra, $\hat{\boldsymbol{F}}(\boldsymbol{\theta})$, can be obtained simply because J_{MAP} is quadratic in the spectra $\boldsymbol{F}$. If the latter are replaced by $\hat{\boldsymbol{F}}(\boldsymbol{\theta})$ in J_{MAP}, we obtain a partially optimised function for minimisation, which now only depends explicitly on the positions:

$$J_{\mathrm{MAP}}^{\dagger}(\boldsymbol{\theta}) = J_{\mathrm{MAP}}\big[\boldsymbol{\theta}, \hat{\boldsymbol{F}}(\boldsymbol{\theta})\big]. \tag{9.102}$$

This criterion is minimised by a sequential search for the planets, as in the CLEAN algorithm. Figure 9.19 shows the maps of $J_{\mathrm{MAP}}^{\dagger}$ obtained for a single planet as a function of the prior information used. It is clear that the constraints of positivity and smoothness imposed on the spectra significantly improve estimates of the position of the planet, by discrediting (Fig. 9.19 right) positions compatible with the data (Fig. 9.19 left and centre) but corresponding to highly *chaotic* spectra.

Figure 9.20 shows the estimated spectrum of an Earth-like planet. As expected, spectral regularisation avoids noise amplification and has a beneficial effect on the estimate.

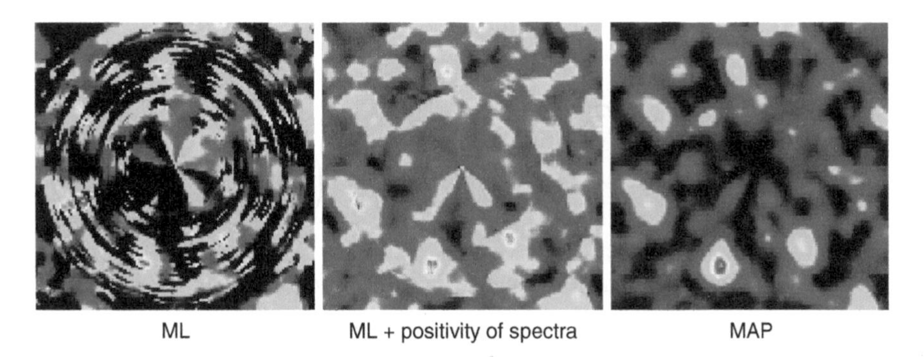

Fig. 9.19 Likelihood maps for the position of a planet. *Left*: Likelihood alone. *Centre*: Likelihood under the constraint that spectra are positive. *Right*: MAP, i.e., likelihood penalised by a spectral regularisation criterion. The true position of the planet is at the bottom, slightly to the left, and clearly visible on the right-hand image. From Mugnier, L., Thiébaut, E., Belu, A., in *Astronomy with High Contrast Imaging III*, EDP Sciences, Les Ulis, 2006, and also Thiébaut, E., Mugnier, L., Maximum a posteriori planet detection with a nulling interferometer, in Proceedings IAU Conf. 200, Nice, 2006

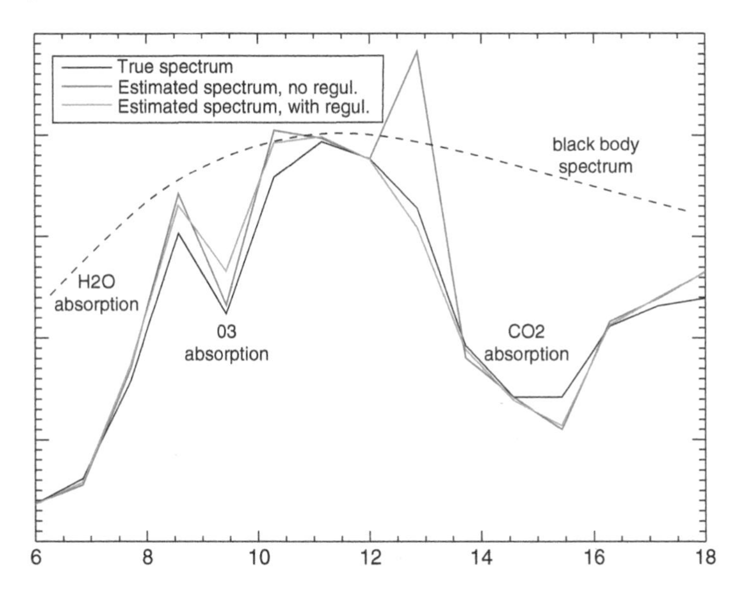

Fig. 9.20 Reconstructed spectrum at the estimated position of the planet. *Red*: Without regularisation. *Green*: With regularisation. *Black*: True spectrum of the Earth. Same reference as for Fig. 9.19

Problems

Note. These exercises refer also to the subjects treated in Appendixes A and B.

9.1. In each of the three examples, use the Fourier transform on the left to deduce the one on the right. Fourier transform is indicated by an arrow:

$$\frac{\sin x}{x} \to \pi \Pi(\pi s), \qquad \left(\frac{\sin x}{x}\right)^2 \to \Lambda(\pi s),$$
$$\mathrm{e}^{-x^2} \to \sqrt{\pi}\mathrm{e}^{-\pi^2 s^2}, \qquad \delta(ax) \to \frac{1}{|a|},$$
$$\mathrm{e}^{-x^2/2\sigma^2} \to \sqrt{2\pi}\sigma\mathrm{e}^{-2\pi^2\sigma^2 s^2}, \quad \mathrm{e}^{\mathrm{i}x} \to \delta\left(s - \frac{1}{2\pi}\right).$$

9.2. Using the convolution theorem, give an expression for

$$\mathrm{e}^{-ax^2} \star \mathrm{e}^{-bx^2}.$$

(Problem 6.15 is an application of this result.)

Answer. The following property is used: the Fourier transform of a convolution product is the pointwise product of the Fourier transforms of the factors in that product. This property is very useful in numerical calculations, the algorithms for multiplication being so much more rapid. Linearity and similarity properties are also used in the following deduction:

$$\begin{aligned} F(x) &= \mathrm{e}^{-ax^2} \star \mathrm{e}^{-bx^2} \\ &= \mathrm{FT}\left[\mathrm{FT}\left[\mathrm{e}^{-ax^2}\right] \cdot \mathrm{FT}\left[\mathrm{e}^{-bx^2}\right]\right] \\ &= \mathrm{FT}\left[\sqrt{\frac{\pi}{a}}\mathrm{e}^{-x^2/a}\sqrt{\frac{\pi}{b}}\mathrm{e}^{-x^2/b}\right] \\ &= \frac{\pi}{\sqrt{ab}}\mathrm{FT}\left[\exp\left(-\frac{a+b}{ab}x^2\right)\right], \end{aligned}$$

so that, finally,

$$F(x) = \frac{\pi^{3/2}}{\sqrt{a+b}}\exp\left(-\frac{ab}{a+b}x^2\right).$$

9.3. Prove the Wiener–Khinchine theorem: The autocorrelation and spectral density are a Fourier pair. Apply it to the case of a time-dependent electric current $i(t)$, and its Fourier transform $\tilde{i}(\nu)$, where ν is a temporal frequency, and thus interpret the theorem physically in terms of power.

9.4. Compare the FTs of the convolution and the correlation products of two functions.

9.5. A signal $X(t)$, whose spectrum has bounded support, passes through a linear filter which has instantaneous response $F(t)$. Let $Y(t)$ be the output signal. $X(t)$ and $Y(t)$ are sampled, giving sequences of results $\{X_i\}$ and $\{Y_i\}$. Show that

$$\{Y_i\} = \{I_i\} \star \{X_i\}.$$

How can the sequence $\{I_i\}$ be found?

9.6. Filtering by Running Mean (Smoothing). Let $x(t) = x_s + n(t)$ be a signal, where x_s is a constant to be determined, and $n(t)$ is the stationary random noise affecting the measurement, such that $\langle n(t)\rangle = 0$. Define

$$y_T(t) = \frac{1}{T}\int_t^{t+T} x(\theta)\,d\theta,$$

which is called the *running mean* of $x(t)$. Put $y_T(t)$ into the form of a convolution, and deduce that it is the result of a linear filtering of $x(t)$. Determine the transfer function $\tilde{H}(f)$ of this filter, and its equivalent bandpass Δf defined by

$$\Delta f = \int_{-\infty}^{+\infty} |\tilde{H}(f)|^2\,df.$$

Answer. An expression equivalent to the running mean is given, up to a translation by $T/2$, by

$$y_T(t) = x(t) \star \frac{1}{T}\Pi\left(\frac{t}{T}\right).$$

The running mean is therefore the result of linear filtering by the filter

$$\tilde{H}(f) = \mathrm{sinc}(Tf),$$

and so, by Parseval's theorem,

$$\Delta f = \int \left|\frac{1}{T}\Pi\left(\frac{t}{T}\right)\right|^2 dt = \frac{1}{T}.$$

9.7. Find the autocorrelation function $R(\tau)$ of a Gaussian process after filtering by the linear filter

$$\tilde{H}(\nu) = \frac{1}{1+(2\pi RC\nu)^2}.$$

9.8. Consider a stochastic process $\mathbf{x}(t)$. Define a new process $\mathbf{x}_T(t)$, referred to as the *estimate of the mean of* $\mathbf{x}(t)$, by

$$x_T(t) = \frac{1}{2T}\int_{t-T}^{t+T} x(t)\,dt.$$

Show that, when $\mathbf{x}(t)$ is stationary,

$$\langle x_T(t)\rangle = \langle x(t)\rangle = \eta,$$

and

$$\sigma_{\mathbf{x}_T}^2 = \frac{1}{4T^2} \int_{-T}^{+T} \int_{-T}^{+T} C(t_1 - t_2)\, \mathrm{d}t_1\, \mathrm{d}t_2,$$

where

$$C(t_1 - t_2) = R(t_1 - t_2) - \eta^2.$$

Express $\sigma_{\mathbf{x}_T}^2$ as a function of $t_1 - t_2 = \tau$.

Apply this result to the filtering process in Problem 9.7 and deduce a simple analytic form for its variance. Graph the quantity $\sigma_{\mathbf{x}_T}^2(\tau)$ and give a simple interpretation.

Apply these results to the following astrophysical problem, frequently encountered in Fourier transform spectroscopy, speckle interferometry, the study of solar oscillations, and other areas. An astronomer wishes to measure the spectral density of a process $\mathbf{x}(t)$, which he knows to be normal and centred. Give a practical procedure for doing this. What value of T should be chosen so that the estimate of the spectral density $S_T(\nu)$ will lie within a previously chosen confidence interval?

9.9. Optimal Sampling Rate and Shannon's Theorem. In a turbulent fluid, viscous dissipation gives a wave number cutoff κ_{M} to the spectrum of spatial frequencies in the turbulence (see Chap. 2). Assuming the medium to be isotropic, what geometrical arrangement of temperature sensors should be set up in order to be able to reconstruct by interpolation the full temperature distribution of the medium?

A sinusoidal plane temperature wave, with wave number κ_0 in a given direction, is superposed upon the turbulence. How is the spectral density modified?

9.10. Shannon Optimal Sampling Rate. The aim is to determine to great accuracy the period T_{p} of the Vela optical pulsar PSR 0833-45 (Fig. 7.9), knowing that it is close to 89 ms. The signal is received by a photomultiplier and then sampled at the frequency f_{S}. The dark current of the detector produces Poissonian noise, with a white spectral density up to $f_{\mathrm{c}} \gg f_{\mathrm{p}}$, where $f_{\mathrm{p}} = 1/T_{\mathrm{p}}$. What is the optimal value of f_{S}? Determine the signal-to-noise ratio if the measurement lasts for a time T (assuming availability of all the required quantities: the magnitude of the pulsar, the transmission and quantum efficiency of the setup, the aperture of the telescope, and the noise current).

Show that f_{S} must be modified if information is sought concerning the shape of the periodic signal, and explain how it must be modified. What would be the effect of such a change on the signal-to-noise ratio of the PSD?

9.11. Digitisation and Truncation. Consider two centred normal processes $\mathbf{x}(t)$ and $\mathbf{y}(t)$, with the same variance σ^2, whose correlation is given by

$$\langle x(t)\, y(t) \rangle = \rho\sigma^2.$$

This is equivalent to the joint probability distribution

$$\text{prob}\{X \leq \mathbf{x} < X + \mathrm{d}X,\ Y \leq \mathbf{y} < Y + \mathrm{d}Y\}$$
$$= \frac{1}{2\pi\sigma^2(1-\rho^2)^{1/2}} \exp\left[-\frac{X^2 + Y^2 - 2\rho XY}{2\sigma^2(1-\rho^2)}\right].$$

Let a_i and b_i be samples of $x(t)$ and $y(t)$ taken at regular intervals t_i. Instead of digitising x and y, only their signs are recorded:

$$a_i = \begin{cases} +1 \text{ if } & x(t) \geq 0, \\ -1 \text{ if } & x(t) < 0, \end{cases}$$

and similarly for b_i with reference to $y(t)$. This *digitisation on one bit* is extremely rapid. Show that the correlation between the new discrete random variables a_i and b_i is given by

$$\langle a_i\, b_i \rangle = \frac{2}{\pi} \arcsin \rho,$$

so that, for weak correlation $\rho \ll 1$,

$$\langle a_i\, b_i \rangle \sim \frac{2\rho}{\pi} \propto \langle x(t)\, y(t) \rangle.$$

Assume that the spectral densities of $x(t)$ and $y(t)$ have bounded support (filtered signals). Let $R(\tau)$ be the autocorrelation of the processes **a** and **b**, and $S(\nu)$ their cross-spectrum (the FT of R). Show that the signal-to-noise ratio for an estimate of S is only a factor $2/\pi$ less than the signal-to-noise ratio that would be obtained by measuring the original processes **x** and **y** over the same period of time. *Real time digital correlators* use this property, referred to as *clipping*, to increase their speed of calculation.

9.12. The whole visible face of the Sun (integrated flux) is observed with a high-resolution spectrometer (see Sect. 8.3.3), in the wing of an absorption line of the photosphere. The received intensity is highly-dependent on the Doppler shift of the line, which in turn depends on the line-of-sight component of the mean velocity. Once the Doppler effects due to the various relative motions of the Sun and the Earth have been corrected for, any non-zero residue is interpreted as a large scale oscillation of the Sun's surface. A period of 160 min is observed.

What is the optimal sampling of the signal? At non-polar latitudes on Earth, only daytime observation is possible. What effect does this truncation have on the observed spectrum? Show that it will contain *artificial* or *ghost lines*. If, in addition, randomly occurring cloud formations sometimes prevent sampling, what will be their effect on the PSD? This could be simulated on a personal computer. (See Grec G. et al., Nature **288**, 541, 1980, for a description of such observations.)

9.13. The spectral density of an image of *photospheric solar granulation*, observed in white light, is given in Fig. 9.21. The frame has dimensions 13 × 13 arcsec. What is the *spatial frequency resolution* obtained in the figure? The calculation

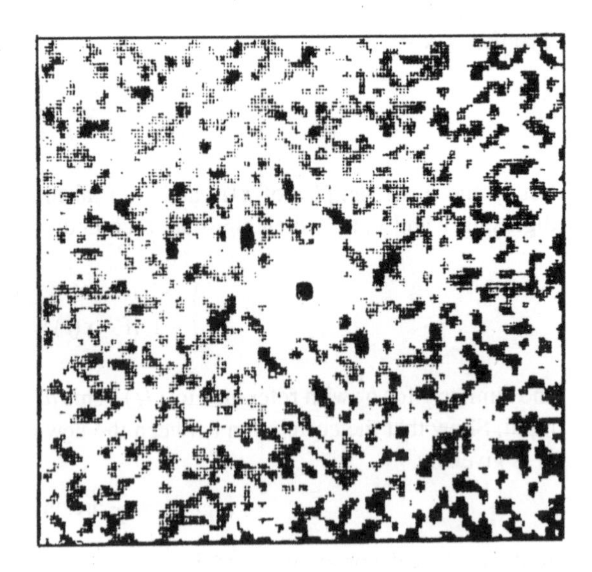

Fig. 9.21 Spectral density of solar granulation. From Aime C. et al., Astr. Ap. **79**, 1, 1979. With the kind permission of Astronomy and Astrophysics

has been simplified by replacing each granulation cell by a Dirac spike. How can the calculated two-dimensional PSD be interpreted, and in particular, the apparent hexagonal structure at low frequency and the noise at high frequency?

9.14. Given the brightness of the sky background around 500 nm (see Sect. 2.3), determine the number of counts resulting from the sky background given by each pixel of a CCD (quantum efficiency $\eta = 0.6$), observing a field of 2×2 arcsec through a 3.6 m telescope, whose transmission is assumed to be 100%. What is the limiting magnitude of a galaxy which could be detected against this background, with a reasonable degree of confidence, in an observation lasting one hour? (Be careful not to confuse the sky background with the fluctuation of this background.) What would be gained by observing over one month, or one year?

Answer. According to Sect. 2.3.1, the sky has magnitude $m_\mathrm{V} = 22$ arcsec^{-2}, which implies a sky background flux per square arcsec given by

$$F_\mathrm{c} = 3.92 \times 10^{-8} \cdot 10^{-22/2.5} = 6.21 \times 10^{-17}\,\mathrm{W\,m^{-2}\,\mu m^{-1}\,arcsec^{-2}}.$$

The number of counts due to the sky background is

$$n_\mathrm{c} = F_\mathrm{c} S \Omega t/(hc/\lambda),$$

where S is the collecting area of the telescope, and hence $S = 1.8^2\pi$ m^2, t is the integration time, Ω is the solid angle subtended at each pixel, namely 4 arcsec^2, and hc/λ is the energy of each photon.

The number of photons required from the galaxy in order to give a reasonable confidence interval (3σ) is

$$n_g = 3\sqrt{F_c S \Omega t/(hc/\lambda)},$$

implying a flux from the galaxy of

$$\begin{aligned} F_g &= n_g(hc/\lambda)/St \\ &= 3\sqrt{(hc/\lambda)\Omega F_c/St} \quad \mathrm{W\,m^{-2}\,\mu m^{-1}}. \end{aligned}$$

Note that the smallest detectable flux varies as $t^{-1/2}$.

For $t = 3\,600$ s, $F_g = 1.56 \times 10^{-19}\,\mathrm{W\,m^{-2}\,\mu m^{-1}}$, which corresponds to a magnitude $m_V = 28.5$. For $t = 1$ month, a factor of $\sqrt{24 \times 30} = 26.8$ is gained, corresponding to 3.5 magnitudes. For $t = 1$ year, only a further 1.35 magnitudes is gained. This does assume that the noise remains stationary over such long periods.

9.15. A photoconducting infrared detector is placed at the focus of a 4 m telescope, with a field of view of 10 arcsec, in the photometric N band, of spectral width 0.5 μm. Two terms contribute to the received signal. Firstly, the flux of photons emitted by the star, and secondly, the emission of the Earth's atmosphere, with emissivity $\varepsilon(\lambda)$. How can $\varepsilon(\lambda)$ be evaluated (Sect. 2.3)?

Taking $\varepsilon(\lambda) = 0.5$ for the site under consideration, calculate the thermal atmospheric background noise. What is the limiting magnitude m_N imposed by this background for an observation of duration 1 s, or 1 hr? Suggest ways of improving the observation.

9.16. Lock-in Detection. A procedure for extracting a signal from noise. In order to extract a signal, drowned out by noise, any available information concerning the signal, such as positivity, periodicity, and so on, should be put to use, with a view to reducing the final uncertainty. Detection is particularly simple if both the frequency and the phase of the signal are known. In Fig. 9.22, the radiation $\mathbf{x}(t)$ is periodically modulated by a rotating disk (or some equivalent device), with frequency ω and

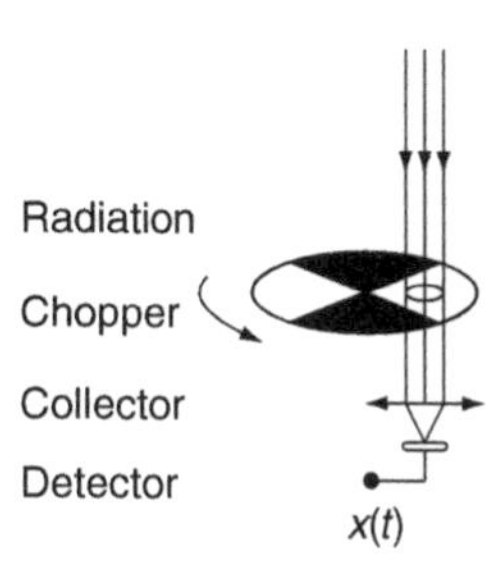

Fig. 9.22 Signal modulated by a rotating disk

phase ϕ. The signal measured is then

$$X_0(t) = n(t) + x(t)F(t),$$

where $n(t)$ is the detector noise, $F(t)$ the periodic, and in general non-sinusoidal, function characterising the *chopper*.

The following operations are then carried out, using either analogue electronic or digital techniques:

- Filtering, $X_1(t) = X_0(t) \star H(t)$, where $H(t)$ is a narrow passband filter, $\Delta\omega \ll \omega$, centred on ω.
- Multiplication, $X_2(t) = X_1(t) \cos(\omega t + \phi')$.
- Low-frequency filtering, for example, by running mean,

$$X_3(t) = X_2(t) \star \Pi(t/T), \quad \text{with} \quad T \gg 2\pi/\omega.$$

Show that this sequence of operations only leaves the part of the noise $n(t)$ with power spectrum in $\Delta\omega$, and then uses the phase information. In the case of white noise, estimate the output signal-to-noise ratio. If $\mathbf{x}(t)$ is random (e.g., a Poissonian process for the arrival of photons), what is the effect on the measurement?

Chapter 10
Sky Surveys and Virtual Observatories

Astronomers have always catalogued the objects they observe and study. In ancient times, the Greek astronomer Hipparchos (between 161 and 127 BC) and the Chinese astronomers Shi Shen and Gan De (in the Warring States period, around 500 BC) built up the first systematic catalogues of the celestial regions accessible from their parts of the world. Tycho Brahe did likewise, followed by surveys of non-stellar objects, such as Messier's.[1] At the end of the nineteenth century, *sky surveys* marked the beginning of the great modern catalogues. Computers completely transformed this landscape in the last decade of the twentieth century, by vastly increasing the volumes of data that could be stored, while improving accessibility, addressability, and communications through a range of automated systems. In this chapter, we shall consider some of the results of this revolution. In addition to the development of *statistical astrophysics* and systematic *whole sky surveys* across all wavelengths, we will also discuss the advent of *virtual observatories* which bring together all available maps and data about a specific object, to assist in solving whatever astrophysical problem is under investigation.

10.1 Statistical Astrophysics

Modern astronomy is characterised by a genuine flood of data, growing exponentially. There are several reasons for this rapid evolution. The first, rather obvious, stems from growing instrumental capacities in terms of telescope size (see Chap. 5), number of pixels on 2D detectors (see Chap. 7), and resolution and spectral range of spectrometers (see Chap. 8). Another, more subtle reason is that, in several areas of astrophysics, the desired information, whether it concern a phenomenon on the cosmic scale or a particular class of objects, can now only be firmly established

[1]Charles Messier (1730–1817) was a French astronomer who catalogued nebulas that still carry his initial M even today.

P. Léna et al., *Observational Astrophysics*, Astronomy and Astrophysics Library,
DOI 10.1007/978-3-642-21815-6_10, © Springer-Verlag Berlin Heidelberg 2012

by analysing a sample containing thousands up to tens of millions of individual objects. Only these large samples suffice to reduce the effects of dispersion or lack of resolution in order to isolate some very dilute effect. Very ambitious observing programmes called *large surveys* often involve dedicated instruments as well as general purpose instruments on large telescopes. They are carried out over months or years by international consortiums, bringing together dozens of researchers.

Galactic astronomy and modern cosmology are two disciplines that depend heavily on this type of observation. And recently, scientists interested in the more remote regions of the Solar System have also become active in the development of sky surveys, in their search for objects beyond the orbit of Neptune.

The study of the Galaxy, e.g., stellar populations, structure, and components of the interstellar medium, has given rise to many exhaustive surveys, particularly in the centimetre and millimetre wave radio regions. In other wavelength regions, satellites observing in the infrared and at high energies have supplemented these surveys. Figure 1.7 illustrates the exceptional spectral coverage obtained in this way.

Cosmology has generated some very extensive programmes, and is continuing to do so. In each endeavour, whether it be the identification of large-scale structures as instantiated by clusters and associations of galaxies, the study of the evolution of the first galaxies in the remote universe, witness to the processes involved in assembly of the first structures, or the mapping of dark matter by weak gravitational lensing effects, the images of millions of galaxies must be acquired and analysed in order to identify statistically significant trends. In this context, the need for a statistical dimension covering very large numbers of objects is well exemplified by attempts to determine cosmological parameters as accurately as possible.

The search for rare objects is rather different but nevertheless involves quite analogous needs in terms of the amount of information that must be acquired and analysed. In this case, the desired property is not shared in some tenuous way by millions of individual sources, but is particular feature of a rarefied population that can only be tracked down within a huge population of superficially similar objects by identifying some specific signature. The search for quasars, planetary nebulas, brown dwarfs, free-floating planets,[2] and trans-Neptunian objects are all the subject of such ambitious programmes.

One particular type of large survey is the class of exhaustive sky surveys, i.e., covering the whole celestial sphere in a certain wavelength region. This kind of survey was first attempted in the visible at the end of the nineteenth century with the ambitious international undertaking known as the *Carte du ciel*.[3] The aim was to produce a complete photographic atlas of the sky in the two hemispheres,

[2]These are planet-like objects with lower masses than brown dwarfs, viz., around $13M_{\text{Jupiter}}$, but which do not appear to be gravitationally bound to any star.

[3]This remarkable enterprise involved astronomers the world over. The decision was taken at an international conference in Paris in 1887, delegating the work to 18 observatories around the world, each equipped with the same model of refracting telescope built by the Henry brothers at the Paris Observatory. Work continued for three quarters of a century and the photographic archives thereby generated are today of inestimable scientific value.

comprising no fewer than 22 000 photographic plates. Unfortunately, it was never quite finished, but was taken up again with more success at the beginning of the second half of the twentieth century by the *Palomar Optical Sky Survey* (POSS). This particularly effective survey was carried out using the very wide field telescope invented by Bernhardt Schmidt. It was then supplemented by equivalent surveys carried out in the southern hemisphere from the vantage point of Australia and Chile. A little later, radioastronomy was to catch up, e.g., with the Parkes–MIT–NRAO (PMN) survey, closely followed by infrared astronomy with the IRAS satellite in 1983, which mapped the sky in the far infrared. And high energies have not been neglected. With surveys like the one carried out by ROSAT in the X-ray region in 1992, there is scarcely any wavelength region in which the whole sky has not already been surveyed at the time of writing (2007). Astrometric surveys are also extremely important, with the Hipparcos mission in 1989, and the even more impressive GAIA mission, planned for 2012, which should investigate a billion stars (see Chap. 4).

Concerning the analysis of data generated in this way, a large survey is characterised by several parameters, of which the most important are the field covered, the sensitivity (or depth), the completeness, and the confusion limit:

- **Field.** This is the area covered by the large survey. Some surveys cover the whole sky, or just one of the two hemispheres. The field must be extensive to ensure that the results are not dominated by specific effects of concentration or rarity. The term *cosmic variance* is generally used in observational cosmology to qualify this cause of uncertainty.
- **Sensitivity or Depth.** This is the magnitude or flux of the faintest unambiguously detected objects. This is often a criterion of the kind flux $> n\sigma$, where σ is the uncertainty in the flux measurement. The depth of the survey generally depends directly on the exposure time in a given direction. The survey must be deep to ensure that reliable statistics can be obtained for particularly remote or intrinsically faint objects. Depending on the objects under investigation, a compromise is always sought between sensitivity and field, since the total time allocated to such a programme must remain within reasonable bounds of months or years.
- **Completeness.** This is closely related to the last feature. It is defined as the apparent magnitude (or flux) beyond (or below) which only a given fraction (usually 50%) of the relevant objects is effectively detected. A conventional method for determining this completeness limit is to introduce simulated data into the real data in a realistic way and then to determine the fraction effectively detected by the algorithm. A survey can by complete, in the case where almost no member of the relevant class of objects has been omitted.
- **Confusion Limit.** If several sources are located in the element of spatial resolution of the observing instrument, the counts obtained will be incorrect because the apparent position and flux of the sources will be modified by the mixing with generally fainter sources. This is known as *source confusion*, and the resulting measurement error is referred to as *confusion noise*. This problem first appeared in radioastronomy, where antenna sidelobes are significant.

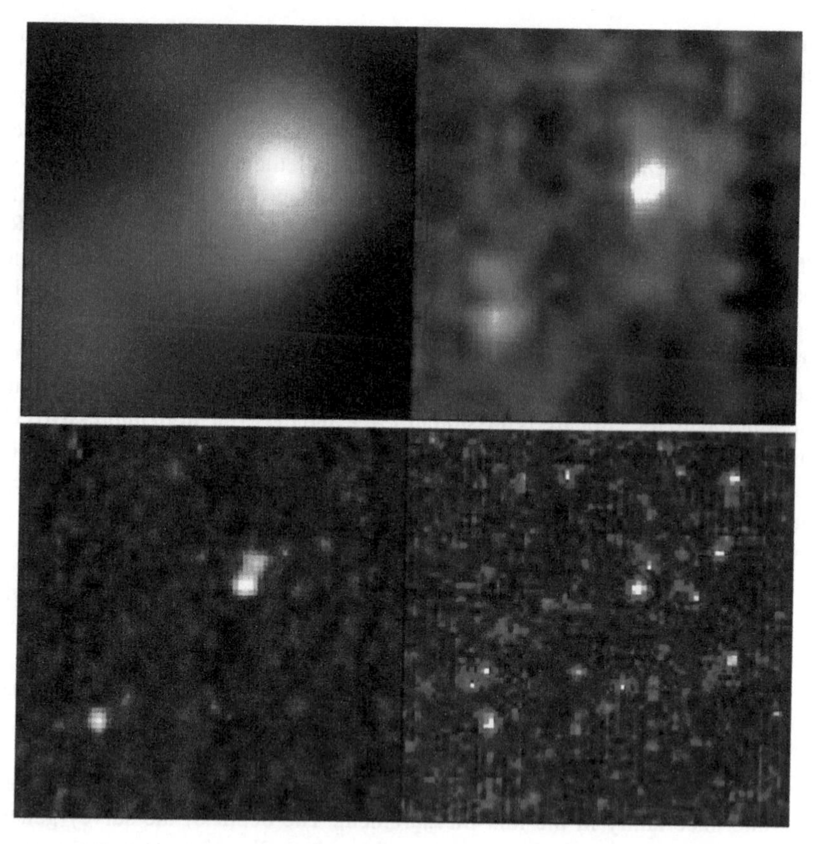

Fig. 10.1 Source confusion and spatial resolution in the images of a field of $6.6' \times 6.6'$ observed in the infrared and containing galaxies. *Top*: ISO mission, $\lambda = 170\ \mu$m and Spitzer mission, $\lambda = 160\ \mu$m. *Bottom*: $\lambda = 70\ \mu$m, $\lambda = 24\ \mu$m. G. Lagache et al. ARAA **43**, 727, 2005, with the kind permission of the publisher

The infrared region can also be sensitive to this limit, whenever the sensitivity becomes high while the angular resolution remains modest. This is exemplified by infrared space observatories like IRAS (1983), ISO (1995), and Spitzer (2003), where the telescopes, with diameters less than one metre, provide angular resolutions in the range between a few arcsec and one arcmin, while the sensitivities achieved are excellent owing to the absence of thermal background in space (see Fig. 10.1).

10.2 Large Sky Surveys

The period from 1980 to 2007 has certainly been the golden age for large sky surveys. A series of such programmes led to major changes in our understanding, particularly in cosmology, which until then had been restricted to largely theoretical

Table 10.1 Some large sky surveys currently available or in progress (2007)

Name	Means	λ	Field
Photometric sky surveys			
APM	Measuring machine	Visible	Southern hemisphere
SDSS	Camera (ground)	Visible	10 000 deg^2
CFHT Legacy	Camera (ground)	Visible	$410 + 170 + 4$ deg^2
IRAS	Satellite (1983)	12, 25, 60, 100 μm	Southern hemisphere
2MASS	Camera (ground)	Near infrared	Whole sky
DENIS	Camera (ground)	Near infrared	Whole sky
Einstein	Satellite (1979)	X ray	Whole sky
XMM	Satellite (1999)	X ray	5 deg^2
Spectroscopic sky surveys			
CfA2	Ground	Visible	17 000 deg^2
DEEP2	Ground	Visible	3.5 deg^2
SDSS	Ground	Visible	10 000 deg^2
VVDS wide	Ground	Visible	16 deg^2
VVDS deep	Ground	Visible	1.5 deg^2
ZCOSMOS	Ground	Visible	1.7 + 1.0 deg^2
Cosmic microwave background surveys			
COBE	Satellite (1989)	1.25 μm–5 mm	Whole sky
BOOMERANG	Balloon	25–412 GHz	Southern hemisphere
WMAP	Satellite (2001)	22–90 GHz	Whole sky
PLANCK	Satellite (2008)	30–857 GHz	Whole sky

developments, owing to the lack of sufficient observational data. This golden age is set to continue for a few more years at least, with ever more ambitious projects. Without trying to be exhaustive, we shall discuss here the large photometric, spectroscopic, and cosmic microwave background surveys summarised in Table 10.1.

The list is impressive, but reflects the spectacular progress in cosmology over the years 1995–2005, including the discovery that the expansion of the Universe is actually accelerating, with the accompanying dark energy hypothesis, confirmation of inflation, confirmation of the flat universe model, measurements of most cosmological parameters with ever increasing accuracy, evidence for dark matter, and so on.

Large sky surveys generally use dedicated systems either in the form of a specialised instrument at the focus of a telescope that can be used for other purposes, or as a complete telescope/instrument system custom built for the task at hand. In the latter case, this system may be spaceborne.

10.2.1 Sky Surveys at Visible Wavelengths

Schmidt Sky Surveys and Their Digitisation

To work effectively, these surveys must be able to access a wide field in one go, while nevertheless keeping a good angular resolution. In the optical region, the large photographic plate set up at the focus of a Schmidt telescope served this purpose well for several decades. The surveys made with these instruments from 1950 to 2000 nevertheless have a rather poor quantum efficiency, limiting them to magnitudes below 19 to 21, depending on the colour, and poor photometric quality owing to the non-linearity of the photographic emulsion. Despite these limitations and their age, these surveys are still the only ones providing access to the whole sky at resolutions of the order of a few arcsec, so they remain genuinely useful. The main drawback with them, namely that they could not directly provide digital information suitable for mass processing and archiving, is no longer a problem, thanks to systematic digitisation of the plates.

The three large Schmidt surveys were:

- SERC/AAO (Siding Spring Telescope, Australia) for the southern hemisphere ($\delta < 0°$). The two series of plates are denoted by J (blue) and I, from 1974 to 1987, and R from 1990 to 2000.
- POSSI/POSSII (Palomar Schmidt Telescope): E (red) and O (blue) for $\delta > -20°$, from 1950 to 1958. J (blue), F (red), N (band I) for $\delta > 0°$, from 1987 to 2000. The field of a POSS plate is in general $5° \times 5°$ with mean angular resolution 3 arcsec.
- ESO: R (red) for $\delta < -20°$ from 1978 to 1990.

These surveys were distributed to observatories the world over in the form of copies on large format photographic paper, filling enormous amounts of shelving and generally accompanied by superposable transparent sheets containing multiwavelength information about identified sources.

Today, the Schmidt sky surveys have been made available by digitising the photographic plates. This was done using specialised machines which explore the plates point by point with high mechanical accuracy, thereby translating their grey levels into digitised pixels.

- The *Automatic Plate Scanner* (APS) at the University of Minnesota produced the POSSI E and O plate digitisations. The resulting catalogue contains 200 million objects (north) from the first epoch along with a catalogue of galaxies. All the recorded objects correspond to double detections in two colours.
- The *Proper Motion Machine* (PMM) was implemented by the US Naval Observatory (Washington, USA). The POSSI and POSSII measurements of the AAO-J, SERC, and ESO were used to produce the USNO catalogue, containing 520 million objects. Two rough magnitudes (R and B) are given for these objects. This is a reference catalogue for astrometry.
- The *Automated Plate Measurer* (APM) is in Cambridge (UK). This machine digitised the AAO-R, SERC-J, and POSSI plates. A galaxy catalogue well known to cosmologists was produced first. In 2007, two whole-sky catalogues of the northern and southern hemispheres are available.
- In France, MAMA is a high accuracy measuring machine at the Paris Observatory, which has been used for specific programmes rather than exhaustive coverage of sky surveys.

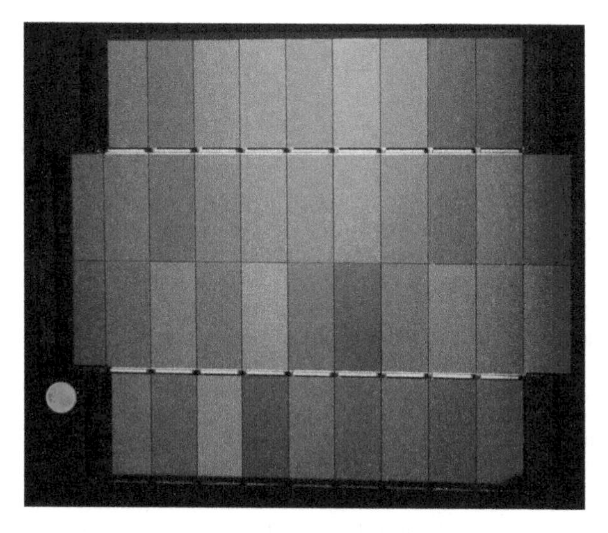

Fig. 10.2 Focal plane of the MEGACAM camera, showing the array of 40 thinned $4\,612 \times 2\,048$ pixel CCDs made by EEV (UK). Source SApCEA

Large CCD Sky Surveys

With the advent of CCD detectors (see Chap. 7) around 1980, the photographic plate was gradually superseded, although it was only during the 1990s that surveys could finally be produced with any real efficiency. It had to be possible to form very large area arrays of CCD detectors, developed specifically for this purpose. The whole area had to be sensitive without loss at edges, using rear face connections to keep the sides free, and highly accurate juxtaposition techniques had to be developed. A good example is the MEGACAM camera (see Fig. 10.2), developed by the astrophysics department of the French atomic energy authority (CEA). This simultaneously operates, at low temperatures, an array of 40 CCD detectors each with $4\,612 \times 2\,048$ pixels, manufactured by the British company EEV. This camera is set up at the primary focus of the CFH telescope (CFHT) (Hawaii) behind a dedicate focal reducer, comprising a sophisticated combination of very large lenses and a mobile image stabilisation component. An excellent image quality is achieved over almost the whole of the impressive field of view of $1° \times 1°$.

Deep Wide-Field Photometric Surveys

These are used to probe regions exceeding the characterise size of the large scale structures in the Universe ($\approx$60 Mpc), and their measurements are thus barely affected by cosmic variance. One classic use of these surveys is to measure the angular correlation function. They can also be used to select sub-samples of rare objects in the Universe.

As an illustration, let us consider the observational means and corresponding performance of deep wide-field surveys made with the CFH telescope in Hawaii:

- The first such survey, the Canada–France Deep Field (CFDF) survey, was carried out from 1996 to 1997 using the UH8K camera, an array of 8 CCDs with a field of view of 29 arcmin × 29 arcmin. This survey covered 1 deg^2 and included over 100 000 objects, down to the detection limit $I_{AB} = 25.5$.
- The CFH12K camera, an array of 12 CCDs with a field of view of 42 arcmin × 28 arcmin, was then used to carry out the VIMOS VLT Deep Survey (VVDS) from 1999 to 2000, which covered 16 deg^2 to a depth of $I_{AB} = 24.8$.
- Since 2004, the MEGACAM camera has been producing a survey with an even wider field of view. The CFHT Legacy Survey (CFHT-LS) has taken five years and a half years. The aim is to image 170 deg^2 of the sky (the Wide component) in 5 colours (u*, g$'$, r$'$, i$'$, z$'$), up to magnitude $I_{AB} = 25.5$, and 4 deg^2 (the Deep component) up to magnitude $I_{AB} = 28.3$. CFHT-LS also includes a Very Wide component in the plane of the ecliptic to seek out trans-Neptunian objects.

Among the most important large sky surveys, we should mention the *Sloan Digital Sky Survey* (SDSS) carried out at the Apache Point Observatory in New Mexico (USA) since 1998. This CCD survey, which covers quarter of the northern sky around the galactic pole (10 000 deg^2) and part of the galactic plane, includes 5 colours (u$'$, g$'$, r$'$, i$'$, z$'$) using the filters of the US Naval Observatory (USNO) photometric system, and reaches magnitude 23 in r$'$. A dedicated 2.5 m telescope was designed to acquire 3 × 3 deg^2 images, using a 5 × 6 array of 2 048 × 2 048 CCDs operating in continuous scanning mode. More than 40 Tbytes of data will be generated. A subset of around 1 Tbyte will include 1 million spectra, together with positions and mini images in each colour for more than 100 million objects.

Ultra-Deep Field Photometric Surveys

The Hubble Space Telescope (HST) was equipped from 2002 with the *Advanced Camera for Surveys* (ACS), which could make images with a field of view of 202 × 202 arcsec^2 and a pixel size of 0.049 arcsec. Before the failure that put it out of action at the beginning of 2007, it was indeed a remarkable tool for observing the faintest objects, by virtue of the unequalled resolution and contrast of its images. Two deep photometric surveys have been particularly spectacular, namely the *Hubble Deep Field North* and *South* (HDF-N and HDF-S). Objects were detected up to magnitudes $I_{AB} = 27.6$, although the field of view remained limited to 4 × 4 arcmin^2. Hence only 2 500 galaxies were observed over these two fields. The angular resolution of the images obtained with the HST was unique for studying the morphology of galaxies and spectral shifts $z > 1$, until the advent around 2005 of the first near-infrared surveys using adaptive optics, which are in the process of changing this situation.

Spectroscopic Sky Surveys

It is particularly in observational cosmology that the large spectroscopic surveys have been conceived as indispensable supplements to the large imaging surveys. This is because they provide a way to access the third dimension, i.e., distance, by measuring the spectral shift toward the red of light from receding galaxies.

Note that multicolour photometric surveys contain information that can be treated as a very low resolution spectrum, and can be used to estimate the distance to galaxies by examining photometric redshifts. To obtain sufficiently accurate measurements of z, one must cover the broadest possible spectral region with a large number of narrow bandpass filters.

The COMBO-17 survey is a deep field survey with imaging over 0.78 deg^2, made using the *Wide Field Imager* (WFI) camera installed on the 2.2 m MPG/ESO telescope at La Silla in Chile. Images are obtained in 17 optical bands, including 12 narrow filters, covering the spectral region 300–900 nm. The COMBO-17 sample contains 25 000 photometric spectral shifts.

Compared with the use of spectroscopic shifts, the measurement is less robust and less accurate. The error is estimated to be 10 times as great. For this reason, true spectroscopic surveys soon came to the fore.

Since it is not feasible to measure this shift efficiently galaxy by galaxy, even at the low spectral resolutions required, multi-object spectroscopic methods flourished. These fall into two main families: multi-slit spectrographs and optical fibre spectrographs in medusa mode (see Sect. 8.3).

In the nearby Universe ($z < 0.1$), two large spectroscopic surveys have acquired a considerable amount of data since 2001:

- **Two Degree Field Galaxy Redshift Survey (2dFGRS).** This is an Anglo-Australian project, using the 3.9 m telescope of the Anglo-Australian Observatory, with the *Two Degree Field* (2dF) multi-object spectrograph, where 400 optical fibres are automatically arranged by a robot to measure the target objects (see Chap. 8). These objects are selected automatically from digitised photographic plates of the APM Galaxy Survey. This survey covers 2 000 deg^2 and contains at total of 230 000 z measurements on as many galaxies.
- **Sloan Digital Sky Survey (SDSS).** Apart from its camera (see above), the SDSS has a fibre-fed multi-object spectrograph able to acquire 640 spectra at the same time. The aim of this survey, in progress at the time of writing (2011), is to obtain the spectra of a million galaxies.

At redshifts z higher than 0.3, there is still no spectroscopic sample of the size of the 2dFGRS and SDSS surveys. To illustrate the difficulty involved at high z, note that the flux observed from a galaxy at $z = 4$ is 6 000 times fainter than the flux from a similar galaxy located at $z = 0.1$. The compromise between the size of the observed field of view, depth of observations, and fraction of objects targeted generally leads to a selection of sources established purely on the basis of a limiting apparent magnitude.

Surveys carried out between 1994 and 2000 were made using telescopes in the 4 m category and multi-object spectrographs with multiplex gain in the range 50–70. The broadest spectroscopic sample contains 2 000 galaxies with magnitudes $R < 21.5$ for the CNOC2 *Galaxy Redshift Survey*, which is 100 times less than the size of samples made in the near Universe. The *Canada–France Redshift Survey* (CFRS) made statistical analyses up to $z = 1.3$, thanks to a selection of faint objects. The results established by CFRS are spectacular, but the sample contains fewer than 250 galaxies in each age range.

Telescopes in the 8 m category have made much deeper surveys. The K20 spectroscopic survey was carried out with the Very Large Telescope (VLT) in Chile, using the FORS1 and FORS2 spectrographs which have multiplex gains of 19 and 52, respectively. This survey measured the redshift z of 550 objects. However, it was a combination of multi-object spectrographs (MOS) with very high multiplex gain and intensive use of 8 m telescopes which, in 2007, produced spectroscopic surveys containing several tens of thousands of galaxies with redshifts up to $z \approx 6$.

Two such large, deep-field spectroscopic surveys are underway at the time of writing (2007), namely, DEEP2 and the VIMOS-VLT Deep Survey (VVDS). DEEP2 uses the DEIMOS spectrograph of the Keck II telescope. DEIMOS is a slit MOS, obtaining 75 spectra simultaneously with spectral resolution $R = 4\,000$. DEEP2 will measure the spectra of 65 000 galaxies at $z > 0.7$.

The VVDS project is based at the VLT and uses the *VIsible Multi-Object Spectrograph* (VIMOS). This spectrograph can observe up to 1 000 spectra simultaneously at low resolution ($R = 200$). The aim is to acquire 150 000 galaxy redshifts up to $z = 6$. The field of the VLT is 30 arcmin, this being split into four subfields, each imaged spectrally through a slit mask and a grism on a CCD array. Each mask is a thin sheet of aluminium in which a high power laser has cut narrow slits. Simultaneous spectral analysis of 1 000 objects is possible with this instrument.

10.2.2 Infrared Sky Surveys

The first infrared survey worthy of the name was carried out from space. This was the almost complete coverage of the sky by the IRAS satellite (1983) in four bands, viz., 12, 25, 60, and 100 μm, with a resolution of about 1 arcmin. The sky was scanned continuously by placing the satellite on a polar orbit held perpendicular to the direction of the Sun, with the telescope pointing to the local vertical. The excellent sensitivity achieved here is due to the fact that the whole setup, including telescope and detectors, was cooled. Over the past 15 years, the quality of the final product has been greatly improved by three complete reprocessing operations.

Ground-based surveys in the near infrared only became feasible at the beginning of the 1990s, with the advent of detectors of appreciable size (256×256 pixels). The two main programmes of this kind were 2MASS and DENIS (see Table 10.1):

- **2 μm All Sky Survey (2MASS).** This is a whole-sky survey in the spectral bands J, H, and K of the near infrared, carried out using two purpose-built telescopes, one in each hemisphere. The limiting magnitudes were 17, 16.4, and 15.6, respectively.

- **Deep Near Infrared Southern Sky Survey (DENIS).** This is a European survey of the southern sky in the spectral bands I, J, and K, reaching limiting magnitudes of 18.5, 16.5, and 14, respectively. The survey was carried out between 1997 and 2003, using the general purpose 1 m telescope at the ESO in La Silla, Chile, which was allocated to this project on a full time basis. The survey is available in the form of images and a catalogue of point sources, supplemented by a specific catalogue for the Magellanic Clouds.

10.3 A Virtual Observatory

At the beginning of the twenty-first century, it became desirable to develop ways of exploiting this extraordinary wealth of survey data, so that all multiwavelength information about a region or object could be put together simultaneously in order to extract the maximum possible astrophysical value. This idea has now become a reality for any research scientist around the world. Thanks to a set of standardised tools, any scientist can obtain access to several of these databases through a computer terminal, superposing information in a uniform manner in the form of maps. This is the notion of *virtual observatory*, now a priority development in many countries and the subject of international coordination.

Hundreds of terabits (10^{12} bits) of observation data, corresponding to several thousand billion pixels are generated by ground-based and spaceborne observatories, and at ever increasing rates. Of course, this information, stored in digital form in huge databases, is exploited by the scientists responsible for producing it, but often only to extract some rather narrow set of information, while the full content of these databases would allow others to find a wealth of other information, e.g., concerning other, previously ignored sources, nevertheless worthy of analysis in some different context. Making this data available to a wider community would certainly avoid the unnecessary duplication of costly observation nights. Observatories and space agencies have understood this simple fact and most offer this service today to the broader scientific community, providing free access to their archives after generally rather short periods of one or two years, during which exclusive access is maintained for those who made the observations.

There is now a clear desire to take a further step and provide the possibility of reconstructing the sky virtually, at all wavelengths. Online access to databases held in observatories around the world, in a user friendly format that can be directly exploited scientifically, will require the help of new information technologies and up-to-date research and analysis tools that are fast and simple to use. The aim of the European project *Astrophysical Virtual Observatory* (AVO), born in 2001, is precisely to provide scientific research with such tools. Supported by the European Commission, this project is piloted by the ESO with several European partners.

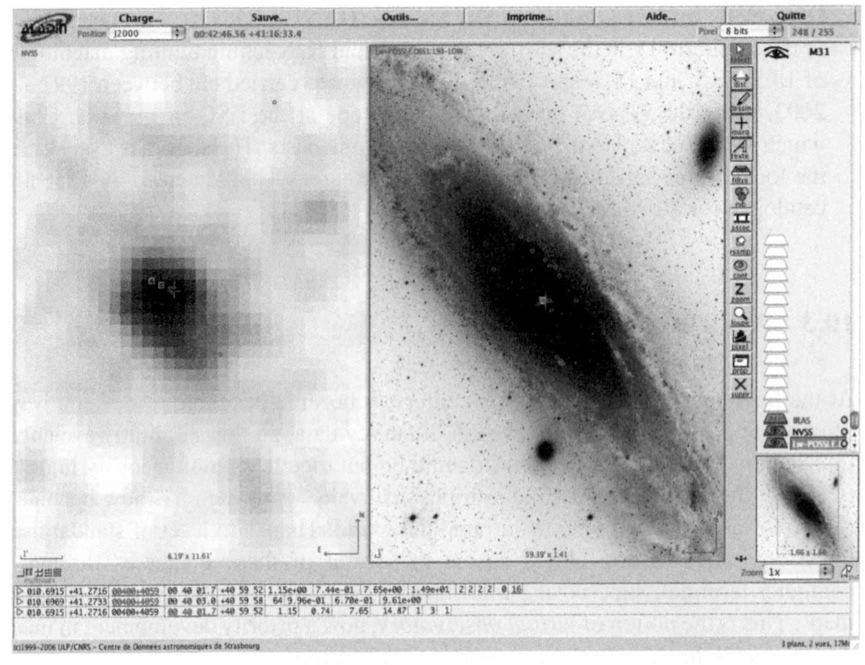

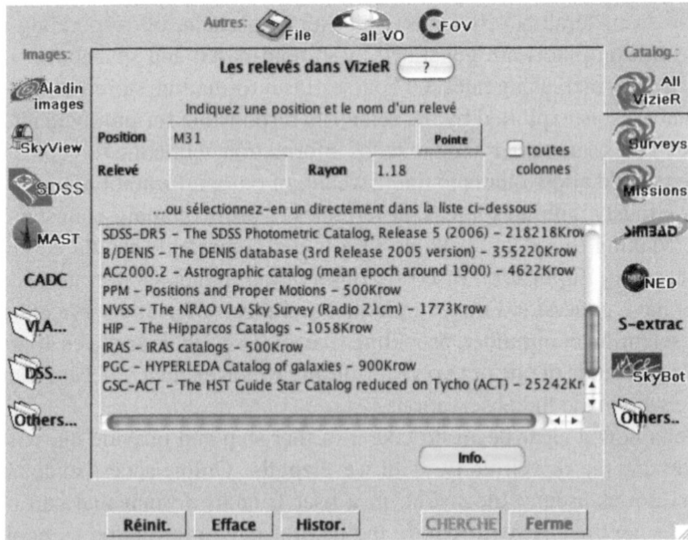

Fig. 10.3 *Upper*: Example of what is possible with the Aladin software at the CDS. Comparing images produced by large sky surveys, in this case, IRAS 100 μm (*left*) and POSS visible (*right*), and superposing object identification and photometry from other catalogues (here 2MASS). Measurement values appear at the bottom. *Lower*: User interface for selecting image database or catalogue

The American counterpart of the AVO, also born in 2001, is the *US National Virtual Observatory* (NVO).[4]

Among those developing the AVO, two French institutes are playing an exemplary role. One is the *Centre de données astronomiques de Strasbourg* (CDS), run by the CNRS and the Louis Pasteur university, which specialises in archiving, and the other is the TERAPIX data centre (*Traitement Elémentaire, Analyse et Réduction des PIXels*) based at the *Institut d'Astrophysique de Paris*, and run by the CNRS and the University of Paris 6, working upstream by processing the huge amounts of data from these large sky surveys in a fully automated way.

For many years now, the CDS has been a reference in the gathering, identification, and archiving of astronomical and bibliographical data. Its unique SIMBAD database can be accessed online. The CDS has also pioneered the notion of virtual observatory with the development of the ALADIN system, which can already be used to visualise and superpose images from the main large imaging sky surveys on a personal computer, identifying catalogued sources and supplying the corresponding photometric and astrometric data (see the example in Fig. 10.3).

The TERAPIX center was set up to exploit data provided by the MEGACAM (visible) and WIRCAM (infrared) CCD cameras operating at the focal point of the 3.6 m Canada–France–Hawaii (CFH) telescope. It was in particular the very ambitious CFHT Legacy Survey which originally mobilised a significant fraction of the efforts at this centre. Since then, its missions have been extended and TERAPIX now processes data from other very large sky surveys.

[4] www.us-vo.org/about.cfm.

Appendix A
Fourier Transforms

The reviews given in Appendices A and B are intended to help with the understanding of Chaps. 6, 7, and 9, and to introduce the notation used throughout the book. Although they have been written with a certain minimum of rigour, mathematical proof and systematic formalism were not the main aim. Examples have been chosen as close as possible to the themes of the book.

A.1 Definitions and Properties

A.1.1 Definitions

Given a function $f(x)$, where $x \in \mathbb{R}$, the *Fourier transform* of $f(x)$ is the function $\tilde{f}(s)$, where $s \in \mathbb{R}$, and

$$\tilde{f}(s) = \int_{-\infty}^{\infty} f(x) \mathrm{e}^{-2\mathrm{i}\pi s x} \, \mathrm{d}x \ .$$

The functions f and $\tilde{f}$ are said to form a Fourier pair, and we sometimes write

$$\tilde{f} \rightleftarrows f \quad \text{or} \quad \tilde{f} = \mathrm{FT}[f] \ .$$

The function $\tilde{f}(s)$ exists if the function $f(x)$ is bounded, integrable and has only a finite number of maxima, minima, and discontinuities. This does not necessarily imply that the transform of $\tilde{f}$ is f. For the Fourier transformation to be reciprocal, i.e.,

$$f(x) = \int_{-\infty}^{\infty} \tilde{f}(s) \mathrm{e}^{2\mathrm{i}\pi s x} \, \mathrm{d}s \ ,$$

it is sufficient that f should be square-integrable, that is, that the integral

$$\int_{-\infty}^{\infty} |f(x)|^2 \, \mathrm{d}x$$

should exist. The definition of the FT can be extended to distributions. The FT of a distribution is not necessarily square-integrable.

The functions f and $\tilde{f}$ may be either real or complex.

P. Léna et al., *Observational Astrophysics*, Astronomy and Astrophysics Library,
DOI 10.1007/978-3-642-21815-6, © Springer-Verlag Berlin Heidelberg 2012

Generalisation

It is possible to generalise the FT to several dimensions, if f is defined on $\mathbb{R}^n$ (i.e., is a function of n real variables). Let $\boldsymbol{r}, \boldsymbol{w} \in \mathbb{R}^n$. Then

$$\tilde{f}(\boldsymbol{w}) = \int_{-\infty}^{\infty} f(\boldsymbol{r}) \mathrm{e}^{-2\mathrm{i}\pi \boldsymbol{r}\cdot\boldsymbol{w}} \, \mathrm{d}\boldsymbol{r} \; .$$

A Simple Interpretation

If $f(t)$ is a function of time, $\tilde{f}(s)$ represents its content of *temporal frequencies*. Similarly, if $f(\boldsymbol{r})$ is defined on $\mathbb{R}^2$, representing a two-dimensional space, the function $\tilde{f}(\boldsymbol{w})$, where $\boldsymbol{w} \in \mathbb{R}^2$, represents its content of *spatial frequencies*.

A.1.2 Some Properties

Linearity

$$\mathrm{FT}\,[af] = a\mathrm{FT}\,[f]\,, \quad a = \text{const.} \in \mathbb{R}\;,$$

$$\mathrm{FT}\,[f+g] = \mathrm{FT}\,[f] + \mathrm{FT}\,[g]\;.$$

Symmetry and Parity

Symmetry considerations are useful in the study of FTs. Let $P(x)$ and $Q(x)$ be the even and odd parts of $f(x)$

$$f(x) = P(x) + Q(x)\;.$$

Then

$$\tilde{f}(s) = 2\int_0^{\infty} P(x)\cos(2\pi xs)\,\mathrm{d}x - 2\mathrm{i}\int_0^{\infty} Q(x)\sin(2\pi xs)\,\mathrm{d}x\;.$$

We also have the trivial relation

$$\tilde{f}(0) = \int_{-\infty}^{\infty} f(x)\mathrm{d}x\;,$$

Table A.1 Symmetry of Fourier pairs

$f(x)$	$\tilde{f}(s)$
Real and even	Real and even
Real and odd	Imaginary and odd
Imaginary and even	Imaginary and even
Complex and even	Complex and even
Complex and odd	Imaginary and odd
Real, arbitrary	Real part even, imaginary part odd
Imaginary, arbitrary	Imaginary part even, real part odd

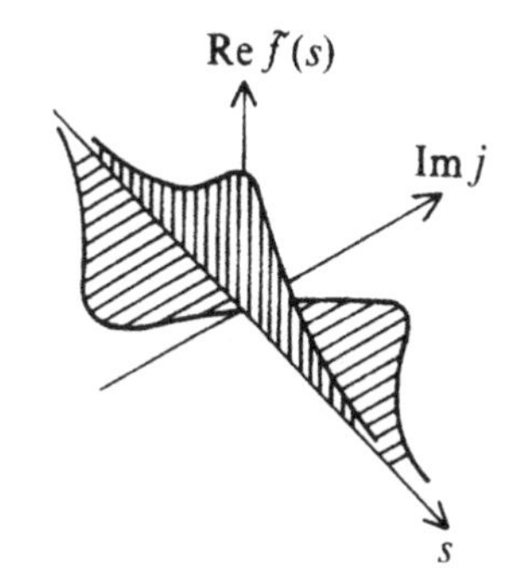

Fig. A.1 Representation of the Fourier Transform $\tilde{f}(s)$ of an arbitrary real function $f(x)$, showing the real and imaginary parts

which says that the zeroth order moment corresponds to the zero frequency. This result leads to Table A.1. Figure A.1 gives a simple representation of the transform $\tilde{f}(s)$.

Similarity

The similarity relation is written

$$f(ax) \rightleftarrows \frac{1}{|a|}\tilde{f}\left(\frac{s}{a}\right) , \qquad a = \text{constant} \in \mathbb{R} .$$

The dilatation of a function causes a contraction of its transform. This easily-visualised property is useful in understanding why the transform of a function with very compact support (i.e., which is non-zero on only a small region), has a very extended transform. In analysing temporal frequencies, we find that an impulse of short duration contains a very wide spectrum of frequencies, i.e., the shorter the impulse, the higher the frequencies its transform contains. This is the classic relation for the spectrum of a wave packet, according to which our knowledge of the properties of a signal cannot be arbitrarily precise simultaneously in both time and frequency.

Translation

The translation of a function gives

$$f(x-a) \rightleftarrows \mathrm{e}^{-2\mathrm{i}\pi a s}\, \tilde{f}(s)\ .$$

A translation in one space is a phase rotation in the transform space.

Derivative

It follows directly from the definition of a Fourier pair that

$$\frac{\mathrm{d}f(x)}{\mathrm{d}x} \rightleftarrows (2\pi \mathrm{i} s)\tilde{f}(s)\ , \qquad \frac{\mathrm{d}^n f(x)}{\mathrm{d}x^n} \rightleftarrows (2\pi \mathrm{i} s)^n\, \tilde{f}(s)\ .$$

A.1.3 Important Special Cases in One Dimension

Box Function

The box function, written $\Pi(x)$, is defined by

$$\Pi(x) = 1 \quad \text{for} \quad x \in \left]-\frac{1}{2}, +\frac{1}{2}\right[\ ,$$

$$\Pi(x) = 0 \quad \text{for} \quad x \in \left]-\infty, -\frac{1}{2}\right] \text{ or } \left[\frac{1}{2}, +\infty\right[\ ,$$

and shown in Fig. A.2. We have likewise, for the box of width $a > 0$,

$$\Pi\left(\frac{x}{a}\right) = 1 \quad \text{for} \quad x \in \left]-\frac{a}{2}, +\frac{a}{2}\right[\ ,$$

$$\Pi\left(\frac{x}{a}\right) = 0 \quad \text{for} \quad x \in \left]-\infty, -\frac{a}{2}\right] \text{ or } \left[\frac{a}{2}, +\infty\right[\ .$$

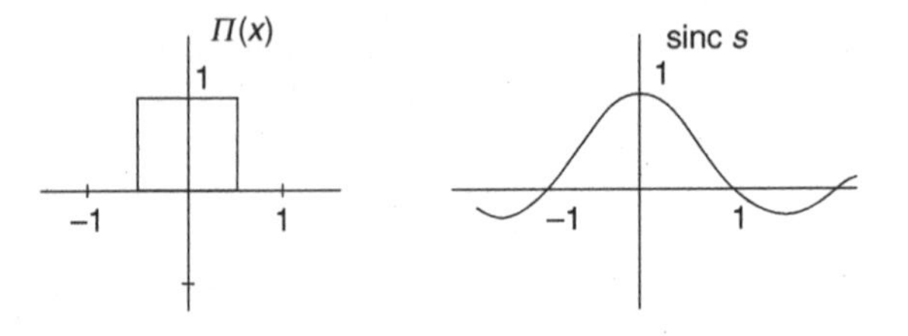

Fig. A.2 The box function and its Fourier transform

The FT of the box function $\Pi(x)$ is written

$$\operatorname{sinc} s = \frac{\sin(\pi s)}{\pi s} \rightleftarrows \Pi(x) ,$$

and, using the similarity relation,

$$\Pi\left(\frac{x}{a}\right) \rightleftarrows |a|\operatorname{sinc}(as) .$$

The Dirac Distribution

The Dirac distribution $\delta(x)$, also known as the Dirac delta function, is not strictly speaking a function. It is defined by

$$\delta(x) = \int_{-\infty}^{\infty} \mathrm{e}^{2\mathrm{i}\pi s x}\, \mathrm{d}s .$$

Its Fourier transform is thus 1 on the whole interval $]-\infty, \infty[$.

The Dirac Comb

This distribution is constructed by periodic repetition of the Dirac distribution. It is written $\sqcup\!\sqcup(x)$, and sometimes called the Shah function (after the Cyrillic character which it resembles):

$$\sqcup\!\sqcup(x) = \sum_{n=-\infty}^{+\infty} \delta(x-n) .$$

This distribution has the remarkable property of being identical to its FT, i.e.,

$$\sqcup\!\sqcup(x) \rightleftarrows \sqcup\!\sqcup(s) .$$

It is also called the *sampling function*, because of the following property, illustrated in Fig. A.3:

$$\sqcup\!\sqcup(x)\, f(x) = \sum_{n=-\infty}^{+\infty} f(n)\delta(x-n) .$$

This may be intuitively understood as follows: starting from a continuous function $f(x)$, the operator reads a number (here infinite) of discrete values of $f(x)$, which we consider as being *samples* from $f(x)$. The term *sample* is understood here in a meaning close to the familiar one. (See also Appendix B and Sect. 9.1.)

This function can also be used as replication operator (Fig. A.3)

$$\sqcup\!\sqcup(x) \star f(x) = \sum_{n=-\infty}^{+\infty} f(x-n) ,$$

where $\star$ denotes convolution (see below).

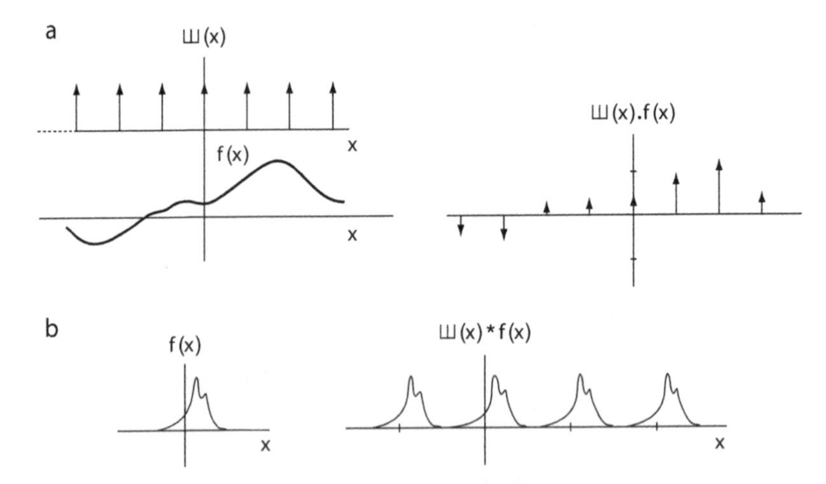

Fig. A.3 (**a**) Sampling of a function $f(x)$ by a Dirac comb. (**b**) Replication of a function $f(x)$ by convolution with a Dirac comb

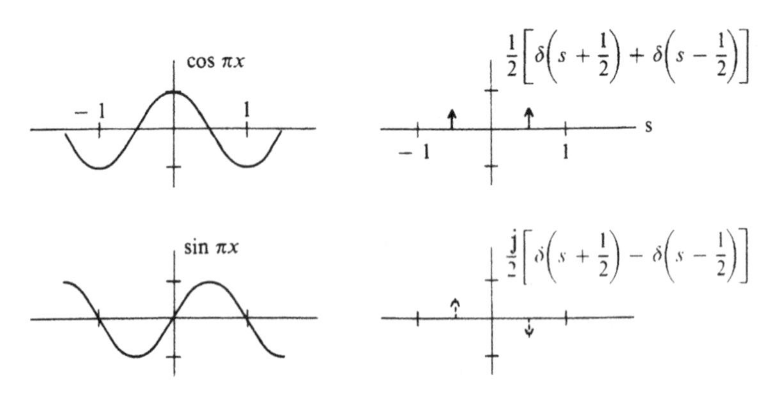

Fig. A.4 Fourier transforms of trigonometric functions (*dashed arrows* indicate imaginary quantities)

This distribution plays an important role in the study of signal sampling, for example, when digitising, but also in the study of periodic structures such as interferometer antennas, or the lines of diffraction gratings.

Trigonometric Functions

As they are not square integrable, trigonometric functions do not have FTs in function space. The Fourier transforms nevertheless exist in the form of distributions (Fig. A.4).

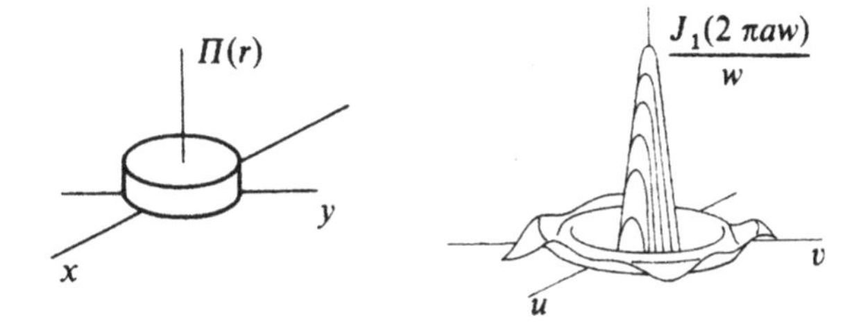

Fig. A.5 Fourier transform of the two-dimensional axisymmetric box function

$$\cos \pi x \rightleftarrows \frac{1}{2}\left[\delta\left(s+\frac{1}{2}\right)+\delta\left(s-\frac{1}{2}\right)\right] \quad \text{symmetric},$$

$$\sin \pi x \rightleftarrows \frac{\mathrm{i}}{2}\left[\delta\left(s+\frac{1}{2}\right)-\delta\left(s-\frac{1}{2}\right)\right] \quad \text{antisymmetric}.$$

A.1.4 Important Special Cases in Two Dimensions

Box Function in Two Dimensions

This function is defined (Fig. A.5) as a constant inside the unit circle and zero outside, with $r^2 = x^2 + y^2$:

$$\Pi\left(\frac{r}{2}\right) = \begin{cases} 1 & r < 1, \\ 0 & r \geq 1. \end{cases}$$

The FT of this function can be written, with $w^2 = u^2 + v^2$,

$$\Pi\left(\frac{r}{2}\right) \rightleftarrows \frac{J_1(2\pi w)}{w},$$

where $J_1(x)$ is the Bessel function of order 1. The similarity relation implies

$$\Pi\left(\frac{r}{2a}\right) \rightleftarrows a\frac{J_1(2\pi a w)}{w}, \quad a > 0.$$

Dirac Distribution in Two Dimensions

This distribution is defined by

$$\delta(x, y) = \delta(\boldsymbol{r}) = \iint_{\text{plane}} \mathrm{e}^{2\mathrm{i}\pi \boldsymbol{r}\cdot\boldsymbol{w}}\, \mathrm{d}\boldsymbol{w},$$

and its FT takes the value 1 on the whole plane $\boldsymbol{w} \in \mathbb{R}^2$.

Two-dimensional Sampling Function

This distribution is constructed by repetition of the two-dimensional Dirac function in the plane:

$$\text{Ш}(x, y) = \sum_{m=-\infty}^{+\infty} \sum_{n=-\infty}^{+\infty} \delta(x - m, y - n) \ .$$

It is identical to its own Fourier transform

$$\text{Ш}(x, y) \rightleftarrows \text{Ш}(u, v) \ .$$

Pairs of Frequently Used Fourier Transforms

Figures A.6 and A.7 show graphically some frequently used Fourier pairs in one and two dimensions, respectively.

Note in particular the one-dimensional Gauss function

$$\mathrm{e}^{-\pi x^2} \rightleftarrows \mathrm{e}^{-\pi s^2} \ ,$$

and in two dimensions

$$\mathrm{e}^{-\pi r^2} \rightleftarrows \mathrm{e}^{-\pi w^2} \ ,$$

which are preserved under Fourier transformation. The similarity relation leads to

$$\exp\left[-\pi\left(\frac{x}{a}\right)^2\right] \rightleftarrows |a| \exp\left[-\pi(as)^2\right] \ .$$

A.1.5 Important Theorems

Convolution

The *convolution* of two functions (or distributions) is defined by the integral

$$h(x) = f(x) \star g(x) = \int_{-\infty}^{+\infty} f(u) g(x - u) \mathrm{d}u \ .$$

The convolution can be considered as a linear transformation determined by $g(x)$ and applied to $f(x)$. This transformation represents the behaviour of many physical systems which impose a linear operation on the input signal, represented by $f(x)$, and lead to the output signal $h(x)$ (Fig. A.8).

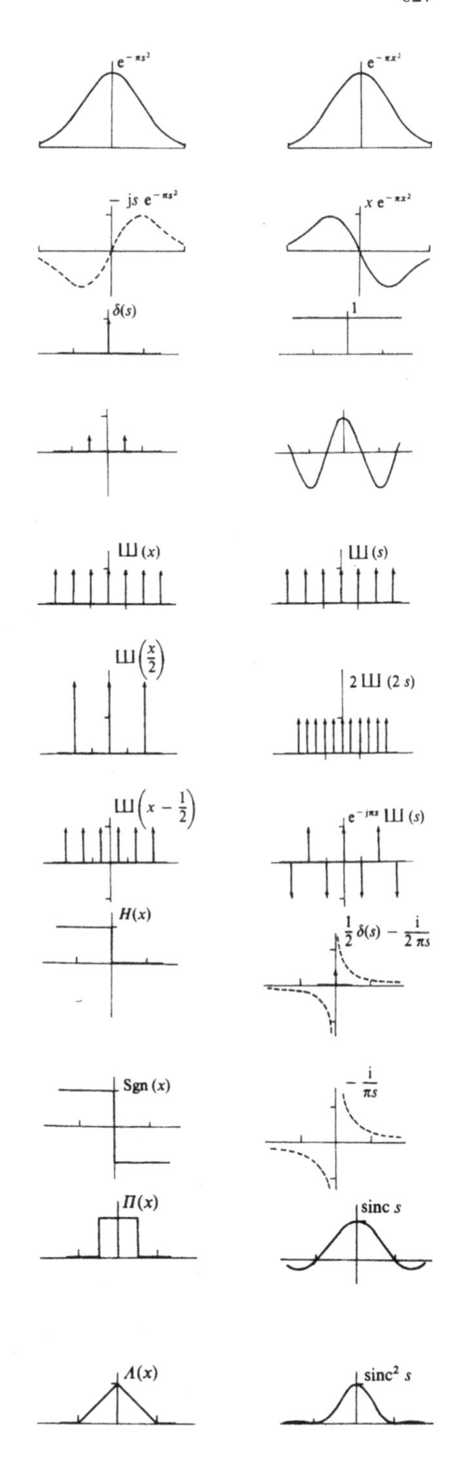

Fig. A.6 Common Fourier pairs. From Bracewell R.N., *The Fourier Transform and Its Applications*, McGraw-Hill, New York, 1965. With the kind permission of the McGraw-Hill Book Co

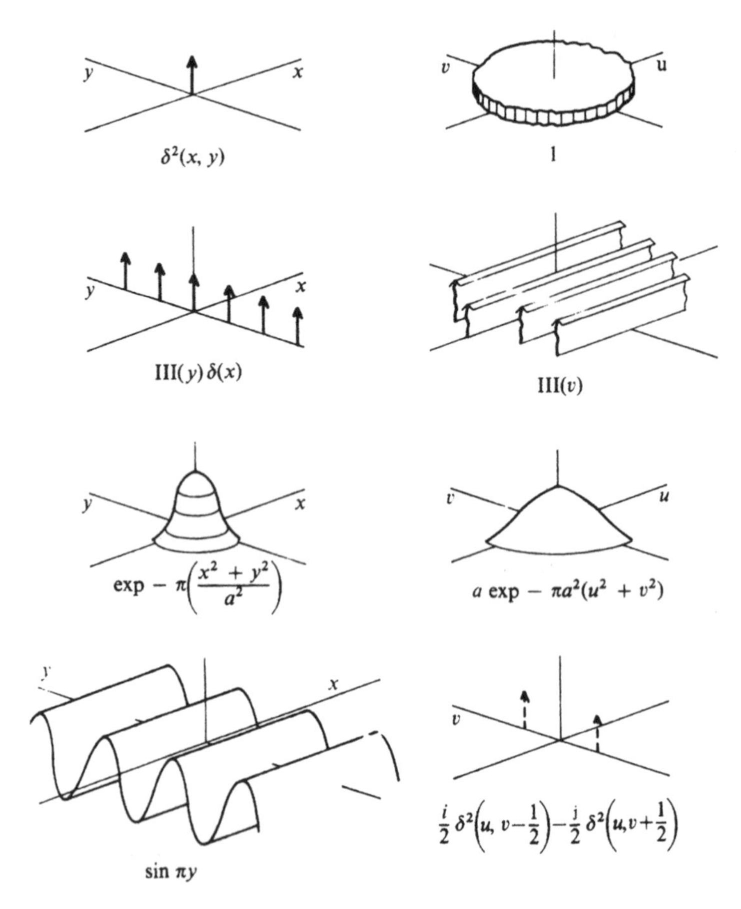

Fig. A.7 Common Fourier pairs in two dimensions. From Bracewell R.N., *The Fourier Transform and Its Applications*, McGraw-Hill, New York, 1965. With the kind permission of the McGraw-Hill Book Co. Note that the FT of $\sin(\pi y)$ takes purely imaginary values

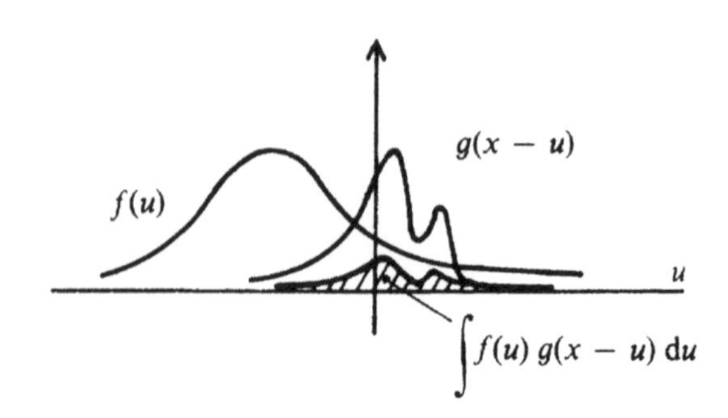

Fig. A.8 Convolution of two functions $f(u)$ and $g(u)$

The Fourier transform acts on convolutions in a remarkable way

$$f(x) \rightleftarrows \tilde{f}(s) ,$$

$$g(x) \rightleftarrows \tilde{g}(s) ,$$

$$h(x) = f(x) \star g(x) \rightleftarrows \tilde{f}(s)\tilde{g}(s) = \tilde{h}(s) .$$

The convolution of two functions (or distributions) is transformed into the pointwise product of their FTs. The convolution product is commutative, associative and distributive over addition. All results mentioned here extend to higher dimensions. For example,

$$H(\boldsymbol{r}) = F(\boldsymbol{r}) \star G(\boldsymbol{r}) = \iint F(\boldsymbol{\rho})G(\boldsymbol{r} - \boldsymbol{\rho})\mathrm{d}\boldsymbol{\rho} ,$$

$$\tilde{H}(\boldsymbol{w}) = \tilde{F}(\boldsymbol{w})\tilde{G}(\boldsymbol{w}) .$$

Correlation

The *correlation product* or *cross- correlation* of two real functions (or distributions) is defined by

$$k(x) = \int_{-\infty}^{+\infty} f(u)g(u + x)\mathrm{d}u ,$$

and its interpretation is straightforward, with the help of Fig. A.9. Note that it differs from the convolution only by a change of sign in the argument of the second function. There is no universally accepted notation for the correlation product. We can write

$$k = C_{fg} \quad \text{or} \quad k = f \otimes g .$$

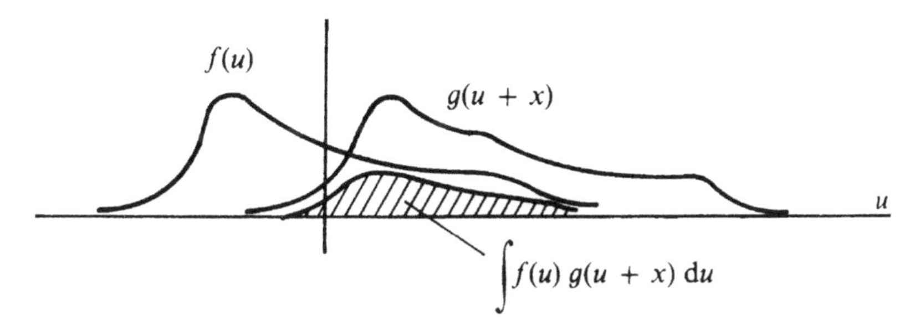

Fig. A.9 The value of the correlation function $k(x) = f(x) \otimes g(x)$, calculated for a particular value of x, is equal to the area *shaded*

If f and g are complex, the correlation product is defined by

$$f \otimes g = \int_{-\infty}^{+\infty} f^*(u) g(u+x) \mathrm{d}u \,,$$

where the asterisk denotes complex conjugation.

The *autocorrelation* of a function (or distribution) is given by

$$k(x) = \int_{-\infty}^{+\infty} f(u) f(u+x) \mathrm{d}u \,.$$

Parseval's (Rayleigh's) Theorem

This theorem says that a function and its FT have the same norm

$$\int |f(x)|^2 \,\mathrm{d}x = \int \left|\tilde{f}(s)\right|^2 \mathrm{d}s = \int f f^* \,\mathrm{d}x \int \tilde{f} \tilde{f}^* \,\mathrm{d}s \,.$$

Similarly, for two functions (or distributions),

$$\int f g^* \,\mathrm{d}x = \int \tilde{f} \tilde{g}^* \,\mathrm{d}s \,.$$

Spectral Density

The *spectral density*, or *power spectral density* of the function $f(x)$ is the quantity

$$S_f(s) = \left|\tilde{f}(s)\right|^2 \,,$$

and $\tilde{f}(s)$ is generally called the *spectrum* of $f(x)$. These terms are sometimes confused. In optics or spectroscopy, what is commonly called the spectrum is in fact a spectral density (see Chap. 5). The spectral density is also very commonly called the *power spectrum*.

Finally, we call the quantity

$$\left|\tilde{f}(s)\right| = \left\{\left[\mathrm{Re}\,\tilde{f}(s)\right]^2 + \left[\mathrm{Im}\,\tilde{f}(s)\right]^2\right\}^{1/2}$$

the *amplitude spectrum*, and the quantity

$$\arg \tilde{f}(s) = \arctan\left(\frac{\mathrm{Im}\,\tilde{f}(s)}{\mathrm{Re}\,\tilde{f}(s)}\right)$$

the *phase spectrum*. We collect these definitions together for convenience:

$$\begin{array}{ll} \tilde{f}(s) & \text{spectrum}\ , \\ \left|\tilde{f}(s)\right| & \text{amplitude spectrum}\ , \\ \arg \tilde{f}(s) & \text{phase spectrum}\ , \\ \left|\tilde{f}(s)\right|^2 & \text{spectral density, power spectrum}\ . \end{array}$$

Autocorrelation (Wiener–Khinchine) Theorem

The Fourier transform of the autocorrelation of $f(x)$ is the power spectrum of $f(x)$

$$\begin{array}{cc} f(x) & f(x) \otimes f(x) \\ \downarrow\uparrow & \downarrow\uparrow \\ \tilde{f}(s) & \left|\tilde{f}(s)\right|^2 \end{array}$$

The power spectrum $|\tilde{f}(s)|^2$ is a real positive quantity. The phase information in the complex function $f(x)$, which is contained in the real and imaginary parts of $\tilde{f}(s)$, is lost in the operation of autocorrelation.

Cross-Power Spectrum and Cross-Correlation

If the functions (or distributions) $f(x)$ and $g(x)$ have FTs $\tilde{f}(s)$ and $\tilde{g}(s)$, respectively, the *cross-power spectrum* or, better, the *cross-spectral density*, is defined by

$$S_{fg}(s) = \tilde{f}(s)\tilde{g}^*(s)\ ,$$

so that $S_{fg}(s) = S^*_{gf}(s)$; and the *cross-correlation* of f and g was defined above. Then we find the relation

$$S_{gf}(s) \rightleftarrows C_{fg}(x)\ .$$

A.2 Physical Quantities and Fourier Transforms

The FT proves to be invaluable for the representation of physical quantities. It can be used to isolate, in a complicated signal, simple (sinusoidal) components which are easy to handle, represent and interpret. However, it should be borne in mind that no information whatever is lost or gained when we go to the Fourier transform space.

Physical quantities, whether represented by functions or by distributions, are restricted in certain ways, and this allows us to develop some of the above ideas.

Support of $f(x)$

The Fourier transform of $f(x)$ is defined if $f(x)$ is defined on a support $]-\infty, +\infty[$. A physical quantity is generally only known on a *bounded support*. For example, a signal $f(t)$ which is a function of time is only known on a finite time interval $[-T/2, T/2]$. It is clearly possible to extend $f(t)$ outside this interval by setting

$$f(t) = 0 \quad \text{for} \quad t \in \left]-\infty, -\frac{T}{2}\right[\quad \text{or} \quad \left]\frac{T}{2}, +\infty\right[.$$

(This is an example of *windowing*.) Then we can calculate the FT and find the spectrum without ambiguity if $f(t)$ is known at all points of the interval, or rather, if $f(t)$ is known from an infinitely dense sampling. If on the other hand $f(t)$ is unspecified outside the interval $[-T/2, T/2]$, we can give no precise information about its spectrum without further assumptions (for example, concerning periodicity, or square-integrability, and the like). The method of windowing described here can be generalised. For example, the width T of the window can be left undetermined, in the case of a function of time. This Fourier analysis on a variable and translatable window (*adaptive Fourier analysis*) is known as *time–frequency analysis*. A powerful tool is the *Wigner–Ville representation*.

Negative Frequencies

The function $\tilde{f}(s)$ is also defined on the support $]-\infty, +\infty[$, and therefore refers to negative frequencies. Is it possible to give some physical interpretation of negative frequencies? A physical quantity is real, and all we need to know its FT on the whole of $]-\infty, +\infty[$ are the values taken by the real and imaginary parts of the FT on $[0, +\infty[$, extending these symmetrically and antisymmetrically, respectively. The function $\tilde{f}(s)$ is then *Hermitian* (even real part, odd imaginary part). Its Fourier transform $f(x)$ is real. Suppose we wish to carry out a numerical filtering on $f(x)$, which amounts to multiplying its FT by a certain function $\tilde{h}(s)$. We carry out the operation of pointwise multiplication of $\tilde{f}(s)$ and $\tilde{h}(s)$ on the whole support $]-\infty, +\infty[$, before taking the inverse FT. This is shown in the following diagram.

$$\begin{array}{ccc}
f(x) \quad \text{real} & \overrightarrow{\text{FT}} & \tilde{f}(s) \text{ calculated on } [0, \infty] \\
\downarrow & & \downarrow \\
\text{Filtering operation} & & \tilde{f}(s) \quad \text{completed} \\
\text{(i.e., convolution)} & & \text{on } [0, -\infty] \\
\downarrow & & \downarrow \\
f(x) \star h(x) & \overleftarrow{\text{FT}} & \tilde{f}(s)\tilde{h}(s)
\end{array}$$

Power of a Signal

Let $f(x)$ be some signal, which may be complex; for example, a complex electric field whose amplitude and phase are functions of time.

We define the *instantaneous power* of the signal as

$$p(x) = f(x) f^*(x) .$$

When the variable x is time and f is a current or a voltage, this definition agrees with the normal usage.

The *average power* over an interval X centred on x_0 is defined as

$$P(x_0, X) = \frac{1}{X} \int_{x_0 - X/2}^{x_0 + X/2} f(x) f^*(x) \, \mathrm{d}x .$$

The integral of the instantaneous power is the *energy* associated with the signal

$$\int_{-\infty}^{+\infty} p(x) \, \mathrm{d}x .$$

If the signal is zero outside the interval X, the integral generally converges without problems. If this is not the case, the signal may have an infinite energy, although its average power is finite. For example, the signal

$$f(x) = \cos(2\pi x)$$

has average power

$$\langle P \rangle = \frac{1}{2} ,$$

although its energy is infinite.

The average power of a signal with support $[-\infty, +\infty]$ is defined by the limit

$$\langle P \rangle = \lim_{X \to \infty} \frac{1}{X} \int_{-X/2}^{+X/2} f(x) f^*(x) \, \mathrm{d}x ,$$

if it exists.

Although the terms energy and power only agree with their usual physical meanings for signals which are functions of time, they are universally used in the senses defined above.

Power Spectrum of a Signal

Finite Energy Signal

If $f(x)$ is a signal, we can calculate $\tilde{f}(s)$, and the quantity

$$S_f(s) = \tilde{f}(s)\tilde{f}^*(s) = \left|\tilde{f}(s)\right|^2 ,$$

already defined as the spectral density of $f(x)$ is also called the *power spectrum* of $f(x)$. $S_f(s)$ has the dimensions of energy per frequency interval (spectral energy density) and not a power. The designation as the power spectrum or *power spectral density* (PSD) is nevertheless standard.

The total energy of the signal is clearly

$$\int_{-\infty}^{+\infty} S_f(s)\,\mathrm{d}s ,$$

if the integral exists.

Infinite Energy Signal

In this case, $\tilde{f}(s)$ cannot be calculated. Nevertheless, the autocorrelation function of $f(x)$ may exist, given by

$$C_{ff}(x) = \lim_{X\to\infty} \frac{1}{X} \int_{-X/2}^{+X/2} f(u) f(x+u)\,\mathrm{d}u ,$$

and the Wiener–Khinchine theorem gives the spectral density of the signal $f(x)$

$$S_f(s) = \mathrm{FT}\left[C_{ff}(x)\right] .$$

Here $S_f(s)$ is an energy, or a power per frequency interval, which is consistent with previous definitions and makes it appropriate to use the term power spectral density. For example, if we consider a signal in the form of a voltage $v(t)$, the PSD is measured in $\mathrm{V}^2\ \mathrm{Hz}^{-1}$, and the amplitude spectrum in $\mathrm{V\ Hz}^{-1/2}$.

Cross-Spectral Density and Cross-Correlation

The definitions and remarks above apply equally to the combination of two functions $f(x)$ and $g(x)$.

For example, in the case of a signal of finite energy, the *cross-spectral density* is written

$$S_{fg}(s) = \tilde{f}(s)\tilde{g}^*(s) ,$$

and the *cross-correlation* is

$$C_{fg}(x) = \int_{-\infty}^{+\infty} f(u)g(u+x)\mathrm{d}u .$$

If the energies are infinite, the cross-correlation is the limit

$$C_{fg}(x) = \lim_{X\to\infty} \frac{1}{X} \int_{-X/2}^{+X/2} f(u)g(u+x)\,\mathrm{d}u ,$$

if it exists, and the cross-spectral density is defined by

$$S_{fg}(s) = \mathrm{FT}\left[C_{fg}(x)\right] .$$

A.3 Wavelets

By Fourier analysis, a signal can be decomposed into periodic functions of infinite support. This is a global transformation which is well-suited to signals with little or no evolution in time, the main example being stationary or quasi-stationary signals. Nevertheless, a physical signal is always defined on a bounded support, its emission being locally defined in time. The most important information is often contained in its non-stationarity (its beginning and end). The same comment could be made about an image bounded in space, or other types of signal, as well as the time signals referred to above. A partial solution is provided by the windowing technique mentioned earlier, which replaces a definition on an infinite support by a definition of the signal including a minimum number of significant coefficients. In a word, it *compresses* the signal in the most efficient way possible.

Two methods have been specifically developed to represent and analyse this type of signal. They are relevant here, because they are more and more frequently used in astronomy, to process data as varied as observed spectra, images, temporally variable emissions such as solar flares and periodic bursts of a pulsar, and many others. In these methods the time variable can, of course, be replaced by a space coordinate, or several (the two coordinates of an image), or indeed any other variable, such as the wavelength of a spectrum.

The *time–frequency representation* takes into account the time development of the frequency content of a signal. It gives the instantaneous frequency of the signal and the time dependence of this instantaneous frequency.

The *wavelet transform*, or *time–scale representation*, describes the development of a signal relative to some observation scale. Wavelets are oscillating functions, localised in time, unlike the sinusoidal functions of the Fourier transform. They all have the same shape, differing only in the instant of their appearance and in their duration. Wavelets of short duration and small amplitude thus represent very localised components of the signal, which would be missed upon a larger scale examination.

These two methods are particularly useful for classification and morphological analysis, which are all-important in astronomy. A signal (for example, the image of a type of galaxy) will thus have a unique signature on the wavelet transform of the image of a cluster. The relevant information is concentrated, in such a way that, in the Fourier transform, a particular frequency, drowned out by the rest of the signal, will stand out as a spike in its spectrum.

Appendix B
Random Processes and Variables

Noise phenomena, due to the fluctuations of a thermodynamic system or the quantum nature of interactions, impose a practical and theoretical limit on any measurement. Some basic mathematical notions are useful, given the importance of this subject in the book (but see the note at the beginning of Appendix A).

B.1 Random Variables

Definition of a Random Variable (r.v.)

Consider a process (for example, dice-throwing or absorption of a photon by the photoelectric effect), which has a number of possible outcomes ζ. A pre-assigned rule associates a quantity $\mathbf{x}(\zeta)$ with each outcome. (The r.v. is denoted $\mathbf{x}$, not to be confused with the notation for a vector $\boldsymbol{x}$. The difference should be clear from the context. Naturally, a vector can also be a random variable, but we shall not introduce any specific notation to cover this case.) $\mathbf{x}(\zeta)$ is called a *random variable* of probability $P(\zeta)$, where $P(\zeta)$ is the probability of getting the result ζ (Fig. B.1).

We denote by $\{\mathbf{x} \leq x\}$ the set of outcomes for which the random variable $\mathbf{x}$ has a value no larger than a number x. This set may be empty, or contain just one, or several, or all the outcomes, and each of these cases is called an *event*. Each event, in turn, has a *probability* associated with it, so that probability becomes a positive definite function on the set of events, satisfying certain rules to be stated below.

A real random variable is defined when the set $\{\mathbf{x} \leq x\}$ is an event for any real number x. It is also required that the probabilities of the events $\{\mathbf{x} = -\infty\}$ and $\{\mathbf{x} = +\infty\}$ should be zero.

A complex random variable is defined by a process which associates with each outcome ζ a complex number

$$\mathbf{z}(\zeta) = \mathbf{x}(\zeta) + \mathrm{i}\mathbf{y}(\zeta) ,$$

P. Léna et al., *Observational Astrophysics*, Astronomy and Astrophysics Library,
DOI 10.1007/978-3-642-21815-6,

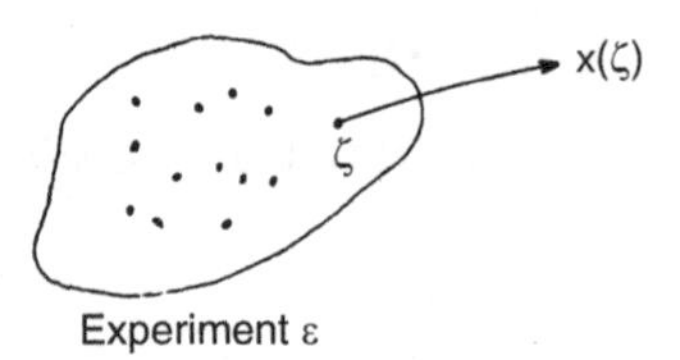

Fig. B.1 An experiment ε formed by the set of outcomes ζ (here discrete), with each of which is associated the value $\mathbf{x}(\zeta)$ of the r.v. $\mathbf{x}$

in such a way that the functions $\mathbf{x}$ and $\mathbf{y}$ are themselves real random variables. Unless otherwise stated, all random variables in the following are real.

Distribution Function

Given a real number x, the set $\{\mathbf{x} \leq x\}$ of all outcomes ζ such that $\mathbf{x}(\zeta) \leq x$, is an *event*. The probability $P\{\mathbf{x} \leq x\}$ of this event is a number depending on x, and therefore a function of x, which we denote by $F_{\mathbf{x}}(x)$, or simply $F(x)$ if there is no risk of ambiguity regarding the r.v. it refers to. Hence,

$$F_{\mathbf{x}}(x) = P\{\mathbf{x} \leq x\} .$$

This is called the *distribution function* (not to be confused with distributions) of the r.v. $\mathbf{x}$. It has the following properties:

$$F(-\infty) = 0 , \qquad F(+\infty) = 1 ,$$

it is a non-decreasing function of x, i.e.,

$$F(x_1) \leq F(x_2) \qquad \text{for } x_1 < x_2 ,$$

and $F(x)$ is right continuous, i.e.,

$$F(x^+) = F(x) .$$

Probability Density

The derivative of the distribution function, in the sense of distributions,

$$f_{\mathbf{x}}(x) = \frac{\mathrm{d}F_{\mathbf{x}}(x)}{\mathrm{d}x}$$

is called the *density* (or density function, or sometimes *frequency*) of the r.v. $\mathbf{x}$. (The theory of distributions makes differentiation possible even when the function is not continuous. The derivative is then not a function, but a distribution.)

If $F(x)$ is continuous and differentiable everywhere except at a countable number of values of x, the r.v. is said to be continuous. Attributing arbitrary positive values

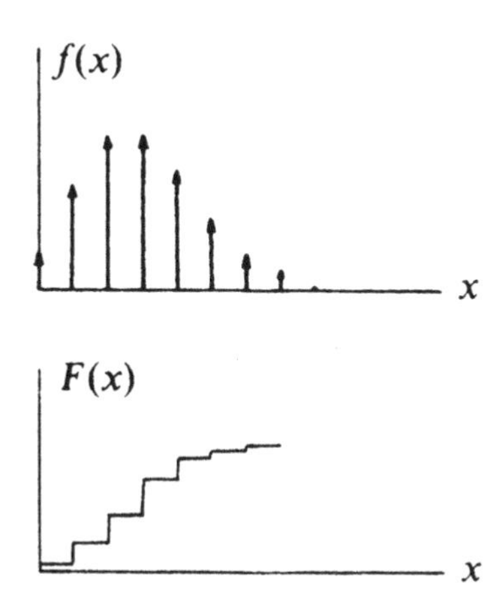

Fig. B.2 Probability density function $f(x)$ and corresponding Poisson ($a = 3$) distribution function $F(x)$

to $f(x)$ at the set of values of x (of measure zero) at which F is not differentiable, we can treat $f(x)$ as a function defined for all x.

If there are a discrete number of outcomes, $F(x)$ has a staircase appearance, as shown in Fig. B.2. The density is then

$$f(x) = \sum_i p_i \delta(x - x_i) ,$$

a succession of Dirac distributions of amplitude p_i, where p_i is the probability of outcome i. We obtain

$$\left.\frac{\mathrm{d}F(x)}{\mathrm{d}x}\right|_{x=x_i} = \left[F(x_i) - F(x_i^-)\right]\delta(x - x_i) .$$

The probability density is normalised to unity:

$$\int_{-\infty}^{+\infty} f(x)\,\mathrm{d}x = F(+\infty) = 1 .$$

Expectation Values

The expectation of any function ϕ of the continuous r.v. $\mathbf{x}$ of density f is written

$$E\{\phi(x)\} = \int_{-\infty}^{+\infty} \phi(x) f(x)\,\mathrm{d}x .$$

Similarly, if $\mathbf{x}$ is a discrete r.v., taking values x_n with probabilities p_n, then

$$E\{\phi(x)\} = \sum_n \phi(x_n) P_n .$$

In physics, the expectation value of ϕ is called the ensemble average and denoted $\langle \phi \rangle$.

Besides having other useful properties, the expectation value is the moment generating operator for the probability measure.

Moments

Any probability measure is completely described by either its distribution function or its density function, but these both contain more information than we can easily apprehend. A small number of global parameters, called the *moments* of the distribution, already provide the essential content of the model.

For any whole positive number k, the moment of order k of the r.v. $\mathbf{x}$ is defined by $\mu'_k = E\{\mathbf{x}^k\}$, and the central moment of order k by $\mu_k = E\{(\mathbf{x} - E\{\mathbf{x}\})^k\}$.

We always have $\mu_0 = 1$. Knowing all the moments (an infinite number) of a distribution function amounts to knowing the distribution function $F(x)$.

Mean

The *mean value* of a continuous random variable $\mathbf{x}$ is

$$\eta = E\{\mathbf{x}\} = \int_{-\infty}^{+\infty} x f(x)\,\mathrm{d}x \ ,$$

where $f(x)$ is the density of the variable. If $\mathbf{x}$ is a discrete random variable, taking the values x_n with probability p_n, then

$$E\{\mathbf{x}\} = \sum_n x_n p_n \ .$$

The quantity η gives information about how the density $f(x)$ is centred. We can also define the *most probable value* of x, which is that value x_1 such that $f(x_1)$ is a maximum.

The *median* of $\mathbf{x}$ is that value x_m such that

$$P\{\mathbf{x} \le x_\mathrm{m}\} = F(x_\mathrm{m}) = 1/2 \ .$$

Variance

The *variance* or *dispersion* is defined as the central moment of order two

$$\mu_2 = E\left\{(\mathbf{x} - \eta)^2\right\} = \int_{-\infty}^{+\infty} (x - \eta)^2 f(x)\,\mathrm{d}x \ .$$

This quantity, usually denoted σ^2, describes how the r.v. is concentrated about its mean η. Its positive square root is called the *standard deviation* or *root mean square deviation* (rms). It follows that

$$\sigma^2 = E\left\{\mathbf{x}^2 - 2\mathbf{x}\eta + \eta^2\right\} = E\left\{\mathbf{x}^2\right\} - 2\eta E\left\{\mathbf{x}\right\} + \eta^2 = E\left\{\mathbf{x}^2\right\} - \eta^2 ,$$

which gives the important relation

$$\sigma^2 = E\left\{\mathbf{x}^2\right\} - \left(E\left\{\mathbf{x}\right\}\right)^2 .$$

Characteristic Function

The *characteristic function* $\Phi(\omega)$ of a random variable $\mathbf{x}$ is the expectation value of the new random variable $\exp \mathrm{i}\omega\mathbf{x}$ defined in terms of $\mathbf{x}$. Hence

$$\Phi(\omega) = E\left\{\exp \mathrm{i}\omega\mathbf{x}\right\} = \int_{-\infty}^{+\infty} \mathrm{e}^{\mathrm{i}\omega x} f(x)\, \mathrm{d}x .$$

This is just the Fourier transform of $f(x)$

$$\Phi(\omega) \rightleftarrows f(x) .$$

Similarly, if $\mathbf{x}$ is a discrete random variable

$$\Phi(\omega) = \sum_k \mathrm{e}^{\mathrm{i}\omega x_k} P\left\{\mathbf{x} = x_k\right\} .$$

Some Important Random Variables

It can be shown that, given a function $G(x)$ such that

$$G(-\infty) = 0 , \quad G(+\infty) = 1 , \quad \lim_{x\to\infty} G(x) = 1 ,$$

$$G(x_1) \le G(x_2) \quad \text{if} \quad x_1 < x_2 ,$$

$$G(x^+) = G(x) ,$$

there exists a random process and a random variable defined on it with distribution function $G(x)$.

The functions $G(x)$ below thus define random variables. Indeed, they provide a good approximation to the random behaviour of certain physical quantities.

Normal or Gaussian Distribution

The probability density of the normal distribution is given by

$$f(x) = \frac{1}{\sigma\sqrt{2\pi}} \exp\left[-\frac{(x-\eta)^2}{2\sigma^2}\right] = N\left(x;\eta,\sigma^2\right) .$$

The distribution has two parameters: η (position) and σ (scale).

The corresponding distribution function is

$$F(x) = \int_{-\infty}^{x} f(u)\,\mathrm{d}u = 0.5 + \mathrm{erf}\left(\frac{x-\eta}{\sigma}\right) ,$$

where $\mathrm{erf}(x)$ denotes the *error function*, which can be found in tables of functions or programmed.

The parameter η is the mean and σ^2 is the variance of the distribution. The odd moments about the mean are zero, and the even moments about the mean are given by

$$\mu_{2r} = \frac{(2r)!}{2^r r!}\sigma^{2r} .$$

The two parameters η and σ completely specify the normal distribution. The width half way up to the peak, called the *full width at half maximum* or FWHM, is given by

$$\mathrm{FWHM} = 2\sigma\sqrt{2\ln 2} \approx 2.3548\,\sigma .$$

The r.v. $\mathbf{y} = (\mathbf{x}-\eta)/\sigma$ is a normally distributed r.v. with mean zero and variance one. It is called the *reduced normal* r.v.

We shall see later that a great number of physical phenomena lead to quantities which are distributed normally. It is useful to be able to give the interval $[-Q, +Q]$ such that the reduced r.v. should lie with probability almost 1 in that interval:

$$P\left\{\left|\frac{\mathbf{x}-\eta}{\sigma}\right| < Q\right\} = 1-\alpha .$$

The parameter Q determines the $(1-\alpha)$ *confidence interval*. Table B.1 gives the values of Q for various values of α.

We deduce that a r.v. distributed as $N(x;\eta,\sigma)$ has probability 99.7% of lying in the interval $[\eta-3\sigma, \eta+3\sigma]$ during a trial. The 99.7% confidence level thus corresponds to the three sigma rule.

Table B.1 Confidence intervals for the normal distribution

$(1-\alpha)$ [%]	50.0	68.3	90.0	95.0	95.4	99.0	99.7	99.9
Q	0.6745	1	1.64	1.96	2	2.5758	3	3.2905

Poisson Distribution

The Poisson distribution gives the probability of finding k events in a certain interval (an interval of time, or more generally, a range), when the events are independent of each other, unlimited in number, and occur at a constant rate in time or with a constant spacing over the range (space, for example). The probabilities of the various values of the discrete r.v. $\mathbf{x}$, which lie in $\{0, 1, 2, 3, \ldots\}$, are given by

$$P\{\mathbf{x} = k\} = \mathrm{e}^{-a} \frac{a^k}{k!} .$$

The single parameter a is called the *parameter* of the distribution. The probability density is given by

$$f(x) = \mathrm{e}^{-a} \sum_{0}^{\infty} \frac{a^k}{k!} \delta(x - k) ,$$

an infinite series of Dirac distributions. The mean and variance of the Poisson distribution are both equal to a.

χ^2 Distribution

A random variable has a χ^2 distribution if its probability density is

$$f(x) = \begin{cases} \dfrac{1}{2\Gamma(n/2)} \left(\dfrac{x}{2}\right)^{n/2-1} \mathrm{e}^{-1/2} & \text{for } x > 0 , \\ 0 & \text{for } x < 0 . \end{cases}$$

There is only one parameter, n, which is strictly positive and called the *degree of freedom*. Note that if $n = 3$, we obtain the Maxwell distribution in the kinetic theory of gases. The Γ function is tabulated and programmable. The mean and variance of the r.v. are n and $2n$, respectively.

The main feature of the χ^2 distribution is as follows: if $\{\mathbf{x}_i, i = 1, 2, \ldots, n\}$ are n independent random variables with the reduced normal distribution, then the r.v. χ defined by

$$\chi^2 = \sum_{i=1}^{n} \mathbf{x}_i^2$$

has a χ^2 distribution with n degrees of freedom.

Central Limit Theorem

Let $\mathbf{x}_1, \mathbf{x}_2, \mathbf{x}_3, \ldots, \mathbf{x}_n$ be a sequence of independent random variables with the same density $f_{\mathbf{x}}(x)$, with mean η and variance σ^2. it can be shown that their arithmetic mean

$$\overline{\mathbf{x}} = \frac{1}{n}\sum_{1}^{n} \mathbf{x}_i$$

has mean μ and variance σ^2/n. It thus tends to μ when n tends to infinity (law of large numbers). Furthermore, the central limit theorem (from a German term meaning that it is of central importance in probability theory) states that this variable behaves in the limit as a normal r.v. with mean μ and variance σ^2/n. This result barely depends on the precise nature of the density f. It is only required to have a well-defined mean and variance. There are less restrictive versions of these conclusions, e.g., when the distributions of the $\mathbf{x}_i$ are different but all have a well-defined mean μ_i and variance σ_i^2. The variable $\overline{\mathbf{x}}$ then behaves as a normal r.v. with mean η and variance σ^2 given by

$$\eta = \frac{1}{n}\sum \eta_i \quad \text{and} \quad \sigma^2 = \frac{1}{n^2}\sum \sigma_i^2 \ .$$

As $n \to \infty$, the density $f_{\overline{\mathbf{x}}}(x)$ of $\overline{\mathbf{x}}$ tends to the normal distribution

$$f(x) = \frac{1}{\sigma\sqrt{2\pi}} \exp\left[-\frac{(x-\eta)^2}{2\sigma^2}\right] .$$

An intuitive idea of the proof of the theorem is obtained by noting that the density $f_{\overline{\mathbf{x}}}(x)$ is the n-fold convolution

$$f_{\overline{\mathbf{x}}}(x) = f_1(x) \star f_2(x) \star \cdots \star f_n(x) \ .$$

But the convolution of an increasing number of positive functions, each of which tends to zero at $\pm\infty$, tends to a Gaussian. This is easy to visualise for an n-fold convolution of box functions $\Pi(x)$.

The central limit theorem is very useful in physics, where the statistical behaviour of a system often results from the accumulation of a large number of independent effects.

B.2 Random or Stochastic Processes

Definition

Consider an experiment having a set of possible outcomes ζ, where $P(\zeta)$ is the probability. A real or complex function of time t denoted by $\mathbf{x}(t,\zeta)$ is assigned to each outcome. The family of functions thus created is called a stochastic process (abbreviated to s.p. in the following). ζ belongs to the set of outcomes of the experiment, and t is a real number in $]-\infty, +\infty[$. The notation $\mathbf{x}(t)$ will be used to represent the process. This notation can be considered as representing four different aspects of the process:

(a) A family of functions depending on time (indexed by ζ).
(b) A particular function of time (ζ being fixed).
(c) A random variable (at fixed t, for a set of trial outcomes ζ).
(d) A number (at fixed t, and for fixed ζ).

Aspect (c) is clearly a necessary condition if we hope to speak of a random process.
The following three examples should motivate these definitions:

- The experiment consists in choosing a molecule of a gas and the quantity $\mathbf{x}(t)$ is the component of its velocity along a given axis. This component varies in time.
- The experiment consists in choosing an oscillator from the production line. The output voltage of the oscillator can be written

$$\mathbf{x}(t,\zeta) = \mathbf{a}(\zeta)\sin\omega_{\zeta}t \ ,$$

  where amplitude and frequency depend on the oscillator. The sequence of output voltages is a stochastic process. Note that in this case the function $\mathbf{x}(t)$ is completely determined for any given outcome, i.e., for any choice of oscillator.
- The experiment consists in throwing a die, and

$$\mathbf{x}(t) = \begin{cases} \sin t & \text{for heads ,} \\ \mathbf{x}(t) = 2t & \text{for tails .} \end{cases}$$

  The function representing $\mathbf{x}(t)$ is random, but completely specified after each throw.

Figure B.3 illustrates the various ways of thinking about random processes. $\mathbf{x}(t)$ is assumed to be real.

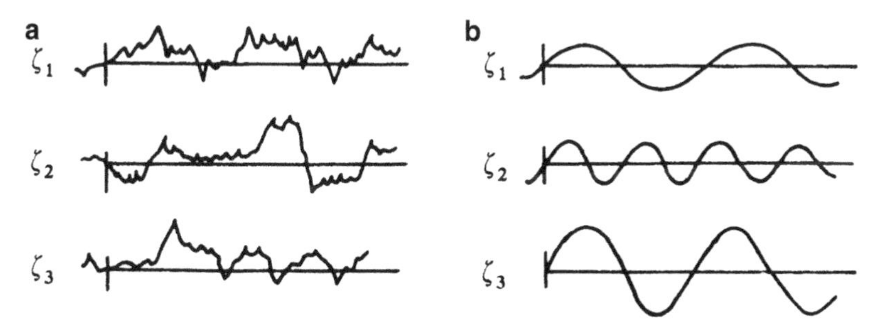

Fig. B.3 Different representations of a continuous stochastic process $\mathbf{x}(t)$. (**a**) $\mathbf{x}(t,\zeta)$ is a random function of t. (**b**) $\mathbf{x}(t,\zeta)$ is a determinate function of t

Distribution Function

The *first order distribution function* of the stochastic process $\mathbf{x}(t)$ is

$$F(x;t) = P\{\mathbf{x}(t) \leq x\} ,$$

the probability of the event $\{\mathbf{x}(t) \leq x\}$, consisting of all the outcomes ζ such that, at time t, the functions $\mathbf{x}(t, \zeta)$ do not exceed the value x.

The corresponding *density function* is

$$f(x,t) = \frac{\partial F(x;t)}{\partial x} .$$

The *second order distribution function* of the stochastic process $\mathbf{x}(t)$ is defined as follows. For two times t_1 and t_2, the two r.v.s $\mathbf{x}(t_1)$ and $\mathbf{x}(t_2)$ are used to write

$$F(x_1, x_2; t_1, t_2) = P\{\mathbf{x}(t_1) \leq x_1, \mathbf{x}(t_2) \leq x_2\} .$$

The corresponding density function is

$$f(x_1, x_2; t_1, t_2) = \frac{\partial^2 F(x_1, x_2; t_1, t_2)}{\partial x_1 \partial x_2} .$$

Mean of a Stochastic Process

The *mean* of a stochastic process is the expectation value

$$\eta(t) = E\{\mathbf{x}(t)\} ,$$

and is generally time dependent.

Autocorrelation and Autocovariance of a Stochastic Process

The *autocorrelation* of the stochastic process $\mathbf{x}(t)$ is the expectation of the product of the random variables $\mathbf{x}(t_1)$ and $\mathbf{x}(t_2)$, viz.,

$$R(t_1, t_2) = E\{\mathbf{x}(t_1)\mathbf{x}(t_2)\} = \int_{-\infty}^{+\infty} x_1 x_2 f(x_1, x_2; t_1, t_2) \mathrm{d}x_1 \mathrm{d}x_2 .$$

The *autocovariance* of $\mathbf{x}(t)$ is the mathematical expectation of the product of the random variables $\mathbf{x}(t_1)$ and $\mathbf{x}(t_2)$, viz.,

$$C(t_1, t_2) = E\{[\mathbf{x}(t_1) - \eta(t_1)]\,[\mathbf{x}(t_2) - \eta(t_2)]\} .$$

It follows immediately that

$$C(t_1, t_2) = R(t_1, t_2) - \eta(t_1)\eta(t_2) ,$$

$$\sigma^2_{\mathbf{x}(t)} = C(t, t) = R(t, t) - \eta^2(t) ,$$

and the autocorrelation is equal to the autocovariance for a stochastic process with zero mean. The autocovariance is centred and the autocorrelation is not.

Stationary Processes

A particularly simple case, which often arises in physics, is one in which the various moments of the r.v. $\mathbf{x}(t)$ are invariant under time translation. The time origin is thus arbitrary, and although the process is time dependent, it is conserved in a probabilistic sense.

A stochastic process $\mathbf{x}$ is said to be *stationary* if the random variables

$$\mathbf{x}(t) \quad \text{and} \quad \mathbf{x}(t + \varepsilon)$$

have the same statistics, for any ε. A weaker definition, which is often sufficient in practice, requires the mean of the s.p. to be time independent and the autocorrelation to depend only on the interval $\tau = t_2 - t_1$ and not on t_2:

$$E\{\mathbf{x}(t)\} = \eta = \text{constant} , \quad E\{\mathbf{x}(t + \tau)\mathbf{x}(t)\} = R(\tau) .$$

Spectral Density

The *spectral density* or *power spectrum* of a stationary s.p. $\mathbf{x}(t)$ is defined to be the Fourier transform $S(f)$, if it exists, of the autocorrelation function $R(\tau)$ of the process.

Examples of Stochastic Processes

Bearing in mind the most common applications, two random processes are described here which use the probability distributions discussed previously, namely the normal distribution and the Poisson distribution.

Normal Process

A process $\mathbf{x}(t)$ is said to be *normal* if, for all $n, t_1, t_2, \ldots, t_n$, the random variables $\mathbf{x}(t_1), \ldots, \mathbf{x}(t_n)$ have a joint normal distribution. The statistics of a normal s.p. are completely determined by its mean $\eta(t)$ and autocovariance $C(t_1, t_2)$. Since

$$E\{\mathbf{x}(t)\} = \eta(t) , \qquad \sigma^2_{\mathbf{x}(t)} = C(t, t) ,$$

the first order density of the s.p. is given by

$$f(x,t) = \frac{1}{\sqrt{2\pi C(t,t)}} \exp\left\{-\frac{[x-\eta(t)]^2}{2C(t,t)}\right\} .$$

If the s.p. is stationary, it follows that

$$E\{\mathbf{x}(t)\} = \eta , \qquad \sigma_{\mathbf{x}}^2 = C(0) .$$

The autocorrelation $R(\tau)$ and the autocovariance $C(\tau)$ of the s.p. are not determined a priori. There are infinitely many normal stochastic processes, with the same mean η and variance $C(0)$, which differ in the function $C(\tau)$ for $\tau \neq 0$.

The power spectrum of this process is the Fourier transform of the autocorrelation function,

$$S(f) = \mathrm{FT}[R(\tau)] ,$$

if it exists (Wiener–Khinchine theorem).

Poisson Process

Consider an interval $[0, T]$ within which n points are chosen at random. The probability of finding exactly k points within the sub-interval $[t_1, t_2]$ of $[0, T]$ is given by

$$P\{k \text{ in } [t_1, t_2]\} = C_n^k p^k (1-p)^{n-k} = \frac{n!}{k!(n-k)!} p^k (1-p)^{n-k} ,$$

where

$$p = \frac{t_2 - t_1}{T} .$$

Suppose we have

$$n \gg 1 \quad \text{and} \quad \frac{t_2 - t_1}{T} \ll 1 , \quad n \to \infty , \; T \to \infty , \; \frac{n}{T} = \text{constant} .$$

Then, putting $n/T = \lambda$, the probability of finding k points in an interval of length $t_2 - t_1$, placed anywhere along the time axis, is

$$P\{k \text{ in } (t_2 - t_1)\} = \mathrm{e}^{-\lambda(t_2 - t_1)} \frac{[\lambda(t_2 - t_1)]^k}{k!} .$$

We thus obtain a set of point events randomly distributed in time, the probabilities associated with each interval being independent, provided that the intervals do not intersect.

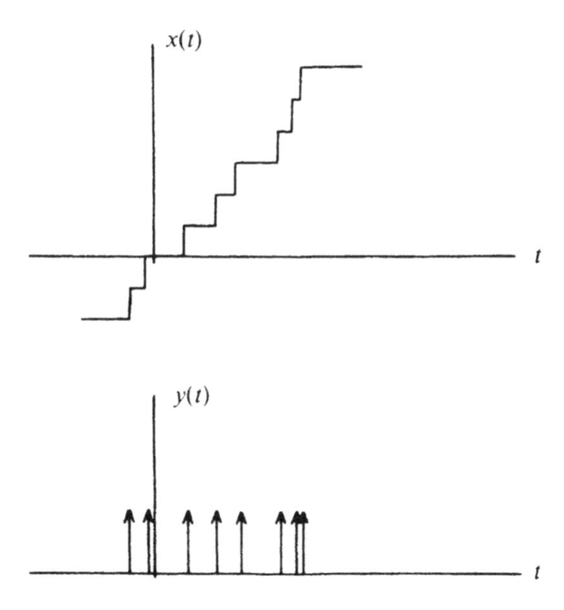

Fig. B.4 $\mathbf{x}(t)$ is a Poisson process and $\mathbf{y}(t)$ is a Poisson impulse process

We now define the *Poisson process* $\mathbf{x}(t)$ in the following way: we set $\mathbf{x}(0) = 0$ and $\mathbf{x}(t_2) - \mathbf{x}(t_1)$ equal to the number of points in the interval $[t_1, t_2]$. The family of functions resulting from this definition is called the Poisson process (Fig. B.4). At a given time t, the constructions $\mathbf{x}(t)$ constitute a random variable with a Poisson distribution of parameter λt.

It follows immediately that

$$E\{\mathbf{x}(t_\mathrm{a}) - \mathbf{x}(t_\mathrm{b})\} = \lambda(t_\mathrm{a} - t_\mathrm{b}) ,$$

$$E\left\{[\mathbf{x}(t_\mathrm{a}) - \mathbf{x}(t_\mathrm{b})]^2\right\} = \lambda^2 (t_\mathrm{a} - t_\mathrm{b})^2 + \lambda(t_\mathrm{a} - t_\mathrm{b}) .$$

As an example of a Poisson process, consider a distant star emitting photons which are collected by a telescope. A photoelectric detector (see Chap. 7) gives an impulse at the arrival of each photon. These impulses are registered by a counter, and the counter reading constitutes a Poisson process $\mathbf{x}(t)$.

Poisson Impulses

Consider the process $\mathbf{y}(t)$,

$$\mathbf{y}(t) = \sum_i \delta(t - t_i) ,$$

formed from a series of Dirac distributions at the random instants t_i described in the last section. The process can be regarded as the limit as ε tends to zero of the stochastic process

$$\mathbf{z}(t) = \frac{\mathbf{x}(t+\varepsilon) - \mathbf{x}(t)}{\varepsilon} ,$$

where $\mathbf{x}(t)$ is the Poisson s.p. It is therefore the derivative of $\mathbf{x}$ in the sense of distributions. Then

$$\mathbf{y}(t) = \frac{\mathrm{d}\mathbf{x}(t)}{\mathrm{d}t} = \lim_{\varepsilon \to 0} \mathbf{z}(t) ,$$

and in the stationary case

$$E\{\mathbf{y}(t)\} = \lambda , \quad R(\tau) = \lambda^2 + \lambda\delta(\tau) , \quad C(\tau) = \lambda\delta(\tau) .$$

The spectral density $S(f)$ of the process $\mathbf{y}(t)$ is the Fourier transform of $R(\tau)$

$$S(f) = \lambda^2\delta(f) + \lambda .$$

Apart from an impulse at the origin, this spectral density consists of a term independent of the frequency, and is said to have a *white spectral density*.

Transformation of a Process

Consider a process $\mathbf{x}(t)$, and a rule associating a new function $\mathbf{y}(t)$ with each function $\mathbf{x}(t)$. The statistics of $\mathbf{y}(t)$ can, in general, be deduced from those of $\mathbf{x}(t)$. There are many applications, when we regard $\mathbf{x}(t)$ as the input of a system S, and $\mathbf{y}(t)$ as its output (Fig. B.5). The output s.p. depends partly on the input s.p. and partly on the properties of the system. One important application is filtering (see Sect. 9.1.3).

Example

Consider a stationary and central normal s.p., and the transformation

$$\mathbf{y}(t) = \mathbf{x}^2(t) .$$

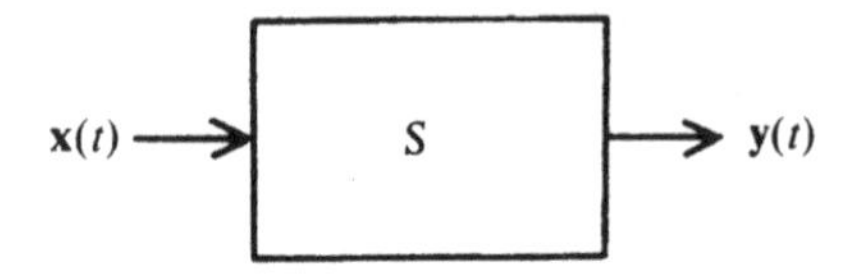

Fig. B.5 Transformation of a process by a system S

The density of $\mathbf{x}(t)$ is given by

$$f_{\mathbf{x}}(x) = \frac{1}{\sigma\sqrt{2\pi}} \mathrm{e}^{-x^2/2\sigma^2} ,$$

and it can be shown that the density of the s.p. $\mathbf{y}(t)$, which is not a normal s.p., is

$$f_{\mathbf{y}}(y) = \frac{1}{\sqrt{2\pi R(0) y}} \exp\left[-\frac{y}{2R(0)}\right] H(y) ,$$

where $R(\tau)$ is the autocorrelation of $\mathbf{x}(t)$ and H is the Heaviside step function.

The autocorrelation of the s.p. $\mathbf{y}(t)$ can be deduced from the relation

$$E\{\mathbf{x}^2(t+\tau)\mathbf{x}^2(t)\} = E\{\mathbf{x}^2(t+\tau)\}E\{\mathbf{x}^2(t)\} + 2E^2\{\mathbf{x}(t+\tau)\mathbf{x}(t)\} ,$$

which gives

$$R_{\mathbf{y}}(\tau) = R^2(0) + 2R^2(\tau) .$$

The variance of $\mathbf{y}(t)$ is then

$$\sigma_{\mathbf{y}}^2 = E\{\mathbf{y}^2(t)\} - E^2\{\mathbf{y}(t)\} = 2R^2(0) .$$

Ergodicity of a Stochastic Process

The *ergodicity* of an s.p. bears on the question of whether, starting from a single outcome $\mathbf{x}(t, \zeta)$ of an experiment, it is possible to determine the statistical properties of the stochastic process $\mathbf{x}(t)$. Put another way, are the mean, variance, autocorrelation, and so on, calculated from a single outcome $x(t)$, identical to the same quantities evaluated for the entire set of outcomes as we have defined them above?

It is intuitively clear that such a result can only hold if the process is stationary and if $x(t)$ takes all possible values in the course of time. Not all stochastic processes will satisfy this condition. Consider the example of the oscillators given above; the variance of the s.p. is not equal to the variance of the voltage measured for a particular outcome.

In a physical process, the time variation of the variable is easily accessible using some measuring device. On the other hand, it is generally difficult to measure a large number of independent outcomes of the s.p. at the same instant t_0. Ergodicity is therefore a property which facilitates statistical analysis. (*Brownian motion* is an example of an ergodic process: it makes no difference whether we take the time average of the velocity of one molecule, or the mean over the whole set of molecules at a given time. The velocity of any molecule will range over all possible values in the course of time.)

Definition. A stochastic process $\mathbf{x}(t)$ is ergodic if the averages taken over time are equal to the averages taken over the ensemble, that is, the expectation values.

We can establish criteria for the ergodicity of a stationary s.p. by formulating the notion of a time average. Define

$$x_T = \frac{1}{T} \int_0^T x(t)\,\mathrm{d}t \;,$$

$$\hat{R}_T(\tau) = \frac{1}{T} \int_0^T x(t)x(t+\tau)\,\mathrm{d}t \;.$$

Then $\mathbf{x}_T$ and $\hat{\mathbf{R}}_T(\tau)$ are respectively a random variable and a stochastic process, whose values are determined for each outcome ζ. We wish to establish conditions for the following to hold:

$$\lim_{T\to\infty} x_T = \eta = E\{\mathbf{x}(t)\} \;,$$

$$\lim_{T\to\infty} \hat{R}_T(\tau) = R(\tau) = E\{\mathbf{x}(t)\mathbf{x}(t+\tau)\} \;.$$

The question is closely related to the following: starting from a physical measurement made over a finite time T and giving the values of one outcome $x(t)$ of the stochastic process during this interval, is it possible to obtain significant information about the expectation values characterising the s.p., such as the mean, variance, autocorrelation, and so on?

The expressions x_T and $\hat{R}_T(\tau)$ do not always converge to the corresponding expectation values. This point will be discussed in more detail later (see Sect. B.3), when measurements of random physical quantities are considered.

Filtering

Frequency Filtering

Consider the linear operation

$$y(t) = \int_{-\infty}^{+\infty} x(t-\theta)h(\theta)\,\mathrm{d}\theta$$

applied to a stationary process $\mathbf{x}(t)$, where $h(\theta)$ is a given function. This convolution is called *filtering* of the process $\mathbf{x}(t)$ by the filter h:

$$y(t) = x(t) \star h(t) \;.$$

From the convolution theorem

$$\tilde{y}(s) = \tilde{x}(s)\tilde{h}(s) \;,$$

where

$$\tilde{h}(s) = \int_{-\infty}^{+\infty} h(t)\mathrm{e}^{2\mathrm{i}\pi st}\,\mathrm{d}t$$

is called the *transfer function* of the filter and is generally complex. If we set $x(t) = \delta(t)$, we obtain

$$y(t) = h(t) \ .$$

This implies that the function $h(t)$ is the *response* of the filter to an instantaneous impulsive input signal. It is also called the *impulsive filter response*.

The function $|\tilde{h}(s)|^2$ is the transfer function for the spectral power, or power, of the filter. It is real and positive, but only gives information about the modification of the amplitude spectrum of the process by the filter. It reveals nothing about modifications to the phase.

Time Filtering

When a process $\mathbf{x}(t)$ is measured over an interval $[-T/2, T/2]$, and assumed to be zero outside this interval, this is equivalent to considering a second process $\mathbf{y}(t)$ such that

$$\mathbf{y}(t) = \Pi\left(\frac{t}{T}\right)\mathbf{x}(t) \ .$$

Such a situation is typical of a physical measurement, where a signal can only be measured during some finite time T. The Fourier transform of the above expression is

$$\tilde{y}(s) = \tilde{x}(s) \star T\,\mathrm{sinc}(Ts) \ ,$$

so that multiplication by the box function in time amounts to a convolution in frequency space. This convolution constitutes a blurring of the spectrum of the original process due to the fact that the signal is only measured for a finite time. It should be noted in particular that all information about frequencies below $1/T$ is lost.

B.3 Physical Measurements and Estimates

In this section we discuss physical problems for which the rigorous mathematical theory of probability was developed. In real physical situations, a continuous or discrete quantity is only ever measured finitely often or over a finite time. It is in the nature of things that the results of repeating the same experiment should be different and random in character. As seen in Sect. 7.2, this fluctuation or *noise* is an apparently unavoidable feature of all physical measurement.

The physicist studying some physical phenomenon, for example a quantity x varying in time, can only do so over some finite period of time T. He can then evaluate the mean, variance, or other attributes, for the resulting discrete or continuous measurements. The same situation would arise if he measured a quantity y (for example, a pressure) a certain number of times. Each measurement would give a different result, and he could once again evaluate a mean and a variance. It could then be said that we are studying the statistics of the phenomenon $x(t)$ or the quantity y. The change in terminology, from speaking of probabilities to speaking of *statistics*, is an indication that a priori we do not know whether or not $x(t)$ or y have the properties of a process or a random variable.

Measurements provide an *instance* or an *observation* of the underlying phenomenon. Statistical calculation (of mean, variance, and so on), carried out on the values thus obtained, provides *estimators* of the mean and variance of the underlying process $\mathbf{x}(t)$ or random variable $\mathbf{y}$, insofar as these can be supposed to constitute a correct representation of the physical phenomena. These estimators are themselves fluctuating quantities; a further set of measurements leads to a new value of the estimators. The problem facing the statistician is to evaluate the validity of these estimators with a view to describing some underlying reality.

As physical measurement always involves estimation, it is actually an abuse of language to speak of measuring the moments of a variable or a stochastic process. This abuse is so widespread that we shall also indulge in it, having drawn attention to the problem.

B.3.1 An Example of Estimation: The Law of Large Numbers

Consider a random variable $\mathbf{x}$ with mean η and variance σ^2, and an experiment E_n consisting of the n-fold repetition of an identical experiment. The latter leads to n variables $\mathbf{x}_1, \mathbf{x}_2, \ldots, \mathbf{x}_n$, from which the sample mean can be calculated:

$$\overline{\mathbf{x}} = \frac{1}{n}(\mathbf{x}_1 + \mathbf{x}_2 + \cdots + \mathbf{x}_n) .$$

We know that

$$E\{\overline{\mathbf{x}}\} = \eta , \qquad \sigma^2_{\overline{\mathbf{x}}} = \frac{\sigma^2}{n} ,$$

and it can be shown that

$$P\{\eta - \varepsilon < \overline{\mathbf{x}} < \eta + \varepsilon\} \geq 1 - \frac{\sigma^2}{n\varepsilon^2} .$$

The experiment E_n is now carried out once, giving a set of values $x_1, x_2, \ldots, x_n$. The quantity

$$x = \frac{1}{n}(x_1 + x_2 + \cdots + x_n)$$

is considered to be a reliable estimate of the mean η, for we can be practically certain that the inequality

$$\eta - \varepsilon < x < \eta + \varepsilon$$

will hold. Put another way, if we were to repeat the experiment a very large number of times, the quantity x would satisfy these inequalities 99% of the time, provided

$$1 - \frac{\sigma^2}{n\varepsilon^2} = 0.99 .$$

We see that, compared to the rigorous results of probability theory, statistical conclusions suffer from a certain fuzziness.

B.3.2 Estimating the Moments of a Process

Consider a stationary ergodic process $\mathbf{x}(t)$, with mean η and autocorrelation $R_{\mathbf{x}}(t)$. The physical measurement only provides information concerning a single instance of this process, and only during the time interval $[0, T]$. From the measurement, the quantities

$$x_T = \frac{1}{T}\int_0^T x(t)\,\mathrm{d}t ,$$

$$\hat{R}_T(\tau) = \frac{1}{T}\int_0^T x(t)x(t+\tau)\,\mathrm{d}t ,$$

can be calculated. A priori, these quantities are not equal to η and $R_{\mathbf{x}}(t)$. They can be regarded as instances of a random variable and a stochastic process, respectively, resulting from a transformation of the s.p. $\mathbf{x}(t)$. Indeed, these quantities are defined for any given outcome. The physical measurement gives a single value for each and these values are called *estimators* of η and $R_{\mathbf{x}}(t)$.

An estimator is said to be *unbiased* if

$$E\{\mathbf{x}_T\} = \eta , \qquad E\{\hat{\mathbf{R}}_T(\tau)\} = R_{\mathbf{x}}(\tau) .$$

We can show that, under certain convergence restrictions, the quantities x_T and $\hat{R}_T(\tau)$ tend to η and $\hat{R}_{\mathbf{x}}(\tau)$, respectively, as T tends to infinity (see Chap. 9).

Mean

It can be shown that the variance of the estimate $\mathbf{x}_T$ is given as a function of T by the expression

$$\sigma_{\mathbf{x}_T}^2 = \frac{1}{T}\int_{-\infty}^{+\infty} \left[R_{\mathbf{x}}(\tau) - \eta^2\right]\left[\Pi\left(\frac{\tau}{T}\right) \star \Pi\left(\frac{\tau}{T}\right)\right] \mathrm{d}\tau .$$

Application

Let $\mathbf{x}(t)$ be a stationary r.v. with zero mean ($\eta = 0$) and constant spectral density (i.e., white)

$$S_{\mathbf{x}}(f) = S(0) = \text{constant} ,$$

filtered by a low-pass filter

$$h(f) = \Pi\left(\frac{f}{2f_{\mathrm{c}}}\right) .$$

After filtering, the process $\mathbf{y}(t)$ has variance

$$\sigma_{\mathbf{y}}^2 = R(0) - \eta^2 = R(0) = \int_{-\infty}^{+\infty} S_{\mathbf{y}}(f)\,\mathrm{d}f = 2f_{\mathrm{c}}S(0) .$$

Let us now calculate the variance of the estimate $\mathbf{y}_T$ made on the interval $[0, T]$:

$$\begin{aligned}
y_T &= \frac{1}{T}\int_0^T y(t)\,\mathrm{d}t , \\
\sigma_{\mathbf{y}_T}^2 &= \frac{1}{T}\int_{-\infty}^{+\infty} R_{\mathbf{y}}(\tau)\left[\Pi\left(\frac{\tau}{T}\right) \star \Pi\left(\frac{\tau}{T}\right)\right]\mathrm{d}\tau \\
&= \frac{1}{T}\left|\mathrm{FT}[R_{\mathbf{y}}(\tau)] \star T[\mathrm{sinc}(Tf)]^2\right|_{f=0} \\
&= \frac{1}{T}\int_{-\infty}^{+\infty} S(f)\,|\mathrm{sinc}(Tf)|^2\,T\,\mathrm{d}f ,
\end{aligned}$$

noting that

$$\Pi\left(\frac{\tau}{T}\right) \star \Pi\left(\frac{\tau}{T}\right) = \Lambda\left(\frac{\tau}{T}\right) , \quad \mathrm{FT}[R_{\mathbf{y}}(\tau)] = S(f) .$$

If $T \gg 1/f_{\mathrm{c}}$, a condition that is easily satisfied, $\sigma_{\mathbf{y}_T}^2$ is approximated by

$$\sigma_{\mathbf{y}_T}^2 \simeq \frac{1}{T}S_{\mathbf{y}}(0) ,$$

and hence

$$\sigma_{\mathbf{y}_T}^2 \simeq \frac{\sigma_{\mathbf{y}}^2}{2f_{\mathrm{c}}T} .$$

The *estimate* $\mathbf{y}_T$ of the mean of the s.p. has indeed a variance tending to zero as T increases.

Autocorrelation

In a similar way, the estimator of the autocorrelation function is taken to be

$$\hat{R}(\tau) = \frac{1}{T}\int_0^T x(t)x(t+\tau)\,\mathrm{d}t\ .$$

It can be shown that this estimator is unbiased and that its variance

$$\mathrm{Var}\{\hat{\mathbf{R}}(\tau)\} = E\left\{|\hat{\mathbf{R}}(\tau) - R_{\mathbf{x}}(\tau)|^2\right\}$$

tends to zero, when T tends to infinity. For large τ

$$\mathrm{Var}\{\hat{\mathbf{R}}(\tau)\} \simeq \frac{1}{T}\int_{-\infty}^{+\infty} R_{\mathbf{x}}^2(\tau)\,\mathrm{d}\tau\ .$$

When the process has white spectral density and is limited to frequencies $f < f_{\mathrm{c}}$, it follows that

$$\mathrm{Var}\{\hat{\mathbf{R}}(\tau)\} \leq \frac{R_{\mathbf{x}}^2(0)}{f_{\mathrm{c}}T}\ .$$

Spectral Density

Can the quantity

$$\frac{1}{T}\left|\int_0^T x(t)\mathrm{e}^{-2\pi\mathrm{i}ft}\,\mathrm{d}t\right|^2$$

be regarded as an unbiased estimator for the spectral density $S_{\mathbf{x}}(f)$ of the process $\mathbf{x}(t)$? This could only be the case if

$$\lim_{T\to\infty}\frac{1}{T}\left|\int_0^T x(t)\mathrm{e}^{-2\pi\mathrm{i}ft}\,\mathrm{d}t\right|^2 = S_{\mathbf{x}}(f)$$

and, moreover, the variance of the estimator tends to zero as T tends to infinity. In general, these conditions are not met, the estimator is biased, and care is needed over the limit as T becomes large. We refer to more specialised treatments for further discussion.

The power spectrum can also be estimated by measuring the signal through a filter $\tilde{g}(f)$ of width Δf, centred on the frequency f_0. Then

$$\hat{S}(f_0) = \text{estimator of } S_{\mathbf{x}}(f_0) = \frac{1}{T}\int_0^T |y(t)|^2\,\mathrm{d}t$$

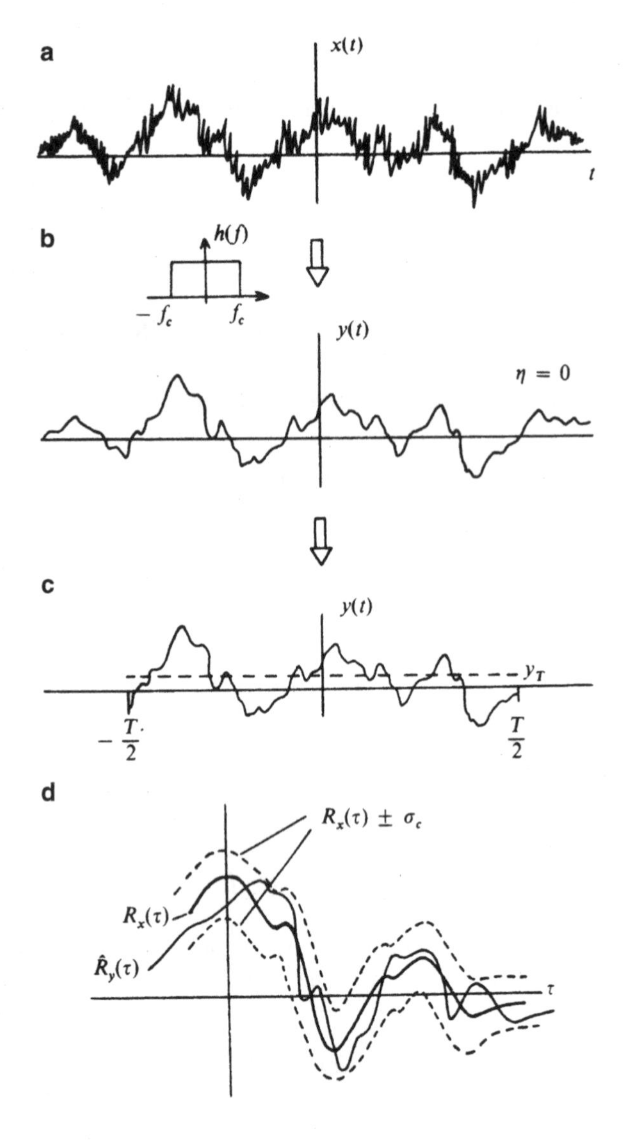

Fig. B.6 Estimation of the mean and power spectrum of a stochastic process $\mathbf{x}(t)$. (**a**) An instance $x(t,\zeta)$ of the process. (**b**) The process $\mathbf{y}(t)$ obtained from $\mathbf{x}(t)$ by the low-pass filtering $h(f)$, with cutoff frequency f_c. The high frequencies have disappeared. (**c**) Physical measurement of $\mathbf{y}(t)$ over a time T. (**d**) Autocorrelation function $R_{\mathbf{x}}(\tau)$ of the process $\mathbf{x}(t)$, and autocorrelation function $\hat{R}_{\mathbf{y}}(\tau)$ of the measurement, the *dashed curves* giving the confidence interval σ_c. The mean $\eta = 0$ of the s.p. $\mathbf{x}(t)$, and the estimate y_T of this mean ($y_T \neq 0$), have also been indicated

with

$$\tilde{y}(f) = \tilde{x}(f) \,.$$

It can then be shown that

$$\mathrm{Var}\{\hat{\mathbf{S}}(f_0)\} = \frac{S_{\mathbf{x}}^2(f)}{T\Delta f} \,.$$

Figure B.6 illustrates the above discussion.

Appendix C
Physical and Astronomical Constants

The vast majority of physical and astronomical constants required in this book can be found in Allen C.W., *Astrophysical Quantities* (AQ). A few of the most frequently used quantities are quoted here.

Speed of light	c	$2.999\,792\,458 \times 10^{8}$ m s^{-1}
Gravitational constant	G	6.670×10^{-11} N m^{2} kg^{-2}
Planck's constant	h	$6.626\,20 \times 10^{-34}$ J s
Electron charge	e	$1.602\,192 \times 10^{-19}$ C
Electron mass	m_{e}	$9.109\,56 \times 10^{-31}$ kg
Boltzmann's constant	k	$1.380\,62 \times 10^{-23}$ J K^{-1}
Proton mass	m_{p}	$1.672\,661 \times 10^{-27}$ kg
Rydberg constant	R_{H}	$109\,677.576$ cm^{-1}
Zeeman shift	$\Delta\nu/B$	$1.399\,61 \times 10^{10}$ Hz T^{-1}
Wavelength associated with 1 eV	λ_0	$1.239\,85 \times 10^{-6}$ m
Frequency associated with 1 eV	ν_0	$2.417\,965 \times 10^{14}$ Hz
Temperature associated with 1 eV		11 604.8 K
Astronomical unit	AU	$1.495\,979 \times 10^{11}$ m
Parsec	pc	$3.085\,678 \times 10^{16}$ m
		3.261 633 light yr
Mass of Sun	$M_{\odot}$	1.989×10^{30} kg
Radius of Sun	$R_{\odot}$	$6.959\,9 \times 10^{8}$ m
Luminosity of Sun	$L_{\odot}$	3.826×10^{26} W

P. Léna et al., *Observational Astrophysics*, Astronomy and Astrophysics Library,
DOI 10.1007/978-3-642-21815-6,

Blackbodies

Stefan–Boltzmann constant	$\sigma = 5.669\,56 \times 10^{-8}$ W m^{-2} K^{-4}	
Planck's law	$B_\lambda = 2hc^2\lambda^{-5}/(e^{hc/k\lambda T} - 1)$	[W m^{-2} m^{-1} sr^{-1}]
	$B_\lambda = c_1\lambda^{-5}/(e^{c_2/\lambda T} - 1)$	[W m^{-2} m^{-1} sr^{-1}]
	$c_1 = 2hc^2 = 1.191\,07 \times 10^{-16}$	[W m^2 s^{-1}]
	$c_2 = hc/k = 1.438\,83 \times 10^{-2}$	[m K]
	$B_\nu = 2h\nu^3c^{-2}/(e^{h\nu/kT} - 1)$	[W m^{-2} Hz^{-1} sr^{-1}]
	$N_\lambda = 2c\lambda^{-4}/(e^{c_2/\lambda T} - 1)$	[photons m^{-2} m^{-1} sr^{-1}]
	$N_\nu = 2\nu^2c^{-2}/(e^{h\nu/kT} - 1)$	[photons m^{-2} Hz^{-1} sr^{-1}]

Appendix D
Tables of Space Missions

Space missions at all wavelengths, as mentioned in the book.

Mission	Objective	Launch
Apollo	Lunar missions	1963–1972
AXAF (Advanced X-ray Astrophysics Facility)	X-ray observatory	1999
Boomerang	Balloon studying the diffuse cosmological background	1998
Cassini–Huygens	Exploration of Saturn	1997
CHANDRA (AXAF)	X-ray observatory	1999
Clementine	Exploration of the Moon	1994–1994
Corot (COnvection ROtation and planetary Transits)	Search for exoplanets and asteroseismology	2006
Cos-B (Cosmic ray Satellite)	γ-ray observatory	1975–1982
Darwin	Search for exo-Earths	2010
Einstein	X-ray observatory	1979
FUSE (Far Ultraviolet Spectroscopic Explorer)	UV observatory	1999
Gaia	Astrometry	2010
Galex (Galaxy Evolution Explorer)	UV Observatory	2003
Glast (Gamma-ray Large Area Space Telescope)	γ-ray observatory	2008
Granat	X-ray observatory	1989–1998
GRO (Gamma-Ray Observatory)	γ-ray observatory	1991–2000
HEAO (High Energy Astronomy Observatories)	High energy observatories	1979–1981
Hipparcos–Tycho	Astrometry	1989–1993
HST (Hubble Space Telescope)	Visible–IR observatory	1990
Integral (International Gamma-Ray Astrophysics Laboratory)	X and γ observatory	2002
IRAS (Infrared Astronomical Satellite)	IR observatory	1983

(continued)

P. Léna et al., *Observational Astrophysics*, Astronomy and Astrophysics Library,
DOI 10.1007/978-3-642-21815-6,

(continued)

Mission	Objective	Launch
ISO (Infrared Space Observatory)	IR observatory	1995–1998
IUE (International Ultraviolet Explorer)	UV observatory	1978-1996
JWST (James Webb Space Telescope)	Visible–IR observatory	2013
KAO (Kuiper Airborne Observatory)	Visible–IR observatory	1993
Kepler	Search for exo-Earths	2009
LISA (Laser Interferometer Space Antenna)	Gravitational wave detection	2018
Maxim (Microarcsecond X-ray Imaging Mission)	X-ray observatory	2020
OAO (Orbiting Astronomical Observatories)	UV observatories	1976
ODIN	Submillimetre observatory	2001
Opportunity	Exploration of Mars	2003
OSO8 (Orbiting Solar Observatory)	γ-ray observation of the Sun	1975-1978
Planck	Exploration of diffuse cosmological background	2007
Pronaos (Projet National d'AstronOmie Submillimétrique)	Submillimetre balloon	1994
Rosat (ROentgen SATellite X-ray observatory)	X-ray observatory	1992–1999
Rosetta	Study of a comet	2004
Rosita	X-ray observatory	?
Sas2 (Small Astronomy Satellite)	γ-ray observatory	1970
Sigma	γ-ray observatory	1989
SIM (Space Interferometry Mission)	X-ray observatory	2013
Skylab	Space station	1973–1979
SOFIA (Stratospheric Observatory For Infrared Astronomy)	IR and submillimetre observatory	2004
Spitzer	IR observatory	2003
STARS	Stellar photometry	2002
SUZAKU	X-ray observatory	2005
TPF (Terrestrial Planet Finder)	Search for exo-Earths	2020
UHURU (SAS-1)	X-ray observatory	1970–1973
Ulysses	Exploration of the Sun	1990
Viking	Exploration of Mars	1975–1976
Voyager	Exploration of the Solar System	1977
WMAP (Wilkinson Microwave Anisotropy Probe)	Exploration of diffuse cosmological background	2001
XEUS (X-Ray Evolving Universe Spectroscopy Mission)	X-ray observatory	2010
XMM–Newton (X-ray Multi-Mirror Mission)	X-ray observatory	1999

Appendix E
Webography

E.1 Main Earth-Based Telescopes

Site	Name	Type and diameter	Nationality	URL
Radio and submillimetre				
Australia	Australia Telescope Compact Array	Radio (6 m × 22 m)	Australia	http://www.narrabri.atnf.csiro.au/
Australia	Mopra Antenna	Radio (22 m)	Australia	http://www.atnf.csiro.au/
Australia	Parkes Observatory	Radio (64 m)	Australia	http://www.parkes.atnf.csiro.au/
Chile	Atacama Large Millimeter/Submillimeter Array (ALMA)	Europe, Japan, USA, Chile	mm (0.3–9.6)	http://science.nrao.edu/alma/index.shtml
England	Cambridge Ryle Telescope	Radio (8 × 13 m)	England	http://www.mrao.cam.ac.uk/telescopes/ryle/
England	Jodrell Bank Observatory	Radio (76 m)	England	http://www.jb.man.ac.uk/
England	Cosmic Anisotropy Telescope	mm (3 × 0.7 m)	England	http://www.mrao.cam.ac.uk/telescopes/cat/
France	Nançay Radio Observatory	Radio (100 m)	France	http://www.obs-nancay.fr/
Germany	Effelsberg Radio Telescope	Radio (100 m)	Germany	http://www.mpifr-bonn.mpg.de/div/effelsberg/index_e.html

(continued)

P. Léna et al., *Observational Astrophysics*, Astronomy and Astrophysics Library,
DOI 10.1007/978-3-642-21815-6,

(continued)

Site	Name	Type and diameter	Nationality	URL
Holland	Westerbork Synthesis Radio Telescope	Radio (14 × 25 m)	Holland	http://www.astron.nl/radio-observatory/public/public-0
India	Giant Metrewave Radio Telescope	Radio (30 × 45 m)	India	http://gmrt.ncra.tifr.res.in/
Japan	Nobeyama Radio Observatory	Radio (45 m)	Japan	http://www.nro.nao.ac.jp/indextop-e.html
Mexico	Large Millimeter Telescope	mm (50 m)	Mexico, USA	http://www.lmtgtm.org/
Puerto Rico	Arecibo Observatory	Radio (305 m)	USA	http://www.naic.edu/
Sweden	Onsala Space Observatory	mm (20 and 25 m)	Sweden	http://www.oso.chalmers.se/
USA	Five College Radio Astronomy Observatory	Radio (14 m)	USA	http://donald.phast.umass.edu/~fcrao/
USA	Green Bank	Radio (45 and 100 m)	USA	http://www.gb.nrao.edu/
USA	Hat Creek Radio Observatory	Radio (6 × 6 m)	USA	http://www.hcro.org/
USA	Owens Valley Radio Observatory	Radio (6 × 10.4 m)	USA	http://www.ovro.caltech.edu/
USA	Very Large Array	Radio (27 × 25 m)	USA	http://www.vla.nrao.edu/
USA (Hawaii, Mauna Kea)	Caltech Submillimeter Observatory	mm (10.4 m)	USA	http://www.submm.caltech.edu/cso/
USA (Hawaii, Mauna Kea)	James Clerk Maxwell Telescope	mm (15 m)	Canada, UK, Holland	http://www.jach.Hawaii.edu/JCMT/
Optical and infrared				
Australia	Anglo-Australian Observatory	Optical (3.9 m)	Australia, UK	http://www.aao.gov.au/
Australia	Mount Stromlo and Siding Spring Observatories	Optical (1.85 and 2.3 m)	Australia	http://msowww.anu.edu.au/observing/
Chile (Cerro Panchón) and USA (Hawaii)	Gemini Observatory	Optical–IR (8 m)	USA, UK, Canada, Chile, Argentina, Brazil	http://www.gemini.edu/

(continued)

(continued)

Site	Name	Type and diameter	Nationality	URL
Chile (Cerro Tololo)	Cerro Tololo Interamerican Observatory	optique (1, 1.5 et 4m)	USA	http://www.ctio.noao.edu/ctio.html
Chile (La Silla)	La Silla–ESO Facilities	Optical (1.5, 2.2, 3.5, and 3.6 m)	Europe	http://www.ls.eso.org/
Chile (Cerro Paranal)	Very Large Telescope	Optical (4 × 8 m)	Europe	http://www.eso.org/public/teles-instr/vlt.html
Chile	European Extremely Large Telescope (E-ELT)	Optical (42 m)	Europe	http://www.eso.org/projects/e-elt/
England	Cambridge Optical Aperture Synthesis Telescope	Optical (4 × 0.4 m)	England	http://www.mrao.cam.ac.uk/telescopes/coast/
France	Observatoire de Haute-Provence	Optical (1.2, 1.5, and 1.93 m)	France	http://www.obs-hp.fr/
France	Observatoire Midi-Pyrenées	Optical (1 and 2 m)	France	http://www.obs-mip.fr/
Italy	Loiano Telescopes	Optical (0.6 and 1.5 m)	Italy	http://www.bo.astro.it/loiano/
Mexico	Guillermo Haro Observatory	Optical (2.1 m)	Mexico	http://www.inaoep.mx/~astrofi/cananea/
South Africa	South African Astronomical Observatory	Optical (2 m)	South Africa	http://www.saao.ac.za/
Spain (Calar Alto)	Calar Alto Observatory	Optical (3.5 m)	Germany, Spain	www.caha.es
Spain (Canary Islands)	Telescopio Nazionale Galileo	Optical (3.5 m)	Italy	http://www.tng.iac.es/
Spain (Canary Islands)	THEMIS Heliographic Telescope for the Study of the Magnetism and Instabilities on the Sun	Optical (2 m)	France, Italy	http://www.themis.iac.es/

(continued)

(continued)

Site	Name	Type and diameter	Nationality	URL
Spain (Canary Islands)	Gran Telescopio de Canarias	Optical (10 m)	Spain	http://www.gtc.iac.es/
Spain (Canary Islands)	Nordic Optical Telescope	Optical (2.6 m)	Denmark	http://www.not.iac.es/
Spain (Canary Islands)	Isaac Newton Group	Optical (4.2, 2.5, and 1 m)	UK, Holland	http://www.ing.iac.es/
USA (Hawaii, Mauna Kea)	Infra-Red Telescope Facility	IR (3 m)	USA	http://irtfweb.ifa.hawaii.edu
USA	Giant Magellan Telescope (GMT)	Optical (24.5 m)	USA	http://www.gmto.org/
USA	Thirty Meter Telescope (TMT)	Optical (30 m)	USA	http://www.tmt.org/
USA	UK Infra-Red Telescope	IR (3.8 m)	UK	http://www.jach.hawaii.edu/UKIRT/
USA (Hawaii, Mauna Kea)	Air Force Maui Optical Station	Optical (3.7 m)	USA	http://www.fas.org/spp/military/program/track/amos.htm
USA	Apache Point Observatory	Optical (3.5 m)	USA	http://www.apo.nmsu.edu/
USA	Palomar Observatory	Optical (1.2, 1.5, and 5 m)	USA	http://astro.caltech.edu/observatories/palomar/
USA (Hawaii, Mauna Kea)	SUBARU Telescope	Optical (8.3 m)	Japan	http://subarutelescope.org/
USA	Lick Observatory	Optical (3 m)	USA	http://www.ucolick.org/
USA (Hawaii, Mauna Kea)	University of Hawaii Telescope	Optical (2.2 m)	USA	http://www.ifa.hawaii.edu/88inch/88inch.html
USA	Hobby Eberly Telescope	Optical (9 m)	USA, Germany	http://www.astro.psu.edu/het/
USA (Hawaii, Mauna Kea)	Canada–France–Hawaii Telescope	Optical (3.6 m)	Canada, France, USA	http://www.cfht.hawaii.edu/
USA (Hawaii, Mauna Kea)	Keck Observatory	Optical (2 × 10 m)	USA	http://www.keckobservatory.org/
USA	Kitt Peak National Observatory	Optical (0.9, 2.1, and 4 m)	USA	http://www.noao.edu/kpno/

(continued)

(continued)

Site	Name	Type and diameter	Nationality	URL
USA	Large Binocular Telescope Observatory	Optical (2 × 8.4 m)	USA, Italy	www.lbto.org
USA	Mount Wilson Observatory	Optical (1.5, 2.5 m)	USA	http://www.mtwilson.edu/
USA	Multiple Mirror Telescope Observatory	Optical (6 × 1.8 m)	USA	http://cfa-www.harvard.edu/mmt/, http://www.mmto.org/
Gravitational waves and neutrinos				
Antartic	Antarctic Muon and Neutrino Detector Array	mm (13 m × 0.2 m)	USA	http://amanda.uci.edu/
Canada	Sudbury Neutrino Observatory	Neutrinos	Canada, UK, USA	http://www.sno.phy.queensu.ca/
Italy	Virgo Interferometer	Gravitational waves	France, Italy	https://wwwcascina.virgo.infn.it/
USA	Laser Interferometer Gravitational Wave Observatory	Gravitational waves	USA	http://www.ligo.caltech.edu/

E.2 Recent Space Missions

Name	Period of activity	Owners	Main scientific objective and URL
AKARI	2006–2007	Japan, Europe	Photometric and spectroscopic observation in the region 2–180 μm in 13 spectral bands. http://www.astro-f.esac.esa.int/
Cassini–Huygens Mission to Saturn & Titan	1997–present	Europe	Study of Saturn and Titan from the visible to the infrared. http://saturn.jpl.nasa.gov/home/
Chetra X-ray Observatory	1999–present	USA	Observation of X-ray emissions from high-energy phenomena such as supernovas. http://chandra.harvard.edu/
COnvection ROtation et planetary Transits (COROT)	2006–2012 (or beyond)	France	Asteroseismology and search for exoplanets. http://smsc.cnes.fr/COROT/

(continued)

(continued)

Name	Period of activity	Owners	Main scientific objective and URL
COsmic Background Explorer (COBE)	1989–present	USA	IR and microwave measurement of diffuse cosmological background. http://lambda.gsfc.nasa.gov/product/cobe/
Far Ultraviolet Spectroscopic Explorer (FUSE)	1999–2008	USA, France, Canada	Exploration of the Universe with high-resolution spectroscopy in the extreme UV. http://fuse.pha.jhu.edu/
GALaxy Evolution EXplorer (GALEX)	2003–present	USA, Korea, France	UV observation of nearby and remote galaxies. http://www.galex.caltech.edu/
Global Astrometric Interferometer for Astrophysics (GAIA)	~ 2012	Europe	Astrometry of around a billion stars in our Galaxy. http://www.rssd.esa.int/gaia/
High Energy Transient Explorer	2002–present	USA, Japan, France, Brazil, Italy, India	Detection and localisation of gamma-ray bursts. http://space.mit.edu/HETE/
Hubble Space Telescope (HST)	1990–present	USA	UV to IR observatory. http://www.stsci.edu/hst/
INTErnational Gamma-Ray Astrophysics Laboratory (INTEGRAL)	2002–present	Europe	Detection of the highest energy phenomena in the Universe. http://sci.esa.int/science-e/www/area/index.cfm?fareaid=21
Infrared Space Observatory	1995–1998	Europe	Exploration of the IR Universe. http://www.sci.esa.int/iso/
James Webb Space Telescope (JWST)	~ 2014	USA	Optical–IR observatory (successor to HST). http://www.jwst.nasa.gov/
Kepler mission	2009–2014	USA	Detection of exoplanets. http://kepler.nasa.gov/
PLANCK	2009–present	Europe	Probing anisotropies in the diffuse cosmological background. http://www.rssd.esa.int/index.php?project=PLANCK
ROSAT	1990–1999	Germany, UK, USA	Global UV exploration of the Universe. http://heasarc.nasa.gov/docs/rosat/rosgof.html
ROSETTA	2004–present	Europe	Study of the comet 67P/Churyumov–Gerasimenko. http://www.esa.int/esaSC/120389_index_0_m.html

(continued)

(continued)

Name	Period of activity	Owners	Main scientific objective and URL
SOlar and Heliospheric Observatory (SOHO)	1995–present	Europe, USA	Study of the internal structure and atmosphere of the Sun. http://sohowww.nascom.nasa.gov/
Solar TErrestrial RElations Observatory (STEREO)	2006–present	USA	Study of coronal mass ejections from the Sun using 3D data. http://stp.gsfc.nasa.gov/missions/stereo/stereo.htm
Spitzer Space Telescope	2003–present	USA	Imaging and spectroscopy between 3 and 180 microns. http://www.spitzer.caltech.edu/
Stardust	1999–present	USA	Exploration of a comet. http://stardust.jpl.nasa.gov/
Wilkinson Microwave Anisotropy Probe (WMAP)	2001–present	USA	Mapping the diffuse cosmological background to investigate the formation of the first galaxies.http://map.gsfc.nasa.gov/
XMM-Newton	1999–present	Europe	To study the violent Universe, from the neighbourhood of black holes to the formation of the first galaxies. http://www.esa.int/science/xmmnewton, http://xmm.esac.esa.int/

E.3 Databases

Nom	Description	URL
Aladin Sky Atlas	Interactive sky map giving access to digital images of every region of the sky	http://aladin.u-strasbg.fr/
Astrophysical CATalogs support System (CATS) of the Special Astrophysical Observatory	Several astrophysical catalogues, especially in the radio region	http://cats.sao.ru/
Atomic Data for Astrophysics	Data for collisional ionisation and photoionisation, recombination, opacities, etc.	http://www.pa.uky.edu/~verner/atom.html
Besançon Double and Multiple Star Database	Photometric and spectroscopic data on binary and multiple star systems	http://bdb.obs-besancon.fr/

(continued)

(continued)

Nom	Description	URL
Canadian Astronomy Data Centre (CADC)	Storage of ground and space data	http://cadcwww.dao.nrc.ca/
Cassini–Huygens Mission to Saturn & Titan	Data from the Cassini–Huygens mission	http://saturn.jpl.nasa.gov/home/
Catalog of Infrared Observations	Access to published data on more than 10 000 sources observed in the infrared	http://ircatalog.gsfc.nasa.gov
Catalogue of Galactic Supernova Remnants	Catalogue supernova remnants in the Galaxy	http://www.mrao.cam.ac.uk/surveys/snrs/
Center for Earth et Planetary Studies (CEPS)	Several thousand images of the Solar System	http://www.nasm.si.edu/ceps/
Centre de Donnees de la Physique des Plasmas	Data on plasma physics	http://cdpp.cesr.fr/
Centre de Données Astronomiques de Strasbourg	Several thousand astronomical catalogues	http://vizier.u-strasbg.fr/
Digitized Sky Survey at LEDAS	Access to the digitised data of the STScI Digitised Sky Surveys DSS and DSS-II	http://ledas-www.star.le.ac.uk/DSSimage
ELODIE archive	More than 16 000 echelle spectrograms obtained with ELODIE at the Observatoire de Haute Provence	http://atlas.obs-hp.fr/elodie/
ESO Science Archive Facility	Archives from all data obtained with ESO telescopes	http://archive.eso.org/
European Asteroidal Occultation Results	Observations of stellar occultations by asteroids in Europe since 1997	http://www.euraster.net/
European Pulsar Network Data Archive	Measurements of pulsar profiles	http://www.mpifr-bonn.mpg.de/old_mpifr/div/pulsar/data/
Extrasolar Planets Encyclopedia	Catalogue of all extrasolar planets discovered up to now	http://exoplanet.eu/
Galactic Classical Cepheid Database	List of all known population I Cepheids in our Galaxy	http://www.astro.utoronto.ca/DDO/research/cepheids/
High Energy Astrophysics Science Archive Research Center	Archives from several high energy space missions	http://heasarc.gsfc.nasa.gov/
Hypercat	Interface for acces to several extragalactic databases	http://www-obs.univ-lyon1.fr/hypercat/

(continued)

(continued)

Nom	Description	URL
INTErnational Real-time MAGnetic Observatory NETwork	Data and information relating to magnetic observatories	http://www.intermagnet.org/
InfraRed Science Archive	Data from NASA's infrared and millimetre missions	http://irsa.ipac.caltech.edu/
Italian Database of Astronomical Catalogues	Around 280 catalogues developed since 1984 by the Astronet group in Italy	http://db.ira.cnr.it/dira2/
LEicester Database and Archive Service	Access to data from high energy space missions like EXOSAT and ROSAT	http://ledas-www.star.le.ac.uk/
Libraries of stellar spectra	Catalogue of stellar spectra	http://www.ucm.es/info/Astrof/invest/actividad/spectra.html
McMaster Cepheid Photometry and Radial Velocity Archive	Photometric and radial velocity data from galactic and extragalactic Cepheids	http://www.physics.mcmaster.ca/Cepheid/
Minor Planet Center	Astrometric data and orbits of comets and small objects	http://minorplanetcenter.net/
Multimission Archive at STScI	A wide range of astronphysical archives, mainly in the optical, near-infrared, and UV regions	http://archive.stsci.edu/mast.html
NASA National Space Science Data Center	36 Tb of digital data from 440 NASA space missions	http://nssdc.gsfc.nasa.gov/
NASA/IPAC Extragalactic Database	Access to a large amount of published extragalactic data	http://nedwww.ipac.caltech.edu/
NCSA Astronomy Digital Image Library	Collection of astronomical images in FITS format	http://www.dlib.org/dlib/october97/adil/10plante.html
Open Cluster Database	Information on more than 100 000 stars belonging to about 500 open clusters	http://www.univie.ac.at/webda/
POLLUX database of stellar spectra	High-resolution optical stellar spectra	http://pollux.graal.univ-montp2.fr/
SDSS	Access to data from the SDSS extragalactic sky survey	http://www.sdss.org
SkyCat	Tool developed at the ESO for visualising images and accessing astronomical catalogues	http://archive.eso.org/skycat/
TOPbase at CDS: The Opacity Project	Energy levels of the most abundant elements in astrophysics	http://cdsweb.u-strasbg.fr/topbase/topbase.html

(continued)

(continued)

Nom	Description	URL
UK Astronomy Data Centre	Selection of data obtained by Earth-based telescopes in the UK	http://casu.ast.cam.ac.uk/casuadc
VizieR Catalogue Service	Access to a complete library of published astronomical catalogues	http://vizier.u-strasbg.fr/
Washington Double Star Catalog	Main astrometric database of double and multiple star systems	http://ad.usno.navy.mil/wds/wds.html
Wide Field Plate Database	Descriptive information about wide field photographic data stored in several places around the world	http://www.skyarchive.org/
XMM–Newton catalogue interface	Public interface providing access to the catalogue of sources observed by XMM–Newton	http://xcatdb.u-strasbg.fr/

E.4 Journals

Name	URL
Acta Astronomica	http://acta.astrouw.edu.pl/
Annual Reviews of Astronomy and Astrophysics	http://www.annualreviews.org/journal/astro
ASP Conference Series	http://www.astrosociety.org/pubs/cs/confseries.html
Astronomical Journal	http://iopscience.iop.org/1538-3881
Astronomy & Geophysics	http://www.blackwellpublishing.com/journals/AAG/
Astronomy and Astrophysics	http://www.aanda.org/
Astrophysical Journal	http://iopscience.iop.org/0004-637X/
Astrophysical Journal Supplement	http://iopscience.iop.org/0067-0049/
Astrophysics and Space Science	http://www.wkap.nl/journalhome.htm/0004-640X
Cambridge University Press	http://www.cup.cam.ac.uk
Classical and Quantum Gravity	http://www.iop.org/Journals/cq/
ESO Publications	http://www.eso.org/gen-fac/pubs/
IAU publications	http://www.iau.org/science/publications/iau/
Icarus	http://icarus.cornell.edu
Journal of Astronomical Data	http://www.vub.ac.be/STER/JAD/jad.htm
Journal of Cosmology and Astroparticle Physics	http://www.iop.org/EJ/journal/JCAP/
Journal of Optics A : Pure and Applied Optics	http://www.iop.org/Journals/oa
Journal of the British Astronomical Association	http://www.britastro.org/journal/
Monthly Notices of the Royal Astronomical Society	http://www.blackwellpublishing.com/journal.asp?ref=0035-8711
Nature	http://www.nature.com/

(continued)

(continued)

Name	URL
Observatory Magazine	http://www.ulo.ucl.ac.uk/obsmag/
Optics Express	http://www.opticsexpress.org/
Physical Review Letters	http://prl.aps.org/
Publications of the Astronomical Society of Australia	http://www.publish.csiro.au/nid/138.htm
Publications of the Astronomical Society of Japan	http://pasj.asj.or.jp/
Publications of the Astronomical Society of the Pacific. Electronic Edition	http://www.journals.uchicago.edu/PASP/journal/
	http://pasp.phys.uvic.ca/
Science	http://www.sciencemag.org/

E.5 Bibliographical Research

Name	URL
Astrophysics Data System	http://adswww.harvard.edu/
E-print archive in Astrophysics (astro-ph, arXiv)	http://arxiv.org/archive/astro-ph/

E.6 Image Sources

Name	Description	URL
APM Galaxy Survey	About 2 million galaxies	http://www.nottingham.ac.uk/~ppzsjm/apm/apm.html
Astronomy Picture of the Day	A different astronomical image for every day of the week	http://antwrp.gsfc.nasa.gov/apod/astropix.html
Big Bear Solar Observatory	Images of the Sun at several wavelengths, updated each day	http://www.bbso.njit.edu/
CCD Images of Messier Objects	Images of Messier objects obtained with several instruments	http://zebu.uoregon.edu/messier.html
Center for Earth and Planetary Studies	A large collection of images of planets in the Solar System	http://www.nasm.si.edu/ceps/
Color Stereo Photos of Mars from the 2004 Rovers	Images of the surface of Mars	http://astro.uchicago.edu/cosmus/projects/marsstereo/

(continued)

(continued)

Name	Description	URL
Digital Archive of Historical Astronomy Pictures	Collection of historical images	http://www42.pair.com/infolund/bolaget/DAHAP/
ESO Photo Gallery. Astronomical Images	Images obtained with ESO telescopes	http://www.eso.org/outreach/gallery/astro/
Galaxy Zoo	Project for classifying a large number of nearby galaxies	http://www.galaxyzoo.org/
Hubble Heritage Project	Large collection of images obtained with the Hubble Space Telescope	http://heritage.stsci.edu/
Hubble Space Telescope picture gallery	Images from the Hubble Space Telescope	http://hubblesite.org/gallery/
Infrared Space Observatory Science Gallery	Images from the ISO satellite	http://iso.esac.esa.int/science/
IPAC Image Gallery	Images from several infrared space missions	http://coolcosmos.ipac.caltech.edu/image_galleries/missions_gallery.html
JPL Planetary Photojournal	Images of the Solar System	http://photojournal.jpl.nasa.gov/
Messier Pages at SEDS	Images of the Messier catalogue	http://www.seds.org/messier/
Multiwavelength Atlas of Galaxies	Images of several nearby galaxies at several wavelengths	http://astronomy.swin.edu.au/staff/gmackie/atlas/atlas_edu.html
Multiwavelength Milky Way	Images of the Milky Way at several wavelengths	http://mwmw.gsfc.nasa.gov/
NASA JSC Digital Image Collection	Images from several NASA missions	http://images.jsc.nasa.gov/
NOAO Image Gallery	Images obtained with the NOAO telescope	http://www.noao.edu/image_gallery/
NSSDC Photo Gallery	Collection of images classified in terms of the type of astrophysical object	http://nssdc.gsfc.nasa.gov/photo_gallery/photogallery.html
Radio Galaxy and Quasar Images	Images of extragalactic radio sources obtained with the VLA	http://www.cv.nrao.edu/~abridle/images.htm
Two Micron All Sky Survey (2MASS) Image Gallery	Images from the 2MASS survey	http://www.ipac.caltech.edu/2mass/gallery/

E.7 Education

Name	Description	URL
Associations and Organisations		
Astronomical Society of the Pacific	Education in astronomy	http://www.astrosociety.org/
European Association for Astronomy Education	Association with the aim of encouraging and propagating astronomy	http://www.eaae-astro.org/
Hands On Universe	Educational programme for students to learn about the Universe	http://www.handsonuniverse.org/
Educational Sites		
A website for popularization of Astrophysics and Astronomy	Several basis lessons in astronomy	http://www.astrophysical.org/
A knowledge for extragalactic astronomy and cosmology	Documents around the theme of extragalactic astronomy	http://nedwww.ipac.caltech.edu/level5/
Astrocentral	Popularisation	http://www.astrocentral.co.uk
Atlas of the Universe	Overview of the Universe on different scales	http://www.atlasoftheuniverse.com/
Constellations and Their Stars	A page for each constellation	http://www.astro.wisc.edu/~dolan/constellations/
Curious About Astronomy? Ask an Astronomer	Site for asking professional astronomers questions	http://curious.astro.cornell.edu/
Falling into a Black Hole	Some information about the physics of black holes	http://casa.colorado.edu/~ajsh/schw.shtml
Imagine the Universe	Astronomical site for children	http://imagine.gsfc.nasa.gov/
Infrared Astronomy Tutorial	On the subject of infrared astronomy	http://coolcosmos.ipac.caltech.edu/
MAP Introduction to Cosmology Page	Introduction to cosmology	http://map.gsfc.nasa.gov/m_uni.html
NASA Spacelink	NASA's outreach site	http://www.nasa.gov/audience/foreducators/index.html
Ned Wright's Cosmology Tutorial	Lectures on cosmology	http://www.astro.ucla.edu/~wright/cosmolog.htm
sci.astro Frequently Asked Questions	Answers to frequently asked questions	http://sciastro.astronomy.net/
Science Fiction Stories Using Good Astronomy	Introduction to astrophysics through science fiction	http://www.astrosociety.org/education/resources/scifi.html
Space Time Travel	Illustrations by simulations of predictions from general relativity	http://www.spacetimetravel.org

(continued)

(continued)

Name	Description	URL
StarChild	Basic lessons for children	http://starchild.gsfc.nasa.gov/
Windows to the Universe	General astronomy site	http://www.windows.ucar.edu/
History of astronomy		
Astronomiae Historia	General history of astronomy	http://www.astro.uni-bonn.de/~pbrosche/astoria.html
Galileo Project	Information about Galileo	http://www.rice.edu/Galileo/

E.8 Computing and Astronomy

Description	URL
Work environment	
Basic UNIX commands	http://mally.stanford.edu/~sr/computing/basic-unix.html
Word processing	
LaTeX	http://www.latex-project.org/
An introduction to LaTeX	http://www.maths.tcd.ie/~dwilkins/LaTeXPrimer/
LaTeX mathematical symbols	http://web.ift.uib.no/Fysisk/Teori/KURS/WRK/TeX/symALL.html
Text editors (programming)	
Emacs	http://www.gnu.org/software/emacs/
An introduction to Emacs	http://www2.lib.uchicago.edu/~keith//tcl-course/emacs-tutorial.html
IDL under Emacs	http://idlwave.org/
Data processing software	
Matlab	http://mathworks.com
Octave (Free version of Matlab)	http://www.gnu.org/software/octave/
IDL	http://www.ittvis.com/index.asp
An introduction to IDL	http://nstx.pppl.gov/nstx/Software/IDL/idl_intro.html
Library of IDL routines astronomy	http://idlastro.gsfc.nasa.gov/contents.html
GDL (Free version of IDL)	http://gnudatalanguage.sourceforge.net/
IRAF	http://iraf.noao.edu/
Python	http://www.python.org/
TOPCAT	http://www.star.bris.ac.uk/~mbt/topcat/
Graphics	
Gnuplot	http://www.gnuplot.info/
SuperMongo	http://www.astro.princeton.edu/~rhl/sm/

E.9 Resources

Name	URL
Astronomical Internet Resources	http://www.stsci.edu/resources/
Astronomy resources on the Internet	http://www.cv.nrao.edu/fits/www/astronomy.html
AstroWeb at CDS: yellow-page services	http://cdsweb.u-strasbg.fr/astroweb.html
WebStars	http://heasarc.gsfc.nasa.gov/docs/www_info/webstars.html

Appendix F
Acronyms

For improved readability, most of these acronyms will be given in capital letters.

AAT	Anglo-Australian Telescope
ACS	Advanced Camera for Surveys
AIC	Achromatic Interferential Coronagraphy
AIPS	Astronomical Image Processing System
ALMA	Atacama Large Millimeter Array
a.m.	*ante meridiem*
AMANDA	Antartic Muon And Neutrino Detector Array
AMBER	Astronomical Multi-Beam Combiner
ANTARES	Astronomy with a Neutrino Telescope and Abyss environmental Research
ASIC	Application Specific Integrated Circuit
AVO	Astronomical Virtual Observatory
AXAF	Advanced X-ray Astronomical Facility
BAT	Burst Alert Telescope
BATSE	Burst And Transient Source Experiment
BDL	Bureau des Longitudes
BIB	Blocked Impurity Band
BIH	Bureau International de l'Heure
BIMA	Berkeley Illinois Maryland Array
BIPM	Bureau International des Poids et Mesures
CAD	Computer Aided Design
CARMA	Combined Array for Research in Millimeter-wave Astronomy
CAT	Cherenkov Array at Themis
CCD	Charge Coupled Device
CDA	Centre de Données Astronomiques (Strasbourg)
CEA	Commissariat à l'Énergie Atomique (France)
CFHT	Canada–France–Hawaii Telescope
CFRS	Canada–France Redshift Survey
CHARA	Center for High Angular Resolution Astronomy

P. Léna et al., *Observational Astrophysics*, Astronomy and Astrophysics Library,
DOI 10.1007/978-3-642-21815-6, © Springer-Verlag Berlin Heidelberg 2012

CIA	Coronographe Interférométrique Achromatique
CIAXE	Coronographe Interférométrique Achromatique Axé
CIB	Cosmic Infrared Background
CID	Charge Injection Device
CINECA	Consorzio Interuniversitario per il Calcolo Automatico dell'Italia Nord Orientale
CLV	Center-to-Limb Variation
CMA	Channel Multiplier Array
CMB	Cosmic Microwave Background
CMOS	Complementary Metal Oxide Semiconductor
CNES	Centre National d'Études Spatiales (France)
CNOC	Canadian Network for Observational Cosmology
CNRS	Centre National de la Recherche Scientifique (France)
COAST	Cambridge Optical Array Synthesis Telescope
COB	Cosmic Optical Background
COBE	COsmic Background Experiment
COMBO	Classifying Objects by Medium Band Observation
COROT	Convection, Rotation, et Transits planétaires
CRIRES	Cryogenic High Resolution Infrared Echelle Spectrograph
CSO	CalTech Submillimeter Observatory
DARA	Deutsche Agentur für Raumfahrt Angelegenheiten (Germany)
DAT	Digital Audio Tape
DEIMOS	Deep Imaging Multi-Object Spectrograph
DENIS	Deep Near Infrared Survey
DIMM	Differential Monitor
DIVA	Deutsches Interferometer für Vielkanalphotometrie und Astrometrie
DLR	Deutsches Luft- und Raumfahrt
DORIS	Détermination d'Orbite et Radiopositionnement Intégré par Satellite
DQE	Detector Quantum Efficiency
DRO	Direct Read Out
DSB	Double Side Band
DUMAND	Deep Underwater Muon And Neutrino Detector
DVR	Direct Voltage Readout
EBCCD	Electron Bombarded Charge Coupled Device
EBS	Electron Bombarded Silicon
ECU	European Currency Unit
EGO	European Gravitation Observatory
EGRET	Energetic Gamma-Ray Experiment Telescope
ELT	Extremely Large Telescope
E-ELT	European Extremely Large Telescope
ESA	European Space Agency
ESO	European Southern Observatory
ESRF	European Synchrotron Research Facility

ET	Ephemeris Time
EUV	Extreme Ultra Violet
EVN	European VLBI Network
FAME	Full Sky Astrometric Mapping Explorer
FET	Field Effect Transistor
FFT	Fast Fourier Transform
FGRS	(2D) Field Galaxy Redshift Survey
FIRAS	Far InfraRed Astronomical Spectrometer
FIRST	Far InfraRed Space Telescope
FK	Fundamental Katalog
FOC	Faint Object Camera (Hubble)
FORS	Focal Reducer and Spectrograph
FP	Fabry–Perot
FT	Fourier Transform
FTS	Fourier Transform Spectrometer
FU	Flux Unit
FUSE	Far Ultraviolet Spectroscopic Explorer
FWHM	Full Width at Half Maximum
GAIA	Global Space Astrometry
GALLEX	Gallium Experiment
GHz	Gigahertz
GI2T	Grand Interféromètre à deux Télescopes
GLAO	Ground Layer Adaptive Optics
GLAST	Gamma-ray Large Area Space Telescope
GLONASS	GLObal'naya NAvigatsionnaya Sputnikovaya Sistema
GMRT	Giant Meter Radio Telescope
GNO	Gallium Neutrino Observatory
GPS	Global Positioning System
HALCA	Highly Advanced Laboratory for Communications and Astronomy
HARPS	High Accuracy Radial velocity Planet Searcher
HCRF	Hipparcos Celestial Reference Frame
HDF	Hubble Deep Field
HEAO	High Energy Astronomical Observatory
HEB	Hot Electron Bolometer
HEMT	High Electron Mobility Transistor
HESS	High Energy Stereoscopic System
HETG	High Energy Transmission Spectrometer
HIPPARCOS	HIgh Precision PARallax COllecting Satellite
HPBW	Half Power Beam Width
HRS	High Resolution Spectrograph (Hubble Space Telescope)
HST	Hubble Space Telescope
HURA	Hexagonal Uniform Redundant Array
IAU	International Astronomical Union
IBC	Impurity Band Conduction
IBIS	Imager on Board Integral

ICRF	International Celestial Reference Frame
ICRS	International Celestial Reference System
IDL	Interactive Data Language
IDRIS	Institut du Développement et des Ressources en Informatique Scientifique (France)
IF	Intermediate Frequency
INSU	Institut National des Sciences de l'Univers (France)
IOTA	Infrared Optical Telescope Array
IPCS	Image Photon Counting System
IR	Infrared
IRAC	Infrared Camera
IRAF	Image Reduction and Analysis Facility
IRAM	Institut de RadioAstronomie Millimétrique (France)
IRAS	InfraRed Astronomy Satellite
ISGRI	Integral Soft Gamma Ray Imager
ISI	Infrared Spatial Interferometer
ISM	Interstellar Medium
ISO	Infrared Space Observatory
ISS	International Space Station
IUE	International Ultraviolet Explorer
JCMT	James Clerk Maxwell Telescope
JD	Julian Day
JSET	Junction Supraconducting Effect Transistor
JWST	James Webb Space Telescope
KAO	Kuiper Airborne Observatory
KI	Keck Interferometer
LAMOST	Large Meter Optical Spectroscopy Telescope
LBT	Large Binocular Telescope
LGS	Laser Guide Star
LIDAR	LIght Detection And Ranging
LIGO	Laser Interferometer Gravitational wave Observatory
LISA	Large Interferometric Space Antenna
LO	Local Oscillator
LOFAR	Low Frequency Array
LS	Least Squares
LSR	Local Standard of Rest
LTE	Local Thermodynamic Equilibrium
MACAO	Multi Application Curvature Adaptive Optics
MAMA	Machine À Mesurer pour l'Astronomie (France)
MAP	See WMAP
MAXIM	Micro Arcsecond X-ray Imaging Mission
MCAO	Multi Conjugate Adaptive Optics
MEM	Maximum Entropy Method
MEMS	Micro Electro Mechanical System
MHD	Magnetohydrodynamic

MIDAS	Münich Interactive Data Analysis System
MIDI	Mid-Infrared Interferometric Instrument
MIRA	Mitaka Optical InfraRed Interferometer
MIS	Metal Insulator Semiconductor
MISTRAL	Myopic Edge-Preserving Image Restoration Method, with Application to Astronomical Adaptive-Optics-Corrected Long-Exposure Images
ML	Maximum Likelihood
MMT	MultiMirror Telescope
MOS	Metal Oxide Semiconductor
MPG	Max-Planck Gesellschaft
MRO	Magdalena Ridge Observatory
MTF	Modulation Transfer Function
MURA	Modified Uniform Redundant Array
NACO	NAOS-Conica
NAOS	Nasmyth Adaptive Optics System
NASA	National Aeronautic and Space Administration (USA)
NASDA	National Air and Space Development Agency (Japan)
NBS	National Bureau of Standards
NEC	Noise Equivalent Charge
NEF	Noise Equivalent Flux
NEMO	Neutrino Ettore Majorana Observatory
NEP	Noise Equivalent Power
NESTOR	Neutrino Extended Submarine Telescope with Oceanographic Research
NGC	New General Catalog
NGS	Natural Guide Star
NICMOS	Near Infrared Camera for Multi-Object Spectrograph
NOAO	National Optical Astronomy Observatory (USA)
NPOI	Navy Prototype Optical Interferometer
NRAO	National RadioAstronomy Observatory (USA)
NSF	National Science Foundation
NSO	National Solar Observatory
NTT	New Technology Telescope
OAO	Orbiting Astronomical Observatory
OECD	Organisation for Economic Cooperation and Development
OHANA	Optical Hawaiian Array for Nanoradian Astronomy
OVLA	Optical Very Large Array
OVRO	Owens Valley Radio Observatory
PACS	Photodetector Array Camera and Spectrometer (Herschel mission)
PAPA	Precision Analog Photon Address
PC	Photoconductor
p.d.	Potential difference
PIAA	Phase Induced Amplitude Apodization
PICSIT	Pixelated CsI Telescope (Integral mission)

PM	PhotoMultiplier
p.m.	*post meridiem*
PMM	Precision Measuring Machine
POSS	Palomar Optical Sky Survey
POST	POlar Stratospheric Telescope
ppb	Parts per billion
ppm	Parts per million
PSD	Power Spectral Density
PSD	Pulse Shape Discrimination
PSF	Point Spread Function
PSR	Pulsar
PSS	Palomar Sky Survey
QSO	QuasiStellar Object (quasar)
RADAR	RAdio Detection and Ranging
RENATER	Réseau National de la Recherche et de la Technologie (France)
RF	Radiofrequency
RGS	Resolution Grating Spectrometer (X-ray mission)
RISC	Reduced Instruction Set Computer
rms	Root mean square
ROSAT	Roentgen Space Astronomical Telescope
ROSITA	Roentgen Survey with an Imaging Telescope on the ISS
SAGE	Russian American Gallium Experiment
SALT	South African Large Telescope
SAO	Smithsonian Astronomical Observatory
SCIDAR	SCIntillation Detection and Ranging
SDSS	Sloane Digital Sky Survey
SEC	Secondary Electron Conduction
SERC	Science and Engineering Research Council (UK)
SIM	Space Interferometry Mission
SIN	Superconductor Insulator Normal
SINFONI	Spectrograph for Integral Field Observation in the Near Infrared
SIS	Superconductor Insulator Superconductor
SIT	Silicon Intensified Target
SKA	Square Kilometer Array
SMA	SubMillimeter Array
SNO	Sudbury Neutrino Observatory
SNR	SuperNova Remnant
SNU	Solar Neutrino Unit
SODAR	Sonic Detection And Ranging
SOFIA	Stratospheric Observatory for Far Infrared Astronomy
SOHO	Space HelioSpheric Observatory
SPAN	Space Physics Analysis Network
SPI	Spectrometer on Integral (γ-ray mission)
SPIRE	Spectral and Photometric Imaging Receiver
SPOT	Satellite Probatoire d'Observation de la Terre

SQUID	Superconducting Quantum Interference Device
SSB	Single Side Band
SSPM	Solid State PhotoMultiplier
STJ	Superconducting Tunnel Junction
STP	Standard Temperature Pressure
TAI	Temps Atomique International
TCP	Transmission Control Protocol
TDRSS	Transmission Data and Relay Satellite System
THz	Terahertz
TIMMI	Thermal Infrared MultiMode Instrument
TMT	Thirty Meter Telescope
TPF	Terrestrial Planet Finder
TV	Television
UCAC	USNO CCD Astrograph Catalog
UHF	UltraHigh Frequency
UKIRT	United Kingdom InfraRed Telescope
USNO	United States Naval Observatory
UT	Universal Time
UTC	Universal Time Coordinated
UV	UltraViolet
UVES	UltraViolet Echelle Spectrograph
VESPA	Versatile Spectrometer Assembly
VIMOS	Visible Multiple Object Spectrograph
VISIR	VLT Imager and Spectrometer in the InfraRed
VLA	Very Large Array
VLBI	Very Long Baseline Interferometry
VLT	Very Large Telescope
VLTI	Very Large Telescope Interferometer
VSOP	VLBI Space Observatory Program
VVDS	VIMOS VLT Deep Survey
WFPC	Wide Field Planetary Camera
WMAP	Wilkinson Microwave Anisotropy Probe
XMM	X-ray Multimirror Mission
ZIMPOL	Zurich Imaging Polarimeter
ZOG	Zero Order Grating

Bibliography

Since this book is intended as a reference, only a very limited number of books or papers have been cited, and then explicitly. The present bibliography, like the webography, supplements these references with books and review articles considered to be of fundamental importance. They are organised here by theme, with headings and subheadings that do not necessarily correspond to chapter titles in the book. We have not included atlases, which can be found in the CFHT bibliography below.

The general bibliographic sources to follow can be consulted by a highly efficient search using key words. These sources sometimes require a password, which must be obtained through a university library:

- **Base enterBooks.com.** Bibliographic search for books, with access in French, English, or Spanish. Online address: www.enterbooks.com.
- **Libraries of the European Southern Observatory (ESO).** An exceptional collection of books and professional review. Online address: http://libhost.hq.eso.org:8088/uhtbin/webcat.
- **SAO/NASA Database: Astrophysics Data System — Centre de données astronomiques de Strasbourg (CDS).** Access to all scientific journals for the professional astronomer. Online address: http://cdsads.u-strasbg.fr.
- **Bibliographic Data at the Centre national de la recherche scientifique (CNRS) — Institut national des sciences de l'univers (INSU).** Access to sources and open archives (HAL, ArXiv). Online address: http://biblioplanets.inist.fr.

Since the tools and methods of astronomical observation are changing fast, many useful references do not feature in didactic works, but appear rather in conference proceedings. The latter provide up-to-date results and original contributions, but do not always have the synoptic value of a textbook, nor the same level of refereeing as in a peer-reviewed journal. However, it is worth listing some particularly important organisations, to be cited generically under the more detailed subject headings below:

- **Experimental Astronomy.** A good refereed review, published by Springer and bringing together publications on experimental aspects of astronomical observation.
- **European Astronomical Society Publication Series (EAS).** This series began in 2001 and can be accessed online at www.eas-journal.org.
- **Astronomical Society of the Pacific Conference Series (ASP).** All titles published since 2003 are available as e-books at the address www.aspbooks.org.
- **Society of Photographic Instrumentation Engineers (SPIE).** Professional and commercial society founded in the USA in 1955. Only the acronym is used today. Among other things, the SPIE Digital Library contains a section entitled *Astronomy and Astronomical Optics*, bringing together a large number of conference proceedings at very different levels but

P. Léna et al., *Observational Astrophysics*, Astronomy and Astrophysics Library,
DOI 10.1007/978-3-642-21815-6, © Springer-Verlag Berlin Heidelberg 2012

well up to date, dealing with astronomical observation both on the ground and from space, telescopes, their instrumentation, and so on. The sections entitled *Electronic Imaging and Processing* and *Optics and Electro-optics* may also provide useful information. Online address: spiedigitallibrary.aip.org/.

General References

Annual Review of Astronomy and Astrophysics. Annual Reviews Series. Annual volume published since 1963, containing a dozen review articles, generally excellent. The state of the art in observational techniques and methods is often described. This reference is denoted by ARAA (volume, page, year) in this book and in the bibliographic references below.

The Astronomy & Astrophysics Review, Springer, Berlin. European equivalent of the previous, published since 1989, edited by Th. Courvoisier in the form of several volumes each year. This contains few review articles on observation in astronomy.

The Astronomy & Astrophysics Encyclopædia, Maran S.P. Ed., Van Nostrand, New York, 1992. An excellent synopsis, with detailed discussions and bibliography for all astrophysical themes. Also discusses the main observational tools.

Encyclopedia of Astronomy and Astrophysics, Murdin, P. Ed., 4 vol., Institute of Physics, London, 2001. High-quality, up-to-date professional encyclopedia, available at http://eaa.crcexpress.com.

ALLEN C.W., Astrophysical Quantities, Cox, A. Ed. Springer, 4th edn., 2000. Defines key quantities and provides numerical values of useful constants. This is a basic reference for any astronomer, denoted by AQ in this book.

HARWIT M., Astrophysical Concepts, Springer, 2000. An excellent reference book, presenting the main physical concepts used in astrophysics.

HEYVAERTS, J., Astrophysique. Étoiles, univers et relativité, Dunod, Paris, 2006. An excellent astrophysics manual at MSc and doctoral levels, with exercises.

LANG K.R., Astrophysical Formulae, Astronomy & Astrophysics Library, Springer, 3rd edn., 1999, 2 volumes, reprinted in 2006. A compendium of basic physical tools used in the main astrophysical problems. This standard reference is denoted by AF in this book.

PECKER J.C., SCHATZMAN E., Astrophysique générale, Masson, Paris, 1959. Standard but sometimes out-of-date reference work. The section on observation gives a good description of conventional astronomy in the 1960s.

Historical Works

HAWKING, S., On the Shoulders of Giants, Running Press, 2003. The main foundational texts of astronomy.

HOCKEY, T. et al. The Biographical Encyclopedia of Astronomers, Springer, 2 volumes, 2007. This book contains more than a thousand biographical descriptions of astronomers, from historical times to the present day.

LANG K.R., GINGERICH O., A Source Book in Astronomy and Astrophysics 1900–1975, Harvard University Press, 1979. Collection of key papers, reporting particularly on the considerable progress made in observational methods and associated discoveries.

VERDET J.P., Astronomie et astrophysique, Larousse, Paris, 1993. Collection of key papers, with comments and historical context.

WALLERSTEIN G., OKE J.B., The First 50 Years at Palomar, 1949–1999, Another View: Instruments, Spectroscopy, and Spectrophotometry in the Infrared, ARAA **38**, 113–141, 2000.

Most annual volumes of the ARAA begin with an article in which an eminent astronomer, usually American, gives a personal account of his or her life of research and contributions. These portrayals thus form part of the rich history of twentieth century astronomy.

Terminology

HECK A., Star Briefs 2001: A Dictionnary of Abbreviations, Acronyms, and Symbols in Astronomy and Related Space Sciences, Kluwer, 2000..

HOPKINS J., Glossary of Astronomy and Astrophysics, University of Chicago Press, 2nd edn., 1980. Rather physically oriented introduction to English and American terminology.

LANG K.R., A Companion to Astronomy and Astrophysics: Chronology and Glossary with Data Tables, Springer, 2006.

Desktop Publishing in Astronomy and Space Sciences, Heck A. Ed., World Scientific Publishing, Singapore, 1992.

HECK A., HOUZIAUX L. Future Professional Communication in Astronomy, Académie royale de Belgique, 2007. Detailed analysis of the way publication has evolved with the advent of digital communications.

Research Policy in Astronomy

HECK A. Star Guides 2001: A Worldwide Directory of Organizations in Astronomy, Related Space Sciences and Other Related Fields, Kluwer, 2003. An invaluable practical guide to all centres of research and observation, with references, addresses, and so on.

Cosmic Vision 2015–2025, European Space Agency 2008, www.esa.int/esaCP/index.html. Conclusions concerning orientations and choices for the ESA's astronomical space missions in the specified decade.

Towards a Strategic Plan for European Astronomy, Astronet European Network, 2006. Document available at www.astronet-eu.org, and also published as a book (see next item). The result of a wide-ranging consultation on the aims and means of both ground-based and spaceborne European astronomy.

DE ZEEUW T., A Science Vision for European Astronomy, Astronet, 2007.

WOLTJER L., Europe's Quest for the Universe, EDP, Paris, 2006. A comprehensive overview of the rebirth of European astronomy in the second half of the twentieth century, both on the ground and in space.

OECD Science, Technology, and Industry Outlook, OECD, Paris, 2010. Annual review of the main trends in public policy in member countries of the OECD.

UNESCO Science Report 2010, UNESCO, Paris, 2010. Global picture of scientific development in the world, and in particular in astronomy.

New Worlds, New Horizons in Astronomy and Astrophysics, National Academies (USA), National Academic Press, 2010. Guidelines for Earth-based and spaceborne astronomy in the USA, prepared every ten years.

JAXA Vision — JAXA 2025. Japan Aerospace Exploration Agency, Tokyo, 2005, available online at www.jaxa.jp/about/2025/index_e.html. Prospects for the Japanese space programme.

Les sciences spatiales: Adapter la recherche française aux enjeux de l'espace, Académie des sciences (France), Report no. 30, ed. by Puget J.L., EDP Sciences, Paris, 2010. Overview of the development of space research in France and Europe.

Long Range Plan 2010 for Canadian Astronomy and Astrophysics, National Research Council of Canada, www.casca.ca/lrp2010/index.php, 2010. Guidelines for Earth-based and spaceborne astronomy in Canada, prepared every ten years.

Observation in Astrophysics

Modern Technology and Its Influence in Astronomy, Wall J.V. & Boksenberg A. Eds., Cambridge University Press, New York, 1990. Collective work describing trends in imaging and radiation detection.

Compendium of Practical Astronomy, Roth G.D. Ed., Vols. I, II, and III, Springer-Verlag, New York, 1994. Extremely comprehensive textbook built up from individual contributions with many references. The first volume is more specifically devoted to observation, its tools, and its techniques.

BRADT H., Astronomy Methods, Cambridge University Press, 2004.

HARWIT M., Cosmic Discovery: The Search, Scope, and Heritage of Astronomy, MIT Press, 1984. Excellent book, showing in particular how observational capabilities have conditioned astrophysical discovery.

KITCHIN C.R., Astrophysical Techniques, Taylor & Francis, 4th edn., 2003.

Exploring the Cosmic Frontier: Astrophysical Instruments for the 21st Century, Lobanov A.P. et al (Eds.), Springer, 2008.

WILSON T.L., ROHLFS K., HITTEMEISTER S., Tools of Radio Astronomy, 5th edn., Astronomy and Astrophysics Library, Springer-Verlag, New York, 2009.

SMITH R.C., Observational Astrophysics, Cambridge University Press 1995. This book 'borrowed' the title of the original book by LENA, P., published by Springer in 1988.

SPIE Digital Library. Instrumentation in Astronomy, CDP27. Two CDROMs presenting a set of particularly significant papers in this area, appearing in the SPIE volumes, 2002.

Astronomical Photometry

BESSELL M.S., Standard Photometric Systems, ARAA **43**, 293–336, 2005.

BUDDING E. & DEMIRCAN O., Introduction to Astronomical Photometry, 2nd edn., Cambridge University Press, 2007.

STERKEN C. & MANFROID J., Astronomical Photometry: A Guide, Kluwer, 1992. Clear and comprehensive reference.

TINBERGEN J., Astronomical Polarimetry, Cambridge University Press, 1996.

Terrestrial Atmosphere

Structure and Properties

VALLEY S., Handbook of Geophysics and Space Environnement, McGraw-Hill, New York, 1965. A good general reference for all properties and numerical values associated with the atmosphere and space environment of the Earth.

HOUGHTON J.T., The Physics of the Atmosphere, Cambridge University Press, 1977. Excellent introductory monograph on the physics and dynamics of the atmosphere.

HUMPHREYS W.J., Physics of the Air, Dover Publishing Inc., 1964. Old textbook (1920), but in a more recent edition that contains a good elementary description of many meteorological and atmospheric phenomena.

SALBY M.L. Fundamentals of Atmospheric Physics, International Geophysics Series, Vol. 61, Academic Press, New York, 1996. Excellent didactic book on the fundamental properties of the Earth atmosphere.

Atmospheric and Interplanetary Radiation

GOODY R.M., Atmospheric Radiation, Cambridge University Press, 1977.

McCORMAC B.M., The Radiating Atmosphere, Reidel, Dordrecht, 1971.

REACH F.E. & GORDON J.L., The Light of the Night Sky, Reidel, Dordrecht, 1973.

Optics of Cosmic Dust, Videen G., Kocifaj M. (Eds.), NATO Workshop, Kluwer, 2002.

Interplanetary Dust, Grün E., Gustafson B.A.S., Dermott S., et al. (Eds.), Springer, 2001.

Scattering in the Atmosphere (Selected Papers on), S.P.I.E. **MS 07**, 1989. A collection of fundamental articles on atmospheric scattering.

SPIE Digital Library. See the volumes of conference proceedings in the section on *Atmospheric Sciences*.

Transmission of Electromagnetic Radiation

Atlases of the solar spectrum obtained with very high resolutions also identify telluric lines.

The Infrared Handbook. US Govt. Print. Off., Washington, 1979. Although outdated, this remains an excellent reference for theory and practice in the region 1–1 000 μm.

MIGEOTTE M., NEVEN L., SWENSSON J., Le Spectre Solaire de 2.8 à 23.7 μm, Mém. Soc. Roy. Sc. Liège **2**, 1957.

DELBOUILLE L., ROLAND G., BRAULT J., TESTERMAN L., Spectre solaire de 1 850 à 10 000 cm^{-1}, Université de Liège, 1981.

WALLACE L., HINKLE K., LIVINGSTON W., An Atlas of the Photospheric Spectrum (735–1 123 nm), National Solar Observatory, Tucson, 1994.

Atmospheric Turbulence

The *Journal of Atmospheric Sciences* contains research directly relating to atmospheric properties. For many other references, see also under the heading *Adaptive Optics and Radioastronomy*.

LESIEUR M., Turbulence in Fluids, 4th edn., revised and updated, Springer, 2007.

LUMLEY J.L. & PANOFSKY H.A., The Structure of Atmospheric Turbulence, Interscience, Paris, 1964.

Astronomical Sites on Earth

Astronomical Site Evaluation in the Visible and Radio Range, Vernin J., Benkhaldoun Z., Muñoz-Tuñon C. Eds., ASP Series vol. 266, 2002.

ARENA — Large Astronomical Infrastructures at CONCORDIA. Prospects and Constraints for Antarctic Optical & Infrared Astronomy, Candidi M., Epchtein N. Eds., EAS Series vol. 25, EDP, Paris, 2007.

A Vision for European Astronomy and Astrophysics at the Antarctic station CONCORDIA Dome C in the next decade 2010–2020. The ARENA Consortium, Epchtein N. Ed., arena.unice.fr, 2009.

Dome C Astronomy & Astrophysics Meeting, Giard M., Casoli F., Paletou F. (Eds.), EAS Series, vol. 14, EDP, 2005.

The Protection of Astronomical and Geophysical Sites, Kovalevsky J. Ed., Editions Frontières, Gif-sur-Yvette, 1992.

ABAHAMID A., VERNIN J., BENKHALDOUN Z., Seeing, outer scale of optical turbulence, and coherence outer scale at different astronomical sites using instruments on meteorological balloons, Astron. Astrophys. **422**, 1123, 2004. Interesting article for site comparisons.

GARSTANG R.H., The status and prospects for ground-based observatory sites, ARAA **27**, 19–40, 1989.

Spaceborne Observation

Space Missions

This is not the place to give an exhaustive bibliography of spaceborne observation. The reader can consult the initial reports and feasibility studies published by the European Space Agency and other large space agencies (USA, Japan, China, Russia, etc.) whenever a new mission is under study or new guidelines are being established. See the webography for more details.

SPIE Digital Library. See the volumes of conference proceedings in the sections entitled *Astronomy, Astronomical Instrumentation, Space Technologies*.

Moon-Based Observation

Towards a World Strategy for the Exploration and Utilization of our Natural Satellite, European Space Agency **SP-1170**, 1994.

Astrophysics from the Moon, Mumma M.J. & Smith H.J. Eds., AIP Proceedings **207**, Amer. Inst. Phys., 1990.

Spatial and Temporal Reference Systems

AUDOIN C., GUINOT B., LYLE S. (transl.), The Measurement of Time: Time, Frequency, and the Atomic Clock, Cambridge University Press, 2001. This book discusses the ideas and techniques involved in time measurement and the establishment of time scales.

GREEN R.M., Spherical Astronomy, Cambridge University Press, 1985. Very good and up-to-date elementary textbook, with vector formulations and taking into account relativistic effects.

KOVALEVSKY J., Modern Astrometry, 2nd edn., Astronomy and Astrophysics Library, Springer-Verlag, 2002. Recent, general work covering in some detail the many aspects of astrometry, such as image formation, instrumentation, and data compression.

KOVALEVSKY J. & SEIDELMAN P.K., Fundamentals of Astrometry, Cambridge University Press, 2004. For advanced students. The most recent textbook on current astrometry, including the most recent conventions of the International Astronomical Union.

MIGNARD F., LES échelles de temps, Introduction aux Ephémérides Astronomiques, Connaissance des temps, Chap. 2, EDP Sciences. Updated annually in Connaissance des temps.

MURRAY C.A., Vectorial Astrometry, Adam Hilger, Bristol, 1983. This book deals above all with the relevant concepts in a very general relativistic framework which makes it difficult to follow. However, the reader's efforts will be rewarded by the quality of the contents.

VANIER J. and AUDOIN C., The Quantum Physics of Atomic Frequency Standards, Adam Hilger, Bristol, 1989. The most comprehensive book on the physics of frequency standards. High level.

WALTER H.G., SOVERS O.J., Astrometry of Fundamental Catalogues, Springer, 2000. Monograph discussing the underlying ideas and construction of fundamental catalogues, with a detailed presentation of FK5, the Hipparcos Catalogue, and the ICRF with the VLBI technique.

The Hipparcos Mission, **SP-1111**, European Space Agency, 1989. Three volumes published for the launch of Hipparcos, describing the instrument, preparation and contents of the initial catalogue, and compression methods.

The Hipparcos Catalogue, **SP-1200**, European Space Agency, 1997. Final publication of the Hipparcos Catalogue. Volume 1 specifies the properties of the catalogue and volume 3 presents all the data processing techniques.

Reference Frames in Astronomy and Geophysics, Kovalevsky J., Mueller I.I. & Kolaczek B. Eds., Kluwer Academic Publishers, 1989. About twenty synoptic papers written by the best specialists, making a real pedagogical effort. The work covers celestial and terrestrial reference systems, the relationships between these systems, and also temporal reference systems.

High-Accuracy Timing and Positional Astronomy, van Paradijs J. & Maitzen H.M. Eds. Lecture Notes in Physics **418**, Springer-Verlag, New York, 1993. Excellent didactic book resulting from the European Astrophysical Doctoral Network (EADN).

GAIA: A European Space Project, ESA Publication Series **2**, O. Bienaymé & C. Turon Eds., EDP 2002. Report from a Les Houches School on the GAIA mission and its scientific objectives.

GAIA: Composition, Formation and Evolution of the Galaxy, ESA-SCI(2000)4, 2000. Official document containing the scientific proposal for the GAIA mission, and providing a lot of information about the current situation and issues in stellar and galactic physics.

Telescopes

BAARS J.W.M., The Paraboloidal Reflector Antenna in Radio Astronomy and Communication. Theory and Practice, Springer, 2007.

BELY P., The Design and Construction of Large Optical Telescopes, Springer, 2003. Excellent introductory work on new telescopes in the twenty-first century.

DANJON A. & COUDER A., Lunettes et Télescopes, Blanchard, Paris, 1983. Re-edition of an excellent classical textbook on visible observation.

KING H.C., The History of the Telescope, Dover Publish. Inc., New York, 2nd edn. 1977. Excellent historical introduction.

KITCHIN C., Solar Observing Techniques, Springer, 2002. Telescopes and instruments for observing the Sun.

LEMAITRE G., Astronomical Optics and Elasticity Theory: Active Optics Methods, Springer, 2010. Basic principles of active optics.

Astronomical Optics (selected papers on), Schroeder D. Ed., **MS 73**, 1993. A collection of fundamental historical articles on optical astronomy (telescopes, interferometers, turbulence, etc.), from 1920 to the present day.

ROHLFS K. & WILSON T.L., Tools of Radio Astronomy, Springer, 2004. Telescopes and instrumentation used in radio astronomy.

WILSON R.W., Reflecting Telescope Optics I. Basic Design, Theory and Its historical Developments, 2nd revised edn., Springer, 2007.

WILSON R.W., Reflecting Telescope Optics II. Manufacture, Testing, Alignement, Modern techniques, Springer, 2001. Two fundamental works dealing with the design and construction of optical telescopes.

SPIE Digital Library. See the volumes of conference proceedings in the sections entitled *Astronomical Instrumentation, Astronomy*.

Optics and Image Formation

BORN M. & WOLF E., Principles of Optics, Pergamon Press, Oxford, 6th edn., 1980. Fundamental reference book, especially for the formalism of geometric or wave optics, the complete theory of diffraction, etc. Also contains the general principles of interferometers, e.g., Michelson, Fabry–Perot, particularly useful in the ultraviolet, visible, infrared, and submillimetre regions.

BRUHAT G. & KASTLER A., Optique, Masson, Paris, 1954, 6th edn. dating from 1965. Key textbook, although sometimes out of date.

GOODMAN J.W., Introduction to Fourier Optics, 3rd edn., Roberts & Co., 2005. Excellent reference.

HARBURN G., TAYLOR C.A. & WELBERRY T.R., Atlas of Optical Transforms, Bell, 1975. Useful depictions of Fourier transforms.

HARVEY A.F., Coherent Light, Wiley, New York, 1970. A particularly clear discussion of issues relating to radiation coherence.

HECHT E., Optics: International Edition, Pearson Education, 4th edn. 2003. Excellent book with a very comprehensive didactic introduction to optics.

MERTZ L., Transformations in Optics, Wiley, New York, 1965. Original work, with many direct references to astronomy.

SCHNEIDER P., EHLERS J. & FALCO E.E., Gravitational Lenses, Springer, 1992, reprinted 1999.

Handbook of Optics, vols. I and II, Bass M. Ed., McGraw-Hill, 2001. Reference book discussing all features of image formation, light sources, and light detectors used in astronomy.

Guided Optics

AGRAWAL G., Nonlinear Fiber Optics, 3rd edn., Academic Press, 2006.

BUCK J.A., Fundamentals of Optical Fibers, Wiley, New York, 1995. Reference work on all aspects of optical fibres.

HEACOX W.D. & CONNES P., Astronomy & Astrophysics Review **3**, 169, 1992.

YOUNG M., Optics and Lasers, Fibers and Optical Waveguides, Springer, 2000.

Image Formation in Turbulent Media

GOODMAN J.W., Statistical Optics, Wiley, New York, 1985. The main reference book for imaging in random media and partial coherence properties of radiation.

RODDIER F., The Effects of Atmospheric Turbulence in Optical Astronomy, Progress in Optics **XIX**, 281, 1981. Old paper but of fundamental importance, clearly describing the formalism and the main results needed to understand the effects of the atmosphere on light propagation.

TATARSKI V.I., Wave Propagation in Turbulent Media, Dover Publ. Inc., New York, 1961. Standard work on the physics of wave propagation in random media.

WOOLF N.J., High Resolution Imaging from the Ground, ARAA **20**, 367, 1982. Motivations and techniques leading to the construction of very large ground-based telescopes in the 1980s.

Adaptive Optics

Adaptive Optics for Astronomy, Alloin D.M. & Mariotti J.-M. Eds., NATO-ASI Series **423**, 1994.

Adaptive Optics in Astronomy, Roddier F. Ed., Cambridge University Press, 1999. Basic work for the study of adaptive optics.

BECKERS J., Adaptive Optics for Astronomy: Principles, Performances, and Applications, ARAA **31**, 13–62, 1993. A review written early in the development of adaptive optics.

HUBIN N., MAX C.E., WIZINOWICH P., Adaptive Optics Systems, SPIE, 2008. A compendium of the state of the art in this technique, as applied to astronomy.

TYSON R.K., Principles of Adaptive Optics, Academic Press, 1991.

SPIE Digital Library. See the volumes of conference proceedings under the heading *Adaptive Optics*.

High Dynamic Range Imaging

Optical techniques for direct imaging of exoplanets, C. R. Acad. Sc. 8, 273–380, Aime C. Ed., 2007.

Astronomy with High Contrast Imaging II, Aime C. Ed., European Astronomical Society Publ., vol. 12, 2004.

BRACEWELL R.N., Detecting non-solar planets by spinning infrared interferometer, Nature **274**, 780, 1978. The original idea for nulling interferometry.

JACQUINOT P., ROIZEN-DOSSIER B., Prog. Opt. 3, 31, 1964. Fundamental historical paper.

LYOT B., ZEIT. F. Astrophys. **5**, 73, 1932. Key historical paper.

Interferometry

Radio Interferometry

CORNWELL T.J. & PERLEY R.A., Radio Interferometry: Theory, Techniques, Applications, American Society of the Pacific Conference Series **19**, 1991.

HALL P.J. Ed., The Square Kilometer Array: An Engineering Perspective, Springer, 2004.

KELLERMANN K.I., MORAN J.M., The development of high-resolution imaging in radioastronomy, ARAA **39**, 457–509, 2001.

THOMSON A.R., MORAN J.M. & SWENSON G.W., Interferometry and Synthesis in Radioastronomy, Wiley, New York, 1986. Standard work for the study of imaging in radioastronomy.

WILSON T.L. & HÜTTEMEISTER S., Tools of Radio Astronomy. Problems and Solutions, Springer, 2005.

Subarcsecond Radioastronomy, Davis R.J. & Booth R.S. Eds., Cambridge University Press, 1993.

Optical Interferometry

ALLOIN D.M. & MARIOTTI J.M., Diffraction-Limited Imaging with Large Telescopes, NATO ASI Series **274**, Kluwer Academic Publishers, Dordrecht, 1989. Lectures from a summer school, providing a didactic introduction at the beginnings of adaptive optics and optical interferometry.

FELLI M. & SPENCER R.E., Very Long Baseline Interferometry: Techniques and Applications, Reidel, Dordrecht, 1989.

GLINDEMAN A., Principles of Stellar Interferometry, Astronomy and Astrophysics Library, Springer, 2011. Standard reference for optical interferometry, and especially the setting up of the European VLT.

HANBURY-BROWN R., Photons, Galaxies, and Stars (selected papers on), Indian Academy of Science, Bangalore, 1985. Collection of original papers on optical interferometry by photon correlation.

LABEYRIE A., LIPSON S.G., NISENSON P., An Introduction to Optical Stellar Interferometry, Cambridge University Press, 2006.

LAWSON P., Principles of Long-baseline Stellar Interferometry, Wiley Blackwell, 2010.

QUIRRENBACH A., Optical Interferometry, ARAA **39**, 353–401, 2001.

SAHA S.K., Aperture Synthesis: Methods and Applications to Optical Astronomy, Astronomy and Astrophysics Library, Springer, 2010.

SHAO M. & COLAVITA M.M., Long Baseline Optical and Infrared Stellar Interferometry, ARAA **30**, 457–498, 1992.

The Power of Optical & Infrared Interferometry: Recent Scientific Results, Paresce F., Richichi A., Delplancke F. & Chelli A. (Eds.), Springer, 2008.

SPIE Digital Library. See the volumes of conference proceedings under the heading *Astronomical Instrumentation*.

Detectors and Receivers

General Features

COHEN-TANNOUDJI C., DUPONT-ROC J., GRYNBERG G., Photons et atomes. Introduction l'électrodynamique quantique, EDP, Paris, 1987.

COHEN-TANNOUDJI C., DUPONT-ROC J., GRYNBERG G., Processus d'interaction entre photons et atomes, EDP, Paris, 1996.

GRYNBERG G., ASPECT A., FABRE C., Introduction to Quantum Optics: From the Semi-Classical Approach to Quantized Light, Cambridge University Press, 2010.

HARRIS A.I., Coherent Detection at Millimetric Waves and Their Application, Nova Science Publ., Conmack, 1991.

KITTEL C., Introduction to Solid State Physics, Wiley, 8th edn., 2004. Essential source for the basics of solid state physics, relevant to most radiation detectors in all energy regions.

LAMARRE J.M., Photon noise in photometric instruments at far infrared and submillimeter wavelengths, Appl. Opt. **25**, 870–876, 1986.

LANDAU L., LIFSHITZ E.M., Statistical Physics, 3rd edn., Part I, Course on Theoretical Physics, Vol. 5, Butterworth–Heinemann, 1980.

PITAEVSKII L.P., LIFSHITZ E.M., Statistical Physics, Part 2, Course on Theoretical Physics, Vol. 9, Butterworth–Heinemann, 1980.

LONGAIR M., High Energy Astrophysics, vol. I, Particles, Photons and Their Detection, Cambridge University Press, 1992. Very comprehensive textbook, including a section on the ultraviolet to radiofrequency regions and associated techniques. Good bibliography.

MACKAY C.D., Charge coupled devices in astronomy, ARAA **24**, 255–283, 1986.

MATHER J., Bolometer noise: Non-equilibrium theory, Appl. Opt. **21**, 1125, 1982. Detailed analysis of thermal noise.

OLIVIER B.M., Thermal and quantum noise, Proc. IEEE **53**, 436, 1965.

RIEKE G.H., Detection of Light from the Ultraviolet to the Submillimeter, Cambridge University Press, 1994. Synoptic view of light detection, written by an astronomer for researchers.

SZE S.M., Physics of Semiconductor Devices, Wiley, New York, 1981. This can be considered as the basic reference for all aspects of semiconductor physics.

Radiofrequencies

PHILIPS T.G. & WOODY D.P., Millimeter and submillimeter wave receivers, ARAA **20**, 285, 1982.

ROBINSON B., Frequency allocation: The first forty years, ARAA **37**, 65–96, 1999.

Infrared and Submillimeter

Millimetric Astronomy, Les Houches, Winter Workshop, March 1990, Nova science Publ., New York, 1991. Millimetre and submillimetre detectors.

The Infrared and Electro-optical Systems Handbook, Accetta & Shumaker Executive Eds., copubl. ERIM and SPIE, 1993. Eight volumes.

The Infrared Handbook. US Govt. Print. Off., Washington, 1985. Although somewhat out of date, still an excellent reference work for theory and practice in the region 1–1 000 μm.

LOW F.J., RIEKE G., GEHRZ R.D., The beginning of modern infrared astronomy, ARAA **45**, 43–65, 2007. Up-to-date discussion by the pioneers in this area.

RIEKE G., Infrared dectector arrays for astronomy, ARAA **45**, 77–115, 2007.

WOLSTENCROFT R.D. & BURTON W.B., Millimeter & Submillimeter Astronomy, Kluwer Academic Publishers, New York, 1988.

Visible and CCD Detectors

Astronomy with Schmidt-Type Telescopes, Capaccioli M. Ed., Reidel, Dordrecht, 1984. General review of the uses of photographic plates and the photography of faint objects.

BUIL C., CCD Astronomy: Construction and Use of an Astronomical CCD Camera, Willmann–Bell, 1991.

HOWELL S.B., Handbook of CCD Astronomy, Cambridge University Press, 2006.

JANESICK J.R., Scientific Charge-Coupled Devices, SPIE, Bellingham, 2001.

McLEAN I.S., Modern Instrumentation in Astronomy, Springer-Verlag, New York, 1989. High quality textbook on CCDs.

Ultraviolet

BOWYER S., DRAKE J.J., VENNES S., Extreme ultraviolet astronomy, ARAA **38**, 231–288, 2000.

X-Ray Astronomy

BEIERSDORFER P., Laboratory X-ray astrophysics, ARAA **41**, 343–390, 2003.

BRANDT H.V.D., OHASHI T. & POUNDS K.A., X-ray astronomy missions, ARAA **30**, 391–427, 1992.

GIACCONI R. & GURSKY H., X-Ray Astronomy, Reidel, Dordrecht, 1974.

PAERELS F.B.S., KAHN S.M., High resolution X-ray spectroscopy with Chandra and XMM–Newton, ARAA **41**, 291–342, 2003.

TRÜMPER J., ROSAT, Physica Scripta **T7**, 209–215, 1984. Detailed description of the X-ray astronomy mission (Röentgen Satellite).

Gamma-Ray Astronomy

DEBERTIN K. & HELMER R.G., Gamma-Ray and X-Ray Spectrometry with Semiconductor Detectors, Elsevier Science Publishers, North-Holland, 1988.

Towards a Major Atmospheric Cerenkov Detector for TeV Astro/particle Physics, Fleury P. & Vacanti G. Eds., Éditions Frontières, Gif-sur Yvette, 1992.

GILMORE G., HEMINGWAY J., Practical Gamma-Ray Spectrometry, John Wiley & Sons, 1995.

Semiconductor Radiation Systems, K. Iniewski, CMOS Emerging Technologies Inc., Vancouver 2009.

KNOLL G.F., Radiation Detection and Measurement, Wiley, 1999. A standard reference.

LECOQ P., ANNENKOV, A., GEKTIN, A., KORZHIK, M., PEDRINI, C., Inorganic Scintillators for Detector Systems, Springer, 2006.

LUTZ G., Semiconductor Radiation Detector Device Physics, Springer, 1999.

RODNYI P., Physical Processes in Inorganic Scintillators, CRC Press, New York, 1997.

SPIELER H., Semiconductor Detector Systems, Oxford University Press 2005.

Cosmic Rays

SCHLICKEISER R., Cosmic Ray Astrophysics, Springer, 2002.

Neutrino Astronomy

ANSELMANT P. et al., Phys. Lett. B **285**, 1992. Contains a discussion of solar neutrinos and their detection.

BAHCALL J.N., Highlights in Astrophysics: Concepts and Controversies, Shapiro S. & Tevkolsky S. Eds., Wiley, New York, 1985.

BAHCALL J.N. & DAVIS R., Essays in Nuclear Astrophysics, Cambridge University Press, 1982.

Gravitational Astronomy

Gravitational Radiation, Deruelle N. & Piran T. Eds., Cours de l'École de Physique Théorique des Houches, North Holland, 1983.

HAKIM R., Gravitation relativiste, InterÉditions, 1994.

SCHNEIDER P., EHLERS J. & FALCO E.E., Gravitational Lenses, Springer-Verlag, Heidelberg, 1991.

VINET J.-Y., Optical detection of gravitational waves, Compte-Rendus Acad.Sc. **8**, 2007.

Spectrometry

Precision Spectroscopy in Astrophysics, Santos N.C., Pasquini L., Correia A.C.M., Romaniello M. (Eds), Springer, 2008.

Next Generation Wide-Field Multi-Object Spectroscopy, Brown M.J., Dey A. (Eds.), PASP 2002.

AERTS C., CHRISTENSEN-DALSGAARD J., KURTZ D.W., Asteroseismology, Springer, 2008.

CHAFFEE F. & SCHROEDER D., Astronomical Applications of Echelle Spectroscopy, ARAA **14**, 23, 1976.

CONNES P., Astronomical Fourier Spectroscopy, ARAA **8**, 209, 1970. Comprehensive overview of Fourier spectroscopy.

JAMES J., Spectrograph Design Fundamentals, Cambridge University Press, 2007.

KITCHIN C.R., Optical Astronomical Spectroscopy, Institute of Physics, London, 1995.

TENNYSON J., Astronomical Spectroscopy. An Introduction to the Atomic and Molecular Physics of Astronomical Spectra, Imperial College Press, London, 2005.

SPIE Digital Library. See the volumes of conference proceedings in the section entitled *Astronomical Instrumentation*.

Signal Processing

TASSI Ph., Méthodes statistiques, Economica, Paris, 1985. Clear synopsis.

ANDREWS H.C. & HUNT B.R., Digital Image Restoration, Prentice Hall, 1977. Basic techniques of information restoration in image processing.

BIJAOUI A., Image et information: Introduction au traitement numérique des images, Masson, Paris, 1981.

BLANC-LAPIERRE A. & FORTET R., Théorie des fonctions aléatoires, Masson, Paris, 1953.

CLARKE R.J., Transform Coding of Images, Academic Press, Orlando, 1985. Standard reference for signal processing.

DAS P.K., Optical Signal Processing, Springer-Verlag, New York, 1991. Basics of optical signal processing.

KENDALL M. & STUART A., The Advanced Theory of Statistics, Vols. 1 and 2, Ch. Griffin & Co. Ltd., London & High Wycombe, 1977, 1979. Standard reference book for statistical treatment of data.

KENDALL M., STUART A. & ORD J.K., The Advanced Theory of Statistics, Vol. 3, Ch. Griffin & Co. Ltd., London & High Wycombe, 1983. Third and final part of this standard reference. Half the volume is devoted to time series.

LEVINE B., Fondements théoriques de la radiotechnique statistique, Vols. I, II, and III, Editions Mir, Moscow 1973, 1973, 1976. A remarkable, didactic introductory text, but unfortunately out of print.

MAX J., Méthodes et techniques de traitement du signal et applications aux mesures physiques, Masson, Paris, 4th edn., 1985–1987. This can be considered as a basic reference on noise-affected measurement.

NARAYAN R. & NITYANANDA R., Maximum entropy image restoration in astronomy, ARAA **24**, 127–170, 1986.

PAPOULIS A., Probability, Random Variables, and Stochastic Processes, McGraw-Hill, New York, 3rd edn., 1991. Standard reference for random variables and stochastic processes, including proofs and applications to physics.

PUETTER R.C., GOSNELL T.R., YAHIL A., Digital image reconstruction: Deblurring and denoising, ARAA **43**, 139–194, 2005.

STARCK J.-L., MURTAGH F., Astronomical Image and Data Analysis, Springer, 2006.

Inverse Problems

TITTERINGTON D.M., General structure of regularization procedures in image reconstruction, Astron. & Astrophys. **144**, 381–387.

DEMOMENT G., Image reconstruction and restoration: Overview of common estimation structures and problems, IEEE Trans. Acoust. Speech Signal Process **37** (12), 2024–2036, December 1989.

Bayesian Approach to Inverse Problems, Idier J., Ed., ISTE, London, 2008

MUGNIER L.M., BLANC A., Idier J., Phase diversity: A technique for wavefront sensing and for diffraction-limited imaging, In Hawkes P., ed., Advances in Imaging and Electron Physics, vol. 141, Chap. 1, pp. 1–76, Elsevier, 2006.

MUGNIER L.M., LE BESNERAIS G., MEIMON S., Inverse problems in optical imaging through atmospheric turbulence, In Idier J., Ed., Bayesian Approach for Inverse Problems, Chap. 10. ISTE, London, 2008.

STIGLER, S.M., Gauss and the invention of least squares, Annals of Statistics **9** (3), 465–474, May 1981.

Data Processing and Archiving

Astronomy from Large Data Bases II, Heck A. & Murtagh F. Eds., European Southern Observatory Proc. **43**, Garching, 1992.

Statistical Challenges in Modern Astronomy, Babu G.J. & Feigelson E.D. Eds., Springer-Verlag, New York, 1993

Mathematics

BRACEWELL R.M., The Fourier Transform and Its Applications, McGraw-Hill, New York, 1965. An invaluable book for its wide range of exercises relating to astronomy and physics.

JAFFARD S., MEYER Y., RYAN R.D., Wavelets: Tools for Science and Technology, Society for Industrial Mathematics, 2001.

RODDIER F., Distributions et transformation de Fourier, Ediscience, Paris, 1971.

SCHWARTZ L., Mathematics for the Physical Sciences, Dover, 2008. Fundamental work for understanding optics and signal processing in astronomy.

Error, Bias, and Uncertainties in Astronomy, Jaschek C. & Murtagh F. Eds., Cambridge University Press, New York, 1990.

Index

AB magnitude, 107
Abbe's condition, 200
Aberration
  optical, 179
  Seidel, 180
  spherical, 180
Absorption
  coefficient of, 443
  spectrum, 443
Abundance, 6, 453
Acceptance of a fibre, 263
Accretion, 21
Acoustic spectrometer, 498
Active optics, 181, 191
Adaptive Fourier analysis, 632
Adaptive optics, 240
  extreme, 254
  ground layer, 253
  multi-conjugate, 252
  multi-object, 254
  NAOS, 536
ADU, 538
Aerosols, 76
  scattering by, 59
AIC, 308
Air mass, 62
Air shower, 71, 410
Airglow, 50
Airy
  function, 224
  profile, 247
Aladdin detector, 376
Aladin database, 24, 616
Aliasing, 522
Alkali halide, 413
Allocating frequencies, 76
Altazimuth mount, 178, 192
AMANDA, 426
Amateur astronomy, 29
AMBER instrument, 289
Amplitude detector, 325
Analogue-to-digital converter (ADC), 519
Anamorphic factor, 472
Anastigmatic optics, 177
Angel's cross, 314
Anglo-Australian Telescope (AAT), 190, 479
Angular resolution, 23
Anisoplanicity, 244
Annihilation, 445
Anomaly, South Atlantic, 83
Antarctic, 76
ANTARES, 426, 429
Antenna temperature, 388
Anticoincidence devices, 414
Aperture, 12
Aperture supersynthesis, 277
Aperture synthesis, 227, 276
APEX telescope, 501
APM digitisation, 610
Apodisation, 113, 225, 299, 305, 505
Application specific integrated circuits (ASIC), 527
APS digitisation, 610
Arecibo radiotelescope, 186
Array of detectors, 365
Artificial intelligence, 78
Artificial reference star, 250
Astigmatism, 180, 242
Astrometry, 276
  VLBI, 149
Astrophysical Virtual Observatory (AVO), 615
Astrophysics Data System (ADS), 37
Atacama Large Millimeter Array (ALMA), 74, 282

P. Léna et al., *Observational Astrophysics*, Astronomy and Astrophysics Library,
DOI 10.1007/978-3-642-21815-6, © Springer-Verlag Berlin Heidelberg 2012

Atmospheric
  constituents, 41
    minor, 41
  degradation of image, 228
  refraction, 61
  scattering, 58
  transmission, 53
  turbulence, 62
Attenuation, 263
Autocorrelation, 630, 631
  of stochastic process, 646
Autocovariance of stochastic process, 646
Avalanche photodiode, 248, 363
AXAF mission, 201
Azimuthal angle, 132

Baffle, 193
Baikal telescope, 428
Band gap, 345
BAT, 207
Bayes estimator, 553
Bayes principle, 553
BEAR, 488
Bias, 517, 545, 547, 552, 555
BIB detector, 377
Bimorphic mirror, 245
Bin, 337, 457
Binning mode, 373
Blackbody
  fluctuations, 100
  radiation, 99
Blaze, 466
  angle, 466
  wavelength, 466
Blocked impurity band detector, 377
Boiling, 70
Bolometer, 383
  germanium, 383
  hot electron, 383, 385
  superconducting, 385
  X-ray, 386
BOOMERANG mission, 197
Bootstrap method, 559
Bose–Einstein statistics, 104
Bouguer's line, 109
Box function, 622
  in two dimensions, 625
Bracewell interferometer, 313
Bragg
  diffraction, 206, 493
  lens, 206
  reflection, 498
  spectrometer, 491
Bragg–Fresnel lens, 201
Brightness, reduced, 97
Bunching, 104
Bure interferometer, 281, 533
Bureau des Longitudes, 157
Bureau International de l'Heure, 164
Bureau of Standards (US), 161

Cadmium telluride, 419
Caesium clocks, 158, 161
Calar Alto telescope, 190, 251
Calendar, 170
Calibration, 110
  angular, 110
  energy, 110
  gamma, 120
  infrared, 114
  radiofrequency, 110
  spectral, 110, 117
  time, 110
  ultraviolet, X, 117
  visible, 114
Calorimeter, 355
Caltech Submillimeter Observatory, 385
Camera obscura, 202
Capella, 294
Carbon dioxide, 44
Carcinotron, 396
CARMA array, 281
Carte du ciel, 606
Cartesian coordinates, 129
CASIMIR instrument, 501
Cassegrain focus, 178
Cat's-eye mirror, 488
Catalogue
  FK4, 146
  FK5, 24, 146
  GSC, 146
  Hipparcos, 146, 152
  PPM, 146
  SAO, 146
  Tycho, 146, 152
  UCAC, 146
Cauchy distribution, 573
CCD
  astrograph catalog, 154
  controller, 367, 372
  format, 371
  frame transfer, 368, 539
  full frame, 368
  intensified, 361
  low light level, 372
  readout noise, 370

thinned, 368
time delay integration, 549
windowing mode, 373
for X-ray detections, 372
CDS, 37
CELESTE, 411
Celestial sphere, 176
Central limit theorem, 643
Centre–limb darkening, 124
Cerenkov radiation, 71
CFH telescope, 190, 611
Chacaltaya site, 46
Chajnantor site, 73
Chandra mission, 493
Characteristic function of random variable, 641
Charge loss, 416
Cherenkov air shower, 411
Cherenkov effect, 353
Cherenkov radiation for neutrino detection, 424
Cherenkov telescope, 410
Chopping, 545
Chopping secondary mirror, 192
Chronometry, 110
CID detector, 366
Circadian cycle, 44
Citations, 37
CLEAN algorithm, 270
Clementine mission, 87
Clocks, stability of, 284
Cloud, 72
CMOS detector, 373
COBE satellite, 121, 198
Coded mask, 202, 541
Coding, image, 203
Coherence
area of, 235, 236
complex degree of, 101
order n, 103
étendue, 211, 214, 333
length, 102
of radiation, 100
spatial, 101
temporal, 101
time, 101
Collimating mirror, 465
Colour indices, 108
Coma, 180, 242
Compton observatory, 207
Compton scattering, 202, 352, 412
Compton telescope, 407
Concordia Station, 77
Conduction band, 345
Confidence interval, 518, 642
Confusion noise, 607
CONICA, 536
Continuous spectrum, 443
Contrast, 54
Convective instability, 41
Convergence of an estimator, 554
Convolution, 626
Cooper pairs, 349
CORALIE, 474
CORAVEL, 474
Coronagraph
AIC, 308
Guyon, 308
Kasdine–Spergel, 306
visible nulling, 309
Coronagraphic interferometer, 311
Coronagraphy, 298, 299
Coronography, 60
COROT mission, 532, 538
Correlation product, 629
Correlator, 275
Cosmic background radiation, 121
Cosmic infrared background, 98
Cosmic microwave background, 98
Cosmic optical background, 98
Cosmic rays, 5, 79, 85, 420
on CCDs, 513
Cosmic variance, 607
Cosmological background, gravitational, 436
Cosmological horizon, 9
Cosmological window, 82
COSMOS, 358
Cost of equipment, 33
Coudé focus, 178
Counting, 360
photon, 361
Coupler, 389
Coupling efficiency, 263
Crab Nebula, 205
Cross talk, 363
Cross-correlation, 629, 635
Cross-power spectrum, 631
Cross-spectral density, 631, 635
Cross-validation, 525
CSO telescope, 396
Current–voltage characteristic, 401
Curvature of focal plane, 183
Curve of growth, 453
Cutoff wavelength, 263

Dark charge, 546
Dark current, 359, 371, 512, 545
Dark signal, 324

Darwin mission, 296
Data, 35
Data bank, 26
Data compression, 26, 206, 527
Data-processing software, 26
Decision theory, 551
Declination, 133
Deconvolution, 459
Defocus, 242
Deformable mirror, 245
Degeneracy factor, 104, 334
Degree of confidence, 515
Degree of freedom, 643
Densified pupils, 291
Densitometry, 330
Depletion zone, 401
Detection
  coherent, 325
  of gamma rays, 412
  of gravitational waves, 431
  incoherent, 326
  limit, 515
  lock-in, 603
  of radiofrequencies, 387
  submillimetre, 326, 343, 355, 387
  ultraviolet, 344, 368
  video, 392
  visible, 344
  X rays, 328, 344
Detector
  amplitude, 325
  array, 365
  CMOS, 373
  format, 371, 382
  hybrid, 373
  multichannel, 328
  quadratic, 325
  single channel, 326
Detector quantum efficiency (DQE), 356
Diatomic molecule, 447
Differential measurement, 56
Diffraction, 210
  Bragg, 206, 493
  Fraunhofer, 216
  Fresnel, 231, 321
  at infinity, 216
Diffuse cosmic background, 82
Digital correlator, 601
Digitisation, 519
  step, 520
Dilution refrigerator, 384
Diode, avalanche, 363
Dirac
  comb, 623
  distribution, 623
  distribution in two dimensions, 625
Direct voltage readout, 381
Dirty beam, 269
Dirty map, 269
Discharge tube, 111
Discrimination of weak source, 53
Dispersion
  corrector, 62
  of a grating, 464
  in optical fibre, 264
  of random variable, 640
Distortion, 180
χ^2 distribution, 643
Distribution function
  of random variable, 638
  of stochastic process, 646
Dome C, 77
Dopant, 346
Doppler
  effect, 450
  imaging, 451
  lines, 453
  profile, 453
  width, 454
Double sideband reception (DSB), 394
DUMAND, 425
Dynamic range, 519

E-ELT, 194
Earth's atmosphere, 40
EB-CCD tube, 248, 362, 363
Ebert–Fastie mounting, 470
Echellogram, 468
Ecliptic, 136
  frame, 136
  inclination angle, 136
Edison tasimeter, 375
Effelsberg radiotelescope, 188
EGRET instrument, 408
Einstein effect, 452
Einstein observatory, 199, 494
Einstein ring, 519
Electromagnetic spectrum, 4
Electrometer, 350, 359
Electron bombarded silicon (EBS) tube, 362
Electronic camera, 360
Electronic transitions, 444, 445
  fine structure, 446
  hyperfine structure, 446
Electronographic tube, 360
Emission spectrum, 443
Emissivity, 112

Emittance, 96
Emulsion
  density of, 358
  nuclear, 350, 360
End-to-end model, 532
Ensemble average, 640
Entropy of data, 527
Epitaxy, 403
  molecular, 402
Equation of time, 160
Equatorial
  frame, 134
  mount, 178
Equinox, 135
Ergodicity
  of stochastic process, 651
  of turbulence, 64
Error function, 642
ESRF, synchrotron, 201
Estimation, 515, 550, 654
  error, 551
Estimator, 512, 551
  biased, 568
  convergence, 554
  efficient, 564
  of mean, 656
  minimax, 554
  reliable, 571
  robust, 572
  unbiased, 555, 655
Etendue, 94, 211, 215, 457, 472
Ethics, 37
Euler angles, 137
Event, 637
EVN, 285
Exhaustive statistic, 561
EXOSAT mission, 493
Exosat satellite, 201, 365
Exozodi, 314
Expectation value, 639
Exposure time calculator, 532
External scale of turbulence, 64, 230
Extinction, 109
  interstellar, 109
Extrasolar planet, 296
Extreme adaptive optics, 254
Extremely large telescopes, 33

Fabry–Pérot etalon, 490
Factor analysis, 528
Faint Object Camera, 362
Fano factor, 417
Fast Fourier transform, 501
Fermat's principle, 183
FFT, 501
Fibre
  graded-index, 263
  multimode, 263
  single-mode, 263
  step-index, 263
Field
  curvature, 180
  fully illuminated, 180
  of a telescope, 177
  of view, 12
  vignetted, 180
Field derotator, 192
Field effect transistor (FET), 367
FIFI instrument, 379
Figure of merit, 462
Filling factor, geometrical, 329
Filtering
  frequency, 652
  online, 524
  optimal, 523
  of stochastic process, 652
  time, 653
Finesse of a spectrometer, 490
Fisher information, 570
  matrix, 571
Fisher–Neyman criterion, 561
Fizeau interferometry, 257, 291
Flat field, 548
Flop, 27
Fluctuations, 332
Fluorescence, 50
Fluoride glasses, 263
Flux unit, 96
Flux, monochromatic, 95
Focal
  ratio, 179
  scale, 179
Format of detector, 371, 382
Fourier pairs, 627
Fourier transform, 619
Frame
  change of, 138
  ecliptic, 136
  equatorial, 134
  galactic, 137
  horizontal, 131
  hour, 133
Fréchet inequality, 564
Free spectral interval, 468, 490
Frequency change, 494
Frequency doubler, 396
Frequency of random variable, 638

Frequency standards, 161
  atomic, 159
  primary, 164
Fresnel lens, 201
Fried parameter, 233, 322
Fringe tracker, 264, 289
Fringe visibility, 102, 227
Full width at half maximum (FWHM)
  of normal distribution, 642
Fundamental catalogues, 145
Fundamental plane, 130, 144
FUSE mission, 195

GAIA, 87, 155, 607
  characteristics, 156
  ICRF, 156
Galactic
  background, 57
  centre, 137
  frame, 137
  pole, 137
GALEX mission, 195
Galileo time, 166
Gallium Neutrino Observatory (GNO), 422
Gamma
  background noise, 514
  burst telescope, 206
  burster, 207
  bursts, 20, 21, 206, 438
  collimator, 407
  detection, 412
  function, 643
  imaging, 202
  instruments, 406
  spectrometry, 412
  telescopes, 201
  transmission, 82
Gaunt factor, 118
Gauss–Markov theorem, 574
Gaussian distribution, 642
Gemini telescopes, 190
Generalised inverse, 581
Geocorona, 52
Geodesics, 183
Geodesy, space, 142
Geometric optics, 177
Geometric reference system, 147
Germanium, 346
Giant Meterwave Radio Telescope (GMRT), 74
GIRAFFE spectrometer, 479
Gladstone's relation, 68
GLAST mission, 408
GMRT array, 278
GPS, 285
  satellite network, 284
  time, 166
Graham, Mount, 190
Grain of photographic plate, 328, 350, 513
Grating, 463
  blazed, 466
  dispersion of, 464
  echelle, 466
Gravitational
  lens, 184
  optics, 175
  telescope, 183
  waves, 9
    detection, 431
    telescope, 431
Greenbank radiotelescope, 501
Gregorian calendar, 171
Gregorian reform, 171
Grism, 469
Ground layer adaptive optics, 253
GSPC, 364
Gunn effect, 396
Gurney–Mott theory, 350

Hadley cells, 72
HALCA mission, 286
Half power beam width, 222
Harmonic generator, 396
HARPS, 475
  spectroscopy, 474
Hecht's equation, 416, 440
HEGRA instrument, 412
Height, 132
Heliosphere, 83
Helium, 445, 450, 496
HEMT, 397, 398
Herschel mission, 87, 190, 198, 480, 501
Hertzsprung–Russell diagram, 109
HESS observatory, 411
Heterodyne detection, 393
Heterodyne interferometry, 260
  infrared, 287
Heterojunction, 348
HETG, 493
HIFI, 385
High dynamic range imaging, 298
High Energy Stereoscopic System (HESS), 75
High resolution spectrograph (HRS), 473
Hipparcos
  catalogue, 146
  frame, 151

HCRF, 155
mission, 152, 607
satellite, 87
Hodges–Lehmann median, 573
Holographic etching, 468
Homodyne interferometry, 261
Horizontal frame, 131
Horn, 388
Hot electron bolometer (HEB), 383, 385, 401
Hot electron mixer, 400
Hour angle, 133
Hour frame, 133
HPBW, 222
Hubble Space Telescope (HST), 195, 612
HURA mask, 204
Hybrid array, 374
Hybrid detector, 373
Hybrid mapping, 273
Hydrogen
atom, 444
maser, 284
molecular, 447
Hydrogen-like atom, 444, 445
Hygrometer, spectral, 56
Hyperfine structure, 445, 446
Hypersensitised photographic plate, 350
Hypertelescope, 291

IBIS instrument, 408
IBIS/ISGRI instrument, 541
Ice, 448
IceCube Neutrino Observatory, 426, 428
ICRF, 148
ICRS, 148
Image
agitation, 228
coding, 203
geometric, 177
intensifier, 360
long-exposure, 232
quality, 182
sampling, 222
short-exposure, 234
slicer, 473, 480
Image–object relation, 218
Impartiality, principle of, 553
Impurity, 346
Inertial range, 64
Infrared
detection, 339
imaging, 234
interferometry, 287
photometry, 108, 112
spectrometry, 463, 486, 487, 490
telescopes, 192
transmission of, 46, 53, 73, 81
Instrumental
MTF, 268
profile, 483
spectral, 456
signature, 544
Integral field spectroscopy (IFS), 461, 479
INTEGRAL mission, 206, 408, 541
spectrometer, 418
Integration time, 518
Intensity interferometry, 335
Intensity standards, 110
Interference filter, 463
grid, 463
Interferogram, 482
Interferometer
intensity, 320
Michelson, 102
spectral, 481
Interferometry, 256
Bracewell, 313
coronagraphic, 311
differential, 238
direct, 257, 259
fibre, 292
Fizeau, 257, 291
heterodyne, 260
heterodyne infrared, 287
homodyne, 261
intensity, 335
Michelson, 259
nulling, 296, 311
pupil plane, 228
radio, 256
space-based, 294
speckle, 236
VLBI, 149, 282
X-ray, 297
Intermediate frequency (IF), 394
Internal scale of turbulence, 64
International Astronomical Union, 29, 148, 159, 169
Interplanetary dust, 80
Inversion
layer, 40
subsidence, 72
temperature, 72
Ionosphere, 44
Ionospheric layers, 44
IRAC instrument, 376
IRAM millimetre interferometer, 533
IRAS mission, 197, 381, 607, 614

Irradiance, 95
IRS, 501
ISGRI, 206, 420
ISI interferometer, 287
ISO mission, 381, 608
Isoplanatic
angle, 237
patch, 236
Isoplanicity, 219
IUE satellite, 25, 195, 467

Jackknife method, 560
Jansky, 96
JCMT radiotelescope, 186, 400
Jensen inequality, 559
Jet stream, 66
Johnson noise, 382
Joule–Thomson refrigerator, 374
Julian
calendar, 170
date, 170
day, 169
Junction, 348
JWST mission, 87, 195

Kamiokande, 424
Karhunen–Loève polynomials, 243, 528
Kasdine–Spergel coronagraph, 306
Keck interferometer, 289
Keck telescopes, 190, 193
Kepler mission, 296
Kinematic reference system, 147
Kirchoff–Fresnel relation, 217
Kitt Peak, 73, 190
Klein–Nishina formula, 352
Klystron, 396
Km3Net telescope, 431
Kolmogorov turbulence, 64, 243
Kuiper Airborne Observatory (KAO), 47, 196

La Palma site, 190
La Silla site, 73, 190
Labview, 531
Lagrange equation, 183
Lagrange points, 87
Lallemand, A., 360
Lambert's law, 100
LAMOST telescope, 191
Large optical mirrors, 189
construction of, 191
Laser guide star, 248, 250, 480
Laser Interferometer Space Antenna (LISA), 435
Law of large numbers, 551, 654
LBT, 190, 191, 194, 264, 291
Lehmann–Scheffe theorem, 563
Lepton, 7
Lidar, 52
Light interference, 547
LIGO experiment, 432
Likelihood function, 554
Line of sight, 176
Line spectrum, 443
Lithium fluoride, 493
Lithium niobate, 498
Littrow condition, 466
Lobe of radiotelescope, 389
Local oscillator, 284, 389, 394
Local Standard of Rest (LSR), 451
Local thermodynamic equilibrium (LTE), 55, 104
LOFAR, 280
Lorentz profile, 49, 455
Loss function, 552
Low light level (L3-CCD), 372
Luminosity, 97
monochromatic, 97
Lunar occultation, 225
Lyot coronagraph, 302

MACAO, 247
Mach's principle, 148
Magellan telescope, 194
Magnetograph, 452, 481
Magnetosphere, 79
Magnitude, 104
AB, 107
absolute, 107
bolometric, 107
photometric, 106
STMAG, 107
systems, 106
MAMA, 358, 610
Maser amplifier, 398
Maser line, 454
Mauna Kea site, 55, 72, 73, 190
MAXIM mission, 298
Mean intensity, 96
Mean of stochastic process, 646
Mean squared error, 568
Mean value, 640
Median of random variable, 640
MEGACAM, 371, 611
MEMS, 245

Mercury iodide, 420
Meridian, astronomical, 132
Mesopause, 41
Mesosphere, 41
Meteorites, 6
Meton cycle, 170
Michelson interferometer, 102, 259
  spectral, 481
Microchannel plate, 360, 365
Microdensitometer, 358
Microlens array, 480
MIDI instrument, 289
Mie scattering, 59
MilliCrab, 544
Minimax principle, 554
Minimax risk, 554
Minimum variance bound, 566, 568
Mixer, 349, 355, 387, 394, 399
  hot electron, 400
  noise temperature of, 391
  SIS, 349
Mixing ratio, 42
Modulation transfer function
  detector, 328
Modulation transfer function (MTF), 182, 220
Modulation, rapid, 57
Molecular
  aggregates, 448
  laser, 396
  transitions, 446
Molière scattering, 408
Moments, 640
Moon, 87
  laser telemetry, 148
MOS capacitance, 366
Most probable value, 640
Multi-object adaptive optics, 254
Multi-object spectrometer, 461
Multi-object spectroscopy (MOS), 475
Multichannel detector, 328
Multiplex
  advantage, 486
  gain, 486

Nadir, 132
Nançay radiotelescope, 186, 280
NAOS, 536
Nasmyth focus, 178
National Virtual Observatory (NVO), 617
Natural guide star, 480
Negative frequencies, 632
NEMO telescope, 429
NESTOR telescope, 428
Neutrino detection, 420
  using Cherenkov radiation, 424
  using chlorine, 421
  using gallium, 422
  using heavy water, 423
  using indium, 423
NICMOS detector, 376
Nodding, 545
Noise, 511, 512
  background, 339, 512
    gamma, 514
    thermal, 513
  detector, 512
  diode, 111
  Gaussian, 517
  generator, 111
  Johnson, 382
  photon, 334
  pointing, 540
  quantum, 334, 336
  readout, 357, 370
  scintillation, 515
  seismic, 434
  shot, 334
  signal, 514
  temperature, 335, 390, 402, 497
  thermal, 340
  transmission, 514
  white, 584
Noise equivalent charge (NEC), 357
Noise equivalent flux (NEF), 339
Noise equivalent power (NEP), 339, 383
Non-linearity, 546
Normal distribution, 642
Normal process, 647
Notch filter mask, 304
NTT telescope, 190, 192, 381
Nuclear
  lines, 447
  transitions, 445
Nucleosynthesis, 447
Nulling fringe mode, 291
Nulling interferometry, 296, 311
Numerical aperture, 179, 263
Nutation, 134, 137
Nyquist formula, 342

OAO satellite, 195
Observation
  daytime, 59
  from space, 77
  in situ, 11
  sites, 71

Observing block, 537
Observing system, 12
  astrophysical model, 531
  computing model, 531
  detector, 530
  electrical model, 531
  environmental model, 531
  mechanical model, 530
  optical model, 530
  thermal model, 530
Obukhov's law, 64
ODIN mission, 398
Offset signal, 57
OH radical, 55
OHANA, 292
On–off technique, 57
Open acess, 36
Opticad, 530
Optical absorption, 347
Optical depth, 443
Optical fibre, 263
Optical telescopes, 189
Optimal filtering, 26
Orbit
  high circular, 86
  low equatorial, 86
Oversampling, 179, 522
Oxygen, molecular, 89
Ozone, 44

Pair creation, 202, 351, 353
Pair effect, 408
Palomar Optical Sky Survey (POSS), 607
Palomar, Mount, 73, 190
Paraboloid, 177
Parallax, 152
Parametric amplifier, 398
Paranal site, 73, 190
Parent population, 550
Parkes–MIT–NRAO (PMN) survey, 607
Parseval's theorem, 630
Partial correction, 247
Partial ordering, 552
Peltier refrigerator, 359
Phase closure, 239, 271
Phase mask, 304
Phase noise, 239
Phonons, 347, 413, 448
Phoswich detector, 415
Photocathode, 345
Photochemical effect, 350
Photoconductor
  extrinsic, 346
  intrinsic, 345
  stressed, 347
Photocurrent, 337
Photodetector Array Camera and Spectrometer (PACS), 379, 480, 501
Photoelectric effect, 202
Photographic plate, 357
Photography, 350
Photoionisation of gas, 350
Photometric quality, 73
Photometry
  gamma-ray, 201
  infrared, 108, 112
  radiofrequency, 110
  ultraviolet, 117
  visible, 114
  X-ray, 118
Photomultiplier (PM), 359
  solid- state (SSPM), 379
Photon
  correlation, 261
  counting camera, 362
  properties of, 13
  statistics, 103
Photovoltaic effect, 347
Pico Veleta radiotelescope, 189
PICSIT, 206
Pierre Auger observatory, 75
Piston, 242
Piston effect, 66
Pixel, 328, 510
  dead, 549
  frequency, 373
  hot, 549
Planck mission, 87, 198, 385
Planck's law, 99
Planetary boundary layer, 66
PlanetQuest, 295
Plasma
  frequency, 50
  ionospheric, 46, 50, 62
PMM digitisation, 610
Point estimation, 551
Point source response function, 222
Point spread function (PSF), 219, 222, 224, 580
Pointing noise, 540
Poisson
  distribution, 643
    parameter of, 643
  impulse, 649
  process, 648
  statistics, 104
Polarisation, 24, 97

Pollution
light, 76
radiofrequency, 76, 88
Polycyclic aromatic hydrocarbons, 448
Positronium, 447
Power spectral density (PSD)
of random variable, 630
of signal, 634
Power spectrum, 647
Precession, 133, 134
Precipitable water, quantity of, 43
Predisperser, 468, 491
Primary mirror, 177
Primary standard, 106
Principal components, 528
Probability, 637
Probability density, 638
Proleptic calendar, 171
PRONAOS balloon, 47
Proper motion, 133, 135, 145, 146
Proportional counter, 364, 493
Prouhet–Thué–Morse sequence, 314
Publications, 34
Pulse shape discrimination (PSD), 415
Pupil, 216, 218
circular, 223
disconnected, 226
function, 217
masking, 273
non-redundant, 277
redundant, 277
Pyrometry, 112

Quadratic detector, 325, 342
Quantisation of EM field, 332
Quantum efficiency, 337, 344
Quantum noise, 334, 336
compression, 339
Quantum well, 348
Quantum yield, 350
Quasar, 148, 155
Quasi-monochromatic radiation, 101
Quasi-optical mode, 380
Quenouille method, 559

Radiation belts, 83
Radiation pattern, 113
Radiative transfer, 94
Radio dish, 185
Radiofrequency
detection, 335, 339, 387
photometry, 110
spectrometry, 494
telescopes, 185
transmission, 74
Radiometry, 94
Radiotelescope, 185
Ramo's theorem, 416, 440
Random process, 644
Random variable, 637
reduced normal, 642
Rao–Blackwell theorem, 562
Rao–Cramér inequality, 564
Raster scanning, 186, 328
Ray tracing, 179, 181, 182
Rayleigh
resolution criterion, 224
scattering, 58
star, 251
theorem, 630
unit, 51
Rayleigh–Jeans law, 99
Readout
noise, 357
rate, 381
Real-time data handling, 26
Realisation, 550
Recycling of light, 432
Reddening, 108
Redundancy of data, 527
Reflection grating spectrometer (RGS), 468, 494
Refracting telescope, 189
Refractive index of air, 61
Reliability, 78
Remote sensing, 10
Resistor, thermal noise of, 341
Resonant cavity, 388
Reynold's number, 63
Richardson number, 67
Richardson's law, 359
Right ascension at Greenwich, 142
Risk, 552
Roddier and Roddier coronagraph, 302
Root mean square (rms) deviation, 641
ROSAT mission, 201, 607
ROSITA mission, 372
Rotation lines, 445, 446
Rotation of field, 192
Rotation of the Earth, 132, 134, 141, 165
Rotation–vibration lines, 445, 446
Rouan coronagraph, 304
Rowland mounting, 470
Ruby maser, 399
Running mean, 339, 599

Rydberg formula, 444
Rydberg maser, 399

SALT telescope, 191
Sample, 550, 623
Sampling, 520
  function, 623
  rate, 520
Saturation threshold, 324
Scale height, 40
Schmidt sky surveys, 610
Schmidt telescope, 183
Schottky diode, 396, 401
Scidar, 66
Scintillation, 228, 515
  interplanetary, 237
  interstellar, 237
Scintillator, 413
  conversion yield of, 413
Second (time), 158, 162
Secondary electron conduction tube (SEC), 362
Secondary standard, 106
Seeing, 23
  angle, 234
Self-coherent camera, 309
Semi-classical model of light, 332
Semiconductor detector, 415
Sensitised photographic plate, 350
Sensitivity of detection, 18
Shack–Hartmann analysers, 244
Shah function, 623
Shannon's theorem, 521
Shear, wind, 66
Shift-and-add, 236
Shot noise, 334
SI units, 162
Sidelobes, 113
Sidereal angle, 141
  Greenwich, 142
Signal
  average power of, 633
  energy of, 633
  instantaneous power of, 633
  power spectrum of, 634
Signal-to-noise ratio (S/N), 512, 516, 518
  in spectrometry, 459
Silicates, 448
Silicon, 346
Silicon intensified target (SIT) tube, 362
SIMBOL-X mission, 202
Similarity relation, 621
Sinc function, 622, 623
SINFONI, 480
Single sideband (SSB) reception, 394
Single-channel detector, 326
Single-mode propagation, 263
SIS junction, 349, 400
SIS mixer, 349
SKA, 279
Sky
  background
    magnitude of, 52
    subtraction of, 57
  chopping, 57
  coverage, 250
  noise, 57, 68
  subtraction, 538
Sky survey, 608
  CCD, 611
  CFDF, 612
  CFHT-LS, 612
  completeness, 607
  confusion limit, 607
  cosmic microwave background, 609
  DEEP2, 614
  DENIS, 614
  depth, 607
  2dFGRS, 613
  ESO, 610
  field, 607
  HDF-N and -S, 612
  infrared, 614
  2MASS, 614
  photometric, 609, 612
  POSSI/POSSII, 610
  Schmidt, 610
  SDSS, 612, 613
  SERC/AAO, 610
  spectroscopic, 609, 613
  VVDS, 612, 614
Slit mask, 475
Smearing, 228
Snapshot mode, 273
SOFIA telescope, 47, 197, 501
SOHO, 87
Solar
  aureole, 60
  corona, 60
  eclipse, 60
  flares, 83, 122, 280
  granulation, 601
  modulation, 85
  spectrum, 121
  wind, 5, 83
Solid transitions, 447
Solid-state imagers, 365

Soviet–American Gallium Experiment (SAGE), 422
Space Interferometry Mission, 295
Space telescopes, 194
Space–time frames, 24
Spallation, 86
Spark chamber, 410, 513
Spatial
  cutoff frequency, 222
  filter, 220, 226, 263
  filtering by turbulence, 233
  frequencies, 620
  modes, 241
  reference systems, 129
Specific intensity, 95
  of the Sun, 114
Speckle, 236, 334
  cancellation, 310
Speckle interferometry, differential, 238
Spectral
  coverage, 15
  element, 457
  lamp, 117
  line, 444
    equivalent width, 444, 449
    flanks of, 444
    intensity, 449
    polarisation, 450
    profile, 449
    wings of, 444
  mask, 475
  multiplexing, 482
  resolution, 456
  width, 101
Spectral density
  estimator of, 657
  of random variable, 630
  of stochastic process, 647
  white, 650
Spectroheliogram, 481, 482
Spectroheliograph, 461, 481
Spectrometer, 455
  acoustic, 498
  autocorrelation, 499
  Bragg, 491
  Fabry–Pérot, 489
  fibre optic, 477
  finesse of, 490
  Fourier transform, 481, 488
  grating, 463
  heterodyne, 494
  holographic Fourier, 485
  imaging, 455, 460
  interferometric, 462
  multi-object, 461, 475
  multichannel, 460, 497
  resonance, 503
  robot arm, 479
  sequential, 460
Spectrometry
  gamma, 412
  heterodyne, 503
  infrared, 463, 490
  integral field, 471, 479
  radiofrequency, 494
  submillimetre, 501
  ultraviolet, 463
  visible, 490
  X-ray, 491
Spectrophotometry, 94
Spectrum, 630
  amplitude, 630
  phase, 631
SPHERE, 307
Spherical coordinates, 129
Spin–orbit coupling, 445
SPIRE, 488
Spitzer mission, 117, 198, 376, 501
SQUID, 385
Standard deviation, 641
Stationary process, 647
Statistical contrast, 586
Statistics, 550
Stenope, 202
STMAG, 107
Stochastic process, 644
  transformation of, 650
Stokes parameters, 99, 500
Storage capacity, 381
Stratopause, 41
Stratosphere, 40, 41
Stratospheric balloon, 197
Strehl ratio, 247
Structure constant
  of refractive index, 68, 230
  of temperature, 66
Structure function, 65
  of phase, 230
Sub-pupil, 226
Subaru telescope, 190
Submillimetre
  detection, 326, 339, 343, 355, 387
  spectrometry, 501
  telescope, 192, 196
  transmission of, 53
Submillimetron project, 397
Sudbury Neutrino Observatory (SNO), 423
Sun, mean displacement of, 142

Super-Kamiokande, 424
Super-resolution, 238, 319
Supercomputers, 27
Superconducting tunnel junction (STJ), 349
Supernova remnant (SNR), 494
Supernova SN 1987A, 425
Surface wave, 499
SWIFT satellite, 207
Synchrotron radiation, 119
Synchrotron sources, 120
System temperature, 390
Systematic error, 555

Taylor turbulence, 69
TDRSS, 87
Telluric absorption bands, 45
Temperature
  antenna, 388
  brightness, 116
  colour, 124
  noise, 335, 390, 402, 497
  potential, 67
  system, 390
Temporal frequencies, 620
Temporal reference system, 157
Temporal width, 101
TERAPIX, 617
Thermal emission of atmosphere, 55
Thermal noise of resistor, 341
Thermosphere, 41
Thinned CCD, 368
Thirty Meter Telescope (TMT), 33, 194
Thomson scattering cross-section, 350
Throughput, 94, 211, 457, 472
  advantage, 485
  of a spectrometer, 484
Time
  apparent solar, 159
  atomic, 161
  civil, 161
  Coordinated Universal (UTC), 164
  dynamical, 167
  Ephemeris (ET), 167
  International Atomic (TAI), 163
  local sidereal, 140
  mean solar, 159
  Terrestrial Dynamical (TDT), 169
  Universal (UT), 161
Time delay integration, 549
Time scales, 157
Time variability, 20
Time–frequency analysis, 632
TIMMI instrument, 381
Tip–tilt, 242
  correction, 244
  fluctuation, 244
Tololo site, 190
Transfer equation, 442
  inversion of, 443
Transparency function, 203
Triad, local, 132
Triangulation, localisation by, 207
Triplet corrector, 181
Tropopause, 40
Troposphere, 40, 63
Truncated mean, 572
Truncation, 600
Turbulence
  effect on images, 228
  frozen, 69
  homogeneous, 64
  ionospheric, 70
  spatial filtering by, 233
Turbulent dissipation, 63

UCAC, 154
UKIRT telescope, 190
Ultraviolet
  detection, 335, 344, 368
  photometry, 117
  spectrometry, 463
  telescopes, 195
  transmission of, 46
Ultraviolet Visible Echelle Spectrograph (UVES), 468
Undersampling, 522
Uniform minimum variance unbiased (UMVU) estimator, 553
URA mask, 205

Valence band, 345
Van Allen belts, 83
Varactor, 398
Variance of random variable, 640
Vega, 114
Venus, spectrum of, 487
Vernal equinox, 134, 145
Vertical, local, 131
Very Large Array (VLA), 74, 277, 278, 391
Very Large Telescope (VLT), 190, 193
Very long baseline interferometry (VLBI), 148, 260, 282
VESPA correlator, 500
Vibrations, man-made, 76
Video detection, 392

Vidicon tube, 361
VIMOS, 476
VIRGO experiment, 432
Virtual instrument, 529
Virtual observatory, 615
Virtual reality, 78
Viscosity, 63
Visible
  detection, 335, 344
  imaging, 234
  nulling coronagraph, 309
  photometry, 114
  spectrometry, 490
  telescopes, 189, 193, 194
  transmission of, 73
VISTA, 190
VLBA, 285, 286
VLTI, 288
Voigt profile, 455
Volume holographic grating, 469
VSOP mission, 286

Wave number, 96
Wave packet, 621
Wavefront
  analyser, 244
  measurement of, 241
  perturbations of, 229
  sensor, 537
Waveguide, 262, 388
Wavelets, 635
Weber's experiment, 431
Weighting telescope, 207
White paper, 33
White spectral density, 650
Wien's law, 99
Wiener–Khinchine theorem, 598, 631
Wiener–Kolmogorov filter, 523
Wigner–Ville representation, 632
Windowing, 632
Windowing mode of CCD, 373
Winsorized mean, 572
Wire chamber, 351
WMAP mission, 123
Wolter telescope, 200

X-ray
  detection, 328, 344
  photometry, 118
  spectrometry, 491
  telescopes, 199
  transmission, 82
Xing Long, 190
XMM–Newton mission, 201, 468, 493

Young slits, 102

Zeeman effect, 450, 451, 455, 481
Zelenchuskaya site, 73, 190
Zemax, 530
Zenith, 132
  distance, 45
Zentralen Grenzversatz, 518
Zernike polynomials, 181, 241
Zernike theorem, 211
Zernike–Van Cittert theorem, 213
Zodiacal
  emission, 80
  light, 80, 314
  nebula, 80

Lightning Source UK Ltd.
Milton Keynes UK
UKOW07n1334050516

273613UK00001B/23/P